VOLUME ONE HUNDRED AND SEVENTEEN

Current Topics in DEVELOPMENTAL BIOLOGY

Essays on Developmental Biology, Part B

CURRENT TOPICS IN DEVELOPMENTAL BIOLOGY

"A meeting-ground for critical review and discussion of developmental processes"
A.A. Moscona and Alberto Monroy (Volume 1, 1966)

VOLUME ONE HUNDRED AND SEVENTEEN

CURRENT TOPICS IN DEVELOPMENTAL BIOLOGY

Essays on Developmental Biology, Part B

Edited by

PAUL M. WASSARMAN
Department of Developmental and Regenerative Biology
Icahn School of Medicine at Mount Sinai
New York, USA

ELSEVIER

AMSTERDAM • BOSTON • HEIDELBERG • LONDON
NEW YORK • OXFORD • PARIS • SAN DIEGO
SAN FRANCISCO • SINGAPORE • SYDNEY • TOKYO
Academic Press is an imprint of Elsevier

AP

Academic Press is an imprint of Elsevier
50 Hampshire Street, 5th Floor, Cambridge, MA 02139, USA
525 B Street, Suite 1800, San Diego, CA 92101-4495, USA
The Boulevard, Langford Lane, Kidlington, Oxford OX5 1GB, UK
125 London Wall, London, EC2Y 5AS, UK

First edition 2016

ISBN: 978-0-12-801382-3
ISSN: 0070-2153

For information on all Academic Press publications visit our website at http://store.elsevier.com/

CONTENTS

CONTRIBUTORS

Bipul R. Acharya
Division of Cell Biology and Molecular Medicine, Program in Membrane Interface Biology, Institute for Molecular Bioscience, The University of Queensland, Brisbane, Queensland, Australia

Megan Addison
The Francis Crick Institute, Mill Hill Laboratory, London, United Kingdom

Claire Anderson
Department of Cell & Developmental Biology, University College London, London, United Kingdom

M. Azim Surani
Wellcome Trust/Cancer Research UK Gurdon Institute; Department of Physiology, Development and Neuroscience, University of Cambridge, Cambridge, United Kingdom

Hao Chang
Department of Molecular Biology and Genetics; Howard Hughes Medical Institute, Johns Hopkins University School of Medicine, Baltimore, Maryland, USA

Hélène Chanut-Delalande
Centre de Biologie du Développement, Université de Toulouse, UPS; CNRS, UMR5547, Toulouse, France

Alys M. Cheatle Jarvela
Entomology Department, University of Maryland, College Park, Maryland, USA

Michelle M. Collins
Department of Developmental Genetics, Max Planck Institute for Heart and Lung Research, Bad Nauheim, Germany

Justin Crocker
Janelia Research Campus, Howard Hughes Medical Institute, Ashburn, Virginia, USA

Eric H. Davidson
Division of Biology and Biological Engineering, California Institute of Technology, Pasadena, California, USA

David P. Doupé
Department of Genetics, Harvard Medical School, Boston, Massachusetts, USA

Nicolas Fossat
Embryology Unit, Children's Medical Research Institute, Westmead; Discipline of Medicine, Sydney Medical School, University of Sydney, Sydney, New South Wales, Australia

Giuseppe Fusco
Department of Biology, University of Padova, Padova, Italy

Ufuk Günesdogan
Wellcome Trust/Cancer Research UK Gurdon Institute; Department of Physiology, Development and Neuroscience, University of Cambridge, Cambridge, United Kingdom

Sarah J.L. Graham
Department of Physiology, Development and Neuroscience, University of Cambridge, Cambridge, United Kingdom

Thierry Hoquet
Faculté de philosophie, Université Jean Moulin Lyon 3, Lyon Cedex 02, France

Chi Kin Ip
Embryology Unit, Children's Medical Research Institute, Westmead; Discipline of Medicine, Sydney Medical School, University of Sydney, Sydney, New South Wales, Australia

Uwe Irion
Max-Planck-Institut für Entwicklungsbiologie, Tübingen, Germany

Steven L. Klein
Department of Anatomy and Regenerative Biology, The George Washington University School of Medicine and Health Sciences, Washington, DC, USA

Lucie Laplane
UMR 8590 (CNRS, IHPST-Université Paris I Panthéon-Sorbonne, ENS), Paris, France

Eveline S. Litscher
Department of Developmental and Regenerative Biology, Icahn School of Medicine at Mount Sinai, New York, USA

David A.F. Loebel
Embryology Unit, Children's Medical Research Institute, Westmead; Discipline of Medicine, Sydney Medical School, University of Sydney, Sydney, New South Wales, Australia

Terry Magnuson
Department of Genetics; Lineberger Comprehensive Cancer Center, University of North Carolina, Chapel Hill, North Carolina, USA

David R. McClay
Department of Biology, Duke University, Durham, North Carolina, USA

Andrew P. McMahon
Department of Stem Cell Biology and Regenerative Medicine, Eli and Edythe Broad Center for Regenerative Medicine and Stem Cell Research, Keck School of Medicine of the University of Southern California, Los Angeles, California, USA

Douglas A. Melton
Department of Stem Cell and Regenerative Biology, Harvard Stem Cell Institute, Harvard University, Cambridge, Massachusetts, USA

Alessandro Minelli
Department of Biology, University of Padova, Padova, Italy

Marek Mlodzik
Department of Developmental and Regenerative Biology, Graduate School of Biomedical Sciences, Icahn School of Medicine at Mount Sinai, New York, USA

Sally A. Moody
Department of Anatomy and Regenerative Biology, The George Washington University School of Medicine and Health Sciences, Washington, DC, USA

Jeremy Nathans
Department of Molecular Biology and Genetics; Howard Hughes Medical Institute; Department of Neuroscience; Department of Ophthalmology, Johns Hopkins University School of Medicine, Baltimore, Maryland, USA

Christiane Nüsslein-Volhard
Max-Planck-Institut für Entwicklungsbiologie, Tübingen, Germany

Virginie Orgogozo
Institut Jacques Monod, CNRS, UMR 7592, Univ Paris Diderot, Sorbonne Paris Cité, Paris, France

Virginia E. Papaioannou
Department of Genetics and Development, Columbia University Medical Center, New York, USA

François Payre
Centre de Biologie du Développement, Université de Toulouse, UPS; CNRS, UMR5547, Toulouse, France

Norbert Perrimon
Department of Genetics, Harvard Medical School; Howard Hughes Medical Institute, Boston, Massachusetts, USA

Isabelle S. Peter
Division of Biology and Biological Engineering, California Institute of Technology, Pasadena, California, USA

Leslie Pick
Entomology Department, University of Maryland, College Park, Maryland, USA

Serge Plaza
Centre de Biologie du Développement, Université de Toulouse, UPS; CNRS, UMR5547, Toulouse, France

Thomas Pradeu
ALYSAI, CNRS, UMR5164, University of Bordeaux, Bordeaux, France

Ella Preger-Ben Noon
Janelia Research Campus, Howard Hughes Medical Institute, Ashburn, Virginia, USA

Karine Prévot
Université Paris Ouest Nanterre-La Défense, Ireph, Nanterre Cedex, France

Jesse R. Raab
Department of Genetics; Lineberger Comprehensive Cancer Center, University of North Carolina, Chapel Hill, North Carolina, USA

Mirana Ramialison
Australian Regenerative Medicine Institute; Systems Biology Institute Australia, Monash University, Clayton, Victoria, Australia

Amir Rattner
Department of Molecular Biology and Genetics, Johns Hopkins University School of Medicine, Baltimore, Maryland, USA

Valentine Reynaud
Institut de Recherches Philosophiques de Lyon (IRPhiL), Centre de la recherche Eugène Chevreul, Lyon, France

Janet Rossant
Program in Developmental and Stem Cell Biology, Hospital for Sick Children, University of Toronto, Toronto, Ontario, Canada

John S. Runge
Curriculum for Genetics and Molecular Biology; Department of Genetics; Lineberger Comprehensive Cancer Center, University of North Carolina, Chapel Hill, North Carolina, USA

Noriyuki Satoh
Marine Genomics Unit, Okinawa Institute of Science and Technology Graduate University, Onna, Okinawa, Japan

Hans R. Schöler
Department of Cell and Developmental Biology, Max Planck Institute for Molecular Biomedicine, Münster, Germany

Trudi Schupbach
Department of Molecular Biology, Princeton University, Princeton, New Jersey, USA

Matthew P. Scott
Carnegie Institution for Science, Washington, D.C.; Stanford University School of Medicine, Stanford, California, USA

Ajeet Pratap Singh
Max-Planck-Institut für Entwicklungsbiologie, Tübingen, Germany

Jonathan M.W. Slack
Department of Biology and Biochemistry, University of Bath, Bath, United Kingdom; Department of Genetics, Cell Biology and Development, University of Minnesota, Minneapolis, Minnesota, USA

Sergei Y. Sokol
Department of Developmental and Regenerative Biology, Icahn School of Medicine at Mount Sinai, New York, USA

Davor Solter
Siriraj Center of Excellence in Stem Cell Research, Mahidol University, Bangkok, Thailand; Max Planck Institute of Immunobiology and Epigenetics, Freiburg, Germany

Philippe Soriano
Department of Developmental and Regenerative Biology, The Tisch Cancer Institute, Icahn School of Medicine at Mount Sinai, New York, USA

Allan C. Spradling
Department of Embryology, Carnegie Institution for Science, Baltimore, Maryland, USA

Didier Y.R. Stainier
Department of Developmental Genetics, Max Planck Institute for Heart and Lung Research, Bad Nauheim, Germany

Claudio D. Stern
Department of Cell & Developmental Biology, University College London, London, United Kingdom

David L. Stern
Janelia Research Campus, Howard Hughes Medical Institute, Ashburn, Virginia, USA

Patrick P.L. Tam
Embryology Unit, Children's Medical Research Institute, Westmead; Discipline of Medicine, Sydney Medical School, University of Sydney, Sydney, New South Wales, Australia

Paul A. Trainor
Stowers Institute for Medical Research, Missouri; Department of Anatomy and Cell Biology, University of Kansas Medical Center, Kansas City, Kansas, USA

Harish N. Vasudevan
Department of Developmental and Regenerative Biology, The Tisch Cancer Institute, Icahn School of Medicine at Mount Sinai, New York, USA

Michel Vervoort
Institut Jacques Monod, CNRS, UMR 7592, Univ Paris Diderot, Sorbonne Paris Cité, Paris, France

Yanshu Wang
Department of Molecular Biology and Genetics; Howard Hughes Medical Institute, Johns Hopkins University School of Medicine, Baltimore, Maryland, USA

Paul M. Wassarman
Department of Developmental and Regenerative Biology, Icahn School of Medicine at Mount Sinai, New York, USA

Gary M. Wessel
Department of Molecular and Cellular Biology, Brown University, Providence, Rhode Island, USA

Eric Wieschaus
Molecular Biology Department, HHMI, Princeton University, Princeton, New Jersey, USA

Emilie Wilkie
Embryology Unit; Bioinformatics Unit, Children's Medical Research Institute, Westmead, New South Wales, Australia

David G. Wilkinson
The Francis Crick Institute, Mill Hill Laboratory, London, United Kingdom

Lewis Wolpert
Cell and Developmental Biology, University College London, London, United Kingdom

Hugh R. Woodland
School of Life Sciences, University of Warwick, Coventry, United Kingdom

Guangming Wu
Department of Cell and Developmental Biology, Max Planck Institute for Molecular Biomedicine, Münster, Germany

Alpha S. Yap
Division of Cell Biology and Molecular Medicine, Program in Membrane Interface Biology, Institute for Molecular Bioscience, The University of Queensland, Brisbane, Queensland, Australia

Jennifer Zanet
Centre de Biologie du Développement, Université de Toulouse, UPS; CNRS, UMR5547, Toulouse, France

Kenneth S. Zaret
Institute for Regenerative Medicine, Epigenetics Program, Department of Cell and Developmental Biology, Perelman School of Medicine, University of Pennsylvania, Philadelphia, Pennsylvania, USA

Magdalena Zernicka-Goetz
Department of Physiology, Development and Neuroscience, University of Cambridge, Cambridge, United Kingdom

PREFACE

ESSAYS ON DEVELOPMENTAL BIOLOGY-2016

These volumes commemorate the 50th or "golden" anniversary of the founding of *Current Topics in Developmental Biology* (CTDB) in the summer of 1966 by Aron Moscona (1921–2009) and Alberto Monroy (1913–1986) with Academic Press as publisher. In 1966 Moscona and Monroy anticipated the changing tide of developmental biology, from a classical to a more molecular science, and intended that CTDB be a venue for the change. From the beginning CTDB was to serve as "a meeting ground for critical review and discussion of developmental processes". More than 115 volumes later, still published by Academic Press (Elsevier), CTDB continues to strive to serve this purpose for developmental biologists.

The idea of publishing these essays came about in late 2013 when I realized that 2016 would be the 50th anniversary of the CTDB series. The object was to assemble essays contributed by a large number of outstanding developmental biologists around the world working on different aspects of development with a variety of model organisms. To achieve this goal more than 120 research scientists were contacted individually by e-mail in January 2014 and invited to contribute to this project. The response was surprisingly good with about 90 invitees agreeing to submit essays by September 15, 2015. Many flattering comments accompanied the acceptances, including "I think it is very appropriate to have an anniversary edition of what has been a very influential series over the years" and "CTDB is a review series that I appreciate. It has been, and is, important for my field of developmental biology". When the late Walter Gehring received my invitation and list of contributors he wrote back immediately saying, "After looking at your impressive list of contributors, I have to join your gang."

In the end more than 70 essays were submitted for publication by some of the most accomplished and well-known developmental biologists in the world. The essays appear here in alphabetical order according to the invited author's last name, not according to topic or on some other basis. Just a glance at the Table of Contents reveals the wide breadth of research topics and variety of model organisms discussed. Many essays relate the history of a particular area of research and/or personal experiences in research, whereas others are quite philosophical about a specific area of research and/or research in general. The authors were encouraged to speculate and in many instances have done so. The essays are not necessarily intended to be comprehensive reviews, however, they provide a window onto the very rich landscape of contemporary research in developmental biology.

I'm grateful to all the authors who devoted a great deal of time and effort preparing their essays. From a personal point of view, it's been a great pleasure to interact via e-mail with so many outstanding developmental biologists during the past 18 months or so. They tolerated my frequent reminders ("nagging") about the essays and, in a few instances, with my pleas to contribute to these volumes of CTDB. Despite a few sleepless nights worrying about who might not deliver an essay on time, overall this turned out to be a gratifying experience. My goal was to assemble 50th anniversary volumes of CTDB that would reflect the current state of developmental biology and be useful to students and investigators for years to come. It's up to you the reader to decide whether this goal was achieved. I can't help but wonder what Moscona and Monroy would think about the essays and the state of developmental biology presented in these 50th anniversary volumes of CTDB.

Finally, I want to thank Zoe Kruze and Sarah Lay at Elsevier, Oxford, and my wife, Eveline Litscher, for their forbearance, advice, and assistance during the course of this lengthy project.

We fondly remember four outstanding developmental biologists, Professors Eric H. Davidson (1937–2015; California Institute of Technology, Pasadena, California), Walter J. Gehring (1939–2014; Biozentrum, University of Basel, Switzerland), Julian H. Lewis (1946–2014; Cancer Research, London Research Institute, England), and Yoshiki Sasai (1962–2014; RIKEN Center for Developmental Biology, Japan) who agreed to be contributors to Essays on Developmental Biology-2016, but died unexpectedly in 2014 or 2015. Our condolences to their families and friends.

Paul M. Wassarman
Series Editor
Current Topics in Developmental Biology

CHAPTER ONE

Epigenetic Regulation by ATP-Dependent Chromatin-Remodeling Enzymes: SNF-ing Out Crosstalk

John S. Runge[*,†,‡], **Jesse R. Raab**[†,‡], **Terry Magnuson**[†,‡,1]
*Curriculum for Genetics and Molecular Biology, University of North Carolina, Chapel Hill, North Carolina, USA
†Department of Genetics, University of North Carolina, Chapel Hill, North Carolina, USA
‡Lineberger Comprehensive Cancer Center, University of North Carolina, Chapel Hill, North Carolina, USA
[1]Corresponding author: e-mail address: terry_magnuson@med.unc.edu

Contents

Abstract

Cells utilize precise mechanisms to access genomic DNA with spatiotemporal accuracy. ATP-dependent chromatin-remodeling enzymes (also known simply as "remodelers") comprise a specialized class of enzymes that is intimately involved in genomic organization and accessibility. Remodelers selectively position nucleosomes to either alleviate chromatin compaction or achieve genomic condensation locally, based on a multitude of cellular signals. By dictating nucleosome position, remodelers control local euchromatic and heterochromatic states. These activities govern the accessibility of regulatory regions like promoters and enhancers to transcription factors, RNA polymerases, and coactivators or -repressors. As studies unravel the complexities of epigenetic topography, evidence points to a chromatin-based interactome where regulators interact competitively, cooperatively, and/or codependently through physical and functional means. These types of interactions, or crosstalk, between remodelers raise important questions for tissue development. Here, we briefly review the evidence for remodeler interactions and argue for additional studies examining crosstalk.

Current Topics in Developmental Biology, Volume 117
ISSN 0070-2153
http://dx.doi.org/10.1016/bs.ctdb.2015.10.009

1. INTRODUCTION

The hypothesis that the epigenome contains a readable language of DNA methylation, posttranslational modifications to unstructured histone tails, and multi-dimensional chromatin boundary elements has dramatically improved the understanding of chromatin regulation by stimulating careful analysis of previously unappreciated epigenetic details (Strahl & Allis, 2000). Regulation of gene subsets by epigenetic features can dictate cellular transitions (i.e., proliferation, potency, and differentiation) and drive *de novo* tissue development with distinct epigenomic signatures (Kundaje et al., 2015). Within the context of the epigenome, remodelers have emerged as critical regulators of developmental programs (Clapier & Cairns, 2009).

There are four major subfamilies of remodelers (SWI/SNF, INO80, ISWI, and CHD). All are considered derivatives of the SNF2 ATPase family (Fig. 1). Each remodeling complex contains a single catalytic ATPase that utilizes ATP hydrolysis to achieve the energy-intensive modulation of chromatin (Fig. 1, parentheses). Notably, several remodelers utilize multiple ATPases via mutually exclusive relationships (Fig. 1; SWI/SNF, NURD). Mutual exclusivity is not limited to the ATPase subunit, as some remodeling families contain accessory positions within a complex that can be filled by mutually exclusive proteins (SWI/SNF). Additionally, remodelers can utilize the same accessory subunits across family and subfamily classifications (Fig. 1). The fact that remodeler components form distinct biochemical

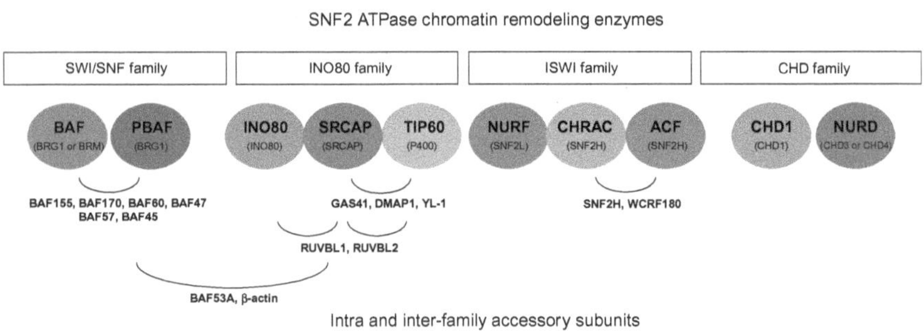

Figure 1 SNF2 ATPase chromatin-remodeling enzymes. Representative diagram of ATP-dependent chromatin-remodeling families. Families are designated by white boxes. Specific remodelers are represented as circles. Catalytic ATPases for each remodeling complex are indicated in parentheses. Subunits that incorporate into multiple complexes are identified by curved lines. Notably, the ISWI family includes additional complexes not depicted here (NoRC, RSF, WICH). (A color version of this figure is located at the end of this chapter.)

complexes on the basis of ATPase inclusion and accessory subunit incorporation is one of the most intriguing topics in remodeler biology. The compositional similarities between remodelers may implicate the complexes in direct relationships via cooperative or competitive means. Likewise, it is probable that variation in composition between remodelers is indicative of functional heterogeneity across remodeler subfamilies. However, limited knowledge on the biological functionality, locus-specific assembly, and regulatory crosstalk of remodelers exists to-date. These topics are of central importance to the biomedical community given the broad implications for understanding gene regulation through epigenetic mechanisms.

Remodeler ATPases are engines to the nuclear machines. They are archetypal subunits around which each remodeling complex is assembled, and are evolutionarily conserved across eukaryotes indicating their essential contribution to life. They hydrolyze hundreds of ATP molecules per minute to facilitate nucleosome sliding, histone variant exchange, and promoter clearance on a genome-wide scale (Holstege et al., 1998). They also contain additional functional domains that direct their function. For example, the predominant ATPase containing subunit of SWI/SNF, BRG1 (encoded by the *SMARCA4* gene in humans), contains a carboxy-terminal bromodomain that targets acetylated lysines on exposed histone tails, thereby promoting activation of transcription (Dutta et al., 2014). As each remodeler ATPase contains an essential SNF2 ATPase domain as well as conserved domains that promote specific activities of the remodelers, they are the subject of intense investigation.

In addition to the ATPase, each complex also contains multiple protein subunits that facilitate the assembly and action of each remodeling machine. The BPTF subunit of the NURF complex in the ISWI subfamily, for example, deposits trimethylation on lysine 4 residues of histone 3 (H3K4me3), a hallmark of chromatin priming that establishes local euchromatin and recruits transcriptional activators (Guenther, Levine, Boyer, Jaenisch, & Young, 2007). Deposition of H3K4me3 allows the NURF complex to dramatically alter the chromatin state of specific targets through both chromatin remodeling and posttranslational modification of histones (Wysocka et al., 2006). This bridging of chromatin remodeling and chromatin-modifying activities is common in chromatin-remodeling complexes. In comparison to the H3K4me3 deposition by the NURF complex, the NURD complex in the CHD subfamily contains multiple histone deacetylases (HDAC1, HDAC2, HDAC3, HDAC8) that enable the complex to deacetylate histone tails and compact nucleosome arrays (Denslow & Wade, 2007).

The counteracting activities of the NURF and NURD complexes are representative of differing functions between remodelers as well as the importance of non-ATPase subunits in directing remodeler function. The canonical compositions and activities of each remodeler subfamily have been reviewed extensively and biochemically derived maps of complex members have been defined (Bartholomew, 2014; Conaway & Conaway, 2009; Gerhold, Hauer, & Gasser, 2015; Kadoch & Crabtree, 2015; Koster, Snel, & Timmers, 2015). Yet we still know surprisingly little about the biological function of many proteins that copurify with the complexes. Because each remodeler contains numerous accessory subunits that have undefined biological roles and it is likely that each confers a functional advantage to the complex(es) they associate with, we project that current knowledge on remodeler compositions and interactions underestimates the diversity of activities within remodeler–class epigenetic regulation *in vivo*.

New evidence indicates that remodelers colocalize to a large proportion of the mammalian genome and directly compete or cooperate (Morris et al., 2014). Studies on remodeler codependencies indicate that remodelers may antagonize one another for promoter occupancy and locus control in yeast and mammals (Parnell, Schlichter, Wilson, & Cairns, 2015), similar to the counteracting remodeling exhibited by the NURF and NURD complexes. Additionally, extensive studies have revealed unexpected stochasticity in noncanonical assemblies of remodeler components in developmental and pathological contexts (de la Serna, Carlson, & Imbalzano, 2001; Lessard et al., 2007; Lickert et al., 2004; Sawa, Kouike, & Okano, 2000; Staahl et al., 2013). To gain a comprehensive understanding of remodeler–class regulatory activities, it is necessary to examine current data that implicate remodelers in an interdependent and dynamic network genome-wide. As we learn more about the abundance and significance of interactions between remodelers, the consequences for gene regulation, cell differentiation, and tissue development are immense. Such functional crosstalk between remodelers is the focus of this essay.

2. REMODELER LOCALIZATION GENOME-WIDE

Despite longstanding evidence from genetic experiments that remodelers perform-specific functions throughout the genome (Bultman et al., 2000; Hargreaves & Crabtree, 2011), there are new indications that remodelers engage in dynamic functional interactions. This unexpected result has shaped recent outlooks on chromatin biology in development by challenging biochemically derived molecular structures and predictions on remodeler functionality. These insights reveal genomic crosstalk occurs on an unprecedented scale.

Extensive work has identified that the SWI/SNF complex localizes to active genomic elements (Euskirchen et al., 2011; Ho, Jothi, et al., 2009). In one particular study, multiple subunits of the SWI/SNF complex were mapped genome-wide (Euskirchen et al., 2011). In addition, all interactions associated with the SWI/SNF complex were analyzed by mass spectrometry. Together, these data showed the full genome-wide and proteome-wide interactions carried out by the canonical SWI/SNF complex. While this work provides a valuable baseline for the genomic occupancy and the interacting partners of the SWI/SNF complex, variable subunits within the complex have not been mapped. Future work analyzing the various biochemically distinct forms of SWI/SNF and other chromatin-remodeling complexes is needed to understand the functional and phenotypic diversity of these remodelers.

The localization of SWI/SNF components to regulatory positions throughout the genome aligned with previous findings that targeted mutations in remodelers cause dramatic phenotypes. Murine embryos lacking the Brg1 subunit of SWI/SNF fail to complete the perimplantation stage of development, for example (Bultman et al., 2000). However, investigations of remodelers from different families have confounded the assumption that mutagenic phenotypes arise in part by unique genomic localization. ChIP paired with high-throughouput sequencing (ChIP-seq) for the catalytic subunits of SWI/SNF, ISWI, and CHD subfamilies revealed that these distantly related remodelers colocalize across the genome with a high frequency (Morris et al., 2014). Intriguingly, disruption of the complexes led to unique chromatin organization phenotypes according to nucleosome-mapping techniques. Therefore despite colocalization, remodelers perform nonredundant roles in regulating genomic packaging. The regulation of this activity does not appear to be at the level of genomic occupancy, leaving a major unanswered question of how remodelers that bind similar regions and presumably modulate overlapping genes sets exhibit different phenotypes. Resolution of DNA binding by remodelers may obscure interpretations given the inability to accurately distinguish subtle variations using current technologies. One exciting alternative is that remodelers engage with local epigenetic cues after binding chromatin, resulting in dynamic regulation of remodelers despite similar targeting preferences.

One interesting scenario to evaluate crosstalk between remodelers would be to test the regulatory activities of closely related complexes. Examining the genomic localization preferences and gene expression control by complexes that are biochemically similar would clarify the biological functionality of biochemically discrete remodelers. For example, the ARID1A and ARID2 subunits of SWI/SNF do not simultaneously appear in biochemical extractions of the SWI/SNF complex *in vitro* (Nie et al., 2000). Each ARID (AT-rich interacting domain) associates with discrete configurations of SWI/SNF, known as BAF (ARID1A) and PBAF (ARID2), and is required for promoter targeting by the intact complex (Chandler et al., 2013). It is likely that ARID1A and ARID2 target discrete loci to facilitate unique cellular programs for the two SWI/SNF configurations. However, given the co-occupancy of genomic sites by remodelers from different families, it is possible that ARID1A- and ARID2-containing complexes also co-occupy binding sites. ARID1A and ARID2 may target the same loci at different times, utilizing SWI/SNF subunits to target similar features but unique ARID subunits to achieve different regulatory results. We hypothesize that given their overlapping, yet distinct subunit compositions, BAF and PBAF exhibit both competition and cooperation with one another in their function as regulators of gene expression (Fig. 2).

Currently, genomic colocalization by remodelers is best attributed to transient interactions at sites of nucleosome-depleted chromatin (Euskirchen et al., 2011; Morris et al., 2014; Nagaich, Walker, Wolford, & Hager, 2004). We propose that colocalization may serve a more specific purpose.

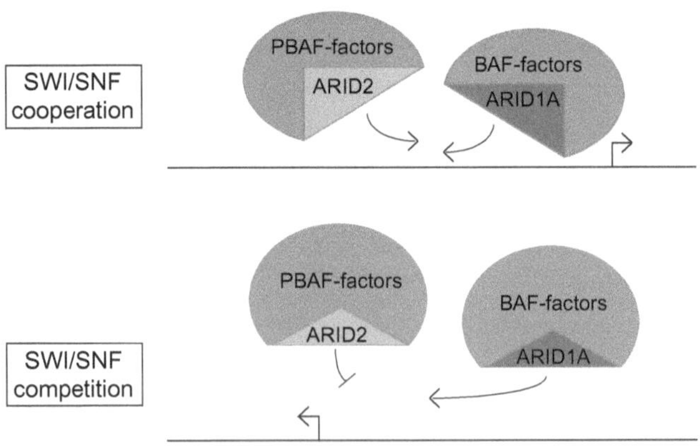

Figure 2 SWI/SNF interactions genome-wide. Putative functional interactions between SWI/SNF subcomplexes. Illustrations demonstrate competitive or cooperative genomic binding and gene regulation by SWI/SNF complexes that share regulatory components. (A color version of this figure is located at the end of this chapter.)

Subunit sharing between independent remodeling complexes could contribute to genomic targeting preferences by remodelers, where chromatin targeting by remodelers is dependent on direct cooperativity between counteracting or coacting remodelers. Moreover, we predict that observed colocalization reflects that remodelers conduct separate regulatory functions throughout the genome and therefore allows cells to personalize their expression profiles based on local stimuli and epigenetic cues.

3. ACCESSORY SUBUNITS IN REMODELER ACTIVITIES

Given the mutual exclusivity of some remodeler factors and the similar localization patterns of evolutionarily divergent remodelers, the most elusive feature of remodelers to-date pertains to their functional composition *in vivo*. From glycerol gradient purifications and mass spectrometry, each complex has been stringently purified to identify stable associates. SWI/SNF subfamily members are comprised of 8–15 subunits, INO80 subfamily members have up to 18 subunits, while ISWI and CHD have fewer ranging between 3–4 and 1–11 subunits, respectively (Clapier & Cairns, 2009). Identifying if and when functional variations of these canonical complex compositions exist *in vivo* has been more challenging.

In addition to compositional heterogeneity, some remodelers like the INO80 and SWR complexes exhibit structural modularity (Watanabe et al., 2015). Each complex exists in two states, one elongated form and a compressed form. Presumably, these different forms represent captured states of the complexes mid-remodeling, in accordance with the "hinge" hypothesis where the remodeler stretches to target the nucleosome then bends to embrace and mobilize it (Tosi et al., 2013; Watanabe et al., 2015). The extent of structural modularity within remodeler families is not known, despite several crystal structures that indicate nucleosome targeting is a predominant feature of remodeling complexes (Dürr et al., 2006; Grüne et al., 2003; Hauk, McKnight, Nodelman, & Bowman, 2010; Kasten, Clapier, & Cairns, 2011). Likewise, there is currently limited knowledge on the functionality of biochemically defined complexes. High-resolution genomic mapping of SWR and INO80 complexes in yeast identified nucleosomal targeting preferences of 20 subunits from the two complexes in total (Yen, Vinayachandran, & Pugh, 2013). But each complex has been typically studied from the perspective of only one or two subunits in functional settings, leaving many of the accessory subunits with limited characterization. The gap in our understanding of the functional assemblage of complexes, particularly in higher eukaryotes or single cells, has left the roles of the accessory subunits largely undefined.

Several studies have recently provided intriguing hypotheses on the utility of lesser-known subunits. One well-supported hypothesis is that unique combinations of SWI/SNF subunits create distinct complexes, each with a different functional role. Namely, the degree of heterogeneity within SWI/SNF suggests accessory subunits perform essential and diverse functions. For example, SWI/SNF factor BAF53A (encoded by *ACTL6A* in humans) is replaced by BAF53B (encoded by *ACTL6B*) upon differentiation of neuronal progenitors to neurons in mammals (Lessard et al., 2007). The subunit "switch" is essential for cells to complete cell lineage specification. Genetic mutations in BAF53B lead mice to acquire defects in synaptic plasticity and long-term memory suggesting BAF53B is systematically essential for the activities of mature neurons (Vogel-Ciernia et al., 2013). Similar phenotypes have been observed in hematopoietic and embryonic tissue types where constitutive BAF53A expression is required for repopulating cells but insufficient for differentiation (Krasteva et al., 2012; Lu et al., 2015).

Identifying the activities of accessory subunits could be key to unlocking remodeler specification *in vivo*. In several developmental systems, evidence suggests specific SWI/SNF assemblies drive cell lineage choices in development (Ho, Jothi, et al., 2009; Ho, Ronan, et al., 2009; Lickert et al., 2004; Sawa et al., 2000; Staahl et al., 2013) or cell lineage abandonment in malignancy (Kadoch & Crabtree, 2013). This evidence implicates accessory subunits as central players in remodeling activities. In development, subunit-exchanges by discrete remodelers are dictated by expression. The mechanism is likely more complex in cancer where remodeler components are ubiquitously expressed. Mutations in a single complex member likely allow chromatin-remodeling activity by other forms of the complex in ways that are important for tumor formation and cancer proliferation. To understand physiology and pathophysiology, the roles of subunits that exist in multiple remodeling complexes and families should be clarified. We advocate for common subunits to be extensively characterized in order to clarify their role in remodeler–class crosstalk.

Intriguingly, BAF53A stably associates with many SNF2 remodelers. SWI/SNF, INO80, TIP60, and SRCAP complexes all contain BAF53A as a core subunit (Kadoch & Crabtree, 2015). Therefore assigning developmental phenotypes to any single BAF53A-containing remodeler is problematic because genetic manipulation of BAF53A would affect numerous activities of the other remodelers. Notably, BAF53A has been extensively studied in the context of the INO80 complex. *In vitro*, BAF53A has been shown to be required for efficient DNA-dependent ATPase activity, histone targeting, and nucleosome mobility of the INO80 complex (Chen et al., 2011; Chen, Conaway, & Conaway, 2013). In addition, BAF53A forms a distinct structural module within the INO80 complex composed of β-actin and the actin-related protein ACTR8. This module has a high affinity for the helicase-SANT-associated (HSA) domain within the INO80 ATPase indicative of a stable and evolutionarily conserved interaction (Szerlong et al., 2008). SWI/SNF ATPase BRG1 also contains an HSA domain with BAF53A preference. Loss of either BAF53A alone or the BAF53A-β-actin-ACTR8 module is not required for the remodeling activities of the INO80 complex, but greatly enhances them (Chen et al., 2011; Szerlong et al., 2008). Therefore, it is likely that BAF53A acts in concert with the additional remodeler subunits to maximize the biological efficiency for nucleosome eviction and placement, perhaps by increasing the targeting efficiency of the complex to specific epigenetic features where remodeling is necessary. In such a scenario, the subunit "switch" to BAF53B may facilitate a different targeting preference for SWI/SNF thereby enabling

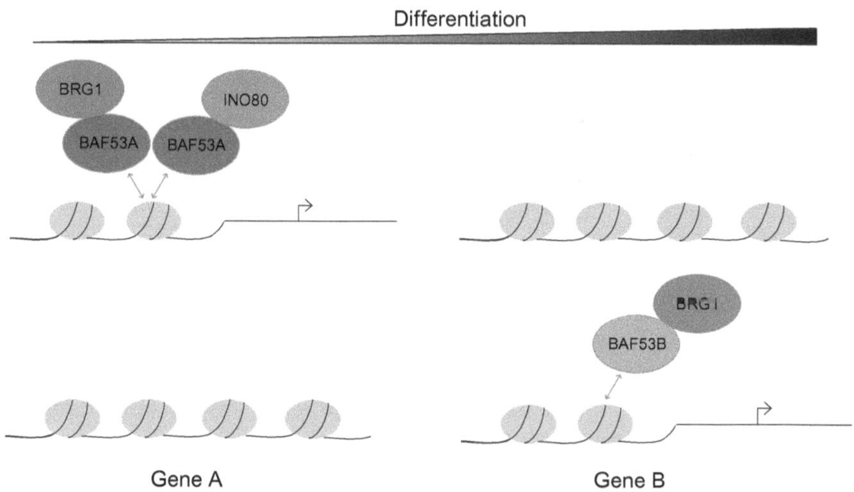

Figure 3 Role of accessory subunits in remodeler localization. Predicted remodeler localization preferences based on subunit composition. We hypothesize that BAF53A-containing complexes utilize the histone-binding subunit to target similar epigenetic features. Similarly, we hypothesize that subunit switching from BAF53A to BAF53B in SWI/SNF in cell lineage specification serves to relocalize SWI/SNF to sites that require expression at specific cell lineage timepoints. (A color version of this figure is located at the end of this chapter.)

expression of differentiation-specific genes (Fig. 3). Little is known regarding the protein domains within BAF53A but clarification of its ability to be modified structurally or functionally may provide insight into its incorporation into specific remodeling complexes and its preference for histone epitopes. While the exact functions for BAF53A in the other SNF2 remodelers have not been clarified, it is probable that BAF53A serves a unifying feature among BAF53A-containing SNF2 remodelers that makes them functionally cooperative and competitive.

Other subunits are shared between remodelers. The helicases RUVBL1 and RUVBL2 (RUVBLs) exist in INO80, TIP60, and SRCAP yet their remodeling-related functions are largely unknown, due in part to the fact that the RUVBLs promiscuously participate in nonremodeling activities through the Fanconi Anemia complex (Rajendra, Garaycoechea, Patel, & Passmore, 2014) and in snoRNP assembly (Bizarro et al., 2014). Assembly of INO80 and TIP60 has been shown to be dependent on the RUVBLs (Jha, Gupta, Dar, & Dutta, 2013; Jónsson et al., 2004). In addition, RUVBLs enhance the DNA-binding ability of YY1, a transcription factor that canonically associates with INO80 (López-Perrote et al., 2014). This evidence suggests that the RUVBL proteins are required members of remodeler assemblies that exist to enhance binding efficiency and specification at epigenetic docking sites, similar to BAF53A. As with BAF53A, understanding the recruitment and action of accessory subunits like the RUVBLs within remodeling assemblies will be essential to clarifying their precise function in the context of chromatin remodeling and gene regulation.

Recent data on remodeler interactions, cooperation, and functional heterogeneity of remodeling complexes suggest there is likely a diversity of remodeler compositions *in vivo*. Emerging genome-wide studies of remodeler assemblies will improve our understanding of the range of functional compositions that remodelers form. We anticipate that intrafamily and intercomplex variation exists to provide cells maximal plasticity to regulate gene expression rapidly and effectively. Biochemical purifications of remodelers inform our understanding of remodeler compositions and are informative at the cell population level. However, at an individual locus and a single point of time, the combinatorial assembly of proteins from a diverse set of subunits may yield unique function on a locus-specific scale. Highlighting the different functions of unique compositions of chromatin-remodeling complexes will be key to understanding the many roles of these complexes in gene regulation.

4. CONCLUSION

Faithful regulation of chromatin allows cells to maintain control over nucleosome position and gene expression. Because remodelers intimately interact with the genome to participate in chromatin regulation, their study is essential to understanding the dynamic epigenetic landscape as it pertains to genomic organization, integrity, and gene expression. As we currently understand only a fraction of the complexity that exists to modulate the epigenomic landscape *in vivo*, the study of remodelers, their interactions, and their components will provide important steps to clarify emergent questions of genomic regulation. The critical role of these complexes in development and disease highlights the necessity of understanding crosstalk within, and between, chromatin-remodeling families.

REFERENCES

Bartholomew, B. (2014). ISWI chromatin remodeling: One primary actor or a coordinated effort? *Current Opinion in Structural Biology*, *24*, 150–155.

Bizarro, J., Charron, C., Boulon, S., Westman, B., Pradet-Balade, B., Vandermoere, F., et al. (2014). Proteomic and 3D structure analyses highlight the C/D box snoRNP assembly mechanism and its control. *The Journal of Cell Biology*, *207*, 463–480.

Bultman, S., Gebuhr, T., Yee, D., La, M. C., Nicholson, J., Gilliam, A., et al. (2000). A Brg1 null mutation in the mouse reveals functional differences among mammalian SWI/SNF complexes. *Molecular Cell*, *6*, 1287–1295.

Chandler, R., Brennan, J., Schisler, J., Serber, D., Patterson, C., & Magnuson, T. (2013). ARID1a-DNA interactions are required for promoter occupancy by SWI/SNF. *Molecular and Cellular Biology*, *33*, 265–280.

Chen, L., Cai, Y., Jin, J., Florens, L., Swanson, S., Washburn, M., et al. (2011). Subunit organization of the human INO80 chromatin remodeling complex: An evolutionarily conserved core complex catalyzes ATP-dependent nucleosome remodeling. *The Journal of Biological Chemistry*, *286*, 11283–11289.

Chen, L., Conaway, R., & Conaway, J. (2013). Multiple modes of regulation of the human Ino80 SNF2 ATPase by subunits of the INO80 chromatin-remodeling complex. *Proceedings of the National Academy of Sciences of the United States of America*, *110*, 20497–20502.

Clapier, C., & Cairns, B. (2009). The biology of chromatin remodeling complexes. *Annual Review of Biochemistry*, *78*, 273–304.

Conaway, R., & Conaway, J. (2009). The INO80 chromatin remodeling complex in transcription, replication and repair. *Trends in Biochemical Sciences*, *34*, 71–77.

de la Serna, I. L., Carlson, K., & Imbalzano, A. (2001). Mammalian SWI/SNF complexes promote MyoD-mediated muscle differentiation. *Nature Genetics*, *27*, 187–190.

Denslow, S., & Wade, P. (2007). The human Mi-2/NuRD complex and gene regulation. *Oncogene*, *26*, 5433–5438.

Dürr, H., Flaus, A., Owen-Hughes, T., & Hopfner, K. (2006). Snf2 family ATPases and DExx box helicases: Differences and unifying concepts from high-resolution crystal structures. *Nucleic Acids Research*, *34*, 4160–4167.

Dutta, A., Gogol, M., Kim, J., Smolle, M., Venkatesh, S., Gilmore, J., et al. (2014). Swi/Snf dynamics on stress-responsive genes is governed by competitive bromodomain interactions. *Genes & Development*, *28*, 2314–2330.

Euskirchen, G., Auerbach, R., Davidov, E., Gianoulis, T., Zhong, G., Rozowsky, J., et al. (2011). Diverse roles and interactions of the SWI/SNF chromatin remodeling complex revealed using global approaches. *PLoS Genetics*, *7*, e1002008.

Gerhold, C., Hauer, M., & Gasser, S. (2015). INO80-C and SWR-C: Guardians of the genome. *Journal of Molecular Biology*, *427*, 637–651.

Grüne, T., Brzeski, J., Eberharter, A., Clapier, C., Corona, D., Becker, P., et al. (2003). Crystal structure and functional analysis of a nucleosome recognition module of the remodeling factor ISWI. *Molecular Cell*, *12*, 449–460.

Guenther, M., Levine, S., Boyer, L., Jaenisch, R., & Young, R. (2007). A chromatin landmark and transcription initiation at most promoters in human cells. *Cell*, *130*, 77–88.

Hargreaves, D., & Crabtree, G. (2011). ATP-dependent chromatin remodeling: Genetics, genomics and mechanisms. *Cell Research*, *21*, 396–420.

Hauk, G., McKnight, J., Nodelman, I., & Bowman, G. (2010). The chromodomains of the Chd1 chromatin remodeler regulate DNA access to the ATPase motor. *Molecular Cell*, *39*, 711–723.

Ho, L., Jothi, R., Ronan, J., Cui, K., Zhao, K., & Crabtree, G. (2009). An embryonic stem cell chromatin remodeling complex, esBAF, is an essential component of the core pluripotency transcriptional network. *Proceedings of the National Academy of Sciences of the United States of America*, *106*, 5187–5191.

Ho, L., Ronan, J., Wu, J., Staahl, B., Chen, L., Kuo, A., et al. (2009). An embryonic stem cell chromatin remodeling complex, esBAF, is essential for embryonic stem cell self-renewal and pluripotency. *Proceedings of the National Academy of Sciences of the United States of America*, *106*, 5181–5186.

Holstege, F., Jennings, E., Wyrick, J., Lee, T., Hengartner, C., Green, M., et al. (1998). Dissecting the regulatory circuitry of a eukaryotic genome. *Cell*, *95*, 717–728.

Jha, S., Gupta, A., Dar, A., & Dutta, A. (2013). RVBs are required for assembling a functional TIP60 complex. *Molecular and Cell Biology*, *33*, 1164–1174.

Jónsson, Z., Jha, S., Wohlschlegel, J., & Dutta, A. (2004). Rvb1p/Rvb2p recruit Arp5p and assemble a functional Ino80 chromatin remodeling complex. *Molecular Cell*, *16*, 465–477.

Kadoch, C., & Crabtree, G. (2013). Reversible disruption of mSWI/SNF (BAF) complexes by the SS18-SSX oncogenic fusion in synovial sarcoma. *Cell*, *153*, 71–85.

Kadoch, C., & Crabtree, G. R. (2015). Mammalian SWI/SNF chromatin remodeling complexes and cancer: Mechanistic insights gained from human genomics. *Science Advances*, *1*(5), e1500447. http://dx.doi.org/10.1126/sciadv.1500447. http://dx.doi.org/10.1126/sciadv.1500447.

Kasten, M., Clapier, C., & Cairns, B. (2011). SnapShot: Chromatin remodeling: SWI/SNF. *Cell*, *144*, 310.e1.

Koster, M., Snel, B., & Timmers, H. (2015). Genesis of chromatin and transcription dynamics in the origin of species. *Cell*, *161*, 724–736.

Krasteva, V., Buscarlet, M., Diaz-Tellez, A., Bernard, M., Crabtree, G., & Lessard, J. (2012). The BAF53a subunit of SWI/SNF-like BAF complexes is essential for hemopoietic stem cell function. *Blood*, *120*, 4720–4732.

Kundaje, A., Meuleman, W., Ernst, J., Bilenky, M., Yen, A., Heravi-Moussavi, A., et al. (2015). Integrative analysis of 111 reference human epigenomes. *Nature*, *518*, 317–330.

Lessard, J., Wu, J., Ranish, J., Wan, M., Winslow, M., Staahl, B., et al. (2007). An essential switch in subunit composition of a chromatin remodeling complex during neural development. *Neuron*, *55*, 201–215.

Lickert, H., Takeuchi, J., Von, B. I., Walls, J., McAuliffe, F., Adamson, S., et al. (2004). Baf60c is essential for function of BAF chromatin remodelling complexes in heart development. *Nature*, *432*, 107–112.

López-Perrote, A., Alatwi, H., Torreira, E., Ismail, A., Ayora, S., Downs, J., et al. (2014). Structure of Yin Yang 1 oligomers that cooperate with RuvBL1-RuvBL2 ATPases. *The Journal of Biological Chemistry*, *289*, 22614–22629.

Lu, W., Fang, L., Ouyang, B., Zhang, X., Zhan, S., Feng, X., et al. (2015). Actl6a protects embryonic stem cells from differentiating into primitive endoderm. *Stem Cells*, *33*, 1782–1793.

Morris, S., Baek, S., Sung, M., John, S., Wiench, M., Johnson, T., et al. (2014). Overlapping chromatin-remodeling systems collaborate genome wide at dynamic chromatin transitions. *Nature Structural & Molecular Biology*, *21*, 73–81.

Nagaich, A., Walker, D., Wolford, R., & Hager, G. (2004). Rapid periodic binding and displacement of the glucocorticoid receptor during chromatin remodeling. *Molecular Cell*, *14*, 163–174.

Nie, Z., Xue, Y., Yang, D., Zhou, S., Deroo, B., Archer, T., et al. (2000). A specificity and targeting subunit of a human SWI/SNF family-related chromatin-remodeling complex. *Molecular and Cellular Biology*, *20*, 8879–8888.

Parnell, T., Schlichter, A., Wilson, B., & Cairns, B. (2015). The chromatin remodelers RSC and ISW1 display functional and chromatin-based promoter antagonism. *elife*, *4*, e06073.

Rajendra, E., Garaycoechea, J., Patel, K., & Passmore, L. (2014). Abundance of the Fanconi anaemia core complex is regulated by the RuvBL1 and RuvBL2 AAA+ ATPases. *Nucleic Acids Research*, *42*, 13736–13748.

Sawa, H., Kouike, H., & Okano, H. (2000). Components of the SWI/SNF complex are required for asymmetric cell division in C. elegans. *Molecular Cell*, *6*, 617–624.

Staahl, B., Tang, J., Wu, W., Sun, A., Gitler, A., Yoo, A., et al. (2013). Kinetic analysis of npBAF to nBAF switching reveals exchange of SS18 with CREST and integration with neural developmental pathways. *The Journal of Neuroscience*, *33*, 10348–10361.

Strahl, B., & Allis, C. (2000). The language of covalent histone modifications. *Nature*, *403*, 41–45.

Szerlong, H., Hinata, K., Viswanathan, R., Erdjument-Bromage, H., Tempst, P., & Cairns, B. (2008). The HSA domain binds nuclear actin-related proteins to regulate chromatin-remodeling ATPases. *Nature Structural & Molecular Biology*, *15*, 469–476.

Tosi, A., Haas, C., Herzog, F., Gilmozzi, A., Berninghausen, O., Ungewickell, C., et al. (2013). Structure and subunit topology of the INO80 chromatin remodeler and its nucleosome complex. *Cell*, *154*, 1207–1219.

Vogel-Ciernia, A., Matheos, D., Barrett, R., Kramár, E., Azzawi, S., Chen, Y., et al. (2013). The neuron-specific chromatin regulatory subunit BAF53b is necessary for synaptic plasticity and memory. *Nature Neuroscience*, *16*, 552–561.

Watanabe, S., Tan, D., Lakshminarasimhan, M., Washburn, M., Hong, E., Walz, T., et al. (2015). Structural analyses of the chromatin remodelling enzymes INO80-C and SWR-C. *Nature Communications*, *6*, 7108.

Wysocka, J., Swigut, T., Xiao, H., Milne, T., Kwon, S., Landry, J., et al. (2006). A PHD finger of NURF couples histone H3 lysine 4 trimethylation with chromatin remodelling. *Nature*, *442*, 86–90.

Yen, K., Vinayachandran, V., & Pugh, B. (2013). SWR-C and INO80 chromatin remodelers recognize nucleosome-free regions near +1 nucleosomes. *Cell*, *154*, 1246–1256.

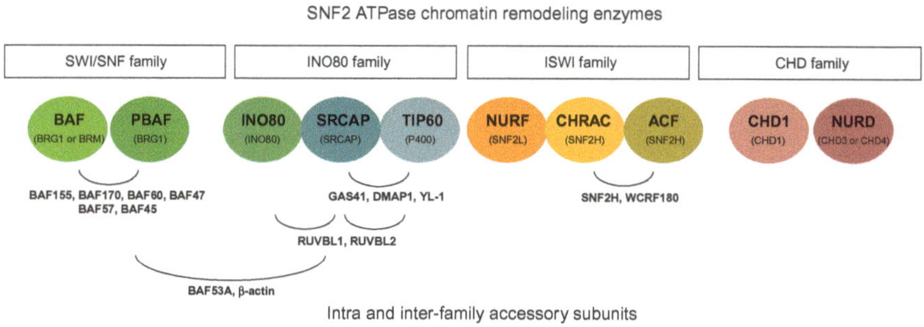

John S. Runge *et al.*, Figure 1 SNF2 ATPase chromatin-remodeling enzymes. Representative diagram of ATP-dependent chromatin-remodeling families. Families are designated by white boxes. Specific remodelers are represented as circles. Catalytic ATPases for each remodeling complex are indicated in parentheses. Subunits that incorporate into multiple complexes are identified by curved lines. Notably, the ISWI family includes additional complexes not depicted here (NoRC, RSF, WICH).

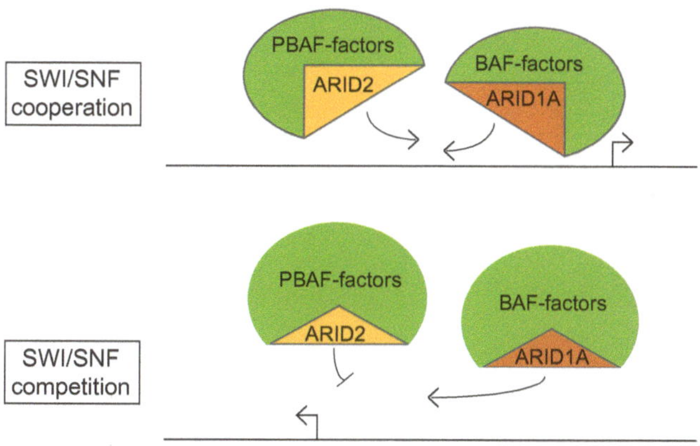

John S. Runge *et al.*, Figure 2 SWI/SNF interactions genome-wide. Putative functional interactions between SWI/SNF subcomplexes. Illustrations demonstrate competitive or cooperative genomic binding and gene regulation by SWI/SNF complexes that share regulatory components.

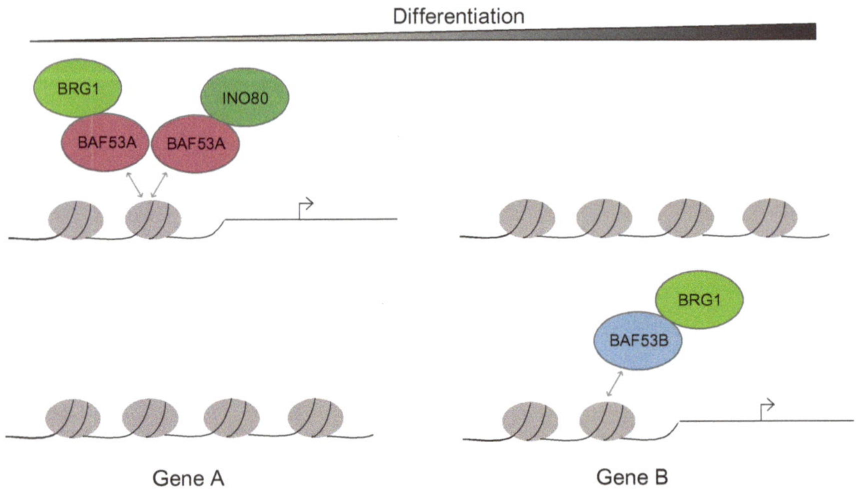

John S. Runge *et al.*, Figure 3 Role of accessory subunits in remodeler localization. Predicted remodeler localization preferences based on subunit composition. We hypothesize that BAF53A-containing complexes utilize the histone-binding subunit to target similar epigenetic features. Similarly, we hypothesize that subunit switching from BAF53A to BAF53B in SWI/SNF in cell lineage specification serves to relocalize SWI/SNF to sites that require expression at specific cell lineage timepoints.

CHAPTER TWO

Sea Urchin Morphogenesis

David R. McClay[1]
Department of Biology, Duke University, Durham, North Carolina, USA
[1]Corresponding author: e-mail address: dmcclay@duke.edu

Contents

Abstract

In the sea urchin morphogenesis follows extensive molecular specification. The specification controls the many morphogenetic events and these, in turn, precede patterning steps that establish the larval body plan. To understand how the embryo is built it was necessary to understand those series of molecular steps. Here an example of the historical sequence of those discoveries is presented as it unfolded over the last 50 years, the years during which major progress in understanding development of many animals and plants was documented by CTDB. In sea urchin development a rich series of experimental studies first established many of the phenomenological components of skeletal morphogenesis and patterning without knowledge of the molecular components. The many discoveries of transcription factors, signals, and structural proteins that contribute to the shape of the endoskeleton of the sea urchin larva then followed as molecular tools became available. A number of transcription factors and signals were discovered that were necessary for specification, morphogenesis, and patterning. Perturbation of the transcription factors and signals provided the means for assembling models of the gene regulatory networks used for specification and controlled the subsequent morphogenetic events. The earlier experimental information informed perturbation experiments that asked how patterning worked. As a consequence it was learned that ectoderm provides a series of patterning signals to the skeletogenic cells and as a consequence the skeletogenic cells secrete a highly patterned skeleton based on their ability to genotypically decode the localized reception of several signals. We still do not understand the complexity of the signals received by the skeletogenic cells, nor do we understand in detail how the genotypic information shapes the secreted skeletal biomineral, but the current knowledge at least outlines the sequence of events and provides a useful template for future discoveries.

Current Topics in Developmental Biology, Volume 117
ISSN 0070-2153
http://dx.doi.org/10.1016/bs.ctdb.2015.11.003

ABBREVIATIONS

BMP bone morphogenetic protein
EMT epithelial–mesenchymal transition
FGF fibroblast growth factor
SC skeletogenic cells
VEGF vascular endothelial growth factor

The reviews published in the CTDB volumes record 50 years of remarkable advances in our understanding of how embryos work. The tools available, the understanding of molecular information and pathways, the evolutionary relationships between different animal forms, and our understanding of morphogenesis have gone from relatively primitive sketches to, in many cases, a detailed in-depth understanding, and many unsuspected mechanisms were revealed along the way. At the same time, new questions that were not even dreamed of 50 years ago were addressed, while others remain as puzzles. My goal in this essay is to provide a perspective, especially for young readers, on morphogenesis of the sea urchin embryo, the model organism that has been the focus of attention of my laboratory for almost that long.

The sea urchin was designated as a deuterostome by early systematists, because during gastrulation, the mouth is the second opening to form in the embryo. That may seem like an odd trait to hang together an entire phylogenetic branch of the animal kingdom, but to the credit of those systematists, DNA sequence analysis upheld that phylogenetic relationship and as a consequence the sea urchin embryo model continues to be used both for a glimpse at possible evolutionary underpinnings of vertebrate development and for the relative simplicity of morphogenesis that characterizes sea urchin gastrulation. That simplicity also facilitates approaches for the experimentalist to access molecular processes guiding morphogenetic movements. The early analyses of gastrulation in this indirectly developing embryo took advantage of its transparency and also took advantage of advances in microscopy starting in the middle of the twentieth century. Descriptions of the epithelial–mesenchymal transition (EMT) that characterized ingression of the skeletogenic mesenchyme cells, and descriptions of cell shape changes and motility that characterized invagination of the archenteron provided a backdrop for our current approaches aimed at understanding how these processes work at a molecular level.

To get at those mechanisms, however, the field desperately needed tools and molecules. I remember, for example, the summer of 1969 when students returning from Woods Hole broadly dispersed the new technology of sodium dodecyl sulfate protein gel electrophoresis to research campuses around the country (Laemmli, 1970). Suddenly, researchers went from being able to resolve almost nothing without first purifying abundant proteins, to an ability of resolving hundreds of proteins easily, thereby providing experimental access to low-abundance proteins. Shortly thereafter, Milstein and colleagues introduced a technology for producing monoclonal antibodies and that enabled developmental biologists to identify and follow valuable markers of cells and tissues (Milstein & Lennox, 1980). These occurred just prior to the revolutionary advances in molecular biology over two decades that made it increasingly easy to identify and produce reagents that could specifically perturb individual molecules in a system. With those advances, it became possible to explore morphogenesis mechanistically. Many other technologies were introduced and with each new advance our ability to explore mechanism gained new dimensions. Molecular genetics, for example, provided a growing stockpile of genes that are universally involved in aspects of morphogenesis and patterning.

Gustafson and Wolpert (1961) and Wolpert (1971) described the behavior of cells as the sea urchin embryo underwent gastrulation. It provided a good description of ingression (an "EMT") of skeletogenic cells (SCs) that was further detailed in a later paper by Katow and Solursh (1981). They showed that the SCs, like all mesenchyme cells that originate as epithelial cells, progressively become motile, change shape, change polarity, move through a basement membrane, and deadhere from the epithelium in order to enter the interstitial spaces of the embryo. At the time, those papers appeared there were no molecular markers nor were there reagents to characterize the causal changes driving these EMT processes, but it was possible to isolate the cells that later underwent the EMT. That enabled us to show that during the time of EMT, the SCs dramatically alter their adhesive properties suggesting a large change in molecular components of the membranes of the cells (Fink & McClay, 1985; McClay, 1979). Once it became possible to identify adhesion molecules, it was realized that cadherins, integrins, and other adhesion molecules are highly conserved (Takeichi, 1988), and moreover, at the time of EMT, the membrane massively remodels expression of these molecules enabling the cells to release themselves from the epithelium and to change behavior and now move along an extracellular matrix as they position themselves for their later role in biomineralization of the embryonic skeleton (Miller & McClay, 1997).

The molecular identification of adhesion molecules opened the door to many other questions. It was obvious that the EMT was a well-coordinated morphogenetic movement that required an enormous number of molecular changes, but it remained to be shown how that event was controlled, and it also remained to be understood how conserved the mechanism might be. By the early 1990s, it was clear that the proteins conferring function to cells had to be coordinately produced, and by then, the central dogma was well in the grasp of molecular biologists. However, it still was not clear how complex morphogenetic movements might be controlled. Sure, all the components were encoded in the DNA, and regulatory mechanisms of RNA production and protein synthesis were subjects of intense scrutiny, but it was not clear how all of that could be assembled into an apparatus to control the many component functions that constituted the EMT. That concern sent my laboratory in a very different direction. We retooled and began a two-decade-long adventure into systems biology in order to build gene regulatory networks. We appreciated the fact that a complex cellular event such as an EMT required the coordinate control of many transcription factors. Consequently, we collaborated with Davidson at Caltech and with others in the sea urchin community to explore how transcription factors operate (in our case) to set the stage for morphogenesis. A large number of transcription factors were identified, and those that were exclusive to the SCs were eventually identified. Each was perturbed and the effect of that perturbation on the others in the SCs was noted (Davidson, 2006; Davidson, McClay, & Hood, 2003; Davidson et al., 2002). From that analysis by 2008, it was possible to build a fairly comprehensive gene regulatory network model of SC specification, and shortly thereafter, models for specification of most other early embryonic cell lineages were deduced (Li, Materna, & Davidson, 2013; Oliveri, Tu, & Davidson, 2008; Peter & Davidson, 2011).

With a gene regulatory network in hand, one could ask which transcription factors were responsible for controlling the EMT. Were all SC transcription factors involved or were there subcircuits? To address this question, we designed rather simple quantitative assays to examine onset of motility, cell shape change, cell polarity change, degradation of the basement membrane for invasion into the interstitial space (the blastocoel), and deadhesion. A number of transcription factors were perturbed in the assays, and we learned that there are a number of subcircuits involved in the EMT. No single transcription factor controlled all EMT processes though several were involved in multiple subcircuits (Saunders & McClay, 2014). Identification of the controls of each EMT process enabled us to examine those processes in greater detail. As an example, we found three transcription factors involved in production of the hole in the basement membrane. This enabled us to screen for protease candidates that might be responsible for making that hole. We found three membrane-tethered matrix metalloproteases that are transcriptionally activated by that subcircuit. When the production of the proteases was perturbed, the invasive hole was compromised. Of importance, we began with the cell biology to identify the question, built a gene regulatory network circuit that appeared to control that process, and returned to the cell biology with an ability to penetrate deeper into causal mechanism.

Of course for the EMT process, as with detailed analyses of most morphogenetic processes, there is much that remains to be learned. Each of the other component processes of the EMT can now be explored in greater detail given at least an elementary understanding of the control circuits that direct the cell biology. For example, as mentioned above, the plasma membrane of the SC precursors contains cadherin. At the time of EMT, the cadherin is lost and the membrane exhibits nascent mesenchymal properties through insertion of integrins and neuroglial cell adhesion molecule. Somehow that massive membrane remodeling must take place and be controlled by the specification circuits. We know a little about transcription factors involved (*snail* and *twist* and others), but much remains to be learned since it appears that membrane trafficking on a massive scale removes the epithelial membrane and replaces it with membrane components suitable for a mesenchymal lifestyle. Additionally, as the nascent mesenchymal cells emerge into the blastocoel, they begin extending long thin filopodia that appear to be used as sensors for the patterning process these cells will soon experience (Miller, Fraser, & McClay, 1995).

Other morphogenetic processes in the sea urchin remain to be analyzed to the same depth as the EMT, and like that process, an even deeper understanding is within reach. Invagination of the archenteron, for example, has been well described (Ettensohn, 1984a, 1984b, 1985; Hardin, 1987, 1988; McClay, Morrill, & Hardin, 1990), the cell types that participate have been delineated, and a series of gene regulatory networks of the gut have been published and are constantly being updated (Annunziata et al., 2014; Peter & Davidson, 2011). As yet, however, the invagination process has not been analyzed in mechanistic detail at a molecular level. There are ideas of how the invagination works and biomechanical properties have been measured, but the causal molecular sequences remain to be addressed (Davidson, Oster, Keller, & Koehl, 1999). The tools for that analysis are available and are available for other morphogenetic movements such as coelomic pouch formation, and pigment cell EMT and a mesenchyme-to-epithelial transition as these cells populate the aboral ectoderm. Other movements such as homing of the primordial germ cells to the coelomic pouches, formation of the stomodaeum, and fusion of the foregut with the stomodaeum also are within range of mechanistic analysis. In each case, to better understand how a process works, it will be useful to have knowledge of the gene regulatory network underlying that process, of the component cell biological processes that contribute to it, and of the patterning inputs that localize that process within the embryo. These latter processes, together called "pattern formation," are processes that control placement and provide inputs governing the organization of three-dimensional structures. To illustrate how pattern formation fits into morphogenesis, the following section continues its focus on SCs to illustrate advances in understanding how pattern formation contributes to construction of the larval skeleton in the sea urchin.

1. SKELETAL PATTERNING

Pattern formation is a component of morphogenesis that also has enjoyed enormous mechanistic advance since CTDB was started. Since the metazoan is a three-dimensional organism built by organizing millions of cells into many shapes and complex structures, there is a deep curiosity among developmental biologists as to how those shapes and structures emerge, a process often referred to as "patterning." The same curiosity exists for plants, though much of the patterning in plants is very different than mechanisms used by animals. The problem is difficult for several reasons. First, the patterning mechanisms tend to be fleeting and sequential. Second, unicellular model organisms do not contribute and cells in culture contribute only to a very limited extent to understanding patterning mechanisms. Thus, much of what is learned must be addressed *in vivo*. Third, there are many barriers to study of patterning *in vivo* especially in mammals, where embryo patterning can be studied for only short periods of time outside the uterus. Nevertheless, many advances have been made. To illustrate some of the advances and the ways in which they occurred, this section will follow advances in understanding how the three-dimensional shape of the sea urchin larval skeleton emerges. Other excellent examples that could be followed include anterior–posterior patterning in many animals, limb patterning in flies and vertebrates, vulval organization, and neural patterning in *Caenorhabditis elegans*. For all of these, the sequence of discovery is similar over the past 50 years, so the choice of the sea urchin skeleton allows one to gain an understanding of how the patterning component of morphogenesis has reached today's level of understanding and will illustrate the holes that remain to be filled.

The pattern of the calcium carbon skeleton as seen in the sea urchin *Lytechinus variegatus* 24 h after fertilization is similar to an artist's easel and the name comes from an ancient Roman reading desk which also looks like an easel. Sixty-four SCs are responsible for production of that entire structure, and as seen above, SCs emerge into the blastocoel as a consequence of an EMT at a stage called the "mesenchyme blastula" stage about 9.5 hpf (hours postfertilization). The biology of the system is favorable during skeletogenesis since the 64 SCs have a known origin, a known gene regulatory network, and much about their biology has been discovered. To build the skeleton, another simplifying component is that during the 24 h skeletogenic period, the 64 SCs that initiate skeletogenesis do not divide until several days later when the larva begins to feed. The SCs are easily viewed inside the transparent embryo so their behavior following EMT is well documented. The SCs first form a ring around the floor of the blastocoel, then fuse to form a syncytium, and finally, a group of cells on either side collect to form lateral clusters of SCs. Within those clusters, skeletogenesis is initiated.

In patterning, prior to any molecular knowledge, there were a number of ideas about how an embryo might construct a three-dimensional shape. The issue became the nature and complexity of the information necessary for building that shape. Early ideas were based on experimental observations. Horstadius (1939), for example, examined phenotypes of sea urchin embryo recombinants, and based on his observations, he proposed a double gradient of morphogens that could provide information of enough complexity to explain his results. His idea suggested that patterning of the embryo was based on cells somehow interpreting morphogen concentrations at any position in the embryo. There were a number of problems with this idea but it started the conversation. Wolpert (1971), in a CTDB chapter, later used the French flag as an analogy to model positional information based on gradients. The idea was that somehow cells could detect their position along a gradient and as a consequence would then use that information to pattern the structure. That model also had conceptual problems, but provided at least a conceptual possibility. A succession of gradient ideas including Turing reaction–diffusion models (Turing, 1953) and others provided the field with many ideas about how patterning might occur, but without detailed molecular information, each of these ideas could be argued but remained largely untested in an actual morphogenetic situation.

Given the many ideas, how were patterning principles explored? In the case of the sea urchin embryo, a series of experiments conducted prior to knowing any molecular information established several properties of the system: it was learned that any one of the 64 SCs could reside at any location along the skeleton. That indicated that each SC was essentially equal to the other 63, and it meant that patterning information had to have been provided from somewhere else for each cell to execute its unique contribution to skeletal pattern (Ettensohn, 1990, 1999; Ettensohn & Sweet, 2000; Guss & Ettensohn, 1997; Malinda & Ettensohn, 1994). An experiment with Nickel, which blocks calcium channels, established that at least some patterning information comes from the ectoderm (Armstrong, Hardin, & McClay, 1993; Armstrong & McClay, 1994; Hardin & Armstrong, 1997; McClay, Armstrong, & Hardin, 1992). In embryo recombinants, if the SCs were treated with Nickel prior to the recombination, the skeletons were normal. If, on the other hand, only the ectoderm was treated with Nickel prior to the recombination, the skeleton was radialized, or lacked right–left and dorsal–ventral positional information. In other words, the ectoderm provided important patterning input to the SCs which executed skeletogenesis using those instructions. The pattern that was executed by SCs was genotypically appropriate to the SC DNA. If SCs of one species were inserted into the blastocoel of a second species, the detailed skeleton pattern was genotypically appropriate to the transplanted SCs even though they required host ectodermal information telling them where to produce the skeleton. That meant that information on position within the blastocoel was dictated by ectoderm (positional information), and once informed of that position, the SCs produced a readout of pattern appropriate to that position. Thus, patterning was both autonomous and nonautonomous, and it provided an added benefit by showing that positional information signals were highly conserved between species. Since position could be appropriately decoded by SCs, despite the fact that the ectoderm delivering those signals was separated by more than 50 million years since a common ancestor, it was clear that the signaling system was both complex and conserved.

2. ECTODERMAL PATTERNING INPUTS

If, as the evidence suggested, ectoderm provides patterning inputs, how does that component of patterning work? Since the SCs, based on the early results, are equivalent and must somehow interpret the incoming patterning information, transduce it, and produce the spatially correct piece of the skeleton in response, the positional information supplied to the SC must be complex enough for the SC to produce a specific skeletal piece appropriate to the site where it is located. To fully understand the sequence, it will be necessary to understand the ectodermal inputs, how they are transduced, and how the SCs then organize themselves to synthesize the three-dimensional template for the skeletal structure. Here, the essay will document progress on only one of those directions, how the ectoderm sets itself up to provide patterning input to the SCs.

Progress was strongly assisted not only by technical advances but also by work in other systems: a large number of signaling molecules and transcription factors were identified, beginning with the Wieschaus–Nusslein-Volhard screen in *Drosophila* (Nusslein-Volhard & Wieschaus, 1980; Wieschaus, Nusslein-Volhard, & Jurgens, 1984). Those genes, as they were identified, were shown to participate in patterning in many embryos throughout the animal kingdom. A number of signal transduction pathways were discovered and most of those also were discovered to participate in patterning of diverse embryos. Collectively, the many molecular discoveries with all model systems contributing, provided a large number of candidates and led to identification of a number of signaling inputs that patterned the sea urchin ectoderm. In rough historical order, the dorsal–ventral axis was discovered to depend upon Nodal and bone morphogenetic protein (BMP) for specification of those two ectodermal territories (Duboc, Rottinger, Besnardeau, & Lepage, 2004). The Wnt pathway was discovered to provide patterning inputs along the anterior–posterior axis (Logan, Miller, Ferkowicz, & McClay, 1999; Range, Angerer, & Angerer, 2013). Nodal and BMP were utilized again to establish the right–left asymmetry (Duboc, Rottinger, Lapraz, Besnardeau, & Lepage, 2005). It remained to discover how the ectoderm responded to those signals to set up a system for supply of positional information for skeletogenesis.

Skeletal patterning cues began to surface with the discovery that the SCs expressed receptors for vascular endothelial growth factor (VEGF) and for fibroblast growth factor (FGF), and the ectoderm expressed VEGF and FGF from two small lateral patches of ectodermal cells immediately over the site where skeletogenesis began (Duloquin, Lhomond, & Gache, 2007; Rottinger et al., 2008). Knockdown of each signal strongly impaired skeletogenesis. Those data indicated that VEGF and FGF have a role in skeletal patterning but among the questions remaining was the question as to how VEGF and FGF were secreted from such specific regions of the ectoderm. That question was answered when it was learned that the two lateral sites of growth factor production were at the intersection of two bands of ectoderm. The ciliary band became localized at the boundary between the dorsal and ventral ectodermal territories governed by BMP and Nodal signaling. An orthogonal band of ectoderm was specified in response to short-range signaling from the endoderm via Wnt5 (McIntyre, Lyons, Martik, & McClay, 2014; McIntyre, Seay, Croce, & McClay, 2013). This established a circumferential band of ectodermal cells and that band intersected the ciliary band at the site where VEGF and FGF were secreted. Later, at a more anterior site on the ciliary band, VEGF was again expressed and secreted and just beneath that site the growing skeleton branched to initiate the oral skeletal branches that shaped the anterior end of the larva. The patterning was not the consequence of diffusion gradients setting up each territory. The only long-range diffusion in the system was BMP to set up the dorsal half of the ectoderm. All other signals used, at least to set up the sites for VEGF and FGF release, were based on short-range signaling. Further, the two bands of ectoderm appear to participate in all skeletal patterning. All body skeletal rods run parallel to either the ciliary or border ectoderm bands. The sites of VEGF release are the only sites where skeletal branching occurs, and the outgrowth of the oral and postoral rods appears to use FGF and the FGF receptor at the tip of the growing rod (McIntyre et al., 2014, 2013; Rottinger et al., 2008). Many holes remain to be filled, but the general theme of how positional information is supplied to the SCs appears to be within the grasp of understanding.

The next set of phenomena to be solved is to understand how the three-dimensional crystal of calcium carbonate is patterned by the 64 SCs. Assuming that the ectoderm supplies the necessary positional information, what do the SCs do with that information and how do they mold the skeleton? Again, early experiments helped set up the questions to be answered. If a few SCs from another species were inserted into the blastocoel of a host embryo, along with the SCs of the host, a mosaic skeleton was produced in which a local region surrounding the inserted SC was patterned like the skeleton of the donor species (Armstrong et al., 1993; Armstrong & McClay, 1994). In another set of experiments when the 64 SCs were augmented by insertion of up to 200 or more extra SCs, each SC participated in building the skeleton, but the relative contribution of each was commensurately reduced (Ettensohn & McClay, 1988). These outcomes were interpreted to mean that each SC received positional information and produced a portion of skeleton that was delimited by about half the distance from its neighbor. Further, since each SC produces a pattern appropriate to that particular region of the skeleton, the information arriving must be of sufficient complexity to control the specific three-dimensional shape of just that region of skeleton. The precise nature of that information, its transduction, and the means by which the individual SC precisely molds that little piece of skeleton remains unknown, but there are a number of findings to use in those discoveries.

First is the question of how biomineralization occurs. To understand how that works, it is necessary to describe how the SCs relate to each other. After EMT, they migrate within the blastocoel and form a ring of cells and later fuse to form a syncytium (Hodor & Ettensohn, 1998). Two clusters of cells accumulate (by migration) on either side of the ring, immediately beneath the two early sites of VEGF and FGF secretion. Within the cluster, a triradiate calcium carbonate deposit accumulates as the initial seed biomineral (Wilt, 1999, 2005). Recent evidence indicates that the initial secretion is amorphous calcium carbonate which then crystallizes (Knapp, Wu, Mobilia, & Joester, 2012). Rod growth then occurs by sequential addition of more amorphous calcium carbonate at the tips of the rods and it then crystallizes by entrainment to the axes of the existing crystal. Prior to crystallization, SCs somehow define the shape of the local piece of rod. The SC membrane wraps around a cavity to form the template for the amorphous calcium carbonate deposit in much the same way that Schwann cell membranes wrap around neurons (though SC membranes wrap around the cavity only once). The calcium carbonate thus is secreted into an extracellular compartment but that compartment is completely surrounded by a membrane fold. The shape of the cavity produced by the membrane folds decides the shape of the skeleton, and, as above, the region to either side of each SC nucleus is patterned by locally interpreting genotypic information from that nucleus (recalling the outcome of skeletons produced by mixtures of SCs from two species). Consequently, as the rods of the skeleton grow out, each nucleus and the membrane to either side of it programs the shape of the membrane cavity and supplies amorphous calcium carbonate to the space. In addition, a number of proteins called spicule matrix proteins provide added structural support, and perhaps supply part of the genotypic patterning components to the crystal (Urry, Hamilton, Killian, & Wilt, 2000; Wilt, 2005). A recent RNA-seq analysis suggests the number of spicule matrix proteins could be rather large, though how their distribution might contribute to the pattern of the skeleton is unknown (Rafiq, Cheers, & Ettensohn, 2012; Rafiq, Shashikant, McManus, & Ettensohn, 2014).

This essay has focused on how one three-dimensional structure is formed and how discoveries unfolded that led to our current understanding of that structure. Each of those discoveries was preceded by advances in technology that provided the tools, and in each case, discoveries in other embryos added information that was of value for the sea urchin. In thinking about the above account, one must realize that while a lot has been learned even more remains to be discovered in this system, and the same is true for other model patterning system. In the sea urchin larval skeleton as in most other systems, the highly localized cellular conversations are of the greatest interest. Those conversations are initiated by the ectoderm that establishes two orthogonal bands of cells that provide patterning information. How that information is released, and its complexity remains to be discovered. The SCs that receive the ectodermal positional information then must interpret it and produce an output that precisely molds the calcium carbonate they secrete into the correct shape. The details of how all of that works also remains to be discovered in much the same way that in the vertebrate limb, there is much information known about positioning the humerus in the upper arm, for example, but the details of how that bone is structured and how muscle origins and insertions are established along that limb remain to be discovered. It will be fascinating to see what the CTDB reports over the next 50 years to illuminate details of these and other morphogenetic patterning systems.

REFERENCES

Annunziata, R., Perillo, M., Andrikou, C., Cole, A. G., Martinez, P., & Arnone, M. I. (2014). Pattern and process during sea urchin gut morphogenesis: The regulatory landscape. *Genesis*, *52*, 251–268.

Armstrong, N., Hardin, J., & McClay, D. R. (1993). Cell-cell interactions regulate skeleton formation in the sea urchin embryo. *Development*, *119*, 833–840.

Armstrong, N., & McClay, D. R. (1994). Skeletal pattern is specified autonomously by the primary mesenchyme cells in sea urchin embryos. *Developmental Biology*, *162*, 329–338.

Davidson, E. (2006). *The regulatory genome: Gene regulatory networks in development and evolution*. San Diego: Academic Press/Elsevier.

Davidson, E. H., McClay, D. R., & Hood, L. (2003). Regulatory gene networks and the properties of the developmental process. *Proceedings of the National Academy of Sciences of the United States of America*, *100*, 1475–1480.

Davidson, L. A., Oster, G. F., Keller, R. E., & Koehl, M. A. (1999). Measurements of mechanical properties of the blastula wall reveal which hypothesized mechanisms of primary invagination are physically plausible in the sea urchin Strongylocentrotus purpuratus. *Developmental Biology*, *209*, 221–238.

Davidson, E. H., Rast, J. P., Oliveri, P., Ransick, A., Calestani, C., Yuh, C. H., et al. (2002). A genomic regulatory network for development. *Science*, *295*, 1669–1678.

Duboc, V., Rottinger, E., Besnardeau, L., & Lepage, T. (2004). Nodal and BMP2/4 signaling organizes the oral-aboral axis of the sea urchin embryo. *Developmental Cell, 6*, 397–410.

Duboc, V., Rottinger, E., Lapraz, F., Besnardeau, L., & Lepage, T. (2005). Left-right asymmetry in the sea urchin embryo is regulated by nodal signaling on the right side. *Developmental Cell, 9*, 147–158.

Duloquin, L., Lhomond, G., & Gache, C. (2007). Localized VEGF signaling from ectoderm to mesenchyme cells controls morphogenesis of the sea urchin embryo skeleton. *Development, 134*, 2293–2302.

Ettensohn, C. A. (1984a). Mechanisms of epithelial invagination. *Quarterly Review of Biology, 60*, 289–307.

Ettensohn, C. A. (1984b). Primary invagination of the vegetal plate during sea urchin gastrulation. *American Zoologist, 24*, 571–588.

Ettensohn, C. A. (1985). Gastrulation in the sea urchin embryo is accompanied by the rearrangement of invaginating epithelial cells. *Developmental Biology, 112*, 383–390.

Ettensohn, C. A. (1990). The regulation of primary mesenchyme cell patterning. *Developmental Biology, 150*, 261–271.

Ettensohn, C. A. (1999). Cell movements in the sea urchin embryo. *Current Opinion in Genetics & Development, 9*, 461–465.

Ettensohn, C. A., & McClay, D. R. (1988). Cell lineage conversion in the sea urchin embryo. *Developmental Biology, 125*, 396–409.

Ettensohn, C. A., & Sweet, H. C. (2000). Patterning the early sea urchin embryo. *Current Topics in Developmental Biology, 50*, 1–44.

Fink, R. D., & McClay, D. R. (1985). Three cell recognition changes accompany the ingression of sea urchin primary mesenchyme cells. *Developmental Biology, 107*, 66–74.

Guss, K. A., & Ettensohn, C. A. (1997). Skeletal morphogenesis in the sea urchin embryo: Regulation of primary mesenchyme gene expression and skeletal rod growth by ectoderm-derived cues. *Development, 124*, 1899–1908.

Gustafson, T., & Wolpert, L. (1961). Studies on the cellular basis of morphogenesis in the sea urchin embryo: Directed movements of primary mesenchyme cells in normal and vegetalized larvae. *Experimental Cell Research, 24*, 64–79.

Hardin, J. D. (1987). Archenteron elongation in the sea urchin embryo is a microtubule-independent process. *Developmental Biology, 121*, 253–262.

Hardin, J. (1988). The role of secondary mesenchyme cells during sea urchin gastrulation studied by laser ablation. *Development, 103*, 317–324.

Hardin, J., & Armstrong, N. (1997). Short-range cell-cell signals control ectodermal patterning in the oral region of the sea urchin embryo. *Developmental Biology, 182*, 134–149.

Hodor, P. G., & Ettensohn, C. A. (1998). The dynamics and regulation of mesenchymal cell fusion in the sea urchin embryo. *Developmental Biology, 199*, 111–124.

Horstadius, S. (1939). The mechanics of sea urchin development as studied by operative methods. *Biological Reviews, 14*, 132–179.

Katow, H., & Solursh, M. (1981). Ultrastructural and time-lapse studies of primary mesenchyme cell behavior in normal and sulfate-deprived sea urchin embryos. *Experimental Cell Research, 136*, 233–245.

Knapp, R. T., Wu, C. H., Mobilia, K. C., & Joester, D. (2012). Recombinant sea urchin vascular endothelial growth factor directs single-crystal growth and branching in vitro. *Journal of the American Chemical Society, 134*, 17908–17911.

Laemmli, U. K. (1970). Cleavage of structural proteins during the assembly of the head of bacteriophage T4. *Nature, 227*, 680–685.

Li, E., Materna, S. C., & Davidson, E. H. (2013). New regulatory circuit controlling spatial and temporal gene expression in the sea urchin embryo oral ectoderm GRN. *Developmental Biology, 382*, 268–279.

Logan, C. Y., Miller, J. R., Ferkowicz, M. J., & McClay, D. R. (1999). Nuclear beta-catenin is required to specify vegetal cell fates in the sea urchin embryo. *Development, 126*, 345–357.
Malinda, K. M., & Ettensohn, C. A. (1994). Primary mesenchyme cell migration in the sea urchin embryo: Distribution of directional cues. *Developmental Biology, 164*, 562–578.
McClay, D. R. (1979). Surface antigens involved in interactions of embryonic sea urchin cells. In A. Moscona & A. Monroy (Eds.), *Current topics in developmental biology* (pp. 199–214). New York: Academic Press.
McClay, D. R., Armstrong, N. A., & Hardin, J. (1992). Pattern formation during gastrulation in the sea urchin embryo. *Development. Supplement*, 33–41.
McClay, D. R., Morrill, J. D., & Hardin, J. D. (1990). Archenteron morphogenesis in the sea urchin. In J. Gerhart (Ed.), *Symposium of the society for developmental biology: Vol. 49* (pp. 15–30). New York: Wiley-Liss.
McIntyre, D. C., Lyons, D. C., Martik, M., & McClay, D. R. (2014). Branching out: Origins of the sea urchin larval skeleton in development and evolution. *Genesis, 52*, 173–185.
McIntyre, D. C., Seay, N. W., Croce, J. C., & McClay, D. R. (2013). Short-range Wnt5 signaling initiates specification of sea urchin posterior ectoderm. *Development, 140*, 4881–4889.
Miller, J., Fraser, S. E., & McClay, D. (1995). Dynamics of thin filopodia during sea urchin gastrulation. *Development, 121*, 2501–2511.
Miller, J. R., & McClay, D. R. (1997). Characterization of the role of cadherin in regulating cell adhesion during sea urchin development. *Developmental Biology, 192*, 323–339.
Milstein, C., & Lennox, E. (1980). The use of monoclonal antibody techniques in the study of development cell surfaces. *Current Topics in Developmental Biology, 14*, 1–32.
Nusslein-Volhard, C., & Wieschaus, E. (1980). Mutations affecting segment number and polarity in Drosophila. *Nature, 287*, 795–801.
Oliveri, P., Tu, Q., & Davidson, E. H. (2008). Global regulatory logic for specification of an embryonic cell lineage. *Proceedings of the National Academy of Sciences of the United States of America, 105*, 5955–5962.
Peter, I. S., & Davidson, E. H. (2011). A gene regulatory network controlling the embryonic specification of endoderm. *Nature, 474*, 635–639.
Rafiq, K., Cheers, M. S., & Ettensohn, C. A. (2012). The genomic regulatory control of skeletal morphogenesis in the sea urchin. *Development, 139*, 579–590.
Rafiq, K., Shashikant, T., McManus, C. J., & Ettensohn, C. A. (2014). Genome-wide analysis of the skeletogenic gene regulatory network of sea urchins. *Development, 141*, 950–961.
Range, R. C., Angerer, R. C., & Angerer, L. M. (2013). Integration of canonical and noncanonical Wnt signaling pathways patterns the neuroectoderm along the anterior-posterior axis of sea urchin embryos. *PLoS Biology, 11*, e1001467.
Rottinger, E., Saudemont, A., Duboc, V., Besnardeau, L., McClay, D., & Lepage, T. (2008). FGF signals guide migration of mesenchymal cells, control skeletal morphogenesis [corrected] and regulate gastrulation during sea urchin development. *Development, 135*, 353–365.
Saunders, L. R., & McClay, D. R. (2014). Sub-circuits of a gene regulatory network control a developmental epithelial-mesenchymal transition. *Development, 141*, 1503–1513.
Takeichi, M. (1988). The cadherins: Cell-cell adhesion molecules controlling animal morphogenesis. *Development, 102*, 639–655.
Turing, A. M. (1953). The chemical basis for morphogenesis. *Philosophical Transactions of the Royal Society of London (Part B), 237*, 37–72.

Urry, L. A., Hamilton, P. C., Killian, C. E., & Wilt, F. H. (2000). Expression of spicule matrix proteins in the sea urchin embryo during normal and experimentally altered spiculogenesis. *Developmental Biology*, *225*, 201–213.

Wieschaus, E., Nusslein-Volhard, C., & Jurgens, G. (1984). Mutations affecting the pattern of the larval cuticle in *Drosophila melanogaster*: Zygotic loci on the X-chromosome and the fourth chromosome. *Roux's Archives of Developmental Biology*, *193*, 296–307.

Wilt, F. H. (1999). Matrix and mineral in the sea urchin larval skeleton. *Journal of Structural Biology*, *126*, 216–226.

Wilt, F. H. (2005). Developmental biology meets materials science: Morphogenesis of biomineralized structures. *Developmental Biology*, *280*, 15–25.

Wolpert, L. (1971). Positional information and pattern formation. *Current Topics in Developmental Biology*, *6*, 183–224.

CHAPTER THREE

Development of the Mammalian Kidney

Andrew P. McMahon[1]

Department of Stem Cell Biology and Regenerative Medicine, Eli and Edythe Broad Center for Regenerative Medicine and Stem Cell Research, Keck School of Medicine of the University of Southern California, Los Angeles, California, USA

[1]Corresponding author: e-mail address: amcmahon@med.usc.edu

Contents

Abstract

The basic unit of kidney function is the nephron. In the mouse, around 14,000 nephrons form in a 10-day period extending into early neonatal life, while the human fetus forms the adult complement of nephrons in a 32-week period completed prior to birth. This review discusses our current understanding of mammalian nephrogenesis: the contributing cell types and the regulatory processes at play. A conceptual developmental framework has emerged for the mouse kidney. This framework is now guiding studies of human kidney development enabled in part by *in vitro* systems of pluripotent stem cell-seeded nephrogenesis. A near future goal will be to translate our developmental knowledge-base to the productive engineering of new kidney structures for regenerative medicine.

Current Topics in Developmental Biology, Volume 117
ISSN 0070-2153
http://dx.doi.org/10.1016/bs.ctdb.2015.10.010

1. STRUCTURE–FUNCTION: AN OVERVIEW OF THE ANATOMY AND FUNCTION OF THE KIDNEY

The paired kidneys are the central organs of homeostasis for our body systems (Fig. 1A). Around 180 L of blood is filtered per day by the kidneys, accounting for around 20% of cardiac output. Filtering removes metabolic waste products, and kidney action adjusts water, salt, and pH to maintain the homeostatic balance of tissue fluids. The kidneys also regulate blood pressure through the renin–angiotensin–aldosterone system, erythrocyte production through production of erythropoietin, and circulating calcium and phosphate levels, in part through the activation of vitamin D.

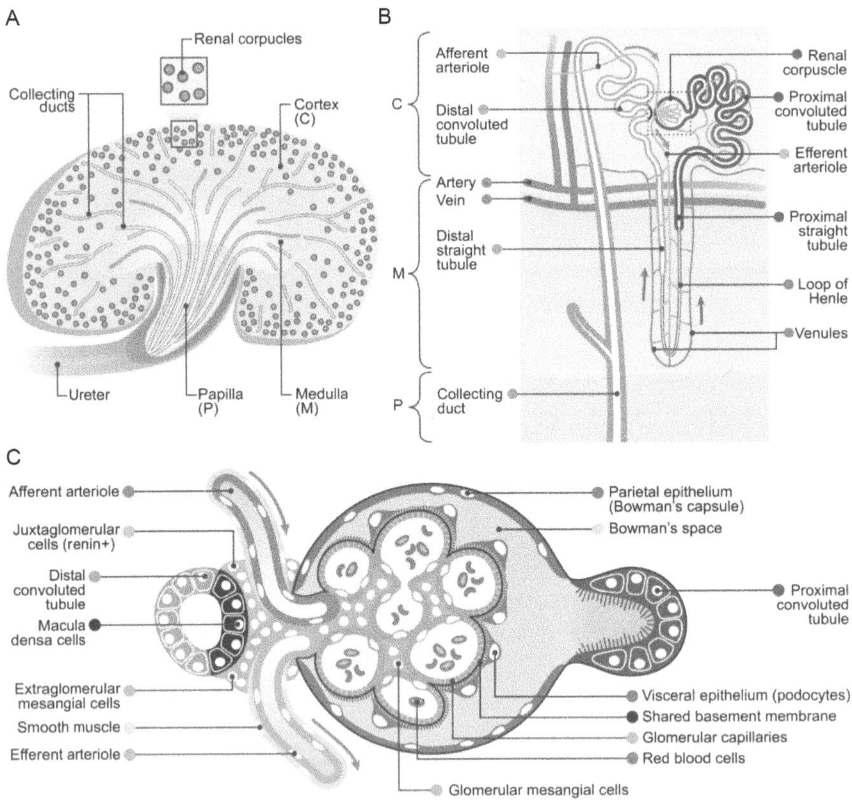

Figure 1 Schematic illustrations of the adult mouse metanephric kidney. (A) Gross regional organization of the adult kidney. (B) Anatomy of the nephron. (C) High-resolution overview of the renal corpuscle and associated structures (region boxed in B). (A color version of this figure is located at the end of this chapter.)

The functional unit of the kidney is the nephron (Fig. 1B). The mouse kidney has 12–16,000 nephrons per kidney, with the numbers varying depending on the strain (Short et al., 2014). Surprisingly, nephron number appears even more variable in man: reports suggest as much as a 10-fold difference in nephron number among individuals with around 1,000,000 nephrons on average per kidney (Bertram, Douglas-Denton, Diouf, Hughson, & Hoy, 2011). This large difference may not be trivial as studies link low nephron number to an elevated risk of kidney disease while higher nephron content in the mouse protects against salt-induced hypertension (Hoy et al., 2011; Puelles et al., 2011; Walker et al., 2012).

Renal filtration takes place within the renal corpuscle at the proximal extent of the nephron (Scott & Quaggin, 2015; Fig. 1A–C). Here, the afferent vasculature forms a convoluted knot of leaky blood vessels, the glomerulus. Mesangial cells, a glomerular variant of vascular-associated pericytes elsewhere in the body, lie closely opposed to the vascular endothelium, maintaining integrity and porosity of the vasculature. Fluid passes from the glomerular vasculature into the interstitial space of the Bowman's capsule, which is lined by podocytes. The podocytes secrete and sit on top of an unusually thick and highly specialized glomerular basement membrane (Miner, 2011). The glomerular basement membrane acts as a filtration barrier, reducing entry of larger molecular weight serum solutes into the nephron (>15 kDa) such as serum albumin (Suh & Miner, 2013).

The podocyte is a spectacular cell. Long, actin-based processes extend from the cell body, each resembling the arms of an octopus. Foot processes radiating from adjacent podocyte arms interdigitate to form, the slit diagrams, narrow openings of around 40 nm that regulate fluid flow into the proximal tubular segment of the nephron (Grahammer, Schell, & Huber, 2013). Given their unique biosynthetic and structural characteristics, podocytes are readily identified by the expression of a number of genes con trolling matrix production, and foot process formation and function (Brunskill, Georgas, Rumballe, Little, & Potter, 2011). Mutations in genes regulating these structural aspects are linked to renal disease (Bierzynska, Soderquest, & Koziell, 2015). Further, podocyte injury has emerged as a key factor in disease progression, most notably diabetic nephropathy, the major cause of end-stage renal disease (Maezawa, Takemoto, & Yokote, 2015).

The renal corpuscle and emerging S1 and S2 segments of the proximal tubule are confined to the outer cortex of the kidney (Fig. 1B). The final segment of the proximal tubule, the S3 segment, extends into the outer medullary domain. A key function of a myriad of membrane-embedded channels and transporters within the proximal tubule epithelial membrane is the reabsorption and recovery of vital small molecules such as glucose, amino acids, and essential minerals. Around 70% of the sodium chloride entering the ultrafiltrate is returned to surrounding efferent vascular capillaries. As a consequence, proximal tubule cells have a distinctive transcriptional signature that reflects their heavy lifting role in membrane transport processes (Brunskill et al., 2008; Thiagarajan et al., 2011). Proximal tubule cells are highly metabolically active and, in part because of this, vulnerable to injury. Acute kidney injury, linked to hypoxic episodes, infection, or nephrotoxic drugs, is one of the leading causes of death in a hospital setting; the proximal tubule is a shared target of these distinct pathological insults.

The loop of Henle extends the renal tubule deep into the medullar before looping back to become the distal tubule segment within the renal cortex (Fig. 1B). The primary role of the loop of Henle is urine concentration, a critical function essential to the radiation and diversification of land-colonizing vertebrates (Sands & Layton, 2014). The distal tubule regulates the return of sodium and calcium to the blood in hormone-directed processes mediated by aldosterone and parathyroid hormone, and participates in water uptake and pH balance (Subramanya & Ellison, 2014).

The macula densa, a specialized distal tubule component, lies just before the convoluted section, close to each glomerulus at the point of entry of afferent arterioles and exit of efferent arterioles (Fig. 1C; Bell, Lapointe, & Peti-Peterdi, 2003). Cells of the macula densa monitor salt levels in the distal tubule, and in response secrete local vasoconstrictors restricting adjacent afferent arterioles, reducing fluid flow into the glomerulus. This is a key feedback system to ensure physiologically important filtration rates. The macula densa also communicates with juxtaglomerular cells (Fig. 1C), a specialized subset of the smooth muscle surrounding the glomerular arterioles, to regulate renin secretion and systemic blood pressure through renin's catalytic action on the angiotensinogen system. Renin release is also triggered by changes in blood pressure in the afferent arteriole.

The distal tubule connects to the collecting duct, a continuous highly arborized epithelial network with a quite distinct origin from the contiguous renal tubule (Fig. 1A and B). The collecting duct system funnels urine from the kidney through the ureter to the bladder. The collecting duct epithelium displays a distinct cortical-medullary axis of branching, growth, and cellular organization, the tubular network focusing into a single (mouse) or multiple (man) renal papillae at its exit point (Fig. 1A). The medullary collecting duct is highly water permeable, facilitating water retention, and is a mosaic of two distinct cell types: abundant principal cells, which are critical for sodium retention, and rarer intercalated cells, which regulate pH through acid–base transporters (Al-Awqati & Gao, 2011; Pearce et al., 2015).

Intermingled between vascular and nonvascular epithelium, closely opposed to their basal surfaces, are interstitial cell types that show varying degrees of smooth muscle characteristics. *Bona fide* contractile smooth muscle surrounds the major blood vessels and the innermost collecting duct epithelium, and contractions propel urine into, and along, the ureter. In addition to specialized renin-secreting interstitial cell types, an interstitial population associated with proximal tubule segments secretes the hormone erythropoietin, stimulating production of red blood cells. Much of the remaining interstitial cells most closely resemble pericyte-like myofibroblasts, a population linked to renal repair and chronic fibrotic kidney disease (Kramann & Humphreys, 2014; Rabelink & Little, 2013). The kidney is also innervated by the sympathetic nervous system, axons tracking vascular pathways to the glomerulus. Sympathetic neurons counter autoregulatory networks responding to decreased blood pressure from excessive fluid loss, constricting afferent arteriole vesicles, and triggering production of renin by juxtaglomerular cells (Grassi, Mark, & Esler, 2015).

2. FROM PROGENITORS TO PRODUCTS: ASSEMBLING THE KIDNEY

2.1 A General Overview

Much of what we understand about kidney development comes from studies of only two mammals: the rat and mouse. Development is comparable in both species, though the mouse is, relatively speaking, 24 h ahead of its rodent cousin. The new genetics made possible by embryo stem cells and gene targeting have led to the primacy of the mouse as the key mammalian model. A broad overview of kidney development in the mouse is followed by a more detailed discussion of developmentally distinct component parts.

The metanephric kidney derives from intermediate mesoderm in temporally and spatially distinct processes (Taguchi et al., 2014; Takasato & Little, 2015). The first event, around embryonic day 8.75 (e8.75), is the emergence of paired nephric ducts at rostral somite levels. These ducts migrate posteriorly, inducing mesonephric tubular structures laterally until reaching the hindlimb level around e10.5 (Fig. 2A). Here, interactions between prespecified metanephric mesenchyme cells and the adjacent

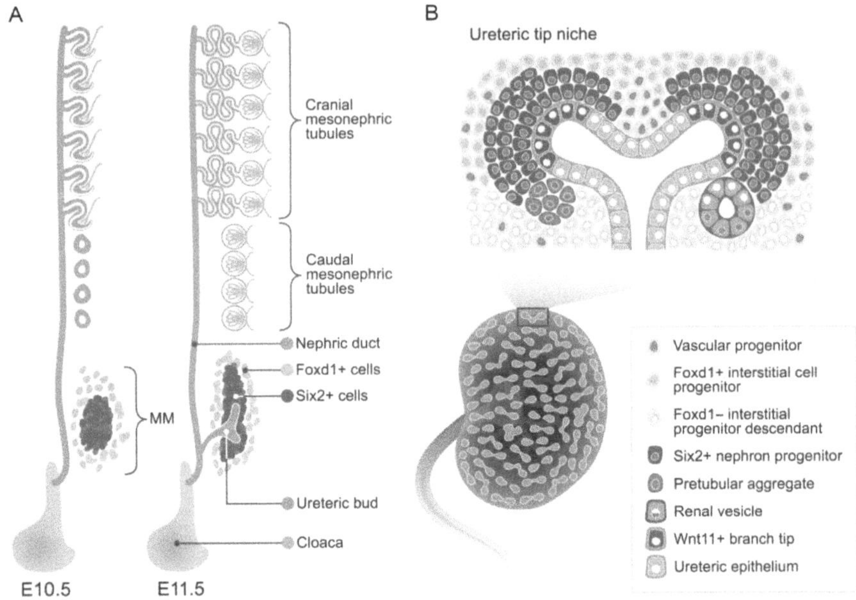

Figure 2 Schematic illustrations of the developing mouse kidney. (A) Overviews of the mesonephric and metanephric kidney anlagen pre-(E10.5) and post-(E11.5) ureteric bud invasion into the metanephric mesenchyme. Anterior is to the top. (B) Cellular and molecular organization of collecting duct (Wnt11+), nephron (Six2+), interstitial (Foxd1+), and vascular progenitors within a ureteric tip niche. (A color version of this figure is located at the end of this chapter.)

nephric duct trigger the dynamic process of kidney development (Combes, Davies, & Little, 2015; Costantini & Kopan, 2010; Kopan, Chen & Little, 2013; Little & McMahon, 2012). The mesenchymal cells induce a single, bilateral, ductal outgrowth, the ureteric bud, which grows into medially positioned metanephric mesenchyme (Fig. 2A). The mesenchyme condenses to encapsulate the growing tip of the ureteric bud and, by e11.5, the ureteric bud has branched to generate a T-like structure (Fig. 2A).

From this time, kidney development takes off (Little, 2015). The ureteric bud-derived epithelium undergoes 12 generations of branching before ceasing 2 days after birth, having established the entire network of the urine trafficking collecting duct system (Short et al., 2014). Initial ingrowth and subsequent branching of the ureteric epithelium is driven by signals supplied by the adjacent capping mesenchyme cells (Figs. 2B and 3A). In turn, the branch tips induce expansion and differentiation of associated capping mesenchyme. At each branching event, a subset of mesenchymal cells forms a tight cluster, the pretubular aggregate, beneath the branch tip. The pretubular aggregate

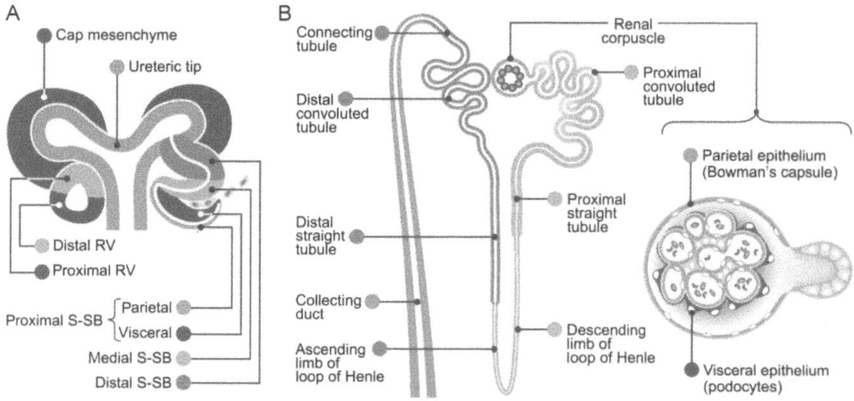

Figure 3 Schematic illustrations of nephron patterning process. (A) Transition of polarized renal vesicle to patterned S-shaped body. (B) Speculative color-encoded map of the cell fate relationship of progenitor subdomains in S-shaped body in (A) to adult nephron anatomy. Verification requires more extensive cell fate-mapping studies. (A color version of this figure is located at the end of this chapter.)

subsequently epitheliarizes to form a small cyst-like renal vesicle. Each renal vesicle gives rise to a nephron. Nephrogenesis continues along with ureteric branching, though approximately half of all nephrons form in a 2-day window after branching ceases (2–4 days after birth; Short et al., 2014). The final nephron number varies between mouse strains, and progenitor depletion studies show that nephron number is dependent on the size of the progenitor pool (Cebrian, Asai, D'Agati, & Costantini, 2014).

Growth, patterning, and morphogenesis over a 24–36-h period transform the renal vesicle into an S-shaped body (Figs. 2B and 3A). The distal-most epithelium of the S-shaped body plumbs into the adjacent collecting duct, generating a patent luminal connection essential for fluid trafficking through the kidney (Kao, Vasilyev, Miyawaki, Drummond, & McMahon, 2012). As the number of nephrons exceeds the number of ureteric branches, multiple nephrons connect to a single branch, or to each other as in the human kidney, generating a cascade of nephrons.

At its proximal-most end, the S-shaped body generates the glomerular cleft. A lengthy period of renal corpuscle assembly initiates with podocyte specification in the proximal epithelium and endothelial cell migration into the cleft (Fig. 3A). Other nephron segments form from the remainder of the S-shaped body. Vasculogenesis and angiogenesis contribute to formation and shaping of the elaborate and highly organized nephron-associated vascular network in conjunction with progressive development of the renal tubule epithelium. Interstitial cell types organize around the vasculature and tubules, and neurons project into the kidney to innervate differentiated target sites.

There is no definitive study on when functional nephrons are first established in the mouse kidney, though some filtering capability is likely by e16.5–e17.5. Given the continual process of recruitment and differentiation of renal vesicles, the late fetal and early neonatal kidney is a mosaic of nephrons at different stages of the maturation, and full kidney function is not established in the mouse until a few weeks after birth.

2.2 The Collecting Duct Network

Abnormal ureteric bud formation often results in a complete failure of kidney development. The fact that this initial event is a particularly vulnerable period in the kidney's developmental program is highlighted by the frequency of unilateral kidney loss in otherwise healthy individuals, estimated at around 1 in 1000 live births from asymptomatic autopsy (Shapiro, Goldfarb, & Ritchey, 2003). Many lines of evidence highlight the critical importance of Gdnf signaling through its Gfra1/Ret receptor complex in these early events (Costantini, 2012). Loss-of-function mutations in any of these components result in complete kidney agenesis (Cacalano et al., 1998; Enomoto et al., 1998; Pichel et al., 1996; Sánchez et al., 1996; Schuchardt, D'Agati, Larsson-Blomberg, Costantini, & Pachnis, 1994). Gdnf is secreted by the metanephric mesenchyme prior to ureteric bud outgrowth, and *Gdnf* expression is maintained in capping mesenchyme throughout the period of active nephrogenesis. Several positive feedback mechanisms, including the upregulation of *Ret* by Gdnf signaling at branch tips, adding robustness to the receptor tyrosine kinase (RTK)-regulated branching process (Costantini, 2012; Pepicelli, Kispert, Rowitch, & McMahon, 1997). Further, Gdnf induces tip-restricted expression of *Wnt11* to maintain normal levels of Gdnf through paracrine Wnt11 action (Fig. 2B; Majumdar, Vainio, Kispert, McMahon, & McMahon, 2003).

Rudimentary buds form in some *Gdnf* and *Ret* mutants suggesting branch-regulating pathways at play. Mouse genetics have highlighted an overlapping role for Fgf7 and Fgf10 from the capping mesenchyme acting through Fgfr2 in the ureteric epithelium (Ohuchi et al., 2000; Qiao et al., 1999; Zhao et al., 2004). As Ret and Fgf receptors encode RTKs, RTK signaling downstream of ligand–receptor engagement is a convergence point for overlapping signaling activities. Both pathways regulate tip-restricted expression of two highly related ETS-domain transcription factors, *Etv4* and *Etv5* (Lu et al., 2009). Together, *Etv4* and *Etv5* are essential for branching (Lu et al., 2009). Sox-domain transcriptional regulators, *Sox8* and *Sox9*, show broadly similar expression to *Etv4* and *Etv5*, and are themselves essential for normal branching outgrowth (Reginensi et al., 2011). *Sox8* and *Sox9* are required for activation of *Etv4/5* and a subset of other tip-expressed Gdnf targets, but *Sox8* and *Sox9* are not required for *Ret* expression (Reginensi et al., 2011).

Tip-specific expression domains are also regulated by retinoic acid (RA) synthesized and released by interstitial progenitors (Rosselot et al., 2010). In ureteric tip explants, RA signaling is required for Gdnf-mediated activation of *Ret* (and most likely Ret-dependent targets) in the ureteric epithelium (Rosselot et al., 2010). Wnt signaling is a third major signaling pathway regulating branch tip programs. Wnt pathway reporters show high endogenous levels of canonical Wnt signaling within branch tips, and Ret expression and branching growth are lost following genetic or biochemical perturbation of the canonical Wnt pathway (Bridgewater et al., 2008; Karner et al., 2010; Marose, Merkel, McMahon, & Carroll, 2008). However, the biologically relevant Wnt ligand(s) have not been identified.

Outgrowth promoting activities are countered by the action of several other pathways (Costantini, 2012). Collectively, their actions ensure the formation of a single ureteric bud of the appropriate size from each nephric duct. As a result, a single ureter will connect each kidney to the bladder. Kidneys with duplex ureters are prone to blockage-related kidney damage with a subsequent loss of kidney function. Factors constraining the outgrowth process act by either limiting the domain of Gdnf expression within the intermediate mesoderm (Slit2/Robo2; Foxc1/Foxc2) or by inhibiting (Bmp4) or attenuating RTK pathway action (Spry1) within the ureteric bud (Basson et al., 2005; Grieshammer et al., 2004; Kume, Deng, & Hogan, 2000; Michos et al., 2004; Wainwright, Wilhelm, Combes, Little, & Koopman, 2015). Importantly, ureteric branching is rescued in *Ret/Gdnf* mutants when *Spry1* action is attenuated: branching now becomes highly dependent upon cap mesenchyme-derived Fgf10 signaling (Michos et al., 2010).

The mouse kidney contains around 3400 branch tips at the cessation of branching 2 days after birth (Short et al., 2014). The continued expression of RTK, Wnt, and RA target genes in branch tips throughout the branching period argues for the continued action of these pathways throughout the branching process. However, the relative importance of each pathway at each stage has not been determined. Genetic analysis also indicates even greater pleiotropy in RTK usage throughout the branching period with evidence for both Met and Egf receptor engagement (Ishibe et al., 2009). New signaling centers may also feed into ongoing branch regulation. Early studies in kidney explants suggested additional signaling from forming nephrons (Saxén, 1987). In support of this model, branching terminates prematurely when nephrogenesis is compromised in *Wnt4* mutants (Stark, Vainio, Vassileva, & McMahon, 1994).

The branch tip harbors the progenitor pool for the entire collecting duct network, and is the mitogenic and motogenic center for branching growth. How each uretic tip-directed pathway controls distinct tip properties is not entirely clear. When Ret signaling, or Etv4 and Etv5 activity, is removed, tip cells rapidly loose tip properties, exit the tip niche, and adopt stalk fates (Chi et al., 2009; Kuure, Chi, Lu, & Costantini, 2010). Thus, competition to remain in the tip niche requires active signaling and, at this level, RTK signaling has an executive function for all tip properties. Gdnf/Ret signaling activates MAP kinase, PI3 kinase, and PLC gamma pathways (Costantini, 2012). Of these, PI3 kinase action is essential for *Etv4* and *Etv5* expression and branching growth (Lu et al., 2009; Tang, Cai, Tsai, Wang, & Dressler, 2002). Genetic removal of MAP kinase pathway components blocks branching, but the ureteric epithelium continues to elongate, highlighting a direct role for MAP kinase signaling in the branching process (Ihermann-Hella et al., 2014). Analysis of *Ctnnd1* (β-catenin) mutants suggests that canonical Wnt signaling maintains the epithelial progenitor pool, as loss of Ctnnd1 results in premature differentiation of progenitor cells (Bridgewater et al., 2008; Marose et al., 2008).

RTK signaling underpins epithelial branching in many organ systems. Modeling suggests that Turing mechanisms can account for early branching patterns in both the lung and the kidney, pointing to similar regulatory processes (Menshykau & Iber, 2013; Menshykau, Kraemer, & Iber, 2012). However, each organ's epithelial network has a distinct shape and form. RTK signaling is likely important not simply in directing branching growth but in directing the branching pattern. Gdnf deficiency is rescued on removing Spry1-mediated antagonism of RTK action, leading ureteric tips to branch extensively; however, their branching pattern is abnormal (Michos et al., 2010). Further, different Fgf ligands stimulate quite distinct modes of epithelial outgrowth: some trigger elongation while others stimulate budding or active branching (Qiao et al., 2001). For the most part, dichotomous branching is the major branching routine throughout kidney development, though trichotomous branching is observed immediately after the T-bud stage (Majumdar et al., 2003). A self-avoidance process, potentially mediated through Bmp7 action, has been implicated in directing spatial outgrowth of each branch tip (Davies, Hohenstein, Chang, & Berry, 2014). Branch angles are not fixed but change over the course of kidney development and likely contribute to the overall shape of the organ (Short et al., 2014).

Cell organization, cell shape, cell division, and cell adhesion have been explored in the context of branching growth. Initial outgrowth of the ureteric bud is preceded by a cuboidal to pseudostratified rearrangement of the epithelium, but this type of remodeling appears restricted to this stage (Chi et al., 2009). Extensive filopodial outgrowth at branch tips is thought to drive branching of many epithelial networks. Filopodial formation is associated with the terminal extension of the nephric duct to the cloaca in a Gdnf/Ret-dependent process that completes the interconnection of the urinary drainage system (Chia et al., 2011). However, there is no evidence of filopodia-directed outgrowth of ureteric epithelial branch tips. Genetic modification of actin-depolymerizing factors indicates that actin remodeling is critical for the branching process (Kuure, Cebrian, et al., 2010; Kuure, Chi, et al., 2010). Further, enhanced epithelial adhesion following removal of MAP-kinase signaling suggests adhesive remodeling is necessary for normal branching (Ihermann-Hella et al., 2014). Altering adhesive properties may also impinge on proliferative programs. Tip growth occurs with a mitosis-associated dispersal of epithelial cells. Following cell division, one of the mitotic products inserts back into the original location of the parent cell while the other inserts a few cell diameters away (Packard et al., 2013). The significance of cell dispersal to the branching process is not understood. However, the process may help in maintaining an active tip environment, facilitating a rapid sorting from the tip niche of cells no longer responsive to tip signals or those committing to differentiated programs.

Importantly, epithelial rearrangements and regulated growth processes continue to shape the ureteric network beyond the branch tips. A Wnt7b-directed epithelial remodeling process is essential for initiating formation of the medullary ureteric architecture around e15.5–16.5 (Yu et al., 2009). Wnt7b secreted by the ureteric epithelium signals to underlying interstitial cells to induce production of additional Wnt signals. Their combined action is thought to bias the plane of cell division within the ureteric epithelium thereby modulating epithelial growth and remodeling to establish the medullary region (Yu et al., 2009). These events are likely to involve extensive rearrangement of the first few branch events of the initial ureteric tree (Short et al., 2014). Wnt9b-mediated planar polarity signaling through the JNK pathway also underlies elongation of the ureteric network (Karner et al., 2009). The critical importance of regulating the plane of cell division for normal epithelial growth is exemplified by a spectrum of cystic kidney diseases that perturb this process (Schnell & Carroll, 2014).

The collecting duct differentiates along a cortico-medullary axis, generating principal and intercalated cells. Principal cells, readily distinguished by the expression of several water transporting membrane channel proteins including *Aqp2/3/4*, are the most abundant differentiated cell type. Intercalated cells come in A- and B-types, displaying distinct ion-exchange proteins, Slc4a1 and Slc26a4, respectively (Al-Awqati & Gao, 2011; Wall, 2005). Notch signaling promotes principal cell fate, whereas a mature intercalated cell fate is dependent on the transcriptional regulator Foxi1 (Blomqvist et al., 2004; Jeong et al., 2009). Although each cell type has a clear, distinct physiological activity, there is evidence for considerable plasticity and interconversion among these cell types, at least in cell culture (Fejes-Tóth & Náray-Fejes-Tóth, 1992).

2.3 The Main Body of the Nephron

The main body of the mouse nephron—from the podocytes of the renal corpuscle to the nephron's interconnection with the collecting duct—derives from a self-renewing, multipotent Six2+ progenitor population established by e10.5, prior to ureteric bud outgrowth (Kobayashi et al., 2008; Fig. 2B). In this, Six2 is essential for maintaining the progenitor state; the absence of Six2 leads to premature and ectopic commitment of the entire progenitor pool to renal vesicle fates by e12.5 (Kobayashi et al., 2008; Self et al., 2006).

Several transcriptional regulators have been identified in the specification or maintenance of the Six2 progenitor pool, including Six1, Hox11 paralogs, Pax2, Wt1, and Sall1 (Little & McMahon, 2012; O'Brien & McMahon, 2014). Genetic analyses suggest nonlinear hierarchical relationships for these regulatory factors. Further, protein–protein interaction studies indicate several of these components complex with Six2 to maintain a nephron progenitor state (Kanda et al., 2014; Park et al., 2012; Xu et al., 2014).

With the exception of the three *Hox11* paralogs (*Hoxa11*, *Hoxc11*, and *Hoxd11*), all the other transcriptional regulatory factors are expressed within the mesonephric kidney (Mugford, Sipilä, Kobayashi, Behringer, & McMahon, 2008). Despite the loss of all *Hox11* paralogs, a condensed mass of Pax2+, Eya1+ mesenchyme is formed at the appropriate axial level for the metanephric kidney anlagen (Wellik, Hawkes, & Capecchi, 2002). Thus, specification of the metanephric mesenchyme is independent of the activity of Hox11 genes. However, the combined activity of Hox11 members is essential for expression of *Gdnf* and other features of the metanephric progenitor program, suggesting that Hox11 paralogs determine metanephric specific kidney outcomes for associated mesenchymal cell derivatives (Wellik et al., 2002). In support of this model, mesonephric tubules ectopically expressing *Hoxd11* adopt metanephric features of regional gene expression (Mugford, Sipilä, Kobayashi, et al., 2008; Mugford, Sipilä, McMahon, et al., 2008).

Ingrowth of a ureteric bud into the metanephric mesenchyme is accompanied by the activation of *Cited1* in uncommitted nephron progenitors (Boyle et al., 2008). Whereas *Six2* expression persists transiently into proximal regions of epitheliarizing nephron precursors, *Cited1* is only expressed within uncommitted nephron progenitors (Mugford, Yu, Kobayashi, & McMahon, 2009; Park et al., 2012). Consequently, *Cited1*'s expression has become the "gold standard" for identifying the nephron progenitor compartment. Surprisingly given the specificity of *Cited1* expression, Cited1 activity is not required for kidney development (Boyle, Shioda, Perantoni, & de Caestecker, 2007). Recent studies examining Six2 levels and cell cycle parameters have identified regional heterogeneity within the nephron progenitor niche (Short et al., 2014). Whether these temporally and spatially distinct progenitor pools play distinct roles in the kidney's developmental program is not clear (Short et al., 2014).

Generating the full complement of nephrons is dependent on balancing expansion of the nephron progenitor compartment with their commitment to epithelial renal vesicles. Quantifying multiple cellular metrics within the developing mouse kidney has provided a broad overview of these events (Combes et al., 2014; Short et al., 2014). The size of the starting nephron progenitor pool at e10.5 is unclear, but a recent analysis has suggested around 2000 cells (Wainwright et al., 2015). Extrapolating using cell cycle estimates from later stages with numbers determined at e11.5 by cell sorting (12,000 progenitors; Kobayashi et al., 2008) and confocal imaging (8000 progenitors, A. Combes and M. Little, personal communication) gives a somewhat larger initiating population of 4000–6800 (A. Combes and M. Little, personal communication). Using these numbers, nephron progenitors undergo a 174- to 600-fold expansion over the course of mouse kidney development to generate more than 1,000,000 progenitors (A. Combes and M. Little, personal communication).

At the level of the mouse tip niche, there is a steady decline in the ratio of nephron progenitors to tip niche until birth when nephrogenesis markedly accelerates (Short et al., 2014). It is unlikely that birth is an actual trigger of these events as kidney development is completed at different times relative to birth in different mammalian species: for example, prior to birth, around 36 weeks in man (Hinchliffe, Sargent, Howard, Chan, & Van Velzen, 1991). All nephron progenitors disappear from the mouse kidney by postnatal days 3–4. Consequently, the population of nephrons generated during the developmental period is the population for life.

Nephron progenitors occupy a distinct position within the tip niche. Progenitors maintain close contact with the tip throughout the branching process and are surrounded by a pool of interstitial progenitors (see later). Given the importance of signaling directed by both the underlying ureteric epithelium and overlying interstitial progenitors, this physical arrangement may have developmental significance. The mechanisms controlling distinct niche progenitor cell layering are not known, though Itga8 in the cap mesenchyme, and its receptor nephronectin on the adjacent ureteric epithelium, are required for interactions promoting Gdnf-dependent ureteric bud outgrowth (Linton, Martin, & Reichardt, 2007; Müller et al., 1997). However, the potential action of this pathway in regulating progenitor niche organization has not been directly addressed. A Sall1-Kif26b pathway has also been linked to adherence of cap mesenchyme to the ureteric bud (Uchiyama et al., 2010).

Fgf, Bmp, Wnt, and Fat signaling all play important roles in regulating nephron progenitors. Fgf9 produced by the ureteric tips, and Fgf9 and Fgf20 in the CM, regulate the proliferative expansion of nephron progenitors through Fgfr1 and Fgfr2 stimulation of RAS and PI3K pathways (Barak et al., 2012; Brown et al., 2011; Poladia et al., 2006). *Bmp7* is expressed broadly in the ureteric epithelium, cap mesenchyme, and nephron derivatives (Dudley & Robertson, 1997). Bmp7 signaling promotes the maintenance and proliferation of progenitors through a JNK mechanism (Blank, Brown, Adams, Karolak, & Oxburgh, 2009; Tomita et al., 2013), and progenitor commitment through Smad-directed processes (Brown et al., 2013).

Wnt9b's action is complex: Wnt9b is essential for both the proliferation and the commitment of progenitor cells (Carroll, Park, Hayashi, Majumdar, & McMahon, 2005; Karner et al., 2011). Both processes require the activity of Ctnnd1, a coactivator in the transcriptional complex formed in response to canonical Wnt signaling (Karner et al., 2011; Park, Valerius, & McMahon, 2007). Application of small-molecule modulators of Ctnnd1 levels suggests that different levels of Wnt signaling may underpin different actions of the canonical pathway: signaling^{low}/β-catenin^{low} proliferation and signaling^{high}/β-catenin^{high} nephron induction (Brown, Muthukrishnan, & Oxburgh, 2015).

Other signaling pathways are also thought to modulate the response to canonical Wnt signaling. Bmp7-directed Smad action is postulated to convert nephron progenitors from a Cited1+/Six2+ progenitor state refractory to Wnt9b/Ctnnd1 inductive signaling to a Cited1−/Six2+ responsive cell state (Brown et al., 2013). Further, atypical cadherin signaling originating from interstitial progenitors may promote adjacent nephron progenitor cell differentiation. Independent studies agree that interstitial cell-derived Fat4 signaling acting through Dchs partners is required to promote progenitor commitment; undifferentiated progenitors accumulate in the absence of Fat4 and Dchs1 (Bagherie-Lachidan et al., 2015; Das et al., 2013; Reginensi et al., 2013). However, there are conflicting views of Yap/Taz and Wnt9b/Ctnnd1 action in these events.

One study focusing on transcriptional targets of Wnt9b's progenitor renewal and differentiation-promoting activities suggests that Fat4/Dchs1-directed nuclear accumulation of Yap/Taz promotes progenitor differentiation in concert with Wnt9/Ctnnd1 (Das et al., 2013). A second study indicates that Fat4/Dchs1 action is independent of Yap/Taz and does not alter canonical Wnt signaling responses but likely has a parallel role in progenitor differentiation (Bagherie-Lachidan et al., 2015). Additional work will be required to resolve these mechanistic differences. Decorin, produced by interstitial progenitors also promotes nephron progenitor differentiation by inhibiting Bmp7 action and thereby sensitizing nephron progenitors to inductive Wnt signaling (Fetting et al., 2014).

Signaling networks connect with distinct transcriptional regulatory programs to control cell state. Here, several groups have reported chromatin immunoprecipitation studies on target interactions of key transcriptional components, including Sall1, Wt1, Six2, and Ctnnd1 (Basta, Robbins, Kiefer, Dorsett, & Rauchman, 2014; Hartwig et al., 2010; Kanda et al., 2014; Park et al., 2012). These studies point to overlapping and unique targets of progenitor maintenance by Sall1 and Six2 (Kanda et al., 2014; Park et al., 2012). Further, they highlight an unexpected molecular interaction between Six2 and canonical Wnt transcriptional complexes on enhancers driving expression of genes regulating progenitor and differentiating cell states mediated by both protein–protein and direct DNA binding interactions (Park et al., 2012). Interestingly, engagement of Ctnnd1 in the complexes is not observed until progenitor commitment. These findings argue against a direct transcriptional role for Ctnnd1-engaged transcriptional complexes in regulating the progenitor state, but support their role in initiating progenitor differentiation (Park et al., 2012).

The first cellular evidence of progenitor commitment is a clustering of cells into a pretubular aggregate beneath ureteric branch tips (Fig. 2B). The cellular mechanisms determining movement and coalescence of this cohort of cells, and their precise origin from within the progenitor pool, are not understood. *De novo* activation of *Fgf8*, *Wnt4*, and *Pax8* distinguishes the pretubular aggregate (Grieshammer et al., 2005; Perantoni et al., 2005; Plachov et al., 1990; Stark et al., 1994). Both Fgf8 and Wnt4 signals are essential for the continued transition of progenitors to an epithelial renal vesicle (Grieshammer et al., 2005; Perantoni et al., 2005; Stark et al., 1994). In this pathway, Fgf8 acts upstream of Wnt4, though Wnt4 is required to maintain both its own and Fgf8 expression (Grieshammer et al., 2005; Perantoni et al., 2005; Stark et al., 1994).

Wnt4 regulates the transition of mesenchymal pretubular aggregates to epithelial renal vesicles (Kispert, Vainio, & McMahon, 1998). In this, the analysis of Wnt4-dependent actions suggests distinct pathways at play. Activation of *Lhx1* in the renal vesicle requires both Wnt4 signaling and Ctnnd1 activity, indicative of canonical Wnt signaling (Park et al., 2007). However, canonical Wnt signaling clearly inhibits epithelial formation (Park et al., 2007), and biochemical and cellular studies point to non-canonical, Ca^{2+}/NFAT-dependent signaling in the epithelial transition (Burn et al., 2011; Tanigawa et al., 2011). The cellular mechanisms underlying epithelial formation have not been determined. Epithelial formation commonly requires cell-to-cell surface interactions mediated by the cadherin family of homophilic cell adhesion molecules. However, the predominant epithelial cadherin, Cdh1, is not present until after renal vesicle formation (Georgas et al., 2009), and Cdh6, which is present within the forming renal vesicle, is not essential for epithelial formation although Cdh6 mutants do display a weak kidney phenotype (Mah, Saueressig, Goulding, Kintner, & Dressler, 2000). Lumen formation within the epithelial vesicle and derived structures is dependent on the action of afadin and nonmuscle myosin action (Yang et al., 2013).

Distinct domains of gene activity segregate the newly formed renal vesicle into proximal and distal domains suggestive of early regional patterning (Fig. 3A; Georgas et al., 2009; Mugford et al., 2009). Conclusive evidence that the observed domains correspond to distinct fates in the adult nephron awaits precise fate-mapping studies from early nephron stages. Notch2 signaling is critical for patterning of proximal fates in renal vesicle derivatives (podocytes and proximal tubule), while graded canonical Wnt signaling has been implicated in specification of the remaining medial and distal segments (Cheng et al., 2007; Lindström et al., 2015). Emerging polarity within renal vesicle derivatives correlates with the proximity of nephron-forming epithelium to the ureteric epithelium: distal markers are activated in cells closest to the ureteric epithelium. Polarizing Wnt signals from the ureteric epithelium (including Wnt9b) are attractive candidates in regulating renal vesicle polarity. However, patterned nephron structures can form when isolated metanephric mesenchyme receives an appropriate inductive signal in the absence of the ureteric epithelium (Saxén, 1987). These disparate results could be explained if renal vesicles self-organize polarity stochastically, while ureteric epithelium-derived signals act to bias and effectively orientate this process.

The cellular dynamics underlying early nephron morphogenesis are not well understood. Growth and dramatic epithelial reorganization accompany the transition of a renal vesicle to an S-shaped body (Fig. 3A): uneven proliferation may factor into distal elongation (Georgas et al., 2009). As there has been no extensive three-dimensional comparison of different stages in the transition to an S-shaped body, it is not clear if there is a reproducible stereotypical process at play. However, by the time, an epithelial connection is established between the distal S-shaped body and the ureteric epithelium, there is considerable molecular heterogeneity evident along the proximo-distal axis of the nascent nephron (Georgas et al., 2009). Limited fate-mapping studies point to some spatial coherence between restricted domains of gene expression in the S-shaped body and adult nephron anatomy (Fig. 3B; Barker et al., 2012; www.gudmap.org).

Proximally, expression of the transcriptional regulators *Mafb*, *Wt1*, and *Lmx1b* marks the onset of podocyte specification and their actions regulate podocyte programs (Miner et al., 2002; Moriguchi et al., 2006; Mundlos et al., 1993; Rohr et al., 2002; Sadl et al., 2002). The midsection of the

S-shaped body is demarcated by expression of genes encoding two Notch ligands, *Jag1* and *Dll1*, and Lfng, a Notch receptor modifying enzyme, suggesting Notch signaling may regulate proximal nephron fates (Leimeister, Schumacher, & Gessler, 2003). Consistent with this view, Notch signaling through the Notch2 receptor regulates podocyte and proximal tubule development (Cheng et al., 2007). As Notch ligands are not freely diffusible, the curvature of the S-shaped body may facilitate Notch ligands produced in the mid-segment contacting Notch2+ responsive cells in the proximal epithelium. Interestingly, Notch2 responsiveness is not conferred by earlier Wnt signaling (Boyle, Kim, Valerius, McMahon, & Kopan, 2011). More likely, Wnt-dependent epitheliarization establishes a tissue context for appropriate spatial signaling to direct the proximal nephron patterning process.

The transcriptional regulators Pou3f3, Hnf1b, Irx1, and Irx2 overlap Dll, Jag1, and Lfng producing domains in the S-shaped body (Blumenfeld, Maury, Chouard, Yaniv, & Condamine, 1991; Heliot et al., 2013; Massa et al., 2013; Nakai et al., 2003). The temporal and spatial overlap in their expression has not been mapped; consequently, the degree of molecular heterogeneity in the midsection of the S-shaped body is unclear. However, Hnf1b does appears to extend more proximally than these other regulatory factors. Both Pou3f3 and Hnf1b are independently required for loop of Henle development. Hnf1b is also required for proximal tubule development, while a Pou3f3-requirement extends distally to the macula densa and distal convoluted tubule (Heliot et al., 2013; Massa et al., 2013; Nakai et al., 2003). The phenotype of *Hnf1b* mutants is more severe than *Pou3f3* mutants: in *Hnf1b* mutants glomeruli connect to the collecting duct through a vestigial epithelial tube, and *Pou3f3*, *Irx1/2*, *Dll1*, *Jag1*, and *Lfng* are all downregulated (Heliot et al., 2013; Massa et al., 2013).

Several transcriptional regulators, including Lef1, Sox9, and Lhx1, extend into the distal-most cells of the S-shaped body (Kobayashi et al., 2005; Mugford et al., 2009; Reginensi et al., 2011). Of these, Lhx1 and Sox9 have been implicated in distal nephron development (Kobayashi et al., 2005; Kumar et al., 2015). Surprisingly, even abnormally patterned S-shaped bodies fuse to the collecting duct suggesting this critical interconnection may not be dependent on distal nephron patterning events (Heliot et al., 2013; Kobayashi et al., 2005; Massa et al., 2013).

Development from the S-shaped body to the functioning nephron sees the emergence of mature cell types coupled to clonal expansion of nephron domains (Barker et al., 2012; Rinkevich et al., 2014). Podocyte maturation has been extensively characterized, driven in part by the clear connection of podocyte biology to kidney disease (Schell, Wanner, & Huber, 2014). However, the regulatory controls driving maturation of other nephron components are less well understood, though there are likely continuing roles for some of the early factors such as Hnf1b in driving late differentiation programs of channel and transporter gene expression in the proximal tubule (Naylor & Davidson, 2014). Development of nephron segments occurs in concert with the development of other kidney structures. For example, the elongation of the loop of Henle is prevented in *Wnt7b* mutants where the medullary ureteric system fails to form (Yu et al., 2009). As in other organ systems, it is likely that function will play a role in regulating tubular maturation; as an example, active fluid transport maintains epithelial polarity of membrane proteins in cells of the adult nephron (Jang et al., 2013).

2.4 The Interstitial Cell Population

The bulk of interstitial cell types, with the exception of smooth muscle lining the renal pelvis and ureter, originate from a self-renewing, multipotent progenitor pool demarcated by the transcriptional regulator Foxd1 (Fig. 2B; Kobayashi et al., 2014). Though definitive clonal mapping has not been possible, distinct Six2+ and Foxd1+ nephron and interstitial progenitor pools, respectively, likely emerge from an Osr1+ progenitor prior to e10.5 (Mugford, Sipilä, McMahon, et al., 2008). At e10.5, a bolus of Six2+ metanephric mesenchyme is surrounded by a halo of Foxd1+ cells. Lineage tracing indicates nephron fate is fixed within Six2+ cells at this time. Interstitial fate is also fixed in most Foxd1+ cells though a small percentage (1–2%) are capable of transitioning to Six2+/Foxd1− nephron progenitors until e11.0 (Kobayashi et al., 2014). Foxd1 is not essential for the interstitial progenitor fate and fate-specifying mechanisms for this progenitor compartment are unclear. A number of transcriptional regulators display a readily identifiable interstitial progenitor prolife in whole kidney *in situ* expression studies (Yu et al., 2012). Of these, Pbx1 plays a role in maintaining the progenitor state blocking premature maturation of the progenitor pool (Hurtado et al., 2015).

Fate mapping shows a continuous commitment of Foxd1+ progenitors to Foxd1− interstitial cell types in parallel with the commitment of neighboring Six2+ nephron progenitors (Fig. 2B; Kobayashi et al., 2014). For the most part, descendants of the Foxd1 pool are closely associated with vascular endothelium forming the mesangial cells of the glomerulus and functionally related pericytes regulating the remainder of the kidney vasculature. Foxd1 descendants also generate smooth muscle and smooth muscle derivatives associated with the arterial system, notably renin-secreting juxtaglomerular cell types (Fig. 1C). The extent to which cellular diversity reflects prepatterning or local responses of a common interstitial cell type to local tissue interactions has not been critically examined, though there is evidence, particularly within arteriole-associated smooth muscle, for regional plasticity.

An interplay between interstitial cells and vascular cells is critical in the vascular knot where mesangial cells support multiple vascular loops, an arrangement vital for normal glomerular function. Both Notch and Pdgfrb signaling are required for mesangial cell programs and, in their absence, the vasculature tuft collapses (Boyle, Liu, & Kopan, 2014; Lin, Sequeira-Lopez, & Gomez, 2014; Lindahl et al., 1998). Further, loss of Foxd1 or Foxd1-derived interstitial progenitors demonstrates a requirement for normal interstitial cell types in forming the renal arterial tree (Sequeira-Lopez et al., 2015). How interstitial cell types interact with the renal tubular component of the nephron is not well understood. Here the shape, size, and close apposition of key cellular players make resolving cell type-specific interactions a particular challenge.

2.5 Vascular and Neural Development

Analysis of organ explants and the tracing of vascular development *in vivo* support multiple routes to vascularizing the mouse kidney (Abrahamson, Robert, Hyink, John, & Daniel, 1998; Herzlinger & Hurtado, 2014). Glomerular vascular progenitors descendant from Osr1+ precursors are present within the initial kidney anlagen, and Flk1+ vascular progenitors collect between bifurcating branch tips within the tip niche at later stages (Fig. 2B; Hyink et al., 1996; Mugford, Sipilä, McMahon, et al., 2008; Mugford et al., 2009). However, vascular ingrowth into the developing kidney is also possible on grafting of kidneys to new tissue sites (Abrahamson et al., 1998; Sariola et al., 1984). While this may suggest distinct developmental origins for the glomerular vasculature and main arterial tree, definitive experiments to settle this issue, without a potential for graft-induced anomalies, have not been possible to date.

The first evidence of nephron-associated vasculature is seen with the convergence of vascular cells at the glomerular cleft of the forming S-shaped body (Fig. 3A). This coincides with, and is likely driven by, Vegfa synthesis in podocyte precursors. In mature podocytes, Vegfa signaling maintains the adjacent glomerular vasculature and promotes endothelial fenestration (Eremina et al., 2003). A second signaling axis essential for the glomerular vasculature is the Cxcl12/Cxcr4 pathway, where Cxcl12 secreted by podocytes and mesangial cells regulates endothelial Cxcr4 activity (Takabatake et al., 2009). Vascular-promoting activities within the glomerulus may be countered by Sem3a inhibition of vascular cell numbers (Reidy et al., 2009). Recent experiments also highlight a role for renal tubule-derived Vegfa in maintenance of the peritubular vasculature (Dimke et al., 2015). Despite the fact that the renal vasculature is critical for normal renal function, the molecular programs establishing renal vasculature architecture and endothelial cell differentiation and function have not been extensively characterized.

Sympathetic innervation of the kidney has attracted considerable interest given the potential for reducing hypertension by surgical deinnervation of the sympathetic input (Grassi et al., 2015). However, though the adult physiological action of sympathetic neural regulation has been extensively studied, the developmental programs generating the neural network have not been comprehensively addressed with modern tracing, fate mapping, and molecular approaches (Ferguson, Ryan, & Bell, 1986; Sariola, Holm, & Henke-Fahle, 1988). Nonrenal studies have highlighted guidance pathways shared by vascular and neural cells in a variety of animal models (Carmeliet & Tessier-Lavigne, 2005). In the kidney, sympathetic axonal processes closely track the main arterial input into the kidney: given precedents in other systems, it seems reasonable to speculate that the physical linkage of arterial and sympathetic systems may reflect coordinate sensing of shared guidance cues.

3. MAN VERSUS MOUSE

How similar is the developing mouse kidney to our own (Cullen-McEwen, Sutherland, & Black, 2016)? Given that the principal drive in studying the mouse is to understand something of us, this is an important and pertinent question. Mutations in genes for many key regulators of the developing mouse kidney associate with congenital anomalies of the urinary tract (CAKUT) syndromes in the patient population (Nicolaou, Renkema, Bongers, Giles, & Knoers, 2015; Vivante, Kohl, Hwang, Dworschak, & Hildebrandt, 2014). Indeed, by taking a set of genes educated by mouse studies to examine patient DNA samples, headway has been made in identifying potential genetic factors underlying developmental anomalies of the human kidney in families with isolated CAKUT (Hwang et al., 2014).

However, there has been little direct comparative molecular analysis of the developing mouse. The human kidney is multilobed, forming 8–15 independent renal calyces, while the mouse kidney makes just one (Treuting & Kowalewska, 2011). This suggests a fundamental difference in ureteric branching and the organization of nephron progenitor pools between the mouse and human kidney at the outset of kidney development, and potentially later differences in remodeling the medullary ureteric architecture. Whereas the mouse ureteric network undergoes dichotomous branching as a standard routine to extend the cortical territory, in the human kidney ureteric branching is reported to be restricted to the first 14–15 weeks of kidney development, the radial extension of the ureteric epithelium continues for many weeks thereafter in the absence of branching (Osathanondh & Potter, 1963). Given the coupling of renal vesicle induction to ureteric branching, and the specific location of renal vesicle formation beneath branch tips, the human observations raise additional questions about the conservation of ureteric growth programs, maintenance of nephron progenitor pools and nephron-forming mechanisms at later developmental stages.

How might similar molecular mechanisms operate at different phases of human kidney development? The observation that different Fgf signals induce elongation versus branching growth mediated through the MAP kinase pathway is especially intriguing in this regard (Ihermann-Hella et al., 2014; Qiao et al., 2001). Arcading is a prevalent mode of hooking up nephrons in the human kidney following the period of branching growth. In this, nephrons interconnect to other nephrons as well as to the ureteric epithelium. Consequently, the connection process is flexible to epithelial origin (Cullen-McEwen et al., 2016). Importantly, the final number of nephrons differs greatly between mouse and man. Does this reflect species differences in the size of the initiating progenitor pool, or different rates of progenitor expansion and commitment? Mechanisms regulating organ size across different species are generally poorly understood; the kidney is no exception.

Detailed and extensive molecular analysis of the mouse kidney by the Genitourinary Molecular Anatomy Project (GUDMAP; www.gudmap.org) has enhanced the understanding of mouse kidney development (McMahon et al., 2008; www.gudmap.org). The high-resolution developmental anatomy produced by GUDMAP participants provides a molecular framework to extend comparative analysis to the human kidney. Efforts are already underway in a human GUDMAP initiative (www.gudmap.org).

Recent studies have revealed some striking differences in *Six1/SIX1* and *Six2/SIX2* regulation within the nephron progenitor compartment of the mouse and human kidney (O'Brien et al., 2016). Given the critical role of Six-factors in maintaining nephron progenitors, variable *SIX* gene activity could contribute to differences in the period of nephrogenesis and final nephron count.

In addressing species differences, a better understanding of the underpinnings of mouse nephrogenesis is essential. A more detailed comparative molecular anatomy of nephron morphogenesis coupled with extensive regional fate mapping can provide a much clearer view of how early domains of gene expression relate to mature structures of the adult kidney (Fig. 3A and B). The cellular complexity of the mammalian kidney is not well understood. Twenty-six cell types have been postulated to form from cap mesenchyme and ureteric bud derivatives (Al-Awqati & Oliver, 2002). However, in the absence of extensive molecular characterization, this figure likely underestimates the number of molecularly distinct differentiated cell derivatives. As an example, detailed studies of proximal tubule development suggest considerable spatial heterogeneity in gene expression patterns (Thiagarajan et al., 2011) though more direct comparative expression studies with multiplex gene expression mapping approaches will be essential to determine the actual extent of cellular diversity.

In the mouse, the majority of nephrons form after birth while in the human kidney nephrogenesis ceases around 36 weeks so that nephrogenesis is complete at birth (Osathanondh & Potter, 1963; Short et al., 2014). Consequently, the final phases of kidney development occur in very different contexts. This begs the question of whether there are species differences regulating these terminal programs and the potential impact of premature birth, now survivable as early as 22–26 weeks? Premature birth is reported to significantly impact kidney function (Sutherland, Ryan, Black, & Kent, 2014). Notably, hypoxia associated with premature birth has damaging effects on multiple organ systems and can influence nephron number in experimental mouse models (Wilkinson et al., 2015).

4. TRANSLATING DEVELOPMENT BIOLOGY: REGENERATIVE APPROACHES TO KIDNEY DISEASE

Given the limits of dialysis, a shortage of kidneys for transplantation and an absence of other therapeutic options, new solutions for treating end-stage renal disease are a high priority. The last few years have seen some important enabling developments with the first reports of directed differentiation of mouse and human pluripotent stem cells to distinct kidney progenitor types (Taguchi et al., 2014; Takasato et al., 2014; Uzarski et al., 2014; Xia et al., 2013). Depending on the induction conditions, cultures can self-organize to generate kidney-like organoids with both ureteric-like and nephron progenitor-like cell types (Takasato et al., 2014). In other procedures, isolated progenitors can be generated independently for each of these lineages (Taguchi et al., 2014; Xia et al., 2013), and nephron progenitor-like cells differentiated into distinct nephron fates (Taguchi & Nishinakamura, 2015). Further, differentiated nephron types can be reprogrammed to earlier nephron progenitor-like states through the ectopic action of a small number of transcriptional regulators associated with normal progenitor programs (Hendry et al., 2013). In addition, *in vitro* culture conditions have been identified that enable the long-term expansion of nephron progenitors while maintaining their competence to initiate nephrogenesis under appropriate conditions (Brown et al., 2015).

Strikingly, all of these advances come from a rigorous application of knowledge gained from studying kidney development in model systems (Taguchi & Nishinakamura, 2015; Takasato & Little, 2015; Vanslambrouck & Little, 2015). Applied developmental biology is opening the door to new approaches to investigate, model, and treat human kidney disease (Little & Takasato, 2015).

ACKNOWLEDGMENTS

I am grateful to Cristy Lytal and Heather Lipari for help with the manuscript and Georgas Designs for production of the figures. Kidney research in my laboratory is supported by grants from the National Institute for Diabetes, Digestive, and Kidney Disease and the California Institute for Regenerative Medicine.

REFERENCES

Abrahamson, D. R., Robert, B., Hyink, D. P., John, P. L. S., & Daniel, T. O. (1998). Origins and formation of microvasculature in the developing kidney. *Kidney International*, *54*, S7–S11.

Al-Awqati, Q., & Gao, X. B. (2011). Differentiation of intercalated cells in the kidney. *Physiology*, *26*(4), 266–272.

Al-Awqati, Q., & Oliver, J. A. (2002). Stem cells in the kidney. *Kidney International*, *61*(2), 387–395.

Bagherie-Lachidan, M., Reginensi, A., Pan, Q., Zaveri, H. P., Scott, D. A., Blencowe, B. J., et al. (2015). Stromal Fat4 acts non-autonomously with Dchs1/2 to restrict the nephron progenitor pool. *Development*, *142*(15), 2564–2573.

Barak, H., Huh, S. H., Chen, S., Jeanpierre, C., Martinovic, J., Parisot, M., et al. (2012). FGF9 and FGF20 maintain the stemness of nephron progenitors in mice and man. *Developmental Cell, 22*(6), 1191–1207.

Barker, N., Rookmaaker, M. B., Kujala, P., Ng, A., Leushacke, M., Snippert, H., et al. (2012). Lgr5+ve stem/progenitor cells contribute to nephron formation during kidney development. *Cell Reports, 2*(3), 540–552.

Basson, M. A., Akbulut, S., Watson-Johnson, J., Simon, R., Carroll, T. J., Shakya, R., et al. (2005). Sprouty1 is a critical regulator of GDNF/RET-mediated kidney induction. *Developmental Cell, 8*(2), 229–239.

Basta, J. M., Robbins, L., Kiefer, S. M., Dorsett, D., & Rauchman, M. (2014). Sall1 balances self-renewal and differentiation of renal progenitor cells. *Development, 141*(5), 1047–1058.

Bell, P. D., Lapointe, J. Y., & Peti-Peterdi, J. (2003). Macula densa cell signaling. *Annual Review of Physiology, 65*(1), 481–500.

Bertram, J. F., Douglas-Denton, R. N., Diouf, B., Hughson, M. D., & Hoy, W. E. (2011). Human nephron number: Implications for health and disease. *Pediatric Nephrology, 26*(9), 1529–1533.

Bierzynska, A., Soderquest, K., & Koziell, A. (2015). Genes and podocytes—New insights into mechanisms of podocytopathy. *Frontiers in Endocrinology, 5*, 226.

Blank, U., Brown, A., Adams, D. C., Karolak, M. J., & Oxburgh, L. (2009). BMP7 promotes proliferation of nephron progenitor cells via a JNK-dependent mechanism. *Development, 136*(21), 3557–3566.

Blomqvist, S. R., Vidarsson, H., Fitzgerald, S., Johansson, B. R., Ollerstam, A., Brown, R., et al. (2004). Distal renal tubular acidosis in mice that lack the forkhead transcription factor Foxi1. *Journal of Clinical Investigation, 113*(11), 1560.

Blumenfeld, M., Maury, M., Chouard, T., Yaniv, M., & Condamine, H. (1991). Hepatic nuclear factor 1 (HNF1) shows a wider distribution than products of its known target genes in developing mouse. *Development, 113*(2), 589–599.

Boyle, S. C., Kim, M., Valerius, M. T., McMahon, A. P., & Kopan, R. (2011). Notch pathway activation can replace the requirement for Wnt4 and Wnt9b in mesenchymal-to-epithelial transition of nephron stem cells. *Development, 138*(19), 4245–4254.

Boyle, S. C., Liu, Z., & Kopan, R. (2014). Notch signaling is required for the formation of mesangial cells from a stromal mesenchyme precursor during kidney development. *Development, 141*(2), 346–354.

Boyle, S., Misfeldt, A., Chandler, K. J., Deal, K. K., Southard-Smith, E. M., Mortlock, D. P., et al. (2008). Fate mapping using Cited1-CreER T2 mice demonstrates that the cap mesenchyme contains self-renewing progenitor cells and gives rise exclusively to nephronic epithelia. *Developmental Biology, 313*(1), 234–245.

Boyle, S., Shioda, T., Perantoni, A. O., & de Caestecker, M. (2007). Cited1 and Cited2 are differentially expressed in the developing kidney but are not required for nephrogenesis. *Developmental Dynamics, 236*(8), 2321–2330.

Bridgewater, D., Cox, B., Cain, J., Lau, A., Athaide, V., Gill, P. S., et al. (2008). Canonical WNT/β-catenin signaling is required for ureteric branching. *Developmental Biology, 317*(1), 83–94.

Brown, A. C., Adams, D., de Caestecker, M., Yang, X., Friesel, R., & Oxburgh, L. (2011). FGF/EGF signaling regulates the renewal of early nephron progenitors during embryonic development. *Development, 138*(23), 5099–5112.

Brown, A. C., Muthukrishnan, S. D., Guay, J. A., Adams, D. C., Schafer, D. A., Fetting, J. L., et al. (2013). Role for compartmentalization in nephron progenitor differentiation. *Proceedings of the National Academy of Sciences of the United States of America, 110*(12), 4640–4645.

Brown, A. C., Muthukrishnan, S. D., & Oxburgh, L. (2015). A synthetic niche for nephron progenitor cells. *Developmental Cell, 34*(2), 229–241.

Brunskill, E. W., Aronow, B. J., Georgas, K., Rumballe, B., Valerius, M. T., Aronow, J., et al. (2008). Atlas of gene expression in the developing kidney at microanatomic resolution. *Developmental Cell*, *15*(5), 781–791.
Brunskill, E. W., Georgas, K., Rumballe, B., Little, M. H., & Potter, S. S. (2011). Defining the molecular character of the developing and adult kidney podocyte. *PloS One*, *6*(9), e24640.
Burn, S. F., Webb, A., Berry, R. L., Davies, J. A., Ferrer-Vaquer, A., Hadjantonakis, A. K., et al. (2011). Calcium/NFAT signalling promotes early nephrogenesis. *Developmental Biology*, *352*(2), 288–298.
Cacalano, G., Fariñas, I., Wang, L. C., Hagler, K., Forgie, A., Moore, M., et al. (1998). GFRalpha1 is an essential receptor component for GDNF in the developing nervous system and kidney. *Neuron*, *21*(1), 53–62.
Carmeliet, P., & Tessier-Lavigne, M. (2005). Common mechanisms of nerve and blood vessel wiring. *Nature*, *436*(7048), 193–200.
Carroll, T. J., Park, J. S., Hayashi, S., Majumdar, A., & McMahon, A. P. (2005). Wnt9b plays a central role in the regulation of mesenchymal to epithelial transitions underlying organogenesis of the mammalian urogenital system. *Developmental Cell*, *9*(2), 283–292.
Cebrian, C., Asai, N., D'Agati, V., & Costantini, F. (2014). The number of fetal nephron progenitor cells limits ureteric branching and adult nephron endowment. *Cell Reports*, *7*(1), 127–137.
Cheng, H. T., Kim, M., Valerius, M. T., Surendran, K., Schuster-Gossler, K., Gossler, A., et al. (2007). Notch2, but not Notch1, is required for proximal fate acquisition in the mammalian nephron. *Development*, *134*(4), 801–811.
Chi, X., Michos, O., Shakya, R., Riccio, P., Enomoto, H., Licht, J. D., et al. (2009). Ret-dependent cell rearrangements in the Wolffian duct epithelium initiate ureteric bud morphogenesis. *Developmental Cell*, *17*(2), 199–209.
Chia, I., Grote, D., Marcotte, M., Batourina, E., Mendelsohn, C., & Bouchard, M. (2011). Nephric duct insertion is a crucial step in urinary tract maturation that is regulated by a Gata3-Raldh2-Ret molecular network in mice. *Development*, *138*(10), 2089–2097.
Combes, A. N., Davies, J. A., & Little, M. H. (2015). Cell-cell interactions driving kidney morphogenesis. *Current Topics in Developmental Biology*, *112*, 467–508.
Combes, A. N., Short, K. M., Lefevre, J., Hamilton, N. A., Little, M. H., & Smyth, I. M. (2014). An integrated pipeline for the multidimensional analysis of branching morphogenesis. *Nature Protocols*, *9*(12), 2859–2879.
Costantini, F. (2012). Genetic controls and cellular behaviors in branching morphogenesis of the renal collecting system. *Wiley Interdisciplinary Reviews. Developmental Biology*, *1*(5), 693–713.
Costantini, F., & Kopan, R. (2010). Patterning a complex organ: Branching morphogenesis and nephron segmentation in kidney development. *Developmental Cell*, *18*(5), 698–712.
Cullen-McEwen, L., Sutherland, M., & Black, M. J. (2016). The human kidney parallels in structure, spatial development and timing of nephrogenesis. In M. H. Little (Ed.), *Kidney development, disease, repair and regeneration* (pp. 27–37): Amsterdam: Elsevier.
Das, A., Tanigawa, S., Karner, C. M., Xin, M., Lum, L., Chen, C., et al. (2013). Stromal–epithelial crosstalk regulates kidney progenitor cell differentiation. *Nature Cell Biology*, *15*(9), 1035–1044.
Davies, J. A., Hohenstein, P., Chang, C. H., & Berry, R. (2014). A self-avoidance mechanism in patterning of the urinary collecting duct tree. *BMC Developmental Biology*, *14*(1), 35.
Dimke, H., Sparks, M. A., Thomson, B. R., Frische, S., Coffman, T. M., & Quaggin, S. E. (2015). Tubulovascular cross-talk by vascular endothelial growth factor A maintains peritubular microvasculature in kidney. *Journal of the American Society of Nephrology*, *26*(5), 1027–1038.

Dudley, A. T., & Robertson, E. J. (1997). Overlapping expression domains of bone morphogenetic protein family members potentially account for limited tissue defects in BMP7 deficient embryos. *Developmental Dynamics*, *208*(3), 349–362.

Enomoto, H., Araki, T., Jackman, A., Heuckeroth, R. O., Snider, W. D., Johnson, E. M., Jr., et al. (1998). GFR alpha1-deficient mice have deficits in the enteric nervous system and kidneys. *Neuron*, *21*(2), 317–324.

Eremina, V., Sood, M., Haigh, J., Nagy, A., Lajoie, G., Ferrara, N., et al. (2003). Glomerular-specific alterations of VEGF-A expression lead to distinct congenital and acquired renal diseases. *Journal of Clinical Investigation*, *111*(5), 707.

Fejes-Tóth, G., & Náray-Fejes-Tóth, A. (1992). Differentiation of renal beta-intercalated cells to alpha-intercalated and principal cells in culture. *Proceedings of the National Academy of Sciences of the United States of America*, *89*(12), 5487–5491.

Ferguson, M., Ryan, G. B., & Bell, C. (1986). Localization of sympathetic and sensory neurons innervating the rat kidney. *Journal of the Autonomic Nervous System*, *16*(4), 279–288.

Fetting, J. L., Guay, J. A., Karolak, M. J., Iozzo, R. V., Adams, D. C., Maridas, D. E., et al. (2014). FOXD1 promotes nephron progenitor differentiation by repressing decorin in the embryonic kidney. *Development*, *141*(1), 17–27.

Georgas, K., Rumballe, B., Valerius, M. T., Chiu, H. S., Thiagarajan, R. D., Lesieur, E., et al. (2009). Analysis of early nephron patterning reveals a role for distal RV proliferation in fusion to the ureteric tip via a cap mesenchyme-derived connecting segment. *Developmental Biology*, *332*(2), 273–286.

Grahammer, F., Schell, C., & Huber, T. B. (2013). The podocyte slit diaphragm—From a thin grey line to a complex signalling hub. *Nature Reviews. Nephrology*, *9*(10), 587–598.

Grassi, G., Mark, A., & Esler, M. (2015). The sympathetic nervous system alterations in human hypertension. *Circulation Research*, *116*(6), 976–990.

Grieshammer, U., Cebrián, C., Ilagan, R., Meyers, E., Herzlinger, D., & Martin, G. R. (2005). FGF8 is required for cell survival at distinct stages of nephrogenesis and for regulation of gene expression in nascent nephrons. *Development*, *132*(17), 3847–3857.

Grieshammer, U., Ma, L., Plump, A. S., Wang, F., Tessier-Lavigne, M., & Martin, G. R. (2004). SLIT2-mediated ROBO2 signaling restricts kidney induction to a single site. *Developmental Cell*, *6*(5), 709–717.

GUDMAP Consortium of Laboratories. (2005–2015). Mouse report data. GenitoUrinary Development Molecular Anatomy Project (GUDMAP), www.gudmap.org, August 2015.

Hartwig, S., Ho, J., Pandey, P., MacIsaac, K., Taglienti, M., Xiang, M., et al. (2010). Genomic characterization of Wilms' tumor suppressor 1 targets in nephron progenitor cells during kidney development. *Development*, *137*(7), 1189–1203.

Heliot, C., Desgrange, A., Buisson, I., Prunskaite-Hyyryläinen, R., Shan, J., Vainio, S., et al. (2013). HNF1B controls proximal-intermediate nephron segment identity in vertebrates by regulating Notch signalling components and Irx1/2. *Development*, *140*(4), 873–885.

Hendry, C. E., Vanslambrouck, J. M., Ineson, J., Suhaimi, N., Takasato, M., Rae, F., et al. (2013). Direct transcriptional reprogramming of adult cells to embryonic nephron progenitors. *Journal of the American Society of Nephrology*, *24*(9), 1424–1434.

Herzlinger, D., & Hurtado, R. (2014). Patterning the renal vascular bed. *Seminars in Cell & Developmental Biology*, *36*, 50–56.

Hinchliffe, S. A., Sargent, P. H., Howard, C. V., Chan, Y. F., & Van Velzen, D. (1991). Human intrauterine renal growth expressed in absolute number of glomeruli assessed by the disector method and Cavalieri principle. *Laboratory Investigation: A Journal of Technical Methods and Pathology*, *64*(6), 777–784.

Hoy, W. E., Hughson, M. D., Diouf, B., Zimanyi, M., Samuel, T., McNamara, B. J., et al. (2011). Distribution of volumes of individual glomeruli in kidneys at autopsy:

Association with physical and clinical characteristics and with ethnic group. *American Journal of Nephrology, 33*(Suppl 1), 15–20.
Hurtado, R., Zewdu, R., Mtui, J., Liang, C., Aho, R., Kurylo, C., et al. (2015). Pbx1-dependent control of VMC differentiation kinetics underlies gross renal vascular patterning. *Development, 142*(15), 2653–2664.
Hwang, D. Y., Dworschak, G. C., Kohl, S., Saisawat, P., Vivante, A., Hilger, A. C., et al. (2014). Mutations in 12 known dominant disease-causing genes clarify many congenital anomalies of the kidney and urinary tract. *Kidney International, 85*(6), 1429–1433.
Hyink, D. P., Tucker, D. C., St John, P. L., Leardkamolkarn, V., Accavitti, M. A., Abrass, C. K., et al. (1996). Endogenous origin of glomerular endothelial and mesangial cells in grafts of embryonic kidneys. *American Journal of Physiology Renal Physiology, 270*(5), F886–F899.
Ihermann-Hella, A., Lume, M., Miinalainen, I. J., Pirttiniemi, A., Gui, Y., Peränen, J., et al. (2014). Mitogen-activated protein kinase (MAPK) pathway regulates branching by remodeling epithelial cell adhesion. *PLoS Genetics, 10*(3), e1004193.
Ishibe, S., Karihaloo, A., Ma, H., Zhang, J., Marlier, A., Mitobe, M., et al. (2009). Met and the epidermal growth factor receptor act cooperatively to regulate final nephron number and maintain collecting duct morphology. *Development, 136*(2), 337–345.
Jang, K. J., Mehr, A. P., Hamilton, G. A., McPartlin, L. A., Chung, S., Suh, K. Y., et al. (2013). Human kidney proximal tubule-on-a-chip for drug transport and nephrotoxicity assessment. *Integrative Biology, 5*(9), 1119–1129.
Jeong, H. W., Jeon, U. S., Koo, B. K., Kim, W. Y., Im, S. K., Shin, J., et al. (2009). Inactivation of Notch signaling in the renal collecting duct causes nephrogenic diabetes insipidus in mice. *The Journal of Clinical Investigation, 119*(11), 3290.
Kanda, S., Tanigawa, S., Ohmori, T., Taguchi, A., Kudo, K., Suzuki, Y., et al. (2014). Sall1 maintains nephron progenitors and nascent nephrons by acting as both an activator and a repressor. *Journal of the American Society of Nephrology, 25*(11), 2584–2595.
Kao, R. M., Vasilyev, A., Miyawaki, A., Drummond, I. A., & McMahon, A. P. (2012). Invasion of distal nephron precursors associates with tubular interconnection during nephrogenesis. *Journal of the American Society of Nephrology, 23*(10), 1682–1690.
Karner, C. M., Chirumamilla, R., Aoki, S., Igarashi, P., Wallingford, J. B., & Carroll, T. J. (2009). Wnt9b signaling regulates planar cell polarity and kidney tubule morphogenesis. *Nature Genetics, 41*(7), 793–799.
Karner, C. M., Das, A., Ma, Z., Self, M., Chen, C., Lum, L., et al. (2011). Canonical Wnt9b signaling balances progenitor cell expansion and differentiation during kidney development. *Development, 138*(7), 1247–1257.
Karner, C. M., Merkel, C. E., Dodge, M., Ma, Z., Lu, J., Chen, C., et al. (2010). Tankyrase is necessary for canonical Wnt signaling during kidney development. *Developmental Dynamics, 239*(7), 2014–2023.
Kispert, A., Vainio, S., & McMahon, A. P. (1998). Wnt-4 is a mesenchymal signal for epithelial transformation of metanephric mesenchyme in the developing kidney. *Development, 125*(21), 4225–4234.
Kobayashi, A., Kwan, K. M., Carroll, T. J., McMahon, A. P., Mendelsohn, C. L., & Behringer, R. R. (2005). Distinct and sequential tissue-specific activities of the LIM-class homeobox gene Lim1 for tubular morphogenesis during kidney development. *Development, 132*(12), 2809–2823.
Kobayashi, A., Mugford, J. W., Krautzberger, A. M., Naiman, N., Liao, J., & McMahon, A. P. (2014). Identification of a multipotent self-renewing stromal progenitor population during mammalian kidney organogenesis. *Stem Cell Reports, 3*(4), 650–662.
Kobayashi, A., Valerius, M. T., Mugford, J. W., Carroll, T. J., Self, M., Oliver, G., et al. (2008). Six2 defines and regulates a multipotent self-renewing nephron progenitor population throughout mammalian kidney development. *Cell Stem Cell, 3*(2), 169–181.

Kopan, R., Chen, S., & Little, M. (2013). Nephron progenitor cells: Shifting the balance of self-renewal and differentiation. *Current Topics in Developmental Biology, 107*, 293–331.

Kramann, R., & Humphreys, B. D. (2014). Kidney pericytes: Roles in regeneration and fibrosis. *Seminars in Nephrology, 34*(4), 374–383.

Kumar, S., Liu, J., Pang, P., Krautzberger, A. M., Reginensi, A., Akiyama, H., et al. (2015). Sox9 activation highlights a cellular pathway of renal repair in the acutely injured mammalian kidney. *Cell Reports, 12*(8), 1325–1338.

Kume, T., Deng, K., & Hogan, B. L. (2000). Murine forkhead/winged helix genes Foxc1 (Mf1) and Foxc2 (Mfh1) are required for the early organogenesis of the kidney and urinary tract. *Development, 127*(7), 1387–1395.

Kuure, S., Cebrian, C., Machingo, Q., Lu, B. C., Chi, X., Hyink, D., et al. (2010). Actin depolymerizing factors cofilin1 and destrin are required for ureteric bud branching morphogenesis. *PLoS Genetics, 6*(10), e1001176.

Kuure, S., Chi, X., Lu, B., & Costantini, F. (2010). The transcription factors Etv4 and Etv5 mediate formation of the ureteric bud tip domain during kidney development. *Development, 137*(12), 1975–1979.

Leimeister, C., Schumacher, N., & Gessler, M. (2003). Expression of Notch pathway genes in the embryonic mouse metanephros suggests a role in proximal tubule development. *Gene Expression Patterns, 3*(5), 595–598.

Lin, E. E., Sequeira-Lopez, M. L. S., & Gomez, R. A. (2014). RBP-J in FOXD1+ renal stromal progenitors is crucial for the proper development and assembly of the kidney vasculature and glomerular mesangial cells. *American Journal of Physiology. Renal Physiology, 306*(2), F249–F258.

Lindahl, P., Hellstrom, M., Kalén, M., Karlsson, L., Pekny, M., Pekna, M., et al. (1998). Paracrine PDGF-B/PDGF-Rbeta signaling controls mesangial cell development in kidney glomeruli. *Development, 125*(17), 3313–3322.

Lindström, N. O., Lawrence, M. L., Burn, S. F., Johansson, J. A., Bakker, E. R., Ridgway, R. A., et al. (2015). Integrated β-catenin, BMP, PTEN, and Notch signalling patterns the nephron. *Elife, 4*, e04000.

Linton, J. M., Martin, G. R., & Reichardt, L. F. (2007). The ECM protein nephronectin promotes kidney development via integrinα 8β1-mediated stimulation of Gdnf expression. *Development, 134*(13), 2501–2509.

Little, M. H. (2015). Improving our resolution of kidney morphogenesis across time and space. *Current Opinion in Genetics & Development, 32*, 135–143.

Little, M. H., & McMahon, A. P. (2012). Mammalian kidney development: Principles, progress, and projections. *Cold Spring Harbor Perspectives in Biology, 4*(5), a008300.

Little, M. H., & Takasato, M. (2015). Generating a self-organizing kidney from pluripotent cells. *Current Opinion in Organ Transplantation, 20*(2), 178–186.

Lu, B. C., Cebrian, C., Chi, X., Kuure, S., Kuo, R., Bates, C. M., et al. (2009). Etv4 and Etv5 are required downstream of GDNF and Ret for kidney branching morphogenesis. *Nature Genetics, 41*(12), 1295–1302.

Maezawa, Y., Takemoto, M., & Yokote, K. (2015). Cell biology of diabetic nephropathy: Roles of endothelial cells, tubulointerstitial cells and podocytes. *Journal of Diabetes Investigation, 6*(1), 3–15.

Mah, S. P., Saueressig, H., Goulding, M., Kintner, C., & Dressler, G. R. (2000). Kidney development in cadherin-6 mutants: Delayed mesenchyme-to-epithelial conversion and loss of nephrons. *Developmental Biology, 223*(1), 38–53.

Majumdar, A., Vainio, S., Kispert, A., McMahon, J., & McMahon, A. P. (2003). Wnt11 and Ret/Gdnf pathways cooperate in regulating ureteric branching during metanephric kidney development. *Development, 130*(14), 3175–3185.

Marose, T. D., Merkel, C. E., McMahon, A. P., & Carroll, T. J. (2008). β-catenin is necessary to keep cells of ureteric bud/Wolffian duct epithelium in a precursor state. *Developmental Biology*, *314*(1), 112–126.

Massa, F., Garbay, S., Bouvier, R., Sugitani, Y., Noda, T., Gubler, M. C., et al. (2013). Hepatocyte nuclear factor 1β controls nephron tubular development. *Development*, *140*(4), 886–896.

McMahon, A. P., Aronow, B. J., Davidson, D. R., Davies, J. A., Gaido, K. W., Grimmond, S., et al. (2008). GUDMAP: The genitourinary developmental molecular anatomy project. *Journal of the American Society of Nephrology*, *19*(4), 667–671.

Menshykau, D., & Iber, D. (2013). Kidney branching morphogenesis under the control of a ligand–receptor-based Turing mechanism. *Physical Biology*, *10*(4), 046003.

Menshykau, D., Kraemer, C., & Iber, D. (2012). Branch mode selection during early lung development. *PLoS Computational Biology*, *8*(2), e1002377.

Michos, O., Cebrian, C., Hyink, D., Grieshammer, U., Williams, L., D'Agati, V., et al. (2010). Kidney development in the absence of Gdnf and Spry1 requires Fgf10. *PLoS Genetics*, *6*(1), e1000809.

Michos, O., Panman, L., Vintersten, K., Beier, K., Zeller, R., & Zuniga, A. (2004). Gremlin-mediated BMP antagonism induces the epithelial-mesenchymal feedback signaling controlling metanephric kidney and limb organogenesis. *Development*, *131*(14), 3401–3410.

Miner, J. H. (2011). Organogenesis of the kidney glomerulus: Focus on the glomerular basement membrane. *Organogenesis*, 7(2), 75–82.

Miner, J. H., Morello, R., Andrews, K. L., Li, C., Antignac, C., Shaw, A. S., et al. (2002). Transcriptional induction of slit diaphragm genes by Lmx1b is required in podocyte differentiation. *The Journal of Clinical Investigation*, *109*(8), 1065–1072.

Moriguchi, T., Hamada, M., Morito, N., Terunuma, T., Hasegawa, K., Zhang, C., et al. (2006). MafB is essential for renal development and F4/80 expression in macrophages. *Molecular and Cellular Biology*, *26*(15), 5715–5727.

Mugford, J. W., Sipilä, P., Kobayashi, A., Behringer, R. R., & McMahon, A. P. (2008a). Hoxd11 specifies a program of metanephric kidney development within the intermediate mesoderm of the mouse embryo. *Developmental Biology*, *319*(2), 396–405.

Mugford, J. W., Sipilä, P., McMahon, J. A., & McMahon, A. P. (2008b). Osr1 expression demarcates a multi-potent population of intermediate mesoderm that undergoes progressive restriction to an Osr1-dependent nephron progenitor compartment within the mammalian kidney. *Developmental Biology*, *324*(1), 88–98.

Mugford, J. W., Yu, J., Kobayashi, A., & McMahon, A. P. (2009). High-resolution gene expression analysis of the developing mouse kidney defines novel cellular compartments within the nephron progenitor population. *Developmental Biology*, *333*(2), 312–323.

Müller, U., Wang, D., Denda, S., Meneses, J. J., Pedersen, R. A., & Reichardt, L. F. (1997). Integrin α8β1 is critically important for epithelial–mesenchymal interactions during kidney morphogenesis. *Cell*, *88*(5), 603–613.

Mundlos, S., Pelletier, J., Darveau, A., Bachmann, M., Winterpacht, A., & Zabel, B. (1993). Nuclear localization of the protein encoded by the Wilms' tumor gene WT1 in embryonic and adult tissues. *Development*, *119*(4), 1329–1341.

Nakai, S., Sugitani, Y., Sato, H., Ito, S., Miura, Y., Ogawa, M., et al. (2003). Crucial roles of Brn1 in distal tubule formation and function in mouse kidney. *Development*, *130*(19), 4751–4759.

Naylor, R. W., & Davidson, A. J. (2014). Hnf1beta and nephron segmentation. *Pediatric Nephrology*, *29*(4), 659–664.

Nicolaou, N., Renkema, K. Y., Bongers, E. M., Giles, R. H., & Knoers, N. V. (2015). Genetic, environmental, and epigenetic factors involved in CAKUT. *Nature Reviews Nephrology*, *11*(12), 720–731.

O'Brien, L. L., & McMahon, A. P. (2014). Induction and patterning of the metanephric nephron. *Seminars in Cell & Developmental Biology*, *36*, 31–38:

O'Brien, L., Guo, Q., Lee, Y. J., Benazet, J. D., Valouev, A., & McMahon, A. P. (2016). Differential regulation of mouse and human nephron progenitors by the Six-family transcriptional regulators (in press).

Ohuchi, H., Hori, Y., Yamasaki, M., Harada, H., Sekine, K., Kato, S., et al. (2000). FGF10 acts as a major ligand for FGF receptor 2 IIIb in mouse multi-organ development. *Biochemical and Biophysical Research Communications*, *277*(3), 643–649.

Osathanondh, V., & Potter, E. L. (1963). Development of human kidney as shown by microdissection. III. Formation and interrelationship of collecting tubules and nephrons. *Archives of Pathology*, *76*, 290–302.

Packard, A., Georgas, K., Michos, O., Riccio, P., Cebrian, C., Combes, A. N., et al. (2013). Luminal mitosis drives epithelial cell dispersal within the branching ureteric bud. *Developmental Cell*, *27*(3), 319–330.

Park, J. S., Ma, W., O'Brien, L. L., Chung, E., Guo, J. J., Cheng, J. G., et al. (2012). Six2 and Wnt regulate self-renewal and commitment of nephron progenitors through shared gene regulatory networks. *Developmental Cell*, *23*(3), 637–651.

Park, J. S., Valerius, M. T., & McMahon, A. P. (2007). Wnt/β-catenin signaling regulates nephron induction during mouse kidney development. *Development*, *134*(13), 2533–2539.

Pearce, D., Soundararajan, R., Trimpert, C., Kashlan, O. B., Deen, P. M., & Kohan, D. E. (2015). Collecting duct principal cell transport processes and their regulation. *Clinical Journal of the American Society of Nephrology*, *10*(1), 135–146.

Pepicelli, C. V., Kispert, A., Rowitch, D. H., & McMahon, A. P. (1997). GDNF induces branching and increased cell proliferation in the ureter of the mouse. *Developmental Biology*, *192*(1), 193–198.

Perantoni, A. O., Timofeeva, O., Naillat, F., Richman, C., Pajni-Underwood, S., Wilson, C., et al. (2005). Inactivation of FGF8 in early mesoderm reveals an essential role in kidney development. *Development*, *132*(17), 3859–3871.

Pichel, J. G., Shen, L., Sheng, H. Z., Granholm, A. C., Drago, J., Grinberg, A., et al. (1996). Defects in enteric innervation and kidney development in mice lacking GDNF. *Nature*, *382*(6586), 73–76.

Plachov, D., Chowdhury, K., Walther, C., Simon, D., Guénet, J. L., & Gruss, P. (1990). Pax8, a murine paired box gene expressed in the developing excretory system and thyroid gland. *Development*, *110*(2), 643–651.

Poladia, D. P., Kish, K., Kutay, B., Hains, D., Kegg, H., Zhao, H., et al. (2006). Role of fibroblast growth factor receptors 1 and 2 in the metanephric mesenchyme. *Developmental Biology*, *291*(2), 325–339.

Puelles, V. G., Hoy, W. E., Hughson, M. D., Diouf, B., Douglas-Denton, R. N., & Bertram, J. F. (2011). Glomerular number and size variability and risk for kidney disease. *Current Opinion in Nephrology and Hypertension*, *20*(1), 7–15.

Qiao, J., Bush, K. T., Steer, D. L., Stuart, R. O., Sakurai, H., Wachsman, W., et al. (2001). Multiple fibroblast growth factors support growth of the ureteric bud but have different effects on branching morphogenesis. *Mechanisms of Development*, *109*(2), 123–135.

Qiao, J., Uzzo, R., Obara-Ishihara, T., Degenstein, L., Fuchs, E., & Herzlinger, D. (1999). FGF-7 modulates ureteric bud growth and nephron number in the developing kidney. *Development*, *126*(3), 547–554.

Rabelink, T. J., & Little, M. H. (2013). Stromal cells in tissue homeostasis: Balancing regeneration and fibrosis. *Nature Reviews. Nephrology*, *9*(12), 747–753.

Reginensi, A., Clarkson, M., Neirijnck, Y., Lu, B., Ohyama, T., Groves, A. K., et al. (2011). SOX9 controls epithelial branching by activating RET effector genes during kidney development. *Human Molecular Genetics*, *20*(6), 1143–1153.

Reginensi, A., Scott, R. P., Gregorieff, A., Bagherie-Lachidan, M., Chung, C., Lim, D. S., et al. (2013). Yap-and Cdc42-dependent nephrogenesis and morphogenesis during mouse kidney development. *PLoS Genetics*, *9*(3), e1003380.

Reidy, K. J., Villegas, G., Teichman, J., Veron, D., Shen, W., Jimenez, J., et al. (2009). Semaphorin3a regulates endothelial cell number and podocyte differentiation during glomerular development. *Development*, *136*(23), 3979–3989.

Rinkevich, Y., Montoro, D. T., Muhonen, E., Walmsley, G. G., Lo, D., Hasegawa, M., et al. (2014). Clonal analysis reveals nerve-dependent and independent roles on mammalian hind limb tissue maintenance and regeneration. *Proceedings of the National Academy of Sciences of the United States of America*, *111*(27), 9846–9851.

Rohr, C., Prestel, J., Heidet, L., Hosser, H., Kriz, W., Johnson, R. L., et al. (2002). The LIM-homeodomain transcription factor Lmx1b plays a crucial role in podocytes. *Journal of Clinical Investigation*, *109*(8), 1073–1082.

Rosselot, C., Spraggon, L., Chia, I., Batourina, E., Riccio, P., Lu, B., et al. (2010). Non-cell-autonomous retinoid signaling is crucial for renal development. *Development*, *137*(2), 283–292.

Sadl, V. S., Jin, F., Yu, J., Cui, S., Holmyard, D., Quaggin, S. E., et al. (2002). The mouse Kreisler (Krml1/MafB) segmentation gene is required for differentiation of glomerular visceral epithelial cells. *Developmental Biology*, *249*(1), 16–29.

Sánchez, M. P., Silos-Santiago, I., Frisén, J., He, B., Lira, S. A., & Barbacid, M. (1996). Renal agenesis and the absence of enteric neurons in mice lacking GDNF. *Nature*, *382*(6586), 70–73.

Sands, J. M., & Layton, H. E. (2014). Advances in understanding the urine-concentrating mechanism. *Annual Review of Physiology*, *76*, 387–409. http://dx.doi.org/10.1146/annurev-physiol-021113-170350.

Sariola, H., Holm, K., & Henke-Fahle, S. (1988). Early innervation of the metanephric kidney. *Development*, *104*(4), 589–599.

Sariola, H., Peault, B., LeDouarin, N., Buck, C., Dieterlen-Lievre, F., & Saxen, L. (1984). Extracellular matrix and capillary ingrowth in interspecies chimeric kidneys. *Cell Differentiation*, *15*(1), 43–51.

Saxén, L. (1987). In *Organogenesis of the kidney: Vol. 19*. Cambridge: Cambridge University Press.

Schell, C., Wanner, N., & Huber, T. B. (2014). Glomerular development—Shaping the multi-cellular filtration unit. *Seminars in Cell & Developmental Biology*, *36*, 39–49:

Schnell, U., & Carroll, T. J. (2014 Nov 13). Planar cell polarity of the kidney. *Experimental Cell Research*. pii: S0014-4827(14)00492-3.

Schuchardt, A., D'Agati, V., Larsson-Blomberg, L., Costantini, F., & Pachnis, V. (1994). Defects in the kidney and enteric nervous system of mice lacking the tyrosine kinase receptor Ret. *Nature*, *367*, 380–383.

Scott, R. P., & Quaggin, S. E. (2015). The cell biology of renal filtration. *The Journal of Cell Biology*, *209*(2), 199–210.

Self, M., Lagutin, O. V., Bowling, B., Hendrix, J., Cai, Y., Dressler, G. R., et al. (2006). Six2 is required for suppression of nephrogenesis and progenitor renewal in the developing kidney. *The EMBO Journal*, *25*(21), 5214–5228.

Sequeira-Lopez, M. L. S., Lin, E. E., Li, M., Hu, Y., Sigmund, C. D., & Gomez, R. A. (2015). The earliest metanephric arteriolar progenitors and their role in kidney vascular development. *American Journal of Physiology. Regulatory, Integrative and Comparative Physiology*, *308*(2), R138–R149.

Shapiro, E., Goldfarb, D. A., & Ritchey, M. L. (2003). The congenital and acquired solitary kidney. *Reviews in Urology*, *5*(1), 2.

Short, K. M., Combes, A. N., Lefevre, J., Ju, A. L., Georgas, K. M., Lamberton, T., et al. (2014). Global quantification of tissue dynamics in the developing mouse kidney. *Developmental Cell*, *29*(2), 188–202.

Stark, K., Vainio, S., Vassileva, G., & McMahon, A. P. (1994). Epithelial transformation of metanephric mesenchyme in the developing kidney regulated by Wnt-4. *Nature, 372*(6507), 679–683.

Subramanya, A. R., & Ellison, D. H. (2014). Distal convoluted tubule. *Clinical Journal of the American Society of Nephrology: CJASN, 9*(12), 2147–2163.

Suh, J. H., & Miner, J. H. (2013). The glomerular basement membrane as a barrier to albumin. *Nature Reviews. Nephrology, 9*(8), 470–477.

Sutherland, M., Ryan, D., Black, M. J., & Kent, A. L. (2014). Long-term renal consequences of preterm birth. *Clinics in Perinatology, 41*(3), 561–573.

Taguchi, A., Kaku, Y., Ohmori, T., Sharmin, S., Ogawa, M., Sasaki, H., et al. (2014). Redefining the *in vivo* origin of metanephric nephron progenitors enables generation of complex kidney structures from pluripotent stem cells. *Cell Stem Cell, 14*(1), 53–67.

Taguchi, A., & Nishinakamura, R. (2015). Nephron reconstitution from pluripotent stem cells. *Kidney International, 87*(5), 894–900.

Takabatake, Y., Sugiyama, T., Kohara, H., Matsusaka, T., Kurihara, H., Koni, P. A., et al. (2009). The CXCL12 (SDF-1)/CXCR4 axis is essential for the development of renal vasculature. *Journal of the American Society of Nephrology, 20*(8), 1714–1723.

Takasato, M., Er, P. X., Becroft, M., Vanslambrouck, J. M., Stanley, E. G., Elefanty, A. G., et al. (2014). Directing human embryonic stem cell differentiation towards a renal lineage generates a self-organizing kidney. *Nature Cell Biology, 16*(1), 118–126.

Takasato, M., & Little, M. H. (2015). The origin of the mammalian kidney: Implications for recreating the kidney *in vitro*. *Development, 142*(11), 1937–1947.

Tang, M. J., Cai, Y., Tsai, S. J., Wang, Y. K., & Dressler, G. R. (2002). Ureteric bud outgrowth in response to RET activation is mediated by phosphatidylinositol 3-kinase. *Developmental Biology, 243*(1), 128–136.

Tanigawa, S., Wang, H., Yang, Y., Sharma, N., Tarasova, N., Ajima, R., et al. (2011). Wnt4 induces nephronic tubules in metanephric mesenchyme by a non-canonical mechanism. *Developmental Biology, 352*(1), 58–69.

Thiagarajan, R. D., Georgas, K. M., Rumballe, B. A., Lesieur, E., Chiu, H. S., Taylor, D., et al. (2011). Identification of anchor genes during kidney development defines ontological relationships, molecular subcompartments and regulatory pathways. *PloS One, 6*(2), e17286.

Tomita, M., Asada, M., Asada, N., Nakamura, J., Oguchi, A., Higashi, A. Y., et al. (2013). Bmp7 maintains undifferentiated kidney progenitor population and determines nephron numbers at birth. *PloS One, 8*(8), e73554.

Treuting, P. M., & Kowalewska, J. (2011). Urinary system. In P. M. Treuting & S. Dintzis (Eds.), *Comparative anatomy and histology: A mouse and human atlas* (pp. 229–251): London, UK: Academic Press.

Uchiyama, Y., Sakaguchi, M., Terabayashi, T., Inenaga, T., Inoue, S., Kobayashi, C., et al. (2010). Kif26b, a kinesin family gene, regulates adhesion of the embryonic kidney mesenchyme. *Proceedings of the National Academy of Sciences of the United States of America, 107*(20), 9240–9245.

Uzarski, J. S., Xia, Y., Belmonte, J. C., & Wertheim, J. A. (2014). New strategies in kidney regeneration and tissue engineering. *Current Opinion in Nephrology and Hypertension, 23*(4), 399–405.

Vanslambrouck, J. M., & Little, M. H. (2015). Direct transcriptional reprogramming to nephron progenitors. *Current Opinion in Genetics & Development, 34*, 10–16.

Vivante, A., Kohl, S., Hwang, D. Y., Dworschak, G. C., & Hildebrandt, F. (2014). Single-gene causes of congenital anomalies of the kidney and urinary tract (CAKUT) in humans. *Pediatric Nephrology, 29*(4), 695–704.

Wainwright, E. N., Wilhelm, D., Combes, A. N., Little, M. H., & Koopman, P. (2015). ROBO2 restricts the nephrogenic field and regulates Wolffian duct–nephrogenic cord separation. *Developmental Biology, 404*(2), 88–102.

Walker, K. A., Cai, X., Caruana, G., Thomas, M. C., Bertram, J. F., & Kett, M. M. (2012). High nephron endowment protects against salt-induced hypertension. *American Journal of Physiology. Renal Physiology*, *303*(2), F253–F258.

Wall, S. M. (2005). Recent advances in our understanding of intercalated cells. *Current Opinion in Nephrology and Hypertension*, *14*(5), 480–484.

Wellik, D. M., Hawkes, P. J., & Capecchi, M. R. (2002). Hox11 paralogous genes are essential for metanephric kidney induction. *Genes & Development*, *16*(11), 1423–1432.

Wilkinson, L. J., Neal, C. S., Singh, R. R., Sparrow, D. B., Kurniawan, N. D., Ju, A., et al. (2015). Renal developmental defects resulting from in utero hypoxia are associated with suppression of ureteric β-catenin signaling. *Kidney International*, *87*(5), 975–983.

Xia, Y., Nivet, E., Sancho-Martinez, I., Gallegos, T., Suzuki, K., Okamura, D., et al. (2013). Directed differentiation of human pluripotent cells to ureteric bud kidney progenitor-like cells. *Nature Cell Biology*, *15*(12), 1507–1515.

Xu, J., Wong, E. Y., Cheng, C., Li, J., Sharkar, M. T., Xu, C. Y., et al. (2014). Eya1 interacts with Six2 and Myc to regulate expansion of the nephron progenitor pool during nephrogenesis. *Developmental Cell*, *31*(4), 434–447.

Yang, Z., Zimmerman, S., Brakeman, P. R., Beaudoin, G. M., Reichardt, L. F., & Marciano, D. K. (2013). *De novo* lumen formation and elongation in the developing nephron: A central role for afadin in apical polarity. *Development*, *140*(8), 1774–1784.

Yu, J., Carroll, T. J., Rajagopal, J., Kobayashi, A., Ren, Q., & McMahon, A. P. (2009). A Wnt7b-dependent pathway regulates the orientation of epithelial cell division and establishes the cortico-medullary axis of the mammalian kidney. *Development*, *136*(1), 161–171.

Yu, J., Valerius, M. T., Duah, M., Staser, K., Hansard, J. K., Guo, J. J., et al. (2012). Identification of molecular compartments and genetic circuitry in the developing mammalian kidney. *Development*, *139*(10), 1863–1873.

Zhao, H., Kegg, H., Grady, S., Truong, H. T., Robinson, M. L., Baum, M., et al. (2004). Role of fibroblast growth factor receptors 1 and 2 in the ureteric bud. *Developmental Biology*, *276*(2), 403–415.

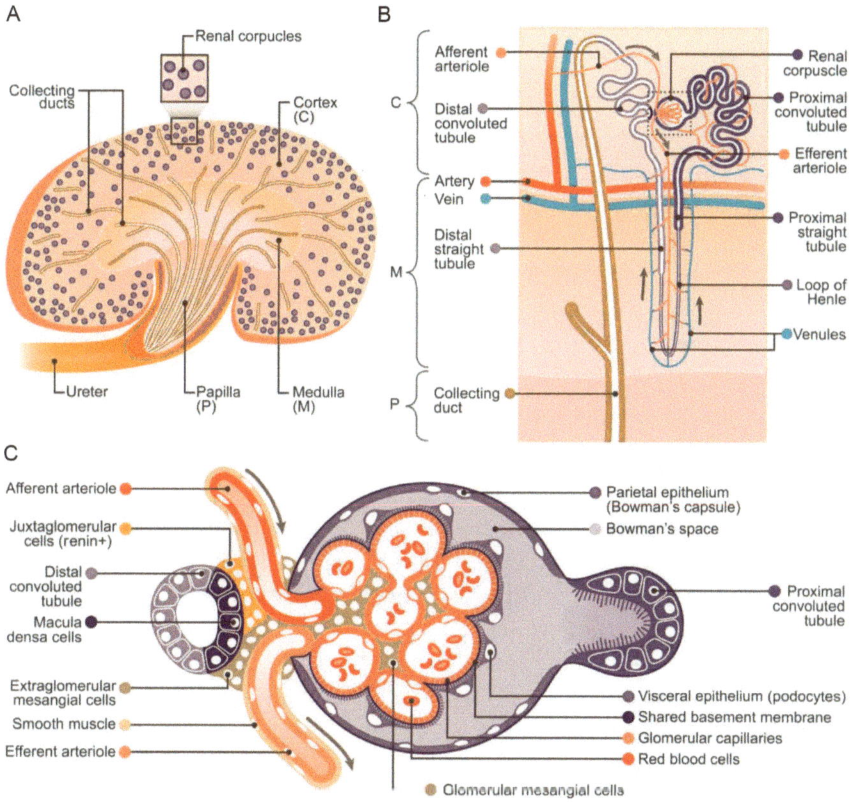

Andrew P. McMahon, Figure 1 Schematic illustrations of the adult mouse metanephric kidney. (A) Gross regional organization of the adult kidney. (B) Anatomy of the nephron. (C) High-resolution overview of the renal corpuscle and associated structures (region boxed in B).

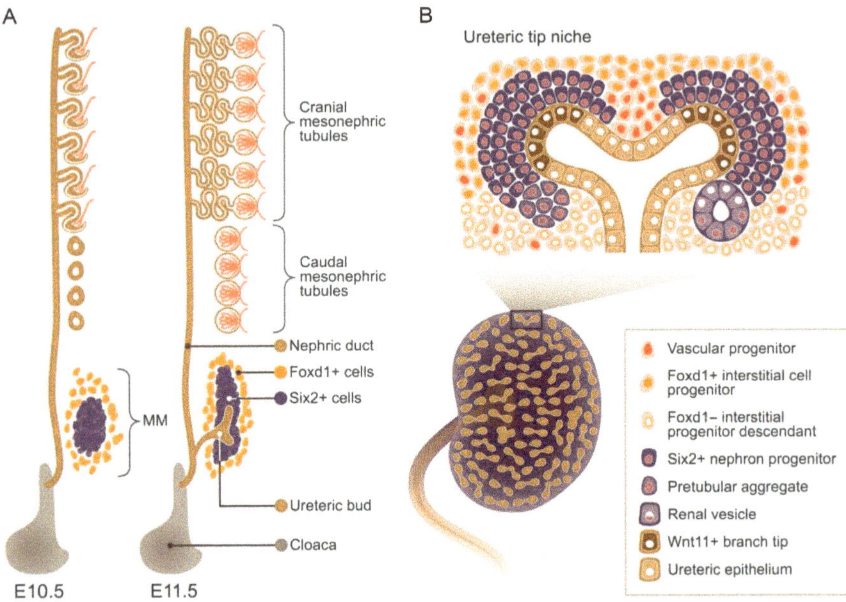

Andrew P. McMahon, Figure 2 Schematic illustrations of the developing mouse kidney. (A) Overviews of the mesonephric and metanephric kidney anlagen pre-(E10.5) and post-(E11.5) ureteric bud invasion into the metanephric mesenchyme. Anterior is to the top. (B) Cellular and molecular organization of collecting duct (Wnt11+), nephron (Six2+), interstitial (Foxd1+), and vascular progenitors within a ureteric tip niche.

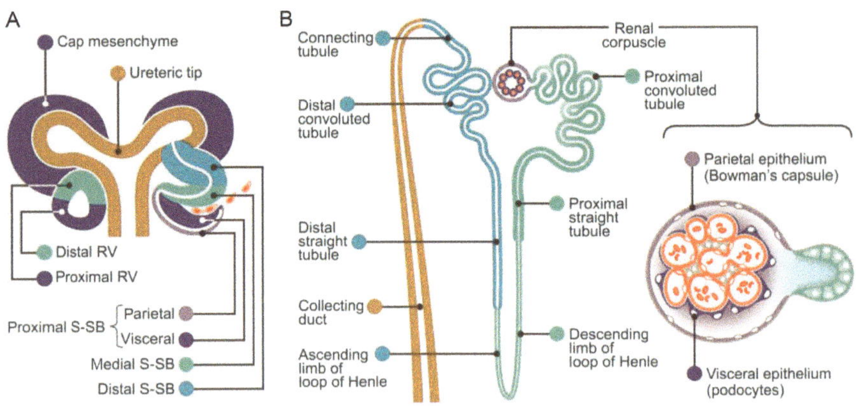

Andrew P. McMahon, Figure 3 Schematic illustrations of nephron patterning process. (A) Transition of polarized renal vesicle to patterned S-shaped body. (B) Speculative color-encoded map of the cell fate relationship of progenitor subdomains in S-shaped body in (A) to adult nephron anatomy. Verification requires more extensive cell fate-mapping studies.

CHAPTER FOUR

Applied Developmental Biology: Making Human Pancreatic Beta Cells for Diabetics

Douglas A. Melton[1]
Department of Stem Cell and Regenerative Biology, Harvard Stem Cell Institute, Harvard University, Cambridge, Massachusetts, USA
[1]Corresponding author: e-mail address: dmelton@harvard.edu

Abstract

Understanding the genes and signaling pathways that determine the differentiation and fate of a cell is a central goal of developmental biology. Using that information to gain mastery over the fates of cells presents new approaches to cell transplantation and drug discovery for human diseases including diabetes.

Developmental biology is a broad inclusive subject that covers formation and maintenance of plants and animals, from the beginning of a new life to its demise in old age. This essay considers one small slice, a sliver, of developmental biology, the formation, function, and utility of pancreatic beta cells.

Why focus on beta cells? One could answer by saying that the beta cell is as interesting as any cell type and in this regard can serve as a test case for understanding how a particular cell type is produced from a fertilized egg. In addition, a strong case for the beta cell derives from its central role in metabolism and diabetes. The beta cell helps regulate cellular metabolism by providing insulin, a hormone that controls the cellular uptake and conversion of sugar to energy. From this perspective, the beta cell helps explain how our bodies use food for immediate energy needs or stores it in fat for times when food is limiting. This places the beta cell at the center of the growing health crisis of diabetes because beta cells are absent, due to an autoimmune destruction, in Type 1 diabetes or dysfunctional in Type 2 diabetes.

Current Topics in Developmental Biology, Volume 117
ISSN 0070-2153
http://dx.doi.org/10.1016/bs.ctdb.2015.11.013

From the perspective of developmental biology, one can put forward two complementary approaches to the study of beta cells. The first is descriptive, delineating the process of making beta cells by describing the genes that come on and go off at each stage of development, the cellular movements, tissue formation, and the histology of insulin-producing cells in the Islet of Langerhans. This would include details on formation of insulin-containing vesicles, the extracellular matrix, neuronal connections, and the special connections of beta cells with blood vessels within the islet. A complete description would show how this impressive cellular machine reads blood sugar concentration every few milliseconds by metabolizing glucose to ATP, closing potassium channels, and elevating intracellular calcium all leading to the fusion of prepacked insulin-containing vesicles with the plasma membrane and release of the hormone. This type of a genetic, anatomical, and physiological description can be likened to a biography, in this case, a "molecular biography" (Shaywitz & Melton, 2005), that can be both satisfying and informative, giving one the sense of truly knowing the beta cell.

A second approach or perspective on studying beta cells is closely aligned with engineering, one that uses any information or means to build a human beta cell. This approach could be called applied developmental biology. The aim here is not to describe what happens in normal development nor be limited by the products or mechanisms of evolution, but rather, to take on the challenge of making human beta cells and doing so at a scale that can provide to new clinical treatments. In the best case, developmental biologists might be able to understand how beta cells are normally made and use that information to make functional human beta cells that can be used to understand and treat a disease that affects hundreds of millions of people.

It may someday be possible to begin with any cell or cell population and convert it directly into enough beta cells to treat a patient. One might be able to begin with easily obtained blood cells or skin cells and manipulate their gene expression with sufficient specificity to convert them all to functional beta cell. This would involve exact control over thousands of genes and while advances in genetic manipulation are advancing at an impressive pace (reviewed by La Russa and Qi, 2015), it is not presently possible to accurately rewire thousands of genes. The transdifferentiation of pancreatic exocrine cells (Li et al., 2014; Zhou, Brown, Kanarek, Rajagopal, & Melton, 2008) or alpha cells (Thorel et al., 2010; Yang, Thorel, Boyer, Herrera, & Wright, 2011) into beta cells shows that this may someday be possible. An alternative approach is to begin with a pluripotent stem cell. The advantages being that stem cells grow indefinitely, providing a virtually unlimited source of material, and they are as yet undifferentiated and therefore they may be more receptive to signals that determine their fate. Simply put, the challenge of direct differentiation *in vitro* is to provide the right signals at the right time to turn pluripotent stem cells into fully functional human beta cells.

In the last few decades, a few model organisms have been very effectively used to describe the normal development of the pancreas. And while there is some variation in the details of frog, zebrafish, mouse, and human pancreatic organogenesis, over all one is struck by the conservation of developmental mechanisms. Dorsal and ventral buds emerge from gut tube endoderm and later rotate and merge to form a multilobed pancreas with exocrine, ductal, and endocrine tissues. These studies identified key early genes and defined the progenitor cell types that have since guided projects aimed at turning stem cells into beta cells. There are many excellent publications by several labs, too many to comprehensively reference here, on the genes and cell types that mark the stages of pancreatic development. A beautiful study combining morphology with gene expression has been published by Jørgensen and colleagues (2007) and there are numerous informative publications on the epigenetic and gene expression changes that accompany beta cell development, some of which are referenced here: Xie, Carrano, and Sander (2015), Wang et al. (2015), Dhawan et al. (2015), Mastracci et al. (2011), Schaffer et al. (2013), Jonsson, Carlsson, Edlund, and Edlund (1994), Pan et al. (2013), Henry and Melton (1998), Hrvatin et al. (2014), Shivdasani (2002), Zorn and Wells (2009), Blodgett, Cura, and Harlan (2014), and Serup (2012). With these guideposts in hand, the task becomes identifying signals, for example, small cell permeable chemicals or growth and differentiation factors, that instruct the cells to move or differentiate from one stage to the next. This general approach is not unique to the pancreas and has been effectively used for other cells, including neurons, and skeletal and heart muscle.

The studies on gene expression and pancreatic development in various vertebrate formed the foundation upon which several groups have attempted to direct the differentiation of human ES cells into the pancreatic lineage (Chen, Borowiak, Sherwood, Kweudjeu, & Melton, 2013; Johannesson et al., 2009; Kroon et al., 2008; Nostro et al., 2015; Rezania et al., 2014; Russ et al., 2015; Shahjalal et al., 2014; Toyoda et al., 2015). Most recently, this approach reached an important juncture by showing that the stem cell-derived beta (sc-beta) cells can show a pattern of gene expression, ultrastructure, and insulin secretion that very closely resemble these features in beta cells found in human cadaveric islets (Pagliuca et al., 2014). While these sc-beta cells are not indistinguishable from human beta cells, they have many of the salient features, most notably glucose-stimulated insulin secretion, that makes them useful for *in vitro* studies and drug screens. Moreover, given the limited number of cadaveric islets that are available for human transplantation, these sc-beta cells can now be explored as a virtually inexhaustible supply of material for human transplantation (Fig. 1).

The present protocols used for directed differentiation of stem cells into sc-beta cells are similar in the sense that they generally begin with a large number (hundreds of millions of stem cells in clusters that are 100–300 μm in diameter grown in suspension; though the cells can also be successfully grown in planar culture) and this number does not change much during the course of the differentiation. While there is division and cell death during the differentiation, the final cell number is typically within two- to fivefold of the starting cell number. One of the great advantages of beginning with stem cells is that they can be relatively easily grown in large quantities before differentiation begins. Consequently, one estimates that a

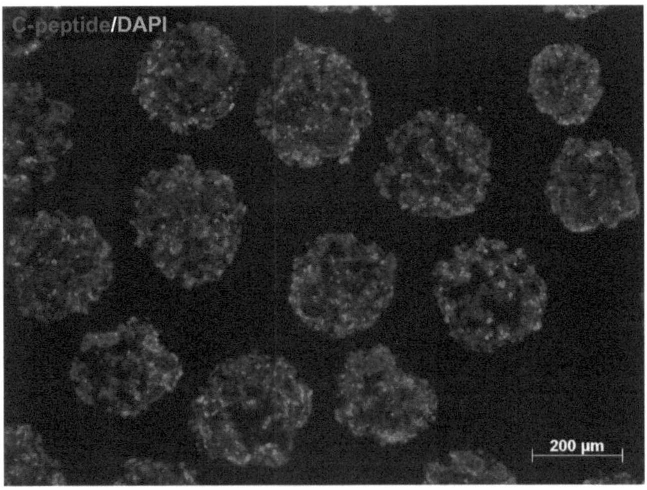

Figure 1 Clusters of stem cell-derived beta cells. Human embryonic stem cells were differentiated *in vitro* using a six step, 30 day protocol described in Pagliuca et al. (2014). This image shows clusters with nuclei stained blue (with DAPI) and insulin stained red (using antibody for C-peptide). Note that the distribution of insulin expressing sc-beta cells is not uniform and only about ¼ of the cells produce insulin. The remaining cells are not fully differentiated and a small percentage produces other endocrine hormones including glucagon and somatostatin (not shown). (A color version of this figure is located at the end of this chapter.)

sufficient number of sc-beta cells ($0.1–1 \times 10^9$) to treat one patient can be grown in less than 1 l of media.

The fact that sc-beta cells can be derived *in vitro* at all, without the requirement for the 3D configuration and cells contacts that accompany normal development, is quite significant. Yet one is mindful of the fact that the efficiency of the process so far described could be improved in several respects. In a typical multistage protocol, the final cell population has about 10–30% sc-beta cells and the remaining cells include a few other endocrine cells (glucagon-producing alpha cells and somatostatin producing delta cells, Pagliuca et al., 2014). However, the majority of the cell population are relatively uncharacterized cells that may be undifferentiated progenitors or other types of unwanted cells. Thus, improving the efficiency, in terms of the percentage of differentiated cells that become beta cells remains an important challenge. In addition, the protocols take 4–5 weeks to complete and the cost of materials is nontrivial.

While the sc-beta cells perform many of the key features of bona fide beta cells, their longevity *in vitro* and *in vivo* has not been determined. Human beta cells divide infrequently and are thought to have a half-life of many months. Similarly, sc-beta cells do not divide at significant rates, but their longevity, exact rate of and capacity for proliferation has not been measured. These parameters will affect their utility for drug screening as well as cell transplantation.

Mature beta cells secrete insulin in a pulsatile fashion (Porksen et al., 1995); it has not yet been determined if sc-beta cells do so as well. And as sc-beta cells are produced *in vitro*, in the absence of other cell types that they would normally in contact during development, most notably endothelial cells (Cao & Wang, 2014), it may be the case that their *in vitro* function lacks other physiological features. The full maturation of sc-beta cells may be possible *in vitro* or may occur shortly after transplantation and contact with other cells of the body.

The most successful cellular transplantations that have rendered diabetic patients insulin independent (reviewed by Vantyghem et al., 2014) have been done with whole (cadaveric) islets. By manipulating the stem cell differentiation protocols, it should be possible to produce all islet endocrine cell types in the same proportion found in human islets. One promising approach to this problem is to use a combination of single cell transcript analysis, combined with "pseudo time" computational analysis (Trapnell et al., 2014), to determine the genes expressed at each step of the process when a cell makes a decision about its fate. In this way, it may be possible to determine which genes need to be turned on and off at various stages of differentiation and then identify small molecules, growth factors, or even genetic manipulations that control that decision. And it should be possible to create islets that contain not only endocrine cells but also mesenchymal/stromal cells, endothelial precursors, and neurons. In this regard, advances in 3D printing of cells onto biological matrices holds enormous promise for making functional "miniorgans" (Murphy & Atala, 2014). This would enable a range of new projects including real-time imaging of cell function (Ben-Zvi & Melton, 2015) and drug screening on reconstructed human tissues, including tissues constructed with a specific genetic composition.

The projects so far described touch on advances and future developments that could lead to functional beta cells, and perhaps islets, that could be used for drug screening. These screens would likely include a search for compounds that increase the survival of beta cells, improve their insulin-secreting function, and revive stressed or damaged beta cells such as those found in Type 2 diabetes. These cells/islets could also be made using a patient's own iPS cells and therefore be used for transplantation in cases where there is no immune component, including Type 2 diabetes and pancreatitis. However, this discussion has bypassed a major challenge for Type 1 diabetes: how to transplant cells into a patient that has a specific immune reaction to their own beta cells.

Two separate, but not mutually exclusive, approaches to protecting transplanted beta cells are being explored. The first is physical protection including microdevices where the cells or small clusters of cells are encapsulated with a biocompatible material that allows the cells to survive and function (reviewed by Weir, 2013). Survival, in an autoimmune context, requires that the material prevent destruction of the beta cells by lymphocyte (primarily T-cell) contact. It is presently unclear whether protection from soluble antibodies directed against beta cells will be a significant problem. For function, the encapsulated beta cells require nutrients (sugar, oxygen) and the ability to secrete insulin across or through the biomaterial. In essence, this comes down to finding a material with holes of the right size to prevent cellular transit and permit transport or diffusion of nutrients and insulin.

A complementary approach aims is to genetically modify sc-beta cells so as to make them opaque or unrecognizable by the immune system. This might be accomplished by a number of genetic alterations that take advantage of CrispR technologies and could be inducible: for example, one could begin with a stem cell that will express alternative MHC molecules in the mature state or blocking molecules such as PDL-1 under control of an inducible promoter (see, for example, Rong et al., 2014). In the longer term, it should it be possible to determine the source (memory T cells) and classes of immune cells that are responsible for the autoimmunity. Once the specific cells responsible for the autoimmunity are identified, it may be possible devise ways to and eliminate them or block their activity by combining genetic modifications with cell transplantation.

In conclusion, this one rather detailed example on the developmental biology of a pancreatic islet is offered to illustrate how an understanding of the normal development of a cell or organ presents interesting challenges in basic developmental biology. This includes asking questions about how decisions are made about cell fates during development, what determines the function and longevity of the final product, and trying to reconstruct the entire process beginning with stem or progenitor cells. And at the same time, the end result has the potential to treat, perhaps even cure diabetes. The combination of asking and answering questions about normal development with the possibility of applying that knowledge to solve a medical problem makes the enterprise a worthwhile goal.

REFERENCES

Ben-Zvi, D., & Melton, D. A. (2015). Modeling human nutrition using human embryonic stem cells. *Cell*, *161*, 12–17.

Blodgett, D. M., Cura, A. J., & Harlan, D. M. (2014). The pancreatic β-cell transcriptome and integrated-omics. *Current Opinion in Endocrinology, Diabetes and Obesity*, *21*(2), 83–88.

Cao, Z., & Wang, X. (2014). The endocrine role between beta cells and intra-islet endothelial cells. *Endocrine Journal*, *61*, 647–654.

Chen, A. E., Borowiak, M., Sherwood, R. I., Kweudjeu, A., & Melton, D. A. (2013). Functional evaluation of ES cell-derived endodermal populations reveals differences between Nodal and Activin A-guided differentiation. *Development*, *140*, 675–686.

Dhawan, S., Tschen, S.-I. I., Zeng, C., Guo, T., Hebrok, M., Matveyenko, A., et al. (2015). DNA methylation directs functional maturation of pancreatic β cells. *The Journal of Clinical Investigation*, *125*, 2851–2860.

Henry, G. L., & Melton, D. A. (1998). Mixer, a homeobox gene required for endoderm development. *Science*, *281*, 91–96.

Hrvatin, S., O'Donnell, C. W., Deng, F., Millman, J. R., Pagliuca, F. W., DiIorio, P., et al. (2014). Differentiated human stem cells resemble fetal, not adult, β cells. *Proceedings of the National academy of Sciences of the United States of America*, *111*, 3038–3043.

Johannesson, M., Ståhlberg, A., Ameri, J., Sand, F. W., Norrman, K., & Semb, H. (2009). FGF4 and retinoic acid direct differentiation of hESCs into PDX1-expressing foregut endoderm in a time- and concentration-dependent manner. *PLoS One*, *4*, e4794.

Jonsson, J., Carlsson, L., Edlund, T., & Edlund, H. (1994). Insulin Promoter Factor 1 is required for pancreas development in mice. *Nature*, *371*, 606–609.

Jørgensen, M. C., Ahnfelt-Rønne, J., Hald, J., Madsen, O. D., Serup, P., & Heckscher-Sørensen, J. (2007). An illustrated review of early pancreas development in the mouse. *Endocrine Reviews*, *28*, 685–705.

Kroon, E., Martinson, L., Kadoya, K., Bang, A., Kelly, O., Eliazer, S., et al. (2008). Pancreatic endoderm derived from human embryonic stem cells generates glucose-responsive insulin-secreting cells in vivo. *Nature Biotechnology*, *26*, 443–452.

La Russa, M. F., & Qi, L. S. (2015). The new state of the art: CRISPR for gene activation and repression. *Molecular and Cellular Biology*, *35*(22), 3800–3809.

Li, W., Nakanishi, M., Zumsteg, A., Shear, M., Wright, C., Melton, D. A., et al. (2014). In vivo reprogramming of pancreatic acinar cells to three islet endocrine subtypes. *Elife*, *3*, e01846.

Mastracci, T. L., Wilcox, C. L., Arnes, L., Panea, C., Golden, J. A., May, C. L., et al. (2011). Nkx2.2 and Arx genetically interact to regulate pancreatic endocrine cell development and endocrine hormone expression. *Developmental Biology*, *359*, 1–11.

Murphy, S. V., & Atala, A. (2014). 3D bioprinting of tissues and organs. *Nature Biotechnology*, *32*, 773–785.

Nostro, M. C., Sarangi, F., Yang, C., Holland, A., Elefanty, A. G., Stanley, E. G., et al. (2015). Efficient generation of NKX6-1+ pancreatic progenitors from multiple human pluripotent stem cell lines. *Stem Cell Reports*, *4*, 591–604.

Pagliuca, F. W., Millman, J. R., Gürtler, M., Segel, M., Van Dervort, A., Ryu, J. H., et al. (2014). Generation of functional human pancreatic β cells in vitro. *Cell*, *159*, 428–439.

Pan, F. C., Bankaitis, E. D., Boyer, D., Xu, X., Van de Casteele, M., Magnuson, M. A., et al. (2013). Spatiotemporal patterns of multipotentiality in Ptf1a-expressing cells during pancreas organogenesis and injury-induced facultative restoration. *Development*, *140*, 751–764.

Porksen, N., Munn, S., Steers, J., Vore, S., Veldhuis, J., & Butler, P. (1995). Pulsatile insulin secretion accounts for 70% of total insulin secretion during fasting. *The American Journal of Physiology*, *269*(3 Pt 1), E478–E488.

Rezania, A., Bruin, J. E., Arora, P., Rubin, A., Batushansky, I., Asadi, A., et al. (2014). Reversal of diabetes with insulin-producing cells derived in vitro from human pluripotent stem cells. *Nature Biotechnology*, *32*, 1121–1133.

Rong, Z., Wang, M., Hu, Z., Stradner, M., Zhu, S., Kong, H., et al. (2014). An effective approach to prevent immune rejection of human ESC-derived allografts. *Cell Stem Cell*, *14*, 121–130.

Russ, H., Parent, A., Ringler, J., Hennings, T., Nair, G., Shvevgert, M., et al. (2015). Controlled induction of human pancreatic progenitors produces functional beta-like cells in vitro. *The EMBO Journal*, *34*(13), 1759–1772.

Schaffer, A. E., Taylor, B. L., Benthuysen, J. R., Liu, J., Thorel, F., Yuan, W., et al. (2013). Nkx6.1 controls a gene regulatory network required for establishing and maintaining pancreatic Beta cell identity. *PLoS Genetics*, *9*, e1003274.

Serup, P. (2012). Signaling pathways regulating murine pancreatic development. *Seminars in Cell and Developmental Biology*, *23*, 663–672.

Shahjalal, H. M., Shiraki, N., Sakano, D., Kikawa, K., Ogaki, S., Baba, H., et al. (2014). Generation of insulin-producing β-like cells from human iPS cells in a defined and completely xeno-free culture system. *Journal of Molecular Cell Biology*, *6*, 394–408.

Shaywitz, D. A., & Melton, D. A. (2005). The molecular biography of the cell. *Cell*, *120*, 729–731.

Shivdasani, R. A. (2002). Molecular regulation of vertebrate early endoderm development. *Developmental Biology*, *249*, 191–203.

Thorel, F., Népote, V., Avril, I., Kohno, K., Desgraz, R., Chera, S., et al. (2010). Conversion of adult pancreatic alpha-cells to beta-cells after extreme beta-cell loss. *Nature*, *464*, 1149–1154.

Toyoda, T., Mae, S.-I., Tanaka, H., Kondo, Y., Funato, M., Hosokawa, Y., et al. (2015). Cell aggregation optimizes the differentiation of human ESCs and iPSCs into pancreatic bud-like progenitor cells. *Stem Cell Research*, *14*, 185–197.

Trapnell, C., Cacchiarelli, D., Grimsby, J., Pokharel, P., Li, S., Morse, M., et al. (2014). The dynamics and regulators of cell fate decisions are revealed by pseudotemporal ordering of single cells. *Nature Biotechnology*, *32*, 381–386.

Vantyghem, M.-C. C., Defrance, F., Quintin, D., Leroy, C., Raverdi, V., Prévost, G., et al. (2014). Treating diabetes with islet transplantation: Lessons from the past decade in Lille. *Diabetes & Metabolism*, *40*, 108–119.

Wang, A., Yue, F., Li, Y., Xie, R., Harper, T., Patel, N. A., et al. (2015). Epigenetic priming of enhancers predicts developmental competence of hESC-derived endodermal lineage intermediates. *Cell Stem Cell*, *16*, 386–399.

Weir, G. C. (2013). Islet encapsulation: Advances and obstacles. *Diabetologia*, *56*, 1458–1461.

Xie, R., Carrano, A. C., & Sander, M. (2015). A systems view of epigenetic networks regulating pancreas development and β-cell function. *Wiley Interdisciplinary Reviews: Systems Biology and Medicine*, *7*, 1–11.

Yang, Y.-P. P., Thorel, F., Boyer, D. F., Herrera, P. L., & Wright, C. V. (2011). Context-specific α- to-β-cell reprogramming by forced Pdx1 expression. *Genes and Development*, *25*, 1680–1685.

Zhou, Q., Brown, J., Kanarek, A., Rajagopal, J., & Melton, D. A. (2008). In vivo reprogramming of adult pancreatic exocrine cells to beta-cells. *Nature*, *455*, 627–632.

Zorn, A. M., & Wells, J. M. (2009). Vertebrate endoderm development and organ formation. *Annual Review of Cell and Developmental Biology*, *25*, 221–251.

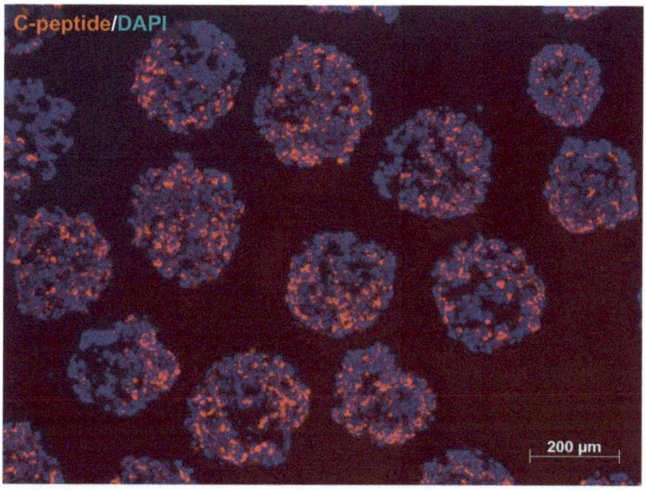

Douglas A. Melton, Figure 1 Clusters of stem cell-derived beta cells. Human embryonic stem cells were differentiated *in vitro* using a six step, 30 day protocol described in Pagliuca et al. (2014). This image shows clusters with nuclei stained blue (with DAPI) and insulin stained red (using antibody for C-peptide). Note that the distribution of insulin expressing sc-beta cells is not uniform and only about ¼ of the cells produce insulin. The remaining cells are not fully differentiated and a small percentage produces other endocrine hormones including glucagon and somatostatin (not shown).

CHAPTER FIVE

The Dishevelled Protein Family: Still Rather a Mystery After Over 20 Years of Molecular Studies

Marek Mlodzik[1]
Department of Developmental and Regenerative Biology, Graduate School of Biomedical Sciences, Icahn School of Medicine at Mount Sinai, New York, USA
[1]Corresponding author: e-mail address: marek.mlodzik@mssm.edu

Contents

Abstract

Dishevelled (Dsh) is a key component of Wnt-signaling pathways and possibly also has other functional requirements. Dsh appears to be a key factor to interpret Wnt signals coming via the Wnt-receptor family, the Frizzled proteins, from the plasma membrane and route them into the correct intracellular pathways. However, how Dsh is regulated to relay signal flow to specific and distinct cellular responses upon interaction with the same Wnt-receptor family remains very poorly understood.

The molecular cloning and initial analysis of *Drosophila dishevelled* (*dsh*), now more than 20 years ago (Klingensmith, Nusse, & Perrimon, 1994), was back then a routine cloning paper of a gene linked to the Wg-signaling pathway. It was shortly followed by the molecular characterization of mouse and *Xenopus* homologs (Sokol, Klingensmith, Perrimon, & Itoh, 1995; Sussman et al., 1994), and a slew of papers addressing the potential roles of Dishevelled proteins (Dsh or Dvl in mammals) in canonical Wnt signaling, Wnt/PCP (planar cell polarity) signaling, and many other contexts from

Current Topics in Developmental Biology, Volume 117
ISSN 0070-2153
http://dx.doi.org/10.1016/bs.ctdb.2015.11.027

Drosophila and *C. elegans* to mice and humans (reviewed in Boutros & Mlodzik, 1999; Wallingford & Habas, 2005; Wynshaw-Boris, 2012), but as of today we still do not really know how Dsh is regulated, how it is "activated" (if such a term is permissible in the Dsh context), and what its functions really are. Its molecular sequence and domain features are highly conserved, suggesting that pooling molecular information from different species should be synergistic; see earlier reviews for the sequences and molecular features of the Dsh/Dvl protein family (Boutros & Mlodzik, 1999; Wallingford & Habas, 2005).

Complications in the functional analyses come from the facts that (1) Dsh proteins usually act maternally (the RNA and protein required for early functions are deposited into the developing eggs by the mother) and (2) issues of redundancy, as there are three Dvls in mice and human, at least two in *Xenopus*, two in *C. elegans*, and >four in zebrafish. The specific problems of functional dissection and associated complications, functional diversity and potential mechanistic insight I will try to discuss below. As this is an essay, it is driven by personal ideas, bias, and attempts at "understanding" Dsh's function(s). It is neither all inclusive nor all data driven (and I also apologize for missing some of the references). I hope it is nonetheless informative and inspiring.

1. THE MANY FUNCTIONS OF DISHEVELLED'S

Understanding the molecular basis and specificity for signal transduction pathways during pattern formation, development, and cell fate specification is of general importance in biology and disease mechanisms. The associated activation of intracellular signaling pathways leads to highly specific responses, ranging from cell proliferation to cell fate induction and terminal differentiation. Erroneous signaling can lead to cellular transformation and ultimately to diseases like cancer. The tight regulation of the "response system(s)" is therefore among the most important elements that govern the development and homeostasis of multicellular higher eukaryotes. Importantly, in response to different upstream signals, cells often use the same signaling molecules for different purposes. Conversely, the activation of the same signaling protein can lead to the activation of distinct downstream pathways and cascades. A well-documented example for the first scenario is the activation of small GTPase of the Ras family in cellular processes. For example, Ras can be activated by different extracellular stimuli and then relay the signal to different effector pathways that can lead for example to activation of transcriptional responses and cytoskeletal rearrangements (Drosten, Lechuga, & Barbacid, 2013). Dsh family members can be considered as a classic example of the second category, with different activation processes/pathways use the same molecule. Yet how the effector proteins know in which context to bind what, and relay specific information remains unclear in many contexts and, certainly, in the case of Dsh it is still unresolved. Nonetheless, the outcome of the Dsh-mediated signaling events *in vivo* is and must be very specific.

Dishevelled (Dsh, Dvl in mammals, where there are three equivalent Dvl genes) is a signaling molecule that functions in distinct contexts and (at least) two signaling pathways (see below), and it appears to have several other biological functions as well. The original *dishevelled* (*dsh*) mutant allele identified in *Drosophila*, *dsh*^{1}, is a viable allele (Fahmy & Fahmy, 1959), with defects in the arrangement of bristles on the body wall and wings, and hence the gene's name. Null alleles turned out to be embryonic lethal, and when lacking both maternal and zygotic function (Perrimon & Mahowald, 1987), and had phenotypes identical to *wingless* (Wg, the *Drosophila* founding member of the Wnt family) and *armadillo* (Arm, *Drosophila* beta-catenin; e.g., Nüsslein-Volhard & Wieschaus, 1980; Perrimon & Mahowald, 1987), the classical embryonic Wg/Wnt-signaling defects in flies. Together, these observations indicated early on that *dsh* has a function in (at least) two biological contexts.

Subsequent work confirmed that Dsh and Dvls are required for the transmission of Wg/Wnt signals in the canonical Wnt pathway and also for signaling during the establishment of PCP (polarity of epithelial cells perpendicular to their apical–basolateral axis and also cellular polarity in mesenchymal cells in several contexts; Adler, 2012; Goodrich & Strutt, 2011; McNeill, 2009; Peng & Axelrod, 2012; Seifert & Mlodzik, 2007; Simons & Mlodzik, 2008; Singh & Mlodzik, 2012; Strutt, 2003; Wallingford, 2006; Wallingford, Fraser, & Harland, 2002; Wang & Nathans, 2007) in vertebrates as well (reviewed in Boutros & Mlodzik, 1999; Wallingford & Habas, 2005). Dsh acts downstream of Fz family receptors in both pathways, although the receptor complexes and Fzs used, and other proteins that associate with Dsh are distinct in the two pathways, raising the question or problem of how a single protein, Dsh (the three mammalian Dvls are equivalent and all have the same potential as *Drosophila* Dsh), downstream of related receptors, specifically activates distinct effector pathways (Axelrod, Miller, Shulman, Moon, & Perrimon, 1998; Boutros, Paricio, Strutt, & Mlodzik, 1998). As this is not all what Dsh/Dvls biologically regulate the situation is more complicated.

Currently, Dsh family members have been linked to the following cellular functions. Besides its two main and best-described functional requirements, (1) downstream regulator of Fz-LPR5/6 receptor complexes in canonical Wg/Wnt signaling and (2) core component of Wnt–Fz PCP signaling, there are several additional biological roles reported: Dsh proteins have been linked to (3) nuclear functions (e.g., Collu et al., 2012; Itoh, Brott, Bae, Ratcliffe, & Sokol, 2005; see also below), although the role of Dsh in the nucleus remains controversial, (4) function in anchoring and/or localizing ciliary basal bodies in multiciliated cells (Park, Mitchell, Abitua, Kintner, & Wallingford, 2008; this function is vertebrate specific as *Drosophila* does not have multiciliated cells), (5) a potential antagonistic function to Notch signaling (Axelrod, Matsuno, Artavanis-Tsakonas, & Perrimon, 1996; Collu et al., 2012), and (6) last not least, a potential role in cell viability, as the *mDvl1,2,3* triple knockout cells are cell lethal ((Wynshaw-Boris, 2012), this potential role cannot be explained by a link to either Wnt pathway and suggest a novel function).

2. MOLECULAR FEATURES AND INTERACTIONS OF DISHEVELLED

The two original Dsh functions in canonical Wnt signaling and Wnt–Fz/PCP signaling have been studied and the longest and several regulatory interactions and domain requirements have been identified. Importantly, the *Drosophila dsh* gene encodes a 623 amino acid protein of 70 kd with no obvious similarities to proteins with catalytic functions, although all its domains and general features are highly conserved (reviewed in Boutros & Mlodzik, 1999; Wallingford & Habas, 2005; see Fig. 1 for cartoon presentation of Dsh domains). Although as a whole the primary sequence of Dsh does not hint at biochemical functions, several domains are highly conserved, giving some clues about its potential molecular functions.

All Dsh genes share three highly conserved domains, an N-terminal DIX (*Di*shevelled-A*x*in) domain, which is also found in Axin and binds to both to the Axin DIX domain or itself (required also for Dsh aggregation; Schwarz-Romond, Merrifield, Nichols, & Bienz, 2005), a central PDZ domain, which interacts with many proteins and also mediates Dsh–Fz interactions (Wong et al., 2003), and a C-terminal DEP domain (based on its original presence in *D*ishevelled-*E*GL-10-*P*leckstrin), which has been linked with the PCP function of Dsh (Axelrod et al., 1998; Boutros et al., 1998;

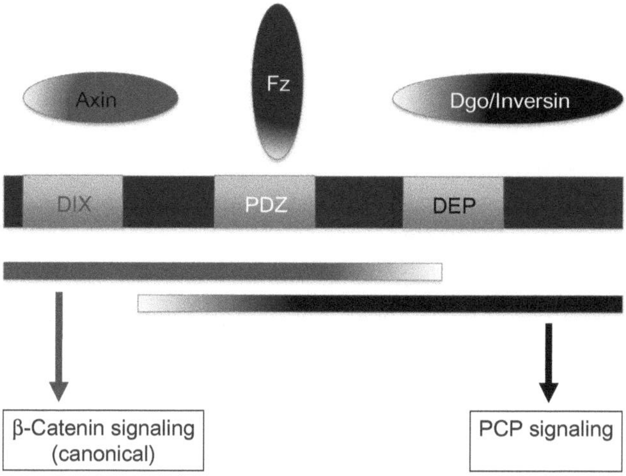

Figure 1 Domain organization of Dishevelled (Dsh) proteins and pathway-specific requirements and interactions. Dsh (in *Drosophila* 623 aa long) is shown in blue with its three main and highly conserved domains, DIX, PDZ, and DEP, labeled accordingly. The color scheme is "red" for canonical Wnt/β-catenin signaling and "black" for Wnt–Fz/PCP signaling. The three domains have their "main" pathway-specific interactors shown in the respective color and the red and black bars below Dsh indicate the approximate extent of requirements for signal relay to the respective pathways. Frizzled receptors directly interact with the PDZ domain and as this interaction is needed for both signaling outcomes it kept in blue. See text for details. (A color version of this figure is located at the end of this chapter.)

Jenny, Reynolds-Kenneally, Das, Burnett, & Mlodzik, 2005) and also physically binds to membrane lipids (Simons et al., 2009; see also review by Wallingford & Habas, 2005 for additional references). Functional studies *in vivo* and in cell culture have suggested that the DIX and PDZ domains and associated sequences (the N-terminal 2/3 of the protein) are critical and sufficient for canonical Wnt signaling, whereas the PDZ and DEP domains and associated protein regions are essential and sufficient for Wnt–Fz/PCP signaling (Boutros et al., 1998; Wallingford & Harland, 2002; Wallingford et al., 2000; reviewed in Boutros & Mlodzik, 1999; Wallingford & Habas, 2005). Consistently, the original dsh^{1} allele and other PCP-specific *dsh* alleles (e.g., dsh^{A3}) are point mutations that map within the DEP domain, the PCP establishment specific region of the protein (Axelrod et al., 1998; Boutros et al., 1998; Penton, Wodarz, & Nusse, 2002). All three conserved domains have been shown to interact with many proteins in several biochemical and yeast two-hybrid assays (Wallingford & Habas, 2005), and strikingly Dsh is currently one of the proteins with the most described molecular interactions. Yet, many of these remain functionally and physiologically unresolved and unclear. So while the domain requirements for the two pathways seem resolved, the dissection of the regulatory input to signaling specificity and pathway-specific protein complex formation remains a challenging task.

The oversimplified view for the Dsh complexes specific to two Wnt pathways looks as follows:

1. For canonical signaling, Dsh forms a complex with Axin (via DIX domain), Fz (via PDZ), and Wnt coreceptor LRP5/6 (arrow in *Drosophila*; via Axins interaction with the coreceptor) and this is triggered by Wnt-mediated (co)receptor activation; these interactions pull Axin out of the beta-catenin "destruction"-complex (reviewed in Clevers & Nusse, 2012) and at the same time the Dsh–Axin multimerization leads to the formation of the so-called signalosomes (Bilic et al., 2007; Zeng et al., 2008); it is likely that this complex also contains other factors from the "destruction"-complex including several of the kinases (Taelman et al., 2010).
2. In Wnt–Fz/PCP signaling, the situation is less clear, although specific interactions are documented. Dsh binds to Fz via the PDZ domain and this membrane-associated protein complex is stabilized through the (acidified) lipid interaction of the DEP domain. The PCP factor Diego (Dgo, Inversin, and Diversin in vertebrates) stabilizes this PCP-specific complex. It is assumed that other proteins are also part of this complex, including Fmi (a.k.a. Stan in *Drosophila*, Celsr in mammals), several kinases, and Rho family GTPases and their activators, including GEFs and Daam1/Formin (see PCP-specific reviews: e.g., Adler, 2012; Goodrich & Strutt, 2011; McNeill, 2009; Peng & Axelrod, 2012; Seifert & Mlodzik, 2007; Simons & Mlodzik, 2008; Singh & Mlodzik, 2012; Strutt, 2003; Wallingford, 2006; Wallingford et al., 2002; Wang & Nathans, 2007). The potential role of the kinases proposed to partake in this will be discussed below.

While these two complexes are the "stable end result" of the involved protein interactions, during PCP establishment Dsh can physically bind other core PCP factors, namely, Van Gogh (Vang, Vangl in vertebrates; a.k.a. Strabismus/Stbm in *Drosophila*) and Prickle (Pk). These interactions should however be less stable and transient, as they serve to antagonize the formation of a Vang–Pk complex on the Fz–Dsh side of planar polarization (see PCP-specific reviews for more details: Adler, 2012; Goodrich & Strutt, 2011; McNeill, 2009; Peng & Axelrod, 2012; Seifert & Mlodzik, 2007;

Simons & Mlodzik, 2008; Singh & Mlodzik, 2012; Strutt, 2003; Wallingford, 2006; Wallingford et al., 2002; Wang & Nathans, 2007). The interactions of Dsh in the proposed other biological contexts are less well defined and will be discussed at the end.

3. KINASES ASSOCIATED WITH Dsh

Dsh proteins contain many potential phosphorylation sites and are known to be heavily phosphorylated. For example, in *Drosophila* Dsh out of the 623 amino acids more than 100 are S/T residues (most of them conserved) and another 11 conserved tyrosines, suggesting that more than 1/6 amino acids is or could be phosphorylated (for sequence alignments, see Boutros & Mlodzik, 1999; Wallingford & Habas, 2005). Actually, many of these sites are indeed phosphorylated as determined by mass spectrometry studies, *in vitro* assays, and *in vivo* confirmations (see, for example, Klein, Jenny, Djiane, & Mlodzik, 2006; Schertel et al., 2013; Serysheva et al., 2013; Singh, Yanfeng, Grumolato, Aaronson, & Mlodzik, 2010; Strutt, Price, & Strutt, 2006; Yanfeng et al., 2011 and others). However, only a few sites have been functionally confirmed as physiologically essential (Singh et al., 2010; Strutt et al., 2006). Many of the S/T residues are functionally clustered and thus a certain level of redundancy is expected, which applies for example to the S/T cluster just N-terminal to the PDZ (Klein et al., 2006; Strutt et al., 2006). In other regions of Dsh however, redundancy is less apparent and phosphorylation has been confirmed, and yet mutations in the specific residues do not appear to be functionally important (Yanfeng et al., 2011). A striking example for this scenario is the sole conserved Y residue within the PDZ domain, which is an F residue in most other PDZs and is phosphorylated in *Drosophila in vivo* (as documented with phospho-specific antibodies). Yet when mutated to a nonphosphorylatable residue the respective mutant Dsh allele fully rescues all detectable functions of the gene in *Drosophila* (Yanfeng et al., 2011).

Despite these complications, several individual residues and S/T-clusters have been identified as functionally critical and the respective kinases have also been shown to act on Dsh and have a Dsh-related function *in vivo*. This applies to an S/T cluster N-terminal to the PDZ domain in Dsh that has been linked with CK1 family phosphorylation PDZ (Klein et al., 2006; Strutt et al., 2006) and a single conserved Y residue within the C-terminal region of the DEP domain (Y473 in *Drosophila*) linked to Abelson (Abl) kinase phosphorylation (Singh et al., 2010). While the Abl–Y473 connection has been shown to push Dsh function toward Wnt–Fz/PCP signaling, the CK1 phosphorylation in the S/T cluster has been linked to both pathways (Klein et al., 2006; Strutt et al., 2006), and there is evidence for redundancy among different CK1 family members, in particular CK1ε/*dco*, CK1α, and CK1γ/*gish* (Gault, Olguin, Weber, & Mlodzik, 2012; Strutt et al., 2006; Zhang et al., 2006). A twist to the CK1ε/*dco* requirement(s) is that it appears that the CK1ε protein is sufficient even as a kinase dead isoform to rescue phosphorylation event, possibly due to complex formation with other CK1 family members or other kinases (Klein et al., 2006; Strutt et al., 2006).

So far, two Dsh phosphorylation-associated kinases presumed specific to canonical Wnt-pathway signaling have been reported. The first, Wnk is an S/T-kinase that lacks the K residue in the standard ATP-binding pocket and hence its name Wnk (standing for "*w*ith *no K*"). There is one family member in *Drosophila*, *dWnk*, and four in mammals (it is a more common kinase subfamily in plants). dWnk was identified in a *Drosophila* cell-based screen for kinases required for Fz-induced Dsh phosphorylation (Serysheva et al., 2013). Accordingly, *dWnk* mutants specifically affect canonical Wg/Wnt signaling *in vivo* and cause a marked reduction of Dsh phosphorylation. However, despite its clear requirement in Wnt signaling, the specific phosphorylation target residues in either Dsh or other coreceptor complex proteins have not yet been identified (Serysheva et al., 2013; Serysheva, Mlodzik, & Jenny, 2014). The second S/T-kinase in this context is Nek2, which was identified in a *Drosophila* gain-of-function screen (Schertel et al., 2013), and has been linked to aspects of cell-cycle regulation previously. Nek2 phosphorylates both, N- and C-terminal S/T residues and is suggested to promote Dsh activity toward canonical signaling (Schertel et al., 2013). The physiological role of Nek2 remains unclear as no mutant phenotypes have been reported yet.

There are several other kinases that have been suggested to act on Dsh or with Dsh, PKC family members, for example, yet in many cases the link to Dsh is still speculative or too vague to allow suggestive functional modes. And, while PKC family members and PKCδ in particular appear involved in Dsh regulation toward the PCP side of signaling (Kinoshita, Iioka, Miyakoshi, & Ueno, 2003), many others are too speculative to be discussed here. One aspect is shared by most of the kinases discussed here and others linked to Dsh, namely that Dsh needs to be first recruited to the membrane, which is mainly mediated by Fz receptor family members, before it interacts with a kinase and gets phosphorylated. It is clear that there remains a lot to be

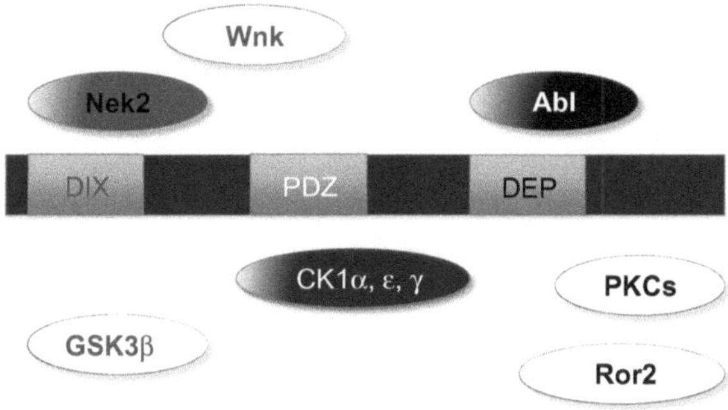

Figure 2 Domain organization of and kinases associate with Dsh function. Color scheme is the same as in Fig. 1. Kinases that are known to directly phosphorylate Dsh are shown in pathway-specific colored solid boxes or in blue (if generally required for Dsh function, e.g., the CK1 family). Some of the more speculative kinases linked to Dsh function and phosphorylation (but not direct phosphorylation of Dsh) are shown as white boxes with pathway-specific colored labels. Only a very limited number of kinases that have been linked to Dsh signaling are shown to reflect a conceptual (over)simplified view. See text for details. (A color version of this figure is located at the end of this chapter.)

discovered at the interplay between kinases and Dsh. See Fig. 2 for a summary of suggested links between specific kinases and Dsh.

4. CELLULAR LEVELS OF Dsh AND REDUNDANCY IN THE MOUSE MODEL

Studies in *Drosophila* suggest that the cellular levels of Dsh are critical for its roles in either pathway or other functions. Strikingly, there are huge amounts of Dsh protein and RNA deposited in the egg during *Drosophila* oogenesis by the mother (Perrimon, Engstrom, & Mahowald, 1989; Perrimon & Mahowald, 1987) and hence the zygotic null *dsh* mutants, originating from a *dsh*+/− mother, survive and look largely normal all the way to third instar larval stages and some even beyond that. This suggests that either the Dsh protein is very stable which has not been addressed thoroughly yet, but seems unlikely as several ubiquitin-linked enzymes and associated proteasome degradation have been lined to Dsh regulation (for example, Chang et al., 2015; de Groot et al., 2014; Madrzak et al., 2015; Strutt, Searle, Thomas-Macarthur, Brookfield, & Strutt, 2013), or that the *dsh* RNA is particularly stable and maintained throughout embryogenesis to later developmental stages. As such, only *maternal-zygotic* double mutant *dsh*$^-$ embryos display the classical Wg/Wnt-signaling canonical defects comparable to *wg* itself or *arm*/β-catenin (Perrimon et al., 1989;

Perrimon & Mahowald, 1987). Nonetheless, at later developmental stages in *Drosophila*, the maintenance of correct Dsh levels is critical for the function of both pathways, canonical Wnt signaling and the Wnt–Fz/PCP pathway.

As such, during larval stages when Wg signaling is critical for the growth and patterning of all imaginal discs and adult tissues, as well as for the subsequent or overlapping process of PCP establishment, too little or too much Dsh is equally deleterious to the tissue and animal. For example, in all assays reported the overexpression of Dsh (driven via components of the Gal4/UAS system; Brand & Perrimon, 1993) is lethal, and similarly reducing dosage of *dsh* very often enhances the lethality of weak Wg-signaling alleles or is a strong suppressor of gain-of-function scenarios in either pathway. The sensitivity to Dsh levels is also observed when a larger than wild-type fraction of Dsh is pushed toward PCP signaling; such scenarios not only cause dominant PCP defects but also cause mild Wg-signaling defects, as the Dsh levels available to canonical Wnt signaling are then too low. A perfect example is the overexpression of Dgo, which locks Dsh in the PCP complex, not only causes gain-of-function PCP defects (Weber, Gault, Olguin, Serysheva, & Mlodzik, 2012), but also leads to loss-of-Wg signaling (Wu, Jenny, Mirkovic, & Mlodzik, 2008). In that context, for example, removing one genomic copy of Dsh (*dsh*+/−) further enhances the canonical Wg-signaling defects, as there is now even less Dsh available for the canonical pathway (while the *dsh*+/− scenario suppresses the PCP gain-of-function defects; Weber et al., 2012; Wu et al., 2008). It is thus surprising that the embryo can easily cope with the huge maternal load of Dsh protein and RNA.

In vertebrate contexts, the levels of Dsh or the sensitivity of the pathways to the protein levels has not been addressed carefully. In *Xenopus*, nevertheless, the protein levels seem to matter as injections of early frog embryos with either wild type of mutant isoforms of xDsh RNA causes strong phenotypes that are readily obtained. The "ease" of such manipulation of Dsh levels and function in *Xenopus* (Itoh & Sokol, 1997; Rothbächer et al., 2000; Wallingford et al., 2000) has been instrumental in shedding light on its functional roles and domain requirements in vertebrate development, which basically confirmed the *Drosophila* data.

The genetic dissection of the Dvl proteins in mice suggests a strict redundancy for canonical signaling and specific roles in the PCP context (Wynshaw-Boris, 2012, see this review by Anthony Wynshaw-Boris for mouse Dvl work-associated references, as these would be too many to list here). In brief, single or double mutants of mDvl1, 2, and 3 display only

Wnt–Fz/PCP-associated defects, and some single mutants are even viable with subtle defects in specific organs like the cochlea. Studies of double knockouts revealed more severe phenotypes (or even novel phenotypes) as compared to single knockouts suggesting overlapping or redundant functions of the *mDvl* genes. The double mutants are generally lethal and associated with PCP-associated developmental defects. As all three Dvl genes are widely expressed, the obvious issue of redundancy is expected to play a significant role here, yet some phenotypes are associated with single mutants and it is not yet clear if all of the single mutant defects are PCP associated. For example, single knockout of mDvl1 leads to viable mice that show social behavior defects and neuronal abnormalities; it remains to be seen how if at all this function can be linked to PCP-associated signaling (Lijam et al., 1997; Wynshaw-Boris, 2012). Nonetheless, largely all of the phenotypes displayed by the mouse *Dvl* mutants appear to be caused by PCP pathway-associated functions, and, importantly, not of the canonical Wnt pathway. These data suggest that the Wnt–Fz/PCP pathway is sensitive to reduction of Dvl levels, but only one *mDvl* allele (of the six present in the genome, two each for the three genes) appears to be sufficient for canonical Wnt-signaling pathway function. This is a rather striking observation and raises many questions on level requirements of the Dsh/Dvl protein family in the individual pathway contexts.

5. A CILIARY FUNCTION OF Dsh

Several quite recent studies have suggested a link between Dsh and ciliogenesis and ciliary positioning. In principle, one can easily link Dsh to ciliary positioning, as it appears to be a key function of Wnt–Fz/PCP signaling to localize cilia to the appropriate region of a cell in vertebrate cells. As such, the Fz–Dsh PCP complex performs (yet not well defined) "functions" to localize the primary cilium to its side near the apical surface of the respective cell, for example, on the posterior side in neural plate cells (e.g., Borovina, Superina, Voskas, & Ciruna, 2010). These data suggest that Dsh is not *per se* required for ciliogenesis (at least its PCP signaling-associated function) as cilia form independently of the Wnt–Fz/PCP core pathway, albeit they are mispositioned. However, the situation is more complicated, as Dsh is found at the base of cilia, associated with the basal body (at least by light microscopy studies) and is also required for apical positioning of the basal bodies (Park et al., 2008). Moreover, in multiciliated cells, every single of the many (often >100) cilia present in such cells does have Dsh localized at its base, suggesting a critical role of Dsh in (at least) apical positioning of cilia and specific trafficking to the ciliary base. In these cells, Dsh levels need to be amplified as the number of cilia and thus basal bodies needs to be massively increased over regular cells with a single, primary cilium. As such the protein levels of the core factors for ciliogenesis and Dsh need to be coordinated (Park et al., 2008).

Additional links to a potential function of Dsh in ciliogenesis come from the observation that one of the key Dsh interactors in *Drosophila*, Dgo, is the homologue of vertebrate Inversin, the product of Nephronophthisis type II gene, which acts in several ciliopathies contexts including kidney (Simons et al., 2005). Inversin has been identified as a key component of the ciliary transition zone protein complexes (Reiter, Blacque, & Leroux, 2012), suggesting that an Inversin/Dgo–Dsh link might serve ciliary functions independent of the Wnt–Fz/PCP contexts.

6. OTHER FUNCTIONS OF Dsh FAMILY MEMBERS?

Of the proposed functions for Dsh family proteins, a direct involvement and function inside the nucleus is controversial and the least well defined. Despite several studies documenting shuttling of Dsh/Dvl into the nucleus and a potential requirement there in canonical Wnt signaling (e.g., Gan et al., 2008; Itoh et al., 2005), and direct interactions with nuclear transcription factors (Barry et al., 2013; Wang et al., 2015), the idea of a nuclear role of Dsh/Dvl has not really caught on. The most recent papers on this subject identify direct interactions between Dvl with members of the YAP/TAZ family (nuclear effectors of the Hippo pathway) (Barry et al., 2013) and FoxK transcription factors (members of the Forkhead transcription factor family) (Wang et al., 2015) in mammalian contexts. As there is no evidence for nuclear Dsh action in *Drosophila*, it is possible that this aspect is an evolutionary addition to the functions of Dsh/Dvl in vertebrates.

Dsh has also been proposed to antagonize Notch signaling both genetically in *Drosophila* (Axelrod et al., 1996) and molecularly in breast cancer cell lines (Collu et al., 2012). While the genetic effects in *Drosophila* can be explained via general signaling pathway interplay, Wnt signaling and the Notch pathway often act antagonistically, the proposed molecular mechanism for Dsh Notch antagonism in breast cancer cells via Dsh/Dvl binding directly to the Su(H)/CSL nuclear effector/transcription factor of the Notch pathway (Collu et al., 2012) is highly intriguing and deserves further attention.

As a final thought, it is worth mentioning that Dsh/Dvl family proteins might also more directly affect cell growth and/or survival, independent of Wg/Wnt signaling. This idea is supported by two general observations. Overexpression of Dsh in *Drosophila* generally causes cell death, and although this has not been "officially" published it is a common observation by several labs, and usually discussed at meetings and conferences informally. Whether it is a physiological effect or an artifact of overexpression remains to be seen. Similarly, the observation that a triple knockout of all three Dvl genes in the mouse appears cell lethal (Wynshaw-Boris, 2012) is also not easily explained via canonical Wnt-signaling effects. As such, it seems that there is still a key role of Dsh/Dvl to be discovered.

7. CONCLUDING REMARKS

The mechanism by which Dsh routes information into the different intracellular pathways and its functional requirements in general may lie in its domain composition (Fig. 1). Using *in vivo* structure–function approaches and mutant rescue experiments, it has been established for quite sometime which regions and which domains of Dsh act in the respective pathways (Axelrod et al., 1998; Boutros et al., 1998), and these observations have been confirmed in vertebrate models (Itoh & Sokol, 1997; Rothbächer et al., 2000; Wallingford & Habas, 2005; Wallingford & Harland, 2002; Wallingford et al., 2000). The molecular nature of the PCP-specific *dsh* alleles further supported these conclusions (Axelrod et al., 1998; Boutros et al., 1998; Penton et al., 2002). The dissection of the distinct domain requirements suggested that a functional understanding is in grasp, yet that was over 15 years ago and we still do not understand Dsh/Dvl regulation much better than that. The addition of regulatory input by kinases has not yet been helpful, possibly because there are so many and because of possible redundancy.

When the domain requirements were established in 1998, it seemed that the lack of knowledge of the respective physical interaction partners was hampering our understanding of *how* Dsh/Dvl signals. We know many interaction partners now, most likely too many, as Dsh/Dvl proteins have pulled down a vast set of molecular interactors, of which many remain questionable. Thus, although we understand the basic principles of signal routing, like the DIX domain interaction with Axin being required for canonical Wnt signaling (e.g., reviewed in Clevers & Nusse, 2012), the PDZ domain interacting with the Fz receptor family members (Axelrod,

2001; Boutros, Mihaly, Bouwmeester, & Mlodzik, 2000; Wong et al., 2003), or the interaction of the DEP domain with membrane lipids and Diego locking Dsh into PCP complexes (Jenny et al., 2005; Simons et al., 2009), we still do not know how these are regulated in a temporal or spatial manner. Most importantly, these signaling events often happen simultaneously in individual cells, e.g., in the *Drosophila* eye (Mlodzik, 1999), and so a tight regulatory mechanism has to be present. At this point, there are still many gaps remaining, too many, in the understanding of Dsh/Dvl function and regulation in the respective Wnt-signaling pathways or other Dsh/Dvl functional requirements. Hopefully, we will not have to wait for another 20 years to have this puzzle solved.

ACKNOWLEDGMENTS

I am most grateful to all my lab members who over the years either directly contributed to these questions or help guide the approaches taken, and my colleague Sergei Sokol who knows more about Dsh than anybody else I know. The work on Dsh in the Mlodzik lab has been recently and continues to be supported by grants from the NIGMS, NEI, and NICHD of the NIH.

REFERENCES

Adler, P. N. (2012). The frizzled/stan pathway and planar cell polarity in the Drosophila wing. *Current Topics in Developmental Biology*, *101*, 1–31.

Axelrod, J., Matsuno, K., Artavanis-Tsakonas, S., & Perrimon, N. (1996). Interaction between Wingless and Notch signaling pathways mediated by Dishevelled. *Science*, *271*, 1826–1832.

Axelrod, J. D. (2001). Unipolar membrane association of Dishevelled mediates Frizzled planar cell polarity signaling. *Genes & Development*, *15*, 1182–1187.

Axelrod, J. D., Miller, J. R., Shulman, J. M., Moon, R. T., & Perrimon, N. (1998). Differential requirement of Dishevelled provides signaling specificity in the Wingless and planar cell polarity signaling pathways. *Genes & Development*, *12*, 2610–2622.

Barry, E. R., Morikawa, T., Butler, B. L., Shrestha, K., de la Rosa, R., Yan, K. S., et al. (2013). Restriction of intestinal stem cell expansion and the regenerative response by YAP. *Nature*, *493*, 106–110.

Bilic, J., Huang, Y. L., Davidson, G., Zimmermann, T., Cruciat, C. M., Bienz, M., et al. (2007). Wnt induces LRP6 signalosomes and promotes dishevelled-dependent LRP6 phosphorylation. *Science*, *316*, 1619–1622.

Borovina, A., Superina, S., Voskas, D., & Ciruna, B. (2010). Vangl2 directs the posterior tilting and asymmetric localization of motile primary cilia. *Nature Cell Biology*, *12*, 407–412.

Boutros, M., Mihaly, J., Bouwmeester, T., & Mlodzik, M. (2000). Signaling specificity by Frizzled receptors in Drosophila. *Science*, *288*, 1825–1828.

Boutros, M., & Mlodzik, M. (1999). Dishevelled: At the crossroads of divergent intracellular signaling pathways. *Mechanisms of Development*, *83*, 27–37.

Boutros, M., Paricio, N., Strutt, D. I., & Mlodzik, M. (1998). Dishevelled activates JNK and discriminates between JNK pathways in planar polarity and *wingless* signaling. *Cell*, *94*, 109–118.

Brand, A. H., & Perrimon, N. (1993). Targeted gene expression as a means of altering cell fates and generating dominant phenotypes. *Development*, *118*, 401–415.

Chang, B., Tessneer, K. L., McManus, J., Liu, X., Hahn, S., Pasula, S., et al. (2015). Epsin is required for Dishevelled stability and Wnt signalling activation in colon cancer development. *Nature Communications*, *6*, 6380.

Clevers, H., & Nusse, R. (2012). Wnt/beta-catenin signaling and disease. *Cell*, *149*, 1192–1205.

Collu, G. M., Hidalgo-Sastre, A., Acar, A., Bayston, L., Gildea, C., Leverentz, M. K., et al. (2012). Dishevelled limits Notch signalling through inhibition of CSL. *Development*, *139*, 4405–4415.

de Groot, R. E., Ganji, R. S., Bernatik, O., Lloyd-Lewis, B., Seipel, K., Sedova, K., et al. (2014). Huwe1-mediated ubiquitylation of dishevelled defines a negative feedback loop in the Wnt signaling pathway. *Science Signaling*, 7, ra26.

Drosten, M., Lechuga, C. G., & Barbacid, M. (2013). Genetic analysis of Ras genes in epidermal development and tumorigenesis. *Small GTPases*, *4*, 236–241.

Fahmy, O. G., & Fahmy, M. J. (1959). New mutants report. *Drosophila Information Service*, *33*, 83–94.

Gan, X. Q., Wang, J. Y., Xi, Y., Wu, Z. L., Li, Y. P., & Li, L. (2008). Nuclear Dvl, c-Jun, beta-catenin, and TCF form a complex leading to stabilization of beta-catenin–TCF interaction. *The Journal of Cell Biology*, *180*, 1087–1100.

Gault, W. J., Olguin, P., Weber, U., & Mlodzik, M. (2012). Drosophila CK1γ, gilgamesh, controls PCP-mediated morphogenesis through regulation of vesicle trafficking. *The Journal of Cell Biology*, *196*, 605–621.

Goodrich, L. V., & Strutt, D. (2011). Principles of planar polarity in animal development. *Development*, *138*, 1877–1892.

Itoh, K., Brott, B. K., Bae, G. U., Ratcliffe, M. J., & Sokol, S. Y. (2005). Nuclear localization is required for Dishevelled function in Wnt/beta-catenin signaling. *Journal of Biology*, *4*, 3.

Itoh, K., & Sokol, S. Y. (1997). Graded amounts of Xenopus dishevelled specify discrete anteroposterior cell fates in prospective ectoderm. *Mechanisms of Development*, *61*, 113–125.

Jenny, A., Reynolds-Kenneally, J., Das, G., Burnett, M., & Mlodzik, M. (2005). Diego and Prickle regulate Frizzled planar cell polarity signalling by competing for Dishevelled binding. *Nature Cell Biology*, 7, 691–697.

Kinoshita, N., Iioka, H., Miyakoshi, A., & Ueno, N. (2003). PKC delta is essential for Dishevelled function in a noncanonical Wnt pathway that regulates Xenopus convergent extension movements. *Genes and Development*, *17*, 1663–1676.

Klein, T. J., Jenny, A., Djiane, A., & Mlodzik, M. (2006). CKIepsilon/discs overgrown promotes both Wnt-Fz/beta-catenin and Fz/PCP signaling in Drosophila. *Current Biology*, *16*, 1337–1343.

Klingensmith, J., Nusse, R., & Perrimon, N. (1994). The *Drosophila* segment polarity gene *dishevelled* encodes a novel protein required for response to the *wingless* signal. *Genes & Development*, *8*, 118–130.

Lijam, N., Paylor, R., McDonald, M. P., Crawley, J. N., Deng, C. X., Herrup, K., et al. (1997). Social interaction and sensorimotor gating abnormalities in mice lacking Dvl1. *Cell*, *90*, 895–905.

Madrzak, J., Fiedler, M., Johnson, C. M., Ewan, R., Knebel, A., Bienz, M., et al. (2015). Ubiquitination of the Dishevelled DIX domain blocks its head-to-tail polymerization. *Nature Communications*, *6*, 6718.

McNeill, H. (2009). Planar cell polarity and the kidney. *Journal of the American Society of Nephrology*, *20*, 2104–2111.

Mlodzik, M. (1999). Planar polarity in the *Drosophila* eye: A multifaceted view of signaling specificity and cross-talk. *The EMBO Journal*, *18*, 6873–6879.

Nüsslein-Volhard, C., & Wieschaus, E. (1980). Mutations affecting segment number and polarity in *Drosophila*. *Nature*, *287*, 795–801.
Park, T. J., Mitchell, B. J., Abitua, P. B., Kintner, C., & Wallingford, J. B. (2008). Dishevelled controls apical docking and planar polarization of basal bodies in ciliated epithelial cells. *Nature Genetics*, *40*(7), 871–879.
Peng, Y., & Axelrod, J. D. (2012). Asymmetric protein localization in planar cell polarity: Mechanisms, puzzles, and challenges. *Current Topics in Developmental Biology*, *101*, 33–53.
Penton, A., Wodarz, A., & Nusse, R. (2002). A mutational analysis of dishevelled in Drosophila defines novel domains in the dishevelled protein as well as novel suppressing alleles of axin. *Genetics*, *161*, 747–762.
Perrimon, N., Engstrom, L., & Mahowald, A. P. (1989). Zygotic lethals with specific maternal effect phenotypes in Drosophila melanogaster. I. Loci on the X chromosome. *Genetics*, *121*, 333–352.
Perrimon, N., & Mahowald, A. P. (1987). Multiple functions of segment polarity genes in Drosophila. *Developmental Biology*, *119*, 587–600.
Reiter, J. F., Blacque, O. E., & Leroux, M. R. (2012). The base of the cilium: Roles for transition fibres and the transition zone in ciliary formation, maintenance and compartmentalization. *EMBO Reports*, *13*, 608–618.
Rothbächer, U., Laurent, M. N., Deardorff, M. A., Klein, P. S., Cho, K. W., & Fraser, S. E. (2000). Dishevelled phosphorylation, subcellular localization and multimerization regulate its role in early embryogenesis. *The EMBO Journal*, *19*, 1010–1022.
Schertel, C., Huang, D., Bjorklund, M., Bischof, J., Yin, D., Li, R., et al. (2013). Systematic screening of a Drosophila ORF library in vivo uncovers Wnt/Wg pathway components. *Developmental Cell*, *25*, 207–219.
Schwarz-Romond, T., Merrifield, C., Nichols, B. J., & Bienz, M. (2005). The Wnt signalling effector Dishevelled forms dynamic protein assemblies rather than stable associations with cytoplasmic vesicles. *Journal of Cell Science*, *118*, 5269–5277.
Seifert, J. R., & Mlodzik, M. (2007). Frizzled/PCP signalling: a conserved mechanism regulating cell polarity and directed motility. *Nature Reviews*, *8*, 126–138.
Serysheva, E., Berhane, H., Grumolato, L., Demir, K., Balmer, S., Bodak, M., et al. (2013). Wnk kinases are positive regulators of canonical Wnt/beta-catenin signalling. *EMBO Reports*, *14*, 718–725.
Serysheva, E., Mlodzik, M., & Jenny, A. (2014). WNKs in Wnt/beta-catenin signaling. *Cell Cycle*, *13*, 173–174.
Simons, M., Gault, W. J., Gotthardt, D., Rohatgi, R., Klein, T. J., Shao, Y., et al. (2009). Electrochemical cues regulate assembly of the Frizzled/Dishevelled complex at the plasma membrane during planar epithelial polarization. *Nature Cell Biology*, *11*, 286–294.
Simons, M., Gloy, J., Ganner, A., Bullerkotte, A., Bashkurov, M., Kronig, C., et al. (2005). Inversin, the gene product mutated in nephronophthisis type II, functions as a molecular switch between Wnt signaling pathways. *Nature Genetics*, *37*, 537–543.
Simons, M., & Mlodzik, M. (2008). Planar cell polarity signaling: From fly development to human disease. *Annual Review of Genetics*, *42*, 517–540.
Singh, J., & Mlodzik, M. (2012). Planar cell polarity signaling: Coordination of cellular orientation across tissues. *Wiley Interdisciplinary Reviews: Developmental Biology*, *1*, 479–499.
Singh, J., Yanfeng, W. A., Grumolato, L., Aaronson, S. A., & Mlodzik, M. (2010). Abelson family kinases regulate Frizzled planar cell polarity signaling via Dsh phosphorylation. *Genes & Development*, *24*, 2157–2168.
Sokol, S. Y., Klingensmith, J., Perrimon, N., & Itoh, K. (1995). Dorsalizing and neuralizing properties of Xdsh, a maternally expressed Xenopus homolog of dishevelled. *Development*, *121*, 3487.
Strutt, D. (2003). Frizzled signalling and cell polarisation in Drosophila and vertebrates. *Development*, *130*, 4501–4513.

Strutt, H., Price, M. A., & Strutt, D. (2006). Planar polarity is positively regulated by casein kinase Iepsilon in Drosophila. *Current Biology*, *16*, 1329–1336.
Strutt, H., Searle, E., Thomas-Macarthur, V., Brookfield, R., & Strutt, D. (2013). A Cul-3-BTB ubiquitylation pathway regulates junctional levels and asymmetry of core planar polarity proteins. *Development*, *140*, 1693–1702.
Sussman, D. J., Klingensmith, J., Salinas, P., Adams, P. S., Nusse, R., & Perrimon, N. (1994). Isolation and characterization of a mouse homolog of the Drosophila segment polarity gene dishevelled. *Developmental Biology*, *166*, 73–86.
Taelman, V. F., Dobrowolski, R., Plouhinec, J. L., Fuentealba, L. C., Vorwald, P. P., Gumper, I., et al. (2010). Wnt signaling requires sequestration of glycogen synthase kinase 3 inside multivesicular endosomes. *Cell*, *143*, 1136–1148.
Wallingford, J. B. (2006). Planar cell polarity, ciliogenesis and neural tube defects. *Human Molecular Genetics*, *15*(Spec No. 2), R227–R234.
Wallingford, J. B., Fraser, S. E., & Harland, R. M. (2002). Convergent extension: The molecular control of polarized cell movement during embryonic development. *Developmental Cell*, *2*, 695–706.
Wallingford, J. B., & Habas, R. (2005). The developmental biology of Dishevelled: An enigmatic protein governing cell fate and cell polarity. *Development*, *132*, 4421–4436.
Wallingford, J. B., & Harland, R. M. (2002). Neural tube closure requires Dishevelled-dependent convergent extension of the midline. *Development*, *129*, 5815–5825.
Wallingford, J. B., Rowning, B. A., Vogeli, K. M., Rothbacher, U., Fraser, S. E., & Harland, R. M. (2000). Dishevelled controls cell polarity during Xenopus gastrulation. *Nature*, *405*, 81–85.
Wang, W., Li, X., Lee, M., Jun, S., Aziz, K. E., Feng, L., et al. (2015). FOXKs promote Wnt/beta-catenin signaling by translocating DVL into the nucleus. *Developmental Cell*, *32*, 707–718.
Wang, Y., & Nathans, J. (2007). Tissue/planar cell polarity in vertebrates: New insights and new questions. *Development*, *134*, 647–658.
Weber, U., Gault, W. J., Olguin, P., Serysheva, E., & Mlodzik, M. (2012). A genetic screen identifies novel planar cell polarity factors in Drosophila. *Genetics*, *191*, 145–162.
Wong, H.-C., Bourdelas, A., Krauss, A., Lee, H.-J., Shao, Y., Wu, D., et al. (2003). Direct binding of the PDZ domain of Dishevelled to a conserved internal sequence in the C-terminal region of Frizzled. *Molecular Cell*, *12*, 1251–1260.
Wu, J., Jenny, A., Mirkovic, I., & Mlodzik, M. (2008). Frizzled–Dishevelled signaling specificity outcome can be modulated by Diego in Drosophila. *Mechanisms of Development*, *125*, 30–42.
Wynshaw-Boris, A. (2012). Dishevelled: In vivo roles of a multifunctional gene family during development. *Current Topics in Developmental Biology*, *101*, 213–235.
Yanfeng, W. A., Berhane, H., Mola, M., Singh, J., Jenny, A., & Mlodzik, M. (2011). Functional dissection of phosphorylation of Disheveled in Drosophila. *Developmental Biology*, *360*, 132–142.
Zeng, X., Huang, H., Tamai, K., Zhang, X., Harada, Y., Yokota, C., et al. (2008). Initiation of Wnt signaling: Control of Wnt coreceptor Lrp6 phosphorylation/activation via frizzled, dishevelled and axin functions. *Development*, *135*, 367–375.
Zhang, L., Jia, J., Wang, B., Amanai, K., Wharton, K. A., Jr., & Jiang, J. (2006). Regulation of wingless signaling by the CKI family in Drosophila limb development. *Developmental Biology*, *299*, 221–237.

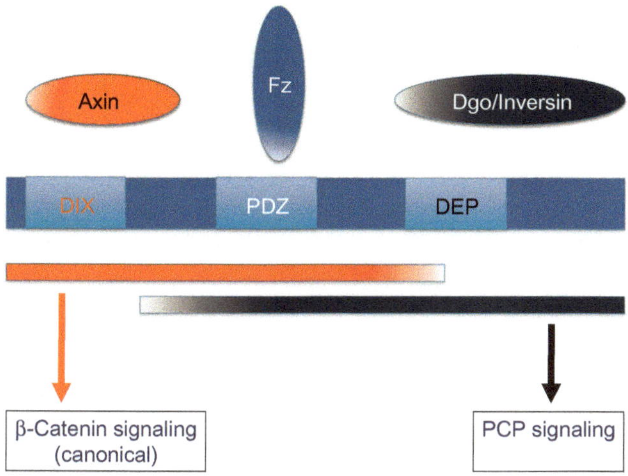

Marek Mlodzik, Figure 1 Domain organization of Dishevelled (Dsh) proteins and pathway-specific requirements and interactions. Dsh (in *Drosophila* 623 aa long) is shown in blue with its three main and highly conserved domains, DIX, PDZ, and DEP, labeled accordingly. The color scheme is "red" for canonical Wnt/β-catenin signaling and "black" for Wnt–Fz/PCP signaling. The three domains have their "main" pathway-specific interactors shown in the respective color and the red and black bars below Dsh indicate the approximate extent of requirements for signal relay to the respective pathways. Frizzled receptors directly interact with the PDZ domain and as this interaction is needed for both signaling outcomes it kept in blue. See text for details.

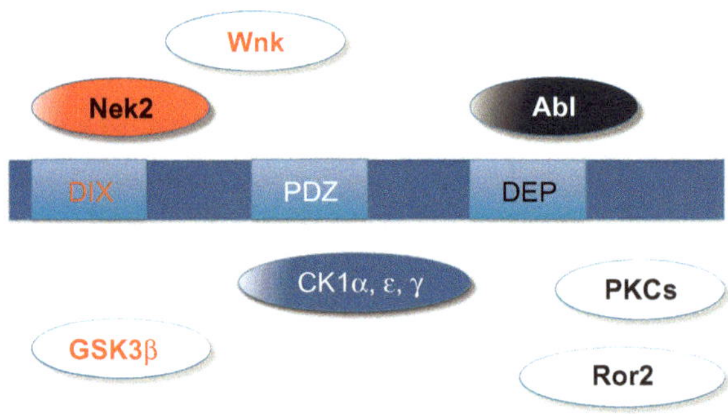

Marek Mlodzik, Figure 2 Domain organization of and kinases associate with Dsh function. Color scheme is the same as in Fig. 1. Kinases that are known to directly phosphorylate Dsh are shown in pathway-specific colored solid boxes or in blue (if generally required for Dsh function, e.g., the CK1 family). Some of the more speculative kinases linked to Dsh function and phosphorylation (but not direct phosphorylation of Dsh) are shown as white boxes with pathway-specific colored labels. Only a very limited number of kinases that have been linked to Dsh signaling are shown to reflect a conceptual (over)simplified view. See text for details.

CHAPTER SIX

When Family History Matters: The Importance of Lineage Analyses and Fate Maps for Explaining Animal Development

Steven L. Klein, Sally A. Moody[1]
Department of Anatomy and Regenerative Biology, The George Washington University School of Medicine and Health Sciences, Washington, DC, USA
[1]Corresponding author: e-mail address: samoody@gwu.edu

Contents

Abstract

Initial interest in understanding how the fertilized egg becomes a multicellular animal suggested two possible answers: either the embryo came from preformed components or it arose through epigenetic processes. Extensive research during the past few decades has identified aspects of development that depend on preformed elements, such as cytoplasmic components and a cell's lineage; it also has identified aspects that depend on epigenetic processes, such as cell interactions and morphogen gradients. These advances have depended on understanding embryonic cell lineage and cell fate. This essay explains how lineage analysis and fate mapping have contributed to our current understanding of embryonic development.

The fundamental question that developmental biologists have asked since the inception of the field is: how is a single cell, the fertilized egg, transformed into a multicellular organism with a complex body plan? Over the field's history, this question has been asked at many different levels of

Current Topics in Developmental Biology, Volume 117
ISSN 0070-2153
http://dx.doi.org/10.1016/bs.ctdb.2015.10.011

complexity, from organisms to germ layers to tissues to cells to molecules. The answers that prevailed at each time depended upon the available technologies and methodologies. Because so much of our current understanding relies on knowing a cell's lineage and fate, in this essay we consider the fundamental philosophical questions that the first lineage analyses and fate maps set out to address, and we review some of the seminal studies that analyzed cell lineage and cell fate. We hope to instill an appreciation of the fact that knowing a founder cell's lineage and fate is an essential component of our understanding of the processes that control animal development.

First, we need some definitions. Often the terms "cell lineage" and "cell fate" are used interchangeably. But, the two terms define different aspects of an embryonic cell's journey from zygote to mature animal. A cell's lineage is its complete family tree (think www.ancestry.com) and defines its ancestral relationship to every other cell in the embryo. In contrast, a cell's fate is what it will become in the end (of a developmental process or an experiment) (Fig. 1). A lineage map defines every cell division, starting with a defined founder cell (often either the zygote or an identifiable progenitor such as a blastomere, micromere, macromere, teloblast, or mother cell) and continuing down through the generation of daughter cells, granddaughter cells, etc., until a specific cell fate is assumed (Fig. 1A). A lineage analysis can produce a fate map that illustrates all the cell divisions that occur to produce a particular structure. For example, in the sea urchin, *S. purpuratus*, the 16-cell mesomeres each divide two more times to produce the progenitors of the aboral ectoderm (Wray, 1999). However, as we will discuss later, complete lineage maps are feasible only in animals that have a small number of cells and/or a small number of cell divisions between the founder cells and their final products. On the other hand, fate maps have been made in a variety of animals using several different labeling techniques. They can be constructed by labeling either groups of cells or individual cells (the latter of which are correctly called *clonal* fate maps). In either case, the marked cells are followed through morphogenetic movements and traced to their germ layer or organ anlage without tracking specific cell divisions, and these fates are projected back onto the progenitor cells (Fig. 1B). While lineage maps and fate maps are constructed in two different ways and can provide unique types of information, in animals in which both are possible, even if only for a portion of the development period, the information often concurs (Fig. 1). Importantly, both approaches have been critical for elucidating essential features of embryonic development, including where, when, and how cell fates are determined.

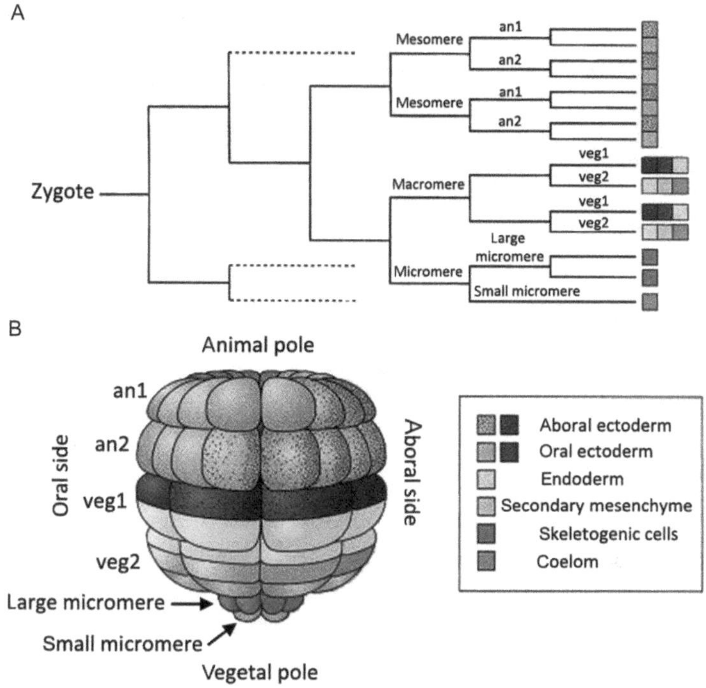

Figure 1 Lineage analysis (A) and fate map (B) of the 60-cell embryo of the sea urchin, *S. purpuratus*. (A) The embryonic cell lineage is a map of each cell division and the ultimate fate of each blastomere. For simplicity, the lineage of only one of the four-cell blastomeres is shown. The lineages of the other blastomeres (dotted lines) have very similar patterns. (B) The left side of the 60-cell embryo with the origins of the various embryonic cell fates (key on the right side) projected onto the blastomeres. *Modified from Wray (1999); see this reference for details.* (A color version of this figure is located at the end of this chapter.)

1. THE ADVENT OF ANALYZING CELL FATES

The question of how the adult form arises from an embryo was considered by Greek philosophers over 2500 years ago (reviewed in Adelman, 1966). One school believed that there was no change in the world, only constancy. They argued, therefore, that the components of the adult were present in the fertilized egg from the beginning, and development was a process that merely brought these components to their recognizable adult form. They called this "preformation." The other school believed that there was only change in the world. They argued that the embryo consisted of components that were qualitatively different from those of the adult and that external forces acted to gradually transform the embryonic components into the adult form. They called this "epigenesis."

These two opposing explanations for how the embryo developed remained at the forefront of investigation for many centuries. Soon after the formulation of the Cell Theory by Schleiden and Schwann (in 1838/1839), embryologists came to realize that the egg was, in fact, a single cell and that the funny little furrows in the egg that formed after fertilization were, in fact, cleavage planes that divided the egg cell into daughter cells (reviewed in Wilson, 1925). The recognition that the embryo was structured from its earliest stages, i.e., composed of cells and cleavage furrows, provided an opportunity to address whether the embryo was preformed or formed by epigenetic processes. This realization motivated investigators to determine whether there was any relationship between the early cells and cleavage furrows, and the later-forming germ layers and organ anlage. Accordingly, they began to visually follow cell divisions from the fertilized egg to germ layers and organ rudiments in a wide variety of embryos. The giants of these classical investigations (e.g., C.O. Whitman, E.B. Wilson, T.H. Boveri, E.G. Conklin) used direct observation of transparent invertebrate embryos (e.g., various species of leeches, ascidians, ascaris, mollusks, and echinoderms). These embryos were large enough to visualize cellular detail, abundant enough to compare results across many individuals, had regular and reproducible cell divisions, and consisted of a small number of cells that rapidly developed into an identifiable fate (germ layer or organ rudiment). These features enabled embryologists to follow the progression of cell divisions from initial cleavages to identifiable progeny (e.g., Fig. 1). These careful examinations of the genealogy of embryonic structures made two consistent observations: (1) the first few cleavage planes demarcated the embryo's cardinal axes and (2) the cleavage pattern could predict the later contribution of the lineage to a specific germ layer and/or organ (reviewed in Moritz & Sauer, 1996; Wilson, 1925). Since specific blastomere progenitors of germ layers and organs could be identified, these results were interpreted to support the preformation hypothesis that the adult form was delineated in the egg.

This conclusion was strongly supported by the observation that the eggs of some animals contained a region of identifiable cytoplasm that was partitioned into a specific lineage that gave rise to a particular tissue type. For example, Conklin (1905) traced the crescent of yellow cytoplasm from the ascidian egg through subsequent cell divisions and eventually into the somites, the myogenic precursors. This observation suggested that a preformed component within the yellow cytoplasm of the egg (a maternally derived mRNA or protein, e.g., Jeffrey, 1985; Swalla & Jeffrey, 1995) was responsible for the formation of muscle cells. These and similar direct observations in other animals championed the idea that preformation plays a significant role in embryonic development. Support for preformation was also provided by several experimental manipulations. For example, transferring the yellow cytoplasm to other cells that normally did not produce muscle caused them to ectopically express muscle proteins (Whittaker, 1980). The search for "preformed" factors has led to our current appreciation of the critical role that inherited maternal molecules play in the formation of the cardinal axes and the determination of some tissues, such as primordial germ cells and endoderm (reviewed in Davidson, 1990; Haston & Reijo-Pera, 2007; White & Heasman, 2008).

However, the preformation interpretations derived from descriptive lineage analyses began to be challenged by the advent of "experimental embryology," which used perturbations rather than observation alone to test the validity of a hypothesis. In animals large enough and sturdy enough to withstand manual poking and prodding, investigators ablated cells to see if the remaining cells changed their fate and rearranged the positions of cells in the embryo to see if they acquired different fates. Critical experiments in frog (by Roux in 1888) and in sea urchins (by Driesch in 1891) supported the idea that embryonic pattern was not always "fixed" by preformed cytoplasmic components (reviewed in Spemann, 1967; Wilson, 1925). In both species, the destruction of one cell of the two-cell embryo still resulted in a normal, completely formed, albeit smaller, embryo. These results could only be explained if the remaining cell had the ability to "regulate" its cell fate, i.e., to change its potential in response to the new environmental conditions. Thus, the field had to reach a compromise: the dichotomy between preformation and epigenesis was overly simplistic, and the development of individual organisms must utilize both processes. Some developmental events depend on preformed components (rebranded as "mosaic" development), whereas others depend on interactions with the external environment or between embryonic components (rebranded as "regulative" development).

The acceptance that both mosaic and regulative mechanisms could be involved in specifying cell fate spurred the community to elucidate where and when each process was used. Cells were ablated to test the response of the remaining cells, they were transplanted to new locations to test whether their fate was already determined, and they were placed in culture to test their potential when developing alone or in the presence of other cells. It was immediately realized that to critically evaluate these experimental perturbations, the field needed techniques that could identify the origins of the cells being manipulated and ways to accurately fate map the embryos of other kinds of animals that are not transparent and/or consist of too many cells to individually track each cell division.

2. CREATING FATE MAPS

At the end of the nineteenth century, investigations into the relative roles of mosaic versus regulative development turned to vertebrates. The embryos of fish and amphibians were particularly useful because they develop external to the mother, are large enough to be experimentally manipulated, and are easily cultured in the laboratory. A central goal of the day was to understand the cell movements that produced the primary germ layers during gastrulation. However, vertebrate embryologists could not perform the elegant cell lineage analyses being done in the marine invertebrates because the cells of their embryos were opaque, and did not cleave in regular patterns at gastrulation stages. Thus, new techniques were sought that would allow the gastrulation movements of cells to be traced.

2.1 Fate Mapping Using Vital Dyes

This problem was solved in the amphibian embryo by Vogt (reviewed in Spemann, 1967), who impregnated agar with a "vital," i.e., nontoxic, dye (Nile Red), and placed several small slivers of them along the midline of the surface of the blastula. The dye would transfer to the cells underlying the agar sliver, and the colored cells could be observed in the living embryo through normal, unperturbed gastrulation movements. Vogt's fate map of the blastula midline cells showed that cells in the vegetal hemisphere moved into the gastrula's interior to become the endoderm, the cells in the animal hemisphere spread around the surface to become the ectoderm, and the cells in the equatorial zone moved between the interior and exterior layers to become the mesoderm. This map illustrated regular and reproducible fields of blastula cell fates for each germ layer, which proved invaluable for interpreting several subsequent experimental studies. For example, it showed that the transplanted dorsal lip of the blastopore, the famous "Organizer," was presumptive mesoderm, whereas the induced secondary nervous system arose from ventral, nonneural ectoderm (reviewed in Spemann, 1967). This indicated that the nervous system was not "preformed" in the embryo, but rather, likely arose via interactions between embryonic parts. This epigenetic (or regulative) process was the first experimental demonstration of tissue "induction" (then called "primary embryonic induction"). Thus, fate maps were invaluable for establishing the idea that communication between different parts of the embryo can lead to the formation of a new cell type, and they ultimately led to decades of experiments that have identified the specific signaling pathways that are responsible for neural induction.

Decades later, Ray Keller expanded Vogt's approach by applying vital dye patches systematically across the entire surface of the newly established laboratory amphibian, *Xenopus laevis* (Keller, 1975, 1976). Other amphibians, including those used by Vogt and Spemann, could only be studied during their natural breeding season, which limited the time during which experiments could be performed. In contrast, it was discovered that the very peculiar South African clawed frog, *Xenopus*, could breed in the laboratory throughout the year, basically "on demand" (see Gurdon & Hopwood, 2000, for a historical account of the development of *Xenopus* as a laboratory model for developmental biology). Keller's comprehensive fate maps covered all superficial and deep regions of the embryo, followed cell movements to much later points in organ formation, and indicated which structures were already represented in the blastula and which depended on later interactions. Because this work had a major focus on the source of the different types of mesoderm, it provided important support for the experimental evidence of Peter Nieuwkoop that mesoderm fate is induced in the equatorial cells of the blastula by interactions between the precursors of the endoderm (vegetal cells) and precursors of the ectoderm (animal cells) (reviewed in Nieuwkoop, 1997). Together, the fate maps of Keller and the tissue recombination experiments of Nieuwkoop enabled the important discoveries of the transcription factors and signaling pathways responsible for formation of the mesoderm germ layer, as well as its patterning into axial, paraxial, lateral plate, and ventral tissue domains.

The ability to trace the movement of cells through gastrulation to their germ layer and then into their specific organs was soon recognized to be of critical importance for any study of how cells are specified into particular types of cells, tissues, and organs. Transplanting cells to new environments in the embryo or into culture to test whether they maintained their original fate or changed fate in response to the new environment became a critical experimental test of fate commitment (and still is: see Davidson, Camus, & Tam, 1999; Gallagher, Hainski, & Moody, 1991; Heasman, Wylie, Hausen, & Smith, 1984 as examples). To accomplish this, a fate map is required, but the vital dye-infused agar approach is not optimal for other animals. The signal fades with time making it difficult to analyze later time points of slower growing animals, and it diffuses away from the original mark limiting the cellular resolution of the map. Superior marking techniques were necessary to improve the resolution of fate mapping, particularly in order to expand its use to two important amniote animal models: chick and mouse embryos.

A number of different marking techniques have been introduced over the years, such as carbon crystals, India ink, and quantum dots. However, a real breakthrough was the introduction of lipophilic fluorescent dyes (see http://tools.thermofisher.com/content/sfs/manuals/mp00282.pdf). These were introduced in the 1980s for marking individual neurons and tracing their axonal projections. In the hands of developmental biologists, they soon became favored for fate mapping because they are simple to detect by their bright fluorescence, they intercalate into cell membranes to illuminate the entire cell, for the most part do not diffuse from a labeled cell to an unlabeled cell and are long-lasting (months to years) in both living and fixed preparations. They can be applied to much smaller patches of cells (~10 cells in many modern studies), thus greatly improving the resolution of fate maps, and spots of different fluorophores can be combined in the same embryo, which makes mapping more efficient. These dyes have been used extensively to map the origins of specific organs from the epiblast and primitive streak of both chick and mouse (e.g., Garcia-Martinez & Schoenwolf, 1993; Hatada & Stern, 1994; Quinlan, Williams, Tan, & Tam, 1995; Sulik et al., 1994; Wilson & Beddington, 1996). They also can be applied much later in development to track the movement of cells during organogenesis, thereby extending fate mapping well beyond the formation of the germ layers. However, one must consider that the small patches of initially labeled cells are potentially heterogeneous, i.e., not necessarily lineally/clonally related. For example, DiI labeling of small patches of cells along the border of the neural plate has produced very significant information about the intermixing of precursors of the neural crest and various cranial placodes (e.g., Pieper, Eagleson, Wosniok, & Schlosser, 2011; Streit, 2002), but without single-cell labeling, it still is not clear whether a single border cell gives rise to both neural crest and placode cells.

2.2 Fate Mapping by Making Chimeras

Another early technique for mapping cell origins was the construction of chimeras. In this case, a small bit of tissue from an animal that carries a marker is transplanted into another animal devoid of that marker. An important early approach was transplanting tissue from a darkly pigmented species into a lightly pigmented related species. This technique, called heteroplastic grafting, was used extensively by Ross Harrison and his students in amphibians to distinguish between graft and host tissue (Harrison, 1935). It also was critical for the discovery of the Organizer. Dorsal lip mesoderm from a lightly pigmented salamander was transplanted into the ventral region of a darkly pigmented host. Because the "induced" neural tube was darkly pigmented, Spemann and Hilde Mangold could conclude that it was derived from the host ventral epidermis and not from the transplanted dorsal lip (reviewed in Spemann, 1967). But, pigmentation is not always sufficient to confidently identify the origin of a single cell.

A major breakthrough was achieved by the discovery that the cells of the Japanese quail carry a nucleolar marker that is very distinct from that of chick embryos and that tissue can be transplanted between their embryos (Le Douarin, 1973; Le Douarin & Barq, 1969; Teillet, Ziller, & Le Douarin, 2008). Using the appropriate histological preparation, it is very easy to distinguish whether a single cell arose from the host or the transplanted tissue. This led to an explosion of fine, microsurgical transplantation experiments that mapped many tissues, including the epiblast (e.g., Garcia-Martinez, Alvarez, & Schoenwolf, 1993), neural crest (e.g., Le Douarin & Dupin, 2012), and craniofacial muscles (e.g., Noden, 1986).

Other approaches, however, were needed for animals that do not have natural markers. In mouse and zebrafish embryos, for example, a number of genetic marking approaches were developed. In particular, there was an explosion in fate mapping the origin of postgastrulation tissues in mouse by transplanting small tissue bits from lines that carried an indelible "reporter," such as LacZ or green fluorescent protein (GFP) (Soriano, 1999; Srinivas et al., 2001), into a wild-type embryo (reviewed in Davidson et al., 1999). Thus, in several different animals, chimera-based fate mapping has had an enormous impact on understanding cell fate determination in embryos in which cell lineage analyses and single-cell fate maps are not feasible. However, these results must be carefully interpreted because transplanting tissues perturb the normal course of developmental events, such as timing, tissue integration, and necrotic cell death.

3. LINEAGE MAPPING OF IDENTIFIED FOUNDER CELLS

The lineage maps of the marine invertebrates (e.g., Fig. 1) only followed the progeny of blastomeres to an early point in development, when the cells populated a germ layer or embryonic organ rudiment. Extending these maps to fully differentiated adult cells was not feasible in most animals because later development added large numbers of cells. Furthermore, complex morphogenetic or metamorphic cell movements made visualizing every cell division impossible. However, numerous studies of late differentiation events using histological preparations of a timed series introduced the idea that a founder cell may produce different types of progeny whose phenotypes were acquired by the timing and positions of the preceding cell divisions, i.e., its lineage. For example, using electron microscopic preparations, Peter Lawrence mapped the cell divisions of a "mother cell" that produced the hairs and bristles in the insect epidermis and showed that different cell types arise in a precise mitotic order (Lawrence, 1966). These analyses, however, were limited to specific organs that could be easily visualized.

An important breakthrough was made when Sulston and Horvitz (1977) published the postembryonic lineages of each terminally differentiated adult cell in *C. elegans*. By combining this information with the classically described embryonic lineages (described in Chitwood & Chitwood, 1974; later supplemented by Sulston's group—see Wood, 1999), these postembryonic descriptions provided the first detailed lineage analysis of the entirety of the developmental process and showed the relationship between each ancestral cell and each adult cell. This approach was possible for a number of reasons. The nematode is transparent from embryonic through adult stages, so every cell can be identified using differential interference contrast microscopy, and every individual worm contains exactly 959 identifiable somatic cells, compared to many millions in vertebrates. Remarkably, these lineage analyses found that each ancestral cell has an invariant lineage, and each cell division is stereotypic and exactly the same across all individuals. With very little biological "noise," therefore, the origin of each differentiated cell could be traced back to its origin from the zygote.

These lineage analyses showed that each ancestral cell undergoes a characteristic number and pattern of cell divisions to produce a fixed collection of progeny. Most clones are composed of a mixture of different cell types, but each type arises at the same branch of the lineage, suggesting that the lineage pattern itself determined a cell's differentiated fate. It appeared that some aspect of an ancestral cell directed the embryo's general features, such as cardinal axes and germ layers, as well as the fate of individual differentiated cells, suggesting that even the adult body plan was preformed by the lineage. However, subsequent single-cell ablations as well as mutant analyses showed that some fate decisions require lineage-specific inheritance of cytoplasmic factors or lineage-specific genes. Other fate decisions depend upon extrinsic positional cues and cell–cell interactions. They demonstrated when and where intrinsic (preformed) versus extrinsic (epigenetic) mechanisms acted during embryonic development and reinforced the fact that elucidating the roles of these factors requires a detailed knowledge of the relationship between ancestral cells and differentiated cells at the level of single cells.

4. CLONAL FATE MAPS

Unfortunately, many of the animals that are used for developmental studies are too large, contain too many cells, and are too opaque to permit the same recording of the history of every cell division. But, the desire to identify the specific ancestral cells of particular differentiated cell types was widespread in the community because it would elucidate how differentiated cell types are generated by the embryo. This was of particular interest in the developing nervous system where anatomical and physiologic studies were revealing an impressive array of very different kinds of neurons. Since specific neuronal types were generated in a lineal manner in *C. elegans*, might the same be true in other animals?

The analysis of the ancestral cell of specific neurons was made possible by a critical technical breakthrough for clonal analyses in opaque embryos: intracellular microinjection of a nontoxic, long-lasting inert label into a single progenitor cell (Weisblat, Sawyer, & Stent, 1978). In the late 1970s, neurobiologists found that a nontoxic plant enzyme, horseradish peroxidase (HRP), could be microinjected or electrophoresed into single neurons. The protein was small enough to rapidly diffuse throughout the cell, but too large to pass through gap junctions and thereby label adjacent cells. After fixing the cell, a simple histochemical staining procedure would reveal the neuron's complex morphology in spectacular detail. This was the cutting-edge technique of the day to relate the electrophysiological properties of a single neuron to its position in the brain and its cellular morphology.

Weisblat and his colleagues were neurobiologists and thus aware of this technique. They cleverly adapted it to label the lineages of single progenitors of the glossiphoniid leech nervous system. The leech species being used in neurobiological studies were particularly amenable because their differentiated neurons are large and stereotypic across individuals, and neurotransmitter types and electrophysiological properties had already been defined for several of them (reviewed in Stent, 1999). The progenitors of the germ layers (macromeres) and of the nervous system (telomeres) are very large, making intracellular microinjections relatively simple to perform. Weisblat et al. (1978) showed that HRP was inherited by the descendants of the injected cell and remained detectable until the progeny differentiated into defined neurons. The label enabled them to follow a lineage from its earliest cell divisions to subsequent stages of terminal differentiation.

Their initial observations confirmed the macromere origin of the germ layers as originally described by Whitman (1987) and significantly extended his description of the teloblast origins of differentiated neurons. Gunter Stent's group published an extensive series of papers that followed a clone from ancestral to differentiated neurons (reviewed in Stent, 1999; Weisblat, Huang, & Isaksen, 1999). Some of the critical discoveries made were that (1) identified neurons in both the central and peripheral nervous systems can be assigned a unique identity based on their position and teloblast origin; (2) each teloblast divides in a sequence that is idiosyncratic and stereotypic with respect to timing, orientation, and asymmetry of the cell division; and (3) there are segmentally iterated patterns of neuronal descent that correspond to five kinship groups. For the most part, these kinship groups give rise to several different types of neurons, similar to nematode and insect lineages. But, interestingly, all serotonin neurons arise from a single kinship group, and all dopamine neurons arise from only two specific kinship groups, suggesting a lineal bias in their development. In addition, this group found that the telomeres are not the only founders of neurons; a few neurons arise from the predominantly mesodermal precursor teloblast, and the supraesophageal ganglion is derived from early cleavage micromeres. These maps were further refined by the introduction of fluorescent tracers that also could not diffuse through gap junctions to adjacent cells (see Stent, 1999). Fluorescent tracers allowed labeling of multiple cell lineages within a single embryo and the combination of lineage mapping with immunofluorescent detection of cell-type-specific markers.

The technique of microinjecting lineage tracers was rapidly adopted by another neurobiologist, Marcus Jacobson, for clonal fate mapping in a vertebrate embryo, *X. laevis* (Hirose & Jacobson, 1979; Jacobson & Hirose, 1981). Like the invertebrates mentioned so far, the fertilized *Xenopus* embryo often (but not always) cleaves in a very regular, stereotyped pattern (Moody, 2000). Although a blastomere fate map had been published using vital dye marking (Nakamura & Kishiyama, 1971), as mentioned above, the staining was not ideal for analyzing the individual cell types produced by the founder cell. Jacobson and his colleague, Giro Hirose, improved upon this map by injecting HRP into the different, identified blastomeres of

2- through 64-cells embryos and fixed them at larval stages to trace their progeny into the different regions of the embryonic nervous system. This created a "clonal fate map," similar to the end stages of the leech maps. They found that a surprisingly large proportion of blastomeres contributed to the CNS (~60% at the 64-cell stage) and that each blastomere populates a characteristic region of the CNS. For example, the anterior-most dorsal blastomeres always populated the ventral brain, whereas the anterior-most ventral blastomeres always populated the dorsal brain. Jacobson (1983) extended this analysis to the surface cells of later cleavages, and observed that by the 512-cell stage there were four defined groups of founder cells on each side of the embryo that contributed to different, restricted CNS domains. This led Jacobson to call these domains "compartments" because of their similarity to those described in fly development (Lawrence, 1975), and to boldly, albeit erroneously, champion a preformed, lineal origin of the vertebrate CNS over inductive interactions (Jacobson, 1982; and see the response by Saxen & Toivonen, 1986).

Subsequent clonal fate maps in *Xenopus* also noted the stereotypic contribution to "positions" within the CNS (Dale & Slack, 1987; Moody, 1987a, 1987b), but the question remained whether particular neuronal types were restricted by their lineage, as shown for nematode and leech neurons. Since the vertebrate embryo has far too many neurons to make a map of "identified" cells, neurons with very characteristic morphologies first were studied. *Xenopus* was very useful for this analysis because terminally differentiated, fully functional neurons are found in the larval CNS, since at this stage the free-living animal needs to swim to escape predation. Jacobson and Hirose (1981) studied the Mauthner neuron, which is extremely easy to identify by its unique morphology and size, and is easy to count because there is only one on each side of the hindbrain. They found that the Mauthner neuron descends from its most common ancestor in only about 80% of cases, indicating that its lineage is biased but not invariant. To extend this finding to other neuron types, the number of primary motoneurons and sensory Rohon-Beard neurons that arise from each blastomere were counted across a large number of embryos (Jacobson, 1983; Jacobson & Moody, 1984; Moody, 1989). These "quantitative" maps showed that primary motoneurons and sensory neurons arise from distinct sets of blastomeres (albeit with some overlap) that varied only slightly between embryos. Statistical analyses supported the interpretation that these two neuronal cell types were "biased" to descend from specific progenitors. To discover, whether the variability in neuronal lineage reflected variation in the partitioning of presumptive cytoplasmic factors due to the imprecision of cleavage planes, or variability in inductive interactions regulated by morphogen gradients, single-cell manipulations were performed (ablations and transplantations), identifying specific neuronal types by location (e.g., retina), morphology (motor neuron and Rohon-Beard neurons), or neurotransmitter expression (hypothalamic dopamine cells and various neurotransmitter subtypes of amacrine cells) (Gallagher & Moody, 2004; Huang & Moody, 1992, 1993, 1995; Moore & Moody, 1999). Like the studies in leech, these papers concluded that a particular neuronal cell type is not produced by a single founder cell, i.e., they are not monoclonal in origin. Instead, there are major and minor progenitors of each neuron type that can be predicted by the position of the lineage in the embryo, and this position biases the type of neuron produced by the founder cell.

It became apparent early on that these *Xenopus* fate maps would be most useful if extended to all organs, not just the nervous system. Two laboratories (Dale & Slack, 1987; Moody, 1987a) concurrently published comprehensive fate maps for the 32-cell stage embryo, and Moody went on to map the earlier cleavage stages as well (Moody, 1987b; Moody & Kline, 1990). These comprehensive fate maps showed the normal fates of the individual blastomeres and, like the nematode and leech, showed that while primary germ layers and embryonic organs also are not monoclonal, they do arise from consistent, identifiable founder cells. Interestingly, this is in contrast to the zebrafish maps that indicate that cleavage stage and blastula clones do not show consistent fate patterns (Kimmel & Law, 1985; Kimmel & Warga, 1987). Because the maps of the 16- and 32-cell *Xenopus* embryos provided finer detail of the origins of important developmental structures (e.g., the Organizer, Bauer, Huang, & Moody, 1994; Vodicka & Gerhart, 1995; the retina, Huang & Moody, 1993), they have been extremely useful in recent studies that target microinjected mRNAs and antisense oligonucleotides into relevant progenitors rather than to the entire embryo. This avoids global and/or nonspecific effects of the injected material, providing specificity to the experimental manipulations.

Of course, blastomere injections initiate clones at very early stages of development and do not allow one to assess later restrictions in fate, as was possible in the nematode and leech lineage maps. So, as researchers got better and better at microinjecting smaller and smaller cells, it became possible to initiate clones after gastrulation using the electroporation approaches that the neurobiologists used to label single neurons. However, it is nearly impossible to label the same cell across different animals because the progenitors are not "identifiable," thus limiting one's ability to assess whether clones are consistent across the population. Nonetheless, this approach has been used in frog, fish, and chick, and demonstrates that even late progenitors produce clones of mixed phenotypes that often intermix with cells from adjacent progenitors (e.g., Bronner-Fraser & Fraser, 1989; Lee, Stainier, Weinstein, & Fishman, 1994; Moody, Chow, & Huang, 2000; Wetts & Fraser, 1988). These studies also emboldened the mammalian embryologists to start microinjecting lineage tracers into cleavage embryos (e.g., Bałakier & Pedersen, 1982). The pioneering work of Roger Pedersen demonstrated that the inner cell mass is established very early (between the 16 and 32 cells), that the epiblast gives rise to all three germ layers, and that there is very little cell mixing of clones at cleavage stages (Pedersen, 1999; Pedersen, Wu, & Bałakier, 1986). These findings were even replicated in human embryos (Mottla et al., 1995). An important concept from the mouse studies was that although transplantation studies showed that any cell of the 32-cell embryo can give rise to all embryonic tissues, in the intact embryo, their fates are already restricted to either trophectoderm or inner cell mass. Thus, family history may direct cell movements and fates, but it can be overridden by external influences in the environment.

However, genetic approaches were needed to create clones in postimplantation mouse embryos because the microinjected tracers did not last long enough. The ability to label single progenitors and thus identify clones at later stages of development had already been elegantly worked out in insects, particularly *Drosophila*, using a variety of methods that rely on mitotic recombination (reviewed in Griffin, Binari, & Perrimon, 2014). In mammals, single progenitors were labeled by infecting them with a retrovirus that expresses a long-lasting tracer molecule, such as LacZ or GFP (Sanes, Rubenstein, & Nicolas, 1986; Soriano & Jaenisch, 1986). These vectors are replication deficient and thus can only infect one cell, and the single integration event indelibly labels all of the progeny of the founder cell. It should be noted that this approach constructs a *clonal* fate map, not a lineage map, because the cell divisions that give rise to the labeled descendants are not recorded. The logic is that the infection rate is so low that statically speaking the marked cells most likely represent a single clone derived from a single founder cell. Other caveats include: this approach cannot precisely control which founder cells are marked, so the same progenitor cannot be identified and labeled across a number of different embryos; and the timing of viral integration cannot be controlled because it depends upon when the founder cell divides. Thus, heterogeneity in the clones, which has often been interpreted to mean that the cell fate has not yet been specified, may be due to variability in the labeling of the founder cells. Nonetheless, there is little doubt that this approach has been essential for describing various aspects of later stages of development. For example, it has enabled significant discoveries in both chick and mouse, such as the separation of somatic and germ line fates (e.g., Soriano & Jaenisch, 1986), the origin of the germ layers (e.g., Carey, Linney, & Pedersen, 1995), the origin of cardiac cell types (e.g., Mikawa & Fischman, 1992), and the heterogeneity of late neural clones (e.g., Walsh & Cepko, 1992).

5. WHAT HAVE WE LEARNED?

Lineage analyses and fate maps have enabled developmental biologists to elucidate the general principles that control embryonic development. They have played an essential role in identifying those developmental processes that depend on "preformed" components, such as inherited maternal factors and lineage relationships, and those that depend on epigenetic processes, such as morphogenetic gradients, cell–cell interactions, and inductions. In recent years, these mapping techniques have been essential in discovering when during development these decisions are made, and in identifying the molecules that regulate the decisions. While complete lineage maps have only been feasible in a few animals, or only in abbreviated parts of a genealogy, they have provided important concepts, such as the importance of a cell division in time and space. They also demonstrate that it is unusual for an embryonic founder cell to give rise to only one cell type; clones typically contain a stereotyped set of cell types. This is in contrast to clones derived from adult stem cells. In addition, by revealing how these processes vary across several types of animals, it becomes possible to explain how the process of fate specification has been modified by evolution to enable different modes of development (e.g., direct vs. indirect) and produce novel features (e.g., neural crest). In this way, fate maps and lineage analyses have provided us the means by which to finally address the dual nature of the question asked by the Greek philosophers: "how are animals formed by their ancestors?"

ACKNOWLEDGMENTS

We would like to acknowledge the significant contributions of the many investigators who over the past century have constructed the lineage and fate maps, and we sincerely apologize to all those whose work we did not specifically cite. We particularly would like to thank

David Weisblat for his helpful discussions about his development of intracellular injection of lineage labels into single cells. We also take this opportunity to acknowledge the contributions of Marcus Jacobson in applying this approach to *Xenopus* and thank him for his efforts to train us when this approach was new. Finally, we thank our recently departed friend, Eric Davidson, for all his insight and advice.

REFERENCES

Adelman, H. B. (1966). *Marcello Malpighi and the evolution of embryology*. Oxford, UK: Oxford University Press.

Bałakier, H., & Pedersen, R. A. (1982). Allocation of cells to inner cell mass and trophectoderm lineages in preimplantation mouse embryos. *Developmental Biology, 90*, 352–362.

Bauer, D. V., Huang, S., & Moody, S. A. (1994). The cleavage stage origin of Spemann's Organizer: Analysis of the movements of blastomere clones before and during gastrulation in *Xenopus*. *Development, 120*, 1179–1189.

Bronner-Fraser, M., & Fraser, S. E. (1989). Cell lineage analysis reveals multipotency of some avian neural crest cells. *Nature, 335*, 161–164.

Carey, F. J., Linney, E. A., & Pedersen, R. A. (1995). Allocation of epiblast cells to germ layer derivatives during mouse gastrulation as studied with a retroviral vector. *Developmental Genetics, 17*, 29–37.

Chitwood, B. G., & Chitwood, M. B. (1974). *Introduction to nematology*. Baltimore, MD: University Park Press.

Conklin, E. G. (1905). Organization and cell-lineage of the ascidian egg. *The Journal of the Academy of Natural Sciences of Philadelphia, 13*, 1–119.

Dale, L., & Slack, J. M. (1987). Fate map for the 32-cell stage of *Xenopus laevis*. *Development, 99*, 527–551.

Davidson, E. H. (1990). How embryos work: A comparative view of diverse modes of cell fate specification. *Development, 108*, 365–389.

Davidson, B. P., Camus, A., & Tam, P. P. L. (1999). Cell fate and lineage specification in the gastrulating mouse embryo. In S. A. Moody (Ed.), *Cell lineage and fate determination* (pp. 491–504). San Diego: Academic Press.

Gallagher, B. C., Hainski, A. M., & Moody, S. A. (1991). Autonomous differentiation of dorsal axial structures from an animal cap cleavage stage blastomere in *Xenopus*. *Development, 112*, 1103–1114.

Gallagher, B. C., & Moody, S. A. (2004). Regulation of primary spinal neuron lineages after deletion of a major progenitor. *Biology of the Cell, 96*, 539–544.

Garcia-Martinez, V., Alvarez, I. S., & Schoenwolf, G. C. (1993). Locations of the ectodermal and nonectodermal subdivisions of the epiblast at stages 3 and 4 of avian gastrulation and neurulation. *The Journal of Experimental Zoology, 267*, 431–446.

Garcia-Martinez, V., & Schoenwolf, G. C. (1993). Primitive-streak origin of the cardiovascular system in avian embryos. *Developmental Biology, 159*, 706–719.

Griffin, R., Binari, R., & Perrimon, N. (2014). Genetic odyssey to generate marked clones in *Drosophila* mosaics. *Proceedings of the National Academy of Sciences of the United States of America, 111*, 4756–4763.

Gurdon, J. B., & Hopwood, N. (2000). The introduction of *Xenopus laevis* into developmental biology: Of empire, pregnancy testing and ribosomal genes. *The International Journal of Developmental Biology, 44*, 43–50.

Harrison, R. G. (1935). Heteroplastic grafting in embryology. *The Harvey Lectures, 1933–1934*, 116–157.

Haston, K. M., & Reijo-Pera, R. A. (2007). Germ line determinants and oogenesis. In S. A. Moody (Ed.), *Principles of developmental genetics* (pp. 150–172). New York: Elsevier.

Hatada, Y., & Stern, C. D. (1994). A fate map of the epiblast of the early chick embryo. *Development, 120*, 2879–2889.
Heasman, J., Wylie, C. C., Hausen, P., & Smith, J. C. (1984). Fates and states of determination of single vegetal pole blastomeres of X. laevis. *Cell, 37*, 185–194.
Hirose, G., & Jacobson, M. (1979). Clonal organization of the central nervous system of the frog. Clones stemming from individual blastomeres of the 16-cell and earlier stages. *Developmental Biology, 71*, 191–202.
Huang, S., & Moody, S. A. (1992). Does lineage determine the dopamine phenotype in the tadpole hypothalamus: A quantitative analysis. *The Journal of Neuroscience, 12*, 1351–1362.
Huang, S., & Moody, S. A. (1993). The retinal fate of *Xenopus* cleavage stage progenitors is dependent upon blastomere position and competence: Studies of normal and regulated clones. *The Journal of Neuroscience, 13*, 3193–3210.
Huang, S., & Moody, S. A. (1995). Asymmetrical blastomere origin and spatial domains of dopamine and neuropeptide Y amacrine cells in *Xenopus* tadpole retina. *The Journal of Comparative Neurology, 360*, 2–13.
Jacobson, M. (1982). Origins of the nervous system in amphibians. In N. C. Spitzer (Ed.), *Current topics in neurobiology* (pp. 44–99). New York: Plenum Press.
Jacobson, M. (1983). Clonal organization of the central nervous system of the frog. III. Clones stemming from individual blastomeres of the 128-, 256- and 512-cell stages. *The Journal of Neuroscience, 3*, 1019–1038.
Jacobson, M., & Hirose, G. (1981). Clonal organization of the central nervous system of the frog. II. Clones stemming from individual blastomeres of the 32- and 64-cell stages. *The Journal of Neuroscience, 1*, 271–284.
Jacobson, M., & Moody, S. A. (1984). Quantitative lineage analysis of the frog's nervous system. I. Lineages of Rohon-Beard neurons and primary motoneurons. *The Journal of Neuroscience, 4*, 1361–1369.
Jeffrey, W. R. (1985). Identification of proteins and mRNAs in isolated yellow crescents of ascidian eggs. *Journal of Embryology and Experimental Morphology, 89*, 275–287.
Keller, R. (1975). Vital dye mapping of the gastrula and neurula of *Xenopus laevis*. I. Prospective areas and morphogenetic movements of the superficial layer. *Developmental Biology, 42*, 222–241.
Keller, R. (1976). Vital dye mapping of the gastrula and neurula of *Xenopus laevis*. II. Prospective areas and morphogenetic movements of the deep layer. *Developmental Biology, 51*, 118–137.
Kimmel, C. B., & Law, R. D. (1985). Cell lineage of zebrafish blastomeres. III. Clonal analyses of the blastula and gastrula stages. *Developmental Biology, 108*, 94–101.
Kimmel, C. B., & Warga, R. M. (1987). Indeterminant cell lineage of the zebrafish embryo. *Developmental Biology, 124*, 269–280.
Lawrence, P. A. (1966). Development and determination of hairs and bristles in the milkweed bug, *Oncopeltus faciatus* (Lygaeidae, Hemiptera). *Journal of Cell Science, 1*, 475–498.
Lawrence, P. A. (1975). The structure and properties of a compartment border: The intersegmental boundary in *Oncopeltus*. *CIBA Foundation Symposium, 0*(29), 3–23.
Le Douarin, N. (1973). A biological cell labeling technique and its use in experimental embryology. *Developmental Biology, 30*, 217–222.
Le Douarin, N., & Barq, G. (1969). Use of Japanese quail cells as "biological markers" in experimental embryology. *Comptes Rendus Hebdomadaires des séances de l'Académie des Sciences. Série D: Sciences Naturelles, 269*, 1543–1546.
Le Douarin, N. M., & Dupin, E. (2012). The neural crest in vertebrate evolution. *Current Opinion in Genetics & Development, 22*, 381–389.
Lee, R. R. K., Stainier, D. Y. R., Weinstein, B. M., & Fishman, M. C. (1994). Cardiovascular development in the zebrafish. II. Endocardial progenitors are sequestered within the heart filed. *Development, 120*, 3361–3366.

Mikawa, T., & Fischman, D. A. (1992). Retroviral analysis of cardiac morphogenesis: Discontinuous formation of coronary vessels. *Proceedings of the National Academy of Sciences of the United States of America*, *89*, 9504–9508.

Moody, S. A. (1987a). Fates of the blastomeres of the 32-cell stage *Xenopus* embryo. *Developmental Biology*, *122*, 300–319.

Moody, S. A. (1987b). Fates of the blastomeres of the 16-cell stage *Xenopus* embryo. *Developmental Biology*, *119*, 560–578.

Moody, S. A. (1989). Quantitative lineage analysis of the origin of frog primary motor and sensory neurons from cleavage stage blastomeres. *The Journal of Neuroscience*, *9*, 2919–2930.

Moody, S. A. (2000). Cell lineage analysis in *Xenopus* embryos. In R. S. Tuan & C. W. Lo (Eds.), *Methods in molecular biology: Developmental biology protocols: Vol. 135* (pp. 1–17). Florida: Humana Press.

Moody, S. A., Chow, I., & Huang, S. (2000). Intrinsic bias and lineage restriction in the phenotype determination of dopamine and neuropeptide Y amacrine cells. *The Journal of Neuroscience*, *20*, 3244–3253.

Moody, S. A., & Kline, M. J. (1990). Segregation of fate during cleavage of frog (*Xenopus laevis*) blastomeres. *Anatomy and Embryology*, *182*, 347–362.

Moore, K. B., & Moody, S. A. (1999). Animal-vegetal asymmetries influence the earliest steps in retinal fate commitment in *Xenopus*. *Developmental Biology*, *212*, 25–41.

Moritz, K. B., & Sauer, H. W. (1996). Boveri's contribution to developmental biology—A challenge for today. *The International Journal of Developmental Biology*, *40*, 27–47.

Mottla, G. L., Adelman, M. R., Hall, J. L., Gindoff, P. R., Stillman, R. J., & Johnson, K. E. (1995). Lineage tracing demonstrates that blastomeres of early cleavage-stage human pre-embryos contribute to both trophectoderm and inner cell mass. *Human Reproduction*, *10*, 384–391.

Nakamura, O., & Kishiyama, K. (1971). Prospective fates of blastomeres at the 32-cell stage of *Xenopus laevis* embryos. *Proceedings of the Japan Academy*, *47*, 407–412.

Nieuwkoop, P. D. (1997). Short historical survey of pattern formation in the endo-mesoderm and the neural anlage in the vertebrates: The role of vertical and planar inductive actions. *Cellular and Molecular Life Sciences*, *53*, 305–318.

Noden, D. M. (1986). Patterning of avian craniofacial muscles. *Developmental Biology*, *116*, 347–356.

Pedersen, R. A. (1999). Cell lineage and cell fate determination in mammals. In S. A. Moody (Ed.), *Cell lineage and fate determination* (pp. 477–489). San Diego: Academic Press.

Pedersen, R. A., Wu, K., & Bałakier, H. (1986). Origin of the inner cell mass in mouse embryos: Cell lineage analysis by microinjection. *Developmental Biology*, *117*, 581–595.

Pieper, M., Eagleson, G. W., Wosniok, W., & Schlosser, G. (2011). Origin and segregation of cranial placodes in *Xenopus laevis*. *Developmental Biology*, *360*, 257–275.

Quinlan, G. A., Williams, E. A., Tan, S. S., & Tam, P. P. (1995). Neuroectodermal fate of epiblast cells in the distal region of the mouse egg cylinder: Implication for body plan organization during early embryogenesis. *Development*, *121*, 87–98.

Sanes, J., Rubenstein, J. L., & Nicolas, J. F. (1986). Use of a recombinant retrovirus to study post-implantation cell lineage in mouse embryos. *The EMBO Journal*, *5*, 3133–3142.

Saxen, L., & Toivonen, S. (1986). Primary embryonic induction. In T. J. Horder, et al. (Ed.), *A history of embryology* (pp. 261–274). Cambridge: Cambridge University Press.

Soriano, P. (1999). Generalized lacZ expression with the ROSA26 Cre reporter strain. *Nature Genetics*, *21*, 70–71.

Soriano, P., & Jaenisch, R. (1986). Retroviruses as probes for mammalian development: Allocation of cells to the somatic and germ cell lineages. *Cell*, *46*, 19–29.

Spemann, H. (1967). *Embryonic development and induction*. New York: Hafner Publishing Company.

Srinivas, S., Watanabe, T., Li, C.-S., William, C. M., Tanabe, Y., Jessell, T. M., et al. (2001). Cre reporter strains produced by targeted insertion of *EYFP* and *ECFP* into the *ROSA26* locus. *BMC Developmental Biology*, *1*, 4. http://www.biomedcentral.com/1471-213X/1/4.

Stent, G. S. (1999). Introduction to the leech. In S. A. Moody (Ed.), *Cell lineage and fate determination* (pp. 173–184). San Diego: Academic Press.

Streit, A. (2002). Extensive cell movements accompany formation of the otic placode. *Developmental Biology*, *249*, 237–254.

Sulik, K., Dehart, D. B., Iangaki, T., Carson, J. L., Vrablic, T., Gesteland, K., et al. (1994). Morphogenesis of the murine node and notochordal plate. *Developmental Dynamics*, *201*, 260–278.

Sulston, J., & Horvitz, H. (1977). Post-embryonic cell lineage of the nematode *Caenorhabditis elegans*. *Developmental Biology*, *56*, 110–156.

Swalla, B. J., & Jeffrey, W. R. (1995). A maternal RNA localized in the yellow crescent is segregated to the larval muscle cells during ascidian development. *Developmental Biology*, *170*, 353–364.

Teillet, M. A., Ziller, C., & Le Douarin, N. M. (2008). Quail-chick chimeras. *Methods in Molecular Biology*, *461*, 337–350.

Vodicka, M. A., & Gerhart, J. C. (1995). Blastomere derivation and domains of gene expression in the Spemann Organizer of *Xenopus laevis*. *Development*, *121*, 3505–3518.

Walsh, C., & Cepko, C. L. (1992). Widespread dispersion of neuronal clones across functional regions of the cerebral cortex. *Science*, *255*, 434–440.

Weisblat, D. A., Huang, F. Z., & Isaksen, D. E. (1999). Cell fate specification in glossiphoniid leech: Macromere, micromere and proteloblasts. In S. A. Moody (Ed.), *Cell lineage and fate determination* (pp. 185–196). San Diego: Academic Press.

Weisblat, D. A., Sawyer, R. T., & Stent, G. S. (1978). Cell lineage analysis by intracellular injection of a tracer enzyme. *Science*, *202*, 1295–1298.

Wetts, R., & Fraser, S. E. (1988). Multipotent precursors can give rise to all major cell types in the frog retina. *Science*, *239*, 1142–1145.

White, J. A., & Heasman, J. (2008). Maternal control of pattern formation in *Xenopus laevis*. *Journal of Experimental Zoology. Part B, Molecular and Developmental Evolution*, *310*, 73–84.

Whitman, C. O. (1987). A contribution to the history of the germ layers in Clepsine. *Journal of Morphology*, *1*, 108–202.

Whittaker, J. R. (1980). Acetylcholinesterase development in extra cells caused by changing the distribution of myoplasm in ascidian embryos. *Journal of Embryology and Experimental Morphology*, *55*, 343–354.

Wilson, E. B. (1925). *The cell in heredity and development*. New York: McMillan.

Wilson, V., & Beddington, R. S. (1996). Cell fate and morphogenetic movement in the late mouse primitive streak. *Mechanisms of Development*, *55*, 79–89.

Wood, W. B. (1999). Cell lineages in *Caenorhabditis elegans* development. In S. A. Moody (Ed.), *Cell lineage and fate determination* (pp. 77–95). San Diego: Academic Press.

Wray, G. A. (1999). Introduction to sea urchins. In S. A. Moody (Ed.), *Cell lineage and fate determination* (pp. 3–9). San Diego: Academic Press.

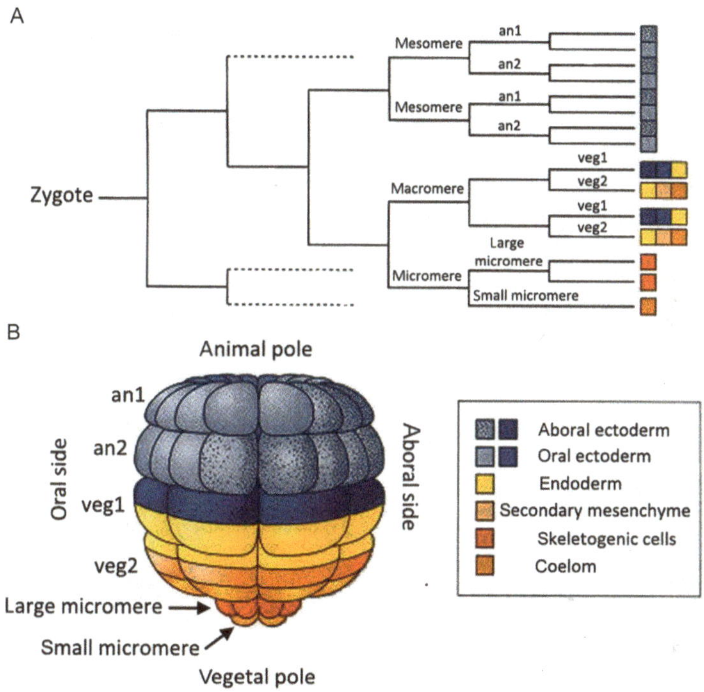

Steven L. Klein and Sally A. Moody, Figure 1 Lineage analysis (A) and fate map (B) of the 60-cell embryo of the sea urchin, *S. purpuratus*. (A) The embryonic cell lineage is a map of each cell division and the ultimate fate of each blastomere. For simplicity, the lineage of only one of the four-cell blastomeres is shown. The lineages of the other blastomeres (dotted lines) have very similar patterns. (B) The left side of the 60-cell embryo with the origins of the various embryonic cell fates (key on the right side) projected onto the blastomeres. *Modified from Wray (1999); see this reference for details.*

CHAPTER SEVEN

Frizzled Receptors in Development and Disease

Yanshu Wang*,†, Hao Chang*,†, Amir Rattner*, Jeremy Nathans*,†,‡,§,1
*Department of Molecular Biology and Genetics, Johns Hopkins University School of Medicine, Baltimore, Maryland, USA
†Howard Hughes Medical Institute, Johns Hopkins University School of Medicine, Baltimore, Maryland, USA
‡Department of Neuroscience, Johns Hopkins University School of Medicine, Baltimore, Maryland, USA
§Department of Ophthalmology, Johns Hopkins University School of Medicine, Baltimore, Maryland, USA
[1]Corresponding author: e-mail address: jnathans@jhmi.edu

Contents

Abstract

Frizzled proteins are the principal receptors for the Wnt family of ligands. They mediate canonical Wnt signaling together with Lrp5 and Lrp6 coreceptors. In conjunction with Celsr, Vangl, and a small number of additional membrane and membrane-associated proteins, they also play a central role in tissue polarity/planar cell polarity (PCP) signaling. Targeted mutations in 9 of the 10 mammalian Frizzled genes have revealed their roles in an extraordinarily diverse set of developmental and homeostatic processes, including morphogenetic movements responsible for palate, ventricular septum, ocular furrow, and neural tube closure; survival of thalamic neurons; bone formation; central nervous system (CNS) angiogenesis and blood–brain barrier formation and maintenance; and a wide variety of processes that orient subcellular, cellular, and multicellular structures relative to the body axes. The last group likely reflects the mammalian equivalent of tissue polarity/PCP signaling, as defined in *Drosophila*, and it includes CNS axon guidance, hair follicle and tongue papilla orientation, and inner ear sensory hair bundle orientation. Frizzled receptors are ubiquitous among multicellular animals and, with other signaling molecules, they very likely evolved to permit the development of the complex tissue architectures that provide multicellular animals with their enormous selective advantage.

Current Topics in Developmental Biology, Volume 117
ISSN 0070-2153
http://dx.doi.org/10.1016/bs.ctdb.2015.11.028

1. INTRODUCTION

Frizzled proteins are the principal receptors for the Wnt family of signaling molecules, and they are found throughout the animal kingdom, including in the most primitive metazoa, but they are not present in plants or in simpler (single cell) eukaryotes, such as yeast (Schenkelaars, Fierro-Constain, Renard, Hill, & Borchiellini, 2015). Frizzled proteins share a common architecture: a conserved extracellular cysteine-rich domain (CRD) is followed by a domain with seven presumptive transmembrane segments. The genetic dissection of Frizzled function began in *Drosophila*, with the characterization of the *frizzled* (*fz*) phenotype by Gubb and Garcia-Bellido (1982). Loss-of-function mutations in *fz* lead to a tissue polarity phenotype, defined as a misorientation of surface structures, such as cuticular bristles and wing hairs, which normally exhibit a precise orientation relative to the body axes. This pathway is generally referred to by the not entirely accurate designation "planar cell polarity" (PCP).

The *Drosophila fz* gene was isolated by Vinson, Conover, and Adler (1989), and the first two vertebrate *Frizzled* homologues were identified several years later (Chan et al., 1992). A search for additional vertebrate *Frizzled* family members using degenerate PCR identified eight family members in mammals, as well as multiple family members in birds and fish (Wang et al., 1996). Two additional mammalian *Frizzled* genes were identified subsequently, bringing the total to 10 (Koike et al., 1999; Wang et al., 1997). The large number of vertebrate *Frizzled* genes together with their receptor-like structure led Wang et al. (1996) to suggest that "If Frizzled proteins act as receptors, that would imply the existence of a corresponding family of ligands. … Currently the only family of ligands known to be of this size and for which no receptors have been identified are the Wnt proteins. The availability of a large number of *Frizzled* genes should facilitate a biochemical test of the possibility that these two families of proteins interact directly." Within a year, that suggestion was confirmed with the demonstration that *Drosophila* Wingless binds to and promotes beta-catenin stabilization via Frizzled2 (Bhanot et al., 1996). These experiments also identified the Frizzled CRD as the site of Wnt binding (Bhanot et al., 1996; Hsieh, Rattner, Smallwood, & Nathans, 1999). The recently determined structure of a Wnt–Frizzled CRD complex shows an unusual ligand–receptor interaction characterized by a relatively small protein–protein interface and a relatively large interface between the CRD and a lipid that is covalently attached to the Wnt (Janda, Waghray, Levin, Thomas, & Garcia, 2012).

The identification of Frizzleds as Wnt receptors implied that Frizzleds act in at least two distinct signaling pathways: tissue polarity/PCP signaling (which may or may not involve a Wnt ligand) and canonical Wnt signaling. The former pathway appears to involve reorganization of the cytoskeleton and is characterized by a small set of asymmetrically localized plasma membrane proteins (including Frizzled) and membrane-associated cytosolic proteins (Goodrich & Strutt, 2011). The latter pathway involves Wnt–Frizzled-dependent disinhibition of a single-pass transmembrane coreceptor (Lrp5/Lrp6 in mammals; arrow in *Drosophila*) leading to inhibition of beta-catenin phosphorylation and proteolysis, with a resultant migration of beta-catenin into the nucleus to regulate transcription in combination with Lef/Tcf transcription factors (MacDonald, Tamai, & He, 2009; Nusse, 2012). A third pathway, the Wnt-calcium pathway, is less well defined and has been suggested to involve a G-protein-based transduction system (Wang & Malbon, 2003).

In this review, we focus on the roles of mammalian *Frizzled* genes *in vivo*, which have been largely defined by studying the effects of targeted mutations in mice. As seen in Fig. 1, a comparison of the amino acid sequences and intron–exon structures among the 10 mammalian Frizzled family members reveals branches consisting of Fz1, Fz2, and Fz7 (subfamily 1); Fz5 and Fz8 (subfamily 2); Fz9 and Fz10 (subfamily 3); Fz4 (subfamily 4); and Fz3 and Fz6 (subfamily 5, the most distant branch). Substantial genetic redundancy has been observed between pairs of Frizzled genes within subfamilies 1, 2, and 5, but, thus far, there appears to be far less redundancy between

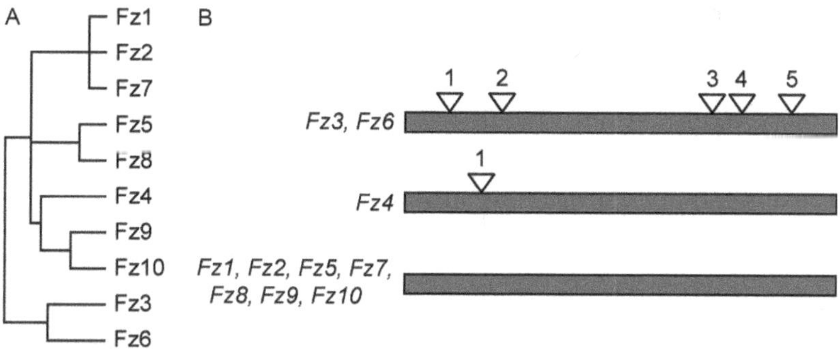

Figure 1 Evolutionary relationships among mammalian Frizzled proteins and genes. (A) Dendrogram showing the relatedness of the 10 mammalian Frizzled amino acid sequences. (B) Intron–exon structures of mammalian *Frizzled* genes, with coding regions depicted schematically as the same length. Only *Fz3*, *Fz4*, and *Fz6* have introns within the coding region. The five coding introns in the *Fz3* and *Fz6* genes are located at identical positions. *From Hua, Chang, Wang, Smallwood, and Nathans (2014).* (A color version of this figure is located at the end of this chapter.)

Frizzled genes in different subfamilies. However, these data should be interpreted cautiously because the redundancy data are still incomplete: many of the 45 possible pairwise combinations of *Frizzled* gene knockouts—especially with *Frizzleds* from different subfamilies—have not been tested. Moreover, in many tissues, there is overlapping expression of multiple *Frizzleds*, as seen, for example, in the spiral ganglion in the inner ear (Shah, Kang, Christensen, Feng, & Kollmar, 2009), so that functional redundancy may involve more than two genes. The patterns of sequence relatedness also correlate with function: Fz3 and Fz6 appear to be devoted largely or exclusively to tissue polarity signaling, whereas Fz4 is devoted largely or exclusively to canonical Wnt signaling. Some mammalian Frizzleds could signal via more than one pathway, as shown for *Drosophila* Frizzled, which signals through both the canonical Wnt and PCP pathways (Bhanot et al., 1999; Bhat, 1998; Müller et al., 1999).

In the text that follows, we have organized the descriptions of *Frizzled* gene function by subfamily. We note that most of the phenotypic analyses in mice were performed on a mixed genetic background.

2. FRIZZLED1, FRIZZLED2, AND FRIZZLED7

Fz1$^{-/-}$ mice show no apparent phenotype, whereas loss of *Fz2* leads to cleft palate and neonatal lethality in ~50% of mice and runting associated with reduced olfactory function in the other ~50% (Yu et al., 2010). Approximately 15% of *Fz7*$^{-/-}$ fetuses have a ventricular septal defect (VSD), a common cardiac anomaly that results in mixing of blood between the left and the right ventricular chambers, while the remaining ~85% of *Fz7*$^{-/-}$ mice have no apparent phenotype except for a kinked tail (Yu, Ye, Guo, & Nathans, 2012). Interestingly, a variety of additional phenotypes appear when *Fz1*, *Fz2*, and *Fz7* mutations are combined (Yu et al., 2010, 2012). 100% of *Fz1*$^{-/-}$;*Fz2*$^{-/-}$ fetuses have a cleft palate (Fig. 2), and approximately two-thirds of *Fz1*$^{-/-}$;*Fz2*$^{-/-}$ fetuses also have cardiac defects—either an isolated VSD or a combination of VSD and double outlet right ventricle, a cardiac anomaly in which the aorta is positioned partially over the right ventricular outflow tract. *Fz1*$^{-/-}$;*Fz7*$^{-/-}$ mice have no additional defects beyond the *Fz7*$^{-/-}$ phenotype except for a gray coat color, suggesting a defect in melanocyte migration and/or development. In contrast, *Fz2*$^{-/-}$;*Fz7*$^{-/-}$ embryos at embryonic day (E)8.5 show a severe loss of convergent extension, fail to close the neural tube, and fail to develop cardiac asymmetry (Fig. 3).

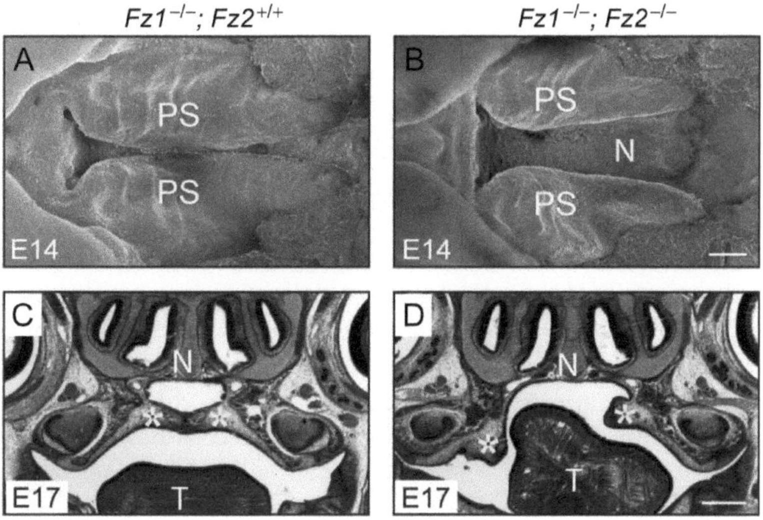

Figure 2 Palate closure defect in $Fz1^{-/-}$;$Fz2^{-/-}$ fetuses. (A and B) Scanning electron microscopy of the palate at E14. PS, palatal shelf. (C and D) Coronal section through the head at E17, stained with hematoxylin and eosin. N, nasal septum; T, tongue; Yellow asterisks, palatal shelves. Scale bars: B, 200 μm; D, 500 μm. *From Yu et al. (2010).* (A color version of this figure is located at the end of this chapter.)

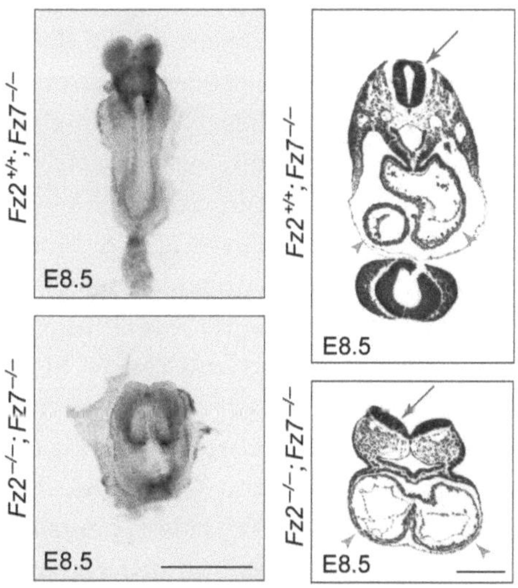

Figure 3 Convergent extension, neural tube closure, and cardiac asymmetry defects in $Fz2^{-/-}$;$Fz7^{-/-}$ embryos. E8.5 embryos are shown intact in a dorsal view (left) and in paraffin sections stained with hematoxylin and eosin (right). Left panels, the $Fz2^{-/-}$;$Fz7^{-/-}$ embryo is shorter and wider than the $Fz2^{+/+}$;$Fz7^{-/-}$ control due to a failure of convergent extension. Right panels, in cross section, the $Fz2^{-/-}$;$Fz7^{-/-}$ embryo shows an open neural tube (red arrows) and symmetric cardiac chambers (green arrowheads). The $Fz2^{+/+}$;$Fz7^{-/-}$ embryo is phenotypically WT. Scale bars: left panels, 1 mm; right panels, 200 μm. *From Yu et al. (2012).* (A color version of this figure is located at the end of this chapter.)

Given that Frizzleds control several signaling pathways (PCP, canonical Wnt, and Wnt-calcium signaling), it is of interest to explore the relationship between signaling pathways affected and developmental phenotypes observed. To this end, Yu et al. (2010, 2012) examined genetic interactions between mutations in *Fz1*, *Fz2*, and *Fz7* and in known PCP genes, such as *Vangl2* [one of the two mammalian homologues of the *Drosophila* core PCP gene *Van Gogh/Strabismus* (*Vang/Stbm*)]. When various combinations of *Fz1*, *Fz2*, or *Fz7* knockout (*KO*) alleles were examined in combination with one copy of the semidominant *Looptail* allele of *Vangl2* ($Vangl2^{Lp/+}$), there was an enhanced frequency of neural tube closure and cleft palate defects. Similarly, the frequency of cardiac defects (mostly VSDs) was enhanced when the $Fz2^{+/-};Fz7^{-/-}$ genotype was combined with one mutant copy of *Wnt3a* (canonical signaling), *Wnt11* (noncanonical signaling), or *Dvl3* (both pathways). While these data show that *Fz1*, *Fz2*, and *Fz7* synergize with other canonical and noncanonical pathway components, they do not definitively indicate which pathway(s) is/are affected by the different *Frizzled* mutations. Palate, ventricular septum, and neural tube closures are known to involve multiple signaling pathways (Biggs, Goudy, & Dunnwald, 2015), and it is therefore plausible that defects in more than one Frizzled-based pathway combine to produce a developmental defect.

In humans, defects in tissue closure processes represent some of the most common congenital anatomic anomalies. Defects in closing the ventricular septum, palate, and neural tube occur, respectively, in ~0.4%, ~0.1%, and ~0.1% of live births, and a defect in closing the ventral urethra occurs in ~0.4% of newborn males (Au, Ashley-Koch, & Northrup, 2010; CDC, 1997; Dolk, Loane, & Garne, 2010). Congenital defects in uterine ligament, ocular furrow, and diaphragm closure are less common. All of these processes involve the directed migration and fusion of tissues, so it is not implausible that they might utilize similar or overlapping cell signaling strategies. In mice, defects in two or more canonical Wnt and/or PCP signaling components from among the *Celsr*, *Dvl*, *Frizzled*, *Lrp*, *Vangl*, and *Wnt* families have been implicated in palate, ventricular septum, neural tube, and ocular furrow closure defects (summarized in Yu et al., 2010). In humans, neural tube closure defects are associated with sequence variants in *VANGL1*, *VANGL2*, *CELSR1*, and *FZD6* in a small fraction of cases (Allache, De Marco, Merello, Capra, & Kibar, 2012; De Marco et al., 2012; Kibar et al., 2009, 2011, 2007). These data suggest that deleterious sequence variants in genes that either code for or control the expression of canonical Wnt and/or PCP signaling components may represent genetic risk factors for a wide variety of tissue closure defects in humans.

3. FRIZZLED5 AND FRIZZLED8

Homozygous deletion of *Fz5* leads to embryonic lethality at midgestation secondary to placental insufficiency (Ishikawa et al., 2001). When the placental phenotype is bypassed by conditionally ablating Fz5 in embryonic but not extraembryonic tissues (*Fz5*$^{CKO/CKO}$*;Sox2-Cre*), the resulting mice survive to adulthood and exhibit interesting phenotypes in the eye and the thalamus (*CKO*, conditional knockout). The ocular phenotype is ~50% penetrant and is characterized by increased cell death in the ventral retina and failure to close the ventral furrow (coloboma), an excess of mesenchymal cells in the vitreous cavity, an excess of astrocytes on the vitreal face of the retina, and persistence of the vascular network within the vitreous cavity (the hyaloid vasculature) into adulthood (Liu & Nathans, 2008). The same ocular phenotype is seen when *Fz5* is selectively ablated in the retina and in retinal astrocytes but not in vascular endothelial cells (ECs; Zhang, Fuhrmann, & Vetter, 2008). In the normal course of ocular development, the hyaloid vasculature regresses as the retinal vasculature matures, leaving an optically clear path between the lens and the retina. In humans, persistence of a hyaloid vascular remnant is a relatively common ocular anomaly (Duke-Elder, 1964; Goldberg, 1997).

In the brain, *Fz5* is expressed selectively in neurons in the parafascicular nucleus (PFN) of the thalamus, and *Fz5*$^{CKO/CKO}$*;Sox2-Cre* mice show a progressive loss of PFN neurons over the first several months of postnatal life (Liu, Wang, Smallwood, & Nathans, 2008). Genetic mosaic experiments in which *Fz5*$^{-/-}$ cells are marked by expression of a reporter show that this loss is cell autonomous. Numerous changes in gene expression in PFN neurons precede neuronal loss, including downregulation of the genes coding for beta-catenin, Wnt4, and Wnt9b, suggesting that a Fz5-dependent positive feedback loop maintains canonical Wnt signaling in the PFN. The physiological role of the PFN is not well understood, but the work of Liu et al. (2008) implicates the PFN in sensorimotor function because loss of PFN neurons in *Fz5*$^{CKO/CKO}$*;Sox2-Cre* mice is accompanied by impaired motor coordination.

Fz8$^{-/-}$ mice have no apparent phenotype. However, loss of both *Fz8* and *Fz5* in embryonic tissues leads to mid-gestational lethality, and loss of one *Fz8* allele in combination with loss of *Fz5* in embryonic tissues leads to 100% penetrance and greater severity of the *Fz5* ocular phenotypes and to a reduction in retinal progenitor proliferation (Liu, Bakeri, Li, & Swaroop, 2012). *Fz8* also exhibits a genetic interaction with *Fz4* in the context of kidney development (Ye, Wang, Rattner, & Nathans, 2011). *Fz4*$^{-/-}$;*Fz8*$^{-/-}$ fetuses show reduced growth of ureteric buds and smaller kidneys, a phenotype similar to that described earlier for *Wnt11*$^{-/-}$ mice (Chi et al., 2004; Majumdar, Vainio, Kispert, McMahon, & McMahon, 2003). The genetic mosaic experiments of Ye, Wang, et al. (2011) revealed strong competition between cells within the ureteric bud, which leads to a dramatic underrepresentation of *Fz4*$^{-/-}$ or *Fz4*$^{-/-}$;*Fz8*$^{-/-}$ cells relative to wild-type cells when both were initially present in roughly equal numbers. Intriguingly, in *Fz4*$^{-/-}$;*Fz8*$^{-/-}$ embryos, in which all cells share the same genotype, the reduction in kidney size is more modest than would be predicted from the almost complete elimination of mutant cells in the mosaic kidney. This last observation suggests the possibility of (1) a feedback signal that enhances cell proliferation and/or survival when overall kidney growth is reduced and/or (2) an interaction between WT and mutant cells that selectively eliminates mutant cells.

4. FRIZZLED9 AND FRIZZLED10

Fz9 has long intrigued human geneticists because it is one of ~20 genes in the chromosome 7 region that is present in a hemizygous state in Williams–Beuren syndrome, a complex multisystem disorder (OMIM 194050). Although *Fz9*$^{-/-}$ mice are viable and fertile, in-depth studies by three research groups have identified: (1) a subtle abnormality in B-cell development (Ranheim et al., 2005), (2) an increase in cell death in the dentate gyrus and a defect in visuo-spatial learning (Zhao et al., 2005), and (3) reduced bone mass and a reduction in newly formed bone in response to fractures, due to impaired osteoblast function (Fig. 4; Albers et al., 2011; Heilmann et al., 2013). Curiously, the bone phenotypes do not appear to be associated with alterations in canonical Wnt signaling, as judged by a comparison of the transcriptomes of *Fz9*$^{-/-}$ and *Fz9*$^{+/+}$ osteoblasts, a cell type in which transcriptome changes in response to canonical Wnt signaling are well defined. This finding contrasts with the well-established role of Lrp5-dependent canonical Wnt signaling in promoting bone formation in mice and humans (Baron & Kneissel, 2013; Joeng, Schumacher, Zylstra-Diegel, Long, & Williams, 2011). Albers et al. (2011) have provided evidence that, in osteoblasts, Fz9 responds to Wnt5a and acts through a noncanonical pathway to control a distinct set of transcripts, including several coding for chemokines.

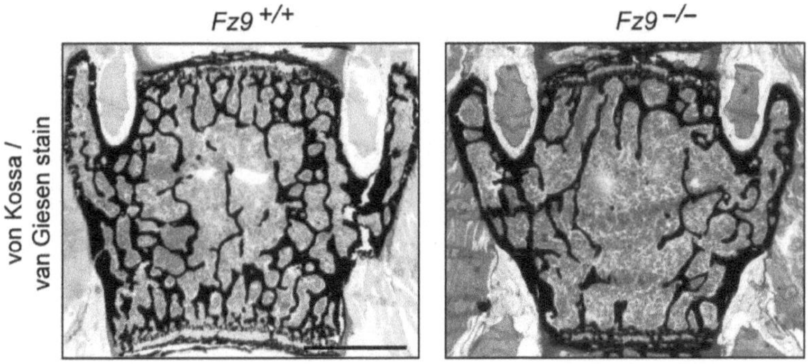

Figure 4 Reduced bone density in *Fz9*$^{-/-}$ mice. The vertebral body shows reduced bone density in a 52-week-old *Fz9*$^{-/-}$ mouse (right) compared to an age- and sex-matched control (left). Scale bar, 250 μm. *From Albers et al. (2011).* (A color version of this figure is located at the end of this chapter.)

Fz10 is the most enigmatic of the Frizzled family members. The *Fz10* gene is expressed widely in the CNS, including in multiple thalamic nuclei (Zhao, Guan, & Pleasure, 2006), but no *Fz10* loss-of-function phenotypes have been reported thus far.

5. FRIZZLED4

Among the Frizzleds that are known or suspected to activate the canonical Wnt pathway, Fz4 is the best understood in terms of its biological function, its interactions with ligands and other receptor components, and its role in human disease. *Fz4*$^{-/-}$ mice exhibit a severe defect in retinal vascularization, progressive hearing loss, and a slowly progressive cerebellar degeneration that is associated with loss of the blood–brain barrier (BBB) in the cerebellum (Wang, Huso, Cahill, Ryugo, & Nathans, 2001; Wang et al., 2012; Xu et al., 2004; Ye et al., 2009). In humans, heterozygosity for loss-of-function mutations in *FZD4* leads to a variable degree of retinal hypovascularization, a syndrome referred to as familial exudative vitreoretinopathy (FEVR; Nikopoulos, Venselaar, et al., 2010; Qin et al., 2005; Robitaille et al., 2002, 2011; Salvo et al., 2015; Toomes et al., 2004).

In WT mice during the first week of postnatal life, ECs emerge from the optic disc and migrate radially outward across the vitreal face of the retina to form a superficial vascular plexus. During the second postnatal week, ECs migrate vertically into the retina and form two additional tiers of capillaries, one immediately proximal and the other immediately distal to the central layer of neuronal cell bodies (the inner nuclear layer;

Fruttiger, 2007). The mature retinal vasculature in mice, as in humans, consists of these three tiers of vessels. In $Fz4^{-/-}$ mice, EC invasion into the retina is aborted after only a short distance, resulting in a complete absence of the two intraretinal capillary beds.

A critical clue to the biochemical mechanism of Fz4 action came from studies of Norrie disease, an X-linked disorder characterized by congenital blindness and slowly progressive hearing loss (Berger & Ropers, 2001). The gene responsible for Norrie disease, *NDP*, codes for a distant relative of the transforming growth factor beta (TGF-beta) family, a dimeric cysteine knot protein with no sequence similarity to Wnts (Berger et al., 1992; Chen et al., 1992). Strikingly, Ndp^{KO} mice (either $Ndp^{-/-}$ females or $Ndp^{-/Y}$ males) exhibit exactly the same retinal phenotype as $Fz4^{-/-}$ mice (Fig. 5), providing strong evidence that the congenital blindness of Norrie disease patients most likely reflects a severe retinal hypovascularization (Luhmann et al., 2005; Rehm et al., 2002). This identity in retinal vascular phenotypes suggested that Norrin (the protein product of the *Ndp* gene) might be a ligand for Fz4. Indeed, this is the case: dimeric Norrin binds a pair of Fz4 CRDs with several nanomolar affinity, it activates canonical Wnt signaling in concert with Fz4 and either Lrp5 or Lrp6 coreceptors, and the Ndp^{KO} retinal vascular phenotype can be fully rescued by artificially stabilizing beta-catenin in developing ECs, thereby circumventing

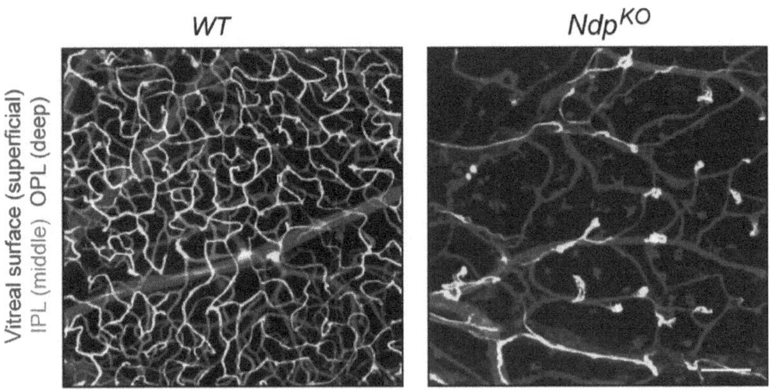

Figure 5 Defective retinal vascularization in Ndp^{KO} mice. Z-stacked confocal images of flat mount adult WT and Ndp^{KO} retinas, with ECs visualized with GS-lectin. The depth of the different vascular structures has been color coded as indicated at left. The Ndp^{KO} retina has only a superficial vascular plexus from which numerous EC clusters invade a short distance into the retina. IPL, inner plexiform layer; OPL, outer plexiform layer. Scale bar, 100 μm. *From Wang et al. (2012).* (A color version of this figure is located at the end of this chapter.)

the signaling defect at the level of ligand and receptor (Xu et al., 2004; Zhou et al., 2014).

Recent structural and biochemical experiments have revealed the precise three-dimensional architecture of the complex between dimeric Norrin and a pair of symmetrically bound Fz4 CRDs and defined the sequence determinants of ligand–receptor recognition (Chang et al., 2015; Ke et al., 2013; Smallwood, Williams, Xu, Leahy, & Nathans, 2007). The dimeric nature of the Norrin–Fz4 CRD complex suggests that Fz4 may normally signal as a dimer, perhaps interacting symmetrically with dimeric Lrp5 or Lrp6 (Liu, Bafico, Harris, & Aaronson, 2003). If a pair of Wnt–Fz complexes is required for signaling, this would create the potential for a nonlinear dependence of signal amplitude on Wnt concentration and for a sharpening of the spatio-temporal boundaries of Wnt action.

Fz4 is expressed in all ECs throughout the body from the earliest times in vascular development through adulthood, and *Ndp* is expressed in selective anatomic regions of the prenatal CNS and in glia throughout the postnatal CNS: Muller glia in the retina, Bergman glia in the cerebellum, and astroglia elsewhere in the brain and spinal cord (Ye, Smallwood, et al., 2011; Ye et al., 2009). Experiments with conditional *Fz4* knockout alleles show that eliminating *Fz4* in ECs produces the same vascular phenotype as global elimination of *Fz4*.

An additional component of the Norrin/Fz4/Lrp signaling complex, Tspan12, was identified in a screen of knockout mice for ocular phenotypes (Junge et al., 2009). Tspan12 has four putative transmembrane segments and is a member of the large and diverse tetraspanin family. In cell culture, Tspan12 acts in conjunction with Fz4 and Lrp5 to specifically enhance Norrin signaling. In humans, *TSPAN12* and *LRP5* mutations are found in a subset of FEVR patients (Nikopoulos, Gilissen, et al., 2010; Poulter et al., 2010, 2012; Salvo et al., 2015). In summary, retinal angiogenesis is controlled by a glial ligand (Norrin) that is sensed by an EC receptor (Fz4), coreceptor (Lrp5), and coactivator (Tspan12).

In the mature retinal and cerebellar vasculatures, Norrin/Fz4 signaling is required for maintenance of the blood–retina barrier and the BBB, respectively (Wang et al., 2012; Ye et al., 2009). This barrier consists of a series of cellular and molecular specializations that distinguishes the CNS and peripheral vasculatures. In CNS ECs, these specializations include the presence of tight junctions, suppression of transcytosis, an absence of fenestrations, and expression of a variety of extrusion pumps and transporters (Daneman & Prat, 2015). The gene expression program that leads to CNS barrier competence is induced and maintained by canonical Wnt signaling (Daneman et al., 2009; Liebner et al., 2008; Stenman et al., 2008; Wang et al., 2012; Zhou et al., 2014).

Experiments in which beta-catenin, Fz4, or Norrin production is either activated or extinguished in the mature vasculature show that CNS barrier integrity is plastic (Liebner et al., 2008; Wang et al., 2012; Zhou et al., 2014). As demonstrated both by expression of molecular markers and by directly assaying the leakiness of CNS vessels, loss of canonical Wnt signaling leads within days to a loss of barrier competence. Conversely, a congenital deficiency in barrier function due to reduced canonical Wnt signaling can be reversed by restoring signaling later in life. Interestingly, the choroid plexus, a highly permeable CNS vascular bed that is the site of cerebrospinal fluid production, shows a cell-autonomous conversion to the barrier competent state (as judged by molecular markers) following EC-specific expression of a stabilized beta-catenin (Zhou et al., 2014). Loss-of-function experiments reveal CNS region-specific variations in ligands (Norrin, Wnt7a, Wnt7b, and possibly other Wnts), receptors (Fz4 and other Frizzleds), and coreceptors (Lrp5 and Lrp6), and the extent of redundancy among these components (Zhou et al., 2014). Taken together, the data reveal a system in which multiple canonical Wnt ligands derived from the CNS parenchyma control vascular barrier integrity.

Recent work on Fz4 signaling has led to the discovery of two additional proteins, Gpr124 and Reck, that function in ECs as Wnt7a- and Wnt7b-specific coactivators of canonical Wnt signaling. Based on its predicted transmembrane topography, Gpr124 had been categorized as a member of the adhesion class of orphan G-protein-coupled receptors. The first link between Gpr124 and vascular biology came with the discovery that *GPR124* transcripts are abundant in human tumor vasculature but rare in quiescent adult vasculature (Carson-Walter et al., 2001; St Croix et al., 2000). The second link came with the discovery that *Gpr124*$^{-/-}$ embryos exhibit anatomically localized defects in CNS angiogenesis that are virtually identical to the defects observed in *Wnt7a* and *Wnt7b* double mutant embryos (Anderson et al., 2011; Cullen et al., 2011; Kuhnert et al., 2010; Stenman et al., 2008). This second observation led Zhou and Nathans (2014) and Posokhova et al. (2015) to test whether Gpr124 promotes signaling by Wnt7a and Wnt7b. Remarkably, in transfected cells Gpr124 stimulated signaling by these two Wnts but had little or no effect on signaling by Norrin or any of the other 17 mammalian Wnts. *In vivo*, various double mutant combinations show that the Wnt7a/Wnt7b/Gpr124/Fz system

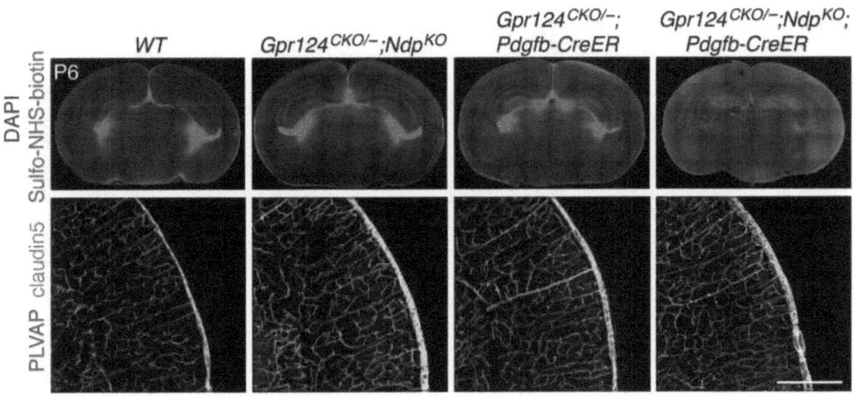

Figure 6 Redundant actions of Gpr124 and Norrin in maintaining BBB integrity. Coronal sections through the cortex and thalamus of postnatal day (P)6 mice with the indicated genotypes. *Pdgfb-CreER* recombines specifically in ECs. 4-Hydroxytamoxifen (4HT) was delivered by intraperitoneal injection at P3–P4. Intravascular Sulfo-NHS-biotin leaks from the choroid plexus in the lateral ventricles in all brain samples (upper panels), but it leaks extensively into the brain parenchyma only in the sample that is deleted for both *Gpr124* and *Ndp* (upper right panel). Intracerebral capillaries in the WT brain (lower left) and in the brains from mice with either one functional copy of *Gpr124* or one functional copy of *Ndp* (lower middle two panels) express claudin5 (a hallmark of CNS vasculature) and suppress plasmalemma vesicle-associated protein (PLVAP; a hallmark of peripheral vasculature). Deletion of both *Gpr124* and *Ndp* reverses this expression pattern (lower right panel). Scale bar, 200 μm. *From Zhou and Nathans (2014).* (A color version of this figure is located at the end of this chapter.)

(which likely signals through Fz4 as well as other EC-expressed Frizzleds) and the Norrin/Tspan12/Fz4 system function redundantly to promote angiogenesis in the embryonic spinal cord and hindbrain and to maintain BBB integrity in many brain regions, including the cerebral cortex (Fig. 6).

The second coactivator, Reck, is a GPI-anchored protein that synergizes with Gpr124 to specifically stimulate Wnt7a/Wnt7b-mediated canonical Wnt signaling in transfected cells (Vanhollebeke et al., 2015). In zebrafish, *Reck* morphants phenocopy *Gpr124* mutants and morphants, and eliminating either Reck or Gpr124 blocks CNS angiogenesis and eliminates EC-specific expression of a reporter for canonical Wnt signaling.

These experiments reveal a general role for canonical Wnt signaling in ECs during embryonic and retinal angiogenesis and BBB development and maintenance. They also reveal the existence of auxiliary receptor proteins that enhance ligand specificity (Fig. 7). The high level of *GPR124* expression in tumor ECs suggests that canonical Wnt signaling in ECs may play a role in tumor angiogenesis.

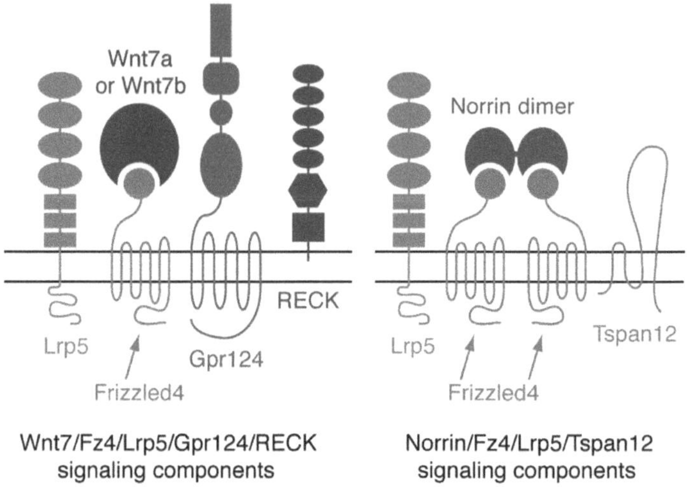

Figure 7 Schematic of ligand and receptor components involved in canonical Wnt signaling in CNS ECs. Horizontal black lines represent the plasma membrane; extracellular space is above and cytoplasm is below. Additional components, such as Lrp6, other Frizzleds, and other Wnts, are also involved, but their roles are not yet well defined. *Modified from Zhou and Nathans (2014).* (A color version of this figure is located at the end of this chapter.)

6. FRIZZLED3 AND FRIZZLED6

Among mammalian Frizzled genes, *Fz6* is the family member that most closely resembles *Drosophila fz*, as determined by the macroscopic appearance of its loss-of-function phenotype (Guo, Hawkins, & Nathans, 2004).

Fz6 is expressed in the epidermis and in hair follicles starting at ~E13. *Fz6*$^{-/-}$ mice are healthy and fertile, with skin and hair follicles that appear microscopically normal. However, at the earliest stages of their development, the orientations of *Fz6*$^{-/-}$ hair follicles are aberrant and in many regions nearly randomized. By contrast, in WT mice, immature follicle orientations generally differ by less than 30° from their final orientations, which are precisely coordinated with the body and limb axes. During the first postnatal week, Fz6$^{-/-}$ follicles gradually reorient to create progressively expanding zones of local order, such as whorls or cruciforms (Fig. 8; Wang, Badea, & Nathans, 2006; Wang et al., 2010). As the number of follicles encompassed by the individual patterns increases and the follicle orientations conform more perfectly to the local geometry of each pattern, the number of patterning centers decreases, a progression that appears to represent a competition between neighboring patterns. In WT mice, a

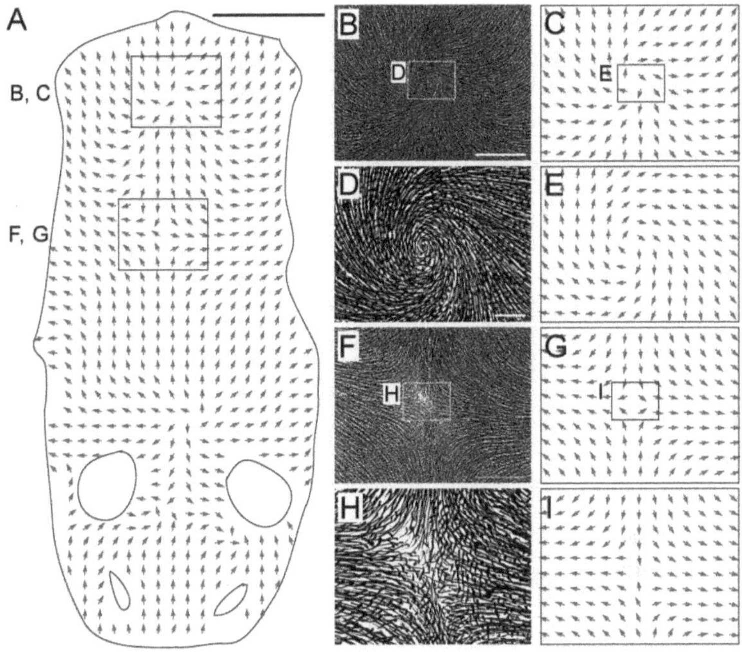

Figure 8 Hair patterns in a $Fz6^{-/-}$;$Vangl2^{Lp/+}$ back skin at P8. (A) Vector map showing hair follicle orientations over the head and back in a flat-mount skin from a P8 $Fz6^{-/-}$;$Vangl2^{Lp/+}$ mouse. Anterior is down; posterior is up. The small ovals are the eye positions; the larger ovals are the ear positions. Four patterning centers are present along the midline. (B–I) The central regions of the two most posterior patterning centers are enlarged in panels B, C, F, and G, and further enlarged in panels D, E, H, and I, as indicated by the rectangles in panels A, B, C, F, and G. Spatial order among follicles is evident at long range (~1 cm) and short range (e.g., between neighboring follicles; tens of microns). Scale bars: A, 1 cm; B and F, 2 mm; D and H, 300 μm. *From Wang, Chang, and Nathans (2010).* (A color version of this figure is located at the end of this chapter.)

more subtle shift in follicle orientations occurs during the same time period, leading to the nearly parallel alignment that characterizes the mature coat.

These and other observations indicate that hair follicle orientation is determined by two fundamentally distinct systems. An embryonic system that requires Fz6 sets up the global pattern of hair follicles orientations, and a postnatal system that is independent of Fz6 refines the initial orientations to minimize differences between neighboring follicles. The refinement system can be modeled with a two-dimensional Ising lattice and a local consensus algorithm of the type that is commonly used to study the behavior of interacting vectors in condensed matter physics (Wang, Badea, et al., 2006). These models are agnostic with regard to molecular and cellular mechanisms, but they capture the global behavior of the system and its responses to various perturbations, such as the juxtaposition of WT and mutant tissue. It is interesting that the wing hairs of *Drosophila* PCP mutants also exhibit large-scale patterns, such as whorls, that are characterized by local order (Taylor, Abramova, Charlton, & Adler, 1998). The similarity in the large-scale patterns formed by PCP-deficient mammalian and insect epidermal structures is especially striking in light of the very different cellular structures involved: a mammalian hair follicle is an elongated cylinder formed by dozens to hundreds of epidermal cells, whereas a *Drosophila* wing hair is an actin-based protrusion that emerges from one side of an epithelial cell (Wong & Adler, 1993).

Hair follicles are associated with auxiliary structures that are polarized in a direction that matches the central follicle: a pair of follicle-derived sebaceous glands, arrector pili muscles, sensory nerve endings, and Merkel cell clusters (epithelial derivatives that are involved in mechanosensation). The largest follicles, which give rise to guard or tylotrich hairs, have all of these structures; the smaller follicles lack Merkel cells. The role of the central follicle in directing the polarized assembly of these different structures was analyzed in *Fz6*$^{-/-}$ mice (Chang & Nathans, 2013). Sebaceous glands, arrector pili muscles, and sensory nerve endings were observed to track the orientation of their associated follicle. Merkel cell clusters on the back skin are normally arranged in a semicircle around the base of the guard hair follicles with the opening of the semicircle pointing toward the anterior. In *Fz6*$^{-/-}$ mice, the Merkel cell clusters are arranged in a closed circle (Fig. 9). These data are consistent with the idea that *Fz6*$^{-/-}$ skin is missing global polarity information and that the circularly symmetric Merkel cell arrangement represents a default response to a nonpolarized environment.

In humans, *FZD6* loss-of-function mutations have been identified in several families with autosomal recessive nail dysplasia (Frojmark et al., 2011; Naz et al., 2012; Wilson et al., 2013). A follow-up histologic and molecular analysis of claw development in *Fz6*$^{-/-}$ mice shows dysplastic growth and downregulation of multiple claw-specific keratin genes in addition to Wnt, bone morphogenetic protein, and Hedgehog genes (Cui et al., 2013). The data imply that Fz6 affects epidermal derivatives in a manner that extends beyond spatial patterning: the simplest explanation is that Fz6 couples directly to the canonical Wnt pathway in nail bed epithelial cells.

Like Fz6, Fz3 appears to signal through the PCP/tissue polarity system, but the *Fz3*$^{-/-}$ phenotype is limited to the nervous system where it controls axon growth and guidance in the forebrain, in cranial and spinal motor

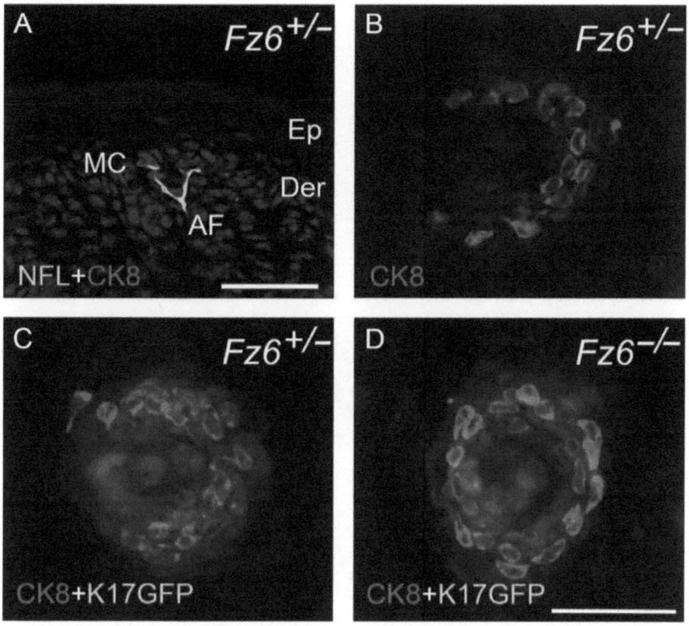

Figure 9 Loss of polarity in the arrangement of Merkel cells in *Fz6*$^{-/-}$ mice. (A) Vertical section of P1 back skin from a phenotypically WT *Fz6*$^{+/-}$ mouse. Merkel cells (MC) are visualized with anticytokeratin8 (CK8) immunostaining and afferent sensory nerve fibers (AF) are visualized with antineurofilament (NFL) immunostaining. Ep, epidermis; Der, dermis. (B–D) Flat mounts of P1 skin in which Merkel cells are visualized with CK8 and the central guard hair follicle is visualized with a *Keratin (K)17-GFP* transgene. Anterior is to the left. The phenotypically WT *Fz6*$^{+/-}$ Merkel cells form a semicircle that opens toward the anterior. The *Fz6*$^{-/-}$ Merkel cells form a closed circle. Scale bar, 50 μm. *From Chang and Nathans (2013).* (A color version of this figure is located at the end of this chapter.)

neurons, in spinal sensory neurons, and in the sympathetic nervous system (Armstrong, Ryu, Chieco, & Kuruvilla, 2011; Fenstermaker et al., 2010; Hua, Jeon, Caterina, & Nathans, 2014; Hua, Smallwood, & Nathans, 2013; Lyuksyutova et al., 2003; Wang, Thekdi, Smallwood, Macke, & Nathans, 2002). In the *Fz3*$^{-/-}$ forebrain, thalamocortical axons grow extensively but never enter the internal capsule, instead projecting inferiorly to innervate the contralateral thalamus (Fig. 10). Similarly, nigrostriatal dopaminergic neurons and brainstem serotonergic axons show extensive misrouting, with many axons growing caudally instead of rostrally. In the WT spinal cord, sensory axons that originate in the dorsal horn cross the midline before turning rostrally toward the brain. In the *Fz3*$^{-/-}$ spinal cord, these sensory axons cross the midline appropriately but appear to lack rostrocaudal cues once their growth cones arrive on the contralateral side.

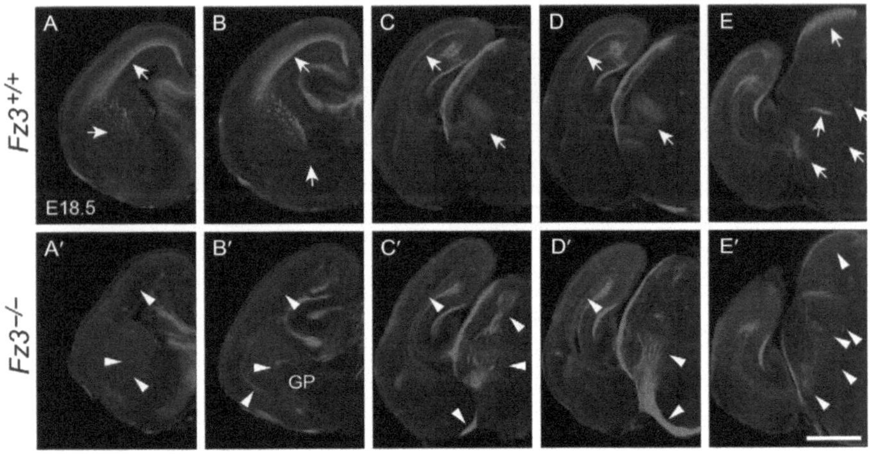

Figure 10 Thalamocortical axon guidance defects in the *Fz3*$^{-/-}$ forebrain at E18.5. Serial coronal sections through the forebrains of E18.5 *Fz3*$^{-/-}$ and *Fz3*$^{+/+}$ fetuses stained with anti-neurofilament antibodies. The midline of each hemibrain image is at the right border of each panel. The sections are arranged from rostral (A and A′) to caudal (E and E′). In the *Fz3*$^{+/+}$ brain, arrows in (A–D) trace the thalamocortical projections originating in the thalamus (D) and projecting via the internal capsule (A–C) to the cortex (A–D). In the *Fz3*$^{-/-}$ brain, thalamic axons track inferiorly with most crossing the midline to the contralateral thalamus (D′) and a minority projecting to the inferior border of the cortex (C′). Additional axon guidance defects are indicated by arrowheads in (A′–E′); see Hua, Jeon, et al. (2014) for details. GP, globus pallidus. Scale bar, 1 mm. *From Hua, Jeon, et al. (2014).* (A color version of this figure is located at the end of this chapter.)

Fz3$^{-/-}$ spinal sensory axons appear to be incapable of finding their appropriate targets in the brainstem and thalamus as determined by behavioral testing of mice that lack Fz3 in all tissues caudal to the neck (*Fz3*$^{CKO/-}$;*Cdx1-Cre*). As adults, these mice show no evidence that thermal or mechanical information from the feet can be transmitted to the brain (Hua, Jeon, et al., 2014). Finally, *Fz3*$^{-/-}$ motor neurons in the dorsal fore- and hindlimbs exhibit a distinctive stalling phenotype in the nerve plexus at the base of the limbs, suggesting that Fz3-dependent signals are required to sense or respond to molecular cues from intermediate targets (Hua et al., 2013).

The developmental roles of Fz3 and Fz6 extend substantially beyond the processes revealed by single gene mutations. *Fz3*$^{-/-}$;*Fz6*$^{-/-}$ embryos reveal redundant functions in neural tube closure and in rotationally orientating the kinocilium/stereociliary bundle complex on the apical face of inner ear sensory hair cells (Wang, Guo, & Nathans, 2006). Additionally, epidermis-specific double knockout of *Fz3* and *Fz6* (in *K14-Cre;*

$Fz3^{CKO/CKO}$;$Fz6^{-/-}$ mice) reveals a redundant role for these genes in orienting the filiform papillae on the dorsal surface of the tongue (Hua, Chang, et al., 2014). Stimulated by the partial redundancy of *Fz3* and *Fz6*, Hua, Chang, et al. (2014) asked whether this redundancy can be extended to other tissues by enlarging the domains of *Fz3* and *Fz6* gene expression. These experiments demonstrated that ubiquitous expression of *Fz3* fully rescues the *$Fz6^{-/-}$* hair follicle orientation phenotype and ubiquitous expression of *Fz6* partially rescues the *$Fz3^{-/-}$* axon guidance phenotype. Thus, the many diverse contexts in which Fz3 and Fz6 act are likely based on a conserved signaling role for these two proteins.

Mutations of *Celsr1*, *Celsr2*, and *Celsr3*, mammalian homologues of the *Drosophila* core PCP gene *Flamingo/Starry night* (*Fmi/Stan*), produce neural tube closure, axon guidance, neuronal migration, inner ear sensory hair cell, and hair follicle phenotypes that closely resemble the phenotypes in *$Fz3^{-/-}$*, *$Fz6^{-/-}$*, and *$Fz3^{-/-}$;$Fz6^{-/-}$* mice (Curtin et al., 2003; Qu et al., 2010; Ravni, Qu, Goffinet, & Tissir, 2009; Tissir, Bar, Jossin, De Backer, & Goffinet, 2005; Zhou et al., 2008). Similarly, *$Vangl2^{Lp/Lp}$* fetuses exhibit defects in inner ear sensory hair cell orientation and neural tube closure that closely resemble the defects observed in *$Fz3^{-/-}$;$Fz6^{-/-}$* fetuses (Kibar et al., 2001; Montcouquiol et al., 2003; Murdoch, Doudney, Paternotte, Copp, & Stanier, 2001; Strong & Hollander, 1949). *$Vangl2^{Lp/Lp}$* fetuses also exhibit a hair follicle orientation defect, but, interestingly, the *$Vangl2^{Lp/Lp}$* follicles are initially oriented perpendicular to the plane of the skin whereas WT and *$Fz6^{-/-}$* follicles are oriented obliquely (Devenport & Fuchs, 2008; Wang et al., 2010). Despite such minor differences, the overall similarities in affected tissues and in polarity defects among *Celsr*, *Vangl*, and *Fz3/Fz6* mutants support the idea that the diverse *$Fz3^{-/-}$*, *$Fz6^{-/-}$*, and *$Fz3^{-/-}$;$Fz6^{-/-}$* phenotypes reflect the activity of a mammalian version of *Drosophila* PCP.

7. CONCLUSION

The past three decades represent the first phase of exploration of the Frizzled family. At least some of the functions of most mammalian Frizzled family members are now known at a descriptive level, and for three family members the signaling pathways in which they participate have been identified. Despite the enormous progress made during this time, our current level of understanding remains rudimentary. To cite three examples: (1) the determinants of Wnt–Frizzled specificity are still unknown, and there is a strong suspicion that additional modulatory/specificity proteins remain to be discovered, (2) the regulation of Frizzled expression, trafficking, and stability is just starting to be explored, with the discovery of R-spondin regulation of Frizzled degradation via the ubiquitin ligase Znrf3 representing a tantalizing appetizer (Hao et al., 2012; Koo et al., 2012), and (3) the role of Frizzled receptors in human disease—both inherited and acquired—is still in its infancy.

Perhaps the greatest challenge is to define the biochemical role of Frizzled proteins in polarity signaling. Tissue polarity has been recognized as an essential attribute of animal development for more than 75 years, with the early microsurgical investigations of polarity in the insect cuticle preparing the intellectual foundations for today's molecular genetic experiments (Lawrence, 1966; Locke, 1959; Wigglesworth, 1940). Tissue polarity signaling also has a connection to the larger world of complex systems, of which biological pattern formation represents one part (Ball, 1999). The local nature of cell–cell signaling suggests that tissue polarity uses conceptually simple "rules" of information transfer that can be modified in various ways to create anatomic diversity. In keeping with this idea, the study of cellular automata demonstrates that large-scale iterative applications of simple rules can generate remarkably complex patterns (Wolfram, 2002). As new mechanistic insights emerge in the field of tissue polarity signaling, it will be of great interest to try to capture the fundamental characteristics of this system in mathematical models, an effort that is already underway (Amonlirdviman et al., 2005; Axelrod & Tomlin, 2011; Wang, Badea, et al., 2006).

ACKNOWLEDGMENTS

Supported by the Howard Hughes Medical Institute, the Ellison Medical Foundation, and the National Eye institute (NIH).

Conflict of interest: The authors have declared that no conflict of interest exists.

REFERENCES

Albers, J., Schulze, J., Beil, F. T., Gebauer, M., Baranowsky, A., Keller, J., et al. (2011). Control of bone formation by the serpentine receptor Frizzled-9. *The Journal of Cell Biology*, *192*(6), 1057–1072.

Allache, R., De Marco, P., Merello, E., Capra, V., & Kibar, Z. (2012). Role of the planar cell polarity gene CELSR1 in neural tube defects and caudal agenesis. *Birth Defects Research. Part A, Clinical and Molecular Teratology*, *94*(3), 176–181.

Amonlirdviman, K., Khare, N. A., Tree, D. R., Chen, W. S., Axelrod, J. D., & Tomlin, C. J. (2005). Mathematical modeling of planar cell polarity to understand domineering nonautonomy. *Science*, *307*(5708), 423–426.

Anderson, K. D., Pan, L., Yang, X. M., Hughes, V. C., Walls, J. R., Dominguez, M. G., et al. (2011). Angiogenic sprouting into neural tissue requires Gpr124, an orphan

G protein-coupled receptor. *Proceedings of the National Academy of Sciences of the United States of America*, *108*(7), 2807–2812.

Armstrong, A., Ryu, Y. K., Chieco, D., & Kuruvilla, R. (2011). Frizzled3 is required for neurogenesis and target innervation during sympathetic nervous system development. *The Journal of Neuroscience*, *31*(7), 2371–2381.

Au, K. S., Ashley-Koch, A., & Northrup, H. (2010). Epidemiologic and genetic aspects of spina bifida and other neural tube defects. *Developmental Disabilities Research Reviews*, *16*(1), 6–15.

Axelrod, J. D., & Tomlin, C. J. (2011). Modeling the control of planar cell polarity. *Wiley Interdisciplinary Reviews. Systems Biology and Medicine*, *3*(5), 588–605.

Ball, P. (1999). *The self-made tapestry: Pattern formation in nature.* Oxford: Oxford University Press. 287 pp.

Baron, R., & Kneissel, M. (2013). Wnt signaling in bone homeostasis and disease: From human mutations to treatments. *Nature Medicine*, *19*(2), 179–192.

Berger, W., Meindl, A., van de Pol, T. J., Cremers, F. P., Ropers, H. H., Doerner, C., et al. (1992). Isolation of a candidate gene for Norrie disease by positional cloning. *Nature Genetics*, *1*(3), 199–203.

Berger, W., & Ropers, H. (2001). Norrie disease. In C. R. Scriver, A. L. Beaudet, W. S. Sly, & D. Valle (Eds.), *The metabolic and molecular bases of inherited disease* (pp. 5977–5985). New York, NY, USA: McGraw-Hill.

Bhanot, P., Brink, M., Samos, C. H., Hsieh, J. C., Wang, Y., Macke, J. P., et al. (1996). A new member of the frizzled family from drosophila functions as a wingless receptor. *Nature*, *382*(6588), 225–230.

Bhanot, P., Fish, M., Jemison, J. A., Nusse, R., Nathans, J., & Cadigan, K. M. (1999). Frizzled and frizzled-2 function as redundant receptors for wingless during drosophila embryonic development. *Development*, *126*(18), 4175–4186.

Bhat, K. M. (1998). Frizzled and frizzled 2 play a partially redundant role in wingless signaling and have similar requirements to wingless in neurogenesis. *Cell*, *95*(7), 1027–1036.

Biggs, L. C., Goudy, S. L., & Dunnwald, M. (2015). Palatogenesis and cutaneous repair: A two-headed coin. *Developmental Dynamics*, *244*(3), 289–310.

Carson-Walter, E. B., Watkins, D. N., Nanda, A., Vogelstein, B., Kinzler, K. W., & St Croix, B. (2001). Cell surface tumor endothelial markers are conserved in mice and humans. *Cancer Research*, *61*(18), 6649–6655.

Centers for Disease Control and Prevention (CDC). (1997). Temporal trends in the incidence of birth defects—United States. *MMWR. Morbidity and Mortality Weekly Report*, *46*(49), 1171–1176.

Chan, S. D., Karpf, D. B., Fowlkes, M. E., Hooks, M., Bradley, M. S., Vuong, V., et al. (1992). Two homologs of the drosophila polarity gene frizzled (fz) are widely expressed in mammalian tissues. *The Journal of Biological Chemistry*, *267*(35), 25202–25207.

Chang, T. H., Hsieh, F. L., Zebisch, M., Harlos, K., Elegheert, J., & Jones, E. Y. (2015). Structure and functional properties of Norrin mimic Wnt for signalling with Frizzled4, Lrp5/6, and proteoglycan. *eLife*, *4*. http://dx.doi.org/10.7554/eLife.06554.

Chang, H., & Nathans, J. (2013). Responses of hair follicle-associated structures to loss of planar cell polarity signaling. *Proceedings of the National Academy of Sciences of the United States of America*, *110*(10), E908–E917.

Chen, Z. Y., Hendriks, R. W., Jobling, M. A., Powell, J. F., Breakefield, X. O., Sims, K. B., et al. (1992). Isolation and characterization of a candidate gene for Norrie disease. *Nature Genetics*, *1*(3), 204–208.

Chi, L., Zhang, S., Lin, Y., Prunskaite-Hyyrylainen, R., Vuolteenaho, R., Itaranta, P., et al. (2004). Sprouty proteins regulate ureteric branching by coordinating reciprocal

epithelial Wnt11, mesenchymal Gdnf and stromal Fgf7 signalling during kidney development. *Development, 131*(14), 3345–3356.
Cui, C. Y., Klar, J., Georgii-Heming, P., Frojmark, A. S., Baig, S. M., Schlessinger, D., et al. (2013). Frizzled6 deficiency disrupts the differentiation process of nail development. *The Journal of Investigative Dermatology, 133*(8), 1990–1997.
Cullen, M., Elzarrad, M. K., Seaman, S., Zudaire, E., Stevens, J., Yang, M. Y., et al. (2011). GPR124, an orphan G protein-coupled receptor, is required for CNS-specific vascularization and establishment of the blood–brain barrier. *Proceedings of the National Academy of Sciences of the United States of America, 108*(14), 5759–5764.
Curtin, J. A., Quint, E., Tsipouri, V., Arkell, R. M., Cattanach, B., Copp, A. J., et al. (2003). Mutation of Celsr1 disrupts planar polarity of inner ear hair cells and causes severe neural tube defects in the mouse. *Current Biology, 13*(13), 1129–1133.
Daneman, R., Agalliu, D., Zhou, L., Kuhnert, F., Kuo, C. J., & Barres, B. A. (2009). Wnt/beta-catenin signaling is required for CNS, but not non-CNS, angiogenesis. *Proceedings of the National Academy of Sciences of the United States of America, 106*(2), 641–646.
Daneman, R., & Prat, A. (2015). The blood–brain barrier. *Cold Spring Harbor Perspectives in Biology*, 7(1), a020412.
De Marco, P., Merello, E., Rossi, A., Piatelli, G., Cama, A., Kibar, Z., et al. (2012). FZD6 is a novel gene for human neural tube defects. *Human Mutation, 33*(2), 384–390.
Devenport, D., & Fuchs, E. (2008). Planar polarization in embryonic epidermis orchestrates global asymmetric morphogenesis of hair follicles. *Nature Cell Biology, 10*(11), 1257–1268.
Dolk, H., Loane, M., & Garne, E. (2010). The prevalence of congenital anomalies in Europe. *Advances in Experimental Medicine and Biology, 686*, 349–364.
Duke-Elder, S. (1964). *System of ophthalmology: Normal and abnormal development* (pp. 764–770). St. Louis: C.V. Mosby.
Fenstermaker, A. G., Prasad, A. A., Bechara, A., Adolfs, Y., Tissir, F., Goffinet, A., et al. (2010). Wnt/planar cell polarity signaling controls the anterior-posterior organization of monoaminergic axons in the brainstem. *The Journal of Neuroscience, 30*(47), 16053–16064.
Frojmark, A. S., Schuster, J., Sobol, M., Entesarian, M., Kilander, M. B., Gabrikova, D., et al. (2011). Mutations in frizzled 6 cause isolated autosomal-recessive nail dysplasia. *American Journal of Human Genetics, 88*(6), 852–860.
Fruttiger, M. (2007). Development of the retinal vasculature. *Angiogenesis, 10*(2), 77–88.
Goldberg, M. F. (1997). Persistent fetal vasculature (PFV): An integrated interpretation of signs and symptoms associated with persistent hyperplastic primary vitreous (PHPV). LIV Edward Jackson Memorial Lecture. *American Journal of Ophthalmology, 124*(5), 587–626.
Goodrich, L. V., & Strutt, D. (2011). Principles of planar polarity in animal development. *Development, 138*(10), 1877–1892.
Gubb, D., & Garcia-Bellido, A. (1982). A genetic analysis of the determination of cuticular polarity during development in Drosophila melanogaster. *Journal of Embryology and Experimental Morphology, 68*, 37–57.
Guo, N., Hawkins, C., & Nathans, J. (2004). Frizzled6 controls hair patterning in mice. *Proceedings of the National Academy of Sciences of the United States of America, 101*(25), 9277–9281.
Hao, H. X., Xie, Y., Zhang, Y., Charlat, O., Oster, E., Avello, M., et al. (2012). ZNRF3 promotes Wnt receptor turnover in an R-spondin-sensitive manner. *Nature, 485*(7397), 195–200.
Heilmann, A., Schinke, T., Bindl, R., Wehner, T., Rapp, A., Haffner-Luntzer, M., et al. (2013). The Wnt serpentine receptor frizzled-9 regulates new bone formation in fracture healing. *PloS One, 8*(12), e84232.

Hsieh, J. C., Rattner, A., Smallwood, P. M., & Nathans, J. (1999). Biochemical characterization of Wnt-frizzled interactions using a soluble, biologically active vertebrate Wnt protein. *Proceedings of the National Academy of Sciences of the United States of America, 96*(7), 3546–3551.

Hua, Z. L., Chang, H., Wang, Y., Smallwood, P. M., & Nathans, J. (2014). Partial interchangeability of Fz3 and Fz6 in tissue polarity signaling for epithelial orientation and axon growth and guidance. *Development, 141*(20), 3944–3954.

Hua, Z. L., Jeon, S., Caterina, M. J., & Nathans, J. (2014). Frizzled3 is required for the development of multiple axon tracts in the mouse central nervous system. *Proceedings of the National Academy of Sciences of the United States of America, 111*(29), E3005–E3014.

Hua, Z. L., Smallwood, P. M., & Nathans, J. (2013). Frizzled3 controls axonal development in distinct populations of cranial and spinal motor neurons. *eLife, 2*, e01482.

Ishikawa, T., Tamai, Y., Zorn, A. M., Yoshida, H., Seldin, M. F., Nishikawa, S., et al. (2001). Mouse Wnt receptor gene Fzd5 is essential for yolk sac and placental angiogenesis. *Development, 128*(1), 25–33.

Janda, C. Y., Waghray, D., Levin, A. M., Thomas, C., & Garcia, K. C. (2012). Structural basis of Wnt recognition by Frizzled. *Science, 337*(6090), 59–64.

Joeng, K. S., Schumacher, C. A., Zylstra-Diegel, C. R., Long, F., & Williams, B. O. (2011). Lrp5 and Lrp6 redundantly control skeletal development in the mouse embryo. *Developmental Biology, 359*(2), 222–229.

Junge, H. J., Yang, S., Burton, J. B., Paes, K., Shu, X., French, D. M., et al. (2009). TSPAN12 regulates retinal vascular development by promoting Norrin- but not Wnt-induced FZD4/beta-catenin signaling. *Cell, 139*(2), 299–311.

Ke, J., Harikumar, K. G., Erice, C., Chen, C., Gu, X., Wang, L., et al. (2013). Structure and function of norrin in assembly and activation of a frizzled 4-Lrp5/6 complex. *Genes & Development, 27*(21), 2305–2319.

Kibar, Z., Bosoi, C. M., Kooistra, M., Salem, S., Finnell, R. H., De Marco, P., et al. (2009). Novel mutations in VANGL1 in neural tube defects. *Human Mutation, 30*(7), E706–E715.

Kibar, Z., Salem, S., Bosoi, C. M., Pauwels, E., De Marco, P., Merello, E., et al. (2011). Contribution of VANGL2 mutations to isolated neural tube defects. *Clinical Genetics, 80*(1), 76–82.

Kibar, Z., Torban, E., McDearmid, J. R., Reynolds, A., Berghout, J., Mathieu, M., et al. (2007). Mutations in VANGL1 associated with neural-tube defects. *The New England Journal of Medicine, 356*(14), 1432–1437.

Kibar, Z., Vogan, K. J., Groulx, N., Justice, M. J., Underhill, D. A., & Gros, P. (2001). Ltap, a mammalian homolog of Drosophila Strabismus/Van Gogh, is altered in the mouse neural tube mutant loop-tail. *Nature Genetics, 28*(3), 251–255.

Koike, J., Takagi, A., Miwa, T., Hirai, M., Terada, M., & Katoh, M. (1999). Molecular cloning of Frizzled-10, a novel member of the Frizzled gene family. *Biochemical and Biophysical Research Communications, 262*(1), 39–43.

Koo, B. K., Spit, M., Jordens, I., Low, T. Y., Stange, D. E., van de Wetering, M., et al. (2012). Tumour suppressor RNF43 is a stem-cell E3 ligase that induces endocytosis of Wnt receptors. *Nature, 488*(7413), 665–669.

Kuhnert, F., Mancuso, M. R., Shamloo, A., Wang, H. T., Choksi, V., Florek, M., et al. (2010). Essential regulation of CNS angiogenesis by the orphan G protein-coupled receptor GPR124. *Science, 330*(6006), 985–989.

Lawrence, P. A. (1966). Gradients in the insect segment: The orientation of hairs in the milkweed bug Oncopeltus fasciatus. *The Journal of Experimental Biology, 44*, 607–620.

Liebner, S., Corada, M., Bangsow, T., Babbage, J., Taddei, A., Czupalla, C. J., et al. (2008). Wnt/beta-catenin signaling controls development of the blood–brain barrier. *The Journal of Cell Biology, 183*(3), 409–417.

Liu, G., Bafico, A., Harris, V. K., & Aaronson, S. A. (2003). A novel mechanism for Wnt activation of canonical signaling through the LRP6 receptor. *Molecular and Cellular Biology, 23*(16), 5825–5835.

Liu, C., Bakeri, H., Li, T., & Swaroop, A. (2012). Regulation of retinal progenitor expansion by frizzled receptors: Implications for microphthalmia and retinal coloboma. *Human Molecular Genetics, 21*(8), 1848–1860.

Liu, C., & Nathans, J. (2008). An essential role for frizzled 5 in mammalian ocular development. *Development, 135*(21), 3567–3576.

Liu, C., Wang, Y., Smallwood, P. M., & Nathans, J. (2008). An essential role for Frizzled5 in neuronal survival in the parafascicular nucleus of the thalamus. *The Journal of Neuroscience, 28*(22), 5641–5653.

Locke, M. (1959). The cuticular pattern in an insect Rhodnius prolixus Stal. *The Journal of Experimental Biology, 36*, 459–477.

Luhmann, U. F., Lin, J., Acar, N., Lammel, S., Feil, S., Grimm, C., et al. (2005). Role of the Norrie disease pseudoglioma gene in sprouting angiogenesis during development of the retinal vasculature. *Investigative Ophthalmology & Visual Science, 46*(9), 3372–3382.

Lyuksyutova, A. I., Lu, C. C., Milanesio, N., King, L. A., Guo, N., Wang, Y., et al. (2003). Anterior-posterior guidance of commissural axons by Wnt-frizzled signaling. *Science, 302*(5652), 1984–1988.

MacDonald, B. T., Tamai, K., & He, X. (2009). Wnt/beta-catenin signaling: Components, mechanisms, and diseases. *Developmental Cell, 17*(1), 9–26.

Majumdar, A., Vainio, S., Kispert, A., McMahon, J., & McMahon, A. P. (2003). Wnt11 and Ret/Gdnf pathways cooperate in regulating ureteric branching during metanephric kidney development. *Development, 130*(14), 3175–3185.

Montcouquiol, M., Rachel, R. A., Lanford, P. J., Copeland, N. G., Jenkins, N. A., & Kelley, M. W. (2003). Identification of Vangl2 and Scrb1 as planar polarity genes in mammals. *Nature, 423*(6936), 173–177.

Muller, H. A., Samanta, R., & Wieschaus, E. (1999). Wingless signaling in the Drosophila embryo: Zygotic requirements and the role of the frizzled genes. *Development, 126*(3), 577–586.

Murdoch, J. N., Doudney, K., Paternotte, C., Copp, A. J., & Stanier, P. (2001). Severe neural tube defects in the loop-tail mouse result from mutation of Lpp1, a novel gene involved in floor plate specification. *Human Molecular Genetics, 10*(22), 2593–2601.

Naz, G., Pasternack, S. M., Perrin, C., Mattheisen, M., Refke, M., Khan, S., et al. (2012). FZD6 encoding the Wnt receptor frizzled 6 is mutated in autosomal-recessive nail dysplasia. *The British Journal of Dermatology, 166*(5), 1088–1094.

Nikopoulos, K., Gilissen, C., Hoischen, A., van Nouhuys, C. E., Boonstra, F. N., Blokland, E. A., et al. (2010). Next-generation sequencing of a 40 Mb linkage interval reveals TSPAN12 mutations in patients with familial exudative vitreoretinopathy. *American Journal of Human Genetics, 86*(2), 240–247.

Nikopoulos, K., Venselaar, H., Collin, R. W., Riveiro-Alvarez, R., Boonstra, F. N., Hooymans, J. M., et al. (2010). Overview of the mutation spectrum in familial exudative vitreoretinopathy and norrie disease with identification of 21 novel variants in FZD4, LRP5, and NDP. *Human Mutation, 31*(6), 656–666.

Nusse, R. (2012). Wnt signaling. *Cold Spring Harbor Perspectives in Biology, 4*(5). http://dx.doi.org/10.1101/cshperspect.a011163.

Posokhova, E., Shukla, A., Seaman, S., Volate, S., Hilton, M. B., Wu, B., et al. (2015). GPR124 functions as a Wnt7-specific coactivator of canonical beta-catenin signaling. *Cell Reports, 10*(2), 123–130.

Poulter, J. A., Ali, M., Gilmour, D. F., Rice, A., Kondo, H., Hayashi, K., et al. (2010). Mutations in TSPAN12 cause autosomal-dominant familial exudative vitreoretinopathy. *American Journal of Human Genetics, 86*(2), 248–253.

Poulter, J. A., Davidson, A. E., Ali, M., Gilmour, D. F., Parry, D. A., Mintz-Hittner, H. A., et al. (2012). Recessive mutations in TSPAN12 cause retinal dysplasia and severe familial exudative vitreoretinopathy (FEVR). *Investigative Ophthalmology & Visual Science, 53*(6), 2873–2879.

Qin, M., Hayashi, H., Oshima, K., Tahira, T., Hayashi, K., & Kondo, H. (2005). Complexity of the genotype-phenotype correlation in familial exudative vitreoretinopathy with mutations in the LRP5 and/or FZD4 genes. *Human Mutation, 26*(2), 104–112.

Qu, Y., Glasco, D. M., Zhou, L., Sawant, A., Ravni, A., Fritzsch, B., et al. (2010). Atypical cadherins Celsr1-3 differentially regulate migration of facial branchiomotor neurons in mice. *The Journal of Neuroscience, 30*(28), 9392–9401.

Ranheim, E. A., Kwan, H. C., Reya, T., Wang, Y. K., Weissman, I. L., & Francke, U. (2005). Frizzled 9 knock-out mice have abnormal B-cell development. *Blood, 105*(6), 2487–2494.

Ravni, A., Qu, Y., Goffinet, A. M., & Tissir, F. (2009). Planar cell polarity cadherin Celsr1 regulates skin hair patterning in the mouse. *The Journal of Investigative Dermatology, 129*(10), 2507–2509.

Rehm, H. L., Zhang, D. S., Brown, M. C., Burgess, B., Halpin, C., Berger, W., et al. (2002). Vascular defects and sensorineural deafness in a mouse model of Norrie disease. *The Journal of Neuroscience, 22*(11), 4286–4292.

Robitaille, J., MacDonald, M. L., Kaykas, A., Sheldahl, L. C., Zeisler, J., Dube, M. P., et al. (2002). Mutant frizzled-4 disrupts retinal angiogenesis in familial exudative vitreoretinopathy. *Nature Genetics, 32*(2), 326–330.

Robitaille, J. M., Zheng, B., Wallace, K., Beis, M. J., Tatlidil, C., Yang, J., et al. (2011). The role of Frizzled-4 mutations in familial exudative vitreoretinopathy and Coats disease. *The British Journal of Ophthalmology, 95*(4), 574–579.

Salvo, J., Lyubasyuk, V., Xu, M., Wang, H., Wang, F., Nguyen, D., et al. (2015). Next-generation sequencing and novel variant determination in a cohort of 92 familial exudative vitreoretinopathy patients. *Investigative Ophthalmology & Visual Science, 56*(3), 1937–1946.

Schenkelaars, Q., Fierro-Constain, L., Renard, E., Hill, A. L., & Borchiellini, C. (2015). Insights into Frizzled evolution and new perspectives. *Evolution & Development, 17*(2), 160–169.

Shah, S. M., Kang, Y. J., Christensen, B. L., Feng, A. S., & Kollmar, R. (2009). Expression of Wnt receptors in adult spiral ganglion neurons: Frizzled 9 localization at growth cones of regenerating neurites. *Neuroscience, 164*(2), 478–487.

Smallwood, P. M., Williams, J., Xu, Q., Leahy, D. J., & Nathans, J. (2007). Mutational analysis of Norrin-Frizzled4 recognition. *The Journal of Biological Chemistry, 282*(6), 4057–4068.

St Croix, B., Rago, C., Velculescu, V., Traverso, G., Romans, K. E., Montgomery, E., et al. (2000). Genes expressed in human tumor endothelium. *Science, 289*(5482), 1197–1202.

Stenman, J. M., Rajagopal, J., Carroll, T. J., Ishibashi, M., McMahon, J., & McMahon, A. P. (2008). Canonical Wnt signaling regulates organ-specific assembly and differentiation of CNS vasculature. *Science, 322*(5905), 1247–1250.

Strong, L. C., & Hollander, W. F. (1949). Hereditary loop-tail in the house mouse accompanied by imperforate vagina and with lethal craniorachischisis when homozygous. *The Journal of Heredity, 40*, 329–334.

Taylor, J., Abramova, N., Charlton, J., & Adler, P. N. (1998). Van Gogh: A new Drosophila tissue polarity gene. *Genetics, 150*(1), 199–210.

Tissir, F., Bar, I., Jossin, Y., De Backer, O., & Goffinet, A. M. (2005). Protocadherin Celsr3 is crucial in axonal tract development. *Nature Neuroscience, 8*(4), 451–457.

Toomes, C., Bottomley, H. M., Scott, S., Mackey, D. A., Craig, J. E., Appukuttan, B., et al. (2004). Spectrum and frequency of FZD4 mutations in familial exudative vitreoretinopathy. *Investigative Ophthalmology & Visual Science, 45*(7), 2083–2090.
Vanhollebeke, B., Stone, O. A., Bostaille, N., Cho, C., Zhou, Y., Maquet, E., et al. (2015). Tip cell-specific requirement for an atypical Gpr124- and Reck-dependent Wnt/beta-catenin pathway during brain angiogenesis. *eLife, 4*. http://dx.doi.org/10.7554/eLife.06489.
Vinson, C. R., Conover, S., & Adler, P. N. (1989). A Drosophila tissue polarity locus encodes a protein containing seven potential transmembrane domains. *Nature, 338*(6212), 263–264.
Wang, Y., Badea, T., & Nathans, J. (2006). Order from disorder: Self-organization in mammalian hair patterning. *Proceedings of the National Academy of Sciences of the United States of America, 103*(52), 19800–19805.
Wang, Y., Chang, H., & Nathans, J. (2010). When whorls collide: The development of hair patterns in frizzled 6 mutant mice. *Development, 137*(23), 4091–4099.
Wang, Y., Guo, N., & Nathans, J. (2006). The role of Frizzled3 and Frizzled6 in neural tube closure and in the planar polarity of inner-ear sensory hair cells. *The Journal of Neuroscience, 26*(8), 2147–2156.
Wang, Y., Huso, D., Cahill, H., Ryugo, D., & Nathans, J. (2001). Progressive cerebellar, auditory, and esophageal dysfunction caused by targeted disruption of the frizzled-4 gene. *The Journal of Neuroscience, 21*(13), 4761–4771.
Wang, Y., Macke, J. P., Abella, B. S., Andreasson, K., Worley, P., Gilbert, D. J., et al. (1996). A large family of putative transmembrane receptors homologous to the product of the Drosophila tissue polarity gene frizzled. *The Journal of Biological Chemistry, 271*(8), 4468–4476.
Wang, H. Y., & Malbon, C. C. (2003). Wnt signaling, Ca2+, and cyclic GMP: Visualizing frizzled functions. *Science, 300*(5625), 1529–1530.
Wang, Y., Rattner, A., Zhou, Y., Williams, J., Smallwood, P. M., & Nathans, J. (2012). Norrin/Frizzled4 signaling in retinal vascular development and blood brain barrier plasticity. *Cell, 151*(6), 1332–1344.
Wang, Y. K., Samos, C. H., Peoples, R., Perez-Jurado, L. A., Nusse, R., & Francke, U. (1997). A novel human homologue of the Drosophila frizzled wnt receptor gene binds wingless protein and is in the Williams syndrome deletion at 7q11.23. *Human Molecular Genetics, 6*, 465–472.
Wang, Y., Thekdi, N., Smallwood, P. M., Macke, J. P., & Nathans, J. (2002). Frizzled-3 is required for the development of major fiber tracts in the rostral CNS. *The Journal of Neuroscience, 22*(19), 8563–8573.
Wigglesworth, V. B. (1940). Local and general factors in the development of 'pattern' in Rhodnius prolixus. *The Journal of Experimental Biology, 17*, 180–200.
Wilson, N. J., Hansen, C. D., Azkur, D., Kocabas, C. N., Metin, A., Coskun, Z., et al. (2013). Recessive mutations in the gene encoding frizzled 6 cause twenty nail dystrophy—Expanding the differential diagnosis for pachyonychia congenita. *Journal of Dermatological Science, 70*(1), 58–60.
Wolfram, S. (2002). *A new kind of science*. Champagne, IL: Wolfram Media, Inc. 1197 pp.
Wong, L. L., & Adler, P. N. (1993). Tissue polarity genes of Drosophila regulate the subcellular location for prehair initiation in pupal wing cells. *The Journal of Cell Biology, 123*(1), 209–221.
Xu, Q., Wang, Y., Dabdoub, A., Smallwood, P. M., Williams, J., Woods, C., et al. (2004). Vascular development in the retina and inner ear: Control by Norrin and Frizzled-4, a high-affinity ligand-receptor pair. *Cell, 116*(6), 883–895.

Ye, X., Smallwood, P., & Nathans, J. (2011). Expression of the Norrie disease gene (Ndp) in developing and adult mouse eye, ear, and brain. *Gene Expression Patterns*, *11*(1–2), 151–155.

Ye, X., Wang, Y., Cahill, H., Yu, M., Badea, T. C., Smallwood, P. M., et al. (2009). Norrin, frizzled-4, and Lrp5 signaling in endothelial cells controls a genetic program for retinal vascularization. *Cell*, *139*(2), 285–298.

Ye, X., Wang, Y., Rattner, A., & Nathans, J. (2011). Genetic mosaic analysis reveals a major role for frizzled 4 and frizzled 8 in controlling ureteric growth in the developing kidney. *Development*, *138*(6), 1161–1172.

Yu, H., Smallwood, P. M., Wang, Y., Vidaltamayo, R., Reed, R., & Nathans, J. (2010). Frizzled 1 and frizzled 2 genes function in palate, ventricular septum and neural tube closure: General implications for tissue fusion processes. *Development*, *137*(21), 3707–3717.

Yu, H., Ye, X., Guo, N., & Nathans, J. (2012). Frizzled 2 and frizzled 7 function redundantly in convergent extension and closure of the ventricular septum and palate: Evidence for a network of interacting genes. *Development*, *139*(23), 4383–4394.

Zhang, J., Fuhrmann, S., & Vetter, M. L. (2008). A nonautonomous role for retinal frizzled-5 in regulating hyaloid vitreous vasculature development. *Investigative Ophthalmology & Visual Science*, *49*(12), 5561–5567.

Zhao, C., Aviles, C., Abel, R. A., Almli, C. R., McQuillen, P., & Pleasure, S. J. (2005). Hippocampal and visuospatial learning defects in mice with a deletion of frizzled 9, a gene in the Williams syndrome deletion interval. *Development*, *132*(12), 2917–2927.

Zhao, C., Guan, W., & Pleasure, S. J. (2006). A transgenic marker mouse line labels Cajal-Retzius cells from the cortical hem and thalamocortical axons. *Brain Research*, *1077*(1), 48–53.

Zhou, L., Bar, I., Achouri, Y., Campbell, K., De Backer, O., Hebert, J. M., et al. (2008). Early forebrain wiring: Genetic dissection using conditional Celsr3 mutant mice. *Science*, *320*(5878), 946–949.

Zhou, Y., & Nathans, J. (2014). Gpr124 controls CNS angiogenesis and blood–brain barrier integrity by promoting ligand-specific canonical Wnt signaling. *Developmental Cell*, *31*(2), 248–256.

Zhou, Y., Wang, Y., Tischfield, M., Williams, J., Smallwood, P. M., Rattner, A., et al. (2014). Canonical Wnt signaling components in vascular development and barrier formation. *The Journal of Clinical Investigation*, *124*(9), 3825–3846.

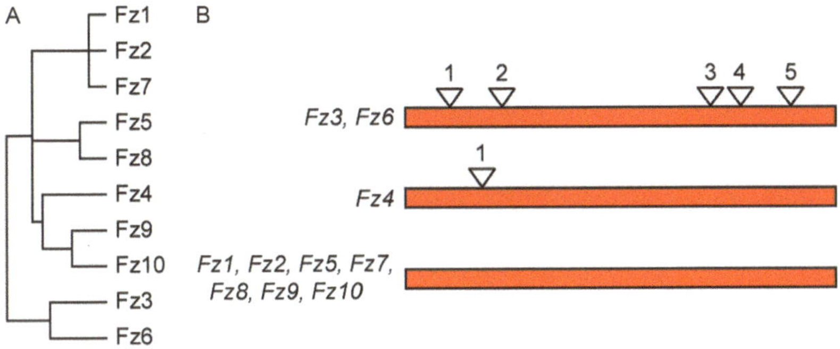

Yanshu Wang *et al.*, Figure 1 Evolutionary relationships among mammalian Frizzled proteins and genes. (A) Dendrogram showing the relatedness of the 10 mammalian Frizzled amino acid sequences. (B) Intron–exon structures of mammalian *Frizzled* genes, with coding regions depicted schematically as the same length. Only *Fz3*, *Fz4*, and *Fz6* have introns within the coding region. The five coding introns in the *Fz3* and *Fz6* genes are located at identical positions. *From Hua, Chang, Wang, Smallwood, and Nathans (2014).*

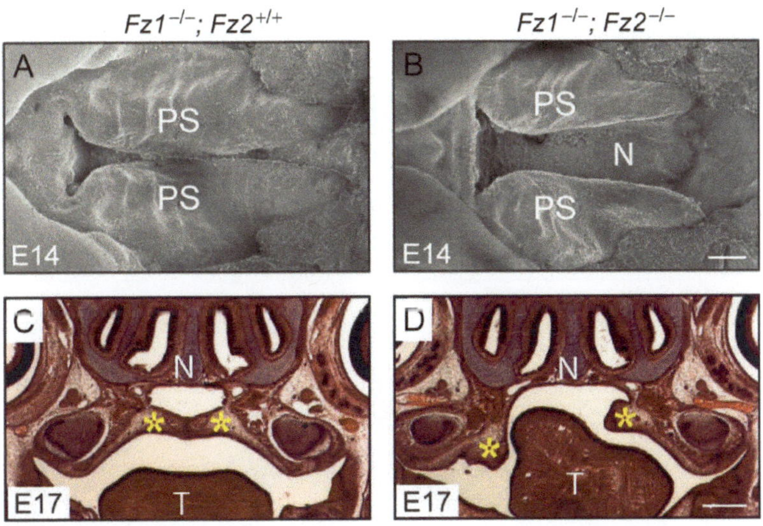

Yanshu Wang *et al.*, Figure 2 Palate closure defect in $Fz1^{-/-}$;$Fz2^{-/-}$ fetuses. (A and B) Scanning electron microscopy of the palate at E14. PS, palatal shelf. (C and D) Coronal section through the head at E17, stained with hematoxylin and eosin. N, nasal septum; T, tongue; Yellow asterisks, palatal shelves. Scale bars: B, 200 μm; D, 500 μm. *From Yu et al. (2010).*

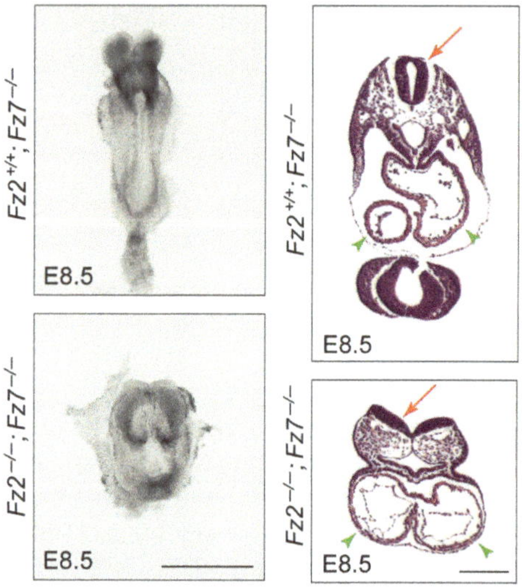

Yanshu Wang *et al.*, Figure 3 Convergent extension, neural tube closure, and cardiac asymmetry defects in *Fz2*$^{-/-}$;*Fz7*$^{-/-}$ embryos. E8.5 embryos are shown intact in a dorsal view (left) and in paraffin sections stained with hematoxylin and eosin (right). Left panels, the *Fz2*$^{-/-}$;*Fz7*$^{-/-}$ embryo is shorter and wider than the *Fz2*$^{+/+}$;*Fz7*$^{-/-}$ control due to a failure of convergent extension. Right panels, in cross section, the *Fz2*$^{-/-}$;*Fz7*$^{-/-}$ embryo shows an open neural tube (red arrows) and symmetric cardiac chambers (green arrowheads). The *Fz2*$^{+/+}$;*Fz7*$^{-/-}$ embryo is phenotypically WT. Scale bars: left panels, 1 mm; right panels, 200 μm. *From Yu et al. (2012).*

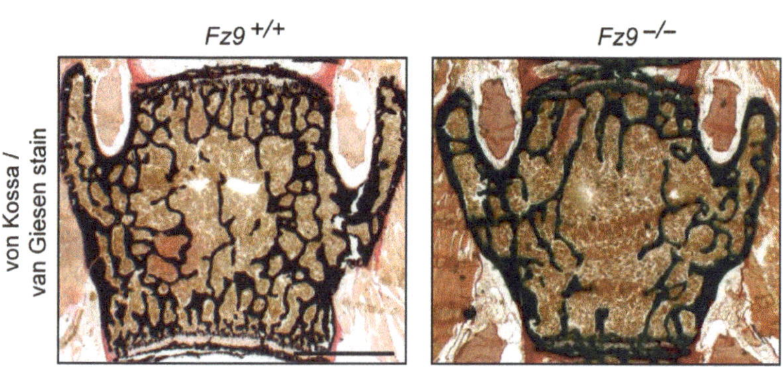

Yanshu Wang *et al.*, Figure 4 Reduced bone density in *Fz9*$^{-/-}$ mice. The vertebral body shows reduced bone density in a 52-week-old *Fz9*$^{-/-}$ mouse (right) compared to an age- and sex-matched control (left). Scale bar, 250 μm. *From Albers et al. (2011).*

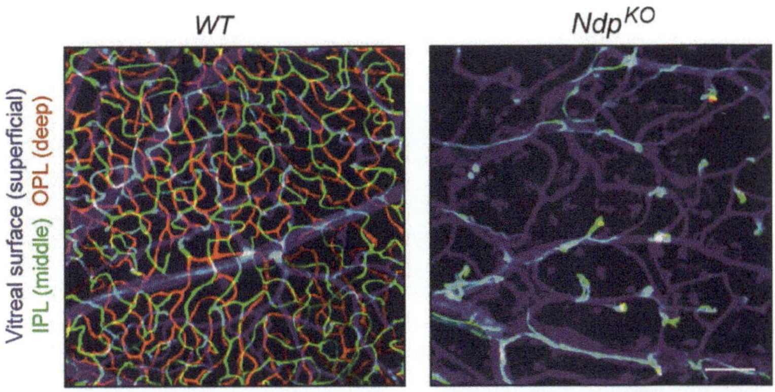

Yanshu Wang *et al.*, Figure 5 Defective retinal vascularization in Ndp^{KO} mice. Z-stacked confocal images of flat mount adult WT and Ndp^{KO} retinas, with ECs visualized with GS-lectin. The depth of the different vascular structures has been color coded as indicated at left. The Ndp^{KO} retina has only a superficial vascular plexus from which numerous EC clusters invade a short distance into the retina. IPL, inner plexiform layer; OPL, outer plexiform layer. Scale bar, 100 μm. *From Wang et al. (2012).*

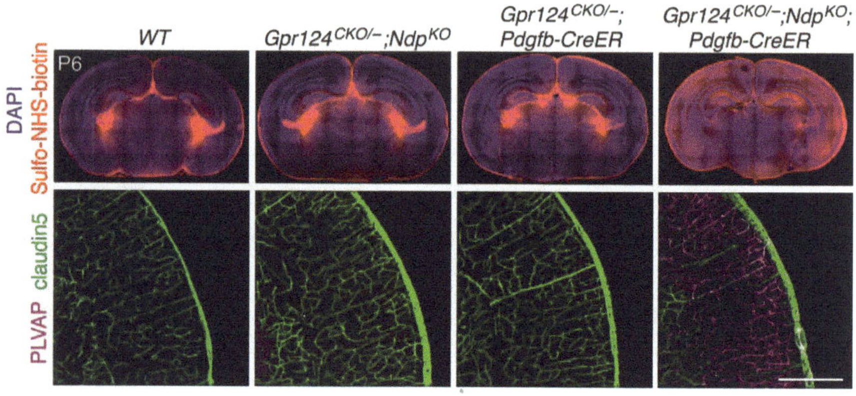

Yanshu Wang *et al.*, Figure 6 Redundant actions of Gpr124 and Norrin in maintaining BBB integrity. Coronal sections through the cortex and thalamus of postnatal day (P)6 mice with the indicated genotypes. *Pdgfb-CreER* recombines specifically in ECs. 4-Hydroxytamoxifen (4HT) was delivered by intraperitoneal injection at P3–P4. Intravascular Sulfo-NHS-biotin leaks from the choroid plexus in the lateral ventricles in all brain samples (upper panels), but it leaks extensively into the brain parenchyma only in the sample that is deleted for both *Gpr124* and *Ndp* (upper right panel). Intracerebral capillaries in the WT brain (lower left) and in the brains from mice with either one functional copy of *Gpr124* or one functional copy of *Ndp* (lower middle two panels) express claudin5 (a hallmark of CNS vasculature) and suppress plasmalemma vesicle-associated protein (PLVAP; a hallmark of peripheral vasculature). Deletion of both *Gpr124* and *Ndp* reverses this expression pattern (lower right panel). Scale bar, 200 μm. *From Zhou and Nathans (2014).*

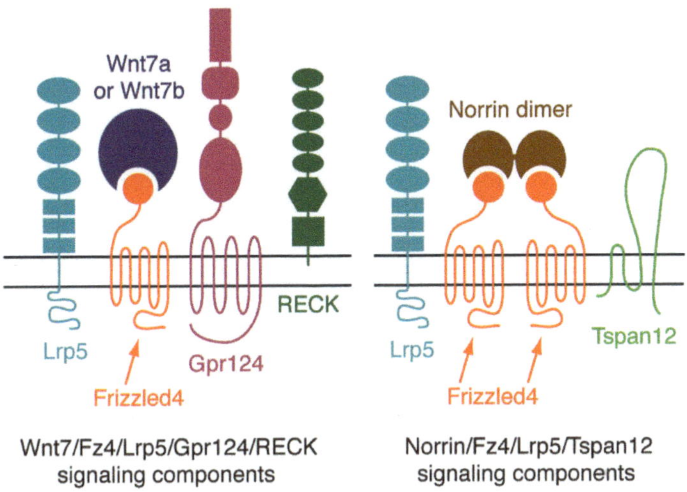

Yanshu Wang *et al.*, Figure 7 Schematic of ligand and receptor components involved in canonical Wnt signaling in CNS ECs. Horizontal black lines represent the plasma membrane; extracellular space is above and cytoplasm is below. Additional components, such as Lrp6, other Frizzleds, and other Wnts, are also involved, but their roles are not yet well defined. *Modified from Zhou and Nathans (2014).*

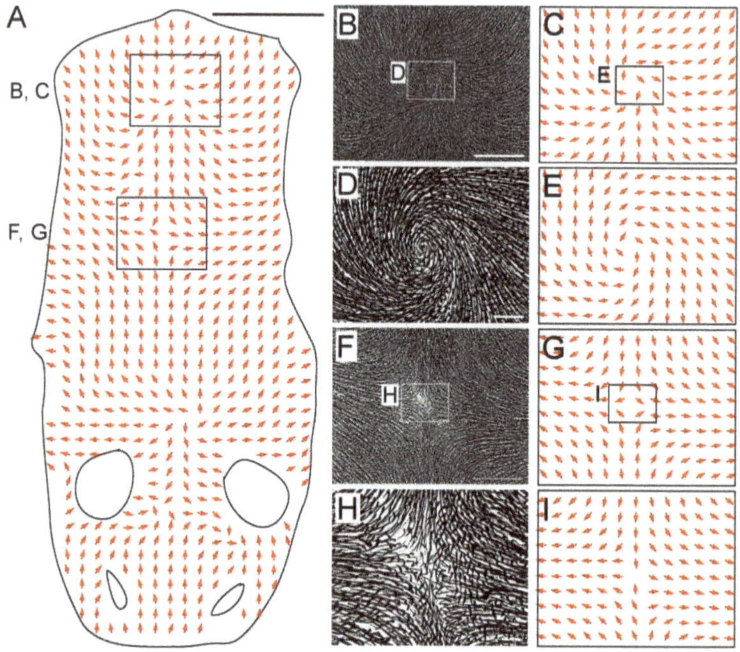

Yanshu Wang *et al.*, Figure 8 Hair patterns in a *Fz6*$^{-/-}$;*Vangl2*$^{Lp/+}$ back skin at P8. (A) Vector map showing hair follicle orientations over the head and back in a flat-mount skin from a P8 *Fz6*$^{-/-}$;*Vangl2*$^{Lp/+}$ mouse. Anterior is down; posterior is up. The small ovals are the eye positions; the larger ovals are the ear positions. Four patterning centers are present along the midline. (B–I) The central regions of the two most posterior patterning centers are enlarged in panels B, C, F, and G, and further enlarged in panels D, E, H, and I, as indicated by the rectangles in panels A, B, C, F, and G. Spatial order among follicles is evident at long range (~1 cm) and short range (e.g., between neighboring follicles; tens of microns). Scale bars: A, 1 cm; B and F, 2 mm; D and H, 300 μm. *From Wang, Chang, and Nathans (2010).*

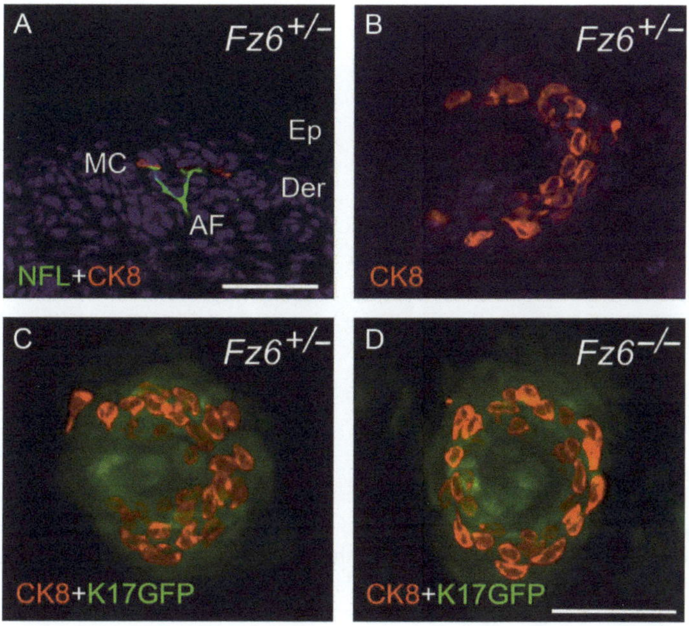

Yanshu Wang *et al.*, Figure 9 Loss of polarity in the arrangement of Merkel cells in *Fz6*$^{-/-}$ mice. (A) Vertical section of P1 back skin from a phenotypically WT *Fz6*$^{+/-}$ mouse. Merkel cells (MC) are visualized with anticytokeratin8 (CK8) immunostaining and afferent sensory nerve fibers (AF) are visualized with antineurofilament (NFL) immunostaining. Ep, epidermis; Der, dermis. (B–D) Flat mounts of P1 skin in which Merkel cells are visualized with CK8 and the central guard hair follicle is visualized with a *Keratin (K)17-GFP* transgene. Anterior is to the left. The phenotypically WT *Fz6*$^{+/-}$ Merkel cells form a semicircle that opens toward the anterior. The *Fz6*$^{-/-}$ Merkel cells form a closed circle. Scale bar, 50 μm. *From Chang and Nathans (2013).*

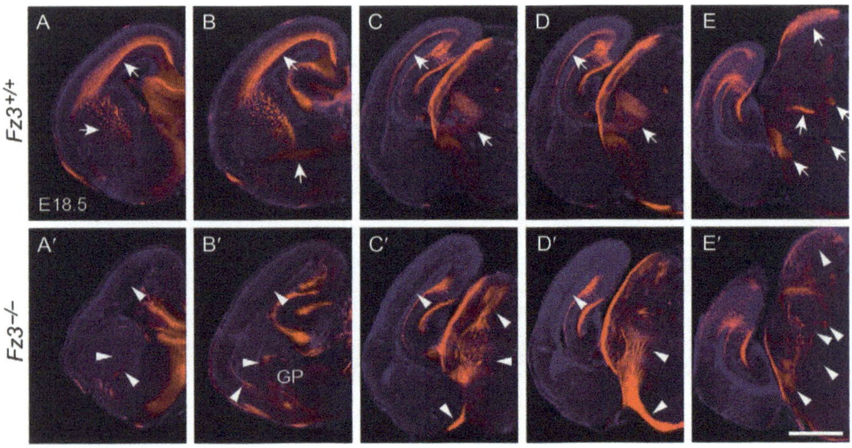

Yanshu Wang *et al.*, Figure 10 Thalamocortical axon guidance defects in the $Fz3^{-/-}$ forebrain at E18.5. Serial coronal sections through the forebrains of E18.5 $Fz3^{-/-}$ and $Fz3^{+/+}$ fetuses stained with anti-neurofilament antibodies. The midline of each hemibrain image is at the right border of each panel. The sections are arranged from rostral (A and A′) to caudal (E and E′). In the $Fz3^{+/+}$ brain, arrows in (A–D) trace the thalamocortical projections originating in the thalamus (D) and projecting via the internal capsule (A–C) to the cortex (A–D). In the $Fz3^{-/-}$ brain, thalamic axons track inferiorly with most crossing the midline to the contralateral thalamus (D′) and a minority projecting to the inferior border of the cortex (C′). Additional axon guidance defects are indicated by arrowheads in (A′–E′); see Hua, Jeon, et al. (2014) for details. GP, globus pallidus. Scale bar, 1 mm. *From Hua, Jeon, et al. (2014).*

CHAPTER EIGHT

The Developmental Genetics of Vertebrate Color Pattern Formation: Lessons from Zebrafish

Uwe Irion, Ajeet Pratap Singh, Christiane Nüsslein-Volhard[1]
Max-Planck-Institut für Entwicklungsbiologie, Tübingen, Germany
[1]Corresponding author; e-mail address: christiane.nuesslein-volhard@tuebingen.mpg.de

Contents

Abstract

Color patterns are prominent features of many animals; they are highly variable and evolve rapidly leading to large diversities even within a single genus. As targets for natural as well as sexual selection, they are of high evolutionary significance. The zebrafish (*Danio rerio*) has become an important model organism for developmental biology and biomedical research in general, and it is the model organism to study color pattern formation in vertebrates. The fish display a conspicuous pattern of alternating blue and golden stripes on the body and on the anal and tail fins. This pattern is produced by three different types of pigment cells (chromatophores) arranged in precise layers in the hypodermis of the fish. In this essay, we will summarize the recent advances in understanding the developmental and genetic basis for stripe formation in the

Current Topics in Developmental Biology, Volume 117
ISSN 0070-2153
http://dx.doi.org/10.1016/bs.ctdb.2015.12.012

zebrafish. We will describe the cellular events leading to the formation of stripes during metamorphosis based on long-term lineage imaging. Mutant analysis has revealed that a number of signaling pathways are involved in the establishment and maintenance of the individual pigment cells. However, the striped pattern itself is generated by self-organizing mechanisms requiring interactions between all three pigment cell types. The involvement of integral membrane proteins, including connexins and potassium channels, suggests that direct physical contacts between chromatophores are involved, and that the directed transport of small molecules or bioelectrical coupling is important for these interactions. This mode of patterning by transmitting spatial information between adjacent tissues within three superimposed cell layers is unprecedented in other developmental systems. We propose that variations in the patterns among *Danio* species are caused by allelic differences in the genes responsible for these interactions.

1. INTRODUCTION

In most animal species, the body is colored. A primary function of colors is easy to comprehend: dark pigments and reflecting structural colors prevent harmful radiation from damaging vital tissues (Brenner & Hearing, 2008). Strikingly, coloration is often displayed in beautiful patterns that are composed of several pigments, as well as various kinds of nanostructures producing a wide range of colors in the skin and its appendages such as bristles, scales, hairs, and feathers. Although the significance of these patterns for the animal is not always obvious, in many instances color patterns have important functions in the communication among individuals of a species, for example, recognition and selection of mating partners, or attraction between many individuals to form large groups. Color patterns are also instrumental in prey–predator interactions by allowing adaptation to the environment, but also serve as deceptive or attractive signals that are recognized by individuals of different species (Protas & Patel, 2008). Often, patterns are highly variable and evolve rapidly, which leads to large diversities in coloration, even within a single genus. In other cases, evolutionary convergence can lead to remarkable similarities in the color patterns of distant genera. In short, color patterns are of high evolutionary relevance as targets of natural as well as sexual selection. Understanding the mechanisms that underlie pigmentation and color pattern formation is an important step toward comprehending the evolution of biodiversity.

In insects, pigments are produced by epidermal cells and shed into the extracellular cuticle, bristles, and hairs. Frequently, color patterns follow morphological landmarks, such as segment boundaries, or wing veins that serve as prepatterns. In these instances, color pattern formation is explained as a readout of positional information in a two-dimensional sheet of cells, singling out epidermal cells to produce the enzymes required to synthesize the pigments (Wittkopp & Beldade, 2009). Well-known developmental signaling systems have been shown to be involved in this mode of pattern formation (Simpson, 2007; Werner, Koshikawa, Williams, & Carroll, 2010). In contrast, vertebrate color patterns are composed of specialized pigment-producing cells that undergo extensive cell movements and cell–cell interactions to form the final pattern. These pigment cells originate from the neural crest, a transient primordium of multipotent cells located at the dorsal neuroectodermal ridge from which progenitor cells migrate out into the periphery to develop a variety of structures and tissues including the peripheral nervous system, glia, and pigment cells (Le Douarin & Dupin, 2003; Le Douarin & Kalcheim, 1999). Birds and mammals have mainly one pigment cell type, the melanocyte, producing the pigment melanin (although in different shades, brownish-black eumelanin and reddish pheomelanin) that is secreted into the skin or the integumentary appendages, feathers, and hairs. Furthermore, birds display carotenoid-based colors and structural colors (Caro, 2013; Roulin & Ducrest, 2013; Uong & Zon, 2010). In contrast, basal vertebrates such as fish, amphibia, and reptiles develop several pigment cell types—chromatophores—producing different colors (Bagnara & Matsumoto, 2007). In fishes in particular, beautiful patterns displaying a wide spectrum of colors arise as multilayered mosaics of chromatophores distributed in the hypodermis of the body, and the epidermis of scales and fins (see a few examples in Figs. 1 and 2).

Although color pattern formation has fascinated scientists since the beginning of modern biology (Darwin, 1871; Poulton, 1890), it is a field that still is dominated by theories rather than detailed knowledge of the molecular, cellular, and developmental events leading to the striking display of different colors in the integument and its appendages of many vertebrate species. Genes affecting the production of pigments and colors have been identified by mutations in many species (Hubbard, Uy, Hauber, Hoekstra, & Safran, 2010); however, long developmental times and increasing body sizes make it technically challenging to study the development of color patterns in any vertebrate. Therefore, it is necessary to concentrate on organisms, in which individuals can be readily observed during the periods when patterning occurs, in which there is both, natural and induced variation, interbreeding is possible and transgenic and imaging methods are applicable. These features are uniquely combined in the zebrafish, *Danio rerio*,

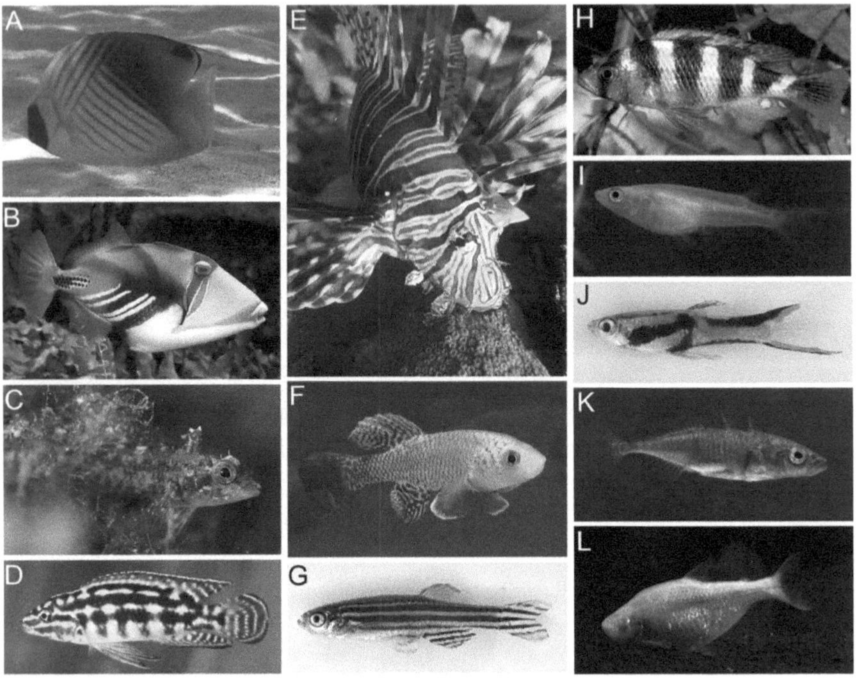

Figure 1 Color pattern diversity in fish. Fish show a great diversity of color patterns mainly by different combinations of stripes and spots. Some patterns are used to attract mates (e.g., in guppies or sticklebacks), others to deter predators (lion fish); cave fish are often blind and not colored. (A) Threadfin Butterfly fish (*Chaetodon auriga*), (B) Picasso fish (*Rhinecanthus aculeatus*), (C) Black-faced blenny (*Tripterygion delaisi*), (D) Checkered Julie (*Julidochromis marlieri*), (E) Common Lionfish (*Pterois spec.*), (F) Turquoise killifish (*Nothobranchius furzeri*), (G) Zebrafish (*Danio rerio*), (H) Cichlid (*Haplochromis latifasciatus*), (I) Japanese rice fish (*Oryzias latipes*), (J) Endler's Guppy (*Poecilia wingei*), (K) Three-spined stickleback (*Gasterosteus aculeatus*), and (L) Mexican cavefish (*Astyanax mexicanus*). Several of these fish are used as model organisms for biomedical research. *Panel (A): Image courtesy of Isabelle Côté. Panel (B): Image source—Wikipedia. Panel (C): Image courtesy of Nico Michiels. Panel (E): Image courtesy of Isabelle Côté. Panel (F): Image courtesy of Nils Hartmann. Panel (H): Image courtesy of Walter Salzburger. Panel (I): Image courtesy of Matthew Harris. Panel (J): Image courtesy of Verena Kottler and Christine Dreyer. Panel (L): Image courtesy of Nicolas Rohner.* (A color version of this figure is located at the end of this chapter.)

owing its name to the striking stereotypic pattern of longitudinal blue and golden stripes on the flanks and on the anal and tail fins (Figs. 2 and 3A) (Kelsh, Harris, Colanesi, & Erickson, 2009; Parichy & Spiewak, 2015; Singh & Nüsslein-Volhard, 2015; Watanabe & Kondo, 2015). The adult striped pattern of zebrafish is composed of melanophores, iridophores, and xanthophores arranged in superimposed layers in the skin

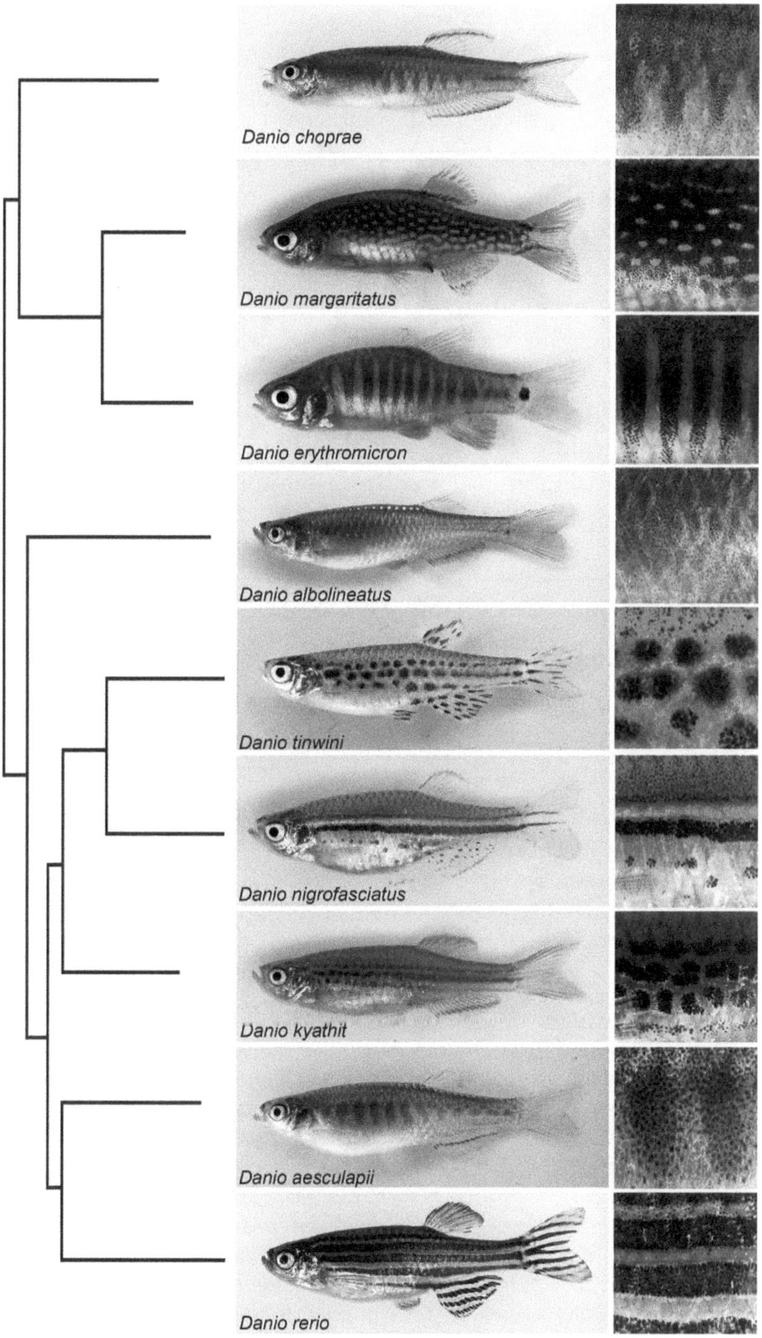

Figure 2 See legend on next page.

(Fig. 3A and B) (Hirata, Nakamura, Kanemaru, Shibata, & Kondo, 2003; Hirata, Nakamura, & Kondo, 2005). Development takes place outside the mother, allowing direct inspection at all stages. Most importantly, an increasing collection of spectacular adult-viable mutants with altered patterns has been accumulated (Table 1). These genes provide the basis of our understanding of the molecular mechanisms of color pattern formation. Novel approaches of long-term imaging of individual fish have been developed, which resulted in a detailed insight into the cellular and molecular background of stripe formation (Frohnhöfer et al., 2013; Singh & Nüsslein-Volhard, 2015; Singh, Schach, & Nüsslein-Volhard, 2014). The color patterns in related *Danio* species are amazingly different (Fig. 2) (McClure, 1999; Parichy, 2007; Parichy & Johnson, 2001); their variation offers a great opportunity to investigate the genetic and developmental basis of color pattern evolution in vertebrates, starting from detailed molecular and cellular investigations in one model species, *D. rerio*.

2. THE STRIPED COLOR PATTERN OF ZEBRAFISH

A common feature of the color patterns of basal vertebrates—fish, amphibians, and reptiles—is the layered organization of pigment cells in the skin, with xanthophores/erythrophores in the outermost layer primarily absorbing short-wavelength light, reflective iridophores/leucophores in the middle layer, and melanophores in the basal layer absorbing light across the full spectrum of wavelengths (Bagnara, Taylor, & Hadley, 1968; Grether, Kolluru, & Nersissian, 2004). Modifications of this standard color-forming unit lead to the particularly rich variation of color patterns in fishes. The striped pattern of the zebrafish body is composed of three distinct types of

Figure 2 Color pattern diversity within the genus *Danio*. The cladogram on the left shows the evolutionary relationship among different *Danio* species. Color patterns vary greatly even between closely related species; they range from almost uniformly distributed melanophores with a posterior red stripe on the trunk of *D. albolineatus*, over dark spots in *D. tinwini*, or stripes and spots on the trunk of *D. nigrofasciatus*, to vertical bars on the trunk of *Danio choprae* and *D. erythromicron* and light spots on a dark background in *D. margaritatus*. *D. rerio* (female) displays sharp boundaries between light and dark stripes. In *D. kyathit*, stripes are broken; *D. aesculapi*, the closest relative of *D. rerio*, has a pattern of irregular vertical bars dissolving into spots. Note the distinct motifs of fin patterning in several *Danio* species such as in *D. choprae* and *D. margaritatus*; the contiguous striped pattern of body and fins as seen in *D. rerio* is an exception. Rightmost panes show enlarged portion of the trunk skin. *Adapted from McCluskey and Postlethwait (2015).* (A color version of this figure is located at the end of this chapter.)

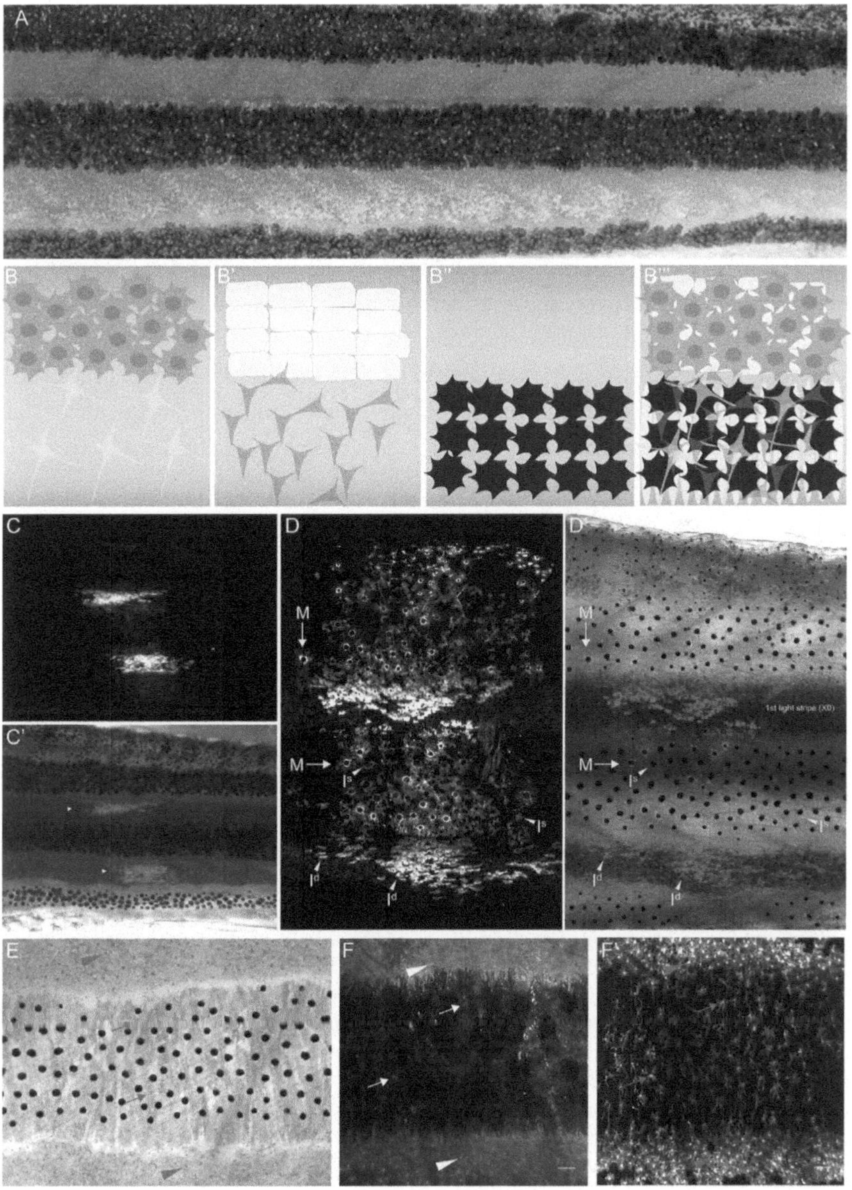

Figure 3 Layered organization of pigment cells in the trunk skin of zebrafish. (A) Close-up view of the lateral trunk of adult zebrafish displaying blue dark stripes and golden light stripes. (B–B‴) Schematic representation of the layered arrangement of pigment cells in the skin of the lateral trunk of zebrafish. The light stripe region consists of dense iridophores (white) covered by dense xanthophores (dark orange). In the stripe

(*Continued*)

pigment cells—xanthophores, iridophores, and melanophores; leucophores are only present in the dorsal and tail fins. Electron microscopic analysis and observation of the zebrafish pigmentation using fluorescent markers reveal distinct differences in the layered organization of these chromatophores between the light and the dark stripe regions (Fig. 3) (Hirata et al., 2003, 2005; Mahalwar, Walderich, Singh, & Nüsslein-Volhard, 2014; Singh et al., 2014). In the light stripe region, compact and bright yellow/orange xanthophores cover a densely packed monolayer of iridophores. Such a layered arrangement imparts a golden iridescent appearance to the light stripes. In the dark stripes, the pigment cells are layered in the following fashion from top to bottom: stellate and faint yellow xanthophores, loose and bluish iridophores, black melanophores, and shiny L-iridophores (Fig. 3B). This arrangement of pigment cells provides a glittering blue coloration with a yellow hue to the dark stripe regions. Whereas xanthophores and iridophores exist in different shapes depending on their position, melanophores of the stripe occur in only one compact shape such that they fill the space almost entirely leaving little room between the cells (Fig. 3C–F) (Frohnhöfer et al., 2013; Hirata et al., 2003, 2005; Mahalwar et al., 2014; Singh et al., 2014). The exact superposition of the differently shaped chromatophores is crucial for the contrast and brightness of the pattern. In the fins, the stripe organization is simpler and essentially composed of alternating melanophores and compact xanthophores. Iridophores do not participate in the fin stripes, but line the fin rays.

Figure 3—Cont'd region, scattered xanthophores (stellate, yellow) and bluish reflective iridophores are present above a compact layer of black melanophores. Beneath melanophores, there is a layer of L-iridophores (white elongated cells in B‴). (C–C′) Overview of a *Tg(sox10:ER^{T2}Cre)*-induced clone (induced on 5 dpf) shows clonally restricted organization of iridophores in the adult skin. Clonally related iridophores exhibit characteristic spreading along the dorsoventral axis. (D–D′) Confocal image of a clone showing detailed organization of labeled pigment cells in the skin with dense iridophores (arrowheads; I^{d}) in the light stripe region, loose iridophores (arrowheads; I^{s}) and melanophores (arrows) in the dark stripe region. (E) Yellow–orange xanthophore pigment in the dark stripe (arrows) and light stripe regions (arrowheads) of the adult skin. Melanophores, compacted due to treatment with epinephrine, display a regular spacing in the dark stripe region. In (F–F′), melanophores are expanded. Xanthophores appear compact and densely packed in the light stripes (arrowheads) and appear stellate and loose in the dark stripe region (arrows). (A color version of this figure is located at the end of this chapter.)

Table 1 Genes Affecting the Color Pattern of Zebrafish

Gene (Protein)	Function	Required for/in	Refs.
Genes affecting one chromatophore type			
albino (Slc45a2) *golden* (Slc24A5) *sparse* (Kit a) *sparse-like* (Kit ligand a) *nacre* (Mitfa)	Solute carrier Solute carrier Receptor Ligand Transcription factor	Melanophores	a, b, c, d, e
pfeffer (Csf1ra)	Receptor	Xanthophores	f, g
shady (Ltk) *rose* (Endothelin receptor Ba) *rose-like* (Endothelin 3) *karneol* (Ece 2) *transparent* (Mpv17)	Receptor Receptor Ligand Enzyme Mitochondrial protein	Iridophores	h, i, j, k, l, m
Genes affecting stripe width and integrity			
leopard (Connexin 41.8)	Gap junction protein	Melanophores, xanthophores	n, o
luchs (Connexin 39.4)	Gap junction protein	Melanophores, xanthophores	p
schachbrett (Tjp 1A, ZO-1)	Tight junction protein	Iridophores	q
seurat (Igsf11)	Cell adhesion protein	Melanophores, xanthophores	r
obelix (Kcnj13)	Inwardly rectifying potassium channel	Melanophores	n, s
dali (Tetraspanin3c)	Transmembrane protein	Unknown	t
asterix (Notch 2)	Receptor	Melanophores	f, m
Genes affecting surrounding tissue			
idefix (Spermidine synthase)	Enzyme	Unknown	m
mau (Aqp3a)	Water channel	Unknown	u
bonaparte (Bnc2)	Zinc finger protein	Unknown	v, w
choker (Meox)	Transcription factor	Myotome	x, y
picasso/hypersensitive (Erbb3b)	Receptor	Dorsal root ganglia, PNS	z

a, Parichy, Rawls, Pratt, Whitfield, and Johnson (1999); b, Lister, Robertson, Lepage, Johnson, and Raible (1999); c, Lamason et al. (2005); d, Dooley, Mongera, Walderich, and Nüsslein-Volhard (2013); e, Dooley, Schwarz, et al. (2013); f, Haffter et al. (1996); g, Parichy, Ransom, Paw, Zon, and Johnson (2000); h, Parichy, Mellgren, et al. (2000); i, Lopes et al. (2008); j, Frohnhöfer, Krauss, Maischein, and Nüsslein-Volhard (2013); k, Krauss, Astrinidis, Fröhnhofer, Walderich, and Nüsslein-Volhard (2013); l, Krauss et al. (2014); m, Frohnhöfer et al. (submitted); n, Maderspacher and Nüsslein-Volhard (2003); o, Watanabe et al. (2006); p, Irion, Frohnhöfer, et al. (2014); q, Fadeev, Krauss, Fröhnhofer, Irion, and Nüsslein-Volhard (2015); r, Eom et al. (2012); s, Iwashita et al. (2006); t, Inoue, Kondo, Parichy, and Watanabe (2014); u, Eskova et al. (in preparation); v, Lang, Patterson, Gordon, Johnson, and Parichy (2009); w, Patterson and Parichy (2013); x, van Eeden et al. (1996); y, Nguyen et al. (2014); and z, Budi, Patterson, and Parichy (2008).

3. STRIPE FORMATION, A SELF-ORGANIZING PROCESS INVOLVING THE INTERACTION OF ALL THREE CELL TYPES

The precise longitudinal orientation of the stripes in adult zebrafish depends on the presence of the horizontal myoseptum serving as a morphological prepattern. In *choker* mutants, which lack the horizontal myoseptum, a meandering pattern of parallel dark and light stripes with random orientations but normal width and composition is formed (Fig. 4) (Frohnhöfer et al., 2013). This indicates that the horizontal myoseptum orients the stripes, whereas stripe formation is a self-organizing process depending on interactions between the pigment cells.

Based on the assumption that iridophores are not necessary for stripe formation, as they are not required in the striped fins, and present in both the dark and light stripes of the body, Turing-type models have been proposed describing stripe formation based on interactions between

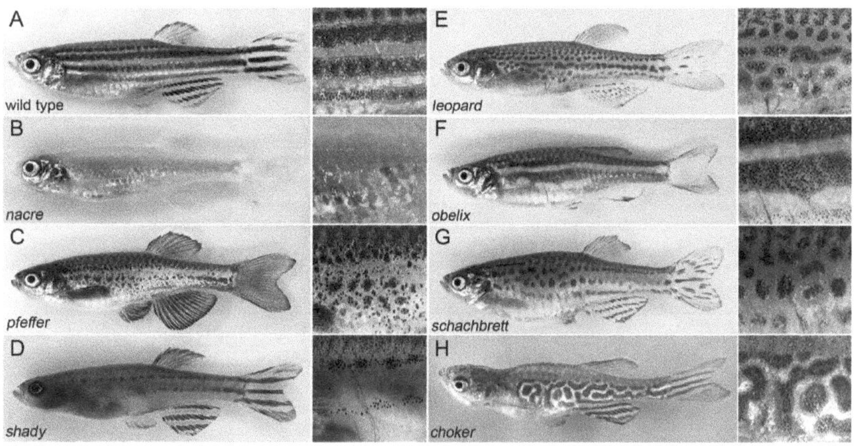

Figure 4 Selected adult-viable pigmentation mutants in zebrafish. (A) Wild-type adult zebrafish male. (B) *nacre* lacks all the neural crest-derived melanophores. (C) *pfeffer* lacks xanthophores, melanophores are reduced in number, and residual dark stripes are discontinuous, displaying a pepper and salt pattern. (D) *shady* mutants lack iridophores and display only two rudimentary melanophore stripes that are broken into spots. Fins are striped. (E) In *leopard*, the stripes break up into spots. (F) *obelix* displays broader and fewer dark stripes and (G) *schachbrett* shows disrupted stripes and spots. (H) In *choker*, which lacks the horizontal myoseptum, a meandering pattern with stripes of normal width is formed. (A color version of this figure is located at the end of this chapter.)

xanthophores and melanophores only (Nakamasu, Takahashi, Kanbe, & Kondo, 2009; Yamaguchi, Yoshimoto, & Kondo, 2007; Yamanaka & Kondo, 2014). This view has been challenged recently by the finding that for the formation of trunk stripes iridophores are absolutely necessary: in mutants lacking iridophores, no dark stripes beyond rudimentary first two stripes are formed in the body, whereas the fins are normally striped (Fig. 4, *shady*) (Frohnhöfer et al., 2013; Haffter et al., 1996; Krauss et al., 2013, 2014; Lopes et al., 2008). In mutants lacking melanophores or xanthophores, residual stripes between iridophores and the remaining cell type are formed in the body, but no stripes are formed in the fins (Fig. 4, *nacre, pfeffer*) (Haffter et al., 1996; Lister et al., 1999; Odenthal et al., 1996). Several mutants display different phenotypes in body and fin stripes. This suggests that models for stripe formation in the body have to be modified substantially to incorporate the recent experimental and genetic data, in particular the dominant role of iridophores (Singh, Fröhnhofer, Irion, & Nüsslein-Volhard, 2015).

Analyses of mutants missing one single pigment cell type have provided crucial insights into pigment cell interdependency and interactions for their development, survival, and stripe pattern formation. Animals depleted in one pigment cell type, such as melanophores in *nacre*, xanthophores in *pfeffer*, or iridophores in *shady*, *rose*, or *transparent*, initiate stripe morphogenesis but the patterns formed by the two remaining cell types are highly abnormal (Fig. 4). The analysis of chimeric animals obtained by blastomere transplantations indicated a cell-autonomous requirement for all these genes, which means that the abnormal behavior of the remaining two cell types observed in the mutants is indirectly caused by the absence of a given cell type. The analysis of double mutants lacking two types of pigment cells revealed that any single chromatophore alone cannot form a pattern; the residual cells evenly cover the zebrafish trunk (Frohnhöfer et al., 2013). This shows that, on its own, each pigment cell type has a tendency to uniformly cover the skin and that the stable stripe pattern emerges due to restrictions imposed by interactions within and between the three pigment cell types; all three chromatophore types have to communicate and interact to form the final striped pattern. Dense iridophores attract xanthophores, locally suppress melanophore survival but promote it at a distance. Compact xanthophores and melanophores exhibit mutual repulsion at a short range, but they are required for melanophore survival at a distance. These observations point to distinct roles of the three pigment cell types in the process of pattern formation. For a thorough understanding of the mechanisms, it is required to investigate the origin of the pigment cells and the process of stripe formation during metamorphosis *in vivo*.

4. DEVELOPMENTAL ORIGIN OF THE THREE PIGMENT CELL TYPES

The larval pigment cells develop directly from the neural crest, delaminate during early development, and migrate to the dorsal, lateral, and ventral aspect of the embryo to form a simple striped larval pattern based on morphological landmarks (Eisen & Weston, 1993; Kelsh et al., 1996). The striped pattern of the adult zebrafish arises considerably later during a metamorphic period that begins about 3 weeks postfertilization and lasts for about 1 month. During this time period, the shape of the fish undergoes a substantial reorganization with the appearance of scales and the adult set of fins replacing the larval fin fold. Stripes sequentially appear in the skin, starting with the first light stripe along the horizontal myoseptum. Lineage tracing has revealed that the adult xanthophores descend directly from larval xanthophores, which evenly cover the body and start to proliferate just before the onset of metamorphosis (Mahalwar et al., 2014; McMenamin et al., 2014). In contrast, adult iridophores and melanophores develop from segmentally reiterated, peripheral nerve-associated multipotent stem cells that have been set apart during early development (Fig. 5A–C) (Budi, Patterson, & Parichy, 2011; Dooley, Mongera, et al., 2013; Singh et al., 2014). Sox10-Cre-induced lineage labeling revealed that stem cells may give rise to glia, peripheral neurons, and all three kinds of chromatophores, predominantly iridophores (Singh et al., 2014). Iridophores emerge along the horizontal myoseptum, proliferate, and spread in the skin (Fig. 5C), whereas melanophore progenitors divide while they migrate along peripheral nerves (Fig. 5B). The melanophores hardly proliferate or migrate after they have reached the skin.

5. FORMATION OF A SERIES OF LIGHT AND DARK STRIPES BY PATTERNED AGGREGATION OF IRIDOPHORES

Stripe morphogenesis begins at the onset of metamorphosis. Thyroid hormone acts as a trigger for metamorphosis and also regulates xanthophore proliferation and differentiation (McMenamin et al., 2014). *In vivo* imaging over extended time periods revealed a leading role of iridophores in stripe

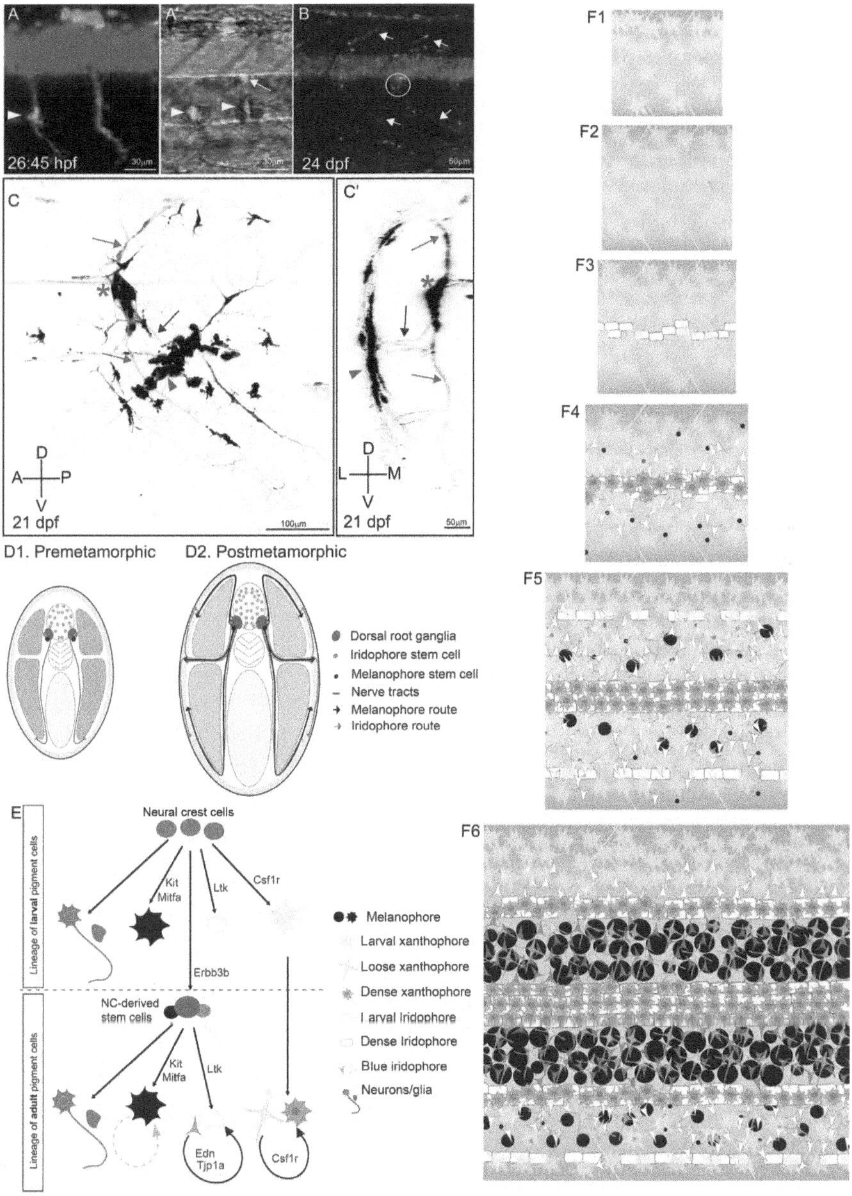

Figure 5 Pigment cell lineage, migration, and stripe pattern formation. (A–A′) During embryogenesis, melanoblasts (Mitfa-positive, green cells) migrate along the primary motor neuron (magenta) and differentiate into larval melanophores (arrowheads in A′). Arrow in (A′) indicates a Mitfa-positive cell that remains at a ventral position of the neural tube, the melanophore stem cell. (B) During metamorphosis, Mitfa-positive

(Continued)

formation. Iridophore progenitors migrate along peripheral nerves through the horizontal myoseptum into the skin where they continue to proliferate and spread, leading to the formation of a contiguous sheet of dense iridophores and the formation of the first light stripe (Singh et al., 2014). Interestingly, there is no clonal restriction of iridophores between segments,

Figure 5—Cont'd melanophore progenitor cells are seen at the dorsal root ganglion (encircled) and along the spinal nerves extending dorsally and ventrally to innervate the lateral skin (arrows). (C–C′) A clone of iridophores (arrowhead) associated with the dorsal root ganglion (asterisk). Blue arrows in (C–C′) indicate the neurons that exit through the horizontal myoseptum; red arrows indicate the neuronal tracts associated with the dorsal and ventral myotomes and the myosepta. (C′) The clone in (C) turned by 90° to show a cross section through the trunk region. Skin is to the left. Abbreviations: D: dorsal, V: ventral, A: anterior, P: posterior, L: lateral, M: medial, hpf: hours postfertilization, and dpf: days postfertilization. (D1–D2) Schematic showing the route followed by melanophore progenitors along the peripheral neurons innervating the skin, whereas iridophores migrate through the horizontal myoseptum and disperse dorsoventrally in the skin. (E) Lineage and the postembryonic source of adult pigment cells. Neural crest cells give rise to several larval cell types including neural tissue and develop directly into the three kinds of larval pigment cells in zebrafish. Adult melanophores, iridophores, and neural tissue are derived from postembryonic stem cells, which are located on the segmentally iterated dorsal root ganglia. Larval xanthophores persist and proliferate at the onset of metamorphosis to generate adult xanthophores. Metamorphic iridophores and xanthophores undergo local proliferation in the skin. In rare cases, pigmented melanophores are seen to divide in the skin (faint circular arrow). A fraction of xanthophores may arise from mixed iridophore clones (faint arrow). Genes required for specific lineages are mentioned in blue along the arrows. (F1–F6) Schematics showing stripe pattern morphogenesis during the transition from the larval stage to the adult stage. (F1) Larval xanthophores (5 dpf) (F2) begin to proliferate at the onset of metamorphosis (2 weeks postfertilization; wpf). (F3) Iridophores appear in the skin through the horizontal myoseptum and organize the first light stripe (~3 wpf). (F4) Melanophores develop from the melanoblasts, which enter the skin through the dorsal and the ventral myotomes (~3–4 wpf). Over the dense iridophores, xanthophores become compact and acquire yellow–orange pigment. Loose iridophores appear at the borders of the dense iridophore region and begin to disperse along the dorsoventral axis. (F5) Melanophores grow in size in the stripe region to form contiguous stripes. Loose iridophores aggregate to organize new light stripes at a distance from the first light stripe (~4–5 wpf). (F6) Xanthophores become stellate over the melanophores and compact in the light stripe regions. Loose iridophores on top of melanophores appear bluish and reflective giving blue color to the dark stripes. Light stripes appear golden due to the presence of silvery dense iridophores covered by yellow–orange xanthophores. (A color version of this figure is located at the end of this chapter.)

and iridophores derived from a single stem cell distribute over two to four segments along the anterior–posterior axis (Fig. 3C and D) (Singh et al., 2014). After the formation of a first light stripe, iridophores continue to proliferate and spread as blue iridophores bidirectionally in the skin toward the dorsal and ventral aspect of the juvenile fish. They form a loose net of cells covering the flank of the fish, in which additional light stripes appear by patterned aggregation of iridophores into the dense form at a given distance from the first light stripe (Figs. 3C and D and 5F1–F6). By cell division, migration, and cell shape transition, iridophores form new light stripes along the dorsoventral axis, prefiguring the region in which dark stripes are formed after the addition of each new light stripe. Consistent with this, mutants lacking iridophores do not add new light or dark stripes beyond a basic pattern of one light stripe (composed of xanthophores only) and two rudimentary dark stripes broken into spots (Frohnhöfer et al., 2013; Krauss et al., 2013, 2014).

At the cellular level, the process of iridophore shape transition and migration appears qualitatively similar to epithelial-to-mesenchymal transitions (EMT) and mesenchymal-to-epithelial transitions (MET). EMT/MET are important cellular shape changes that underlie several events including migration of neural crest cells in the embryo or cancer progression (Kerosuo & Bronner-Fraser, 2012; Polyak & Weinberg, 2009). While it remains to be seen if there are similarities at the levels of gene regulation and cytoskeletal remodeling between EMT/MET and the iridophore behavior, it is interesting to find parallels in terms of cellular behaviors. The recent cloning and characterization of *schachbrett* (German word for checkerboard), a mutant displaying a spotted phenotype (Fig. 4), provide a first clue for the elements of the cytoskeletal machinery that may underlie cell shape transition of iridophores during pattern formation (Fadeev et al., 2015). The *schachbrett* locus encodes Tight junction protein 1a (Tjp1a), a zebrafish homolog of mammalian ZO-1. Tjp1a is specifically expressed in the densely packed iridophores of the light stripe and its expression is lost in loose migratory iridophores. The presence of Tjp1a allows the cells to undergo shape changes from a dense to a loose form during stripe formation. The failure of this change to occur leads to the invasion of iridophores in their dense shape into the dark stripe regions resulting in a spotted phenotype.

While iridophores lay out the first light stripe, melanoblasts, the unpigmented precursors of melanophores, reach the skin in the presumptive dark stripe regions in close association with the peripheral nerves

(Fig. 5B and D) (Budi et al., 2011; Dooley, Mongera, et al., 2013; Singh et al., 2014). In contrast to iridophores, melanophores do not proliferate in the skin but dramatically expand in size to form a contiguous sheet of cells, thus forming the dark stripe (Figs. 3E and F and 5F) (Singh et al., 2014). Hence, the dark and light stripes are formed by rather different cellular behaviors. Xanthophores covering the skin as a coherent sheet start to proliferate at the onset of metamorphosis and undergo marked local reorganization acquiring a flat shape with small cytoplasmic protrusions over the dense iridophores of the light stripes, and a stellate shape with long cytoplasmic protrusions and faint pigmentation upon encountering melanophores (Fig. 3G) (Mahalwar et al., 2014). This leads to the exact superposition of the diverse cell types and shapes required for the proper coloration of the pattern. On their own, iridophores cover the entire body uniformly in the dense form (in *nacre; pfeffer* double mutants), indicating that interactions with xanthophores and melanophores are necessary for the appropriate shape changes during stripe formation (Frohnhöfer et al., 2013; Irion, Frohnhöfer, et al., 2014; Maderspacher & Nüsslein-Volhard, 2003; Nakamasu et al., 2009; Patterson & Parichy, 2013; Yamaguchi et al., 2007). Only iridophores exhibit long distance migration in the skin along the dorsoventral axis (Singh et al., 2014), suggesting that these cells in response to the other two pigment cell types may provide the cues to form new light and dark stripes.

Our present view is that the transition between the loose and dense forms of iridophores prefigures the formation of the striped pattern, which is further refined by short- and long-range interactions between chromatophores across the three layers. We propose that pattern variations among closely related species are caused by changes in the spatial parameters regulating the transitions in chromatophore organization.

6. GENETIC AND MOLECULAR REGULATION OF COLOR PATTERN FORMATION

At the cellular level, the process of stripe formation involves several behaviors such as cell type specification, proliferation and spreading, long-/short-range migration of pigment cells, cell shape transition, and acquisition of appropriate pigmentation. Mutant analysis helps to dissect these processes and provides an insight into the molecular basis of cellular interactions (Table 1).

6.1 Regulation of Pigment Cell Proliferation and Maintenance by Distinct Signaling Systems

The analysis of mutants that specifically lack one type of chromatophore has revealed that different receptor–ligand pairs regulate specification, maintenance, and proliferation of each of the pigment cell types. Most melanophores depend on Kit-signaling; mutations in the receptor (the orthologue of c-Kit, encoded by the gene *sparse*) or the ligand (Kit ligand a, encoded by *sparse-like*) lead to a severe reduction in the number of melanophores (Dooley, Mongera, et al., 2013; Parichy et al., 1999). Leukocyte tyrosine kinase (*shady*) is essential for the specification of iridophores (Lopes et al., 2008). Endothelin signaling is required for dense iridophore development; mutations in endothelin receptor b1 (*rose*) (Parichy, Mellgren, et al., 2000), in endothelin-converting enzyme 2 (*karneol*) (Krauss et al., 2014), or in endothelin 3b (U. I., unpublished) all lead to an absence of dense iridophores in the mutant animals, while loose iridophores are present. Xanthophores depend on colony-stimulating factor signaling; mutations in *pfeffer* (Haffter et al., 1996), which codes for colony-stimulating factor 1 receptor A, lead to the absence of xanthophores (Parichy, Ransom, et al., 2000). Melanophores, in addition, receive unknown cues for maintenance and survival from dense iridophores and xanthophores, whereas the other two cell types develop in a more autonomous manner (Frohnhöfer et al., 2013; Parichy, Ransom, et al., 2000; Walderich, Singh, Mahalwar, & Nüsslein-Volhard, submitted).

Evidence from chimeric animals, where cells of different genetic compositions are juxtaposed, suggests that homotypic cell–cell interactions (interactions between pigment cells of the same kind) regulate pigment cell proliferation and dispersal. Clones of pigment cells that develop in hosts lacking this cell type proliferate more and disperse further compared to similar clones in chimeric animals where the endogenous cells are also present (Walderich et al., submitted). Homotypic competition determines the number, direction of migration, and the individual spacing of pigment cells; it is responsible for the predominant dorsoventral spread of pigment cells of all three kinds during metamorphosis observed in cell lineage experiments (Singh et al., 2014; Walderich et al., submitted). For example, post-embryonic iridophore clones generally contribute to all stripes along the dorsoventral axis, but only to about two segments along the anteroposterior axis (Fig. 3C and D).

Pigment cell shape transitions involved in the formation of the striped organization are regulated by more complex heterotypic cell–cell interactions (interactions between pigment cells of different kind) (Fadeev et al., 2015; Irion, Frohnhöfer, et al., 2014). To understand the molecular nature of these interactions between the chromatophores, it is helpful to analyze mutants, in which all three cell types are present, but nevertheless stripe formation is not normal. These mutants affect stripe integrity or stripe width.

6.2 Formation of Contiguous Boundaries Between the Light and the Dark Stripes

Stripe integrity with straight boundaries between the dark and light stripes is affected in *leopard, luchs, seurat*, and *schachbrett* mutants. All of them are likely to affect contact-dependent cell–cell interactions leading to an expansion of the light stripe regions and an interruption and suppression of the dark stripes. *leopard* and *luchs* code for two different connexins, Cx41.8 and Cx39.4, respectively (Irion, Frohnhöfer, et al., 2014; Watanabe et al., 2006). In both cases, dominant and recessive alleles lead to spotted patterns where light stripe regions invade the dark stripes, whereas the pigment cell composition and layering in the dark and light regions is maintained. Connexins are the subunits of gap junctions, intercellular channels that form by the docking of two hemi-channels from neighboring cells (Kumar & Gilula, 1996; Saez, Berthoud, Branes, Martinez, & Beyer, 2003; Simon & Goodenough, 1998). Both connexins, encoded by *leopard* and *luchs*, are required in melanophores and xanthophores, but not in iridophores (Irion, Frohnhöfer, et al., 2014; Maderspacher & Nüsslein-Volhard, 2003). In double mutants for both connexins, the pattern is completely lost with a reduced number of melanophores appearing as individual cells on a homogeneous background of dense iridophores covered by compact xanthophores. It is suggested that the two connexins form heteromeric gap junctions in and between melanophores and xanthophores resulting in cell–cell communication and ultimately in the instruction of iridophores to undergo a transition from the dense into the loose form characteristic for the dark stripe (Irion, Frohnhöfer, et al., 2014). The mechanism of communication between melanophores and xanthophores on the one hand and iridophores on the other is still unknown. Genetic evidence suggests that in the absence of one of the two connexins homomeric channels can form that still retain some functionality and lead to the formation of spots. Interestingly, in *luchs* the fins are normally striped, whereas in *leopard* no stripes are formed in the fins supporting the notion of a different requirement for fin and body stripes. The short (23 amino acid residues) N-terminal cytoplasmic tail of Cx41.8 is essential for its function. It contains a short motif (ExxxE) predicted to lead to polyamine-sensitivity, suggesting that rectification properties of the gap junctions are important for their function (Watanabe, Watanabe, & Kondo, 2012). Several connexins are known to interact with ZO-1 (Giepmans & Moolenaar, 1998; Nielsen et al., 2003), encoded by the gene affected in *schachbrett* mutants (see above). Because *schachbrett* function is specifically required in iridophores, it has been suggested that different, as yet unknown, connexins might be expressed in these cells and involved in the patterning process (Fadeev et al., 2015).

In *seurat* mutants, Igsf11 (immunoglobulin superfamily member 11), a classical cell adhesion molecule of the immune globulin superfamily, is affected (Eom et al., 2012). Igsf11, which is specifically required in melanophores, mediates adhesive interactions between cells and promotes migration and survival of melanophores. It was suggested that differential adhesive properties of the pigment cells might be required for the formation of the pattern.

6.3 Determination of the Domain of the Light and the Dark Stripe Regions

In *obelix* (aka *jaguar*) mutants, the number of dark stripes is reduced, they are broader, and xanthophores do not adapt their appearance in the presence of melanophores (Haffter et al., 1996). The function of *obelix*, which codes for an inwardly rectifying potassium channel, Kir7.1/Kcnj13, is exclusively required in melanophores (Iwashita et al., 2006; Maderspacher & Nüsslein-Volhard, 2003). It is assumed that dysfunction of the channel leads to a constant depolarization of the plasma membrane in the mutant melanophores resulting in defects in the repulsive signaling between xanthophores and melanophores (Inaba, Yamanaka, & Kondo, 2012). Experiments with isolated cells *in vitro* suggest that *obelix* mutant melanophores are unable to respond appropriately to contact by xanthophores, which in wild-type cells lead to chase-and-run movements (Yamanaka & Kondo, 2014). The dominant phenotype of *obelix* mutations may be caused by haploinsufficiency for the gene (Maderspacher & Nüsslein-Volhard, 2003). Double mutants of *obelix,* and *leopard* or *luchs* show a strong enhancement of the phenotype with a complete loss of the pattern, arguing for a possible link between the two pathways (Irion, Frohnhöfer, et al., 2014; Maderspacher & Nüsslein-Volhard, 2003).

Stripe width is also affected by manipulations of the Notch signaling pathway; melanophore-specific overexpression of the ligand Delta C or the intracellular domain of Notch 1A results in fewer and wider dark stripes

(Hamada et al., 2014). Delta/Notch signaling promotes melanophore survival and it has been shown that melanophores, expressing the Notch 1A and Notch 2 receptors, extend long projections toward xanthophores, which in turn express the ligands Delta C and Delta-like 4. Our own observations show that a specific point mutation in Notch 2, likely leading to the overactivation of the signaling pathway, is causing the dominant *asterix* phenotype, which is also characterized by fewer and wider dark stripes (Haffter et al., 1996).

These genetic and molecular data suggest that the determination of the stripe width, the formation of the straight boundaries between the light and dark stripes, and the patterned aggregation of iridophores, which defines the stripe areas, are all mediated by direct cell contacts among the chromatophores. This mode of pattern formation, occurring over the three distinct cell layers, is very special and quite different from other patterning processes where mainly secreted extracellular signaling molecules and their membrane-bound receptors are involved. However, not all pigment cells in the zebrafish occupy regions that are striped and it is evident that the tissue environment provides additional cues for the behavior of chromatophores in the distinct body regions.

7. THE ROLE OF TISSUE ENVIRONMENT IN REGULATING COLOR PATTERN FORMATION

Pigment cells are present in several regions of the body including the dorsum, trunk hypodermis, head, fins, intestine, and abdominal wall (Hirata et al., 2005), as well as on the epidermis of the scales. However, stripes are only formed in the trunk region, and in the anal and caudal fins, while in other regions the cell types appear singly or mixed in a more or less random pattern. This suggests a role for the tissue environment in providing a context permissible for stripe morphogenesis. So far, we have no clue on the mechanisms determining the different body regions imposing different interaction rules. Recently, based on expression pattern and *gain-of-function* analysis, it has been suggested that the dorsoventral countershading, a pattern of dark dorsum contrasting with a light ventrum that is observed in many vertebrates including zebrafish, is a process that utilizes agouti signaling and that is largely independent of stripe morphogenesis (Ceinos, Guillot, Kelsh, Cerda-Reverter, & Rotllant, 2015).

Mutations in several genes have been identified that are not required in the pigment cells themselves but in other cells to allow the generation of the striped pattern. Some of them encode extracellular ligands for the receptors expressed in the chromatophores, or are responsible for the processing of these ligands (Table 1). These secreted signaling molecules generally affect the respective pigment cells in the entire body, not just the striped regions. Owing to generally low abundances, their distributions are largely unknown.

In other cases, the mechanisms by which the surrounding tissues affect stripe morphogenesis are more complex. In *bonaparte*, the affected gene encodes *basonuclin2*, a nuclear localized zinc finger protein of unknown function (Lang et al., 2009). In the *bonaparte* mutants, the numbers of all the three pigment cell types are highly reduced, and the phenotype is presumably caused by compromised interactions between the chromatophores and the tissue environment (Lang et al., 2009; Patterson & Parichy, 2013). Mutations in the *mau* gene are dominant and cause a spotted pattern with irregular stripe boundaries. *mau* encodes Aquaporin3a, a plasma membrane water channel (Eskova et al., in preparation). Chimeric analyses show that this channel is not required in any of the three pigment cell types but affects an as yet unidentified surrounding tissue. It is not yet clear how a water channel may influence pigment patterning and in which tissue the gene is required. A mutant with fewer dark stripes, but wider light stripes, is *idefix*. The affected gene encodes the enzyme spermidine synthase (U. I., in preparation), which is responsible for the biosynthesis of the polyamine spermidine from putrescine and decarboxylated *S*-adenosyl-methionine. Polyamines, spermidine and its derivative spermine, are important regulators for the gating properties of inwardly rectifying potassium channels (Ficker, Taglialatela, Wible, Henley, & Brown, 1994) and gap junctions (Musa & Veenstra, 2003), underscoring the role of these molecules mediating direct contacts between pigment cells in the patterning process discussed above. We hypothesize that these components modulate the environment such as to allow appropriate function of these channels between pigment cells in all three layers.

An instructive role for the pigment cell environment in determining pigmentation patterns has been demonstrated in mice: dedicated pigment recipient epithelial cells recruit melanocytes to their position in the skin and induce melanin transfer (Weiner et al., 2007). In this context, the mammalian skin has been compared to a coloring book in which the blueprint is created by the pigment recipient cells and the melanocytes fill in the color by transferring melanin (Weiner, Fu, Chirico, & Brissette, 2014). Thus, the tissue environment has an important role in setting up the cues for proper development and organization of the pigment cells in the skin.

The anatomy clearly has an influence on the final stripe pattern on the trunk as exemplified by the analysis of the zebrafish *choker* mutant that lacks the horizontal myoseptum (Frohnhöfer et al., 2013). In fins, which lack muscles that underlie the hypodermis, the stripes have a different pigment cell composition and arrangement, suggesting the mode of pattern formation differs from that in the body. In the epidermis of the scale margins, all three pigment cells are mixing, and the density decreases from dorsal scales to ventral scales, which are devoid of pigment cells. In conclusion, in different body regions the rules of interaction between pigment cells vary, but the molecular basis determining the local differences is largely unknown.

8. EVOLUTION OF COLOR PATTERNS

The color patterns in closely related *Danio* species are amazingly different (Fig. 2) (McClure, 1999; Parichy, 2006; Parichy & Johnson, 2001). The patterns vary in stripe number, width, and orientation, as well as in stripe integrity, ultimately leading to spotted patterns (Fig. 2). In *Danio albolineatus*, almost no pattern is formed, only a short posterior remnant of the central light stripe persists. Strikingly, only in *D. rerio* the stripes in the body and fins are contiguous and the pattern appears periodic along the dorsoventral axis. Other species display repeated vertical bars along the anterior–posterior axis, a frequent motif in fish patterning (see also Fig. 1). In some species, the set of dorsal, anal, and tail fins displays a common theme, quite distinct from that of the body. This pattern variation offers a great opportunity to investigate the genetic and developmental basis of color pattern evolution in vertebrates, starting from detailed molecular and cellular investigations in one model species, *D. rerio*. Viable hybrids between zebrafish and a number of other *Danio* species have been generated; they all show a striped pattern, demonstrating that the zebrafish pattern forming mechanism is dominant (Fig. 6) (Parichy & Johnson, 2001). The hybrids are sterile, thus precluding most genetic experiments; however, exciting technical developments of the recent years, especially next-generation sequencing technologies and the novel possibilities of genome editing with the CRISPR/Cas9 system (Hwang et al., 2013; Irion, Krauss, & Nüsslein-Volhard, 2014; Jinek et al., 2012), allow for the first time to easily expand from model organisms into other species and directly test the function of genes by targeted knockouts. We have already generated mutants in *D. albolineatus* lacking melanophores, xanthophores, or iridophores, by

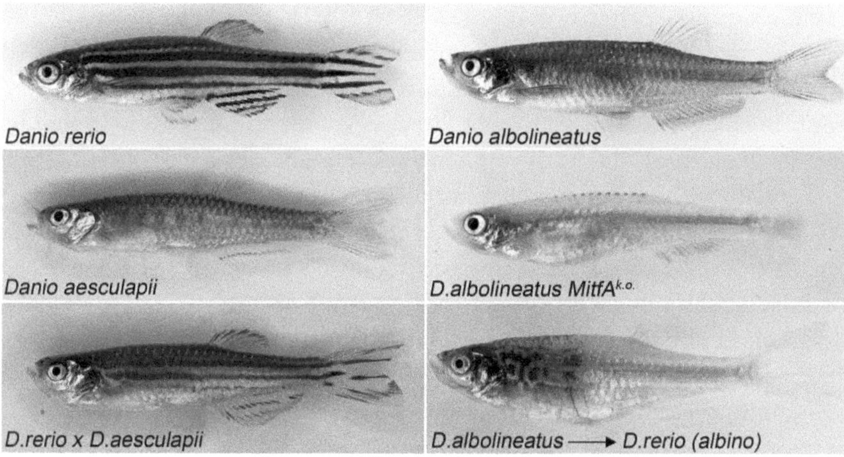

Figure 6 Interspecies hybrid and chimera. Hybrids between *Danio aesculapii* and *Danio rerio* (zebrafish) show a striped pattern very similar to zebrafish; the fins are partially striped. Knockout mutations in *MitfA* in *Danio albolineatus* lead to a loss of melanophores; the patterning, however, is not affected. In chimeric animals obtained by blastomere transplantation, the melanophores derived from *D. albolineatus* can contribute to the stripes of *D. rerio*. (A color version of this figure is located at the end of this chapter.)

knockouts in the genes coding for MitfA, Csf1r, and Ltk (U. I., unpublished). This demonstrates that the signaling pathways for cell specification, proliferation, and maintenance are conserved between zebrafish and *D. albolineatus*. To identify conserved cellular behaviors, heterologous cell transplantations between the different species can be employed. Transplantations of wild-type blastomeres from *D. albolineatus* into *albino* mutant zebrafish embryos result in chimeras in which the (pigmented) *D. albolineatus* donor melanophores integrate into the (unpigmented) stripes of the host animal (Fig. 6), demonstrating that the melanophores from *D. albolineatus* can correctly respond to the patterning cues present in zebrafish, although they would not produce a striped pattern in their normal environment. The next challenge is to identify genes crucial for the pattering process and exchange alleles of candidates between related species using the CRISPR/Cas9 system.

Understanding color pattern formation in birds and mammals presently rests almost entirely on theoretical considerations because their inclusion into eggs or the womb of a gravid female provides very limited imaging possibilities and makes them hardly accessible to experimental manipulations. Although in mammals there is only one pigment cell type, many mammals display patterns composed of two or more colors distributed in vertical stripes or spots. This suggests that there are different types of melanocytes forming the patterns, reminiscent to the most frequent pattern motifs in fish (Watanabe & Kondo, 2012). As fish, amphibians, and reptiles have at least three different pigment cell types, one important question concerns the evolutionary relationship between mammalian melanocytes and the chromatophores of the more basal vertebrates. Iridophores in zebrafish depend on endothelin signaling, as do melanocytes in mammals (Baynash et al., 1994; Frohnhöfer et al., 2013; Krauss et al., 2014; Parichy, Mellgren, et al., 2000). It may not simply have been a case of loss of pigment cell types other than melanophores along the avian/mammalian lineages, but it is conceivable that other pigment cell types acquired the function of melanin production. Stripes in tigers, cats, and zebras may originate from stem cells distributing the melanocytes along the segmentally iterated spinal nerves (Adameyko et al., 2009), possibly explaining their vertical orientation along the dorsoventral body axis, which we also observe in the spreading of the melanophores and iridophores in zebrafish. Although at the level of pigment cells there are profound differences, genes that regulate pigment cell development and color pattern formation are highly conserved between fish, birds, and mammals. Therefore, the studies of zebrafish pigmentation will lay the foundation to understand not only the genetic basis of color pattern variation in fishes but also the evolution of color patterns in other vertebrates.

REFERENCES

Adameyko, I., Lallemend, F., Aquino, J. B., Pereira, J. A., Topilko, P., Muller, T., et al. (2009). Schwann cell precursors from nerve innervation are a cellular origin of melanocytes in skin. *Cell*, *139*, 366–379.

Bagnara, J. T., & Matsumoto, J. (2007). Comparative anatomy and physiology of pigment cells in nonmammalian tissues. In J. J. Nordlund, R. E. Boissy, V. J. Hearing, R. A. King, W. S. Oetting, & J.-P. Ortonne (Eds.), *The pigmentary system: Physiology and pathophysiology* (pp. 11–59). Oxford, UK: Blackwell Publishing.

Bagnara, J. T., Taylor, J. D., & Hadley, M. E. (1968). The dermal chromatophore unit. *The Journal of Cell Biology*, *38*, 67–79.

Baynash, A. G., Hosoda, K., Giaid, A., Richardson, J. A., Emoto, N., Hammer, R. E., et al. (1994). Interaction of endothelin-3 with endothelin-B receptor is essential for development of epidermal melanocytes and enteric neurons. *Cell*, *79*, 1277–1285.

Brenner, M., & Hearing, V. J. (2008). The protective role of melanin against UV damage in human skin. *Photochemistry and Photobiology*, *84*, 539–549.

Budi, E. H., Patterson, L. B., & Parichy, D. M. (2008). Embryonic requirements for ErbB signaling in neural crest development and adult pigment pattern formation. *Development*, *135*, 2603–2614.

Budi, E. H., Patterson, L. B., & Parichy, D. M. (2011). Post-embryonic nerve-associated precursors to adult pigment cells: Genetic requirements and dynamics of morphogenesis and differentiation. *PLoS Genetics*, 7, e1002044.

Caro, T. (2013). The colours of extant mammals. *Seminars in Cell & Developmental Biology*, *24*, 542–552.

Ceinos, R. M., Guillot, R., Kelsh, R. N., Cerda-Reverter, J. M., & Rotllant, J. (2015). Pigment patterns in adult fish result from superimposition of two largely independent pigmentation mechanisms. *Pigment Cell & Melanoma Research*, *28*, 196–209.

Darwin, C. (1871). *The descent of man, and selection in relation to sex*. London: John Murray.

Dooley, C. M., Mongera, A., Walderich, B., & Nüsslein-Volhard, C. (2013). On the embryonic origin of adult melanophores: The role of ErbB and kit signalling in establishing melanophore stem cells in zebrafish. *Development*, *140*, 1003–1013.

Dooley, C. M., Schwarz, H., Mueller, K. P., Mongera, A., Konantz, M., Neuhauss, S. C., et al. (2013). Slc45a2 and V-ATPase are regulators of melanosomal pH homeostasis in zebrafish, providing a mechanism for human pigment evolution and disease. *Pigment Cell & Melanoma Research*, *26*, 205–217.

Eisen, J. S., & Weston, J. A. (1993). Development of the neural crest in the zebrafish. *Developmental Biology*, *159*, 50–59.

Eom, D. S., Inoue, S., Patterson, L. B., Gordon, T. N., Slingwine, R., Kondo, S., et al. (2012). Melanophore migration and survival during zebrafish adult pigment stripe development require the immunoglobulin superfamily adhesion molecule Igsf11. *PLoS Genetics*, *8*, e1002899.

Eskova, A., Garcia-Junco, R., Chauvigné, F., Maischein, H.-M., Ammelburg, M., Cerdà, J., et al. Mutations in aquaporin 3a influence zebrafish pigment patterning through the tissue environment, in preparation.

Fadeev, A., Krauss, J., Fröhnhofer, H. G., Irion, U., & Nüsslein-Volhard, C. (2015). Tight junction protein 1a regulates pigment cell organisation during zebrafish colour patterning. *Elife*, *4*, e06545. http://dx.doi.org/10.7554/eLife.06545.

Ficker, E., Taglialatela, M., Wible, B. A., Henley, C. M., & Brown, A. M. (1994). Spermine and spermidine as gating molecules for inward rectifier K+ channels. *Science*, *266*, 1068–1072.

Frohnhöfer, H. G., Geiger-Rudolph, S., Pattky, M., Meixner, M., Huhn, C., Maischein, H.-M., et al. Spermidine regulates cell-cell interactions during pigment pattern formation in zebrafish, submitted.

Frohnhöfer, H. G., Krauss, J., Maischein, H. M., & Nüsslein-Volhard, C. (2013). Iridophores and their interactions with other chromatophores are required for stripe formation in zebrafish. *Development*, *140*, 2997–3007.

Giepmans, B. N., & Moolenaar, W. H. (1998). The gap junction protein connexin43 interacts with the second PDZ domain of the zona occludens 1 protein. *Current Biology*, *8*, 931–934.

Grether, G. F., Kolluru, G. R., & Nersissian, K. (2004). Individual colour patches as multicomponent signals. *Biological Reviews of the Cambridge Philosophical Society*, *79*, 583–610.

Haffter, P., Odenthal, J., Mullins, M. C., Lin, S., Farrell, M. J., Vogelsang, E., et al. (1996). Mutations affecting pigmentation and shape of the adult zebrafish. *Development Genes and Evolution*, *206*, 260–276.

Hamada, H., Watanabe, M., Lau, H. E., Nishida, T., Hasegawa, T., Parichy, D. M., et al. (2014). Involvement of Delta/Notch signaling in zebrafish adult pigment stripe patterning. *Development*, *141*, 318–324.

Hirata, M., Nakamura, K., Kanemaru, T., Shibata, Y., & Kondo, S. (2003). Pigment cell organization in the hypodermis of zebrafish. *Developmental Dynamics*, *227*, 497–503.

Hirata, M., Nakamura, K., & Kondo, S. (2005). Pigment cell distributions in different tissues of the zebrafish, with special reference to the striped pigment pattern. *Developmental Dynamics*, *234*, 293–300.

Hubbard, J. K., Uy, J. A., Hauber, M. E., Hoekstra, H. E., & Safran, R. J. (2010). Vertebrate pigmentation: From underlying genes to adaptive function. *Trends in Genetics*, *26*, 231–239.

Hwang, W. Y., Fu, Y., Reyon, D., Maeder, M. L., Tsai, S. Q., Sander, J. D., et al. (2013). Efficient genome editing in zebrafish using a CRISPR-Cas system. *Nature Biotechnology*, *31*, 227–229.

Inaba, M., Yamanaka, H., & Kondo, S. (2012). Pigment pattern formation by contact-dependent depolarization. *Science*, *335*, 677.

Inoue, S., Kondo, S., Parichy, D. M., & Watanabe, M. (2014). Tetraspanin 3c requirement for pigment cell interactions and boundary formation in zebrafish adult pigment stripes. *Pigment Cell & Melanoma Research*, *27*, 190–200.

Irion, U., Frohnhöfer, H. G., Krauss, J., Colak Champollion, T., Maischein, H. M., Geiger-Rudolph, S., et al. (2014). Gap junctions composed of connexins 41.8 and 39.4 are essential for colour pattern formation in zebrafish. *Elife*, *3*, e05125.

Irion, U., Krauss, J., & Nüsslein-Volhard, C. (2014). Precise and efficient genome editing in zebrafish using the CRISPR/Cas9 system. *Development*, *141*, 4827–4830.

Iwashita, M., Watanabe, M., Ishii, M., Chen, T., Johnson, S. L., Kurachi, Y., et al. (2006). Pigment pattern in jaguar/obelix zebrafish is caused by a Kir7.1 mutation: Implications for the regulation of melanosome movement. *PLoS Genetics*, *2*, e197.

Jinek, M., Chylinski, K., Fonfara, I., Hauer, M., Doudna, J. A., & Charpentier, E. (2012). A programmable dual-RNA-guided DNA endonuclease in adaptive bacterial immunity. *Science*, *337*, 816–821.

Kelsh, R. N., Brand, M., Jiang, Y. J., Heisenberg, C. P., Lin, S., Haffter, P., et al. (1996). Zebrafish pigmentation mutations and the processes of neural crest development. *Development*, *123*, 369–389.

Kelsh, R. N., Harris, M. L., Colanesi, S., & Erickson, C. A. (2009). Stripes and belly-spots—A review of pigment cell morphogenesis in vertebrates. *Seminars in Cell & Developmental Biology*, *20*, 90–104.

Kerosuo, L., & Bronner-Fraser, M. (2012). What is bad in cancer is good in the embryo: Importance of EMT in neural crest development. *Seminars in Cell & Developmental Biology*, *23*, 320–332.

Krauss, J., Astrinidis, P., Fröhnhofer, H. G., Walderich, B., & Nüsslein-Volhard, C. (2013). transparent, a gene affecting stripe formation in Zebrafish, encodes the mitochondrial protein Mpv17 that is required for iridophore survival. *Biology Open*, *2*, 703–710.

Krauss, J., Frohnhöfer, H. G., Walderich, B., Maischein, H. M., Weiler, C., Irion, U., et al. (2014). Endothelin signalling in iridophore development and stripe pattern formation of zebrafish. *Biology Open*, *3*, 503–509.

Kumar, N. M., & Gilula, N. B. (1996). The gap junction communication channel. *Cell*, *84*, 381–388.

Lamason, R. L., Mohideen, M. A., Mest, J. R., Wong, A. C., Norton, H. L., Aros, M. C., et al. (2005). SLC24A5, a putative cation exchanger, affects pigmentation in zebrafish and humans. *Science*, *310*, 1782–1786.

Lang, M. R., Patterson, L. B., Gordon, T. N., Johnson, S. L., & Parichy, D. M. (2009). Basonuclin-2 requirements for zebrafish adult pigment pattern development and female fertility. *PLoS Genetics*, *5*, e1000744.

Le Douarin, N. M., & Dupin, E. (2003). Multipotentiality of the neural crest. *Current Opinion in Genetics & Development*, *13*, 529–536.

Le Douarin, N. M., & Kalcheim, C. (1999). *The neural crest*. Cambridge, UK: Cambridge University Press.

Lister, J. A., Robertson, C. P., Lepage, T., Johnson, S. L., & Raible, D. W. (1999). nacre encodes a zebrafish microphthalmia-related protein that regulates neural-crest-derived pigment cell fate. *Development, 126*, 3757–3767.

Lopes, S. S., Yang, X., Muller, J., Carney, T. J., McAdow, A. R., Rauch, G. J., et al. (2008). Leukocyte tyrosine kinase functions in pigment cell development. *PLoS Genetics, 4*, e1000026.

Maderspacher, F., & Nüsslein-Volhard, C. (2003). Formation of the adult pigment pattern in zebrafish requires leopard and obelix dependent cell interactions. *Development, 130*, 3447–3457.

Mahalwar, P., Walderich, B., Singh, A. P., & Nüsslein-Volhard, C. (2014). Local reorganization of xanthophores fine-tunes and colors the striped pattern of zebrafish. *Science, 345*, 1362–1364.

McClure, M. (1999). Development and evolution of melanophore patterns in fishes of the genus Danio (Teleostei: Cyprinidae). *Journal of Morphology, 241*, 83–105.

McCluskey, B. M., & Postlethwait, J. H. (2015). Phylogeny of zebrafish, a "model species," within Danio, a "model genus". *Molecular and Biological Evolution, 32*, 635–652.

McMenamin, S. K., Bain, E. J., McCann, A. E., Patterson, L. B., Eom, D. S., Waller, Z. P., et al. (2014). Thyroid hormone-dependent adult pigment cell lineage and pattern in zebrafish. *Science, 345*(6202), 1358–1361.

Musa, H., & Veenstra, R. D. (2003). Voltage-dependent blockade of connexin40 gap junctions by spermine. *Biophysical Journal, 84*, 205–219.

Nakamasu, A., Takahashi, G., Kanbe, A., & Kondo, S. (2009). Interactions between zebrafish pigment cells responsible for the generation of Turing patterns. *Proceedings of the National Academy of Sciences of the United States of America, 106*, 8429–8434.

Nguyen, P. D., Hollway, G. E., Sonntag, C., Miles, L. B., Hall, T. E., Berger, S., et al. (2014). Haematopoietic stem cell induction by somite-derived endothelial cells controlled by meox1. *Nature, 512*, 314–318.

Nielsen, P. A., Baruch, A., Shestopalov, V. I., Giepmans, B. N., Dunia, I., Benedetti, E. L., et al. (2003). Lens connexins alpha3Cx46 and alpha8Cx50 interact with zonula occludens protein-1 (ZO-1). *Molecular Biology of the Cell, 14*, 2470–2481.

Odenthal, J., Rossnagel, K., Haffter, P., Kelsh, R. N., Vogelsang, E., Brand, M., et al. (1996). Mutations affecting xanthophore pigmentation in the zebrafish, Danio rerio. *Development, 123*, 391–398.

Parichy, D. M. (2006). Evolution of Danio pigment pattern development. *Heredity (Edinburgh), 97*, 200–210.

Parichy, D. M. (2007). Homology and the evolution of novelty during Danio adult pigment pattern development. *Journal of Experimental Zoology Part B: Molecular and Developmental Evolution, 308*, 578–590.

Parichy, D. M., & Johnson, S. L. (2001). Zebrafish hybrids suggest genetic mechanisms for pigment pattern diversification in Danio. *Development Genes and Evolution, 211*, 319–328.

Parichy, D. M., Mellgren, E. M., Rawls, J. F., Lopes, S. S., Kelsh, R. N., & Johnson, S. L. (2000). Mutational analysis of endothelin receptor b1 (rose) during neural crest and pigment pattern development in the zebrafish Danio rerio. *Developmental Biology, 227*, 294–306.

Parichy, D. M., Ransom, D. G., Paw, B., Zon, L. I., & Johnson, S. L. (2000). An orthologue of the kit-related gene fms is required for development of neural crest-derived xanthophores and a subpopulation of adult melanocytes in the zebrafish, Danio rerio. *Development, 127*, 3031–3044.

Parichy, D. M., Rawls, J. F., Pratt, S. J., Whitfield, T. T., & Johnson, S. L. (1999). Zebrafish sparse corresponds to an orthologue of c-kit and is required for the morphogenesis of a subpopulation of melanocytes, but is not essential for hematopoiesis or primordial germ cell development. *Development, 126*, 3425–3436.

Parichy, D. M., & Spiewak, J. E. (2015). Origins of adult pigmentation: Diversity in pigment stem cell lineages and implications for pattern evolution. *Pigment Cell & Melanoma Research, 28*, 31–50.

Patterson, L. B., & Parichy, D. M. (2013). Interactions with iridophores and the tissue environment required for patterning melanophores and xanthophores during zebrafish adult pigment stripe formation. *PLoS Genetics, 9*, e1003561.

Polyak, K., & Weinberg, R. A. (2009). Transitions between epithelial and mesenchymal states: Acquisition of malignant and stem cell traits. *Nature Reviews Cancer, 9*, 265–273.

Poulton, E. B. (1890). *The colours of animals, their meaning and use, especially considered in the case of insects*. New York: D. Appleton and Company.

Protas, M. E., & Patel, N. H. (2008). Evolution of coloration patterns. *Annual Review of Cell and Developmental Biology, 24*, 425–446.

Roulin, A., & Ducrest, A. L. (2013). Genetics of colouration in birds. *Seminars in Cell & Developmental Biology, 24*, 594–608.

Saez, J. C., Berthoud, V. M., Branes, M. C., Martinez, A. D., & Beyer, E. C. (2003). Plasma membrane channels formed by connexins: Their regulation and functions. *Physiological Reviews, 83*, 1359–1400.

Simon, A. M., & Goodenough, D. A. (1998). Diverse functions of vertebrate gap junctions. *Trends in Cell Biology, 8*, 477–483.

Simpson, P. (2007). The stars and stripes of animal bodies: Evolution of regulatory elements mediating pigment and bristle patterns in Drosophila. *Trends in Genetics, 23*, 350–358.

Singh, A. P., Fröhnhofer, H. G., Irion, U., & Nüsslein-Volhard, C. (2015). Fish pigmentation. Response to comment on "Local reorganization of xanthophores fine-tunes and colors the striped pattern of zebrafish" *Science, 348*, 297.

Singh, A. P., & Nüsslein-Volhard, C. (2015). Zebrafish stripes as a model for vertebrate colour pattern formation. *Current Biology, 25*, R81–R92.

Singh, A. P., Schach, U., & Nüsslein-Volhard, C. (2014). Proliferation, dispersal and patterned aggregation of iridophores in the skin prefigure striped colouration of zebrafish. *Nature Cell Biology, 16*, 607–614.

Uong, A., & Zon, L. I. (2010). Melanocytes in development and cancer. *Journal of Cellular Physiology, 222*, 38–41.

van Eeden, F. J., Granato, M., Schach, U., Brand, M., Furutani-Seiki, M., Haffter, P., et al. (1996). Mutations affecting somite formation and patterning in the zebrafish, Danio rerio. *Development, 123*, 153–164.

Walderich, B., Singh, A. P., Mahalwar, P., & Nüsslein-Volhard, C. Homotypic cell competition regulates proliferation and tiling of pigment cells during colour pattern formation in the zebrafish, submitted.

Watanabe, M., Iwashita, M., Ishii, M., Kurachi, Y., Kawakami, A., Kondo, S., et al. (2006). Spot pattern of leopard Danio is caused by mutation in the zebrafish connexin41.8 gene. *EMBO Reports*, 7, 893–897.

Watanabe, M., & Kondo, S. (2012). Changing clothes easily: Connexin41.8 regulates skin pattern variation. *Pigment Cell & Melanoma Research, 25*, 326–330.

Watanabe, M., & Kondo, S. (2015). Is pigment patterning in fish skin determined by the Turing mechanism? *Trends in Genetics, 31*, 88–96.

Watanabe, M., Watanabe, D., & Kondo, S. (2012). Polyamine sensitivity of gap junctions is required for skin pattern formation in zebrafish. *Science Reports, 2*, 473.

Weiner, L., Fu, W., Chirico, W. J., & Brissette, J. L. (2014). Skin as a living coloring book: How epithelial cells create patterns of pigmentation. *Pigment Cell & Melanoma Research, 27*, 1014–1031.

Weiner, L., Han, R., Scicchitano, B. M., Li, J., Hasegawa, K., Grossi, M., et al. (2007). Dedicated epithelial recipient cells determine pigmentation patterns. *Cell, 130*, 932–942.

Werner, T., Koshikawa, S., Williams, T. M., & Carroll, S. B. (2010). Generation of a novel wing colour pattern by the wingless morphogen. *Nature*, *464*, 1143–1148.
Wittkopp, P. J., & Beldade, P. (2009). Development and evolution of insect pigmentation: Genetic mechanisms and the potential consequences of pleiotropy. *Seminars in Cell & Developmental Biology*, *20*, 65–71.
Yamaguchi, M., Yoshimoto, E., & Kondo, S. (2007). Pattern regulation in the stripe of zebrafish suggests an underlying dynamic and autonomous mechanism. *Proceedings of the National Academy of Sciences of the United States of America*, *104*, 4790–4793.
Yamanaka, H., & Kondo, S. (2014). In vitro analysis suggests that difference in cell movement during direct interaction can generate various pigment patterns in vivo. *Proceedings of the National Academy of Sciences of the United States of America*, *111*, 1867–1872.

Uwe Irion *et al.*, Figure 1 Color pattern diversity in fish. Fish show a great diversity of color patterns mainly by different combinations of stripes and spots. Some patterns are used to attract mates (e.g., in guppies or sticklebacks), others to deter predators (lion fish); cave fish are often blind and not colored. (A) Threadfin Butterfly fish (*Chaetodon auriga*), (B) Picasso fish (*Rhinecanthus aculeatus*), (C) Black-faced blenny (*Tripterygion delaisi*), (D) Checkered Julie (*Julidochromis marlieri*), (E) Common Lionfish (*Pterois spec.*), (F) Turquoise killifish (*Nothobranchius furzeri*), (G) Zebrafish (*Danio rerio*), (H) Cichlid (*Haplochromis latifasciatus*), (I) Japanese rice fish (*Oryzias latipes*), (J) Endler's Guppy (*Poecilia wingei*), (K) Three-spined stickleback (*Gasterosteus aculeatus*), and (L) Mexican cavefish (*Astyanax mexicanus*). Several of these fish are used as model organisms for biomedical research. *Panel (A): Image courtesy of Isabelle Côté. Panel (B): Image source—Wikipedia. Panel (C): Image courtesy of Nico Michiels. Panel (E): Image courtesy of Isabelle Côté. Panel (F): Image courtesy of Nils Hartmann. Panel (H): Image courtesy of Walter Salzburger. Panel (I): Image courtesy of Matthew Harris. Panel (J): Image courtesy of Verena Kottler and Christine Dreyer. Panel (L): Image courtesy of Nicolas Rohner.*

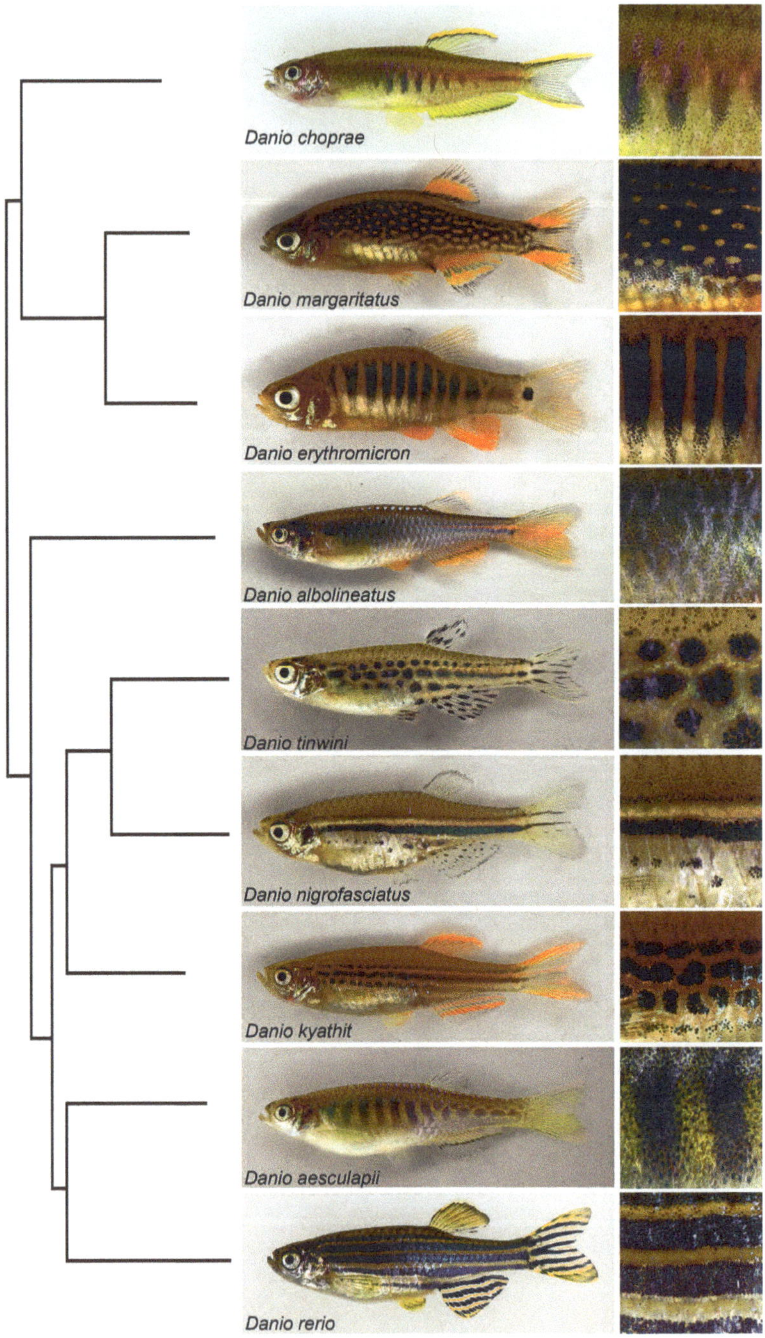

Uwe Irion *et al.*, Figure 2 See legend on opposite page.

Uwe Irion *et al.*, Figure 2 Color pattern diversity within the genus *Danio*. The cladogram on the left shows the evolutionary relationship among different *Danio* species. Color patterns vary greatly even between closely related species; they range from almost uniformly distributed melanophores with a posterior red stripe on the trunk of *D. albolineatus*, over dark spots in *D. tinwini*, or stripes and spots on the trunk of *D. nigrofasciatus*, to vertical bars on the trunk of *Danio choprae* and *D. erythromicron* and light spots on a dark background in *D. margaritatus*. *D. rerio* (female) displays sharp boundaries between light and dark stripes. In *D. kyathit*, stripes are broken; *D. aesculapi*, the closest relative of *D. rerio*, has a pattern of irregular vertical bars dissolving into spots. Note the distinct motifs of fin patterning in several *Danio* species such as in *D. choprae* and *D. margaritatus*; the contiguous striped pattern of body and fins as seen in *D. rerio* is an exception. Rightmost panes show enlarged portion of the trunk skin. *Adapted from McCluskey and Postlethwait (2015).*

Uwe Irion *et al.*, Figure 3 Layered organization of pigment cells in the trunk skin of zebrafish. (A) Close-up view of the lateral trunk of adult zebrafish displaying blue dark stripes and golden light stripes. (B–B‴) Schematic representation of the layered arrangement of pigment cells in the skin of the lateral trunk of zebrafish. The light stripe region consists of dense iridophores (white) covered by dense xanthophores

(*Continued*)

Uwe Irion *et al.*, Figure 3—Cont'd (dark orange). In the stripe region, scattered xanthophores (stellate, yellow) and bluish reflective iridophores are present above a compact layer of black melanophores. Beneath melanophores, there is a layer of L-iridophores (white elongated cells in B‴). (C–C′) Overview of a *Tg(sox10:ER^{T2}Cre)*-induced clone (induced on 5 dpf) shows clonally restricted organization of iridophores in the adult skin. Clonally related iridophores exhibit characteristic spreading along the dorsoventral axis. (D–D′) Confocal image of a clone showing detailed organization of labeled pigment cells in the skin with dense iridophores (arrowheads; I^d) in the light stripe region, loose iridophores (arrowheads; I^s) and melanophores (arrows) in the dark stripe region. (E) Yellow–orange xanthophore pigment in the dark stripe (arrows) and light stripe regions (arrowheads) of the adult skin. Melanophores, compacted due to treatment with epinephrine, display a regular spacing in the dark stripe region. In (F–F′), melanophores are expanded. Xanthophores appear compact and densely packed in the light stripes (arrowheads) and appear stellate and loose in the dark stripe region (arrows).

Uwe Irion *et al.*, Figure 4 Selected adult-viable pigmentation mutants in zebrafish. (A) Wild-type adult zebrafish male. (B) *nacre* lacks all the neural crest-derived melanophores. (C) *pfeffer* lacks xanthophores, melanophores are reduced in number, and residual dark stripes are discontinuous, displaying a pepper and salt pattern. (D) *shady* mutants lack iridophores and display only two rudimentary melanophore stripes that are broken into spots. Fins are striped. (E) In *leopard*, the stripes break up into spots. (F) *obelix* displays broader and fewer dark stripes and (G) *schachbrett* shows disrupted stripes and spots. (H) In *choker*, which lacks the horizontal myoseptum, a meandering pattern with stripes of normal width is formed.

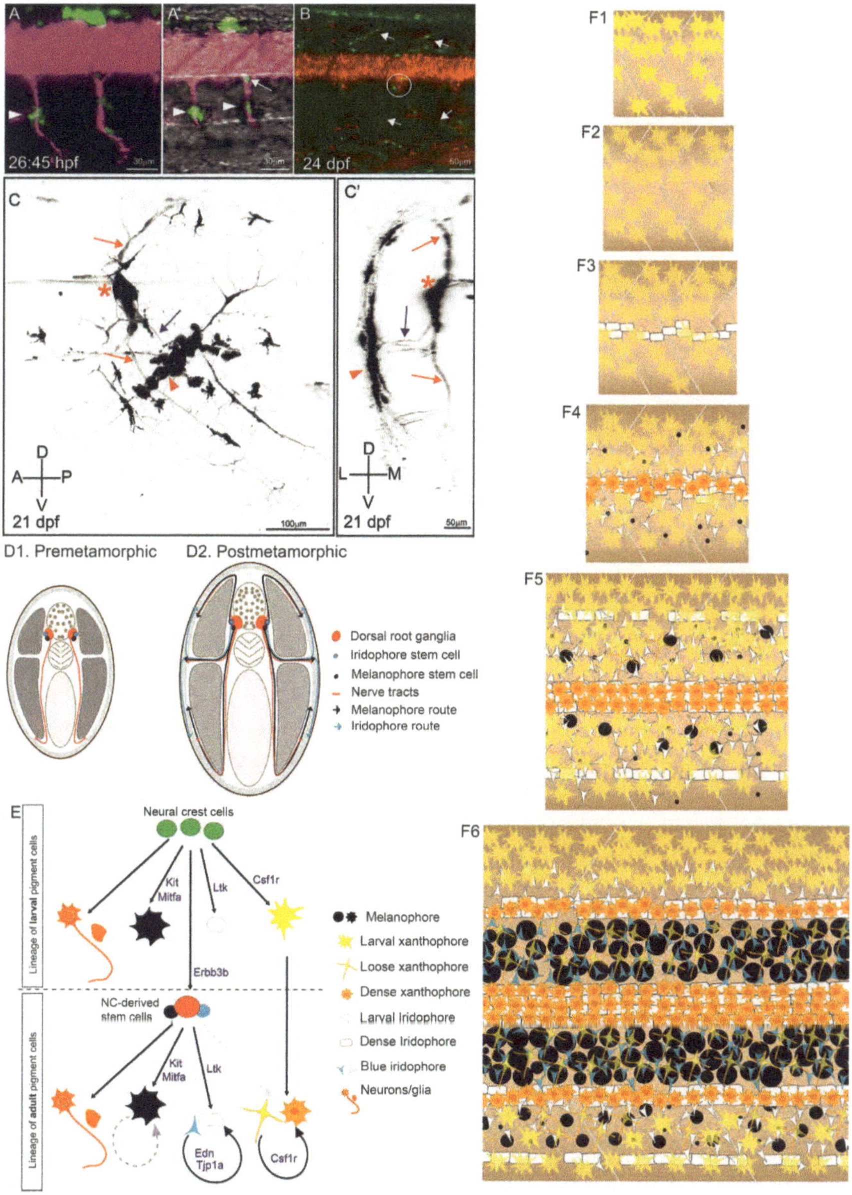

Uwe Irion *et al.*, Figure 5 Pigment cell lineage, migration, and stripe pattern formation. (A–A′) During embryogenesis, melanoblasts (Mitfa-positive, green cells) migrate along the primary motor neuron (magenta) and differentiate into larval melanophores (arrowheads in A′). Arrow in (A′) indicates a Mitfa-positive cell that remains at a ventral position of the neural tube, the melanophore stem cell. (B) During metamorphosis, Mitfa-positive melanophore progenitor cells are seen at the dorsal root ganglion (encircled) and along the spinal nerves extending dorsally and ventrally to innervate the lateral skin (arrows).

(*Continued*)

Uwe Irion *et al.*, Figure 5—Cont'd (C–C′) A clone of iridophores (arrowhead) associated with the dorsal root ganglion (asterisk). Blue arrows in (C–C′) indicate the neurons that exit through the horizontal myoseptum; red arrows indicate the neuronal tracts associated with the dorsal and ventral myotomes and the myosepta. (C′) The clone in (C) turned by 90° to show a cross section through the trunk region. Skin is to the left. Abbreviations: D: dorsal, V: ventral, A: anterior, P: posterior, L: lateral, M: medial, hpf: hours postfertilization, and dpf: days postfertilization. (D1–D2) Schematic showing the route followed by melanophore progenitors along the peripheral neurons innervating the skin, whereas iridophores migrate through the horizontal myoseptum and disperse dorsoventrally in the skin. (E) Lineage and the postembryonic source of adult pigment cells. Neural crest cells give rise to several larval cell types including neural tissue and develop directly into the three kinds of larval pigment cells in zebrafish. Adult melanophores, iridophores, and neural tissue are derived from postembryonic stem cells, which are located on the segmentally iterated dorsal root ganglia. Larval xanthophores persist and proliferate at the onset of metamorphosis to generate adult xanthophores. Metamorphic iridophores and xanthophores undergo local proliferation in the skin. In rare cases, pigmented melanophores are seen to divide in the skin (faint circular arrow). A fraction of xanthophores may arise from mixed iridophore clones (faint arrow). Genes required for specific lineages are mentioned in blue along the arrows. (F1–F6) Schematics showing stripe pattern morphogenesis during the transition from the larval stage to the adult stage. (F1) Larval xanthophores (5 dpf) (F2) begin to proliferate at the onset of metamorphosis (2 weeks postfertilization; wpf). (F3) Iridophores appear in the skin through the horizontal myoseptum and organize the first light stripe (~3 wpf). (F4) Melanophores develop from the melanoblasts, which enter the skin through the dorsal and the ventral myotomes (~3–4 wpf). Over the dense iridophores, xanthophores become compact and acquire yellow–orange pigment. Loose iridophores appear at the borders of the dense iridophore region and begin to disperse along the dorsoventral axis. (F5) Melanophores grow in size in the stripe region to form contiguous stripes. Loose iridophores aggregate to organize new light stripes at a distance from the first light stripe (~4–5 wpf). (F6) Xanthophores become stellate over the melanophores and compact in the light stripe regions. Loose iridophores on top of melanophores appear bluish and reflective giving blue color to the dark stripes. Light stripes appear golden due to the presence of silvery dense iridophores covered by yellow–orange xanthophores.

Uwe Irion *et al.*, Figure 6 Interspecies hybrid and chimera. Hybrids between *Danio aesculapii* and *Danio rerio* (zebrafish) show a striped pattern very similar to zebrafish; the fins are partially striped. Knockout mutations in *MitfA* in *Danio albolineatus* lead to a loss of melanophores; the patterning, however, is not affected. In chimeric animals obtained by blastomere transplantation, the melanophores derived from *D. albolineatus* can contribute to the stripes of *D. rerio*.

CHAPTER NINE

Defining "Development"

Thomas Pradeu*[,1], Lucie Laplane[†], Karine Prévot[‡], Thierry Hoquet[§], Valentine Reynaud[¶], Giuseppe Fusco[||], Alessandro Minelli[||], Virginie Orgogozo[#,1], Michel Vervoort[#]

*ALYSAI, CNRS, UMR5164, University of Bordeaux, Bordeaux, France
[†]UMR 8590 (CNRS, IHPST-Université Paris I Panthéon-Sorbonne, ENS), Paris, France
[‡]Université Paris Ouest Nanterre-La Défense, Ireph, Nanterre Cedex, France
[§]Faculté de philosophie, Université Jean Moulin Lyon 3, Lyon Cedex 02, France
[¶]Institut de Recherches Philosophiques de Lyon (IRPhiL), Centre de la recherche Eugène Chevreul, Lyon, France
[||]Department of Biology, University of Padova, Padova, Italy
[#]Institut Jacques Monod, CNRS, UMR 7592, Univ Paris Diderot, Sorbonne Paris Cité, Paris, France
[1]Corresponding authors: e-mail address: thomas.pradeu@u-bordeaux.fr; orgogozo@ijm.univ-paris-diderot.fr

Contents

Abstract

Is it possible, and in the first place is it even desirable, to define what "development" means and to determine the scope of the field called "developmental biology"? Though these questions appeared crucial for the founders of "developmental biology" in the 1950s, there seems to be no consensus today about the need to address them. Here, in a combined biological, philosophical, and historical approach, we ask whether it is possible and useful to define biological development, and, if such a definition is indeed possible and useful, which definition(s) can be considered as the most satisfactory.

One cannot present a science without at the same time defining its terms.

(Leibniz, New Essays on Human Understanding)

Current Topics in Developmental Biology, Volume 117
ISSN 0070-2153
http://dx.doi.org/10.1016/bs.ctdb.2015.10.012

Insistence on the postulate of complete logical clarification would make science impossible.

(Heisenberg, Physics and Philosophy)

1. INTRODUCTION

How can we define "development"? And, in the first place, why should anyone care about a definition of development? Despite the existence of a well-established and dynamic scientific field called "developmental biology", those two questions are hardly ever raised today (Minelli & Pradeu, 2014). Etymologically, "development" means "unfolding", that is, the progressive unfolding of a preexistent form or preexistent capacities. But very few would endorse today this form of preformationism, which neglects interactions with the environment and stochastic effects (Lewontin, 2000).

In recent years, only a handful of developmental biologists have ventured to define "development" in research papers or books (exceptions include Coen, 1999; Davidson, 2001), and definitions of development are found mainly in textbooks. A frequent textbook definition is "the process through which a single cell—often a fertilized egg—gives rise to a complex multicellular organism" (Wolpert & Tickle, 2011; see also Martinez-Arias & Stewart, 2002; Slack, 1991). Yet this definition has sometimes been criticized for its "adultocentrism" (Bonner, 1965; Minelli, 2003), and the exclusion of unicellular organisms, many forms of asexual reproduction where development starts from a multicellular stage, as well as the phenomena of regeneration and aging. Others have suggested to define development as all the changes that occur in a given organism from its beginning to its death (e.g., Gilbert, 2013; Oyama, 2000; West-Eberhard, 2003), a definition perhaps too broad, as it would include virtually all physiological processes (Pradeu et al., 2011). It seems, therefore, that many developmental biologists would agree with Guirao and Aréchaga (1989) that "the term 'development' is difficult to define", and few try to offer an explicit definition.

But should developmental biologists really care about defining development? The current research field of developmental biology offers an interesting and contrasting situation with regard to this question. In an attempt to better characterize this situation, we asked all the contributors to the present special issue of *CTDB* the two following questions: (i) Do you think that it is important, for your scientific work in developmental biology, to define the word "development"? and (ii) How, according to you, should the notion of "development" be defined? Among the 24 contributors who accepted to answer our minisurvey, 13 of them (54%) consider that they do not need a definition of development, while 11 (46%) believe exactly the opposite. Moreover, these 24 contributors propose diverse definitions that disclose several points of dissension, such as whether development should include regeneration, aging, or the idea of increasing complexity when an organism matures. Of course, the sample of our survey is far too small to draw any definitive conclusion. And yet both our survey and a careful study of past and present literature in developmental biology suggest a lack of consensus about what belongs to "development" or "developmental biology", and what does not.

The term "developmental biology" came to encompass and replace, in intellectual societies and journals, the older labels of "embryology" and "growth" (or "growth and differentiation") at the end of the 1950s (Burian & Thieffry, 2000; Crowe et al., 2015; Gilbert, 1991: ix). In fact, the questions of what development means and how extensive developmental biology should be were crucial for the biologists who, immediately after the Second World War, participated in the construction of the emerging discipline that came to be called "developmental biology". In the Foreword of the very first issue of the *Journal of Embryology and Experimental Morphology* (renamed *Development* in 1986), Dalcq (1953) announces that, with the advent of "the science of development", embryologists witness "an extraordinary extension of their field", in depth (the reach of the "macromolecular level"), space (the progressive establishment of a "common denominator" of morphogenesis, regeneration, asexual reproduction, and embryonic development), and time (the gradual effacement of "the conventional boundaries between the biological phases of life", with a shift of attention from early embryonic phases only to a much broader focus on cytodifferentiation, growth, and even continuous reconstruction throughout life and aging). In a similar vein, in his Introduction to the first issue of the US journal *Developmental Biology*, Weiss (1959) suggests that many seemingly isolated phe nomena studied hitherto by embryologists, plant physiologists, nutritionists, or oncologists, constitute in fact "one continuous spectrum", unified by the "same basic principles" of growth, differentiation, morphogenesis, maturation, aging, regulation, and regeneration, which all together constitute "development".

Because of the remarkable successes of developmental biology from the 1950s to the present day, its current practitioners do not focus as ardently as their predecessors on the question of how to define the term "development" and how to delineate its scope. Importantly, though, this question has not entirely disappeared. For example, the British Society for Developmental Biology recently considered whether its name should change to the

"British Society for Developmental Biology and Stem Cells", illustrating "heated discussion in the developmental biology community about what its relationship should be to the stem cell field" (St Johnston, 2015).

Thus, it appears that there was not, and there still is not, any strong consensus among developmental biologists about both the definition of development and the delineation of the domain of developmental biology. In this essay, our aim is to determine whether today's developmental biologists should, as their predecessors of the 1950s did, seek a definition and a delineation of development. More precisely, we address, in a combined biological, philosophical, and historical approach, the vexing problem of whether it is possible and useful to define biological development, and if such a definition is possible and useful, which definition(s) should be used.

2. SHOULD WE TRY TO DEFINE "DEVELOPMENT"? NO, WE SHOULD NOT

2.1 It is Useless

As illustrated by several sharp answers to our survey, many biologists feel that attempting to define development is simply *useless*. Biologists are generally interested in specific empirical studies, that is, studies of precise mechanisms in a given model organism (e.g., gene expression during embryogenesis in the sea urchin, or Wnt signaling in the formation of the chick's neural crest, or auxin signaling in the formation of a plant embryo root). Therefore, it may seem that general definitions are not useful for the work that really matters, namely, experimentation. If, in practice, developmental biologists can perfectly do their research without ever wondering about what development means, then why should they care?

A precise definition can even be considered as a hindrance for doing practical research in developmental biology. Most scientific concepts remain undefined, implicit, and this is not an obstacle for everyday science (Kuhn, 1962). According to many historians of science, conceptual fuzziness is widespread in science, and even fruitful (Keller, 2002; Rheinberger, 2000). Some degree of imprecision can contribute to foster communication between scientists, both within a scientific domain and between domains. For example, Keller (2002: 181) claims that it is largely because of its malleability that the concept of positional information (Wolpert, 1968) became popular and fruitful, that is, extensively used and discussed until today.

2.2 It is Impossible

A different attitude consists in saying that defining development, as useful as it might be, is in fact *impossible*. In nature, developmental phenomena are so diverse that it would make no sense to pretend that one single definition could bring together all these processes. First, the word "development" can be applied to many organisms across species. Would it not be hazardous to assert that this same word might be applied to plants, microbes, and animals all alike? It seems likely that every general definition of development will inevitably meet with some exceptions (Minelli, 2014). Second, it covers phenomena situated at many different levels (molecular, cellular, and so on). Third, it encompasses a series of generally successive and heterogeneous processes, such as in animals, fertilization, cleavage, gastrulation, organogenesis, and, sometimes, metamorphosis and regeneration. Faced with such a multifaceted diversity, how could developmental biologists offer one single, unifying, definition of development?

Instead of searching for the meaning of development *per se*, one might try to define a series of more specific terms, like "cell differentiation", "morphogenesis", and "organogenesis". For this strategy to be successful, these processes themselves need to be defined. But, to define a word, one must use other words, which themselves might prove difficult to define—a process that could lead to an infinite regress. Potential definitions of "development" are often circumlocutions involving other terms like "morphogenesis", "shape", "irreversibility", "pattern", "complexity", but generally, those terms are themselves particularly difficult to define (see Table 1 in Section 4 below).

3. SHOULD WE TRY TO DEFINE "DEVELOPMENT"? YES, WE SHOULD

3.1 To Delineate Developmental Biology as a Field of Study

A definition of development allows to delineate the *domain* of developmental biology as a scientific field, that is, to determine which living things and processes should be considered as subjects of developmental studies. Now, many biologists might say that they do not care about a definition of development, but, presumably, not so many would say that defining the research field named "developmental biology" is useless. Interestingly, several people who answered our survey remark that biologists often disagree on whether certain processes—e.g., fertilization, regeneration—and certain organisms—e.g., plants—are part of developmental biology. This problem is all but marginal: clarifying what falls under the scope of "developmental biology" is crucial to compare different experimental settings and models, and to determine how each may shed light on the others. As mentioned previously, defining the scope of their field was one of the most pressing issues for the scientists involved in the advent of "developmental biology" in the 1950s. For Weiss (1959), in a time of "progressive splintering of older disciplines into ever smaller and narrower technical specialties", the then emerging field of "developmental biology" aimed at doing exactly the opposite, that is, "to promote the confluence and integration of related, but formerly isolated, lines".

Two examples illustrate why this question is still very important today. The first example concerns the place of plants in developmental biology. The question of whether plants should be included in the domain of embryology was raised as early as the first half of the nineteenth century (e.g., Schleiden, 1848). Including or excluding plants from the study of development changes our vision of development: it questions whether development continues in adulthood (Steeves & Sussex, 1989), whether regeneration is part of development (Birnbaum & Sanchez Alvarado, 2008), and whether development should be defined molecularly (in which case "development" would mean two different things for plants and for animals) or rather at the level of general principles (in which case there may exist a unified meaning of development, adequate for both plants and animals) (Meyerowitz, 2002; Vervoort, 2014).

In a similar vein, an interesting debate today is whether unicellular organisms should be included into the study of development (Love & Travisano, 2013; O'Malley & Dupré, 2007), and if the term "development" means the same thing when applied to multicellular organisms only (a very common practice among developmental biologists), or to unicellular organisms as well. Though textbooks and meetings in developmental biology do not, in general, include unicellular organisms, there has existed for more than 30 years a dynamic field self-identified as "developmental microbiology" (e.g., Peberdy, 1980). Furthermore, multiple observations argue for their inclusion. Several unicellular organisms, such as *Acetabularia* (Dumais, Serikawa, & Mandoli, 2000), social amoebae (Bonner, 2009), and bacteria (Kroos & Maddock, 2003; Peberdy, 1980), can be said to "develop", in the sense that they undergo important and stereotyped morphogenetic changes during their life cycle, including morphogenesis, cellular differentiation, and intercellular signaling (Shimkets, 1999; Straight & Kolter, 2009). A precise definition of development should reach a verdict on whether unicellular organisms are to be included in "developmental biology", and, perhaps most importantly, this would help to determine how studies done on microbes and those done on nonmicrobial entities could inspire each other.

Another similar question, which we mention only briefly here, is whether the notion of development can be applied to cognition and behavior (Gottlieb, 1992; Oyama, Griffiths, & Gray, 2001). Here again, the only way to address such a question is to offer a definition of development.

In addition, an explicit description of what belongs to developmental biology and what does not can be useful from a purely *institutional* point of view. To have a clear view of which journals, scientific societies, scientific meetings, funding opportunities, or academic positions belong to developmental biology, it is necessary to know, if only at a very general level, what development means. In line with Weiss's aim, defining development and delineating the domain of developmental biology can be useful to foster intellectual collaborations and to give rise to a feeling to belong to a shared "community", by bringing together biologists with the same type of questions or approaches, and using the same kind of vocabulary.

3.2 For Considering Development within More Inclusive Processes

In science, whenever you decide to study a process which involves the transformation of something, either through time or space or otherwise, you have to define this something, operationally at least, in order to trace its change. This is the case of development in the context of evolution (e.g., evolution of development) or in the context of ecology (e.g., environmentally induced developmental responses). In other words, when development is viewed from the perspective of a more inclusive process, you need to define it. Current, sometimes pragmatic, available definitions are restricted to specific developmental processes (e.g., gastrulation, or prefoliation, see Section 2.2), which merely allow studying the evolutionary history of these specific processes, but do not allow investigating more general issues, as for instance the role of developmental plasticity in the wider theoretical context of evolutionary models (Fusco & Minelli, 2010).

3.3 For Communication with the Public and Legal Purposes

Even biologists who are skeptical about the need to define development in their own specialized domain would probably agree that defining development could be important for nonbiologists. First, anyone who teaches developmental biology knows that students find definitions extremely useful. Second, many political and legal debates (for example, those related to abortion, assisted reproductive technologies, or the legal status of the fetus) presuppose to have clear notions of "embryo" and "development", for which politicians and judges require the help of biologists. In agreement with Maienschein (2014), we are not suggesting here that our ethical conceptions should be *based* on biological data, but rather that legal decisions should not be in direct conflict with established scientific knowledge, and that, if developmental biologists could provide a clear definition of development and related notions, this would facilitate scientifically informed law making.

3.4 To Overcome Implicit Meanings

Even when left undefined, a concept like "development" often has an implicit meaning, generally rooted in a common background (acquired through education and lab practices) or in intuitions, rather than cautious scientific considerations. Developmental biologists who say that they do not have, and do not need, a definition of development actually do have in mind implicit conceptions of development, which can bias their views and have negative effects on research and discoveries. For example, the idea that development is intrinsically a *construction*, and not destruction, almost certainly slowed down research on the role of apoptosis in development (Baehrecke, 2002; Saunders, 1966). Another telling example concerns the influence of genes on development. That genes play a key role in development is beyond doubt, and an impressive achievement of developmental biology from the 1950s to the present day has been the detailed molecular description of how developmental genes work (Gilbert, 1998; Wolpert, 1994). That being said, the long-prevailing view that development is to a large extent "autonomous" with regard to the environment and genetically determined has oriented research toward "internal" and deterministic components of development, and this implicit assumption led to neglect other, potentially crucial, interactions with "external" factors (Gilbert & Bolker, 2003; Sultan, 2015), such as temperature, predators, symbionts (Gilbert & Epel, 2015), or stochastic processes such as developmental noise (Lewontin, 2000) and somatic mutations during normal development (Campbell, Shaw, Stankiewicz, & Lupski, 2015).

For example, it has been widely assumed that the way immune systems develop and enable hosts to resist diseases was to a large extent dependent on genetic factors. Now, in a striking recent study on human monozygotic and dizygotic twins, Mark Davis' group at Stanford showed that environmental effects are much more important than genetic factors in determining how the adaptive immune system develops and makes the organism resistant to diseases (Brodin et al., 2015). Because of an unquestioned conviction that genes played a major role in the development of the immune system, these environmental factors (especially previous microbial exposures, vaccinations, diet, and dental hygiene) had not been thoroughly investigated.

Importantly, the neglect of environmental factors is to some extent related to the choice of model organisms. Many "classic" model organisms in developmental biology were chosen in part because they were relatively robust under environmental perturbation (Bolker, 1995). In turn, results obtained in these models have tended to reinforce the view that all development is regulated from inside the egg and that "outdoor" experiments were unnecessary (Gilbert & Bolker, 2003).

Thus, it can be argued that most developmental biologists, even those who say they do not need a definition of development, do in fact have implicit conceptions of development, which can affect their practical research. Having a definition of development can help making explicit these otherwise implicit assumptions, and perhaps overcoming them.

4. ANALYSIS OF A FEW POSSIBLE DEFINITIONS OF "DEVELOPMENT"

For those who think that defining development is important, an open question is *how* exactly should the notion of development be defined. Table 1 presents common definitions of development, with their proponents (mostly scientists, though some of them are philosophers or historians), and the potential difficulties they raise.

Obviously, this table is not exhaustive; our aim is simply to open the discussion about possible definitions, on the basis of suggestions regularly found in the literature. Also, this does not entail that anyone interested in defining the notion of "development" would have to adopt a *unique* definition; quite the contrary, it is certainly possible to combine several definitions, among those listed here, or others. Because of the relative malleability of the term "development", one interesting possibility could even be to articulate a "cluster" definition of development, with at its core the processes that everyone considers as part of development (e.g., gastrulation in metazoans or seed germination in spermatophyta), and satellite processes that at least some

Table 1 Some Possible Definitions of Development

Definition	**Proponents**	**Possible Difficulties**
Process through which a single cell gives rise to a complex multicellular organism	Barinaga (1994), Coen (1999), Martinez-Arias and Stewart (2002), and Wolpert and Tickle (2011)	– Excludes several important developmental processes, e.g., regeneration – Excludes many organisms (e.g., uncellulars) and several forms of asexual reproduction – "Adultocentrism" (on these three aspects, see Minelli, 2003)
Elaboration of a new individual form	Amundson (1994), Love (2008), and Pradeu (2014: 15)	– Difficult to define the notions of "new", "shape", or "form"
Change of biological form through time	Slack (2013: 3)	– Difficult to define "form" – Difficult to distinguish development from metabolism
Temporal change of organization along the life cycle	Waddington (1956: 3), Oyama (2000 [1985]), Oyama et al. (2001), Minelli (2003), West-Eberhard (2003: 32), and Gilbert (2013)	– Difficult to define "organization" – Lacks specificity, as it includes too many biological phenomena
Biological readout of the genomically encoded gene networks that determine how the organism is constructed	Davidson (2001) and Peter and Davidson (2015)	– Ignores the role of other factors that can control development, such as hormones, mechanical forces, and environmental factors (Gilbert, 2002)
Irreversible increase of complexity over time	Arthur (1997: 277)	– Difficult to define "complexity" – There are cases where development seems to be associated with a decrease in complexity, in particular during metamorphosis of sessile or parasitic animals (e.g., ascidians, *Sacculina*)
Restriction of possible outcomes (specialization) and increase in levels of organization	Johnson (2001)	– Depends on the level considered

biologists consider as part of development (e.g., fertilization, regeneration). Further investigations will have to determine whether elaborating such a "cluster" definition is possible, and whether it is considered useful by a number of practicing developmental biologists.

5. CONCLUSION

So, should developmental biologists care about defining development? Overall, we believe that arguments in favor of a definition of development prevail. First, it is crucial to determine which subjects belong to "developmental biology" and to distinguish this domain from other biological fields. Second, even when people claim to have no definition of development, they have in mind an implicit conception that impacts their research, highlighting the need to make explicit the implicit. Clarifying the meaning of scientific notions can be a remarkable way to reflect on what we hitherto got right or wrong, and therefore, decide on where to go in the future.

ACKNOWLEDGMENTS

We thank Paul Wassarman for his invitation to contribute to the present special issue of *CTDB*. We also thank Wallace Arthur, Jean Gayon, Alain Ghysen, Scott Gilbert, Paul Griffiths, Jane Maienschein, Michel Morange, Maureen O'Malley, Susan Oyama, and Karola Stotz for their help.

REFERENCES

Amundson, R. (1994). Two concepts of constraint: Adaptationism and the challenge from developmental biology. *Philosophy of Science*, *61*(4), 556–578.

Arthur, W. (1997). *The origin of animal body plans.* Cambridge: Cambridge University Press.

Baehrecke, E. H. (2002). How death shapes life during development. *Nature Reviews. Molecular Cell Biology*, *3*(10), 779–787.

Barinaga, M. (1994). Looking to development's future. *Science*, *266*, 561–564.

Birnbaum, K. D., & Sanchez Alvarado, A. (2008). Slicing across kingdoms: Regeneration in plants and animals. *Cell*, *132*, 697–710.

Bolker, J. (1995). Model systems in developmental biology. *Bioessays*, *17*, 451–455.

Bonner, J. T. (1965). *Size and cycle: An essay on the structure of biology.* Princeton: Princeton University Press.

Bonner, J. T. (2009). *The social amoebae: The biology of cellular slime molds.* Princeton: Princeton University Press.

Brodin, P., et al. (2015). Variation in the human immune system is largely driven by non-heritable influences. *Cell*, *160*, 37–47.

Burian, R., & Thieffry, D. (2000). Introduction to the special issue. 'From embryology to developmental biology'. *History and Philosophy of the Life Sciences*, *3*, 313–323.

Campbell, I. M., Shaw, C. A., Stankiewicz, P., & Lupski, J. R. (2015). Somatic mosaicism: Implications for disease and transmission genetics. *Trends in Genetics*, *31*, 382–392.

Coen, E. (1999). *The art of genes.* Oxford: Oxford University Press.

Crowe, N., et al. (2015). The diversification of developmental biology. *Studies in History and Philosophy of Biological and Biomedical Sciences, 53*, 1–15.

Dalcq, A. M. (1953). Foreword. *Journal of Embryology and Experimental Morphology, 1*(1), 1–4.

Davidson, E. H. (2001). *Genomic regulatory systems: Development and evolution.* San Diego: Academic Press.

Dumais, J., Serikawa, K., & Mandoli, D. F. (2000). *Acetabularia*: A unicellular model for understanding subcellular localization and morphogenesis during development. *Journal of Plant Growth Regulation, 19*(3), 253–264.

Fusco, G., & Minelli, A. (2010). Phenotypic plasticity in development and evolution: Facts and concepts. *Philosophical Transactions of the Royal Society, B: Biological Sciences, 365*, 547–556.

Gilbert, S. F. (1991). Preface. In S. F. Gilbert (Ed.), *A conceptual history of modern embryology* (pp. vii–ix). Baltimore: The Johns Hopkins University Press.

Gilbert, S. F. (1998). Conceptual breakthroughs in developmental biology. *Journal of Biosciences, 23*, 169–176.

Gilbert, S. F. (2002). Book review: Genomic regulatory systems: Development and evolution. *American Journal of Medical Genetics, 108*, 341–342.

Gilbert, S. F. (2013). *Developmental biology* (10th ed.). Sunderland, MA: Sinauer Associates.

Gilbert, S. F., & Bolker, J. A. (2003). Ecological developmental biology: Preface to the symposium. *Evolution and Development, 5*, 3–8.

Gilbert, S. F., & Epel, D. (2015). *Ecological developmental biology* (2nd ed.). Sunderland, MA: Sinauer Associates.

Gottlieb, G. (1992). *Individual development and evolution: The genesis of novel behavior.* New York: Oxford University Press.

Guirao, M., & Aréchaga, J. (1989). Editorial statement. A word of welcome and a changing of the guard. *International Journal of Developmental Biology, 33*, 1.

Johnson, M. H. (2001). Functional brain development during infancy. In Galvin Bremner & Alan Fogel (Eds.), *Blackwell handbook of infant development* (pp. 169–190). Malden, MA: Blackwell.

Keller, E. F. (2002). *Making sense of life.* Cambridge: Harvard University Press.

Kroos, L., & Maddock, J. R. (2003). Prokaryotic development: Emerging insights. *Journal of Bacteriology, 185*, 1128–1146.

Kuhn, T. (1962). *The structure of scientific revolution.* Chicago: University of Chicago Press.

Lewontin, R. (2000). *The triple helix.* Cambridge: Harvard University Press.

Love, A. C. (2008). Explaining the ontogeny of form: Philosophical issues. In S. Sarkar & A. Plutynski (Eds.), *A companion to the philosophy of biology* (pp. 223–247). Malden, MA: Blackwell.

Love, A. C., & Travisano, M. (2013). Microbes modeling ontogeny. *Biology and Philosophy, 28*, 161–188.

Maienschein, J. (2014). *Embryos under the microscope.* Cambridge: Harvard University Press.

Martinez-Arias, A., & Stewart, A. (2002). *Molecular principles of animal development.* Oxford: Oxford University Press.

Meyerowitz, E. M. (2002). Plants compared to animals: The broadest comparative study of development. *Science, 295*, 1482–1485.

Minelli, A. (2003). *The development of animal form.* Cambridge: Cambridge University Press.

Minelli, A. (2014). Developmental disparity. In A. Minelli & T. Pradeu (Eds.), *Towards a theory of development* (pp. 227–245). Oxford: Oxford University Press.

Minelli, A., & Pradeu, T. (2014). Theories of development in biology: Problems and perspectives. In A. Minelli & T. Pradeu (Eds.), *Towards a theory of development* (pp. 1–15). Oxford: Oxford University Press.

O'Malley, M., & Dupré, J. (2007). Size doesn't matter: Towards a more inclusive philosophy of biology. *Biology and Philosophy*, *22*, 155–191.

Oyama, S. (2000 [1985]). *The ontogeny of information*. Durham: Duke University Press.

Oyama, S., Griffiths, P., & Gray, R. (Eds.), (2001). *Cycles of contingency: Developmental systems and evolution*. Cambridge: MIT Press.

Peberdy, J. F. (1980). *Developmental microbiology*. New York: Wiley.

Peter, I., & Davidson, E. H. (2015). *Genomic control process: Development and evolution*. London: Academic Press.

Pradeu, T. (2014). Regenerating theories in developmental biology. In A. Minelli & T. Pradeu (Eds.), *Towards a theory of development* (pp. 15–32). Oxford: Oxford University Press.

Pradeu, T., et al. (2011). The boundaries of development. *Biological Theory*, *6*, 1–3.

Rheinberger, H.-J. (2000). Gene concepts: Fragments from the perspective of molecular biology. In P. Beurton, R. Falk, & H.-J. Rheinberger (Eds.), *The concept of the gene in development and evolution* (pp. 219–239). Cambridge: Cambridge University Press.

Saunders, J. W. (1966). Death in embryonic systems. *Science*, *154*, 604–612.

Schleiden, M. J. (1848). *The plant: A biography*. London: Hippolyte Bailliere.

Shimkets, L. J. (1999). Intercellular signalling during fruiting-body development of *Myxococcus xanthus*. *Annual Review of Microbiology*, *53*, 525–549.

Slack, J. M. W. (1991). *From egg to embryo* (2nd ed.). Cambridge: Cambridge University Press.

Slack, J. M. W. (2013). *Essential developmental biology*. Malden, MA: Blackwell.

St Johnston, D. (2015). The renaissance of developmental biology. *PLoS Biology*, *13*. e1002149.

Steeves, T. A., & Sussex, I. M. (1989). *Patterns in plant development*. Cambridge: Cambridge University Press.

Straight, P. D., & Kolter, R. (2009). Interspecies chemical communication in bacterial development. *Annual Review of Microbiology*, *63*, 99–118.

Sultan, S. E. (2015). *Organism and environment: Ecological development, niche construction, and adaptation*. Oxford: Oxford University Press.

Vervoort, M. (2014). Comparison of animal and plant development: A right track to establish a theory of development? In A. Minelli & T. Pradeu (Eds.), *Towards a theory of development* (pp. 203–217). Oxford: Oxford University Press.

Waddington, C. H. (1956). *Principles of embryology*. New York: Macmillan.

Weiss, P. (1959). Introduction. *Developmental Biology*, *1*(1), 1.

West-Eberhard, M. J. (2003). *Developmental plasticity and evolution*. Oxford: Oxford University Press.

Wolpert, L. (1968). The French flag problem: A contribution to the discussion on pattern development and regulation. In C. H. Waddington (Ed.), *Towards a theoretical biology: Vol. 1* (pp. 125 133). Edinburgh: Edinburgh University Press.

Wolpert, L. (1994). Do we understand development? *Science*, *266*, 571–572.

Wolpert, L., & Tickle, C. (2011). *Principles of development* (4th ed.). Oxford: Oxford University Press.

CHAPTER TEN

Concepts of Cell Lineage in Mammalian Embryos

Virginia E. Papaioannou[1]
Department of Genetics and Development, Columbia University Medical Center, New York, USA
[1]Corresponding author: e-mail address: vep1@cumc.columbia.edu

Contents

Abstract

Cell lineage is the framework for understanding cellular diversity, stability of differentiation, and its relationship to pluripotency. The special condition of *in utero* development in mammals has presented challenges to developmental biologists in tracing cell lineages but modern imaging and cell marking techniques have allowed the gradual elucidation of lineage relationships. Early experimental embryology approaches had limited resolution and relied of suboptimal cell markers and considerable disturbance to the embryos. Transgenic technology introduced genetic markers, particularly fluorescent proteins that, combined with sophisticated imaging modalities, greatly increase resolution and allow clonal analysis within lineages. The concept of cell lineage has also undergone evolution as it became possible to trace the lineage of cells based not only on their physical location or attributes but also on their gene expression pattern, thus opening up mechanistic lines of investigation into the determinants of cell lineage.

1. INTRODUCTION

Where did I come from? Where am I going? And why does it matter? We might all have felt the existential angst of these questions at some point in our lives, but it is really from the point of view, if such existed, of a cell in a

Current Topics in Developmental Biology, Volume 117
ISSN 0070-2153
http://dx.doi.org/10.1016/bs.ctdb.2015.10.013

mammalian embryo that I want to ponder this question and to consider how, over time, imaging and transgenics have honed our ability to trace the lineage of cells during development with increasing precision and allowed us to better understand why and how it matters. As technology has opened new windows into the process of embryogenesis, questions have been framed regarding the interplay between genetics, lineage, and cell behavior, and our concept of lineage analysis has expanded to include determining a mechanistic explanation for lineage patterns. Where things come from can to some extent explain how they got there, but only if we understand the mechanisms through which they reached their destinations. Cell lineage is the framework for understanding causes and mechanisms of cellular diversity, unification of the whole organism, cellular cooperation, stability of the phenotype, and its relationship to pluripotency. It is the thread that holds development together and is arguably the most important concept in developmental biology.

Thus, cell lineage is something every card-carrying developmental biologist thinks about and incorporates into their work. My own pedigree as a developmental biologist includes many cell lineage studies starting with my earliest work in Richard L. Gardner's laboratory. In the best tradition of experimental embryology, Richard had developed a method of introducing cells into preimplantation mammalian embryos that, combined with genetic differences between the introduced cells and the embryo, allowed him to trace the lineage of the introduced cells in a chimeric embryo (Gardner, 1968). Using this method and even more elaborate cut-and-paste micromanipulation procedures in which embryos were taken apart and put back together in different combinations, we carried out studies aimed at tracing the lineage of the earliest differentiated cell types of the mouse blastocyst, the inner cell mass, primitive endoderm, and trophectoderm, in an attempt to determine the origins of postimplantation tissues (Gardner & Papaioannou, 1975; Gardner, Papaioannou, & Barton, 1973; Papaioannou, 1982). Until that time, knowledge of cell lineage in mammalian embryos was inferred from histological studies or by analogy with other experimental organisms like the chick, which were much more straightforward to manipulate. These early studies were far from ideal lineage tracing studies: For one thing, the genetic markers available (isozyme variants) were ubiquitously expressed and developmentally neutral but were impossible to detect at the cellular/spatial level as the tissue was destroyed in the analysis; for another, the micromanipulation method was extremely invasive and the success of the introduction of foreign cells depended entirely on the mammalian embryo's inherent and immense tolerance of abuse and capacity for regulation. In spite of these limitations, the lineage relationships discerned and the fate maps inferred from these experiments provided a foundation of embryonic lineages that has been further refined over the years, but has essentially stood the test of time. This essay chronicles how experiments even with severe limitations can provide useful approximations of lineage relationships, and how the refining of those approximations with methodological innovations has broadened our concept of lineage relationships in the complex mammalian embryo.

2. HOW TO TRACE CELL LINEAGE

Prior to the late eighteenth century, preformationist ideas of embryonic development, in which body parts were preformed in the sperm and only needed to grow larger, did not require any concept of cell lineage. However, long before these ideas were rampant, Aristotle had formulated a theory that the adult existed only potentially in the embryo and that a process of differentiation occurred as development unfolded. It eventually became clear, with the invention and use of the microscope, that there was no homunculus lurking in the sperm and that body parts of a fetus or adult did indeed come from undifferentiated parts of an earlier stage embryo by a process of differentiation, proliferation, and cell movement, hence the need to trace lineage to understand how the final form came into being. Fate maps, the representation of the future fate of cells or tissues projected onto any given embryonic stage, became standard tools for thinking about and describing the dynamic process of differentiation and the destiny of cells during development.

In theory, tracing cell lineage in an embryo is simple: find some means of marking a cell without disturbing the embryo and then keep track of the marked cell and its descendents as the embryo develops to see where they go. In practice, however, this simple goal is fraught with difficulties, particularly in mammalian embryos that are protected from view by the mother's body, presenting a challenge for "seeing where they go." Similarly, the condition "without disturbing the embryo" is tricky. Either the act of marking the cell or the marker itself could potentially disturb the embryo. As for the marker, Anne McLaren proclaimed the attributes of the ideal genetic cell marker for tracking cell lineage in chimeras in her book "Mammalian Chimaeras" (McLaren, 1976). Similar criteria are still applicable to any marker used in cell lineage studies today. Ideally a marker would be (1) ubiquitous, so that any cell population could be marked and traced; (2) cell localized, marking cells and not, for instance, extracellular structures; (3) cell autonomous, so that the mark stayed with the original cell; (4) heritable and (5) stable in mitotic progeny, so that it was not diluted out or lost by cell division; (6) easy to detect, for obvious reasons; and (7) developmentally neutral, such that the presence of the marker did not alter the course of development. Furthermore, for genetic markers, multiple distinguishable forms or alleles would be valuable for tracking more than one cell at a time. McLaren claimed that the ideal marker did not exist, but fortunately that did not stop her or anyone since from developing lineage tracing markers with closer and closer approximations to the ideal.

3. THE CLASSIC EXPERIMENTAL EMBRYOLOGY APPROACH

We can think of mammalian lineage tracing studies as belonging to two different eras: the experimental embryology era and the genetic era. Although the early chimera experiments starting in the 1960s depended on genetic differences to distinguish cells, the methodology was in the tradition of pure experimental embryology—add a cell, take a cell away, move cells around, and see what happens. The goal of many studies was to test not only the fate of cells but also their potential, such that cells might be moved heterotopically (to a different location) and heterochronically (to a different aged embryo), as well as orthotopically (as close as possible to the original location). Even in orthotopic transfers where the goal was to elucidate a cell's normal fate, the extent to which development was disturbed by the procedures, including the time outside the mother's reproductive tract, was impossible to judge as the mammalian embryo's ability to regulate for abuse is staggering and the disturbed course of events may not be the same as the undisturbed course of events even though the end result is a normal embryo. Nonetheless, multiple approaches leading to similar results lent credibility to fate maps that were drawn. As an example, a study by Beddington (1982) probed the fate and potential of different regions of the epiblast by mechanically transferring clumps of marked cells orthotopically and heterotopically in embryos maintained in culture. The cells were labeled with tritiated thymidine, a marker that did not meet the criterion of "easy to detect" as it could only be visualized by autoradiography of sectioned embryos, and it had the disadvantage of dilution with mitosis, thus not meeting the "stable in mitotic progeny" criterion. The culture period no doubt also lent experimental variability, but in spite of these limitations, there was consistency in the results and general agreement with the fate maps of other well-studied amniotes such as chick.

A variation on the cut-and-paste chimeric approach to lineage studies, but still within the experimental embryology tradition, is the application of markers to cells in intact embryos. By leaving the tissues *in situ*, this method reduces disturbance to the embryos, although application of a marker could still affect the behavior of the marked cells. An example of this approach is a series of experiments by Lawson and Pedersen (1987) tracing the lineage of different cell types of the early postimplantation embryo by intracellular injection of horseradish peroxidase and locating the descendents of the cell several days later by enzymatic assay. This particular marker did not appear to adversely affect the cells and although it diluted out with mitosis, it was still easily detectable after 5–6 cell divisions. A feature of this study that came at a high cost in labor was that a single cell was injected and thus only a single clone from a lineage was traced in each embryo. This can be an advantage for precision mapping but also a disadvantage in that even a large number of single cell clones can only provide a sample of the whole picture. Also limiting was the period of invisibility between injecting the marker and harvesting the embryos—a black box between marking and analysis. What was missing was the possibility of seeing exactly how a cell got from one place to the next.

DiI, a lipophilic dye that is incorporated into cell membranes, is an example of a marker that can be followed through many cell divisions due to its intense fluorescence. A study from the Beddington lab using DiI labeling of the pregastrulation distal visceral endoderm inferred by lineage tracing that these cells moved into the position of anterior visceral endoderm (AVE; Thomas, Brown, & Beddington, 1998). The interest in these specific cells came from the expression pattern of the homeobox gene *Hex*, which was seen initially in the distal visceral endoderm and later in the AVE, giving the impression of cell movement. But, as pointed out by these authors, gene expression cannot serve as a lineage marker, hence the need for an actual lineage trace to show the relationship between distal visceral endoderm and AVE. This conjunction of lineage tracing and defining a cell population based on a specific gene expression pattern was a seminal experiment leading to the idea of the AVE, a classical extraembryonic tissue, as a patterning center for the anterior/posterior axis of the embryo and also presaging a new era of lineage tracing based on gene expression.

4. THE RISE OF "APPLIED" GENETIC MARKERS

Experimental embryology and transgenic technology began to come together dramatically for cell lineage studies in the 1990s. The barriers posed by inadequate or cumbersome cell markers and the problem of following cells in real time were answered with the introduction of fluorescent markers into the germ line of mice. Miraculously, the mammalian embryo once again proved to be quite tolerant, not only of the presence of foreign fluorescent proteins in its cells, but also of the various imaging modalities required to visualize the proteins in real time in living embryos in culture. The markers, being the products of transgenes, were indelible, heritable, and easy to follow. The first such transgenic marker, developed by Magda Zernicka-Goetz in Martin Evan's lab was green fluorescent protein (GFP) put under the control of a promoter active in proliferating cells (Zernicka-Goetz et al., 1997). A huge variety of transgenes and transgenic lines have since been fashioned for specific purposes, including, for example, a ubiquitous, nuclear-localized GFP line of mice produced in my lab by Kat Hadjantonakis (Hadjantonakis & Papaioannou, 2004) (Fig. 1), and many

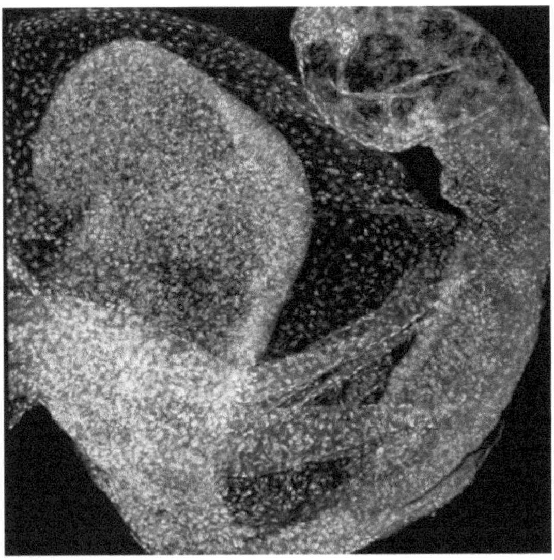

Figure 1 An embryo at midgestation carrying an H2B:eGFP transgene counterstained to highlight cell membranes. *Photo courtesy of Guy Eakin, Anna-Katerina Hadjantonakis, Virginia E. Papaioannou and Richard R. Behringer and reproduced with permission from Cold Spring Harbor Press (Papaioannou & Behringer, 2004).* (A color version of this figure is located at the end of this chapter.)

other, multicolor fluorescent genes under the control of tissue- or cell-specific regulatory elements.

Ubiquitously expressed markers are particularly useful for following cell lineage in chimeras, where marked cells are added to unmarked embryos, as in the work from Patrick Tam's lab establishing fate maps of the primitive streak at different stages of gastrulation (Kinder et al., 1999). Markers that fluoresce under the control of tissue-specific promoters, however, are more useful if one wants to trace cell lineage in undisturbed embryos. But with this type of study, the concept of cell lineage shifts slightly as the definition of the cell type being traced depends on the pattern of gene expression at a particular point in time, rather than on morphology and location. Gene expression is not necessarily a constant, however, and although a particular gene may mark a given cell population at one specific point in development, this may change with time and, in addition, there may be other areas of expression. As elegantly stated in a recent paper, marker gene expression indicates the "state not fate" of cells (Viotti, Foley, & Hadjantonakis, 2014). Only continuous monitoring of individual cells can unequivocally ascertain lineage relationships as the marker will provide a dynamic picture of gene expression changes such that a population of marked cells may grow or decline as cells turn the gene on or off with developmental progress.

Nonetheless, so-called tissue-specific promoters have been extremely valuable in providing insight into cell lineage based on gene expression, with the added benefits of tracking cell behavior. For example, in a study by Shankar Srinivas, Tristan Rodriguez, and others in Rosa Beddingdon's lab, a *Hex:GFP* transgene was used to mark cells of the distal visceral endoderm (Srinivas, Rodriguez, Clements, Smith, & Beddington, 2004). Endogenous *Hex* is expressed in the distal visceral endoderm and later in the AVE, suggesting but not proving that these two areas were linked by lineage. Experiments mentioned previously using DiI as an applied lineage marker supported this idea, but with the *Hex:GFP* transgene, labeled distal visceral endoderm cells could be followed in real time in living embryos. The results showed active migration of the fluorescently labeled cells toward the anterior midline, confirming the lineage of the AVE and corroborating the DiI applied marker study. The potential complications of cessation of expression or of new cells beginning to express the gene were abrogated by constant monitoring of individual fluorescent cells.

Another aspect of this study illustrates the ambiguity about what exactly is being marked by a so-called tissue-specific promoter. It was noted that *Hex* expression in the AVE is not uniform, i.e., there is a mixture of expressing and nonexpressing cells. Similarly, the *Hex:GFP* transgene faithfully recapitulated this salt-and-pepper expression, thus the lineage of only the expressing cells was followed in this experiment. This begs the question of whether the subset of nonexpressing cells of the AVE, which is a morphologically and functionally defined area, have a similar cell lineage, or whether they are derived from another embryonic source.

Patchy gene expression was not an issue in a study from the Hadjantonakis laboratory (Kwon, Viotti, & Hadjantonakis, 2008) in which an entire cell population, the visceral endoderm of the pregastrulation embryo was labeled with an *Afp::GFP* transgene. This study shows the power of lineage tracing an entire population of cells, rather than individual clones or a subset of cells. The results led to an unexpected conclusion at odds with the prevailing models of endoderm development in that they revealed a persistence of the progeny of visceral endoderm cells in areas of the embryos formerly thought to consist entirely of newly formed definitive endoderm from the anterior primitive streak, and further, revealed a contribution of these so-called extraembryonic cells to the fetal gut. In this case it was the perdurance of the fluorescent protein that allowed the recognition of this lineage of cells after the endogenous *Afp* promoter was downregulated. Live imaging then allowed an assessment of the behavior of these cells and the discovery of the intercalation of definitive endoderm by a process of "coordinated egression" rather than displacement of an intact epithelial sheet, as was previously thought.

5. FURTHER REFINEMENTS FOR GENE EXPRESSION LINEAGE TRACING

Rather than putting the cell marker directly downstream of a promoter region of interest, another tactic is to drive Cre recombinase by a specific *cis*-regulatory element and then combine this transgene with a separate Cre-reporter transgene such that a recombination event occurs where Cre is active, causing a marker to be expressed. As this is a heritable genetic change, any cell where the recombination has occurred and all its progeny will express the marker even when gene expression stops. This allows for long term lineage tracing with the caveat that additional cells will be added to the marked lineages as long as activity of the regulatory element of interest continues and wherever else it drives Cre expression. An additional level of control can be added by making the Cre driver inducible, so that the timing of activity of Cre can be experimentally controlled. In this way, cohorts of cells expressing a particular gene (captured by the regulatory element driving that gene) can be marked within particular temporal limits.

Of course each additional experimental manipulation adds complexity and occasionally a limitation to the tracing of undisturbed cell lineages. To wit, the presence of the foreign protein Cre may not always be developmentally neutral (e.g., Lewis, Vasudevan, O'Neill, Soriano, & Bush, 2013; Naiche & Papaioannou, 2007), but when properly controlled, these experimental methods can trace cells into postnatal stages. Conceptually, however, it is important to realize that the transient expression of a particular gene may have little do with the eventual fate of the progeny cells: lineage tracing of cells from a certain location that express a particular gene will show where those cells eventually come to rest, but whether or not that gene played any role in the course of their lineage is a different question altogether. To get at this question, one useful approach is to look in mutant embryos for cell lineage changes due to the loss of the gene.

6. CLONAL ANALYSIS WITHIN CELL LINEAGES

Analysis of the behavior of individual cell clones is related to, but distinct from, cell lineage analysis. Typically, clonal analysis follows all the progeny of a particular cell, but in the context of development, the clonal progeny of a cell from an early embryo may contribute to multiple "lineages" as those lineages segregate during differentiation. In other words, clonal analysis is based simply on the mitotic progeny of a single cell, whilst cell lineage studies are based on following the fate of cells, usually a population of cells, defined by a particular anatomical location and/or gene expression pattern, most commonly after some lineage segregation has occurred. That said, the two concepts are complementary in that every cell has a clonal origin and future (if it is still dividing), and is also part of a cell lineage in terms of its position and differentiated state. Clonal analysis methods are useful for probing the fine grained behavior of individual cells within lineages composed of many cells.

Retrospective analysis of cell clones in genetic mosaics, where a somatic event creates a genetic mutation detectable at the cellular level, has been a classic technique of experimental embryology, particularly in *Drosophila*. Variations on this method have been applied to the mouse using rare somatic mutations in transgenes. An early example was a defective *lacZ* transgene that could be repaired by rare, sporadic interchromosomal recombination events to produce a functional gene that produced a detectable LacZ marker.

This was used, for example, with a ubiquitous promoter to mark random clones to investigate lineage segregation during early embryogenesis and gastrulation (Tzouanacou, Wegener, Wymeersch, Wilson, & Nicolas, 2009). Limitations of this type of retrospective analysis are that the cell of origin is not defined and it can be difficult to distinguish redundant events, particularly when the recombination events are frequent. Thus in this example, analysis was limited to cases where the recombination events were extremely rare, necessitating a large number of specimens to obtain a reasonably comprehensive representation of the patterns of clonal growth. Nonetheless, this study was important in confirming the existence of a bipotential neuromuscular progenitor pool persisting throughout axis elongation.

Fluorescent proteins can also be used for clonal analysis with the advantage that there are many spectral variants that can be simultaneously visualized and distinguished in living embryos. Variations on the theme of clonal analysis have been developed using transgenes with the colorful names of Brainbow, Rainbow and Confetti, which, as their names imply, result in clones of different hues through recombinase-mediated excision and/or inversion of multiple fluorescent protein transgene cassettes to produce recombinant cells with many possible outcomes. Thus, stochastic recombination allows distinction among multiple clones within a population, and multiple clones can be followed within each specimen. Originally developed to study neurons in the brain, they have wide application (Weissman & Pan, 2015). Baggiolini and colleagues used the Confetti transgene, combined with inducible, tissue-specific promoters driving Cre recombinase, to answer a long controversial issue regarding the potency of neural crest cells before and after they leave the neural tube. Using rigorous statistical methods to exclude redundant recombination events, they found that most neural crest cells are multipotent both pre- and postmigration, but that their fates can be heterogeneous, as seen by variation in the range of differentiated fates in different clones (Baggiolini et al., 2015).

7. MUTANT EFFECTS ON CELL LINEAGE

Over the years, a lot of effort has gone into developing the perfect marker for lineage tracing to better understand cell behavior in undisturbed embryos. But one particular kind of disturbance forms the basis of a slightly different concept of cell lineage, that is, investigating the properties of cells with mutations by watching how they interact with normal cells during development. This has long been a staple in the analysis of chimeras to distinguish cell autonomous from nonautonomous mutant effects (e.g., Papaioannou & Gardner, 1979) and of course, determining the tissue-specific effects of any mutation is a standard part of mutation analysis (Papaioannou & Behringer, 2004). But the question here has more to do with competition and whether mutant cells behave differently from wild type cells in the same environment. Thus, the techniques of mosaic analysis can be modified slightly to juxtapose mutant and normal cells with a minimum of disruption to other aspects of development. A beautiful example of this combines the mosaic analysis with double markers system with a linked mutation to mark the two daughter cells of a rare clone with different fluorescent markers and simultaneously produce a homozygous mutation in one of the daughters. This is accomplished through Cre-mediated interchromosomal recombination of two loxP-containing transgenes at homologous chromosomal locations. Originally developed and used to trace neuronal connections (Zong, Espinosa, Su, Muzumdar, & Luo, 2005), the system has great potential for approaching questions of microlineage and the effect of mutations during organogenesis. Frank Costantini's laboratory, for example, has used this method to investigate branching morphogenesis of the ureteric bud during kidney development by following double marked clones of wild type and mutant ureteric bud epithelium with time-lapse imaging *in vitro*. By following differentially marked sister clones with one clone wild type and the other with mutations in the receptor tyrosine kinase gene *Ret* or its target gene, *Etv4,* they showed that Ret/Etv4 signaling regulates cell movements that are part of the basic mechanism of ureteric bud branch formation, independently of effects on cell proliferation (Riccio, Cebrian, Zong, Hippenmeyer, & Costantini, 2016).

8. FINAL THOUGHTS

One of the special challenges of studying mammalian embryology is the flexibility and regulative capacity of the embryo. Because of this ability to absorb and compensate for disruptions, it stands to reason that cell lineages even in undisturbed embryos will not be invariant but that there will be wiggle room at the margins as a tradeoff between flexibility and stability. Nonetheless, the consistency of development implies that lineage history serves as a guiding force in cell fate decisions as development unfolds, so the questions of where do cells come from and where do they go are highly relevant. Lineage studies can link morphology with gene expression patterns and cell fate, converging on the functional role of genes in development and reconciling heterogeneity of gene expression with fate maps. In the future, following the fate of cells and clones of cells with specific, multiple-gene expression profiles will provide an even finer-grained map so that cell lineage studies will not be going out of fashion any time soon.

ACKNOWLEDGMENTS

I would like to thank Kat Hadjantonakis for helpful comments on the manuscript. This work was supported by the Eunice Kennedy Shriver National Institute of Child Health & Human Development of the National Institutes of Health (NIH) grant R37HD033082. The content is solely the responsibility of the author and does not necessarily represent the official views of the NIH.

REFERENCES

Baggiolini, A., Varum, S., Mateos, J. M., Bettosini, D., John, N., Bonalli, M., et al. (2015). Premigratory and migratory neural crest cells are multipotent in vivo. *Cell Stem Cell, 16*(3), 314–322. http://dx.doi.org/10.1016/j.stem.2015.02.017.

Beddington, R. S. (1982). An autoradiographic analysis of tissue potency in different regions of the embryonic ectoderm during gastrulation in the mouse. *Journal of Embryology and Experimental Morphology, 69*, 265–285.

Gardner, R. L. (1968). Mouse chimeras obtained by the injection of cells into the blastocyst. *Nature, 220*, 596–597.

Gardner, R. L., & Papaioannou, V. E. (1975). Differentiation in the trophectoderm and inner cell mass. In M. Balls & A. E. Wild (Eds.), *Early development of mammals* (pp. 107–132). Cambridge: Cambridge Univ. Press.

Gardner, R. L., Papaioannou, V. E., & Barton, S. C. (1973). Origin of the ectoplacental cone and secondary giant cells in mouse blastocysts reconstituted from isolated trophoblast and inner cell mass. *Journal of Embryology and Experimental Morphology, 30*, 561–572.

Hadjantonakis, A.-K., & Papaioannou, V. E. (2004). *BMC Biotechnology, 4*, 33.

Kinder, S. J., Tsang, T. E., Quinlan, G. A., Hadjantonakis, A. K., Nagy, A., & Tam, P. P. (1999). The orderly allocation of mesodermal cells to the extraembryonic structures and the anteroposterior axis during gastrulation of the mouse embryo. *Development, 126*, 4691–4701.

Kwon, G. S., Viotti, M., & Hadjantonakis, A. K. (2008). The endoderm of the mouse embryo arises by dynamic widespread intercalation of embryonic and extraembryonic lineages. *Developmental Cell, 15*, 509–520.

Lawson, K. A., & Pedersen, R. A. (1987). Cell fate, morphogenetic movement and population kinetics of embryonic endoderm at the time of germ layer formation in the mouse. *Development, 101*, 627–652.

Lewis, A. E., Vasudevan, H. N., O'Neill, A. K., Soriano, P., & Bush, J. O. (2013). The widely used Wnt1-Cre transgene causes developmental phenotypes by ectopic activation of Wnt signaling. *Developmental Biology, 379*, 229–234.

McLaren (1976). *Mammalian chimaeras* (p.154). Cambridge: Cambridge University Press.

Naiche, L. A., & Papaioannou, V. E. (2007). *Genesis, 5*, 768–775.

Papaioannou, V. E. (1982). Lineage analysis of inner cell mass and trophectoderm using microsurgically reconstituted mouse blastocysts. *Journal of Embryology and Experimental Morphology, 68*, 199–209.

Papaioannou, V. E., & Behringer, R. R. (2004). *Mouse phenotypes, a handbook of mutation analysis* (p. 226). Cold Spring Harbor, NY: Cold Spring Harbor Press.

Papaioannou, V. E., & Gardner, R. L. (1979). Investigation of the lethal yellow Ay/Ay embryo using mouse chimaeras. *Journal of Embryology and Experimental Morphology, 52*, 153–163.

Riccio, P., Cebrian, C., Zong, H., Hippenmeyer, S., & Costantini, F. (2016). *Ret* and *Etv4* promote directed movements of progenitor cells during renal branching morphogenesis. *PLoS Biology,*(in press).

Srinivas, S., Rodriguez, T., Clements, M., Smith, J. C., & Beddington, R. S. (2004). Active cell migration drives the unilateral movements of the anterior visceral endoderm. *Development, 131*, 1157–1164.

Thomas, P. Q., Brown, A., & Beddington, R. S. P. (1998). Hex: A homeobox gene revealing peri-implantation asymmetry in the mouse embryo and an early transient marker of endothelial cell precursors. *Development, 125*, 85–94.

Tzouanacou, E., Wegener, A., Wymeersch, F. J., Wilson, V., & Nicolas, J. F. (2009). Redefining the progression of lineage segregations during mammalian embryogenesis by clonal analysis. *Developmental Cell, 17*, 365–376.

Viotti, M., Foley, A. C., & Hadjantonakis, A. K. (2014). Gutsy moves in mice: Cellular and molecular dynamics of endoderm morphogenesis. *Philosophical Transactions of the Royal Society of London. Series B, Biological Sciences, 369*, 1657. http://dx.doi.org/10.1098/rstb.2013.0547. (Dec 5, pii: 20130547).

Weissman, T. A., & Pan, Y. A. (2015). Brainbow: New resources and emerging biological applications for multicolor genetic labeling and analysis. *Genetics, 199*(2), 293–306. http://dx.doi.org/10.1534/genetics.114.172510.

Zernicka-Goetz, M., Pines, J., McLean Hunter, S., Dixon, J. P., Siemering, K. R., Haseloff, J., et al. (1997). Following cell fate in the living mouse embryo. *Development, 124*, 1133–1137.

Zong, H., Espinosa, J. S., Su, H. H., Muzumdar, M. D., & Luo, L. (2005). Mosaic analysis with double markers in mice. *Cell, 121*, 479–492.

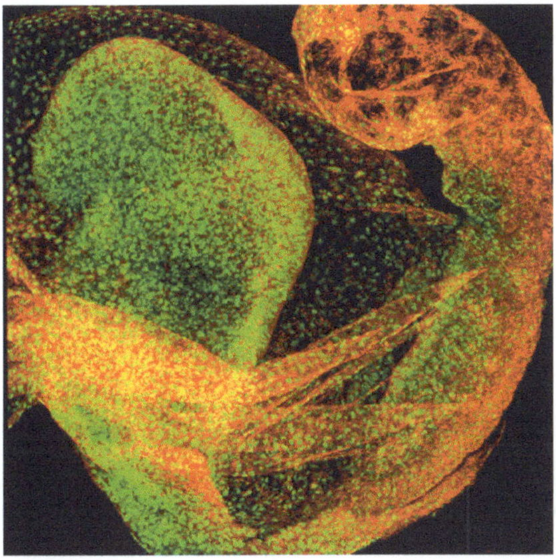

Virginia E. Papaioannou, Figure 1 An embryo at midgestation carrying an H2B:eGFP transgene counterstained to highlight cell membranes. *Photo courtesy of Guy Eakin, Anna-Katerina Hadjantonakis, Virginia E. Papaioannou and Richard R. Behringer and reproduced with permission from Cold Spring Harbor Press (Papaioannou & Behringer, 2004).*

CHAPTER ELEVEN

Small Peptides as Newcomers in the Control of *Drosophila* Development

Jennifer Zanet*[,†], Hélène Chanut-Delalande*[,†], Serge Plaza*[,†,1], François Payre*[,†,1]

*Centre de Biologie du Développement, Université de Toulouse, UPS, Toulouse, France
†Centre de Biologie du Développement, CNRS, UMR5547, Toulouse, France
[1]Corresponding authors: e-mail address: serge.plaza@univ-tlse3.fr; francois.payre@univ-tlse3.fr

Contents

Abstract

Throughout the last century, studies using the fruit fly have contributed to the discovery of many key genetic elements that control animal development. Recent work has shed light on an unexpectedly large number of RNAs that lack the classical hallmarks of protein-coding genes and are thus referred to as noncoding RNAs. However, there is mounting evidence that both mRNA and noncoding RNAs often contain small open reading frames (sORFs/smORFs), which can be translated into peptides. While genome-wide profiling supports a pervasive translation of these noncanonical sORF/smORF/SEP peptides, their functions remain poorly understood. Here, we review recent data obtained in *Drosophila* demonstrating the overlooked role of smORF peptides in the control of development and adult life. Focusing on a few smORF peptides whose functions have been elucidated recently, we discuss the importance of these newly identified regulatory molecules and how they act to regulate the building and function of the whole organism.

Current Topics in Developmental Biology, Volume 117
ISSN 0070-2153
http://dx.doi.org/10.1016/bs.ctdb.2015.11.004

1. INTRODUCTION

Animal development fundamentally relies on the establishment of a complex array of spatiotemporal patterns of gene expression, gradually specifying embryonic territories and executing their differentiation. Therefore, the question of the total number of genes in a given animal specie has been a long-lasting problem in biology. Whole genome sequencing was expected to provide a definitive answer to this question. However, the human genome surprisingly encodes only approximately 25,000 protein-coding genes, i.e., barely four times more than the unicellular yeast. Equally surprising are the modest 13,500 genes in *Drosophila*, a number significantly lower than most conservative estimates based on genetic analyses. Fifteen years after the first genome release, it has now become clear that this protein-centric view of gene content strongly underestimates the number of DNA regions that are transcribed and fulfill important functions for development and physiology. A first milestone was the discovery of hundreds of small (<200 nt) noncoding RNAs, such as miRNAs, siRNAs, or piRNAs, which play regulatory roles in the silencing of genes and transposable elements (reviewed in Moazed, 2009). More recently, genome-wide transcript profiling using RNA-Seq has disclosed the existence of numerous larger RNAs, referred to as long noncoding RNAs (lncRNAs or lincRNAs) since they lack the classical hallmarks of protein-coding genes (Banfai et al., 2012; Cech & Steitz, 2014). Although the functions of lncRNAs remain unknown in the vast majority of cases, there are experimental studies showing key roles of lncRNAs as functional RNAs controlling genome expression at various levels (Guttman & Rinn, 2012).

While RNAs annotated as noncoding cannot produce large-sized proteins, they contain myriads of small open reading frames (smORFs/sORFs) and a growing body of evidence shows that they are often translated into small or short peptides (reviewed in Andrews & Rothnagel, 2014; Pauli, Valen, & Schier, 2015). In addition to lncRNAs, recent work further shows that virtually all classes of RNAs including mRNA (Ingolia et al., 2014), small noncoding RNA (Lauressergues et al., 2015), and antisense transcripts or introns (Prabakaran et al., 2014) can produce smORF-encoded peptides. Even small genomes of bacteria (Storz, Wolf, & Ramamurthi, 2014), mitochondria (Capt, Passamonti, & Breton, 2015; Lee et al., 2015), and viruses (Ingolia et al., 2014; Jaber & Yuan, 2013) have recently proven to encode "new" RNAs translated into peptides. Aside from reaching a full identification of the smORF peptidome, the challenge is now to understand the functions and mode of action of smORF peptides. From a few examples in *Drosophila*, we suggest that smORF peptides may often act as cofactors of protein complexes, constituting an efficient way to control the function of cellular proteins.

2. smORF PEPTIDES: FROM EARLY DISCOVERY TO GENOME-WIDE PROFILING

While there is no genuine definition of smORF peptides (also named, SEP for small encoded peptides or microproteins), it is often considered that they correspond to short proteins inferior to 100 amino acids (aa), translated from a collinear short ORF. Bioactive peptides acting as hormones, e.g., neuropeptides, do not belong to this class since they are generally produced from the post-translational processing of larger precursors. On the other hand, small protein-like ubiquitin (76aa) or antimicrobial peptides (AMPs <100aa), which were previously identified by biochemical or genetic approaches, should be added to the growing family of smORF peptides. A cut-off of approximately 100 codons was often introduced in genome annotations in order to filter off insignificant short ORFs that were expected to occur by chance in any RNA sequence longer than 1000 bp (Dinger, Pang, Mercer, & Mattick, 2008). Similar strategies were used to discriminate whether novel transcripts were coding or noncoding. Thus, every RNA that contains only small ORFs <100aa is generally annotated as noncoding. While there is clear evidence that some RNAs act intrinsically as functional molecules (Guttman & Rinn, 2012), the number of biologically active small proteins/peptides is certainly greatly underestimated (Andrews & Rothnagel, 2014). The picture might be even more complex, given that some RNAs are bifunctional, i.e., act at the RNA level and encode functional peptides (Lauressergues et al., 2015). Therefore, instead of a strict dichotomy between coding and noncoding, eukaryotic genes probably express a gradual spectrum of RNAs, from those encoding proteins, peptides and/or acting as functional RNAs (Pauli et al., 2015).

One of the first smORF peptide was identified by a genetic screen in *Bacillus subtilis* (Sandman, Losick, & Youngman, 1987), showing that the 26aa peptide spoVM is essential for endospore formation. Small peptides were also discovered in plants, such as ENOD40 (van de Sande et al.,

1996) encoding 12 and 24aa peptides implicated in the division of cortical cells and the 36aa Polaris peptide required for root formation (Topping & Lindsey, 1997). Recent studies extended the number of smORF-encoding genes in plants to at least several hundreds (Guillen et al., 2013; Hanada et al., 2013; Hanada, Zhang, Borevitz, Li, & Shiu, 2007). In the same vein, many conserved short ORFs have been identified in the yeast genome, with hundreds of new genes encoding peptides in *S. cerevisiae* (Kastenmayer et al., 2006; Smith et al., 2014).

Genome-wide approaches now further provide definitive evidence that ncRNAs are often translated into small peptides. For example, recent work in zebrafish has identified 190 transcripts encoding small peptides (Bazzini et al., 2014), including Toddler a peptide defining a novel—and evolutionary conserved—signaling pathway (Chng, Ho, Tian, & Reversade, 2013; Pauli et al., 2014). More than 300 smORF peptides were also scored in human cell lines (Ma et al., 2014; Slavoff et al., 2013). In *Drosophila*, 228 out of 274 smORF-containing transcripts analyzed are translated (Aspden et al., 2014). Studies using comparative genomics (Hanada et al., 2013; Ladoukakis, Pereira, Magny, Eyre-Walker, & Couso, 2011), proteomics (Ma et al., 2014; Slavoff et al., 2013), or ribosome profiling coupled to evolutionary conservation (Bazzini et al., 2014; Crappe et al., 2013) all converge to the conclusion that the number of short proteins is far higher than previously suspected (Fig. 1). In the following, we focus on the mechanism of action of smORF peptides in the control of *Drosophila* development and physiology.

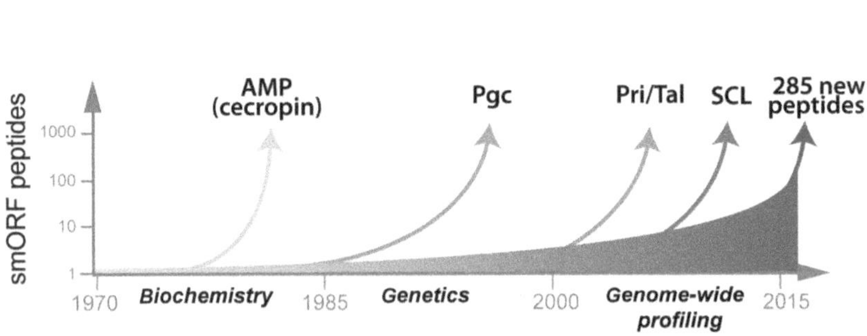

Figure 1 The fast growing family of small peptides in *Drosophila*. Chronological overview of the discovery of short ORF peptides in *Drosophila* and the strategies leading to their discovery. (A color version of this figure is located at the end of this chapter.)

3. THE PGC smORF PEPTIDE PROTECTS THE GERMLINE FROM EMBRYONIC TRANSCRIPTIONAL ACTIVATION

The early stages of *Drosophila* development take place within a syncitial environment. Primordial germ cells are the first cells to individualize and segregate from other embryonic nuclei. Doing so, germ cells escape from global zygotic transcription activation, remaining transcriptionally silent through global repression of RNA polymerase II (RNA PolII) activity. *Polar granule component* (*pgc*) was identified as an atypical RNA specific of nascent germ cells and required for germline development (Nakamura, Amikura, Mukai, Kobayashi, & Lasko, 1996). *pgc* is transcribed in ovaries and transported to the posterior pole of the female gamete (Little, Sinsimer, Lee, Wieschaus, & Gavis, 2015), where germ cell precursors form during early embryogenesis. *pgc* inactivation impairs germ cell migration to the embryonic gonads (Fig. 2A), ultimately leading to agametic adults. The lack of *pgc* induces ectopic expression of somatic genes in germ cells, correlating with premature phosphorylation of the Ser2 residue in the C-terminal domain (CTD) of RNA PolII (Fig. 2B), a post-translational modification required for transcriptional elongation (Martinho, Kunwar, Casanova, & Lehmann, 2004). *pgc* was initially proposed to act as a ncRNA due to the lack of a long ORF with canonical *Drosophila* codon usage. However, the release of genome sequences for 12 *Drosophila* species highlighted the evolutionary conservation of a short ORF encoding a 71aa long peptide. Immunodetection of the Pgc peptide in primordial germ cells provides direct evidence for its translation (Fig. 2B), and a frameshift mutation is sufficient to inactivate *pgc* function *in vivo*. Furthermore, injection of purified Pgc peptide inhibits RNA PolII CTD phosphorylation, both in *ex vivo* and *in vivo* assays, demonstrating that this 71aa peptide is indeed responsible for *pgc* function (Hanyu-Nakamura, Sonobe-Nojima, Tanigawa, Lasko, & Nakamura, 2008; Timinszky, Bortfeld, & Ladurner, 2008). CTD phosphorylation results from the activity of P-TEFb, a protein complex comprising the Cdk9 kinase and a regulatory subunit cyclin T (CycT). Similar to the lack of Pgc, overexpression of P-TEFb induces premature CTD phosphorylation when expressed in germ cells, suggesting that the Pgc peptide may repress P-TEFb activity for the transcriptional quiescence of primordial germ cells. Consistent with this hypothesis, forced Pgc expression in somatic tissues precludes P-TEFb recruitment to transcriptionally active chromatin

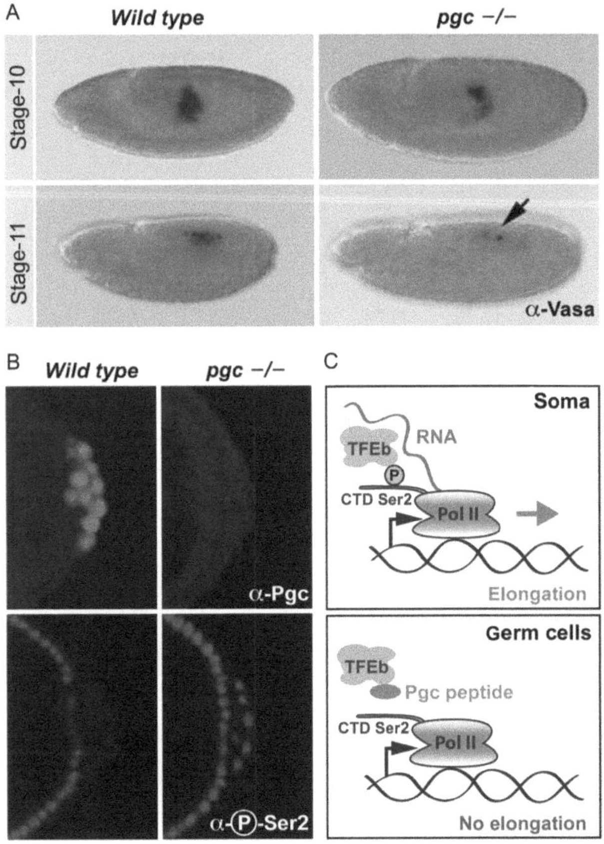

Figure 2 The Pgc peptide controls PolII transcriptional elongation. (A) Defects in germ cell migration in *pgc* mutant embryos, as observed at two successive stages of development during mi-embryogenesis. Migrating germ cells are visualized by immunostaining with an anti-Vasa antibody. (B) Top: Expression of the Pgc peptide detected in polar germ cells in wild-type early embryos and in *pgc* mutants. Bottom: Absence of PolII CTD phosphorylation in normal polar germ cells, whereas precocious PolII CTD phosphorylation is detected in pole cells lacking the Pgc peptide. (C) Model for the action of the Pgc peptide. Binding of the Pgc peptide to the pTFEb protein complex downregulates its kinase activity and counteracts PolII CTD Ser2 phosphorylation that normally occurs in somatic cells. (A color version of this figure is located at the end of this chapter.)

sites and represses CTD phosphorylation (Hanyu-Nakamura et al., 2008). Biochemical analysis of a Pgc recombinant peptide revealed that it adopts an alpha-helical structure that binds to Cdk9, forming a ternary complex with CycT, and leading to the sequestration of P-TEFb (Fig. 2C). Therefore, Pgc represses mRNA transcription during germ cell specification by preventing P-TEFb recruitment to active promoters, leading to the decrease in RNA polII CTD phosphorylation. Transcriptional quiescence is a shared characteristic of primordial germ cells across various species, including worms and *Xenopus* (Nakamura & Seydoux, 2008). For example, although segregated from somatic cells at later stages of development, worm germ cells are maintained transcriptionally inactive by PIE-1, a zinc finger protein that prevents CTD phosphorylation possibly through competitive inhibition (Strome & Lehmann, 2007). Therefore, *Drosophila* has evolved a different mechanism whereby the Pgc peptide binds to, and downregulates, the activity of the PTEF-b protein complex to protect germ cells from general activation of the zygotic genome.

4. CONTROL OF CALCIUM INTRACELLULAR TRAFFICKING BY REGULATORY SMALL PEPTIDES

Bioinformatic analyses of the *Drosophila* genome and ribosome profiling predict hundreds of new genes encoding smORF peptides (Aspden et al., 2014; Ladoukakis et al., 2011). Among these candidates, it has been shown that *pncr003:2L*, a transcript previously annotated as noncoding (Tupy et al., 2005), encodes two highly related peptides (Magny et al., 2013; Savard, Marques-Souza, Aranda, & Tautz, 2006) (see Fig. 3A). During embryogenesis, *pncr003:2L* is expressed in somatic muscles, while it is restricted to the heart in later stages of development and in adults. As deduced from immunodetection *in vivo*, these peptides are translated and accumulated in the sarcoplasmic reticulum (SR) of myocytes (Magny et al., 2013). While *pncr003:2L* loss of genetic function does not lead to detectable alteration in the pattern of muscle formation, mutant flies display severe defects in cardiac muscle function. This includes arrhythmia with perturbations in action potential, as measured by electrophysiology. Analysis of the genetically encoded calcium indicator, GCaMP3, revealed that the absence of *pncr003:2L* increases transient calcium amplitudes and sharpens decay rates, although contractility seems unaffected (Magny et al., 2013). These defects can be rescued by expressing one or the other peptide, but not by a *pncr003:2L* RNA that bears point mutations frameshifting the small ORFs, providing evidence for functional peptides. Although the primary sequence of *pncr003:2L* peptides evolved rapidly across species it nonetheless shares weak similarities with sarcolipin (SLN) and phospholamban (PLN) (Fig. 3A), two peptides previously identified in mammals for their role in regulating the activity of a calcium pump called SERCA, in the heart and smooth muscles (Tada, Kirchberger, & Katz, 1975; Wawrzynow et al.,

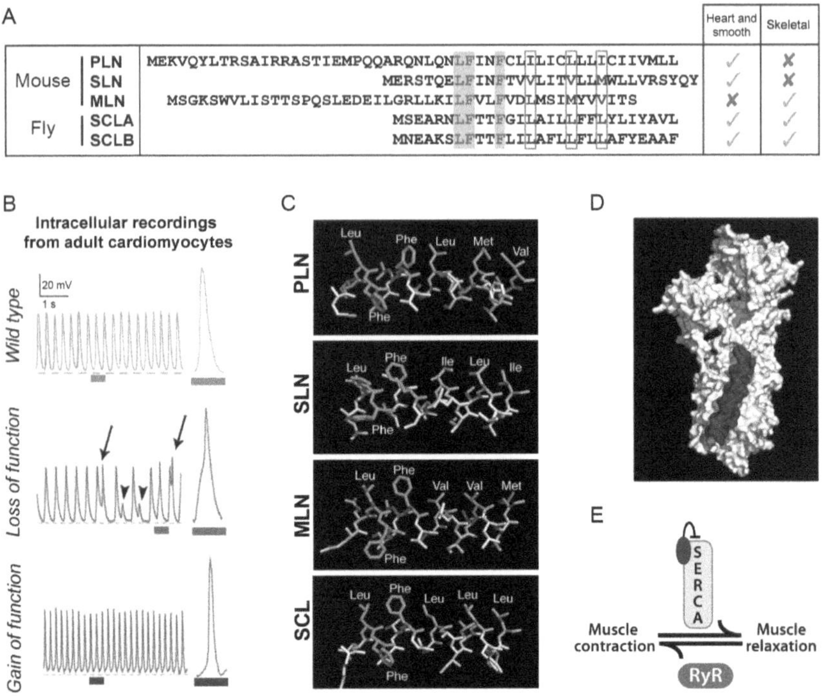

Figure 3 Conservation and function of peptides in muscle physiology. (A) The graph shows sequence alignment of mouse phospholamban (PLN), sarcolipin (SLN), myoregulin (MLN), and *Drosophila* sarcolamban (SCLA and SCLB) peptides, with identical (shaded) and similar (boxed) residues, as well as the type of muscles in which they are expressed. (B) Intracellular recordings from *Drosophila* wild-type (top) and *scl* loss- (middle) or gain-of-function (bottom) adult cardiomyocytes. Arrows indicate double action potential. Arrowheads indicate failed action potential. A sample peak is magnified at the right of each trace. (C) Structure of predicted helices of PLN, SLN, MLN, and SCL peptides, showing a common spatial orientation of conserved residues. (D) The SLN (blue) peptide fits a groove in the structure of the SERCA1 protein (white). (E) SLN peptides bind to the SERCA calcium pump and inhibit its activity, thus decreasing phases of muscle relaxation. (A color version of this figure is located at the end of this chapter.)

1992). Peptides encoded by *pncr003:2L* were thus renamed sarcolamban (SCL) since they also bind to Ca-P60A, the fly ortholog of SERCA (Magny et al., 2013). Expression of a dominant-negative variant of Ca-P60A that inhibits calcium uptake in the SR partially rescues the arrhythmic and transient decay rate phenotypes of *pncr003:2L* mutants, supporting the conclusion that sarcolamban peptides normally downregulate Ca-P60A activity (Fig. 3B). Consistent with an evolutionary role of these small peptides in the control of heart function, overexpression of *Drosophila* SCL peptides, as well as human SLN or PLN peptides, causes cardiac arrhythmia in otherwise wild-type flies (Magny et al., 2013). Recent studies in mice also identified an additional SERCA-binding peptide, myoregulin (MLN), which is specific of skeletal muscles (Anderson et al., 2015). Although SCL, PLN, SLN, and MLN peptides exhibit only a few conserved residues, they all adopt a similar helical conformation, with conserved amino acids positioned in the same spatial orientation (Fig. 3C). As deduced from molecular modeling, these residues form a hydrophobic binding motif directly involved in the interaction with SERCA (Fig. 3D) and their mutation is sufficient to abrogate SERCA binding (Anderson et al., 2015). Despite a weak conservation in primary sequence, these smORF peptides therefore share a common structural organization responsible for the binding to SERCA and, thereby, for the regulation of muscle function and performance from invertebrates to humans (Fig. 3E).

5. MULTIPLE FUNCTIONS OF MLPT/TAL/PRI smORF PEPTIDES IN THE CONTROL OF INSECT DEVELOPMENT

One of the best-understood cases of small peptides as key regulators of animal development is *mille pattes/tarsal-less/polished rice*, a paradigm of apparently noncoding RNA in insects that actually acts through the production of smORF peptides. In the red flour beetle *Tribolium castaneum*, *mille pattes* (*mlpt*) was isolated from a large-scale expression screen because of its specific expression in the posterior growth zone (Savard et al., 2006). This embryonic territory allows the sequential production of posterior segments in species with a so-called short-germ development, which likely corresponds to the ancestral mode of segmentation. Accordingly, *mlpt* RNAi leads to strong segmentation defects, treated embryos displaying a gap-like phenotype (Savard et al., 2006). Ultimately, *mlpt* embryos develop into highly abnormal animals, lacking posterior segments, and showing partial transformation toward thoracic fate with ectopic legs (Fig. 4A). In flies, *tarsal-less* (*tal*) was identified from a forward genetic screen focusing on appendage formation (Galindo, Pueyo, Fouix, Bishop, & Couso, 2007). The initial *tal* mutation leads to abnormal adult legs that lack most tarsal segments (Fig. 4B), whereas loss-of-function alleles are lethal. The same gene was independently identified, and named *polished rice* (*pri*), as it expresses a conserved noncoding RNA showing dynamic tissue-specific embryonic

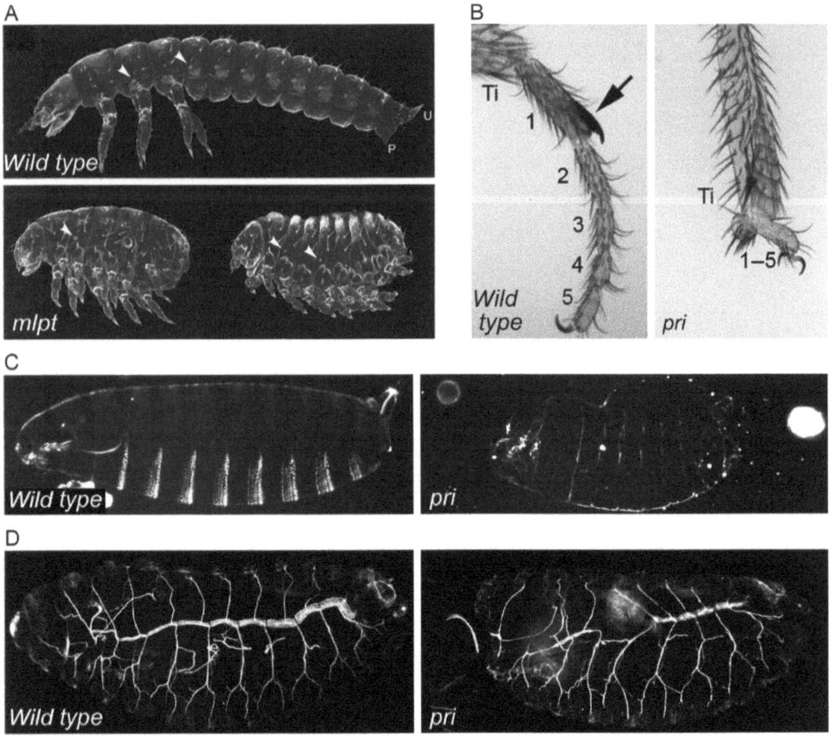

Figure 4 Different functions of *mlpt/tal/pri* during *Tribolium* and *Drosophila* development. (A) Cuticles produced by *Tribolium* first instar larvae. The depletion of *mlpt* induces strong segmentation defects and supernumerary legs. Arrowheads indicate tracheal pits; which are normally specific of the second thoracic and abdominal segments. U, urogomphi; P, pygopodes. (B) *Drosophila* forelegs of adult males showing that inactivation of *pri/tal* leads to lack of terminal tarsal structures (1–5); the tibia (Ti) is not affected. The arrow indicates that the sex comb, a male-specific structure of the first tarsal segment. (C) Cuticles of *Drosophila* larvae highlighting the epidermal differentiation defects observed in *pri* mutant embryos, which totally lack apical denticles and trichomes. (D) Tracheal system of *Drosophila* embryos, as revealed by immunostaining with an anti-Piopio antibody. *pri* mutant embryos display breaks in the main tracheal tube.

expression (Inagaki et al., 2005; Kondo et al., 2007). Reverse genetic analysis revealed essential embryonic functions of this RNA, since mutant embryos die with strong defects in epidermal and tracheal differentiation (Fig. 4C and D). *mlpt/tal/pri* RNA lacks a large-sized ORF and several lines of evidence indicated that its function relies on peptide products. First, it contains four small ORFs that exhibit evolutionary conservation of their coding potential across considerable distances, with all peptides sharing the C-terminal motif

LDPTGnY from dipterans to crustaceans (Galindo et al., 2007; Kondo et al., 2007; Savard et al., 2006). Second, these small ORFs are translated, as seen when fused to GFP in cultured cell assays (Galindo et al., 2007; Kondo et al., 2007). Finally, the introduction of four frameshift mutations is sufficient to inactivate *mlpt/tal/pri*, while a single smORF can substitute for the endogenous products *in vivo* (Galindo et al., 2007; Kondo et al., 2007). All together, these data therefore established that *mlpt/tal/pri*, hereafter referred to as *pri*, encode small peptides required for distinct developmental steps across insect species. How can such tiny peptides (from 10 to 32aa in *Drosophila*) fulfill specific functions to control development?

A striking feature of *pri* embryos is the complete absence of denticles and dorsal hairs, collectively called trichomes, which are epidermal outgrowths that normally represent a prominent morphological trait in insects (Payre, 2004). Previous work has established that the *ovo/shavenbaby* (*svb*) gene plays a pivotal role in epidermal differentiation. It determines the pattern of trichomes during development (Delon, Chanut-Delalande, & Payre, 2003; Frankel et al., 2010; Payre, Vincent, & Carreno, 1999), as well as trichome pattern diversification throughout insect evolution (Frankel et al., 2011; McGregor et al., 2007; Sucena, Delon, Jones, Payre, & Stern, 2003). *svb* encodes a transcription factor that directly activates the expression of a battery of cellular effectors collectively responsible for the formation of trichomes (Chanut-Delalande, Ferrer, Payre, & Plaza, 2012; Menoret et al., 2013). These include factors regulating actin filament polymerization or bundling (Chanut-Delalande, Fernandes, Roch, Payre, & Plaza, 2006) and membrane and cuticle organization (Andrew & Baker, 2008; Fernandes et al., 2010). Although the absence of *pri* results in the lack of trichomes, the expression of *svb* is not altered in *pri* mutant embryos (Galindo et al., 2007; Kondo et al., 2007, 2010). However, in *pri* mutants, Svb target genes are no longer expressed in epidermal cells, suggesting that *pri* is required for the proper activity of the Svb protein (Kondo et al., 2010). Indeed, Svb is first made as a large-sized repressor (1351aa). Pri peptides then trigger a post-translational modification releasing an N-terminal repressor domain, thereby switching Svb into a shorter transcriptional activator (905aa) (Fig. 5). In embryos lacking *pri*, the activating maturation of the Svb protein never occurs, instead, the repressor form of Svb keeps accumulating, preventing the whole program of trichome effector expression (Kondo et al., 2010). Further work has showed that this mechanism also takes place during the formation of adult legs explaining, at least in part, the phenotype observed in *tal* mutants (Pueyo & Couso, 2011).

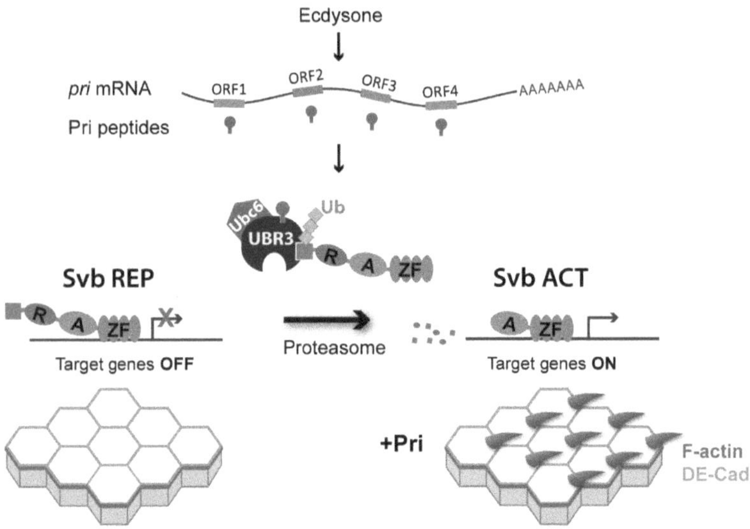

Figure 5 Model of Pri peptide function during epidermal differentiation. The ecdysone signaling pathway directly activates the expression of *pri* and thus times the production of Pri peptides. Pri peptides bind to the E3 ligase Ubr3, enabling it to recognize the N-terminal part of Svb (square box). Together with the E2 conjugating enzyme Ubc6, the Ubr3/Ubc6 complex ubiquitinates the N-terminus of Svb triggering its subsequent recruitment to the proteasome. The proteasome then specifically degrades the N-terminal region of Svb, which contains a repressor (R) domain. This limited degradation releases a shorter form of Svb, which contains an activator (A) and zinc finger (ZF) domains, therefore acting as a transcriptional activator. This matured Svb protein activates the expression of target effectors, collectively responsible for the formation of actin-rich extensions in epidermal cells. (A color version of this figure is located at the end of this chapter.)

A main function of *svb* during embryogenesis is to precisely define the pattern of trichomes (Arif, Kittelmann, & McGregor, 2015; Stern & Frankel, 2013). Several studies have identified a large array of enhancers (Frankel et al., 2010, 2011; McGregor et al., 2007) integrating various upstream regulatory cues to drive *svb* expression in epidermal cells (Payre et al., 1999). For example, Hox factors directly bind to *svb* enhancers and instruct differential expression of *svb* along the anteroposterior axis of *Drosophila* embryos (Crocker et al., 2015) explaining the difference long observed in trichome implantation between thoracic versus more posterior segments. While we now have a wealth of information to understand how *svb* expression builds the pattern of trichomes, the rationale of Pri action to further control the activity of the Svb protein remained poorly understood. A systematic screen of the second chromosome identified a previously uncharacterized gene that, like *pri*, is indispensable for trichome formation

(Chanut-Delalande et al., 2014). This gene called *GstE14/Nopera-Bo* encodes a new component of the biosynthetic pathway leading to ecdysone production (Chanut-Delalande et al., 2014; Enya et al., 2014), the major steroid hormone in insects. Recent work has demonstrated that ecdysone production depends on the integration of various cues including whole body development, tissue growth, nutrients and light, determining the timing of developmental transitions between larval and pupal stages (reviewed in Ou & King-Jones, 2013; Yamanaka, Rewitz, & O'Connor, 2013). Ecdysone is strictly required for *pri* transcription; the Ecdysone nuclear receptor (EcR) directly binds to an enhancer driving *pri* expression in epidermal derivatives (Chanut-Delalande et al., 2014). In response to ecdysone signaling, Pri defines the temporal control of Svb activity throughout development and, thereby, the timing of trichome formation. Uncoupling Pri expression from ecdysone is sufficient to promote untimely epidermal differentiation, i.e., either triggering premature formation of trichomes or keeping the whole program of trichome differentiation on-hold (Chanut-Delalande et al., 2014). Therefore, *svb cis* regulatory elements provide spatial cues to define which cells form trichomes, while Pri peptides act as a developmental timer to determine when trichomes differentiate. These data show that Pri peptides act as a molecular relay of systemic hormonal signaling to synchronize hardwired genetic programs with whole body information, sensing both the internal milieu and the environment.

6. INSIGHTS INTO THE MOLECULAR FUNCTION OF PRI PEPTIDES

Although these studies provided a clear framework for the developmental role of Pri peptides, their molecular mode of action remained elusive. This question was recently tackled by a genome wide RNAi screen in cultured cells, designed to identify cellular factors underlying the activity of Pri peptides for the processing of the Svb transcription factor (Zanet et al., 2015). Unexpectedly, the activating maturation of Svb is not due to endoprotease cleavage but instead, results from a limited degradation by the proteasome. Although the ubiquitin proteasome system is the major machinery involved in eukaryotic protein degradation (Varshavsky, 2012), a growing body of evidence shows that it is also involved in protein processing, as exemplified by transcription factors Ci/Gli, NF-κB, and Stp23/Mga2 (Tian & Matouschek, 2006). Functional dissection of the Svb protein revealed two distinct regions indispensable for Svb processing.

The N-terminal 31aa act as a *pri*-dependent degron, targeting Svb to the proteasome (Zanet et al., 2015). Additional C-terminal domains are required to limit proteasome degradation to the N-terminal repressor domain of Svb, therefore ensuring release of a truncated activator (Fig. 5). Proteins are addressed to the proteasome following the conjugation of polyubiquitin on lysine targets, thanks to the sequential activity of three kinds of enzymes called E1, E2, and E3. E3 ubiquitin ligases generally confer substrate specificity, bridging target proteins with an activated E2 conjugating enzyme that covalently transfers ubiquitin. The N-terminal Svb region contains three evolutionary conserved Lys, which are specifically ubiquitinated following Pri expression (Zanet et al., 2015). The screen identified an E2–E3 complex, Ubc6/Rad6 and Ubr3, which mediates Svb ubiquitination and processing upon Pri expression. Consistent with this conclusion, the lack of Ubr3 prevents Svb maturation *in vivo*, precluding trichome formation both in embryonic and adult epidermal tissues. *In vitro* assays further showed that Pri peptides bind to Ubr3 and Pri/Ubr3 interaction specifically enables Ubr3 binding to the Svb N-terminus. Importantly, this can be achieved using synthetic Pri peptides, while a scrambled peptide of same amino acid composition is devoid of activity (Zanet et al., 2015). In the absence of Pri, Ubr3 remains active nevertheless, still binding and regulating the cellular levels of Diap1 to protect cells from apoptosis (Huang et al., 2014), or the DNA repair factor Ape/1Rrp1 (Meisenberg et al., 2012). Therefore, Pri peptides switch Ubr3 selectivity, possibly via an allosteric conformational change, to regulate Ubr3 binding to Svb and consequently Svb ubiquitin-dependent proteasome processing.

7. CONCLUSIONS AND FUTURE CHALLENGES

The last couple of years have provided compelling evidence for pervasive translation of smORF peptides that likely represent an unexplored reservoir of regulatory functions. Systematic profiling, from viruses to humans, supports that smORF peptides can be produced from various intragenic or extragenic genomic regions, including mRNA, introns, antisense transcripts, noncoding and functional RNAs.

A first challenge is to improve experimental and computational methods to reach a comprehensive identification of functionally relevant smORFs and their associated peptide products. Ribosome profiling now allows the definition of translated regions down to single-nucleotide resolution (Aspden et al., 2014; Bazzini et al., 2014; Ingolia, Brar, Rouskin,

McGeachy, & Weissman, 2013; Ingolia et al., 2014). Furthermore, advances in mass spectrometry should provide direct detection of peptides on a large scale (Prabakaran et al., 2014; Ruiz-Orera, Messeguer, Subirana, & Alba, 2014; Yang et al., 2011). Several bioinformatic packages such as sORF finder (Hanada et al., 2010) and SPADA (Zhou et al., 2013) have been developed to predict translated regions without *a priori* on a minimal size, taking into account evolutionary conservation of the coding potential across species. However, the family of sarcolipin-like smORF peptides shows that selective pressure can apply to the tertiary structure rather than the primary sequence (Fig. 3), hampering the detection of important peptides by sequence-based evolutionary approaches. An additional difficulty is that the tridimensional structure of peptides may also rely on extrinsic parameters such as their interaction with functional partners, stabilizing the spatial conformation of these small molecules.

A second question is to unravel how translating ribosomes discriminate between smORFs and the bulk of noncoding sequences within diverse RNAs. The production of smORF peptides from atypical RNAs such as pri-miRNA (Lauressergues et al., 2015), 12S (Lee et al., 2015), or introns (Smith et al., 2014) shows the existence of noncanonical translation, taking place without the typical hallmarks of translatable RNAs including proper splicing, polyadenylation, and capping. Whether the translation of these peptides requires the same initiation factors and regulators as "classic" proteins remains to be elucidated, and future studies should also define in which cellular compartments this occurs. How initiation codons are selected in the case of polycistronic ORFs, i.e., RNA containing both smORFs and canonical ORFs, or several smORFs-like Pri or SCL, is poorly understood. In the latter cases, there seems to be no obvious signal for internal ribosome entry. Moreover, it is not known whether the respective location of different smORFs along RNA influences their translation efficiency. Therefore, future work on smORF peptides should provide novel insights into the general mechanisms of translation control.

The most intriguing problem is to figure out how smORF peptides fulfill specific functions *in vivo*. It is possible that smORF translation *per se* provides a regulatory function, for example in controlling the intracellular distribution of RNAs or the translation of a downstream ORF, without requiring a specific activity of the translated peptidic products. As discussed above, there is nevertheless clear evidence that smORF peptides may harbor intrinsic regulatory functions. Although the molecular mode of action of smORF peptides has been elucidated in only a handful of cases so far, it is striking to note that they often directly bind and regulate the activity of various classes of proteins and enzymes. Structural studies have shown that protein–protein interactions can involve contacts between short linear protein motifs (Tompa, Davey, Gibson, & Babu, 2014; Van Roey et al., 2014). These motifs, typically less than 10 residues, provide compact interaction interfaces and there is mounting evidence that they mediate important roles in the building of interaction networks and in the functional specification of regulatory factors (Baeza et al., 2015; Buljan et al., 2013). Structurally informed molecular design has shown that small peptide mimics can act as specific inhibitors (Lao et al., 2014), competing with endogenous factors, or rewiring the assembly of protein complexes (Nickerson, Joy, Arora, & Bar-Sagi, 2013). Therefore, smORF peptides are ideally suited to perform a similar function *in vivo*, switching the activity of protein complexes upon binding to a specific target. The evolution of smORF peptides, either from shuffling of existing protein motifs or from the selection of *de novo* sequences, may have provided a large repertoire of regulatory peptides. Future studies should uncover the full spectrum of their molecular functions.

ACKNOWLEDGMENTS

We thank A. Pelissier-Monier, C. Polesello, E. Benrabah, M. Soulard, P. Valenti, and all other members of the lab for invaluable support, and T. Kondo and Y. Kageyama for helpful discussions. The authors would like to dedicate this chapter to the memory of Y. Latapie (1980–2014), whose life was taken by an avalanche in the Pyrenees. We are grateful to Akira Nakamura, Eric N. Olson, Diethard Tautz, and Juan Pablo Couso for their kind permission to reproduce published pictures. This work was supported by grants from the ANR (smORFpep and ChronoNet) and Association pour la Recherche sur le Cancer.

REFERENCES

Anderson, D. M., Anderson, K. M., Chang, C. L., Makarewich, C. A., Nelson, B. R., McAnally, J. R., et al. (2015). A micropeptide encoded by a putative long noncoding RNA regulates muscle performance. *Cell*, *160*(4), 595–606.

Andrew, D. J., & Baker, B. S. (2008). Expression of the Drosophila secreted cuticle protein 73 (dsc73) requires shavenbaby. *Developmental Dynamics*, *237*(4), 1198–1206.

Andrews, S. J., & Rothnagel, J. A. (2014). Emerging evidence for functional peptides encoded by short open reading frames. *Nature Reviews Genetics*, *15*(3), 193–204.

Arif, S., Kittelmann, S., & McGregor, A. P. (2015). From shavenbaby to the naked valley: Trichome formation as a model for evolutionary developmental biology. *Evolution & Development*, *17*(1), 120–126.

Aspden, J. L., Eyre-Walker, Y. C., Phillips, R. J., Amin, U., Mumtaz, M. A., Brocard, M., et al. (2014). Extensive translation of small open reading frames revealed by Poly-Ribo-Seq. *Elife*, *3*, e03528.

Baeza, M., Viala, S., Heim, M., Dard, A., Hudry, B., Duffraisse, M., et al. (2015). Inhibitory activities of short linear motifs underlie Hox interactome specificity in vivo. *Elife*, *4*, e06034.

Banfai, B., Jia, H., Khatun, J., Wood, E., Risk, B., Gundling, W. E., Jr., et al. (2012). Long noncoding RNAs are rarely translated in two human cell lines. *Genome Research*, *22*(9), 1646–1657.

Bazzini, A. A., Johnstone, T. G., Christiano, R., Mackowiak, S. D., Obermayer, B., Fleming, E. S., et al. (2014). Identification of small ORFs in vertebrates using ribosome footprinting and evolutionary conservation. *EMBO Journal*, *33*(9), 981–993.

Buljan, M., Chalancon, G., Dunker, A. K., Bateman, A., Balaji, S., Fuxreiter, M., et al. (2013). Alternative splicing of intrinsically disordered regions and rewiring of protein interactions. *Current Opinion in Structural Biology*, *23*(3), 443–450.

Capt, C., Passamonti, M., & Breton, S. (2015). The human mitochondrial genome may code for more than 13 proteins. *Mitochondrial DNA*, 1–4.

Cech, T. R., & Steitz, J. A. (2014). The noncoding RNA revolution-trashing old rules to forge new ones. *Cell*, *157*(1), 77–94.

Chanut-Delalande, H., Fernandes, I., Roch, F., Payre, F., & Plaza, S. (2006). Shavenbaby couples patterning to epidermal cell shape control. *PLoS Biology*, *4*(9), e290.

Chanut-Delalande, H., Ferrer, P., Payre, F., & Plaza, S. (2012). Effectors of tridimensional cell morphogenesis and their evolution. *Seminars in Cell and Developmental Biology*, *23*(3), 341–349.

Chanut-Delalande, H., Hashimoto, Y., Pelissier-Monier, A., Spokony, R., Dib, A., Kondo, T., et al. (2014). Pri peptides are mediators of ecdysone for the temporal control of development. *Nature Cell Biology*, *16*(11), 1035–1044.

Chng, S. C., Ho, L., Tian, J., & Reversade, B. (2013). ELABELA: A hormone essential for heart development signals via the apelin receptor. *Developmental Cell*, *27*(6), 672–680.

Crappe, J., Van Criekinge, W., Trooskens, G., Hayakawa, E., Luyten, W., Baggerman, G., et al. (2013). Combining in silico prediction and ribosome profiling in a genome-wide search for novel putatively coding sORFs. *BMC Genomics*, *14*, 648.

Crocker, J., Abe, N., Rinaldi, L., McGregor, A. P., Frankel, N., Wang, S., et al. (2015). Low affinity binding site clusters confer hox specificity and regulatory robustness. *Cell*, *160*(1–2), 191–203.

Delon, I., Chanut-Delalande, H., & Payre, F. (2003). The ovo/shavenbaby transcription factor specifies actin remodelling during epidermal differentiation in Drosophila. *Mechanisms of Development*, *120*(7), 747–758.

Dinger, M. E., Pang, K. C., Mercer, T. R., & Mattick, J. S. (2008). Differentiating protein-coding and noncoding RNA: Challenges and ambiguities. *PLoS Computational Biology*, *4*(11), e1000176.

Enya, S., Ameku, T., Igarashi, F., Iga, M., Kataoka, H., Shinoda, T., et al. (2014). A Halloween gene noppera-bo encodes a glutathione S-transferase essential for ecdysteroid biosynthesis via regulating the behaviour of cholesterol in Drosophila. *Scientific Reports*, *4*, 6586.

Fernandes, I., Chanut-Delalande, H., Ferrer, P., Latapie, Y., Waltzer, L., Affolter, M., et al. (2010). Zona pellucida domain proteins remodel the apical compartment for localized cell shape changes. *Developmental Cell*, *18*(1), 64–76.

Frankel, N., Davis, G. K., Vargas, D., Wang, S., Payre, F., & Stern, D. L. (2010). Phenotypic robustness conferred by apparently redundant transcriptional enhancers. *Nature*, *466*(7305), 490–493.

Frankel, N., Erezyilmaz, D. F., McGregor, A. P., Wang, S., Payre, F., & Stern, D. L. (2011). Morphological evolution caused by many subtle-effect substitutions in regulatory DNA. *Nature*, *474*(7353), 598–603.

Galindo, M. I., Pueyo, J. I., Fouix, S., Bishop, S. A., & Couso, J. P. (2007). Peptides encoded by short ORFs control development and define a new eukaryotic gene family. *PLoS Biology*, *5*(5), e106.

Guillen, G., Diaz-Camino, C., Loyola-Torres, C. A., Aparicio-Fabre, R., Hernandez-Lopez, A., Diaz-Sanchez, M., et al. (2013). Detailed analysis of putative genes encoding small proteins in legume genomes. *Frontiers in Plant Science*, *4*, 208.

Guttman, M., & Rinn, J. L. (2012). Modular regulatory principles of large non-coding RNAs. *Nature*, *482*(7385), 339–346.

Hanada, K., Akiyama, K., Sakurai, T., Toyoda, T., Shinozaki, K., & Shiu, S. H. (2010). sORF finder: A program package to identify small open reading frames with high coding potential. *Bioinformatics*, *26*(3), 399–400.

Hanada, K., Higuchi-Takeuchi, M., Okamoto, M., Yoshizumi, T., Shimizu, M., Nakaminami, K., et al. (2013). Small open reading frames associated with morphogenesis are hidden in plant genomes. *Proceedings of the National Academy of Sciences of the United States of America*, *110*(6), 2395–2400.

Hanada, K., Zhang, X., Borevitz, J. O., Li, W. H., & Shiu, S. H. (2007). A large number of novel coding small open reading frames in the intergenic regions of the Arabidopsis thaliana genome are transcribed and/or under purifying selection. *Genome Research*, *17*(5), 632–640.

Hanyu-Nakamura, K., Sonobe-Nojima, H., Tanigawa, A., Lasko, P., & Nakamura, A. (2008). Drosophila Pgc protein inhibits P-TEFb recruitment to chromatin in primordial germ cells. *Nature*, *451*(7179), 730–733.

Huang, Q., Tang, X., Wang, G., Fan, Y., Ray, L., Bergmann, A., et al. (2014). Ubr3 E3 ligase regulates apoptosis by controlling the activity of DIAP1 in Drosophila. *Cell Death and Differentiation*, *21*(12), 1961–1970.

Inagaki, S., Numata, K., Kondo, T., Tomita, M., Yasuda, K., Kanai, A., et al. (2005). Identification and expression analysis of putative mRNA-like non-coding RNA in Drosophila. *Genes to Cells*, *10*(12), 1163–1173.

Ingolia, N. T., Brar, G. A., Rouskin, S., McGeachy, A. M., & Weissman, J. S. (2013). Genome-wide annotation and quantitation of translation by ribosome profiling. *Current Protocols in Molecular Biology*. Chapter 4: Unit 4.18.1–4.18.19.

Ingolia, N. T., Brar, G. A., Stern-Ginossar, N., Harris, M. S., Talhouarne, G. J., Jackson, S. E., et al. (2014). Ribosome profiling reveals pervasive translation outside of annotated protein-coding genes. *Cell Reports*, *8*(5), 1365–1379.

Jaber, T., & Yuan, Y. (2013). A virally encoded small peptide regulates RTA stability and facilitates Kaposi's sarcoma-associated herpesvirus lytic replication. *Journal of Virology*, *87*(6), 3461–3470.

Kastenmayer, J. P., Ni, L., Chu, A., Kitchen, L. E., Au, W. C., Yang, H., et al. (2006). Functional genomics of genes with small open reading frames (sORFs) in *S. cerevisiae*. *Genome Research*, *16*(3), 365–373.

Kondo, T., Hashimoto, Y., Kato, K., Inagaki, S., Hayashi, S., & Kageyama, Y. (2007). Small peptide regulators of actin-based cell morphogenesis encoded by a polycistronic mRNA. *Nature Cell Biology*, *9*(6), 660–665.

Kondo, T., Plaza, S., Zanet, J., Benrabah, E., Valenti, P., Hashimoto, Y., et al. (2010). Small peptides switch the transcriptional activity of shavenbaby during Drosophila embryogenesis. *Science*, *329*(5989), 336–339.

Ladoukakis, E., Pereira, V., Magny, E. G., Eyre-Walker, A., & Couso, J. P. (2011). Hundreds of putatively functional small open reading frames in Drosophila. *Genome Biology*, *12*(11), R118.

Lao, B. B., Drew, K., Guarracino, D. A., Brewer, T. F., Heindel, D. W., Bonneau, R., et al. (2014). Rational design of topographical helix mimics as potent inhibitors of protein–protein interactions. *Journal of the American Chemical Society*, *136*(22), 7877–7888.

Lauressergues, D., Couzigou, J. M., Clemente, H. S., Martinez, Y., Dunand, C., Becard, G., et al. (2015). Primary transcripts of microRNAs encode regulatory peptides. *Nature*, *520*(7545), 90–93.
Lee, C., Zeng, J., Drew, B. G., Sallam, T., Martin-Montalvo, A., Wan, J., et al. (2015). The mitochondrial-derived peptide MOTS-c promotes metabolic homeostasis and reduces obesity and insulin resistance. *Cell Metabolism*, *21*(3), 443–454.
Little, S. C., Sinsimer, K. S., Lee, J. J., Wieschaus, E. F., & Gavis, E. R. (2015). Independent and coordinate trafficking of single Drosophila germ plasm mRNAs. *Nature Cell Biology*, *17*(5), 558–568.
Ma, J., Ward, C. C., Jungreis, I., Slavoff, S. A., Schwaid, A. G., Neveu, J., et al. (2014). Discovery of human sORF-encoded polypeptides (SEPs) in cell lines and tissue. *Journal of Proteome Research*, *13*(3), 1757–1765.
Magny, E. G., Pueyo, J. I., Pearl, F. M., Cespedes, M. A., Niven, J. E., Bishop, S. A., et al. (2013). Conserved regulation of cardiac calcium uptake by peptides encoded in small open reading frames. *Science*, *341*(6150), 1116–1120.
Martinho, R. G., Kunwar, P. S., Casanova, J., & Lehmann, R. (2004). A noncoding RNA is required for the repression of RNApolII-dependent transcription in primordial germ cells. *Current Biology*, *14*(2), 159–165.
McGregor, A. P., Orgogozo, V., Delon, I., Zanet, J., Srinivasan, D. G., Payre, F., et al. (2007). Morphological evolution through multiple cis-regulatory mutations at a single gene. *Nature*, *448*(7153), 587–590.
Meisenberg, C., Tait, P. S., Dianova, I. I., Wright, K., Edelmann, M. J., Ternette, N., et al. (2012). Ubiquitin ligase UBR3 regulates cellular levels of the essential DNA repair protein APE1 and is required for genome stability. *Nucleic Acids Research*, *40*(2), 701–711.
Menoret, D., Santolini, M., Fernandes, I., Spokony, R., Zanet, J., Gonzalez, I., et al. (2013). Genome-wide analyses of shavenbaby target genes reveals distinct features of enhancer organization. *Genome Biology*, *14*(8), R86.
Moazed, D. (2009). Small RNAs in transcriptional gene silencing and genome defence. *Nature*, *457*(7228), 413–420.
Nakamura, A., Amikura, R., Mukai, M., Kobayashi, S., & Lasko, P. F. (1996). Requirement for a noncoding RNA in Drosophila polar granules for germ cell establishment. *Science*, *274*(5295), 2075–2079.
Nakamura, A., & Seydoux, G. (2008). Less is more: Specification of the germline by transcriptional repression. *Development*, *135*(23), 3817–3827.
Nickerson, S., Joy, S. T., Arora, P. S., & Bar-Sagi, D. (2013). An orthosteric inhibitor of the RAS–SOS interaction. *Enzymes*, *34*(Pt. B), 25–39.
Ou, Q., & King-Jones, K. (2013). What goes up must come down: Transcription factors have their say in making ecdysone pulses. *Current Topics in Developmental Biology*, *103*, 35 71.
Pauli, A., Norris, M. L., Valen, E., Chew, G. L., Gagnon, J. A., Zimmerman, S., et al. (2014). Toddler: An embryonic signal that promotes cell movement via Apelin receptors. *Science*, *343*(6172), 1248636.
Pauli, A., Valen, E., & Schier, A. F. (2015). Identifying (non-)coding RNAs and small peptides: Challenges and opportunities. *Bioessays*, *37*(1), 103–112.
Payre, F. (2004). Genetic control of epidermis differentiation in Drosophila. *International Journal of Developmental Biology*, *48*(2–3), 207–215.
Payre, F., Vincent, A., & Carreno, S. (1999). ovo/svb integrates Wingless and DER pathways to control epidermis differentiation. *Nature*, *400*(6741), 271–275.
Prabakaran, S., Hemberg, M., Chauhan, R., Winter, D., Tweedie-Cullen, R. Y., Dittrich, C., et al. (2014). Quantitative profiling of peptides from RNAs classified as noncoding. *Nature Communications*, *5*, 5429.
Pueyo, J. I., & Couso, J. P. (2011). Tarsal-less peptides control Notch signalling through the shavenbaby transcription factor. *Developmental Biology*, *355*(2), 183–193.

Ruiz-Orera, J., Messeguer, X., Subirana, J. A., & Alba, M. M. (2014). Long non-coding RNAs as a source of new peptides. *Elife*, *3*, e03523.

Sandman, K., Losick, R., & Youngman, P. (1987). Genetic analysis of Bacillus subtilis spo mutations generated by Tn917-mediated insertional mutagenesis. *Genetics*, *117*(4), 603–617.

Savard, J., Marques-Souza, H., Aranda, M., & Tautz, D. (2006). A segmentation gene in tribolium produces a polycistronic mRNA that codes for multiple conserved peptides. *Cell*, *126*(3), 559–569.

Slavoff, S. A., Mitchell, A. J., Schwaid, A. G., Cabili, M. N., Ma, J., Levin, J. Z., et al. (2013). Peptidomic discovery of short open reading frame-encoded peptides in human cells. *Nature Chemical Biology*, *9*(1), 59–64.

Smith, J. E., Alvarez-Dominguez, J. R., Kline, N., Huynh, N. J., Geisler, S., Hu, W., et al. (2014). Translation of small open reading frames within unannotated RNA transcripts in Saccharomyces cerevisiae. *Cell Reports*, 7(6), 1858–1866.

Stern, D. L., & Frankel, N. (2013). The structure and evolution of cis-regulatory regions: The shavenbaby story. *Philosophical Transactions of the Royal Society of London. Series B, Biological Sciences*, *368*(1632), 20130028.

Storz, G., Wolf, Y. I., & Ramamurthi, K. S. (2014). Small proteins can no longer be ignored. *Annual Review of Biochemistry*, *83*, 753–777.

Strome, S., & Lehmann, R. (2007). Germ versus soma decisions: Lessons from flies and worms. *Science*, *316*(5823), 392–393.

Sucena, E., Delon, I., Jones, I., Payre, F., & Stern, D. L. (2003). Regulatory evolution of shavenbaby/ovo underlies multiple cases of morphological parallelism. *Nature*, *424*(6951), 935–938.

Tada, M., Kirchberger, M. A., & Katz, A. M. (1975). Phosphorylation of a 22,000-dalton component of the cardiac sarcoplasmic reticulum by adenosine 3′:5′-monophosphate-dependent protein kinase. *Journal of Biological Chemistry*, *250*(7), 2640–2647.

Tian, L., & Matouschek, A. (2006). Where to start and when to stop. *Nature Structural and Molecular Biology*, *13*(8), 668–670.

Timinszky, G., Bortfeld, M., & Ladurner, A. G. (2008). Repression of RNA polymerase II transcription by a Drosophila oligopeptide. *PLoS One*, *3*(6), e2506.

Tompa, P., Davey, N. E., Gibson, T. J., & Babu, M. M. (2014). A million peptide motifs for the molecular biologist. *Molecular Cell*, *55*(2), 161–169.

Topping, J. F., & Lindsey, K. (1997). Promoter trap markers differentiate structural and positional components of polar development in Arabidopsis. *Plant Cell*, *9*(10), 1713–1725.

Tupy, J. L., Bailey, A. M., Dailey, G., Evans-Holm, M., Siebel, C. W., Misra, S., et al. (2005). Identification of putative noncoding polyadenylated transcripts in Drosophila melanogaster. *Proceedings of the National Academy of Sciences of the United States of America*, *102*(15), 5495–5500.

van de Sande, K., Pawlowski, K., Czaja, I., Wieneke, U., Schell, J., Schmidt, J., et al. (1996). Modification of phytohormone response by a peptide encoded by ENOD40 of legumes and a nonlegume. *Science*, *273*(5273), 370–373.

Van Roey, K., Uyar, B., Weatheritt, R. J., Dinkel, H., Seiler, M., Budd, A., et al. (2014). Short linear motifs: Ubiquitous and functionally diverse protein interaction modules directing cell regulation. *Chemical Reviews*, *114*(13), 6733–6778.

Varshavsky, A. (2012). The ubiquitin system, an immense realm. *Annual Review of Biochemistry*, *81*, 167–176.

Wawrzynow, A., Theibert, J. L., Murphy, C., Jona, I., Martonosi, A., & Collins, J. H. (1992). Sarcolipin, the "proteolipid" of skeletal muscle sarcoplasmic reticulum, is a unique, amphipathic, 31-residue peptide. *Archives of Biochemistry and Biophysics*, *298*(2), 620–623.

Yamanaka, N., Rewitz, K. F., & O'Connor, M. B. (2013). Ecdysone control of developmental transitions: Lessons from Drosophila research. *Annual Review of Entomology*, *58*, 497–516.

Yang, X., Tschaplinski, T. J., Hurst, G. B., Jawdy, S., Abraham, P. E., Lankford, P. K., et al. (2011). Discovery and annotation of small proteins using genomics, proteomics, and computational approaches. *Genome Research*, *21*(4), 634–641.

Zanet, J., Benrabah, E., Li, T., Pelissier-Monier, A., Chanut-Delalande, H., Ronsin, B., et al. (2015). Pri sORF peptides induce selective proteasome-mediated protein processing. *Science*, *349*(6254), 1356–1358.

Zhou, P., Silverstein, K. A., Gao, L., Walton, J. D., Nallu, S., Guhlin, J., et al. (2013). Detecting small plant peptides using SPADA (small peptide alignment discovery application). *BMC Bioinformatics*, *14*, 335.

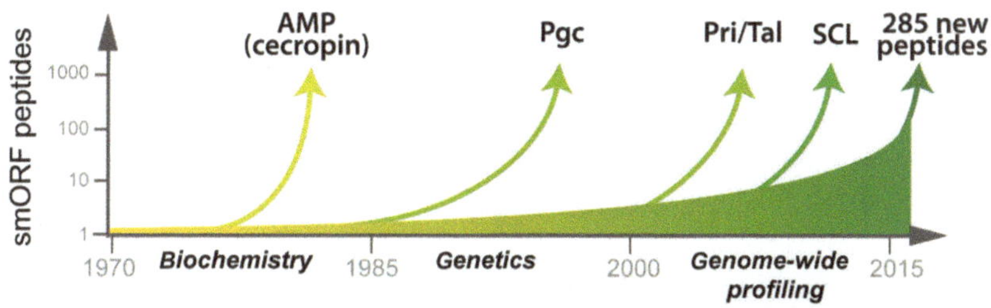

J. Zanet *et al.*, Figure 1 The fast growing family of small peptides in *Drosophila*. Chronological overview of the discovery of short ORF peptides in *Drosophila* and the strategies leading to their discovery.

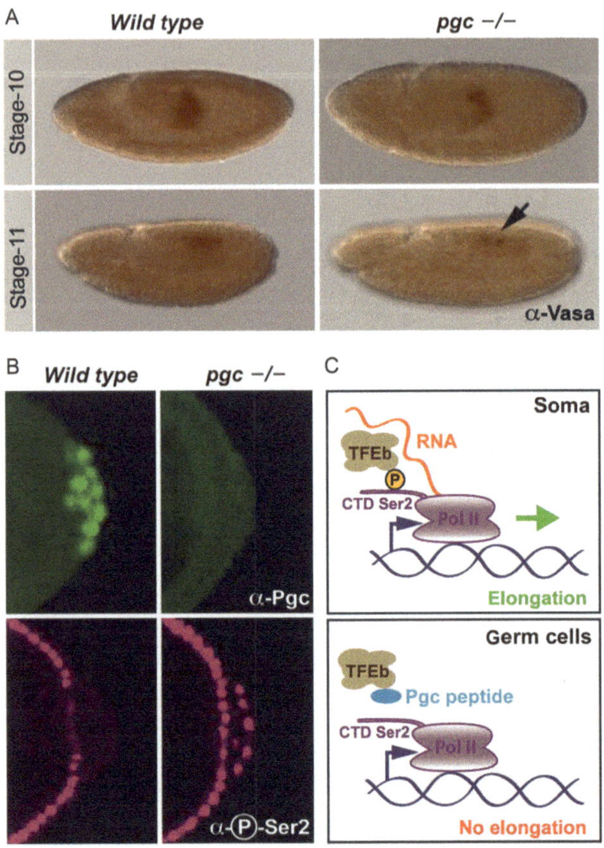

J. Zanet *et al.*, Figure 2 The Pgc peptide controls PolII transcriptional elongation. (A) Defects in germ cell migration in *pgc* mutant embryos, as observed at two successive stages of development during mi-embryogenesis. Migrating germ cells are visualized by immunostaining with an anti-Vasa antibody. (B) Top: Expression of the Pgc peptide detected in polar germ cells in wild-type early embryos and in *pgc* mutants. Bottom: Absence of PolII CTD phosphorylation in normal polar germ cells, whereas precocious PolII CTD phosphorylation is detected in pole cells lacking the Pgc peptide. (C) Model for the action of the Pgc peptide. Binding of the Pgc peptide to the pTFEb protein complex downregulates its kinase activity and counteracts PolII CTD Ser2 phosphorylation that normally occurs in somatic cells.

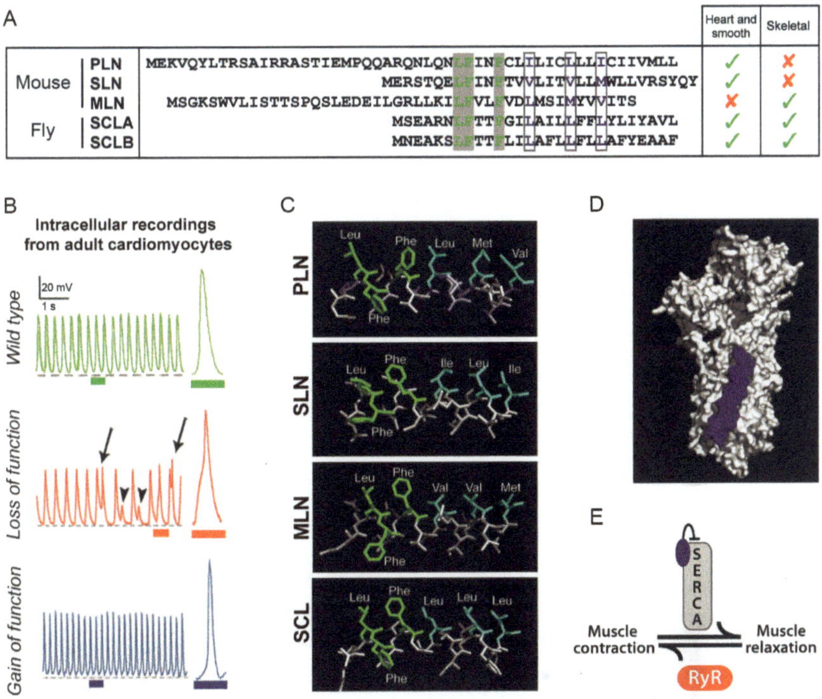

J. Zanet *et al.*, Figure 3 Conservation and function of peptides in muscle physiology. (A) The graph shows sequence alignment of mouse phospholamban (PLN), sarcolipin (SLN), myoregulin (MLN), and *Drosophila* sarcolamban (SCLA and SCLB) peptides, with identical (shaded) and similar (boxed) residues, as well as the type of muscles in which they are expressed. (B) Intracellular recordings from *Drosophila* wild-type (top) and *scl* loss- (middle) or gain-of-function (bottom) adult cardiomyocytes. Arrows indicate double action potential. Arrowheads indicate failed action potential. A sample peak is magnified at the right of each trace. (C) Structure of predicted helices of PLN, SLN, MLN, and SCL peptides, showing a common spatial orientation of conserved residues. (D) The SLN (blue) peptide fits a groove in the structure of the SERCA1 protein (white). (E) SLN peptides bind to the SERCA calcium pump and inhibit its activity, thus decreasing phases of muscle relaxation.

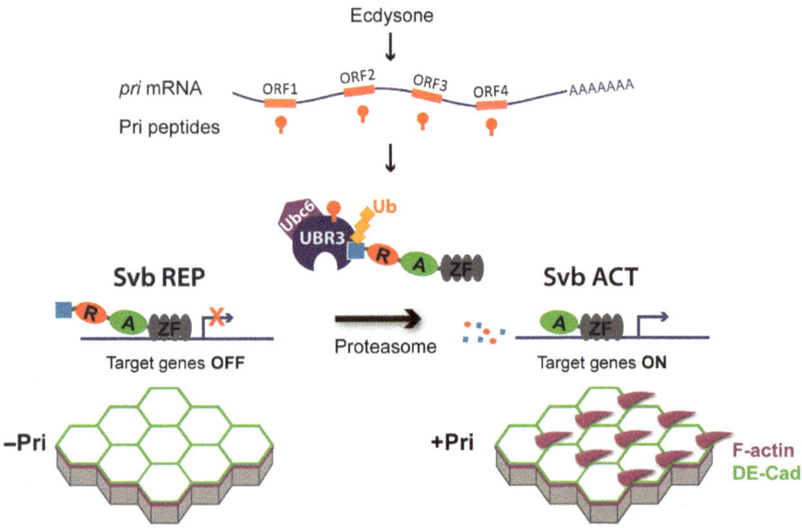

J. Zanet *et al.*, Figure 5 Model of Pri peptide function during epidermal differentiation. The ecdysone signaling pathway directly activates the expression of *pri* and thus times the production of Pri peptides. Pri peptides bind to the E3 ligase Ubr3, enabling it to recognize the N-terminal part of Svb (square box). Together with the E2 conjugating enzyme Ubc6, the Ubr3/Ubc6 complex ubiquitinates the N-terminus of Svb triggering its subsequent recruitment to the proteasome. The proteasome then specifically degrades the N-terminal region of Svb, which contains a repressor (R) domain. This limited degradation releases a shorter form of Svb, which contains an activator (A) and zinc finger (ZF) domains, therefore acting as a transcriptional activator. This matured Svb protein activates the expression of target effectors, collectively responsible for the formation of actin-rich extensions in epidermal cells.

CHAPTER TWELVE

Toward a Systems Understanding of Signaling Pathway Function

David P. Doupé*[,1], Norbert Perrimon*[,†,1]
*Department of Genetics, Harvard Medical School, Boston, Massachusetts, USA
†Howard Hughes Medical Institute, Boston, Massachusetts, USA
[1]Corresponding authors: e-mail address: doupe@genetics.med.harvard.edu; perrimon@receptor.med.harvard.edu

Contents

Abstract

A small number of developmental signaling pathways are used repeatedly throughout development in many different contexts. How these pathways interact with each other and the specific cell context to generate a wide range of appropriate responses remains an important question. The application of genomic and proteomic approaches and imaging at high spatiotemporal resolution are providing answers to this question and revealing new levels of complexity. Here, we discuss pathways as complex networks and examples of how signaling outcomes can be influenced by the temporal nature of the signal, its spatial regulation, and the cell context.

1. INTRODUCTION

When *Current Topics in Developmental Biology* was first published 50 years ago, the biochemical isolation of growth factors had begun to allow the effects of signals on cells to be studied. In the 1980s and 1990s, pioneering genetic screens in model organisms elucidated the major developmental signaling pathways' core components. Subsequent studies showed that these pathways are used repeatedly to perform different developmental, physiological, or

Current Topics in Developmental Biology, Volume 117
ISSN 0070-2153
http://dx.doi.org/10.1016/bs.ctdb.2015.11.005

pathological functions. It soon became clear that relatively few pathways, such as Notch, JAK–STAT, receptor tyrosine kinase, BMP, and Hedgehog were capable of giving rise to diverse cellular responses and that combinatorial signaling and cell context were critical to outcomes (Perrimon, Pitsouli, & Shilo, 2012). In many cases, we now have a good understanding of which pathways are involved in which processes. Today, two critical and interconnected questions in the field remain: (1) how do pathways signal together to generate diverse and robust outcomes, and (2) how does cellular context alter these responses? Over the last decade, a combination of genome-scale approaches and high-resolution imaging are offering answers to these questions and revealing new levels of complexity. Here, we discuss some of the ways in which our understanding of how signaling pathways function to regulate cell fate are changing and leading to a more complex view.

2. PATHWAYS AS COMPLEX NETWORKS

Genetic studies of signaling pathways led to relatively linear textbook views of canonical signaling (Fig. 1A). However, the fact that relatively few pathways can generate such diverse outcomes suggested a more complicated picture. The same pathway can give different outcomes in different cell types, so context plays an important role. Many of the classic examples of developmental signaling, such as the patterning of *Drosophila* ommatidia (Félix & Barkoulas, 2012) and the *Caenorhabditis elegans* vulva (Nagaraj & Banerjee, 2004), involve multiple signaling pathways working together with cross talk between them, so combinatorial signaling adds an additional layer of complexity (Housden & Perrimon, 2014). In addition, work on epidermal growth factor (EGF) and nerve growth factor (NGF) signaling through the mitogen-activated protein kinase (MAPK) pathway made it clear that a single pathway axis can generate different outcomes (Chao, 1992). Simple linear pathways, even acting in different combinations, would be insufficient to explain the diverse outputs generated. Enhancer and suppressor screens in model organisms have identified many additional regulators of signaling pathways, beyond the core components (Fig. 1B). Piecing together a more comprehensive picture of pathway complexity has required the combination of both proteomic approaches, to determine interaction partners of known components, and large-scale functional genomic approaches such as genome-wide RNAi. Functional screens have identified many novel pathway regulators, but in isolation do not distinguish between direct and indirect pathway regulation. Similarly, while proteomic approaches identify many proteins that interact with known pathway components these interactions may not be functional

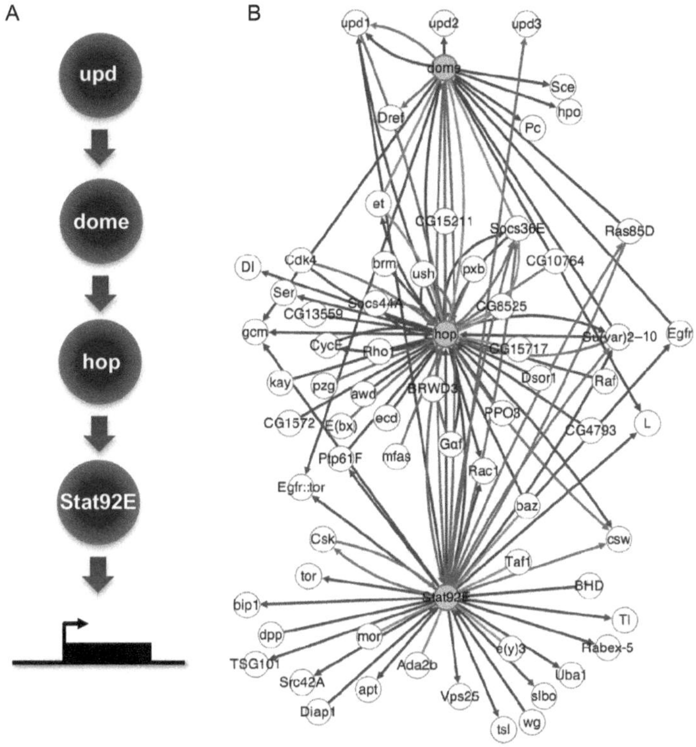

Figure 1 From linear pathways to complex networks. (A) Simple representation of the core Drosophila JAK/STAT pathway: cytokine ligands (upd1, 2, 3) bind to the receptor (dome), which activates JAK (hop) and in turn Stat (Stat92E) to regulate target gene expression. (B) An interaction network for dome, hop, and Stat92E (pink nodes) showing genetic interactions (red and blue lines with arrows) and physical interactions (orange lines). Network generated using EsyN (Bean et al., 2014). (A color version of this figure is located at the end of this chapter.)

regulatory interactions. The two approaches complement one another in characterizing the direct regulatory interactome of a pathway.

Proteomic approaches have allowed interactomes to be derived for several major signaling pathways. Large-scale yeast-two-hybrid approaches (Yu et al., 2008) in combination with siRNA functional analysis have been used to characterize the human MAPK interaction network, identifying a core of over 600 proteins including novel pathway chaperones and scaffolds (Bandyopadhyay et al., 2010). In a similar study across different *Drosophila* cell types, tandem affinity purification mass spectrometry (TAP-MS) and RNAi were used to identify functional RTK–Ras–ERK interactors (Friedman et al., 2011). Many of the novel regulators identified were found to be cell type specific, suggesting that different network wiring offers a mechanism of generating context-specific outcomes. Recent work has also identified the hippo pathway interaction network in *Drosophila* and mammals (Couzens et al., 2013; Kwon et al., 2013; Wang, Li, et al., 2013). This allowed a link to be made between Hippo signaling and vesicle trafficking (Kwon et al., 2013), reflecting the importance of membrane biology and subcellular localization in signaling, as discussed below. In addition to targeted approaches, in which networks are derived by starting from known pathway components and working outward to their interactors, large-scale systematic interaction mapping by mass spectrometry has been performed in fly and mammalian cells (Guruharsha, Obar, & Mintseris, 2012; Guruharsha et al., 2011; Havugimana et al., 2012; Huttlin et al., 2015). While these studies were not focused on signaling, they identified many candidates for novel pathway regulators. Many genes remain without functional annotation and interaction maps allow functional annotations to be proposed based on their interaction partners, including attribution of possible roles in signal transduction. Conservation of interactions between species may indicate conserved functional significance and act as a filter to prioritize candidates for further study.

The combination of interaction and functional regulation from RNAi makes a strong case for a protein being a direct regulator but does not necessarily offer a mechanism. Phosphorylation events are often critical regulators of signaling cascades, but even when an interaction involves a kinase the systematic identification of kinase targets remains a work in progress. Large-scale attempts at mapping kinase targets by phospho-proteomics are beginning to address this and offer insights into the possible branching of signals through phosphorylation targets (Sopko et al., 2014). More generally, as interaction networks of hundreds of proteins are derived for different pathways, the presence of common components offers points for cross talk between pathways in signal transmission.

While the focus of signaling research has largely been on the role of proteins in pathways, nonprotein regulators have begun to take on greater importance, adding to the complexity and offering additional points for diversification of outcomes. As the function of noncoding RNAs has been elucidated in many processes, their roles in signaling networks have become apparent. miRNA regulation has been shown for pathways including Wingless (Silver, Hagen, Okamura, Perrimon, & Lai, 2007) and Hedgehog (Kim, Vinayagam, & Perrimon, 2014). The lin28-let-7 axis has been shown to be critical in a range of stem cell and oncogenic signaling pathways, and let-7 miRNA targets include components of the insulin signaling pathway (Frost & Olson, 2011; Zhu et al., 2011).

The metabolic status of the cell may also influence signaling and there are many examples of intersection between developmental signaling pathways and metabolism, with metabolites acting as regulators and readouts of signaling. For example, the eicosanoid PGE2 has been shown to regulate the Wnt pathway and EGF signaling in hematopoietic and embryonic stem cells, respectively (Goessling et al., 2009; Yun, Lee, Ryu, & Han, 2009), and Akt activity is regulated by TOR in response to amino acids (Saha et al., 2014). Metabolites either individually or collectively can also act as phenotypic sensors of signaling states, for example, the NAD/NADH sensor SoNar has been used as a readout for metabolic screening in cancer drug discovery (Zhao et al., 2015). AMP functions as an energy sensor, indicating low nutrient status by activating AMPK (Zhang et al., 2013), which is capable of cross talk with developmental signaling pathway components such as MEK–ERK via phosphorylation of BRAF (Shen et al., 2013). Physical factors may also have significant effects on signaling outcomes and there is an increasing appreciation of the role of mechanical forces in cellular communication (as reviewed in detail in Miller & Davidson, 2013).

Large interaction networks afford opportunities for context-specific outcomes and cross talk with other pathways. It has become clear that transmission of information through a network, even for a single pathway, is more complex than a linear flow. The differential expression or regulation of network components in distinct cell types offers a possible means to alter the flow of information through pathways in a cell type-specific manner (Kiel, Verschueren, Yang, & Serrano, 2013). Engineering approaches to information flow through a network such as modular response analysis, as opposed to qualitative assessment of inputs and outputs, may allow a better understanding of temporal pathway properties (Bruggeman, Westerhoff, Hoek, & Kholodenko, 2002). Understanding signal transmission in this way requires inputs and outputs of pathways to be measured at different points through the pathway in either real time or time series, and novel imaging approaches are allowing this.

3. TEMPORAL PROPERTIES OF SIGNAL TRANSDUCTION

Dissecting the temporal component of signaling requires tools to detect dynamic readouts and manipulate signaling at multiple levels. The toolset that would be required to do this in a systematic way across the major signaling pathways is very incomplete but, where approaches are available, they have offered new insights into the temporal properties of signaling pathways. Dynamic readouts coupled to inputs that can be adjusted for frequency,

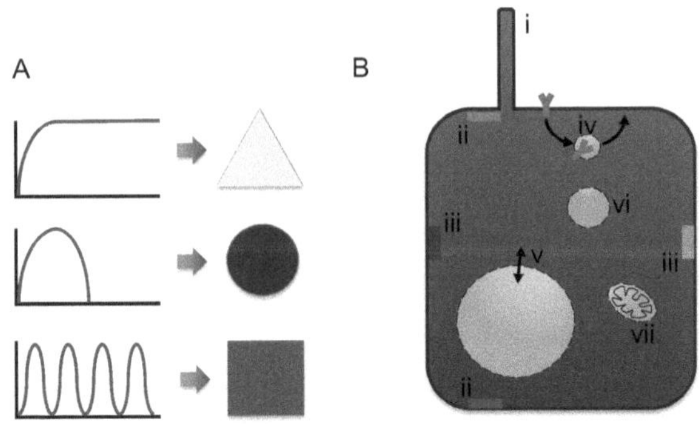

Figure 2 Spatiotemporal properties of signaling pathways. (A) The temporal properties of a single signal (graphs) may determine differential cellular responses (colored shapes). (B) Schematic of a cell showing some possible points of spatial regulation of signal transduction. Cells may localize surface signaling by: extension of short nanotube projections or longer cytonemes (i), apicobasal (ii), and planar (iii) cell polarization. Endocytosis and signaling endosomes (iv), lysosomes (vi), and mitochondria (vii) all have roles in subcellular localization of signaling, while nucleocytoplasmic transport (v) is an important step in many pathways. (A color version of this figure is located at the end of this chapter.)

amplitude, and duration have shown that the nature of the signal can be used to differentially transmit information (Fig. 2A; reviewed in Purvis & Lahav, 2013). In pheochromocytoma 12 (PC12) cells, EGF or NGF treatment results in the near-opposite responses of division and differentiation, respectively (Chao, 1992). Both of these pathways signal through the MAPK–ERK network but the dynamics of ERK activation differ according to the stimulus (Marshall, 1995). EGF triggers transient ERK signaling, whereas NGF causes sustained ERK activity. Differential dynamics of the same pathway effector's activity can therefore result in different cellular outcomes. Modular response analysis using RNAi to knockdown pathway components and phospho-specific antibody staining to detect the activation state of different points in the network has shown that different feedback responses underlie this difference in ERK activation dynamics (Santos, Verveer, & Bastiaens, 2007). EGF activation causes negative feedback within the network and hence transient ERK activation, whereas NGF signaling results in positive feedback and sustained ERK activation. Reversal of these feedback loops is sufficient to reverse the cellular responses to NGF and EGF.

In vivo, manipulating and tracking signaling at sufficient spatiotemporal resolution to study signaling dynamics remains challenging but a growing range of fluorescence-based tools are becoming available (Doupé & Perrimon, 2014). In some cases, subcellular localization of a fluorescently labeled component can be used to detect pathway activity. For example,

GFP tagged STAT translocation to the nucleus when the JAK/STAT pathway is activated (Chen & Reich, 2010). Where this is not possible changes in protein conformations or interactions can sometimes be detected by fluorescence resonance energy transfer (FRET) between donor and acceptor probes fused to signaling components such as MEK and ERK (Burack & Shaw, 2005). An intramolecular FRET probe has been used to successfully measure ERK activity *in vivo* in a *C. elegans* NaCl responsive sensory neuron with high temporal resolution (Tomida, Oda, Takekawa, Iino, & Saito, 2012). In this study, the nature of the upstream signal (NaCl) allowed a microfluidic device to be used to precisely control the signal and visualize the ERK activity response, revealing that cyclic stimulation with a periodicity in the tens of seconds resulted in sustained high levels of ERK activity whereas sustained signals or a shorter signal periodicity did not. The dynamics of an upstream signal rather than just the concentration of the particular signal can therefore be used as a determinant of the cellular response. Optogenetics (reviewed in Toettcher, Voigt, Weiner, & Lim, 2011) has also been used to manipulate Ras–ERK signaling with high temporal resolution demonstrating differential responses to different temporal activations in NIH 3T3 cells (Toettcher, Weiner, & Lim, 2013). Proteomics was used to assess downstream results, offering a nice demonstration of how the genome-scale approaches discussed above can be combined with high-resolution imaging to better understand signal transduction and responses. Dynamic signaling is not a unique feature of the Ras–ERK network. Oscillations in Notch signaling play a critical role in vertebrate somite formation and destabilized transcriptional reporters of Notch signaling have been used to image these oscillations directly (Aulehla et al., 2008; Masamizu et al., 2006). As tools become available to study the temporal properties of other pathways in other contexts, this may prove to be a general theme, providing additional ways to encode information with a finite number of pathways.

4. SPATIAL REGULATION OF SIGNAL TRANSDUCTION

Increasing resolution of signal imaging has also led to an increased understanding of the spatial properties and regulation of signal transduction at the subcellular level (Fig. 2B). In order to regulate gene expression nuclear translocation is an important process in many signaling pathways; for example, STAT6 has been shown to continuously shuttle between nucleus and cytoplasm but to only accumulate in the nucleus when activated (Chen & Reich, 2010). Many organelles have roles in signal transduction;

the mitochondrion is a signaling organelle in apoptosis and is an example of metabolites influencing signaling as mitochondrial reactive oxygen species can affect cellular responses (Chandel, 2014). The lysosome is the site of active TOR and AMPK signaling, representing a key point of intersection between metabolism and signaling (Bar-Peled, Schweitzer, Zoncu, & Sabatini, 2012; Zoncu et al., 2011). Endocytosis and membrane trafficking have particular importance to signal transduction by creating additional sites for signaling, and downregulating or recycling receptors (Di Fiore & von Zastrow, 2014; Gonnord, Blouin, & Lamaze, 2012). Combining RNAi with high-resolution imaging and multiparametric imaging analysis has allowed the complexity of endocytic traffic and its links to many signaling pathways to be revealed (Collinet et al., 2010). G-protein-coupled receptor signaling from endosomes may lead to extended pathway activation and the location of second messenger production is important to the cellular response (Tsvetanova & von Zastrow, 2014). Recycling endosome trafficking of the Notch ligand Delta is important for its function and plays a role in asymmetric fate of *Drosophila* sensory organ precursor (SOP) daughters (Emery et al., 2005). Segregation of endosomes at division can determine the signaling states and hence fates of the resulting cells. Specialized endosomes marked by the presence of Sara have been shown to be responsible for the equal distribution of TGFβ signal components at cell division in the *Drosophila* wing disc (Bökel et al., 2006). Sara endosomes are also critical for asymmetric notch signaling post cell division in both SOP cells and intestinal stem cells in *Drosophila* and in Zebrafish neural precursors (Coumailleau, Fürthauer, Knoblich, & González-Gaitán, 2009; Kressmann, Campos, Castanon, Fürthauer, & González-Gaitán, 2015; Montagne & Gonzalez-Gaitan, 2014).

Even at the plasma membrane, the localization of signaling receptors, ligands, and regulators to particular domains can be important for determining appropriate cellular responses. Apicobasal polarity of neuroblasts, for example, allows the asymmetric segregation of the Notch inhibitor Numb to the ganglion mother cell on division, resulting in biased Notch activity and asymmetric fate (Knoblich, 2010). Planar cell polarity components are responsible for noncanonical Wnt signaling in various development processes (Wansleeben & Meijlink, 2011). In addition to cell polarity, recent work in the *Drosophila* testes has shown that germline stem cells can extend nanotube projections, deforming the membrane of the hub cells that form their niche (Inaba, Buszczak, & Yamashita, 2015). These extensions were found to be strongly enriched for Dpp receptors, allowing highly localized signaling that restricts the ability to self-renew to only the most proximal germline stem cells. On the larger scale of tissues and organs, imaging approaches are also giving new insights into classical questions about positional information. Cytonemes, cellular protrusions that may extend over several cell diameters, have been described in various systems and play important roles in morphogen gradients such as the distribution of Hedgehog in the *Drosophila* wing disc (Bischoff et al., 2013; Kornberg & Roy, 2014). Interpretation of positional information in response to signals is critical for determination of different cell fates in many developmental contexts (Kicheva, Cohen, & Briscoe, 2012; Wolpert, 2011). Many theoretical attempts have been made to model the establishment and maintenance of morphogen gradients and advances in imaging are allowing data to be generated to test these hypotheses, as for the bicoid gradient in the *Drosophila* embryo (Bergmann, Tamari, Schejter, Shilo, & Barkai, 2008; Gregor, Tank, Wieschaus, & Bialek, 2007; Gregor, Wieschaus, McGregor, Bialek, & Tank, 2007). Even a nonprotein gradient, that of retinoic acid in vertebrate hindbrain patterning, has been visualized using a modified retinoic acid receptor fused to a FRET donor and acceptor pair such that retinoic acid binding causes detectable changes in the FRET signal (Shimozono, Iimura, Kitaguchi, Higashijima, & Miyawaki, 2013).

While imaging has helped elucidate many of these processes on a subcellular to tissue scale, genomic and proteomic approaches can also inform studies of spatial aspects of signaling. Networks derived from large-scale proteomic approaches can reflect subcellular organization and where localization is known for a component of a complex, spatial annotation of less well-characterized proteins may be possible (Huttlin et al., 2015). The development of targeted proteomic approaches to profile the proteomes of specific organelles or cell compartments may also prove informative in signaling biology. Techniques such as BioID (Roux, Kim, Raida, & Burke, 2012) and Apex (Rhee et al., 2013), in which a protein-modifying enzyme is fused to a protein or tag that targets it to a specific subcellular localization, allow the proteome of that region of the cell to be identified by isolation of the modified proteins. BioID has already been used to help map the hippo pathway interactome by allowing profiling of proteins not readily extracted by FLAG AP–MS (Couzens et al., 2013). In principle, these approaches could be used both to obtain spatial information about signaling networks and to profile the changes in network component interactions following pathway stimulation. On the tissue, organ or developing embryo scale, recent advances in single-cell sequencing approaches also have potential for mapping differences in the transcriptional outputs of signaling. Two recent studies developed droplet-based single-cell RNA sequencing techniques and applied them to characterize heterogeneous differentiated cell types in the retina, and heterogeneity among differentiating embryonic stem cells (Klein et al., 2015; Macosko et al., 2015). In principle, these approaches could be applied to questions of differential responses to signaling in development on a genome-wide scale at single-cell resolution. Positional information could be inferred retrospectively by imaging the expression patterns of differentially expressed genes *in situ*.

5. THE IMPORTANCE OF CONTEXT

Cell-type specificity is a key feature allowing relatively few pathways to generate the diverse outcomes observed in development. At the level of signal transduction, different cells will express different combinations of network components, allowing different flows of information through the same basic pathway. Altering the abundance of network components that exhibit mutually exclusive interactions with a pathway node, such as CRAF and RIN1 interactions with Ras, has been shown to alter signaling outcomes (Kiel et al., 2013). When modeling information flow within a network, it is therefore important to be aware of the specific cellular context in order to place appropriate constraints on the network. Spatial separation of components to distinct subcellular locations in specific contexts may also need to be considered. As techniques such as RNAseq and proteomics are used more widely, information on the expression of pathway components in specific cell types is increasingly becoming available. However, the need to collate this information and differences in experimental design, sample preparation, and data analysis present challenges to easy interpretation.

Ultimately, signaling effects on cell fate are most commonly mediated at the transcriptional level. The cell type and developmental context critically influence the outcome of signaling events at the level of transcriptional targets. The main effector transcription factors have been identified for the major pathways (Perrimon et al., 2012), but it is clear that their targets vary depending on the cellular context (for example, STAT (Wang, Chen, He, Zhou, & Luo, 2013) and Wnt (Vincent, 2014)). The cell's history and current state of other signaling pathways set a unique ground state upon which a signal can act. There are a number of ways in which context may influence outcomes at the level of transcription itself. Lineage-specific transcription factors (or effectors of other signaling pathways) may recruit signaling pathway effectors, create a permissive chromatin state for their binding, or influence their function when bound (Fig. 3).

Approaches such as ChIPseq and DamID allow the profiling of transcription factor binding and chromatin states on a genome-wide scale and, in combination with expression profiling, can help dissect these mechanisms. Interaction with other transcription factors, which may be effectors of other pathways or lineage-specific factors, can influence both the recruitment to and function at enhancers and promoters across multiple species. Notch signaling, for example, generates different transcriptional responses in *Drosophila* muscle progenitors and hemocytes due to interaction with the lineage-specific factors Twist and Lozenge, respectively (Bernard, Krejci, Housden, Adryan, & Bray, 2010; Terriente-Felix et al., 2013). During vertebrate hematopoietic regeneration, both myeloid and erythroid lineages are generated in response to BMP and Wnt signaling. The effector transcription factors for BMP and Wnt in both lineages have been shown to be SMAD1

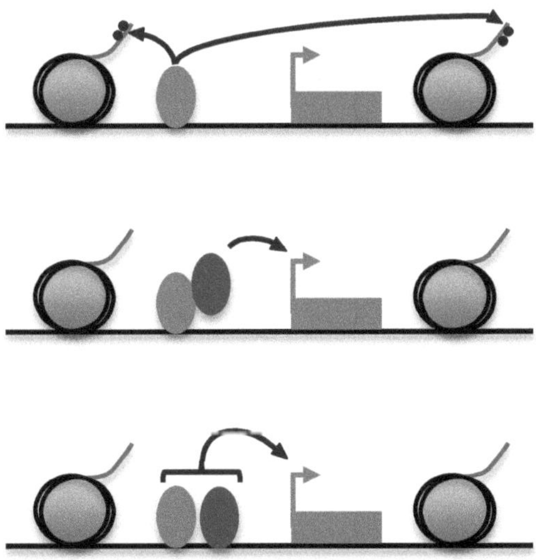

Figure 3 Transcriptional context in signaling. Lineage-specific regulators (or effectors of other pathways) (green) may regulate the targets of signaling pathways by: acting as pioneer factors to open chromatin into a permissive state (top panel, black line is DNA, light blue balls nucleosomes with histone tail extending, dark blue dots represent histone modifications); direct recruitment of effector transcription factors (red) to specific enhancer sites (middle panel); or by cooperative regulation, for example, by recruiting complementary chromatin modifiers or transcriptional apparatus (lower panel). (A color version of this figure is located at the end of this chapter.)

and TCFL2, respectively. However, despite the relatively close relationship between the lineages and the use of the same effector transcription factors, the responses show lineage specificity (Trompouki et al., 2011). In the myeloid lineage, the recruitment of SMAD1 and TCFL2 is mediated by the myeloid-specific transcription factor C/EBPα, whereas in the erythroid lineage GATA1 recruits the same pathway effectors to distinct targets. Interestingly, misexpression of the lineage-specific factor from one lineage in the other is sufficient to result in an exchange of targets.

While in some cases a single lineage-specific factor may act as a critical target determinant, it is likely that in many contexts the combination of factors is key. The binding of transcription factor PU.1 in macrophages and B cells correlates with C/EBPs and EBF1, respectively (Heinz et al., 2010). During macrophage activation, the combination of PU.1 and C/EBPs interact with NFKβ to determine its functional targets (Heinz, Romanoski, Benner, & Glass, 2015; Trompouki et al., 2011). PU.1 may act in this context as a pioneer factor, modifying histones and remodeling nucleosomes to create a permissive chromatin state for the recruitment of subsequent factors (Heinz et al., 2010; Iwafuchi-Doi & Zaret, 2014). In most developmental contexts, multiple signaling pathways may be active at once and a comprehensive understanding on how pathways regulate specific targets in specific cell types will therefore require both the lineage-specific factors and the effector transcription factors of multiple signaling pathways to be taken into account.

6. CONCLUSION

The field of intercellular signaling has developed rapidly and genomics and imaging approaches are offering a wide range of examples of how pathways generate diverse outcomes and how information is robustly transmitted. Complex networks of signaling regulators allow different ways of transmitting information through a single pathway, and provide points of cross talk with different pathways. Spatiotemporal aspects of signaling further increase the complexity of signal transduction and offer additional means to encode information. In addition to differential expression of pathway components, transcriptional context can act as a key determinant of cell type specificity in signaling. In many cases, new approaches are revealing examples of the varied ways in which diverse outcomes can be generated. However, in generating large datasets and describing aspects of signal transduction in ever-greater detail there is a risk of accumulating large quantities of complex information without increasing understanding. It is important to dissect which regulatory processes occur in which context and which are the most functionally important. Moving forward, it will be important to use systematic approaches to generate coherent datasets that can be compared and used together, while anchoring findings to the biological outputs that they regulate. Just as the application of genetics to developmental biology led to the elucidation of many core signaling pathways, in this era of "omics" approaches, collaborative work between cell biology, biochemistry, computational biology, and physics is making substantial contributions to our understanding of signal transduction.

ACKNOWLEDGMENTS

We would like to thank members of the Perrimon lab for helpful discussions. D.P.D. is supported by the Human Frontier Science Program. Work in the Perrimon lab is supported by the NIH and Howard Hughes Medical Institute.

REFERENCES

Aulehla, A., Wiegraebe, W., Baubet, V., Wahl, M. B., Deng, C., Taketo, M., et al. (2008). A beta-catenin gradient links the clock and wavefront systems in mouse embryo segmentation. *Nature Cell Biology*, *10*, 186–193.

Bandyopadhyay, S., Chiang, C., Srivastava, J., Gersten, M., White, S., Bell, R., et al. (2010). A human MAP kinase interactome. *Nature Methods*, 7, 801–805.

Bar-Peled, L., Schweitzer, L. D., Zoncu, R., & Sabatini, D. M. (2012). Ragulator is a GEF for the Rag GTPases that signal amino acid levels to mTORC1. *Cell*, *150*, 1196–1208.

Bean, D. M., Heimbach, J., Ficorella, L., Micklem, G., Oliver, S. G., & Favrin, G. (2014). esyN: Network building, sharing and publishing. *PloS One*, *9*, e106035.

Bergmann, S., Tamari, Z., Schejter, E., Shilo, B. Z., & Barkai, N. (2008). Re-examining the stability of the bicoid morphogen gradient. *Cell*, *132*, 15–17.

Bernard, F., Krejci, A., Housden, B., Adryan, B., & Bray, S. J. (2010). Specificity of Notch pathway activation: Twist controls the transcriptional output in adult muscle progenitors. *Development*, *137*, 2633–2642.

Bischoff, M., Gradilla, A.-C., Seijo, I., Andrés, G., Rodríguez-Navas, C., González-Méndez, L., et al. (2013). Cytonemes are required for the establishment of a normal Hedgehog morphogen gradient in Drosophila epithelia. *Nature Cell Biology*, *15*, 1269–1281.

Bökel, C., Schwabedissen, A., Entchev, E., Renaud, O., & González-Gaitán, M. (2006). Sara endosomes and the maintenance of Dpp signaling levels across mitosis. *Science*, *314*, 1135–1139.

Bruggeman, F. J., Westerhoff, H. V., Hoek, J. B., & Kholodenko, B. N. (2002). Modular response analysis of cellular regulatory networks. *Journal of Theoretical Biology*, *218*, 507–520.

Burack, W. R., & Shaw, A. S. (2005). Live cell imaging of ERK and MEK: Simple binding equilibrium explains the regulated nucleocytoplasmic distribution of ERK. *The Journal of Biological Chemistry*, *280*, 3832–3837.

Chandel, N. S. (2014). Mitochondria as signaling organelles. *BMC Biology*, *12*, 34.

Chao, M. V. (1992). Growth factor signaling: Where is the specificity? *Cell*, *68*, 995–997.

Chen, H.-C., & Reich, N. C. (2010). Live cell imaging reveals continuous STAT6 nuclear trafficking. *Journal of Immunology*, *185*, 64–70.

Collinet, C., Stöter, M., Bradshaw, C. R., Samusik, N., Rink, J. C., Kenski, D., et al. (2010). Systems survey of endocytosis by multiparametric image analysis. *Nature*, *464*, 243–249.

Coumailleau, F., Fürthauer, M., Knoblich, J. A., & González-Gaitán, M. (2009). Directional Delta and Notch trafficking in Sara endosomes during asymmetric cell division. *Nature*, *458*, 1051–1055.

Couzens, A. L., Knight, J. D. R., Kean, M. J., Teo, G., Weiss, A., Dunham, W. H., et al. (2013). Protein interaction network of the mammalian hippo pathway reveals mechanisms of kinase-phosphatase interactions. *Science Signaling*, *6*, rs15.

Di Fiore, P. P., & von Zastrow, M. (2014). Endocytosis, signaling and beyond. *Cold Spring Harbor Perspectives in Biology*, *6*, a016865.

Doupé, D. P., & Perrimon, N. (2014). Visualizing and manipulating temporal signaling dynamics with fluorescence-based tools. *Science Signaling*, *7*, re1.

Emery, G., Hutterer, A., Berdnik, D., Mayer, B., Wirtz-Peitz, F., Gaitan, M. G., et al. (2005). Asymmetric Rab11 endosomes regulate delta recycling and specify cell fate in the Drosophila nervous system. *Cell*, *122*, 763–773.

Félix, M.-A., & Barkoulas, M. (2012). Robustness and flexibility in nematode vulva development. *Trends in Genetics*, *28*, 185–195.

Friedman, A. A., Tucker, G., Singh, R., Yan, D., Vinayagam, A., Hu, Y., et al. (2011). Proteomic and functional genomic landscape of receptor tyrosine kinase and ras to extracellular signal-regulated kinase signaling. *Science Signaling*, *4*, rs10.

Frost, R. J. A., & Olson, E. N. (2011). Control of glucose homeostasis and insulin sensitivity by the Let-7 family of microRNAs. *Proceedings of the National Academy of Sciences of the United States of America*, *108*, 21075–21080.

Goessling, W., North, T. E., Loewer, S., Lord, A. M., Lee, S., Stoick-Cooper, C. L., et al. (2009). Genetic interaction of PGE2 and Wnt signaling regulates developmental specification of stem cells and regeneration. *Cell*, *136*, 1136–1147.

Gonnord, P., Blouin, C. M., & Lamaze, C. (2012). Membrane trafficking and signaling: Two sides of the same coin. *Seminars in Cell & Developmental Biology*, *23*, 154–164.

Gregor, T., Tank, D. W., Wieschaus, E. F., & Bialek, W. (2007). Probing the limits to positional information. *Cell*, *130*, 153–164.

Gregor, T., Wieschaus, E. F., McGregor, A. P., Bialek, W., & Tank, D. W. (2007). Stability and nuclear dynamics of the bicoid morphogen gradient. *Cell*, *130*, 141–152.

Guruharsha, K., Obar, R., & Mintseris, J. (2012). Drosophila protein interaction map (DPiM). *Fly (Austin)*, *6*, 246–253.

Guruharsha, K. G., Rual, J.-F., Zhai, B., Mintseris, J., Vaidya, P., Vaidya, N., et al. (2011). A protein complex network of Drosophila melanogaster. *Cell*, *147*, 690–703.

Havugimana, P. C., Hart, G. T., Nepusz, T., Yang, H., Turinsky, A. L., Li, Z., et al. (2012). A census of human soluble protein complexes. *Cell*, *150*, 1068–1081.

Heinz, S., Benner, C., Spann, N., Bertolino, E., Lin, Y. C., Laslo, P., et al. (2010). Simple combinations of lineage-determining transcription factors prime cis-regulatory elements required for macrophage and B cell identities. *Molecular Cell*, *38*, 576–589.

Heinz, S., Romanoski, C. E., Benner, C., & Glass, C. K. (2015). The selection and function of cell type-specific enhancers. *Nature Reviews. Molecular Cell Biology*, *16*, 144–154.

Housden, B. E., & Perrimon, N. (2014). Spatial and temporal organization of signaling pathways. *Trends in Biochemical Sciences*, *39*, 457–464.

Huttlin, E. L., Ting, L., Bruckner, R. J., Gebreab, F., Gygi, M. P., Szpyt, J., et al. (2015). The BioPlex network: A systematic exploration of the human interactome. *Cell*, *162*, 425–440.

Inaba, M., Buszczak, M., & Yamashita, Y. M. (2015). Nanotubes mediate niche–stem-cell signalling in the Drosophila testis. *Nature*, *523*, 329–332.

Iwafuchi-Doi, M., & Zaret, K. S. (2014). Pioneer transcription factors in cell reprogramming. *Genes & Development*, *28*, 2679–2692.

Kicheva, A., Cohen, M., & Briscoe, J. (2012). Developmental pattern formation: Insights from physics and biology. *Science, 338*, 210–212.
Kiel, C., Verschueren, E., Yang, J.-S., & Serrano, L. (2013). Integration of protein abundance and structure data reveals competition in the ErbB signaling network. *Science Signaling, 6*, ra109.
Kim, K., Vinayagam, A., & Perrimon, N. (2014). A rapid genome-wide microRNA screen identifies miR-14 as a modulator of Hedgehog signaling. *Cell Reports*, 7, 2066–2077.
Klein, A. M., Mazutis, L., Akartuna, I., Tallapragada, N., Veres, A., Li, V., et al. (2015). Droplet barcoding for single-cell transcriptomics applied to embryonic stem cells. *Cell, 161*, 1187–1201.
Knoblich, J. A. (2010). Asymmetric cell division: Recent developments and their implications for tumour biology. *Nature Reviews. Molecular Cell Biology, 11*, 849–860.
Kornberg, T. B., & Roy, S. (2014). Cytonemes as specialized signaling filopodia. *Development, 141*, 729–736.
Kressmann, S., Campos, C., Castanon, I., Fürthauer, M., & González-Gaitán, M. (2015). Directional Notch trafficking in Sara endosomes during asymmetric cell division in the spinal cord. *Nature Cell Biology, 17*, 333–339.
Kwon, Y., Vinayagam, A., Sun, X., Dephoure, N., Gygi, S. P., Hong, P., et al. (2013). The Hippo signaling pathway interactome. *Science, 342*, 737–740.
Macosko, E. Z., Basu, A., Satija, R., Nemesh, J., Shekhar, K., Goldman, M., et al. (2015). Highly parallel genome-wide expression profiling of individual cells using nanoliter droplets. *Cell, 161*, 1202–1214.
Marshall, C. J. (1995). Specificity of receptor tyrosine kinase signaling: Transient versus sustained extracellular signal-regulated kinase activation. *Cell, 80*, 179–185.
Masamizu, Y., Ohtsuka, T., Takashima, Y., Nagahara, H., Takenaka, Y., Yoshikawa, K., et al. (2006). Real-time imaging of the somite segmentation clock: Revelation of unstable oscillators in the individual presomitic mesoderm cells. *Proceedings of the National Academy of Sciences of the United States of America, 103*, 1313–1318.
Miller, C. J., & Davidson, L. A. (2013). The interplay between cell signalling and mechanics in developmental processes. *Nature Reviews. Genetics, 14*, 733–744.
Montagne, C., & Gonzalez-Gaitan, M. (2014). Sara endosomes and the asymmetric division of intestinal stem cells. *Development, 141*, 2014–2023.
Nagaraj, R., & Banerjee, U. (2004). The little R cell that could. *The International Journal of Developmental Biology, 48*, 755–760.
Perrimon, N., Pitsouli, C., & Shilo, B.-Z. (2012). Signaling mechanisms controlling cell fate and embryonic patterning. *Cold Spring Harbor Perspectives in Biology, 4*, a005975.
Purvis, J. E., & Lahav, G. (2013). Encoding and decoding cellular information through signaling dynamics. *Cell, 152*, 945–956.
Rhee, H.-W., Zou, P., Udeshi, N. D., Martell, J. D., Mootha, V. K., Carr, S. A., et al. (2013). Proteomic mapping of mitochondria in living cells via spatially restricted enzymatic tagging. *Science, 339*, 1328–1331.
Roux, K. J., Kim, D. I., Raida, M., & Burke, B. (2012). A promiscuous biotin ligase fusion protein identifies proximal and interacting proteins in mammalian cells. *The Journal of Cell Biology, 196*, 801–810.
Saha, A., Connelly, S., Jiang, J., Zhuang, S., Amador, D. T., Phan, T., et al. (2014). Akt phosphorylation and regulation of transketolase is a nodal point for amino acid control of purine synthesis. *Molecular Cell, 55*, 264–276.
Santos, S. D. M., Verveer, P. J., & Bastiaens, P. I. H. (2007). Growth factor-induced MAPK network topology shapes Erk response determining PC-12 cell fate. *Nature Cell Biology, 9*, 324–330.
Shen, C.-H., Yuan, P., Perez-Lorenzo, R., Zhang, Y., Lee, S. X., Ou, Y., et al. (2013). Phosphorylation of BRAF by AMPK impairs BRAF-KSR1 association and cell proliferation. *Molecular Cell, 52*, 161–172.

Shimozono, S., Iimura, T., Kitaguchi, T., Higashijima, S.-I., & Miyawaki, A. (2013). Visualization of an endogenous retinoic acid gradient across embryonic development. *Nature*, *496*, 363–366.

Silver, S. J., Hagen, J. W., Okamura, K., Perrimon, N., & Lai, E. C. (2007). Functional screening identifies miR-315 as a potent activator of Wingless signaling. *Proceedings of the National Academy of Sciences of the United States of America*, *104*, 18151–18156.

Sopko, R., Foos, M., Vinayagam, A., Zhai, B., Binari, R., Hu, Y., et al. (2014). Combining genetic perturbations and proteomics to examine kinase-phosphatase networks in Drosophila embryos. *Developmental Cell*, *31*, 114–127.

Terriente-Felix, A., Li, J., Collins, S., Mulligan, A., Reekie, I., Bernard, F., et al. (2013). Notch cooperates with Lozenge/Runx to lock haemocytes into a differentiation programme. *Development*, *140*, 926–937.

Toettcher, J. E., Voigt, C. A., Weiner, O. D., & Lim, W. A. (2011). The promise of optogenetics in cell biology: Interrogating molecular circuits in space and time. *Nature Methods*, *8*, 35–38.

Toettcher, J. E., Weiner, O. D., & Lim, W. A. (2013). Using optogenetics to interrogate the dynamic control of signal transmission by the Ras/Erk module. *Cell*, *155*, 1422–1434.

Tomida, T., Oda, S., Takekawa, M., Iino, Y., & Saito, H. (2012). The temporal pattern of stimulation determines the extent and duration of MAPK activation in a Caenorhabditis elegans sensory neuron. *Science Signaling*, *5*, ra76.

Trompouki, E., Bowman, T. V., Lawton, L. N., Fan, Z. P., Wu, D. C., Dibiase, A., et al. (2011). Lineage regulators direct BMP and Wnt pathways to cell-specific programs during differentiation and regeneration. *Cell*, *147*, 577–589.

Tsvetanova, N. G., & von Zastrow, M. (2014). Spatial encoding of cyclic AMP signaling specificity by GPCR endocytosis. *Nature Chemical Biology*, *10*, 1061–1065.

Vincent, J.-P. (2014). Modulating and measuring Wingless signalling. *Methods*, *68*, 194–198.

Wang, H., Chen, X., He, T., Zhou, Y., & Luo, H. (2013). Evidence for tissue-specific JAK/STAT target genes in Drosophila optic lobe development. *Genetics*, *195*, 1291–1306.

Wang, W., Li, X., Huang, J., Feng, L., Dolinta, K. G., & Chen, J. (2013). Defining the protein-protein interaction network of the human hippo pathway. *Molecular & Cellular Proteomics*, *13*, 119–131.

Wansleeben, C., & Meijlink, F. (2011). The planar cell polarity pathway in vertebrate development. *Developmental Dynamics*, *240*, 616–626.

Wolpert, L. (2011). Positional information and patterning revisited. *Journal of Theoretical Biology*, *269*, 359–365.

Yu, H., Braun, P., Yildirim, M. A., Lemmens, I., Venkatesan, K., Sahalie, J., et al. (2008). High-quality binary protein interaction map of the yeast interactome network. *Science*, *322*, 104–110.

Yun, S. P., Lee, M. Y., Ryu, J. M., & Han, H. J. (2009). Interaction between PGE2 and EGF receptor through MAPKs in mouse embryonic stem cell proliferation. *Cellular and Molecular Life Sciences*, *66*, 1603–1616.

Zhang, Y.-L., Guo, H., Zhang, C.-S., Lin, S.-Y., Yin, Z., Peng, Y., et al. (2013). AMP as a low-energy charge signal autonomously initiates assembly of AXIN-AMPK-LKB1 complex for AMPK activation. *Cell Metabolism*, *18*, 546–555.

Zhao, Y., Hu, Q., Cheng, F., Su, N., Wang, A., Zou, Y., et al. (2015). SoNar, a highly responsive NAD+/NADH sensor, allows high-throughput metabolic screening of anti-tumor agents. *Cell Metabolism*, *21*, 777–789.

Zhu, H., Shyh-Chang, N., Segrè, A. V., Shinoda, G., Shah, S. P., Einhorn, W. S., et al. (2011). The Lin28/let-7 axis regulates glucose metabolism. *Cell*, *147*, 81–94.

Zoncu, R., Bar-Peled, L., Efeyan, A., Wang, S., Sancak, Y., & Sabatini, D. M. (2011). mTORC1 senses lysosomal amino acids through an inside-out mechanism that requires the vacuolar H+-ATPase. *Science*, *334*, 678–683.

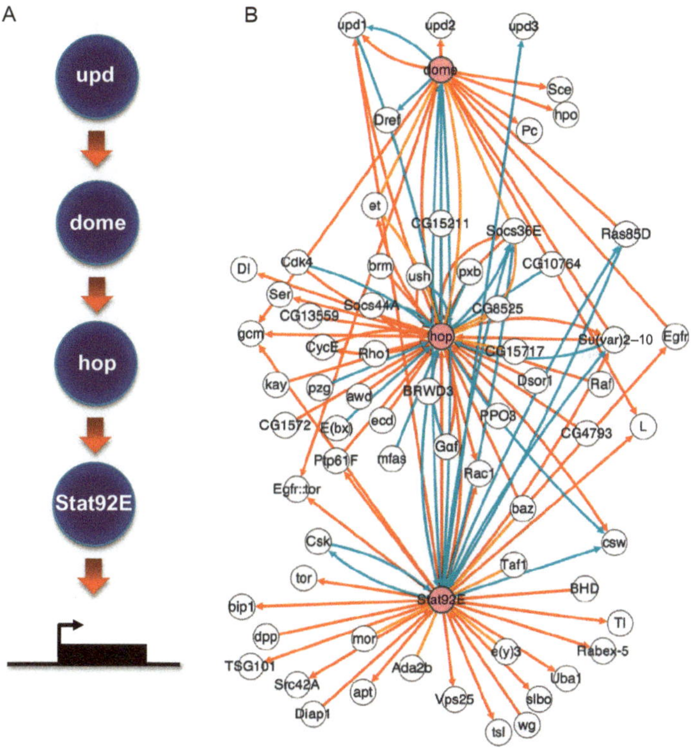

David P. Doupé and Norbert Perrimon, Figure 1 From linear pathways to complex networks. (A) Simple representation of the core Drosophila JAK/STAT pathway: cytokine ligands (upd1, 2, 3) bind to the receptor (dome), which activates JAK (hop) and in turn Stat (Stat92E) to regulate target gene expression. (B) An interaction network for dome, hop, and Stat92E (pink nodes) showing genetic interactions (red and blue lines with arrows) and physical interactions (orange lines). Network generated using EsyN (Bean et al., 2014).

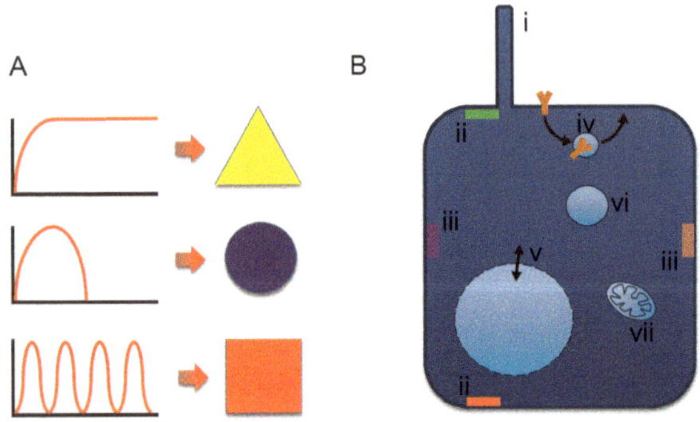

David P. Doupé and Norbert Perrimon, Figure 2 Spatiotemporal properties of signaling pathways. (A) The temporal properties of a single signal (graphs) may determine differential cellular responses (colored shapes). (B) Schematic of a cell showing some possible points of spatial regulation of signal transduction. Cells may localize surface signaling by: extension of short nanotube projections or longer cytonemes (i), apicobasal (ii), and planar (iii) cell polarization. Endocytosis and signaling endosomes (iv), lysosomes (vi), and mitochondria (vii) all have roles in subcellular localization of signaling, while nucleocytoplasmic transport (v) is an important step in many pathways.

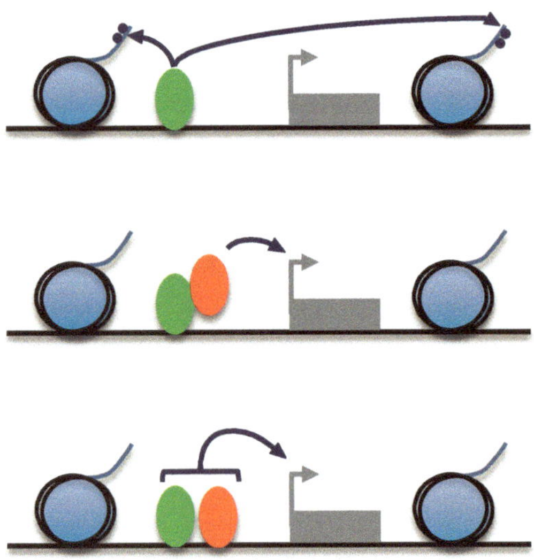

David P. Doupé and Norbert Perrimon, Figure 3 Transcriptional context in signaling. Lineage-specific regulators (or effectors of other pathways) (green) may regulate the targets of signaling pathways by: acting as pioneer factors to open chromatin into a permissive state (top panel, black line is DNA, light blue balls nucleosomes with histone tail extending, dark blue dots represent histone modifications); direct recruitment of effector transcription factors (red) to specific enhancer sites (middle panel); or by cooperative regulation, for example, by recruiting complementary chromatin modifiers or transcriptional apparatus (lower panel).

CHAPTER THIRTEEN

Implications of Developmental Gene Regulatory Networks Inside and Outside Developmental Biology

Isabelle S. Peter[1], Eric H. Davidson[2]
Division of Biology and Biological Engineering, California Institute of Technology, Pasadena, California, USA
[1]Corresponding author: e-mail address: ipeter@caltech.edu

Contents

Abstract

The insight that the genomic control of developmental process is encoded in the form of gene regulatory networks has profound impacts on many areas of modern bioscience. Most importantly, it affects developmental biology itself, as it means that a causal understanding of development requires knowledge of the architecture of regulatory network interactions. Furthermore, it follows that functional changes in developmental gene regulatory networks have to be considered as a primary mechanism for evolutionary process. We here discuss some of the recent advances in gene regulatory network biology and how they have affected our current understanding of development, evolution, and regulatory genomics.

The epistemology of developmental biology has been profoundly affected by the demonstration that gene regulatory networks can provide a causal explanation of the phenomena that have fascinated observers of embryogenesis for a century and a half. The implications spread beyond the field of

[2] Deceased

Current Topics in Developmental Biology, Volume 117
ISSN 0070-2153
http://dx.doi.org/10.1016/bs.ctdb.2015.12.014

development *per se* and affect our views of the function and organization of animal genomes, and inescapably of the process of animal evolution that have given rise to animal development. In this brief essay, we explore some of the consequences as they have altered the orientation of scientific inquiry in this and other domains of animal biology that interlock in the processes and outcome of development.

Why gene regulatory networks: this term denotes the inescapable functional system which transforms the encoded genomic program, the product of 800 million years of animal evolution, into the top level executive operations that control development. Specifically, gene regulatory networks consist of the regulatory genes encoding the transcription factors which control all downstream functions, and the network architecture explicitly indicates the interactions of these genes required for the spatial deployment of their expression. These interactions are of course directly encoded in the A's, G's, C's, and T's of the *cis*-regulatory sequences which determine the transcriptional activity of these regulatory genes. That is, the *cis*-regulatory sequences confer target specificity for each gene: the regulatory genes regulate one another, hence the network constellation. The function of these gene regulatory networks is to generate the specific regulatory states defining the spatial organization of the developing animal. A regulatory state is the sum of functionally active transcription factors (or the sets of mRNAs encoding these). The initial process to be organized in development, as schematically summarized in Fig. 1, is subdivision of the embryo along its axial coordinates into regional regulatory states, thereby initiating the sequence of regional cell fate decisions. Following this, specific regions are assigned regulatory states that endow them with the capability of initiating the formation of particular body parts (i.e., defining body part progenitor fields). Further downstream, these progenitor fields are subdivided by establishment of the regionalized regulatory states required for the development of each component of the body part. As development progresses, cell type specification networks are deployed within all of these regions. At each stage, the inputs for the next stage are the outputs of the networks operating at the immediately preceding stage in that spatial domain. Gene regulatory networks thus provide the mechanism by which the constellation of spatial regulatory states progresses in an automaton-like manner from the initial maternal inputs in the embryo all the way to the termination of development. In our recent work, we provide a comprehensive analysis and interpretation of the operation of the whole developmental control system (Peter & Davidson, 2015).

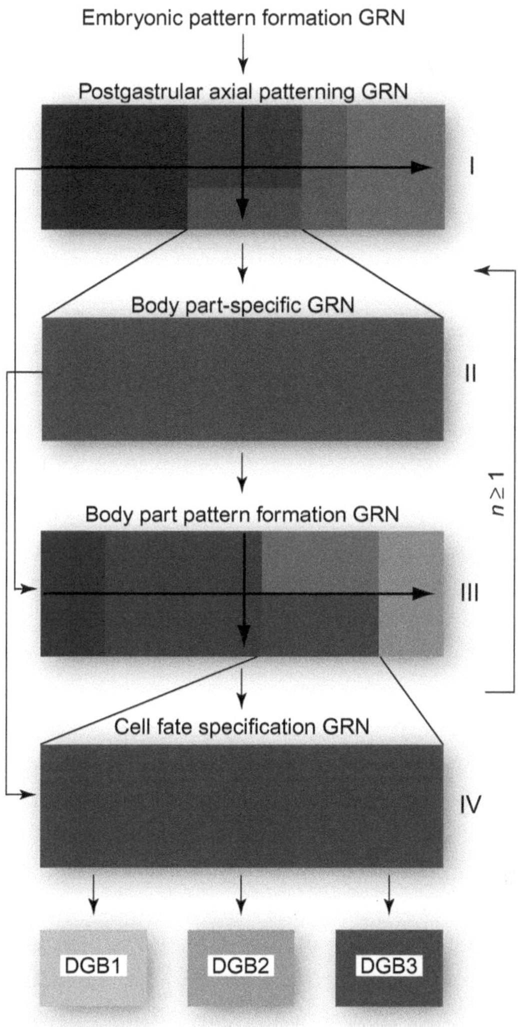

Figure 1 Schematic representation of the hierarchical operation of gene regulatory networks during development. After gastrulation, the progenitor fields for given body parts are specified by expression of regulatory states in given axial positions (I). The regulatory networks operating in given progenitor fields (II) further control the subdivision into progressively defined regulatory state domains, again organized in space according to the embryonic axes (III). This process ultimately leads to the specification of individual cell fates (IV) and the control of cell fate-specific differentiation gene batteries (DGB). Specific network circuitry operates every level of the developmental hierarchy. *Reprinted from Peter and Davidson (2011a) copyright (2011), with permission from Elsevier.* (A color version of this figure is located at the end of this chapter.)

1. IMPLICATIONS FOR CAUSALITY IN DEVELOPMENTAL BIOLOGY

A causal understanding of developmental process means the solution of the mechanism which explains why the phenomena of development occur as they do. To go one step further, solution of "mechanism" means a structure/function analysis couched in terms of the macromolecular apparatus of life. This has never been possible until the regulatory sequence functions of the genome became experimentally accessible at a system level. That is to say, development can only be causally explained in terms of gene regulatory network structure and function.

In the whole twentieth century history of developmental biology, experimental approaches oriented toward causality have been exceedingly rare. A brilliant early exception was Boveri's 1904–1908 polyspermy experiments (Boveri, 1902, 1905, 1907), the object and achievement of which was to demonstrate the requirement for presence of the complete genome in every blastomere in order for (sea urchin) development to take place normally. We learned a great deal in the nature of "if/then" relations among embryonic cells from experimental embryology in the ensuing decades, for example, by studying the effects of blastomere deletions and transplantations. And while this establishes the necessity of given physical organization, it of course provides no access to causal mechanism. Similar demonstration of intertissue or intercellular signaling again defines particular functions and demonstrates their requirement for developmental progression. Similarly, the raise of mutational developmental genetics in the last third of the twentieth century again provided a large collection of necessary elements in the developmental system and told us something about what they are necessary for. Perhaps of more global significance was the emergence from this that the control of development would in some ways devolve from the expression of regulatory genes. All of these advances transcended what still remained the major industry of developmental biology, which was gathering phenomenology, defined ever more elegantly. But looking back from where we are now it is clear that demonstration of necessary relations, whether between signals and morphogenesis, or genes and body part development, or blastomeres and the parts of the embryos, cannot constitute a causal explanation of the developmental process in the above sense and never will. This is because it is inherently impossible to understand a system of interlocking parts by blindly removing individual components and looking at the output of the system. In developmental biology, for example, there is no pathway to mechanistic understanding of how a body part is built the way it is from a catalogue of genes necessary for the process.

The main implication of such reflections is that they define for us what is actually required to achieve a causal explanation in developmental biology. First, we do indeed need to know all the regulatory components operating in the given phase of development, and it is now clear that this information can only be accessed by the tools of molecular biology. For various complex reasons, mostly having to do with network circuitry, many of these components are blind to genetic screens in animal development. Second, we need information on the spatial and temporal patterns of expression of these genes at a system-wide level. Third, we need to identify the specific interactions which constitute the heart of gene regulatory network architecture, and this information can only be acquired through experimental perturbation. Fourth, these regulatory linkages need to be integrated into a model capturing the networks of regulatory interactions which are encoded in the control sequences of the relevant genes. For us a fifth requirement is increasingly fundamental, and that is to be able to address how this interlocking system of regulatory genes executes informational logic transactions at its nodes and in the subcircuits of which the network is composed.

2. WHY SHOULD A DEVELOPMENTAL BIOLOGIST CARE ABOUT GENE REGULATORY NETWORKS?

This question may be superfluous to most readers of this chapter, but is indeed posed not unfrequently. It addresses directly what can be learned from experimentally solving regulatory networks that goes beyond knowing the individual components and their functional necessity for a given developmental process. Why are gene regulatory networks required to provide the detailed causal understanding of the control processes underlying the phenomena of development? A network circuit operating in the context of sea urchin early embryonic mesoderm specification may serve here as an illuminating example. To start with the biological function, this small circuit ensures the spatial specification of mesodermal precursor cells in the context of the rapidly changing cellular geometry of the early embryo. Mesodermal specification is initiated by expression of *glial cell missing* (*gcm*), and continued expression of this gene is required for the specification of the pigment cell fate (Ransick & Davidson, 2006). Thus, continuation of *gcm* is essential for the specification of this cell type, while other cells descending from *gcm*-expressing precursors will give rise to other cell fates and lose *gcm* expression. Both outcomes are the consequence of the same network circuitry. The expression of *gcm* and the specification of almost all mesodermal cell types is initiated in about 14–16 cells, arranged as a ring surrounding the skeletogenic domain. The reason for exactly these and no other cells of the early embryo to become mesodermal is that *gcm* expression at this stage is directly controlled by the transcription factor Su(H) and thus by Delta/Notch signaling, which is activated by the expression of Delta ligand in skeletogenic cells (Ransick & Davidson, 2006). For a few hours, the network circuitry ensures that *gcm* expression is entirely dependent on the Delta/Notch signaling input (Croce & McClay, 2010). Thus, when *gcm*-expressing cells divide such that two rings of cells are formed, only cells of the inner ring continue with the expression of *gcm* and the specification of mesoderm. Cells of the outer ring, more distant from the Delta signal, cease expression of this gene and give rise to the anterior endoderm (Peter & Davidson, 2010, 2011b). Thus, the network circuitry is designed such that this system abolishes mesodermal specification in cells removed from the Delta signaling source. However, a few hours later, skeletogenic cells stop expressing *delta*, and more so, these cells start ingressing into the blastocoel, thus removing the Delta signaling source for the mesodermal cells. At this point, the network circuitry ensures that *gcm* expression becomes independent of the Delta/Notch signaling input. The circuitry active now controls a separate *cis*-regulatory module of *gcm*, which operates using Gcm and Six1/2 as direct activating inputs (Ransick & Davidson, 2012). As the network model in Fig. 2A indicates, this circuit constitutes a double positive feedback loop, where *gcm* expression depends both on its own gene product and on the product of *six1/2*. Expression of *six1/2* in turn is directly regulated by *gatae*, itself a target gene of Gcm. Even though Gcm activates its own expression, if this were to be sufficient for maintenance of *gcm* expression, all cells once turning on *gcm* expression would continue to do so independently of Delta/Notch signaling. Thus, a separate mechanism would have to ensure that *gcm* expression is turned off in endodermal cells. The requirement for both Gcm and Six1/2 has the result that the independence of *gcm* expression from the initial Delta/Notch signaling input is only achieved after several hours, at a time when the change in embryonic geometry and the dislocation of cells demands for signal-independent maintenance of *gcm* expression.

Curiously, a very similar network circuit architecture is used in the development of the mouse heart (Fig. 2B). Here, the regulatory genes building the circuit are *nkx2.5*, *gata4*, and *tbx5* (Herrmann, Groß, Zhou,

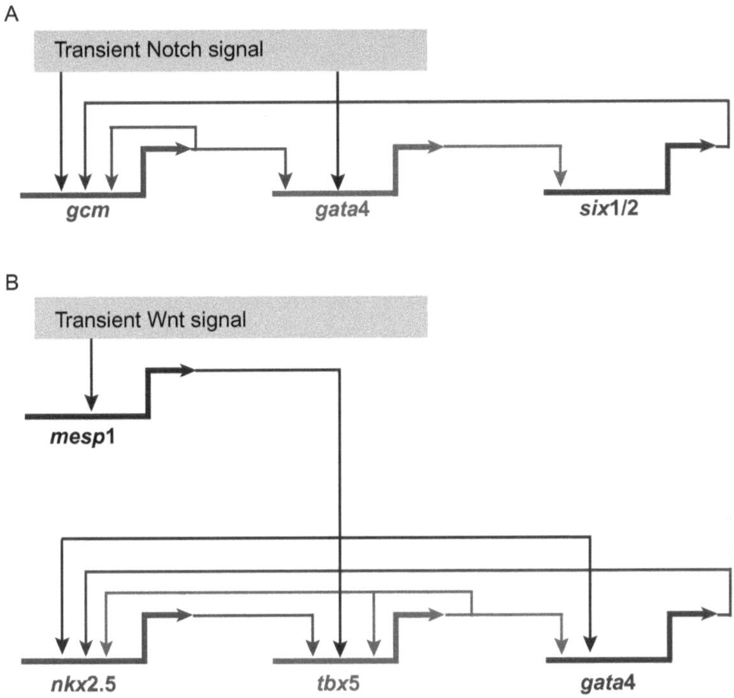

Figure 2 Similar network circuitry operating in different developmental processes. (A) Circuit operating during mesodermal specification in sea urchin embryos. *Gcm* expression is initially activated by Delta/Notch signaling. Two positive feedback loops are responsible for maintaining *gcm* expression after Delta signal is no longer available. Both Gcm and Six1/2 are required for expression of *gcm*, and this AND logic delays stable expression of *gcm* until *six1/2* is expressed. (B) Positive feedback circuit operating in specification of first heart field in mice. The circuit is initially activated by *mesp1* downstream of Wnt signaling. Again, two positive feedback loops maintain expression of *nkx2.5*, mediated by Tbx5 and Gata4, while an additional autoregulatory feedback loop activates expression of *tbx5*. (A color version of this figure is located at the end of this chapter.)

Kestler, & Kühl, 2012). The initial induction of their expression in the first heart field is activated by Mesp1, which is expressed downstream of Wnt signaling in cardiac precursor cells. Again, stable expression of *nkx2.5* is maintained by two positive feedback loops, mediated by Tbx5, expressed downstream of Nkx2.5, and by the product of *gata4*, a target gene of Tbx5. In consequence, the expression of these three regulatory genes is maintained even in the absence of the initial signaling input.

The positive feedback circuit is but one example of circuitry wirings which occur repeatedly in many different developmental gene regulatory networks, as recently reviewed (Peter & Davidson, 2015). The theme remains the same: similar network circuitry may control similar developmental functions, independent of the identity of regulatory genes they are composed of. In fact, the set of regulatory genes deployed in any given developmental context is quite specific. But the interactions among these regulatory genes appear to adhere to a much more universal code of regulatory programing. Thus, the same kind of network circuits, positive feedbacks, coherent or incoherent feedforward loops, or mutual repression circuits, among others, all appear in very different developmental networks, often even at similar positions within the network hierarchy. This leads to the fundamental conclusion that developmental process is indeed encoded in the form of network circuitry, where the constellation of regulatory linkages is the determinant of developmental function. The consequence of this conclusion affects many areas of modern biological sciences. For developmental biology, this means that obtaining understanding of the causality of any developmental process will necessarily require solving the genomic networks controlling the deployment of all genes in the system.

A recent compilation of experimentally solved gene regulatory networks shows the immense progress this field has seen over just the last few years (Peter & Davidson, 2015). At least in part, networks underlying the development of many body parts, such as the heart, limb, eye, and digestive system, have been solved, often covering processes from early specification and patterning of progenitor fields down to the differentiation of the particular cell types these give rise to. Thus, network models are obtained for early endodermal specification, distinction of foregut derivatives such as pancreas and liver, down to the differential specification of pancreatic cell types, including the control of the respective differentiation genes. Similarly, in the vertebrate limb, solved network circuitry illuminates the initial specification of the limb bud, the axes of the developing limb, the anterior/posterior patterning of the autopod, and the interconnection between these processes. Even at this stage, when only fragments of these complex developmental processes are understood at the network level, they provide crucial insights into the design of genomic programs at several levels, from the overall structure of the developmental process (see Fig. 1) down to the individual molecular transactions. They make apparent the powerful impact of the discovery of gene regulatory networks on developmental biology: not only do they uncover the genomic code for development, but they also shift the perception from singular events happening during development to the connectivity of molecular functions responsible for developmental progression.

3. IMPLICATIONS FOR CROSS-SPECIES COMPARATIVE ANALYSIS OF APPARENTLY HOMOLOGOUS DEVELOPMENTAL PROCESSES

The insight that developmental processes are encoded in hierarchically organized gene regulatory networks also offers unique opportunities to directly address genomic sequence alterations underlying evolutionary change. Any difference in morphological structure of a body part must ultimately be the result of a modification of the development process, caused by a functional change of the underlying gene regulatory network. The comparison of gene regulatory networks underlying homologous developmental processes may provide direct access to the mechanisms by which alterations of structure in regulatory networks of common origin has produced morphological change. Such analysis reveals both, shared network features that most likely have been encoded already in the ancestral genome, as well as novel network circuitry that has been acquired during the process of independent evolution. What kind of changes would we expect to find when comparing the regulatory networks underlying homologous developmental processes? They may range from the rewiring of some of the circuitry among the ancestral set of regulatory genes, to the co-option of novel regulatory genes into the network, and maybe even the co-option of entire small circuits, depending in part on the evolutionary distance between divergent species. In addition, the recruitment of differentiation genes at the downstream periphery of networks may lead to the acquisition of novel cellular functions or even of novel cell types within a given developmental context. Obtaining information on conserved as well as novel features in homologous networks will illuminate how network changes have brought about the morphological evolution of the body plan. Sure enough, for the developmental process, both novel and conserved network circuitry may be equally essential, and the degree of conservation will not provide an estimate of the degree of functional importance. For example, wiring among a set of transcription factors within the heart field in vertebrates and flies is remarkably similar, and these regulatory genes all directly control the expression of differentiation genes (Davidson & Erwin, 2006; Peter & Davidson, 2015). However, earlier developmental processes of obvious functional importance, that lead to the specification of the heart field in vertebrates and flies, are controlled by distinct circuitry within the upstream network hierarchy.

Would we be able to distinguish based on network architecture homologous from nonhomologous developmental processes? If so, which features would indicate common ancestry? Current insights indicate that although the expression of similar regulatory gene sets suggests an evolutionary relationship, it does not provide sufficient evidence to indicate a common evolutionary origin of the developmental process *per se*. This is illustrated by the gene regulatory networks encoding the formation of eyes in vertebrates and flies, which are composed of many orthologous regulatory genes, although the developmental processes they contribute to are profoundly different. The large overlap of regulatory gene use in eye development is nevertheless thought to be the outcome of common ancestry. The same set of regulatory genes might have been expressed in photosensitive cell types in the common ancestor (Gehring & Ikeo, 1999). These regulatory genes would thus have acquired, independently in flies and vertebrates, additional functions to give rise to the morphogenetic processes involved in eye development. And sure enough, the regulatory linkages which determine the function of the common regulatory gene set are different in flies and vertebrates, supporting the idea that the gene regulatory networks have been assembled independently and that the resulting developmental processes are nonhomologous. Thus, while sharing of expressed regulatory gene sets might be an indicator of homology at the cell type level, for developmental processes, the architecture of the network, and not the identity of its nodes, will provide the essential information based on which to distinguish related and nonrelated processes.

4. IMPLICATIONS FOR HOW BODY PLANS EVOLVE

Realizing that the genome encodes in its *cis*-regulatory sequences a hierarchical program that computes developmental processes has fundamental consequences on the way we have to think about evolutionary process. Thus, the same type of change of *cis*-regulatory information can have entirely different outcomes, depending on where in the network hierarchy these changes occur. When focusing on changes in *cis*-regulatory activity occurring within the transcriptional control system of individual regulatory genes, the impact of hierarchical position can be directly accessed. Many examples of evolutionary change in regulatory gene use at the downstream periphery of network hierarchy have been described, and they affect post-developmental cellular functions. In mice for example, evolutionary changes in color pattern have been shown to be due to changes in the regulation of *agouti*, a gene encoding an antagonist of a melanocortin receptor (Manceau, Domingues, Mallarino, & Hoekstra, 2011). Expression of Agouti on the ventral side leads to a delay in melanocyte differentiation and to a lighter color in mice. A change of spatial expression of the *agouti* gene to include more dorsal regions has been shown to be responsible for lighter color in mice living in a beach environment, compared to mice living on dark soil. In flies, the species-specific pigmentation patterns of wings have been shown to be the result of *cis*-regulatory changes in the *yellow* pigmentation gene, which lead to the differential spatial expression of this gene in the wing (Gompel, Prud'homme, Wittkopp, Kassner, & Carroll, 2005; Prud'homme et al., 2006). On the other hand, changes in the deployment of regulatory genes higher up in gene regulatory network hierarchy have consequences for the development of entire body parts. Thus, the loss of *pitx1* expression during early limb bud development by evolutionary deletion of the relevant regulatory sequence has been shown to be responsible for the reduction of pelvic fins in sticklebacks (Chan et al., 2010). Similarly, evolutionary changes in beak morphology in Darwin's finches were shown to be the outcome of changes in the expression of signaling molecules such as *bmp4* and *calmodulin*, directly affecting depth/width and length, respectively (Mallarino et al., 2011). Either loss of entire body parts, or the modulation of its dimensions can therefore indeed be the result of evolutionary change in the deployment of individual regulatory genes. However, the evolutionary changes that have led to the acquisition of novel morphological structures are necessarily more complex than that.

In order to give rise to novel body parts or morphological structures, the rewiring within underlying developmental networks must have affected the control of multiple regulatory genes. These changes are likely to have occurred serially, although perhaps within a short time frame, resulting in the sequential assembly of network circuitry. One dramatic example for how changes in gene regulatory network structure have caused the acquisition of a novel morphological feature has been discovered in the echinoderm lineage. While an adult skeleton is a shared feature of all echinoderms, most echinoderm larvae are lacking a skeleton. Larvae of modern sea urchins however do possess a skeleton, which is thought to be formed as a result of the co-option of the entire adult skeletogenic gene regulatory network to operate downstream of an early embryonic patterning circuit (Gao & Davidson, 2008). But even in this case, when the morphological novelty of the embryonic skeleton might represent the co-option of a preexisting network, this could not have occurred without substantial changes in the *cis*-regulatory control regions of multiple regulatory genes. First of all, the regulatory gene set used in the adult and embryonically active skeletogenic networks in the euechinoid *Strongylocentrotus purpuratus* are not absolutely identical. Several regulatory genes used in the embryonic context are not expressed in adult skeletogenic centers, and therefore most likely represent the results of independent co-option events of individual regulatory genes into the skeletogenic network. Furthermore, multiple nodes of the skeletogenic regulatory network respond directly to the double negative gate patterning circuit, which ensures activation of skeletogenic gene expression exclusively in skeletogenic cells, and repression of the same genes everywhere else in the embryo (Oliveri, Tu, & Davidson, 2008). Several regulatory genes of the embryonic skeletogenic network have thus acquired direct transcriptional control by HesC, the transcription factor responsible for excluding expression of the skeletogenic cell fate everywhere but in skeletogenic cells. Operation of the skeletogenic network is permitted in skeletogenic cells because the otherwise ubiquitously expressed *hesc* gene is repressed in these cells by Pmar1. Elements of this same double negative gate patterning circuit have recently been discovered in the cidaroid *Eucidaris tribuloides* (Erkenbrack & Davidson, 2015). Embryos of this species do develop a skeleton as well, although this process is clearly different from the way it operates in *S. purpuratus*. Even though the double negative gate circuit is not functional as such in *E. tribuloides* embryos, the patterning of skeletogenic cells in these embryos does indeed depend on HesC repression of at least some regulatory genes within the skeletogenic network. Thus, we may see, by comparing the regulatory program in these two species, individual steps in the assembly of a network circuit which has resulted in the evolution of a novel morphological feature.

Thus, what so far seemed almost impossible, to rationalize the molecular changes which in deep time have led to the acquisition of novel body parts or the reorganization of entire body plans, becomes accessible in the light of gene regulatory networks. This will require knowledge beyond the expression of particular gene sets, and will involve experimental dissection of developmental regulatory control systems for related developmental processes in species at various crucial positions within the phylogeny. The reward will be to identify parts of networks that are at least functionally conserved, as well as entirely novel circuitry, and to reveal the functional changes in the regulatory genome that have led to the acquisition of novel morphological structures.

5. IMPLICATIONS FOR REGULATORY GENOMICS

Individual *cis*-regulatory modules are relatively small sequence fragments which execute particular information transactions at specific places within a complex developmental process. Each of these regulatory sequences contains arrays transcription factor binding sites that determine its function, but identifying this function is difficult to achieve experimentally and not possible computationally based on sequence alone. For differentiated cell types, experimental identification of *cis*-regulatory functions is facilitated, because many genes are expressed simultaneously and are controlled by the same combination of coexpressed transcription factors. The functional analysis of regulatory sequences of coexpressed differentiation genes, such as for example those controlling the expression of lens crystallins, have shown that they each are composed of binding sites for a subset of the expressed transcription factors (Cvekl & Duncan, 2007). Thus, at the downstream periphery of the network, arrays of differentiation genes in individual cell types are controlled by a common set of transcription factors, and this same combinatorial regulation may persist over long periods of time. However, this is "just" about running a final state of cellular functions. How about all the computational transactions that underlie developmental process? They are driven by *cis*-regulatory modules that may be active for just a short period of time and are controlled by transcription factors that may also not be expressed in that same combination for very long. After all, the expression of transient regulatory states is one of the key drivers of developmental progression. In addition to the temporal restriction of regulatory activity, *cis*-regulatory modules may function in just a part of the embryo, increasing the difficulty of identifying such elements. Since networks themselves are modularly organized, such that individual circuits operate simul taneously and in parallel, regulatory modules active during development often operate based on different inputs. Thus, at higher levels of the network hierarchy, *cis*-regulatory modules tend to be regulated by specific and unique combinations of transcription factors that are different from the regulatory inputs into other *cis*-regulatory modules, even if active in the same cells at the same time.

The statistical interpretation of large-scale genomic analyses of regulatory sequences is designed to recover regulatory design principles common to many *cis*-regulatory modules. Given the ratio in the genome of regulatory genes versus differentiation genes, such analyses will therefore focus on *cis*-regulatory modules operating at the downstream periphery of the network. Yet most of the computation underlying developmental process is encoded in *cis*-regulatory modules active at earlier phases of development, particularly those determining the patterning and organization of a developing embryo. Defining the individual functions of key computational nodes in the network therefore requires knowledge of the architecture of the underlying gene regulatory networks, derived from experimental network analysis. The hierarchical organization of *cis*-regulatory function within developmental networks is of fundamental importance to the control of developmental process, and information on the structure of this regulatory system must therefore be taken into account for interpretation of the regulatory genome.

ACKNOWLEDGMENTS

Many thanks go to Deanna Thomas for preparation of the figures. This work was supported by National Institutes of Health Grant HD-037105.

Commentary. The overall idea and the initial paragraphs of this article have been conceived and written on a sunny fall afternoon, the last day of Eric's life. It thus represents the last product of an incredibly rewarding intellectual collaboration and friendship. Out of respect, these parts of the manuscript were left unchanged. The article was completed on the basis of countless discussions, and therefore possibly in accordance with Eric's views.

REFERENCES

Boveri, T. (1902). Über Mehrpolige Mitosen als Mittel zur Analyse des Zellkerns. *Verhandlungen der Physikalische-medizinischen Gesellschaft zu Würzburg*, *35*, 67–90.

Boveri, T. (1905). *Zellenstudien V. Über die Abhängigkeit der Kerngrösse und Zellenzahl bei Seeigellarven von der Chromosomenzahl der Ausgangszellen*. Jena: Gustav Fischer.

Boveri, T. (1907). *Zellenstudien VI. Die Entwicklung dispermer Seeigeleier. Ein Beitrag zur Befruchtungslehre und zur Theorie des Kerns*. Jena: Gustav Fischer.

Chan, Y. F., Marks, M. E., Jones, F. C., Villarreal, G., Jr., Shapiro, M. D., Brady, S. D., et al. (2010). Adaptive evolution of pelvic reduction in sticklebacks by recurrent deletion of a Pitx1 enhancer. *Science (New York, N.Y.)*, *327*, 302–305.

Croce, J. C., & McClay, D. R. (2010). Dynamics of Delta/Notch signaling on endomesoderm segregation in the sea urchin embryo. *Development (Cambridge, England)*, *137*, 83–91.

Cvekl, A., & Duncan, M. K. (2007). Genetic and epigenetic mechanisms of gene regulation during lens development. *Progress in Retinal and Eye Research*, *26*, 555–597.

Davidson, E. H., & Erwin, D. H. (2006). Gene regulatory networks and the evolution of animal body plans. *Science (New York, N.Y.)*, *311*, 796–800.

Erkenbrack, E. M., & Davidson, E. H. (2015). Evolutionary rewiring of gene regulatory network linkages at divergence of the echinoid subclasses. *Proceedings of the National Academy of Sciences of the United States of America*, *112*, E4075–E4084.

Gao, F., & Davidson, E. H. (2008). Transfer of a large gene regulatory apparatus to a new developmental address in echinoid evolution. *Proceedings of the National Academy of Sciences of the United States of America, 105*, 6091–6096.

Gehring, W. J., & Ikeo, K. (1999). Pax 6: Mastering eye morphogenesis and eye evolution. *Trends in Genetics, 15*, 371–377.

Gompel, N., Prud'homme, B., Wittkopp, P. J., Kassner, V. A., & Carroll, S. B. (2005). Chance caught on the wing: cis-regulatory evolution and the origin of pigment patterns in Drosophila. *Nature, 433*, 481–487.

Herrmann, F., Groß, A., Zhou, D., Kestler, H. A., & Kühl, M. (2012). A Boolean model of the cardiac gene regulatory network determining first and second heart field identity. *PLoS One, 7*, e46798.

Mallarino, R., Grant, P. R., Grant, B. R., Herrel, A., Kuo, W. P., & Abzhanov, A. (2011). Two developmental modules establish 3D beak-shape variation in Darwin's finches. *Proceedings of the National Academy of Sciences of the United States of America, 108*, 4057–4062.

Manceau, M., Domingues, V. S., Mallarino, R., & Hoekstra, H. E. (2011). The developmental role of Agouti in color pattern evolution. *Science (New York, N.Y.), 331*, 1062–1065.

Oliveri, P., Tu, Q., & Davidson, E. H. (2008). Global regulatory logic for specification of an embryonic cell lineage. *Proceedings of the National Academy of Sciences of the United States of America, 105*, 5955–5962.

Peter, I. S., & Davidson, E. H. (2010). The endoderm gene regulatory network in sea urchin embryos up to mid-blastula stage. *Developmental Biology, 340*, 188–199.

Peter, I. S., & Davidson, E. H. (2011a). Evolution of gene regulatory networks controlling body plan development. *Cell, 144*, 970–985.

Peter, I. S., & Davidson, E. H. (2011b). A gene regulatory network controlling the embryonic specification of endoderm. *Nature, 474*, 635–639.

Peter, I. S., & Davidson, E. H. (2015). *Genomic control process, development and evolution.* San Diego, CA: Academic Press/Elsevier.

Prud'homme, B., Gompel, N., Rokas, A., Kassner, V. A., Williams, T. M., Yeh, S. D., et al. (2006). Repeated morphological evolution through cis-regulatory changes in a pleiotropic gene. *Nature, 440*, 1050–1053.

Ransick, A., & Davidson, E. H. (2006). cis-Regulatory processing of Notch signaling input to the sea urchin glial cells missing gene during mesoderm specification. *Developmental Biology, 297*, 587–602.

Ransick, A., & Davidson, E. H. (2012). Cis-regulatory logic driving glial cells missing: Self-sustaining circuitry in later embryogenesis. *Developmental Biology, 364*, 259–267.

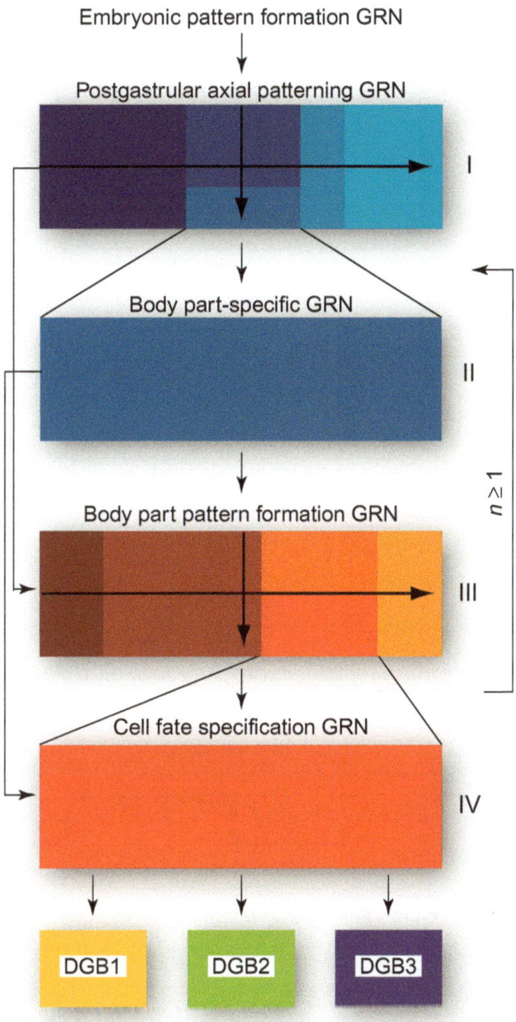

Isabelle S. Peter and Eric H. Davidson, Figure 1 Schematic representation of the hierarchical operation of gene regulatory networks during development. After gastrulation, the progenitor fields for given body parts are specified by expression of regulatory states in given axial positions (I). The regulatory networks operating in given progenitor fields (II) further control the subdivision into progressively defined regulatory state domains, again organized in space according to the embryonic axes (III). This process ultimately leads to the specification of individual cell fates (IV) and the control of cell fate-specific differentiation gene batteries (DGB). Specific network circuitry operates every level of the developmental hierarchy. *Reprinted from Peter and Davidson (2011a) copyright (2011), with permission from Elsevier.*

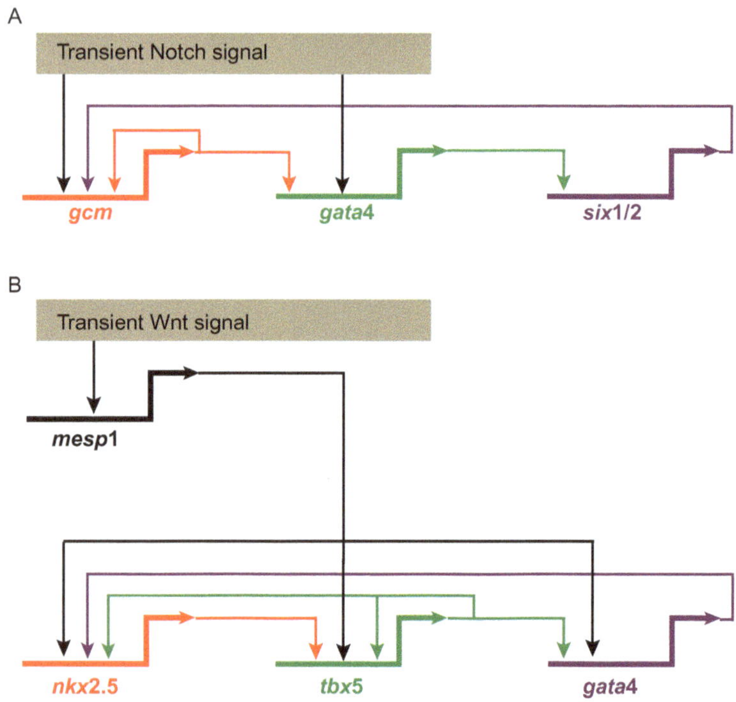

Isabelle S. Peter and Eric H. Davidson, Figure 2 Similar network circuitry operating in different developmental processes. (A) Circuit operating during mesodermal specification in sea urchin embryos. *Gcm* expression is initially activated by Delta/Notch signaling. Two positive feedback loops are responsible for maintaining *gcm* expression after Delta signal is no longer available. Both Gcm and Six1/2 are required for expression of *gcm*, and this AND logic delays stable expression of *gcm* until *six1/2* is expressed. (B) Positive feedback circuit operating in specification of first heart field in mice. The circuit is initially activated by *mesp1* downstream of Wnt signaling. Again, two positive feedback loops maintain expression of *nkx2.5*, mediated by Tbx5 and Gata4, while an additional autoregulatory feedback loop activates expression of *tbx5*.

CHAPTER FOURTEEN

Evo-Devo: Discovery of Diverse Mechanisms Regulating Development

Alys M. Cheatle Jarvela, Leslie Pick[1]
Entomology Department, University of Maryland, College Park, Maryland, USA
[1]Corresponding author: e-mail address: lpick@umd.edu

Contents

Abstract

Evo-devo is a relatively new field focusing on how mechanisms controlling development have changed during evolution. To date, studies of highly conserved members of an animal genetic toolkit have predominated in this field. These toolkit genes were originally identified in a small number of model organisms that do not represent the extant diversity of metazoans. We describe how this research paradigm came to be and what we have learned from it, and then argue that studies of other types of animals and of lineage-specific genes, in both model and nonmodel systems, are needed to expand our understanding of development and evolution.

Current Topics in Developmental Biology, Volume 117
ISSN 0070-2153
http://dx.doi.org/10.1016/bs.ctdb.2015.11.014

1. DISCOVERY OF *HOX* GENES CHANGED THE FIELD OF DEVELOPMENTAL BIOLOGY

The field of developmental genetics began in the early 1900s, based upon a fusion of the disparate fields of embryology and genetics (Gilbert, 2010). Progress in analyzing developmental mechanisms came from transplantation experiments that identified the time and place of cell specification in early embryos and from analysis of mutations—mostly spontaneous mutants—that perturbed development or body patterning. Combining these approaches, Salome Gluecksohn-Waelsch and other forerunners of the field argued that studies of "natural perturbations" of development due to changes in genes controlling this process would lead to an understanding of the rules guiding the development of embryos from single cells into fully functioning organisms. Using different animals systems, scientists began studying these "experiments of nature," examining associated phenotypes and deriving hypotheses about how embryonic development is controlled. Others induced mutations and, using the same genetic logic, studied their phenotypes with the goal of deducing the wild type or endogenous function of these genes.

Perhaps, the most odd among mutants being studied at this time were those causing homeotic transformations, first seen by Bateson in the wild (Bateson, 1894), and later studied in *Drosophila melanogaster* by pioneer Lewis (reviewed in Lewis, 1994, 1997). These mutations resulted in transformation of one body part or segment into a fully formed alternate body part, resulting in phenotypes such as legs replacing antennae. These homeotic transformations were so far from phenotypes being studied in other systems—mouse tail length or coat color, for example—that in a Developmental Biology class attended by one of us (L.P.) taught by Gluecksohn-Waelsch, she pointed out that these were merely bizarre happenings that would have little importance in the broader scheme of developmental biology. It was not until the genes responsible for these bizarre phenotypes were isolated that their broad significance was realized. Wholly unexpectedly, the isolation of what came to be known as *Hox* genes, and the finding that these genes were shared by virtually all animals, in the mid-1980s and early 1990s, led to an explosion in the field of developmental biology.

During this early period of "Homeo-madness" (Marx, 1986), researchers studying animals as diverse as planaria, hydra, mice, and even humans (Garcia-Fernandez, Baguna, & Salo, 1991; McGinnis, Garber, Wirz,

Kuroiwa, & Gehring, 1984; McGinnis, Hart, Gehring, & Ruddle, 1984; Schierwater, Murtha, Dick, Ruddle, & Buss, 1991) were able to readily isolate developmentally important genes based on cross-hybridization to *Drosophila Hox* genes (reviewed in Pick, 2015). The findings that these genes played roles in embryonic patterning in different types of animals and that their chromosomal clustering and colinearity were evolutionarily conserved (see Fig. 1), not only facilitated rapid progress in the field of developmental

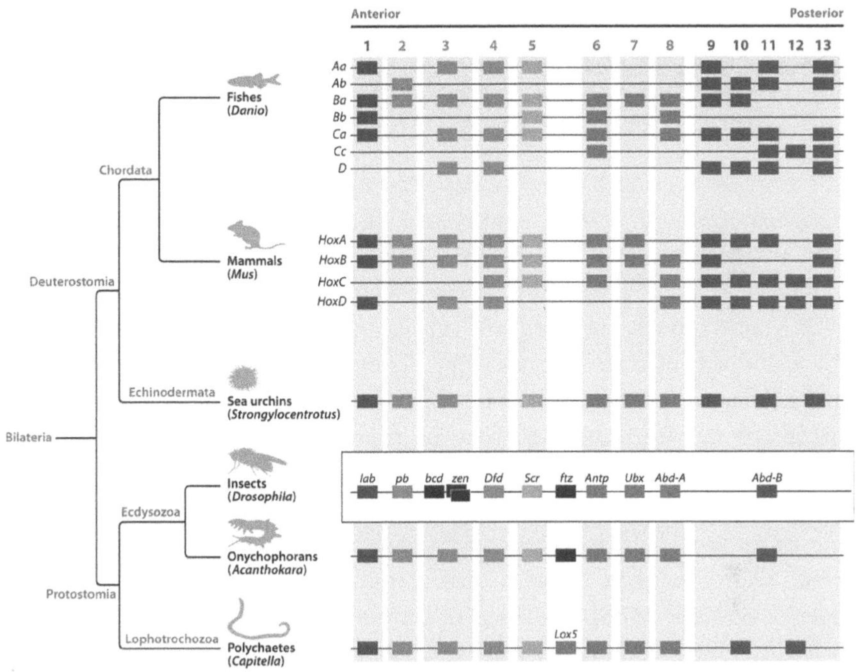

Figure 1 *Hox* gene expression and function have been conserved during animal evolution. *Left*: A cladogram indicating major bilaterian taxa with diagrams showing the basic body plans of selected species. *Right*: Complexes of *Hox* genes from the indicated taxa. The *Drosophila Hox* complex (HOM-C) boxed includes eight homeotic *Hox* genes and three that have taken on novel functions (*bcd, zen,* and *ftz*). Orthology to the *Drosophila Hox* genes is indicated by the color code. The homeotic *Hox* genes are expressed collinearly with their positions in the complex: *lab* is expressed more anteriorly than *pb*, which is expressed more anteriorly than *Dfd*, and so on with *Abd-B* expressed most posteriorly. This colinearity, as well as position of individual *Hox* orthologs within *Hox* complexes, has been conserved during evolution. In Chordata, genome duplication events have generated multiple *Hox* complexes on different chromosomes. Individual paralogs have been lost from some complexes, presumably enabled by redundancy. *Figure from Heffer and Pick (2013).* (A color version of this figure is located at the end of this chapter.)

genetics, it led to a sea-change in the way scientists thought about development. Although it is easy to say in retrospect that this broad conservation of regulatory genes was to be expected, in fact it was not predicted in advance of the finding. It was only in light of evolution, to paraphrase Dobzhansky, that this result could be explained. Animals have evolved from common ancestors and these ancestral mechanisms remain active in extant species of all forms and types. This realization opened up the minds of developmental biologists to the importance of evolution.

2. CONSERVATION OF *HOX* GENES LED TO IDENTIFICATION OF OTHER FAMILIES OF CONSERVED REGULATORY GENES

If genes responsible for the development of ectopic legs and wings in *Drosophila* could control the embryonic development of mammals, pretty much anything was possible. Perhaps, other genes controlling development of *Drosophila*, and of other invertebrate species, were utilized by mammals as well? Many quickly realized the utility of using a conservation-based approach to advance an understanding of development in a variety of organisms, especially vertebrate models which were not amenable to large-scale genetic screens. With the hope that the *Hox* genes were not an odd exception, several groups performed screens to find homologs of known *Drosophila* developmental regulatory genes. For example, orthologs of the *Drosophila* gene *paired* (*prd*), which were known to function in segmentation in *Drosophila*, were identified in the human genome (Burri, Tromvoukis, Bopp, Frigerio, & Noll, 1989). A screen of mouse genomic DNA using the *prd*-family member *gooseberry*, identified *pax1*, which was indeed shown to be expressed in the murine-segmented vertebral column (Deutsch, Dressler, & Gruss, 1988). Similar approaches delivered the vertebrate orthologs of *even-skipped* (*evx1* and *evx2* (Bastian & Gruss, 1990), *achaete-scute* (*mash1* and *mash2*) (Johnson, Birren, & Anderson, 1990), and *Notch* (Coffman, Harris, & Kintner, 1990)), to name only a few. Finding other conserved regulatory genes was more serendipitous. Rijsewjk and colleagues searched for homologs of the *int-1* oncogene by screening a fly genomic library with a probe derived from mouse sequence (Rijsewijk et al., 1987). Not only was a strong match discovered, this gene turned out to be the same as the already well-studied *wg* segment polarity gene. This finding leads to the important realization that *int-1* is a developmental signaling molecule and it was subsequently renamed *Wnt-1*. In a similar example, sequence analysis of *Drosophila Dpp-c*, followed by an unbiased protein database search revealed homology to known mammalian TGF-β genes, which provided valuable insight into the function and mechanism of action by which Dpp regulates developmental processes (Padgett, St Johnston, & Gelbart, 1987). Using homology-based approaches to find and understand the function of genes of interest was so lucrative, it quickly became ubiquitous. Even now that a wide variety of species have sequenced genomes, which would permit broader and more unbiased approaches, many studies still revolve around this now deeply rooted methodology of choosing candidate genes based on assumed homology.

3. A CONSERVED "GENETIC TOOLKIT" CONTROLS THE DEVELOPMENT OF DIVERSE SPECIES

These homology-based approaches led to the finding that developmental regulatory genes encoding transcription factors (TFs) and signaling proteins—from *Hox* TFs to signaling proteins such as Wnt and Hh family members—were utilized by diverse animals and even plants to regulate development. Many of these developmental regulatory genes were identified first in model systems through large genetic screens based upon visible phenotypes that could be readily scored in large numbers of offspring (e.g., Nusslein-Volhard & Wieschaus, 1980). These genes have large effects on body patterning, most often resulting in lethality, infertility, and/ or drastic alteration of body patterns for homozygous loss-of-function alleles. For example, animals homozygous for mutations in segmentation genes in *Drosophila* are missing body segments and die prior to hatching, while mutations in *eyeless* (*Pax6*) result abnormal eye development. Although the evolutionary community was initially skeptical about the significance of these genes, homologs of these genes turned out to be not only present in many species but also responsible for naturally occurring mutation and variation, in keeping with the expectations of early developmental geneticists (see above and Gilbert, 2010).

Two major themes emerged from studies of these conserved genes: First, these genes are regulatory genes (sometimes called master regulatory genes) that are often members of large families of genes encoding proteins with related functions. Second, these genes are utilized, albeit in somewhat different ways, in many, many animal species examined (Fig. 2). As proposed by Sean Carroll, these genes comprise a "genetic toolkit" that has been conserved during evolution, with the implication that the ancestor

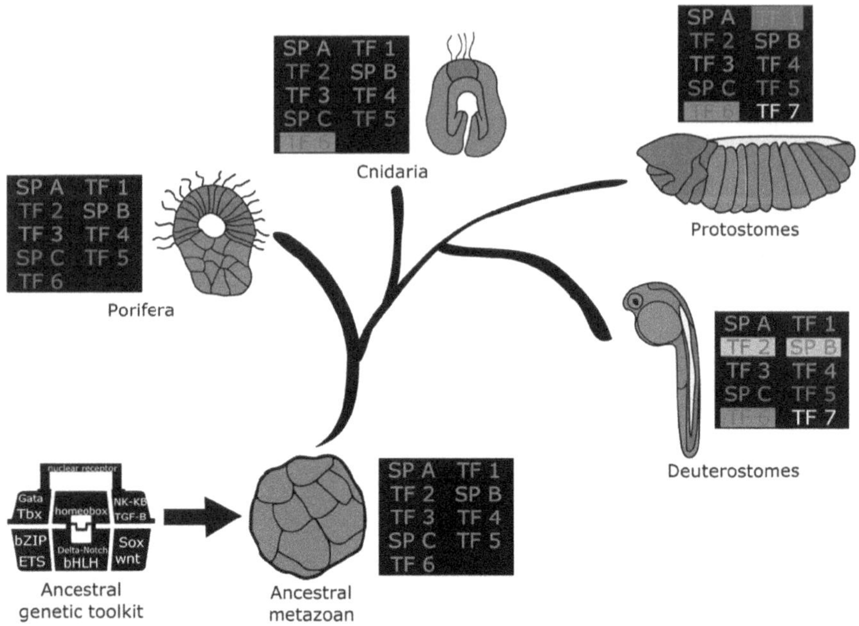

Figure 2 The metazoan genetic toolkit for development. A large body of work demonstrates that the genetic toolkit governing development is highly conserved throughout the animal kingdom (Carroll, 2000). Many of the constituent signaling molecules and transcription factor families are pan-metazoan. Examples of these ancient gene families are pictured in a toolkit (left). TF denotes "transcription factor" and SP denotes "signaling pathway." As animal life became more complex, these toolkit genes were repurposed to allow for the development of new cell types, tissues, and structures, indicated by color changes shared between groups of toolkit genes and a corresponding embryo region (e.g., TF 2, TF 5, TF 6, and SP C families have been changed from ancestral green to derived magenta in Porifera lineage). Importantly, these toolkit genes are often used in more than one time or place in the embryo to help build unique gene networks and allow for greater cell diversity, which is indicated by genes boxed by an additional color representing a second embryo region (for example, TF 1 has cyan font boxed by red in the protostome lineage to indicate use in both the cyan- and red-colored embryo regions). In spite of the impressive reuse and repurposing of these toolkit components, there are also many examples of toolkit genes with deeply conserved developmental functions, indicated by a lack of color change (as examples, TF 3, TF 4, and SP B families are shown in green font in every depicted toolkit). Addition of completely new toolkit factors occurs, but is rare (for example, TF 7 in the bilaterian lineage). (A color version of this figure is located at the end of this chapter.)

of all animals, and most certainly of Bilateria, had a basic version of this toolkit. This toolkit expanded in different metazoan lineages, through gene duplication events and rewiring events that brought the same "old" genes into new processes (reviewed in Carroll, 2000, 2008; Carroll, Grenier, & Weatherbee, 2005). Slowly, the field of evolution and development found common ground with realization that these fields are inextricably intertwined and the new discipline, evolutionary developmental biology (evo-devo), emerged.

The idea of the genetic toolkit found strong support when comparative genomics became feasible at the turn of the millennia. For example, an initial comparison of the *D. melanogaster* and *Caenorhabditis elegans* genomes to available mammalian ESTs demonstrated that nearly all of the signaling genes known to be needed for development were conserved among these three long-diverged species (Rubin et al., 2000). A similar comparison found that 90% of bHLH TFs identified in vertebrates have a protostome ortholog (Ledent & Vervoort, 2001). These studies confirmed that the genetic toolkit originated at least at the base of bilateria. More recent comparisons to basal animals indicate even more ancient toolkit establishment and conservation than was originally imagined. Study of multiple sponge lineages as well as placozoan (*Trichoplax adhaerens*), ctenophore (*Mnemiopsis leidyi*), and cnidarian (*Nematostella vectensis*) genomes revealed that many of the developmental TF families and signaling components considered to be toolkit genes are pan-metazoan or at least pan-eumetazoan (Fortunato, Adamski, & Adamska, 2015; Putnam et al., 2007; Ryan et al., 2013; Srivastava et al., 2008, 2010). Additionally, many toolkit genes were shown to have surprisingly similar functions across taxa—perhaps most famously, *eyeless*, *Pax6*, is responsible for eye development in flies, humans, and many other species despite the large morphological differences between eye types (Halder, Callaerts, & Gehring, 1995; Quiring, Walldorf, Kloter, & Gehring, 1994, reviewed in Callaerts, Halder, & Gehring, 1997). In keeping with this, QTL analyses in systems not amenable to genetic analysis often identify the most likely candidate gene within a region based upon its known role in a model system, or by using model systems to test candidate gene function (e.g., den Hoed et al., 2013). In fact, often toolkit genes form parts of evolutionarily conserved gene networks (Levine & Davidson, 2005). A good example if this comes from detailed gene-regulatory networks for endomesoderm in the sea urchin, *Strongylocentrotus purpuratus*, compared to the newer model system, the sea star, *Patiria miniata* (Hinman & Davidson, 2007; Hinman, Nguyen, Cameron, & Davidson, 2003; McCauley,

Weideman, & Hinman, 2010). Many of the TFs involved are conserved over the 500 million year divergence time between these species, as are their regulatory interactions with each other.

What has been more difficult to understand is how animal life can be so rich in diversity if essentially every animal is built by the same genetic toolkit. The genetic toolkit paradigm concludes that novel phenotypes arise during evolution when toolkit genes experience changes in expression, which allows them to regulate different target genes, or perhaps even the same genes but in a different context or to a different extent. This includes both spatiotemporal expression changes as well as changes in expression level (reviewed in Carroll, Prud'homme, & Gompel, 2008; Stern & Orgogozo, 2009). For example, the highly conserved *hh* pathway—a key member of the genetic toolkit—was implicated in eye degeneration in blind cavefish and also in the adaptive variation observed in jaw morphology in cichlid fish (Roberts, Hu, Albertson, & Kocher, 2011; Yamamoto, Stock, & Jeffery, 2004). Changes in the expression and function of *Hox* genes have been correlated with morphological evolution (reviewed in Pick & Heffer, 2012). For example, *Sex combs reduced* (*Scr*) plays a role in the sexually dimorphic variation in beetle horns and changes in *Ubx* expression levels were implicated in adaptive changes in locomotion in water striders (Refki, Armisen, Crumiere, Viala, & Khila, 2014; Wasik, Rose, & Moczek, 2010). These studies, and many other studies, have demonstrated that rewiring of regulatory interactions, bringing new genes under control of conserved regulators, can result in morphological variations in body patterning.

Much of what we have learned about mechanisms controlling embryonic development has come from studies in tractable model systems such as flies, worms, and mice. Similarly, much of what has been learned about the evolution of these developmental mechanisms has come from applying this knowledge to understand how animals with particular phenotypes compare to known models with respect to developmental processes and/or utilization of toolkit genes. For example, previous work demonstrating that *Ubx* modulates the shape and size of T3 legs in *Drosophila* would have made *Ubx* a prime candidate to examine when trying to understand the unusually proportioned limbs of the water strider. Similarly, the known role of *Scr* in specification of the T1 segment in *Drosophila* immediately suggests this gene as a candidate for regulating modifications of these segments seen in horned beetles. Thus, the detailed knowledge of a variety of developmental processes, gained by the study of model organisms, has propelled the field of evo-devo. However, this also means that the evo-devo community has been somewhat limited to the study of traits that have meaningful model organism cognates.

4. UNDERSTANDING THE ENORMOUS BIODIVERSITY ON OUR PLANET REQUIRES NEW APPROACHES

An estimated 8 million metazoan species currently inhabit the Earth's aquatic and terrestrial environments, a predicted 86% of which have not been characterized (Mora, Tittensor, Adl, Simpson, & Worm, 2011). An understanding of how diversity in cell types, tissues, and organs, as well overall body symmetry, types of germ layers, and developmental strategies are generated during evolution requires an understanding of development in a broad sampling of metazoan taxa. Study of basally branching animal groups, such as Porifera, Cnidaria, and Ctenophora, is especially useful for understanding the origins of these developmental processes. Studies that broadly sample animal taxa from different branches of the tree of life are needed to understand the true pervasiveness and homology of known developmental mechanisms and the extent to which these can be modified (Fig. 3). Yet, the majority of developmental studies have been performed in vertebrate model systems or in the classical model organisms, *D. melanogaster* and *C. elegans*, both of which come from the superphylum Ecdysozoa. This is partly due to the desire to choose models that are likely to yield information relevant to the improvement of human health, thus the emphasis on vertebrate models, and the genetic power of the well-established invertebrate classical model systems. As a result, we have studied a restricted set of conserved genes in a very small sample of the available animal diversity.

As we discuss below, the surprise that a genetic toolkit functions so broadly through the animal kingdom and the success of the study of toolkit members in model organisms has focused studies of development on this class of genes at the expense of studies of non-toolkit members. These non-toolkit genes include a large number of lineage- and species-specific genes that make animals different from each other. In addition, recent comparative studies suggest that functional changes in toolkit ancestors have led to diverse biological roles for similar genes in different taxa. Determining the roles of both toolkit and lineage-specific genes in animals representing extant biodiversity remains an open challenge. Achieving this will require sampling broadly across phylogeny using innovative approaches that assess not only gene expression but also gene function in diverse taxa.

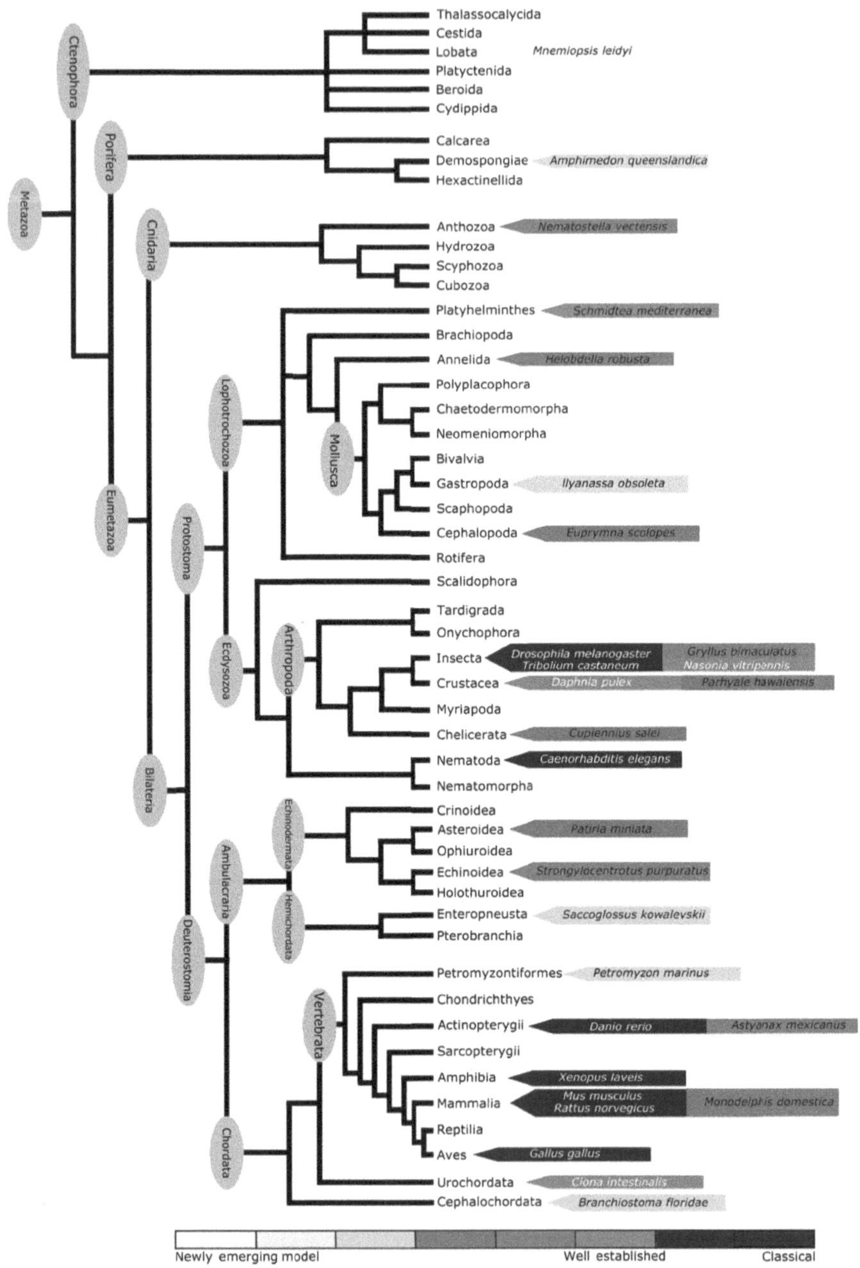

Figure 3 Model systems represent only a tiny subset of metazoan diversity. A cladogram depicting extant metazoan life is on the left. Pink ovals indicate important or relatively better studied clades. Colored flags on the right indicate species within the indicated clade that are being used for developmental or evo-devo work. Flag color indicates

(Continued)

5. LINEAGE-SPECIFIC GENES REMAIN UNDERSTUDIED

The success of studying development based upon a conserved genetic toolkit has left the role of lineage-specific genes largely understudied. Lineage-specific genes, sometimes called "orphans" or "taxonomically restricted genes" depending upon their phylogenetic distribution, are estimated to make up 10–20% of every eukaryotic genome (Khalturin, Hemmrich, Fraune, Augustin, & Bosch, 2009). Because the term "lineage" simply refers to a particular branch on the tree of life (Fig. 3), it can be used to refer to something as broad as "eumetazoans," and therefore include all animals except sponges and ctenophores, or it can be as specific as "primates," and only include apes and monkeys in isolation from other mammals. This means the percentage of genes that are considered to be lineage specific depends greatly on how broad of a range of taxa one is referring to (Fig. 4), although in the literature the term often implies a narrow phylogenetic distribution. The percentage of genes that are species specific, which the narrowest and most specific lineage possible, is thought to be ~1–10% (Wissler et al., 2013; Zhou et al., 2015), which translates to hundreds or even thousands of orphan genes per species. This number can only be estimated when closely related sister species genomes are compared, and so it is difficult to say how universal it may be. In one extreme case, it was found that 71% of ESTs in the genome of the tick, *Amblyomma americanum*, have no homology to anything in the UniProt database (Gibson, Smith, Fuqua, Clay, & Colbourne, 2013). What is less clear is the importance of these orphan genes to development and the evolution of regulatory, and,

Figure 3—Cont'd how well established that species is according to the key at the bottom of the image, ranging from light yellow, minimal resources or lab cultures available, to dark blue, for the best-studied genetic systems—flies, worms, and mice. Model systems were scored based on criteria that represent the richness of available resources (i.e., publically available genome, large research community, and ability to perform sophisticated genetics) and the general amenability of the system for experimentation (i.e., Can the organism be maintained in the lab? Can gametes be obtained year-round? Can embryos be molecularly or genetically perturbed?). Some clades are far better represented by classical and well-established model systems than others, particularly Vertebrata and Ecdysozoa. Lophotrochozoa and Ambulacraria are currently gaining interest and resources, while taxa such as Porifera, Cnidaria, and Ctenophores require much better representation. Essentially, all major clades shown include taxa that are not represented by any experimental system. (A color version of this figure is located at the end of this chapter.)

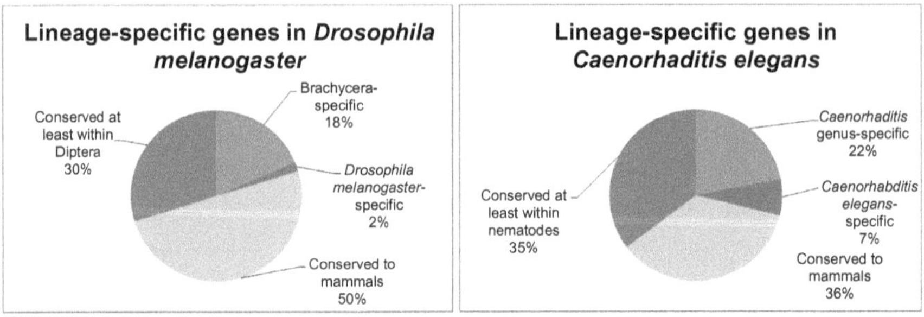

Figure 4 Percentages of lineage-specific genes in model organisms. While a large proportion of genes in both model systems are pan-bilaterian (Rubin et al., 2000) as they are also found in mammals, both also contain a high proportion of genes that are highly taxonomically restricted. 18% of genes in *D. melanogaster* are unique to the Brachycera, a suborder of Diptera. 22% of genes in *C. elegans* are specific to the genus *Caenorhabditis* compared to other nematode species. Both model's gene sets are 1–10% species specific (Wissler, Gadau, Simola, Helmkampf, & Bornberg-Bauer, 2013; Zdobnov et al., 2002; Zhou et al., 2015). (A color version of this figure is located at the end of this chapter.)

by extension, morphological novelty, particularly since lack of homology precludes inference of function for many of these genes. We argue that the importance of these genes to the generation of animal diversity is underestimated and merits greater exploration.

A strict focus on the study of conserved genes identified in a limited number of model organisms will prevent us from fully understanding the evolution of novel body plans and specialized structures. Analysis of the recently sequenced genome of the California two-spot octopus, *Octopus bimaculoides*, illustrates this particularly well (Albertin et al., 2015). It had been thought that whole genome duplication in the cephalopod lineage best explained the many elaborate morphological novelties found in this clade but not other mollusks, such as a greatly expanded complex nervous system and sucker-lined arms that are both chemosensory and dexterous (Hallinan & Lindberg, 2011; Yoshida et al., 2011). Instead, examination of the octopus genome suggests not only that whole genome duplication did not occur, but also that transposons, gene duplication, RNA editing, and the evolution of lineage-specific genes better explain molecular novelty and increased regulatory complexity in this organism. Of the 33,638 predicted protein-coding genes in this species, 3557 are thought to be octopus specific, as they do not align to any known mollusk or cephalopod sequences. Notably, hundreds of octopus-specific genes were specifically expressed in the central nervous system and a few others were limited to the skin, and suckers, where the most striking morphological novelties have evolved in this organism. While they did find fairly standard representation of the developmental toolkit genes compared to other invertebrates, there were also exceptions, such as the C2H2 zinc finger TF family, which has expanded even more extensively in the octopus than in mammals. Transcripts for these novel TFs were found predominantly in embryonic tissue. Overall, these findings suggest that in order to understand how incredible features, such as an advanced CNS and camouflaging skin chromatophores in octopus, arose and develop, we need to understand the roles of novel genes in this process in addition to redeployment of known toolkit genes.

6. EVEN FOR MODEL SYSTEMS, MUCH REMAINS TO BE LEARNED

We can see from the above discussion that there is a world of discovery awaiting us in animals that have not been developed as laboratory model systems. At the same time, established model systems are far from outliving their utility. Modeling of human disease in these models is a major and ongoing endeavor, for example, by generating loss- or gain-of-function alleles of orthologs of genes implicated in human disorders in *Drosophila* (reviewed in Wangler, Yamamoto, & Bellen, 2015). Surprisingly though, the function of a large number of genes in the genomes of even these model systems remains unknown. These genes are generally those that do not mutate to lethal, or otherwise readily evident, phenotypes. They have been annotated in genomes and their expression patterns are often known, but, most frequently, their biological and even biochemical functions are not. In part, this may be because they are not members of large gene families with well-characterized biochemical function such that their functions cannot be predicted by comparison to other family members. A recent query of Flybase revealed that of the 13,918 *Drosophila* genes, 7192 genes are named by annotation ID only. Of these, 1900 lack biological data for any Gene Ontology (GO) category (molecular cell component or biological function). This suggests that there have been no experiments published or even predictions made using InterPro/UniPro. 1323 of these genes are associated with GO terms defined by experimental evidence in at least one category but for some, the only experimental evidence is an expression pattern found in a high-throughput screen and no actual functional data are available. The remainder have been assigned GO terms but these are based on predicted function rather than experimental evidence. These numbers are quite surprising as we tend to think of *Drosophila* as a "saturated" genetic system. What this tells us is that while the genetic tools are sophisticated and deficiencies, transposable element insertions, and other types of mutations do cover the entire genome, a large fraction of *Drosophila* genes have not been studied through loss-of-function genetics to reveal biological roles.

What are these unknown genes? Many will be structural genes, for example, genes encoding products involved in differentiation of fly cuticular structures. Some fraction of these structural genes will surely have taxon or species-specific functions. Others may be regulatory genes with as yet unknown function, whose roles may be unique to specific lineages or may be shared across taxa. These genes may participate in biological processes we have not yet imagined. Thus, continued work on established model systems, particularly by those brave enough to probe the function of these undescribed genes, may open up whole new avenues of research.

7. LIFE BEYOND THE MODEL SYSTEM

The study of embryonic patterning and gene networks in nonmodel systems is in its infancy. We suspect that there are likely whole sets of genes that function in clade-specific manners. Some of these may encode products involved in highly specialized functions of specific species, while others will be regulatory genes whose function was not revealed in genetic screens in model organisms either because these model organisms (fly, worm, and mouse) do not utilize these genes, because their function in the model systems is redundant, or because mutations in these genes do not result in easily observable phenotypes, namely lethality or major changes in patterning. It is important to remember that model systems do not represent the "wild type" of development; they are unique species too that evolved and adapted to a particular niche. Therefore, as discussed above, their developmental processes will have peculiarities that apply only to them, and there will also be other valid ways of achieving the same developmental outcome not observed in these particular species. A well-rounded understanding of development then necessitates continued research in a variety of species.

A good example of this issue involves the diversity of global mechanisms used to determine sex in different taxa. Sex is genetically determined in mammals, while male and female fates are determined environmentally in other taxa. Sex determination in turtles and crocodilians, for example, is controlled by the temperature at which eggs develop. In some fishes, sequential hermaphroditism leads to change from female to male or vice versa over the course of an individuals' life cycle, with social cues playing a role in these sex changes (reviewed in Kobayashi, Nagahama, & Nakamura, 2013; Lorenzi, Earley, & Grober, 2012; Matsumoto & Crews, 2012; Mork, Czerwinski, & Capel, 2014). Model system studies are not adequate for understanding how and why this diversity arose. It is likely that this type of mechanistic divergence is not limited to sex determination and many similar phenomena will be uncovered as the developmental processes of diverse sets of taxa are explored.

Improved tools for studying nontraditional model organisms, along with the growing interest in understanding metazoan evolution have led to the establishment of many new "emergent" model systems from a wider sampling of taxa. The finding that RNA interference functions in a broad range of species to knockdown gene function is a big step forward as it allows for studies of gene function, and not only gene sequence and expression, in animals lacking sophisticated genetics. Similarly, genome editing can generate null alleles without prior knowledge of gene function. Further, genome sequence comparisons, systems biology, and large-scale protein–protein interaction approaches offer ways to identify interacting gene-regulatory networks without *a priori* knowledge of biochemical function. Using these and other approaches, emergent models are being developed using species representing diverse branches of the tree of life (Fig. 3). These emergent models vary greatly in how well developed they are and the types of experiments that can be performed, but overall the steps taken toward developing these systems will likely yield valuable new information about metazoan evolution and development. Yet, it is important to note that there are many large clades of organisms, each containing enormous numbers of species that are still left unrepresented.

Early results from these emergent model systems demonstrate that classical models are not always ideal for understanding the functions of widely conserved genes, or even all genes important to human health. Original comparison of *D. melanogaster* gene sequences with mammalian ESTs revealed approximately 50% orthology (Rubin et al., 2000). This same study demonstrated that ~60% of genes associated with human disease have a fruit fly ortholog. Sequencing of the sea anemone genome (*N. vectensis*) allowed for the inference of gene families present in the eumetazoan ancestor, based upon conservation of the gene between the sea anemone and any one of the following: fly, nematode, human, frog, or pufferfish (Putnam et al., 2007). 72% of these ancestral families were found to be conserved among all living eumetazoan clades: cnidarians, protostomes, and deuterostomes. This work noted that the lineages that include classic model organisms, such as

D. melanogaster and *C. elegans*, must have sustained a large amount of gene loss because a higher percentage (67%) of human genes have a homolog in the sea anemone than either of the aforementioned models (Putnam et al., 2007). Furthermore, *N. vectensis* was found to have orthologs to a higher percentage of human disease genes (68%) compared to *D. melanogaster* (64%) or *C. elegans* (57%) (Maxwell et al., 2014). Although the degree of conservation between classic model organisms and humans is impressive, and as a result many fruitful lines of biomedical research have been performed in such organisms, these comparisons of conservation also demonstrate the need to expand into additional systems where even greater homology, or even just nonoverlapping homology, is present.

We have learned so much from model systems that it sometimes seems that there is little biology left to discover. It is this view that we consider naïve, particularly in light of new genomic data and comparative developmental studies. It is true that so far, examples of clade-specific gene functions are rare, and overall, studies of the genetic toolkit have revealed that many genes regulating embryonic development are conserved in both sequence and function broadly across taxa. This might be due partially to the fact that these toolkit genes were largely identified in the standard model organisms (fly, worm, and mouse), which has led to a "model-system-centric" view of regulatory genes. Yet, the very same close-minded thinking that considers model systems "enough" today applied to the utility of the now well-accepted model systems, not very long ago. At the time, this was actually a long-shot, even though it appears obvious now. We are of the opinion that expanding the variety and diversity of model systems, and especially supporting newly emerging systems with resources, can yield just as high of a pay-off, for both evo-devo and our general understanding of development, as the discovery of conserved *Hox* genes did decades ago.

8. MORE VARIATION THAN EXPECTED: CONSERVED TRAITS CONTROLLED BY DIFFERENT GENES AND OLD GENES TAKE ON NEW FUNCTIONS

One of the more surprising findings from comparative studies of developmental mechanisms is that the traits that are similar in different species are controlled by different genes. This phenomenon, termed Developmental Systems Drift (DSD), was noted first by True and Haag (2001), who observed that morphologically homologous features are often regulated by different regulatory genes (True & Haag, 2001). When a shared feature between taxa is observed—an example is segmentation, which is a defining feature of arthropods—and members of these different taxa share genes (including toolkit genes), a first assumption is that the same genes work in the same way to control the development of this feature. However, this is not the case. There is more variation "beneath the surface" in regulatory gene networks controlling segmentation than there is in the morphology of the segments themselves. For example, the gene *E75A* functions as a pair-rule segmentation gene in *Oncopeltus* (Erezyilmaz, Kelstrup, & Riddiford, 2009). In *Drosophila*, *E75A* is involved in molting and metamorphosis but plays no role in segmentation (Bialecki, Shilton, Fichtenberg, Segraves, & Thummel, 2002). The *Drosophila* pair-rule segmentation genes *even-skipped* and *fushi tarazu* (*ftz*) play no role in segmentation in grasshoppers (Dawes, Dawson, Falciani, Tear, & Akam, 1994; Patel, Ball, & Goodman, 1992) but they are absolutely required for segments to form in *Drosophila*. Interestingly, even different lineage-specific genes can be used to produce similar developmental outcomes. In one example, *C. elegans* and *Caenorhabditis briggsae* evolved hermaphroditism independently, and did so each using a lineage-specific F-box translational regulator, *fog-2* and *she-1*, respectively (Guo, Lang, & Ellis, 2009; Nayak, Goree, & Schedl, 2005; reviewed in Haag, 2009).

Intertwined with the notion that the same trait is controlled by different genes in different species is the fact that old genes have taken on new roles. In the example of convergently evolved hermaphroditism in nematodes, both species use the translational regulator, *GLD-1*, but the gene has opposite effects on spermatogenesis in these two species (Beadell, Liu, Johnson, & Haag, 2011). *Drosophila ftz* is another example of this, switching from an ancestral homeotic gene in basally branching arthropods (Fig. 1) to a pair-rule segmentation gene in *Drosophila* (reviewed in Heffer, Lohr, & Pick, 2011). Another well-known example of both DSD and a gene taking on a new function is *Drosophila bicoid* (*bcd*), which functions as a classic morphogen to promote head formation. While heads form in all insects, *bcd* plays this role in only a small number of highly derived insects, including *Drosophila*, while other mechanisms regulate head development in more basally branching insects (see Lemke et al., 2008). Comparative studies between related species have demonstrated the involvement of rewiring and co-option in the generation of new cells types. In the sea urchin, *glial cells missing* (*gcm*) is expressed in a portion of the developing endomesoderm and helps to initiate a gene network for making pigment cells (Ransick & Davidson, 2006). However, the sea star embryo instead expresses *gcm* in the ectoderm (Hinman & Davidson, 2007). Although it is still unknown what cell type the *gcm*-expressing cells give rise to in this organism, it does not contribute to pigment cell development as there are none in the sea star embryo or larvae. Another gene, *tbr*, which is a top regulatory gene in the sea urchin skeletogenesis gene network, instead works more broadly in the mesoderm of the sea star and has additional expression in the ectoderm (Hinman & Davidson, 2007; McCauley et al., 2010; Oliveri, Carrick, & Davidson, 2002). Overall, these studies suggest that genetic variation among extant species is more extensive than what would be predicted by comparing gene and genome sequences or even by comparing overt morphologies. While morphologies vary due to changes in regulatory gene networks, gene networks can also vary without causing overt changes in morphology.

9. CONCLUDING REMARKS

Through genetic screens in model organisms and candidate gene cloning over the past 30 years, we have "hit development hard" and uncovered many of the genes that control major steps in embryonic development. Established model systems have not outlived their utility—we need more subtle and nuanced approaches to understand how development works in these models, examining gene function for the numerous (>30% of genes) that have been identified by sequence only in these species. At the same time, a whole world of diversity awaits our discovery in nonmodel species. While not discounting its fundamental contribution to understanding of development, we should not be blindsided by the concepts revealed by studies of the shared animal genetic toolkit. New discoveries, based on curiosity-driven, basic research, using diverse animal species, may reveal mechanisms as surprising to us today as the conserved toolkit was to those studying development and evolution before the discovery of *Hox* genes.

ACKNOWLEDGMENTS

We thank the National Institutes of Health (GM113230) for support and Eric Haag for comments on the manuscript. We thank the growing members of the field of Evo-Devo for their stimulating work and amazing discoveries and apologize to those researchers whose work we were unable to cite in this short essay.

REFERENCES

Albertin, C. B., Simakov, O., Mitros, T., Wang, Z. Y., Pungor, J. R., Edsinger-Gonzales, E., et al. (2015). The octopus genome and the evolution of cephalopod neural and morphological novelties. *Nature*, *524*(7564), 220–224.

Bastian, H., & Gruss, P. (1990). A murine even-skipped homologue, Evx 1, is expressed during early embryogenesis and neurogenesis in a biphasic manner. *The EMBO Journal*, *9*(6), 1839–1852.

Bateson, W. (1894). *Materials for the study of variation treated with especial regards to discontinuity in the origin of species*. London: Macmillan.

Beadell, A. V., Liu, Q., Johnson, D. M., & Haag, E. S. (2011). Independent recruitments of a translational regulator in the evolution of self-fertile nematodes. *Proceedings of the National Academy of Sciences of the United States of America*, *108*(49), 19672–19677.

Bialecki, M., Shilton, A., Fichtenberg, C., Segraves, W. A., & Thummel, C. S. (2002). Loss of the ecdysteroid-inducible E75A orphan nuclear receptor uncouples molting from metamorphosis in Drosophila. *Developmental Cell*, *3*(2), 209–220.

Burri, M., Tromvoukis, Y., Bopp, D., Frigerio, G., & Noll, M. (1989). Conservation of the paired domain in metazoans and its structure in three isolated human genes. *The EMBO Journal*, *8*(4), 1183–1190.

Callaerts, P., Halder, G., & Gehring, W. J. (1997). PAX-6 in development and evolution. *Annual Review of Neuroscience*, *20*, 483–532.

Carroll, S. B. (2000). Endless forms: The evolution of gene regulation and morphological diversity. *Cell*, *101*, 577–580.

Carroll, S. B. (2008). Evo-devo and an expanding evolutionary synthesis: A genetic theory of morphological evolution. *Cell*, *134*(1), 25–36.

Carroll, S. B., Grenier, J. K., & Weatherbee, S. D. (2005). *From DNA to diversity: Molecular genetics and the evolution of animal design* (2nd ed.). Malden, MA: Blackwell Science Ltd.

Carroll, S. B., Prud'homme, B., & Gompel, N. (2008). Regulating evolution. *Scientific American*, *298*(5), 60–67.

Coffman, C., Harris, W., & Kintner, C. (1990). Xotch, the Xenopus homolog of Drosophila notch. *Science*, *249*(4975), 1438–1441.

Dawes, R., Dawson, I., Falciani, F., Tear, G., & Akam, M. (1994). *Dax*, a locust Hox gene related to fushi-tarazu but showing no pair-rule expression. *Development*, *120*, 1561–1572.

den Hoed, M., Eijgelsheim, M., Esko, T., Brundel, B. J., Peal, D. S., Evans, D. M., et al. (2013). Identification of heart rate-associated loci and their effects on cardiac conduction and rhythm disorders. *Nature Genetics*, *45*(6), 621–631.

Deutsch, U., Dressler, G. R., & Gruss, P. (1988). Pax 1, a member of a paired box homologous murine gene family, is expressed in segmented structures during development. *Cell*, *53*(4), 617–625.

Erezyilmaz, D. F., Kelstrup, H. C., & Riddiford, L. M. (2009). The nuclear receptor E75A has a novel pair-rule-like function in patterning the milkweed bug, *Oncopeltus fasciatus*. *Developmental Biology*, *334*(1), 300–310.

Fortunato, S. A., Adamski, M., & Adamska, M. (2015). Comparative analyses of developmental transcription factor repertoires in sponges reveal unexpected complexity of the earliest animals. *Marine Genomics*, *24*(Part 2), 121–129.

Garcia-Fernandez, J., Baguna, J., & Salo, E. (1991). Planarian homeobox genes: Cloning, sequence analysis, and expression. *Proceedings of the National Academy of Sciences of the United States of America*, *88*(16), 7338–7342.

Gibson, A. K., Smith, Z., Fuqua, C., Clay, K., & Colbourne, J. K. (2013). Why so many unknown genes? Partitioning orphans from a representative transcriptome of the lone star tick *Amblyomma americanum*. *BMC Genomics*, *14*, 135.

Gilbert, S. F. (2010). *Developmental biology* (9th ed.). Sunderland, MA: Sinauer Associates.

Guo, Y., Lang, S., & Ellis, R. E. (2009). Independent recruitment of F box genes to regulate hermaphrodite development during nematode evolution. *Current Biology*, *19*(21), 1853–1860.

Haag, E. S. (2009). Convergent evolution: Regulatory lightning strikes twice. *Current Biology*, *19*(21), R977–R979.

Halder, G., Callaerts, P., & Gehring, W. J. (1995). Induction of ectopic eyes by targeted expression of the eyeless gene in Drosophila. *Science*, *267*(5205), 1788–1792.

Hallinan, N. M., & Lindberg, D. R. (2011). Comparative analysis of chromosome counts infers three paleopolyploidies in the mollusca. *Genome Biology and Evolution*, *3*, 1150–1163.

Heffer, A., Lohr, U., & Pick, L. (2011). ftz Evolution: Findings, hypotheses and speculations (response to doi:10.1002/bies.201100019). *BioEssays*, *33*(12), 910–918.

Heffer, A., & Pick, L. (2013). Conservation and variation in Hox genes: How insect models pioneered the evo-devo field. *Annual Review of Entomology*, *58*, 161–179.

Hinman, V. F., & Davidson, E. H. (2007). Evolutionary plasticity of developmental gene regulatory network architecture. *Proceedings of the National Academy of Sciences of the United States of America*, *104*(49), 19404–19409.

Hinman, V. F., Nguyen, A. T., Cameron, R. A., & Davidson, E. H. (2003). Developmental gene regulatory network architecture across 500 million years of echinoderm evolution. *Proceedings of the National Academy of Sciences of the United States of America*, *100*(23), 13356–13361.

Johnson, J. E., Birren, S. J., & Anderson, D. J. (1990). Two rat homologues of Drosophila achaete-scute specifically expressed in neuronal precursors. *Nature*, *346*(6287), 858–861.

Khalturin, K., Hemmrich, G., Fraune, S., Augustin, R., & Bosch, T. C. (2009). More than just orphans: Are taxonomically-restricted genes important in evolution? *Trends in Genetics*, *25*(9), 404–413.

Kobayashi, Y., Nagahama, Y., & Nakamura, M. (2013). Diversity and plasticity of sex determination and differentiation in fishes. *Sexual Development: Genetics, Molecular Biology, Evolution, Endocrinology, Embryology, and Pathology of Sex Determination and Differentiation*, 7(1–3), 115–125.

Ledent, V., & Vervoort, M. (2001). The basic helix-loop-helix protein family: Comparative genomics and phylogenetic analysis. *Genome Research*, *11*(5), 754–770.

Lemke, S., Stauber, M., Shaw, P. J., Rafiqi, A. M., Prell, A., & Schmidt-Ott, U. (2008). Bicoid occurrence and Bicoid-dependent hunchback regulation in lower cyclorrhaphan flies. *Evolution & Development*, *10*(4), 413–420.

Levine, M., & Davidson, E. H. (2005). Gene regulatory networks for development. *Proceedings of the National Academy of Sciences of the United States of America*, *102*(14), 4936–4942.

Lewis, E. B. (1994). Homeosis: The first 100 years. *Trends in Genetics*, *10*, 341–343.

Lewis, E. B. (1997). The bithorax complex: The first fifty years. In N. Ringertz (Ed.), *Nobel lectures, physiology or medicine* (pp. 1991–1995). Singapore: World Scientific Publishing Co.

Lorenzi, V., Earley, R. L., & Grober, M. S. (2012). Differential responses of brain, gonad and muscle steroid levels to changes in social status and sex in a sequential and bidirectional hermaphroditic fish. *PLoS One*, 7(12). e51158.

Marx, J. L. (1986). The continuing saga of "homeo-madness". *Science*, *232*(4747), 158–159.

Matsumoto, Y., & Crews, D. (2012). Molecular mechanisms of temperature-dependent sex determination in the context of ecological developmental biology. *Molecular and Cellular Endocrinology*, *354*(1–2), 103–110.

Maxwell, E. K., Schnitzler, C. E., Havlak, P., Putnam, N. H., Nguyen, A. D., Moreland, R. T., et al. (2014). Evolutionary profiling reveals the heterogeneous origins of classes of human disease genes: Implications for modeling disease genetics in animals. *BMC Evolutionary Biology*, *14*, 212.

McCauley, B. S., Weideman, E. P., & Hinman, V. F. (2010). A conserved gene regulatory network subcircuit drives different developmental fates in the vegetal pole of highly divergent echinoderm embryos. *Developmental Biology*, *340*(2), 200–208.

McGinnis, W., Garber, R. L., Wirz, J., Kuroiwa, A., & Gehring, W. J. (1984). A homologous protein-coding sequence in *Drosophila* homeotic genes and its conservation in other metazoans. *Cell*, *37*, 403–408.

McGinnis, W., Hart, C. P., Gehring, W. J., & Ruddle, F. H. (1984). Molecular cloning and chromosome mapping of a mouse DNA sequence homologous to homeotic genes of Drosophila. *Cell, 38*(3), 675–680.

Mora, C., Tittensor, D. P., Adl, S., Simpson, A. G., & Worm, B. (2011). How many species are there on Earth and in the ocean? *PLoS Biology, 9*(8). e1001127.

Mork, L., Czerwinski, M., & Capel, B. (2014). Predetermination of sexual fate in a turtle with temperature-dependent sex determination. *Developmental Biology, 386*(1), 264–271.

Nayak, S., Goree, J., & Schedl, T. (2005). fog-2 and the evolution of self-fertile hermaphroditism in *Caenorhabditis. PLoS Biology, 3*(1), e6.

Nusslein-Volhard, C., & Wieschaus, E. (1980). Mutations affecting segment number and polarity in *Drosophila. Nature, 287*, 795–801.

Oliveri, P., Carrick, D. M., & Davidson, E. H. (2002). A regulatory gene network that directs micromere specification in the sea urchin embryo. *Developmental Biology, 246*(1), 209–228.

Padgett, R. W., St Johnston, R. D., & Gelbart, W. M. (1987). A transcript from a Drosophila pattern gene predicts a protein homologous to the transforming growth factor-beta family. *Nature, 325*(6099), 81–84.

Patel, N. H., Ball, E. E., & Goodman, C. S. (1992). Changing role of even-skipped during the evolution of insect pattern formation. *Nature, 357*(6376), 339–342.

Pick, L. (2015). Hox *genes, evo-devo and the case of the* ftz *gene*. Chromosoma.

Pick, L., & Heffer, A. (2012). Hox gene evolution: Multiple mechanisms contributing to evolutionary novelties. *Annals of the New York Academy of Sciences, 1256*, 15–32.

Putnam, N. H., Srivastava, M., Hellsten, U., Dirks, B., Chapman, J., Salamov, A., et al. (2007). Sea anemone genome reveals ancestral eumetazoan gene repertoire and genomic organization. *Science, 317*(5834), 86–94.

Quiring, R., Walldorf, U., Kloter, U., & Gehring, W. J. (1994). Homology of the eyeless gene of Drosophila to the Small eye gene in mice and Aniridia in humans. *Science, 265*(5173), 785–789.

Ransick, A., & Davidson, E. H. (2006). Cis-regulatory processing of Notch signaling input to the sea urchin glial cells missing gene during mesoderm specification. *Developmental Biology, 297*(2), 587–602.

Refki, P. N., Armisen, D., Crumiere, A. J., Viala, S., & Khila, A. (2014). Emergence of tissue sensitivity to Hox protein levels underlies the evolution of an adaptive morphological trait. *Developmental Biology, 392*(2), 441–453.

Rijsewijk, F., Schuermann, M., Wagenaar, E., Parren, P., Weigel, D., & Nusse, R. (1987). The Drosophila homolog of the mouse mammary oncogene int-1 is identical to the segment polarity gene wingless. *Cell, 50*(4), 649–657.

Roberts, R. B., Hu, Y., Albertson, R. C., & Kocher, T. D. (2011). Craniofacial divergence and ongoing adaptation via the hedgehog pathway. *Proceedings of the National Academy of Sciences of the United States of America, 108*(32), 13194–13199.

Rubin, G. M., Yandell, M. D., Wortman, J. R., Miklos, G. L. G., Nelson, C. R., et al. (2000). Comparative genomics of the eukaryotes. *Science, 287*, 2204–2215.

Ryan, J. F., Pang, K., Schnitzler, C. E., Nguyen, A. D., Moreland, R. T., Simmons, D. K., et al. (2013). The genome of the ctenophore *Mnemiopsis leidyi* and its implications for cell type evolution. *Science, 342*(6164), 1242592.

Schierwater, B., Murtha, M., Dick, M., Ruddle, F. H., & Buss, L. W. (1991). Homeoboxes in cnidarians. *The Journal of Experimental Zoology, 260*(3), 413–416.

Srivastava, M., Begovic, E., Chapman, J., Putnam, N. H., Hellsten, U., Kawashima, T., et al. (2008). The Trichoplax genome and the nature of placozoans. *Nature, 454*(7207), 955–960.

Srivastava, M., Simakov, O., Chapman, J., Fahey, B., Gauthier, M. E., Mitros, T., et al. (2010). The *Amphimedon queenslandica* genome and the evolution of animal complexity. *Nature, 466*(7307), 720–726.

Stern, D. L., & Orgogozo, V. (2009). Is genetic evolution predictable? *Science*, *323*(5915), 746–751.

True, J. R., & Haag, E. S. (2001). Developmental system drift and flexibility in evolutionary trajectories. *Evolution & Development*, *3*, 109–119.

Wangler, M. F., Yamamoto, S., & Bellen, H. J. (2015). Fruit flies in biomedical research. *Genetics*, *199*(3), 639–653.

Wasik, B. R., Rose, D. J., & Moczek, A. P. (2010). Beetle horns are regulated by the Hox gene, Sex combs reduced, in a species- and sex-specific manner. *Evolution & Development*, *12*(4), 353–362.

Wissler, L., Gadau, J., Simola, D. F., Helmkampf, M., & Bornberg-Bauer, E. (2013). Mechanisms and dynamics of orphan gene emergence in insect genomes. *Genome Biology and Evolution*, *5*(2), 439–455.

Yamamoto, Y., Stock, D. W., & Jeffery, W. R. (2004). Hedgehog signalling controls eye degeneration in blind cavefish. *Nature*, *431*(7010), 844–847.

Yoshida, M. A., Ishikura, Y., Moritaki, T., Shoguchi, E., Shimizu, K. K., Sese, J., et al. (2011). Genome structure analysis of molluscs revealed whole genome duplication and lineage specific repeat variation. *Gene*, *483*(1–2), 63–71.

Zdobnov, E. M., von Mering, C., Letunic, I., Torrents, D., Suyama, M., Copley, R. R., et al. (2002). Comparative genome and proteome analysis of *Anopheles gambiae* and *Drosophila melanogaster*. *Science*, *298*(5591), 149–159.

Zhou, K., Huang, B., Zou, M., Lu, D., He, S., & Wang, G. (2015). Genome-wide identification of lineage-specific genes within *Caenorhabditis elegans*. *Genomics*, *106*(4), 242–248.

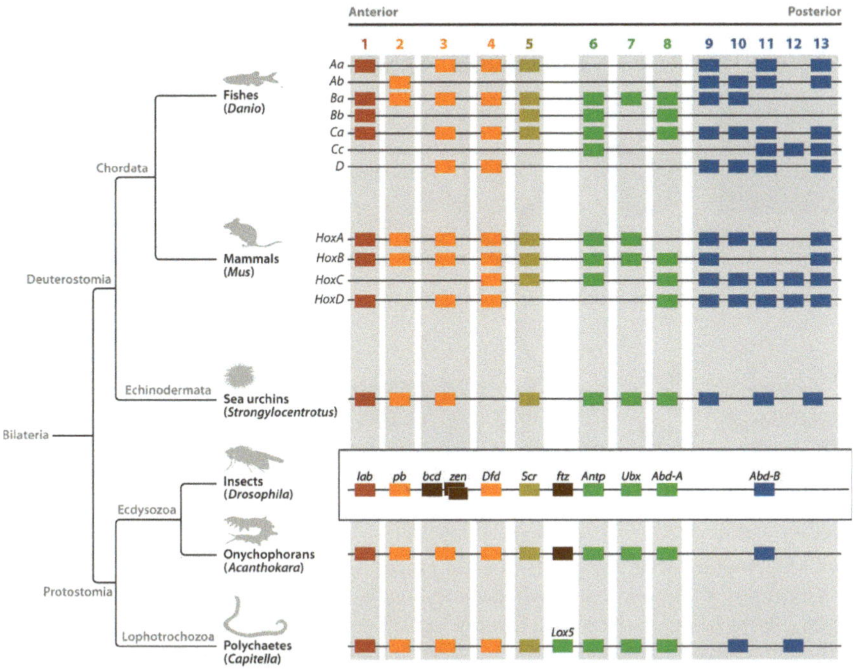

Alys M. Cheatle Jarvela and Leslie Pick, Figure 1 *Hox* gene expression and function have been conserved during animal evolution. *Left*: A cladogram indicating major bilaterian taxa with diagrams showing the basic body plans of selected species. *Right*: Complexes of *Hox* genes from the indicated taxa. The *Drosophila Hox* complex (HOM-C) boxed includes eight homeotic *Hox* genes and three that have taken on novel functions (*bcd*, *zen*, and *ftz*). Orthology to the *Drosophila Hox* genes is indicated by the color code. The homeotic *Hox* genes are expressed colinearly with their positions in the complex: *lab* is expressed more anteriorly than *pb*, which is expressed more anteriorly than *Dfd*, and so on with *Abd-B* expressed most posteriorly. This colinearity, as well as position of individual *Hox* orthologs within *Hox* complexes, has been conserved during evolution. In Chordata, genome duplication events have generated multiple *Hox* complexes on different chromosomes. Individual paralogs have been lost from some complexes, presumably enabled by redundancy. *Figure from Heffer and Pick (2013).*

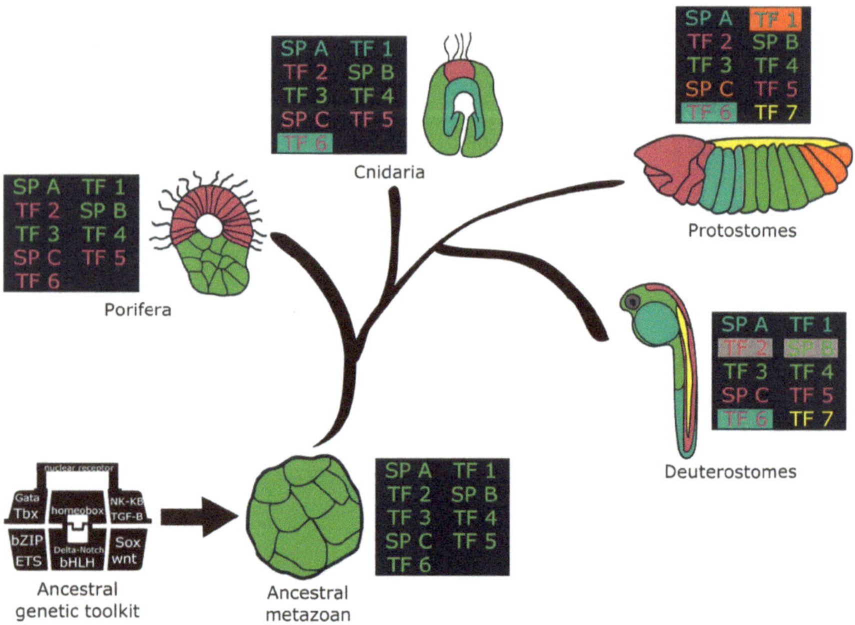

Alys M. Cheatle Jarvela and Leslie Pick, Figure 2 The metazoan genetic toolkit for development. A large body of work demonstrates that the genetic toolkit governing development is highly conserved throughout the animal kingdom (Carroll, 2000). Many of the constituent signaling molecules and transcription factor families are pan-metazoan. Examples of these ancient gene families are pictured in a toolkit (left). TF denotes "transcription factor" and SP denotes "signaling pathway." As animal life became more complex, these toolkit genes were repurposed to allow for the development of new cell types, tissues, and structures, indicated by color changes shared between groups of toolkit genes and a corresponding embryo region (e.g., TF 2, TF 5, TF 6, and SP C families have been changed from ancestral green to derived magenta in Porifera lineage). Importantly, these toolkit genes are often used in more than one time or place in the embryo to help build unique gene networks and allow for greater cell diversity, which is indicated by genes boxed by an additional color representing a second embryo region (for example, TF 1 has cyan font boxed by red in the protostome lineage to indicate use in both the cyan- and red-colored embryo regions). In spite of the impressive reuse and repurposing of these toolkit components, there are also many examples of toolkit genes with deeply conserved developmental functions, indicated by a lack of color change (as examples, TF 3, TF 4, and SP B families are shown in green font in every depicted toolkit). Addition of completely new toolkit factors occurs, but is rare (for example, TF 7 in the bilaterian lineage).

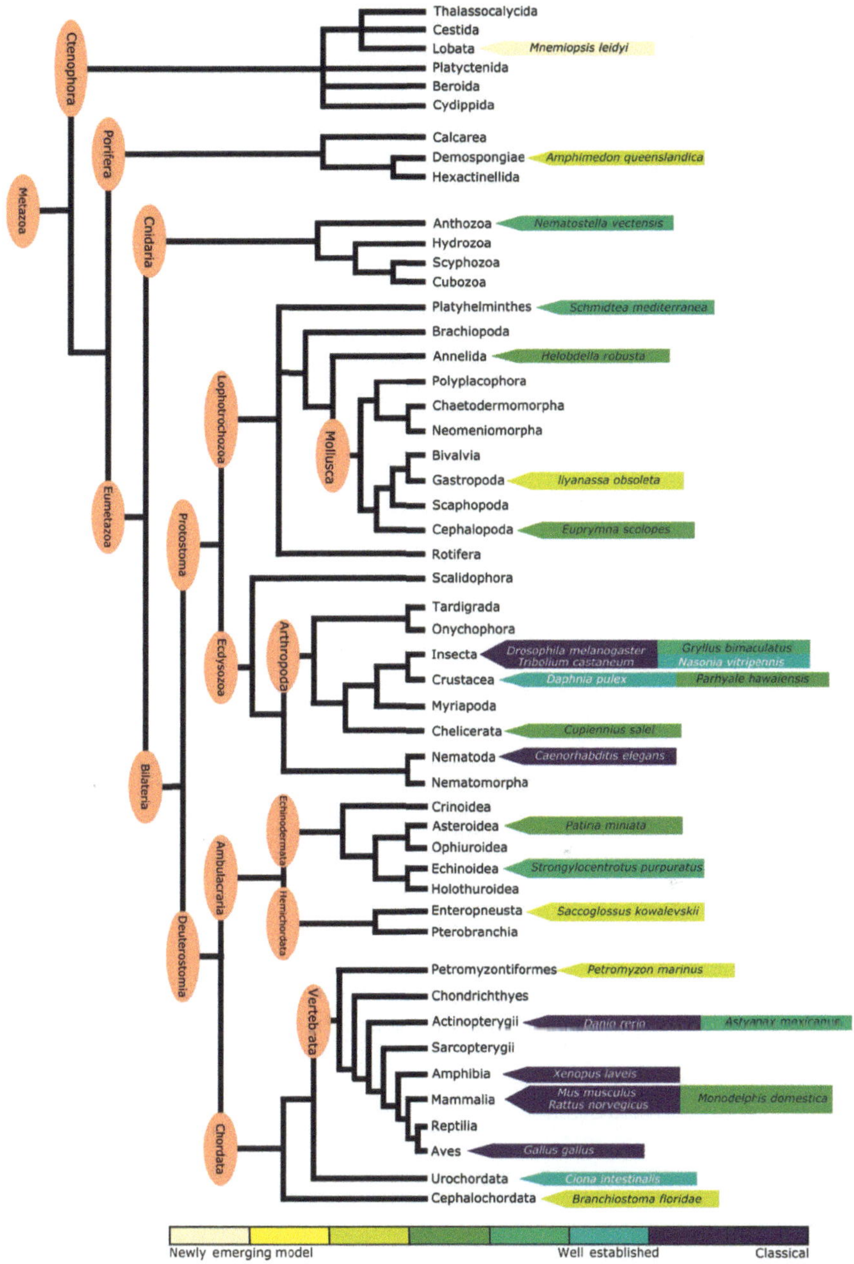

Alys M. Cheatle Jarvela and Leslie Pick, Figure 3 Model systems represent only a tiny subset of metazoan diversity. A cladogram depicting extant metazoan life is on the left. Pink ovals indicate important or relatively better studied clades. Colored flags on the

(Continued)

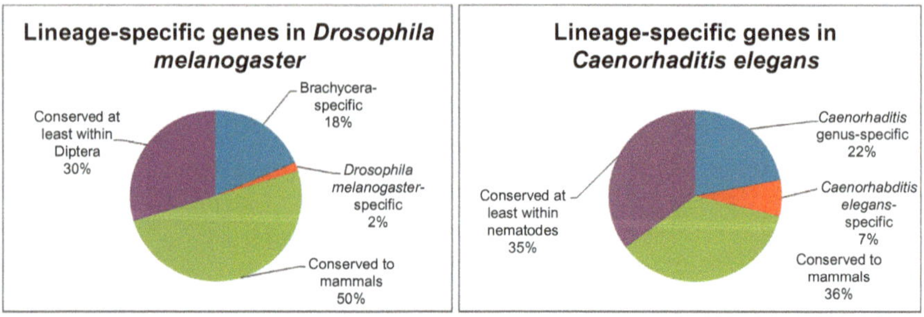

Alys M. Cheatle Jarvela and Leslie Pick, Figure 4 Percentages of lineage-specific genes in model organisms. While a large proportion of genes in both model systems are pan-bilaterian (Rubin et al., 2000) as they are also found in mammals, both also contain a high proportion of genes that are highly taxonomically restricted. 18% of genes in *D. melanogaster* are unique to the Brachycera, a suborder of Diptera. 22% of genes in *C. elegans* are specific to the genus *Caenorhabditis* compared to other nematode species. Both model's gene sets are 1–10% species specific (Wissler, Gadau, Simola, Helmkampf, & Bornberg-Bauer, 2013; Zdobnov et al., 2002; Zhou et al., 2015).

Alys M. Cheatle Jarvela and Leslie Pick, Figure 3—Cont'd right indicate species within the indicated clade that are being used for developmental or evo-devo work. Flag color indicates how well established that species is according to the key at the bottom of the image, ranging from light yellow, minimal resources or lab cultures available, to dark blue, for the best-studied genetic systems—flies, worms, and mice. Model systems were scored based on criteria that represent the richness of available resources (i.e., publically available genome, large research community, and ability to perform sophisticated genetics) and the general amenability of the system for experimentation (i.e., Can the organism be maintained in the lab? Can gametes be obtained year-round? Can embryos be molecularly or genetically perturbed?). Some clades are far better represented by classical and well-established model systems than others, particularly Vertebrata and Ecdysozoa. Lophotrochozoa and Ambulacraria are currently gaining interest and resources, while taxa such as Porifera, Cnidaria, and Ctenophores require much better representation. Essentially, all major clades shown include taxa that are not represented by any experimental system.

CHAPTER FIFTEEN

Making the Mouse Blastocyst: Past, Present, and Future

Janet Rossant[1]
Program in Developmental and Stem Cell Biology, Hospital for Sick Children, University of Toronto, Toronto, Ontario, Canada
[1]Corresponding author: e-mail address: janet.rossant@sickkids.ca

Contents

Abstract

The study of the preimplantation mouse embryo has progressed over the past 50 years from descriptive biology through experimental embryology to molecular biology and genetics. Along the way, the molecular pathways that lead to the establishment of the three cell lineages of the blastocyst have become more clearly understood but the fundamental questions of lineage commitment remain the same as those laid out in early studies. With new tools of genome manipulation, *in vivo* imaging and single-cell analysis, the mouse blastocyst is an excellent model system to understand how organized cell fate decisions are made in a self-organizing developmental context.

1. INTRODUCTION

The last 50 years have encompassed enormous progress in our understanding of the processes of embryonic development in many different model organisms. The genetic programs underlying the initial establishment of different cell lineages have become clearly defined and the complexities of

Current Topics in Developmental Biology, Volume 117
ISSN 0070-2153
http://dx.doi.org/10.1016/bs.ctdb.2015.11.015

morphogenesis and organogenesis are being explored with ever more refined technological tools. And yet, the fundamental questions of development remain the same and still drive our studies today.

Experimental studies of the preimplantation mouse embryo span this 50-year period and have advanced from cellular description to refined lineage modeling and molecular analysis. They were made possible initially by the development of defined culture systems to grow mouse eggs to the blastocyst stage (Brinster, 1963; Whitten & Biggers, 1968), embryo transfer techniques (Mc & Biggers, 1958), and chimera approaches (Mintz, 1965; Tarkowski, 1961). Several questions were posed early on in the field and still are being explored today.

- What is the later fate of the lineages of the blastocyst: trophectoderm (TE) and inner cell mass (ICM), and epiblast (EPI) and primitive endoderm (PE)?
- When are cells restricted to their fate?
- What are the cellular and molecular mechanisms that establish cell fate?

In this essay, rather than provide a broad-based review of the field, I have taken five early papers that addressed some of these issues and asked whether the conclusions stand the test of time, or whether newer technologies have enhanced, modified, or nullified the conclusions. In general, we can still find lessons to learn from the early papers, although maybe not from the rather wordy, discursive style in which they were written!

2. WHAT IS THE LATER FATE OF THE INNER CELL MASS AND TROPHECTODERM?

Gardner, R. L., Papaioannou, V. E., & Barton, S. C. (1973). Origin of the ectoplacental cone and secondary giant cells in mouse blastocysts reconstituted from isolated trophectoderm and inner cell mass. *Journal of Embryology and Experimental Morphology 30*, 561–572. (Gardner, Papaioannou, & Barton, 1973)

In 1968, Richard Gardner published the production of mouse chimeras by injection of inner cell mass (ICM) cells into the blastocyst (Gardner, 1968). Here he took this approach one step further and, by a heroic feat of micromanipulation, generated reconstituted blastocysts, in which the entire ICM was of one genotype and the trophectoderm of another. Using isozymal variants of glucose phosphate isomerase, the most sensitive genetic marker available at the time (Chapman, Ansell, & McLaren, 1972), to distinguish the two genotypes in later tissues, they attempted to define the boundaries of ICM and trophectoderm contributions to the fetus and its extraembryonic membranes. The results were indicative but not clear-cut:

> *"The trophoblast of the implanted embryos was almost exclusively of the trophoblast donor GPI type. The embryos plus membranes were mainly of the ICM-donor type but most also showed a substantial proportion of trophoblast-donor type."*

Papaioannou followed up in 1982 (Papaioannou, 1982) with a more sensitive GPI assay and obtained more convincing evidence of separation of cell fates. However, definitive results required the development of *in situ* genetic markers so that the origin of specific cell types within a complex structure like the placenta or the yolk sac could be identified. Gardner and Johnson generated interspecific chimeras with rat ICMs in mouse blastocysts and used species-specific antibodies to distinguish the resulting chimeras (Gardner & Johnson, 1973). Rossant et al. used *in situ* hybridization against repetitive DNA differences between two species of mice (Rossant, Vijh, Siracusa, & Chapman, 1983) and also used DNA *in situ* hybridization against a large repeat transgene (Rossant & Croy, 1985), thus avoiding issues of potential rat-mouse incompatibilities. All these experiments gave the same clear results; the ICM gave rise to the entire fetus, the extraembryonic yolk sacs, and the allantois, while the trophectoderm produced the trophoblast giant cells, the extraembryonic ectoderm, and the ectoplacental cone, which make up the majority of the placenta (Fig. 1).

The only real caveat to these experiments was that the reconstituted blastocyst might differ in behavior from the undisturbed embryo (Georgiades, Cox, Gertsenstein, Chawengsaksophak, & Rossant, 2007). In particular, the polar trophectoderm overlying the ICM is excluded from a reconstituted blastocyst. With today's refined lineage tracing technologies, we can easily reconfirm the chimera results by direct labeling of the intact blastocyst. For example, infection of the blastocyst with a GFP-expressing lentivirus labels all TE cells and only TE cells and later concept uses show GFP expression restricted to the extraembryonic ectoderm and ectoplacental cone (Georgiades et al., 2007).

3. WHAT IS THE LATER FATE OF THE EPIBLAST AND PRIMITIVE ENDODERM?

Gardner, R. L. & Rossant, J. (1979). Investigation of the fate of 4.5 day post-coitum mouse inner cell mass cells by blastocyst injection. *Journal of Embryology and Experimental Morphology 52*, 141–152. (Gardner & Rossant, 1979)

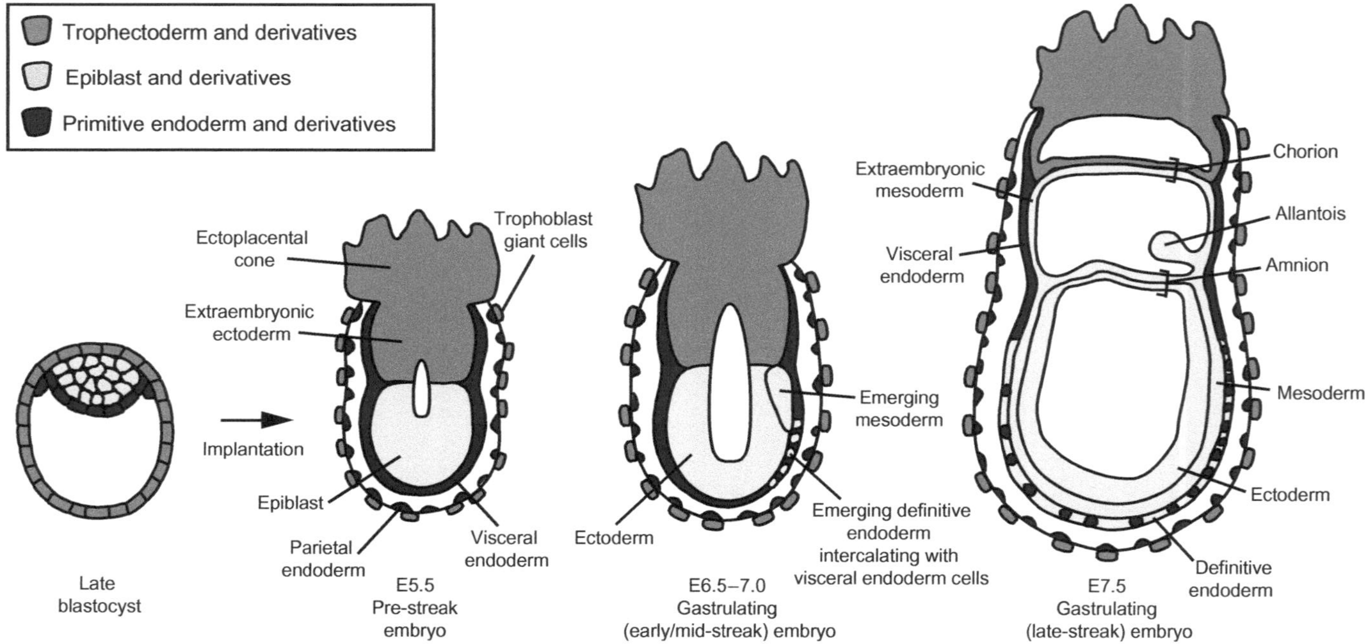

Figure 1 Postimplantation lineage derivatives of the epiblast, primitive endoderm, and trophectoderm. Trophectoderm (red) gives rise to all the later trophoblast lineages that contribute to the developing placenta, as well as to the trophoblast giant cells. Primitive endoderm (green) gives rise to the extraembryonic visceral and parietal endoderm, as well as providing a minor contribution to the developing gut along with the definitive endoderm. The epiblast (blue) gives rise the post-implantation primitive ectoderm, and to the mesoderm and definitive endoderm that arise at gastrulation. Extraembryonic mesoderm arises from the posterior of the primitive streak and forms the allantois and the mesoderm layers of the developing yolk sac, where primitive hematopoiesis occurs. (A color version of this figure is located at the end of this chapter.)

By E4.5, the inner cell mass has clearly segregated into two cell types; a monolayer of cells on the blastocoelic surface called primitive endoderm or hypoblast and an enclosed group of cells called the primitive ectoderm or epiblast (Fig. 1). In this chapter, Gardner and Rossant used cell morphology, without the aid of any molecular markers, to show that rough and smooth cells dissociated from the E4.5 ICM correspond to primitive endoderm and epiblast, respectively. They then injected isolated rough and smooth cells into host blastocysts and showed that the primitive endoderm gave rise to the endoderm but not the mesoderm of the visceral yolk sac, but that the definitive endoderm arose from the epiblast lineage.

They concluded:

> *"The present results demonstrate that morphological differentiation of the ICM in the late blastocyst into two populations of cells is accompanied by the acquisition of distinct cell states, as judged by the criterion of behaviour of the cells following transplantation. We believe* (sic), *furthermore, that they provide an accurate description of the normal lineage of primitive endoderm and ectoderm cells. If this is the case, the former cells give rise to the extra-embryonic endoderm of the conceptus, and the latter to the entire foetus plus extra-embryonic mesoderm."*

While apparently based on belief rather than completely definitive experimental results, these conclusions stood the test of time for many years and were confirmed and extended in chimeras using more sensitive genetic markers.

However, with the advent of GFP-expressing transgenic lines and live-cell imaging, a more nuanced view of the fate of the primitive endoderm has come into play. Incidental observations from tetraploid complementation experiments with embryonic stem cells (Dufort, Schwartz, Harpal, & Rossant, 1998) had shown the persistence of some tetraploid primitive endoderm-derived cells in the definitive endoderm as it forms during gastrulation, but it was not clear whether this occurred in the intact embryo. The Hadjantonakis lab used GFP-lineage specific transgenic lines to show that not all primitive endoderm cells were displaced from the developing definitive endoderm (Kwon, Viotti, & Hadjantonakis, 2008). PE-derived progeny could persist and be observed in the later gut tube. Whether these cells take on definitive endoderm properties or whether they have distinct functions in the gut is still unclear, but the earlier proposed clear distinction between the cell fates of primitive and definitive endoderm is no longer tenable (Fig. 1). Given the similarities in molecular properties of the two lineages, including the dual roles of transcription factors like Sox17 (Artus, Piliszek, & Hadjantonakis, 2011; Kanai-Azuma et al., 2002) and

Foxa2 (Dufort et al., 1998), the boundaries are definitely getting blurred. This has relevance to the early differentiation of endoderm during embryoid body formation from ES cells. Since ES cells rarely contribute to visceral endoderm in chimeras, the identity of this early endoderm layer *in vitro* is not altogether clear. Is it definitive or primitive or maybe a mixed population similar to the early streak stage embryo?

Once the blastocyst is fully expanded and the epiblast and primitive endoderm layers separated, the three blastocyst cell lineages are clearly distinct. Consistent with this stabilization of cell fate, it is possible to isolate and maintain three different lineage-restricted self-renewing progenitor cell lines from the blastocyst: ES cells are epiblast-derived (Boroviak, Loos, Bertone, Smith, & Nichols, 2014), while trophoblast stem cells derive from the TE (Tanaka, Kunath, Hadjantonakis, Nagy, & Rossant, 1998) and XEN cells from the PE (Kunath et al., 2005).

4. WHAT IS THE MECHANISM THAT DRIVES THE FORMATION OF THE ICM AND THE TE?

4.1 The Inside–Outside Hypothesis

Tarkowski, A. K. & Wroblewska, J. (1967). Development of blastomeres of mouse eggs isolated at the 4- and 8-cell stage. *Journal of Embryology and Experimental Morphology 18*, 155–180. (Tarkowski & Wroblewska, 1967)

Early morphological and enzymatic observations had suggested that the mouse egg might contain intrinsic determinants that were segregated during cleavage into specific precursors of later lineages (Dalcq, 1957; Mulnard, 1961). This concept was attractive since it was in accord with early events in many other vertebrate and invertebrate embryos. Tarkowski himself had proposed such a mechanism based on properties of isolated 2-cell blastomeres (Tarkowski, 1961). In this chapter, Tarkowski and Wroblewska studied in detail the *in vitro* development of blastomeres isolated at the 4 and 8-cell stage. They showed that the majority of isolated 8-cell blastomeres formed trophectoderm vesicles, a result not predicted if there were early segregation of ICM and TE lineages based on cytoplasmic determinants. They realized that the predominance of TE structures from isolated blastomeres was predicted from topological considerations if blastocyst formation always occurs after a set number of cell divisions. Not enough cells would be present at the time of cavitation from an isolated 8-cell blastomere to provide any enclosed cells, thus promoting TE vesicle formation. From these results and other observations they then proposed what has come to be known as the "inside–outside hypothesis":

> *"It seems reasonable to assume that the conditions in which external and internal cells find themselves are diametrically different. The internal cells, being completely cut off from the exterior, develop in a micro-environment created by the external cells. In our opinion, the position of a cell in the morula and, a consequence of the position, the different environmental conditions play a decisive role in the differentiation of cells in one of the two directions (trophoblast versus inner mass). For the formation of the inner mass it is necessary that certain blastomeres should become isolated from the exterior before the moment when blastocoelic fluid starts to accumulate between the cells."*

This hypothesis, dating back to 1967, has been extremely influential in all the intervening years and underlies much current thinking on the molecular mechanisms promoting ICM versus TE fate. Although there may be some early predisposition of certain blastomeres to give rise to ICM or TE (Piotrowska-Nitsche, Perea-Gomez, Haraguchi, & Zernicka-Goetz, 2005), the overwhelming evidence suggests that formation of the blastocyst can be achieved as a fully self-organizing system, based on the gradual development of differences between inside and outside cells. In the past few years, focus has been placed on components of the Hippo signaling pathway as candidates driving this segregation (Fig. 2). Sasaki and colleagues first showed that a key coactivator of TEAD transcription factors, YAP, is phosphorylated by the Hippo complex serine–threonine kinase, Lats2, in inside cells of the morula (Nishioka et al., 2009). This leads to Yap's exclusion from the nucleus and prevents TEAD4 activity and expression of the TE-specific transcription factor, Cdx2, in the inside cells. In outside cells, the Hippo pathway is inactive, Yap enters the nucleus and Cdx2 expression is upregulated. Other studies have shown that Cdx2, in turn, represses the expression of the ICM-specific pluripotency genes, *Oct4* and *Nanog* (Strumpf et al., 2005), thus leading to restricted expression of key lineage regulators in the appropriate cell layers by the blastocyst stage. Other players in the Hippo pathway, have also been shown to play necessary roles in ICM/TE segregation, including Neurofibromatosis2 (Nf2) (Cockburn, Biechele, Garner, & Rossant, 2013) and Angiomotin (Hirate et al., 2013). Interestingly, Nf2 protein is inherited from the oocyte: maternal/zygotic loss of function leads to morulae/early blastocysts in which Yap is nuclear in all cells, whether inside or outside (Cockburn et al., 2013). Cdx2 is then expressed in the ICM and transforms those cells into a TE fate. Thus, position-dependent activation of Hippo signaling in inside cells is a prerequisite for formation of a functional ICM.

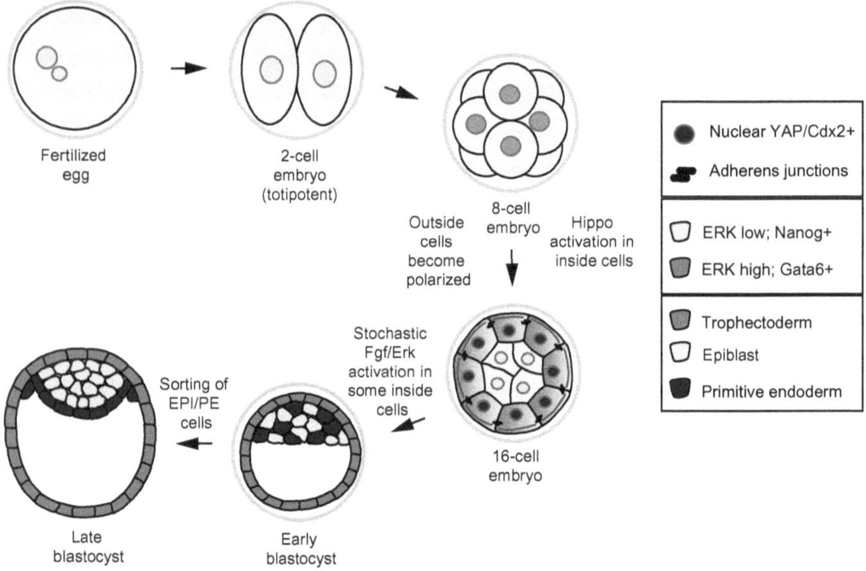

Figure 2 Mechanisms of cell fate specification during preimplantation development. Between the 8- and 16-cell stage, outer blastomeres become polarized and enclose a group of apolar cells. Cell contact and cell polarity lead to differential activation of the Hippo-related Lats kinase in the inside versus the outside cells. Lats phosphorylates Yap in inside cells which causes its exclusion from the nucleus. Hippo signaling is inactive in the outside, polarized cells, Yap is nuclear and activates expression of trophectoderm-specific genes such as *Cdx2*. Absence of *Cdx2* in inside cells allows expression of pluripotency genes and establishment of the inner cell mass. Once the inner cell mass is established at the early blastocyst stage, stochastic differences in the activation of FGF/ERK signaling occur in a mosaic manner within it. Active ERK signaling promotes primitive endoderm formation while absence of ERK signaling promotes epiblast. As the two lineages gradually become specified, they sort out within the ICM so that the primitive endoderm ends up on the blastocoelic surface. (A color version of this figure is located at the end of this chapter.)

4.2 The Polarization Hypothesis

Johnson, M. H. & Ziomek, C. A. (1981). The foundation of two distinct cell lineages within the mouse morula. *Cell 24*, 71–80. (Johnson & Ziomek, 1981)

While activation of Hippo signaling in inside cells clearly drives ICM/TE segregation, it begs the question of how Hippo signaling is constrained to inside cells. A major distinction between inside and outside cells of the morula is that the outside cells show apico-basolateral polarity and eventually form a tight junction coupled epithelium, while enclosed cells lack apical polarity (Fig. 2). Martin Johnson's group in the late 1970s and 1980s carried out extensive cellular studies of the process of polarization, which begins at the 8-cell stage when all cells develop an apical pole. In this chapter and several others (reviewed Johnson & McConnell, 2004), they studied the formation and inheritance of the apical pole during cell divisions from the 8–16 cell stage, mostly using isolated blastomeres and morphology or lectin staining to distinguish the polar region. From these studies, they proposed the "polarization hypothesis":

> *"Two propositions of the polarization hypothesis are ... First, each constituent blastomere of the 8-cell mouse embryo ...becomes structurally and functionally polarized along a radial axis through the embryo.*
> *Second, elements of the cellular polarity established within each blastomere may be conserved during its division to 16 cells. Such a conservation could, with appropriate cleavage planes, generate two cell populations distinct from the moment of their formation as a result of the inheritance of distinctive regions of cytoplasm and membrane."*

This hypothesis has often been stated as an alternative to the inside–outside hypothesis. However, again more recent studies would suggest a more nuanced interpretation. Johnson and colleagues observed that isolated polarized 8-cell blastomeres often divided such that one cell would inherit most of the apical pole and that these kinds of polarized cell divisions were more common than generation of two polarized progeny. The concept that polarized cell divisions generated inside cells in the intact embryo made sense. However, a set of recent papers using detailed live imaging and 3D reconstruction analysis has suggested that some inner cells in the intact embryo at the 16-cell stage arise not by polarized cell division, but by a process of internalization, involving apical constriction, from an original outside position (Anani, Bhat, Honma-Yamanaka, Krawchuk, & Yamanaka, 2014; Samarage et al., 2015; Watanabe, Biggins, Tannan, & Srinivas, 2014). There is no absolute relationship between the angle of cell division and the later internal or external position of the progeny.

However, the idea that the formation of polarized outside and apolar inside cells is key to the segregation of the ICM and TE still holds. Once an outer polarized epithelium is formed it is apparently rare for progeny of inside cells to emerge again into the outer position, although outside cells can still generate more inside progeny at the 16–32 cell stage. Thus, the physical formation of the outer epithelium will provide a mechanism to allow stable generation of an enclosed environment in which Hippo signaling can lead to long-term transcriptional changes in inside cells. Blocking cell adhesion prevents activation of Hippo signaling (Nishioka et al.,

2009; Stephenson, Yamanaka, & Rossant, 2010), while conversely, blocking cell polarization while leaving cells in close contact (Laeno, Tamashiro, & Alarcon, 2013), promotes active Hippo signaling in both inside and outside cells. While the full mechanisms linking cell position, cell–cell contact and cell polarity with ICM/TE fate remain to be determined, elements of both the inside–outside and the polarization hypotheses are still relevant.

5. WHAT ARE THE MECHANISMS DRIVING SEGREGATION OF EPI AND PE WITHIN THE ICM?

Rossant, J. (1975). Investigation of the determinative state of the mouse inner cell mass. II. The fate of isolated inner cell masses transferred to the oviduct. *Journal of Embryology and Experimental Morphology 33*, 991–1001. (Rossant, 1975)

In the absence of lineage-specific markers, the mechanism by which the apparently homogenous ICM of the E3.5 blastocyst is transformed within 24 h into two cell types, distinguished by morphology, position, and fate, was somewhat mysterious. Informed by the fact that the PE formed on the surface of the ICM and by comparison with the inside–outside hypothesis of ICM/TE formation, it was usually assumed that cell position somehow drove this process also. With the primitive tools of direct microdissection of ICMs from blastocysts (prior to the development of immunosurgery (Solter & Knowles, 1975)) and promoting development by transferring them back to the oviduct inside empty zonae pellucidae, Rossant reported on the development of isolated ICMs and concluded:

> *"Although no trophoblast was formed, $3^{1/2}$-day ICMs formed an outer endoderm layer after 1 day in the oviduct, as judged by light and electron microscopical evidence. It is suggested that cell position may be important in endoderm differentiation."*

The conclusion about lack of TE formation is consistent with our knowledge of the restricted potential of the ICM. However, the positional interpretation of the fact that all ICMs produced a full layer of PE on the outside has turned out to be not correct. Thirty years later, the Rossant lab produced evidence that the formation of EPI/PE is a much more interesting and challenging process. With the development of lineage-specific gene markers and technologies to observe expression *in situ* in the early embryo, they were able to show that genes such as *Nanog* and *Gata6*, which were thought to be specific to the EPI and PE, respectively, could already be seen to show variable and complementary "salt and pepper" expression at the E3.5 ICM (Chazaud, Yamanaka, Pawson, & Rossant, 2006). Expression of the markers gradually segregated into the two defined PE and EPI cell layers by E4.5. This suggested that initial activation of lineage specification was position-independent and that cell sorting then generated the two lineages. More detailed and nuanced interpretation of these events has been provided by live-cell imaging of a PE-specific GFP reporter line, showing both cell sorting and death of cells in the wrong position, leading to the final segregation of EPI and PE (Plusa, Piliszek, Frankenberg, Artus, & Hadjantonakis, 2008).

In the 2006 paper from the Rossant lab, they also provided the first clues as the mechanism of the initiation of lineage specification of PE and EPI. In the ICM of embryos mutant for *Grb2*, a key player in signal transduction linking receptor tyrosine kinase signaling to the Ras–MAP kinase pathway, cell numbers were normal but all cells expressed EPI markers and none expressed PE markers (Chazaud et al., 2006). It is now clear from many different studies that FGF4 produced autonomously by cells of the ICM is the driver of the differentiation of the PE (Fig. 2). Blocking FGF/ERK signaling at the blastocyst stage promotes EPI versus PE fate (Nichols, Silva, Roode, & Smith, 2009; Yamanaka, Lanner, & Rossant, 2010) and activating FGF signaling promotes PE versus EPI fate (Yamanaka et al., 2010). Cells within the ICM somehow read out local variation in FGF levels and respond accordingly to generate approximately the right number of EPI versus PE cells. Changes in cellular properties induced by expression of EPI versus PE transcription factors then lead to physical segregation and stabilization of the EPI and PE cell layers. Whether this is entirely a stochastic process or whether the origin of ICM cells from different rounds of cell divisions during cleavage predisposes to EPI or PE fate is somewhat unclear (Morris et al., 2010; Yamanaka, 2011) and the downstream events leading to lineage segregation are not well understood at all.

6. CONCLUSIONS

Building on 50 years of research, the preimplantation mouse embryo is now poised to enter into a new phase as the most powerful system to explore the fundamental processes of lineage development at a single-cell systems level. The entire process can be observed live and imaged *in vitro* in a fully defined culture medium with all the tools of modern imaging and fluorescent tracers. The process of going from one cell to the 100-cell blastocyst with three distinct lineages takes place slowly over 4 days, so that events of lineage specification can be observed evolving in real time. The advent of single-cell transcriptomics allows detailed correlation of gene expression changes with cell behavior. All the tools of mouse genetic manipulation, including newly developed genome editing technologies, are available to explore gene function. New tools of optogenetics promise to allow manipulation of cell signals and responses at the single-cell level in the living embryo. The future is bright!

ACKNOWLEDGMENTS

I thank Eszter Posfai for helpful comments and assistance with the figures. I apologize to the many contributors to our understanding of the formation of the mouse blastocyst whose work was not mentioned in this short personal essay. Much of the author's own work has been supported over the years by the Canadian Institutes of Health Research.

REFERENCES

Anani, S., Bhat, S., Honma-Yamanaka, N., Krawchuk, D., & Yamanaka, Y. (2014). Initiation of Hippo signaling is linked to polarity rather than to cell position in the pre-implantation mouse embryo. *Development*, *141*, 2813–2824.

Artus, J., Piliszek, A., & Hadjantonakis, A. K. (2011). The primitive endoderm lineage of the mouse blastocyst: Sequential transcription factor activation and regulation of differentiation by Sox17. *Developmental Biology*, *350*, 393–404.

Boroviak, T., Loos, R., Bertone, P., Smith, A., & Nichols, J. (2014). The ability of inner-cell-mass cells to self-renew as embryonic stem cells is acquired following epiblast specification. *Nature Cell Biology*, *16*, 516–528.

Brinster, R. L. (1963). A method for in vitro cultivation of mouse ova from TWO-cell to blastocyst. *Experimental Cell Research*, *32*, 205–208.

Chapman, V. M., Ansell, J. D., & McLaren, A. (1972). Trophoblast giant cell differentiation in the mouse: Expression of glucose phosphate isomerase (GPI-1) electrophoretic variants in transferred and chimeric embryos. *Developmental Biology*, *29*, 48–54.

Chazaud, C., Yamanaka, Y., Pawson, T., & Rossant, J. (2006). Early lineage segregation between epiblast and primitive endoderm in mouse blastocysts through the Grb2-MAPK pathway. *Developmental Cell*, *10*, 615–624.

Cockburn, K., Biechele, S., Garner, J., & Rossant, J. (2013). The Hippo pathway member Nf2 is required for inner cell mass specification. *Current Biology*, *23*, 1195–1201.

Dalcq, A. (1957). *Introduction to general embryology*. London: Oxford University Press.

Dufort, D., Schwartz, L., Harpal, K., & Rossant, J. (1998). The transcription factor HNF3beta is required in visceral endoderm for normal primitive streak morphogenesis [in process citation]. *Development*, *125*, 3015–3025.

Gardner, R. L. (1968). Mouse chimaeras obtained by the injection of cells into the blastocyst. *Nature*, *220*, 596–597.

Gardner, R. L., & Johnson, M. H. (1973). Investigation of early mammalian development using interspecific chimaeras between rat and mouse. *Nature: New Biology*, *246*, 86–89.

Gardner, R. L., Papaioannou, V. E., & Barton, S. C. (1973). Origin of the ectoplacental cone and secondary giant cells in mouse blastocysts reconstituted from isolated trophectoderm and inner cell mass. *Journal of Embryology and Experimental Morphology*, *30*, 561–572.

Gardner, R. L., & Rossant, J. (1979). Investigation of the fate of 4.5 day post-coitum mouse inner cell mass cells by blastocyst injection. *Journal of Embryology and Experimental Morphology, 52*, 141–152.
Georgiades, P., Cox, B., Gertsenstein, M., Chawengsaksophak, K., & Rossant, J. (2007). Trophoblast-specific gene manipulation using lentivirus-based vectors. *Biotechniques, 42*, 317–318. 320, 322–315.
Hirate, Y., Hirahara, S., Inoue, K., Suzuki, A., Alarcon, V. B., Akimoto, K., et al. (2013). Polarity-dependent distribution of angiomotin localizes Hippo signaling in preimplantation embryos. *Current Biology, 23*, 1181–1194.
Johnson, M. H., & McConnell, J. M. (2004). Lineage allocation and cell polarity during mouse embryogenesis. *Seminars in Cell & Developmental Biology, 15*, 583–597.
Johnson, M. H., & Ziomek, C. A. (1981). The foundation of two distinct cell lineages within the mouse morula. *Cell, 24*, 71–80.
Kanai-Azuma, M., Kanai, Y., Gad, J. M., Tajima, Y., Taya, C., Kurohmaru, M., et al. (2002). Depletion of definitive gut endoderm in Sox17-null mutant mice. *Development, 129*, 2367–2379.
Kunath, T., Arnaud, D., Uy, G. D., Okamoto, I., Chureau, C., Yamanaka, Y., et al. (2005). Imprinted X-inactivation in extra-embryonic endoderm cell lines from mouse blastocysts. *Development, 132*, 1649–1661.
Kwon, G. S., Viotti, M., & Hadjantonakis, A. K. (2008). The endoderm of the mouse embryo arises by dynamic widespread intercalation of embryonic and extraembryonic lineages. *Developmental Cell, 15*, 509–520.
Laeno, A. M., Tamashiro, D. A., & Alarcon, V. B. (2013). Rho-associated kinase activity is required for proper morphogenesis of the inner cell mass in the mouse blastocyst. *Biology of Reproduction, 89*, 122.
Mc, L. A., & Biggers, J. D. (1958). Successful development and birth of mice cultivated in vitro as early as early embryos. *Nature, 182*, 877–878.
Mintz, B. (1965). Genetic mosaicism in adult mice of quadriparental lineage. *Science, 148*, 1232–1233.
Morris, S. A., Teo, R. T., Li, H., Robson, P., Glover, D. M., & Zernicka-Goetz, M. (2010). Origin and formation of the first two distinct cell types of the inner cell mass in the mouse embryo. *Proceedings of the National Academy of Sciences of the United States of America, 107*, 6364–6369.
Mulnard, J. (1961). Problemes de structure et d'organisation morphogenetique de l'oeuf des Mammiferes. In *Symposium on the germ cells and earliest stages of development (Milan: Fondazione A.Baselli, Instituto Lombardo)* (pp. 639–688).
Nichols, J., Silva, J., Roode, M., & Smith, A. (2009). Suppression of Erk signalling promotes ground state pluripotency in the mouse embryo. *Development, 136*, 3215–3222.
Nishioka, N., Inoue, K., Adachi, K., Kiyonari, H., Ota, M., Ralston, A., et al. (2009). The Hippo signaling pathway components Lats and Yap pattern Tead4 activity to distinguish mouse trophectoderm from inner cell mass. *Developmental Cell, 16*, 398–410.
Papaioannou, V. E. (1982). Lineage analysis of inner cell mass and trophectoderm using microsurgically reconstituted mouse blastocysts. *Journal of Embryology and Experimental Morphology, 68*, 199–209.
Piotrowska-Nitsche, K., Perea-Gomez, A., Haraguchi, S., & Zernicka-Goetz, M. (2005). Four-cell stage mouse blastomeres have different developmental properties. *Development, 132*, 479–490.
Plusa, B., Piliszek, A., Frankenberg, S., Artus, J., & Hadjantonakis, A.-K. (2008). Distinct sequential cell behaviours directing primitive endoderm formation in the mouse blastocyst. *Development, 135*, 3081–3091.
Rossant, J. (1975). Investigation of the determinative state of the mouse inner cell mass II. The fate of isolated inner cell masses transferred to the oviduct. *Journal of Embryology and Experimental Morphology, 33*, 991–1001.

Rossant, J., & Croy, B. A. (1985). Genetic identification of tissue of origin of cellular populations within the mouse placenta. *Journal of Embryology and Experimental Morphology*, *86*, 177–189.

Rossant, J., Vijh, M., Siracusa, L. D., & Chapman, V. M. (1983). Identification of embryonic cell lineages in histological sections of M. Musculus <——> M. caroli chimeras. *Journal of Embryology and Experimental Morphology*, *73*, 179–191.

Samarage, C. R., White, M. D., Alvarez, Y. D., Fierro-Gonzalez, J. C., Henon, Y., Jesudason, E. C., et al. (2015). Cortical tension allocates the first inner cells of the mammalian embryo. *Developmental Cell*, *34*, 435–447.

Solter, D., & Knowles, B. B. (1975). Immunosurgery of mouse blastocyst. *Proceedings of the National Academy of Sciences of the United States of America*, *72*, 5099–5102.

Stephenson, R. O., Yamanaka, Y., & Rossant, J. (2010). Disorganized epithelial polarity and excess trophectoderm cell fate in preimplantation embryos lacking E-cadherin. *Development*, *137*, 3383–3391.

Strumpf, D., Mao, C. A., Yamanaka, Y., Ralston, A., Chawengsaksophak, K., Beck, F., et al. (2005). Cdx2 is required for correct cell fate specification and differentiation of trophectoderm in the mouse blastocyst. *Development*, *132*, 2093–2102.

Tanaka, S., Kunath, T., Hadjantonakis, A. K., Nagy, A., & Rossant, J. (1998). Promotion of trophoblast stem cell proliferation by FGF4. *Science*, *282*, 2072–2075.

Tarkowski, A. K. (1961). Mouse chimaeras developed from fused eggs. *Nature*, *190*, 857–860.

Tarkowski, A. K., & Wroblewska, J. (1967). Development of blastomeres of mouse eggs isolated at the 4- and 8-cell stage. *Journal of Embryology and Experimental Morphology*, *18*, 155–180.

Watanabe, T., Biggins, J. S., Tannan, N. B., & Srinivas, S. (2014). Limited predictive value of blastomere angle of division in trophectoderm and inner cell mass specification. *Development*, *141*, 2279–2288.

Whitten, W. K., & Biggers, J. D. (1968). Complete development in vitro of the pre-implantation stages of the mouse in a simple chemically defined medium. *Journal of Reproduction and Fertility*, *17*, 399–401.

Yamanaka, Y. (2011). Response: Cell fate in the early mouse embryo-sorting out the influence of developmental history on lineage choice. *Reproductive Biomedicine Online*, *22*, 525–527. discussion 528.

Yamanaka, Y., Lanner, F., & Rossant, J. (2010). FGF signal-dependent segregation of primitive endoderm and epiblast in the mouse blastocyst. *Development*, *137*, 715–724.

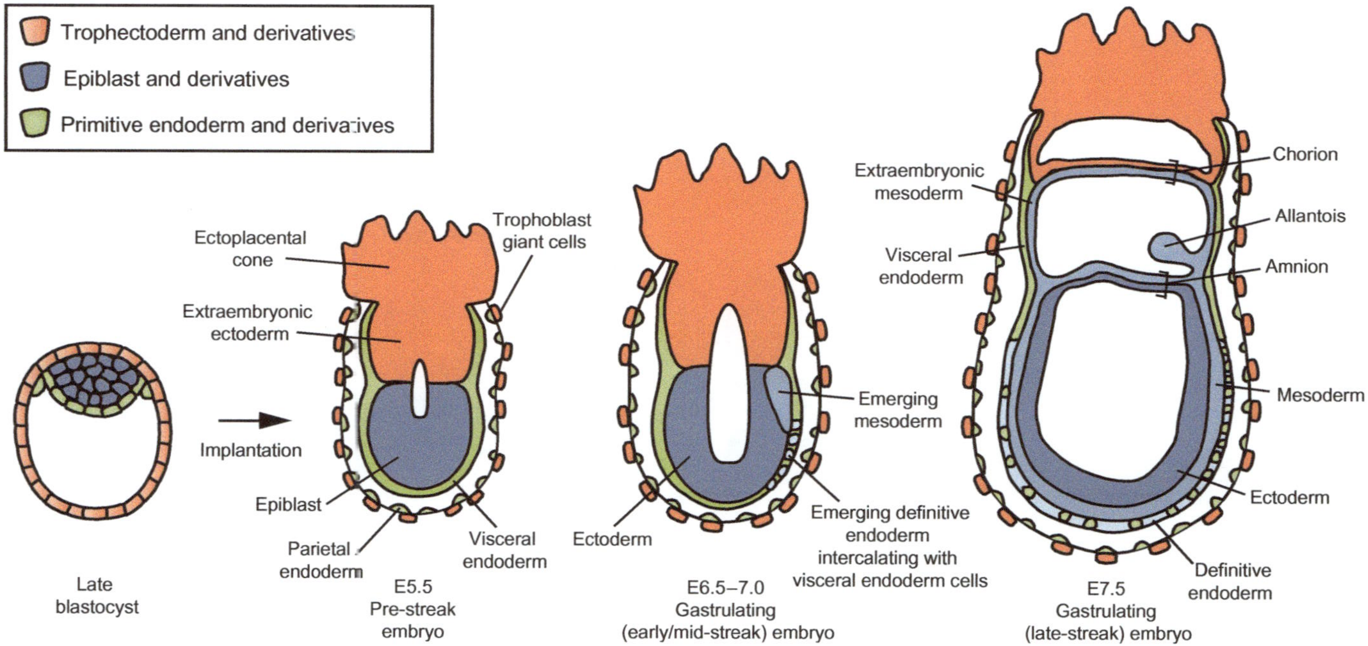

Janet Rossant, Figure 1 Postimplantation lineage derivatives of the epiblast, primitive endoderm, and trophectoderm. Trophectoderm (red) gives rise to all the later trophoblast lineages that contribute to the developing placenta, as well as to the trophoblast giant cells. Primitive endoderm (green) gives rise to the extraembryonic visceral and parietal endoderm, as well as providing a minor contribution to the developing gut along with the definitive endoderm. The epiblast (blue) gives rise the post-implantation primitive ectoderm, and to the mesoderm and definitive endoderm that arise at gastrulation. Extraembryonic mesoderm arises from the posterior of the primitive streak and forms the allantois and the mesoderm layers of the developing yolk sac, where primitive hematopoiesis occurs.

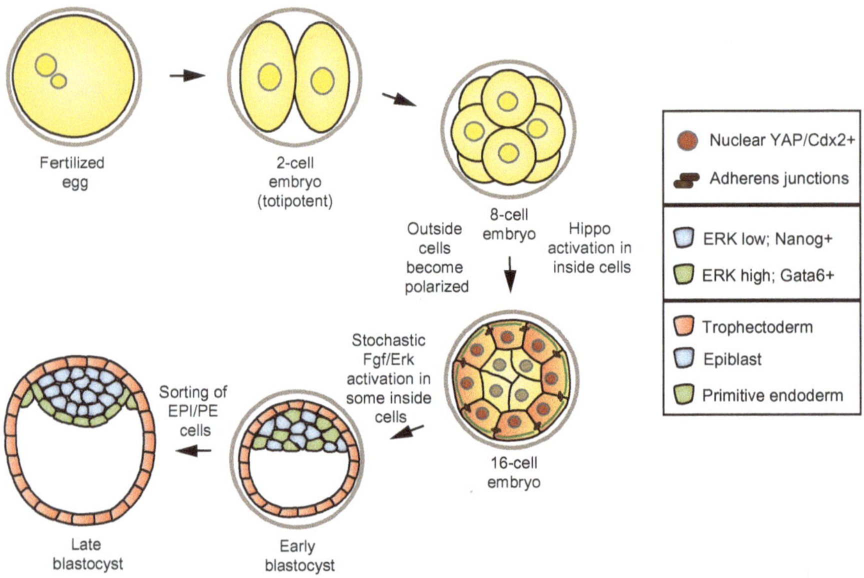

Janet Rossant, Figure 2 Mechanisms of cell fate specification during preimplantation development. Between the 8- and 16-cell stage, outer blastomeres become polarized and enclose a group of apolar cells. Cell contact and cell polarity lead to differential activation of the Hippo-related Lats kinase in the inside versus the outside cells. Lats phosphorylates Yap in inside cells which causes its exclusion from the nucleus. Hippo signaling is inactive in the outside, polarized cells, Yap is nuclear and activates expression of trophectoderm-specific genes such as *Cdx2*. Absence of *Cdx2* in inside cells allows expression of pluripotency genes and establishment of the inner cell mass. Once the inner cell mass is established at the early blastocyst stage, stochastic differences in the activation of FGF/ERK signaling occur in a mosaic manner within it. Active ERK signaling promotes primitive endoderm formation while absence of ERK signaling promotes epiblast. As the two lineages gradually become specified, they sort out within the ICM so that the primitive endoderm ends up on the blastocoelic surface.

CHAPTER SIXTEEN

Two Decades of Ascidian Developmental Biology: A Personal Research Story

Noriyuki Satoh[1]

Marine Genomics Unit, Okinawa Institute of Science and Technology Graduate University, Onna, Okinawa, Japan

[1]Corresponding author: e-mail addresses: norisky@oist.jp; satoh32@gmail.com

Contents

Abstract

Ascidians are the closest relatives of vertebrates. Their utility in experimental embryology has been well recognized because of their simple mode of embryogenesis to form tadpole larvae with a basal chordate body plan. Approximately two decades of research, including decoding of the *Ciona* genome, have promoted ascidians as one of the best systems for exploring genome-wide mechanisms of developmental transcriptional control and chordate evolution.

This chapter provides a personal view of two decades of research in ascidian developmental biology. The progress witnessed in the last 20 years was unimaginable at the time of the field's inception in the mid-1970s.

1. MUSCLE DETERMINANT

One evening in February, 1984, in a room at the Asamushi Marine Biological Station of Tohoku University, Takuya Deno and I had drunk

Current Topics in Developmental Biology, Volume 117
ISSN 0070-2153
http://dx.doi.org/10.1016/bs.ctdb.2015.11.016

too much, and were engaged in a wild discussion as to whether we could identify the molecular nature of muscle determinant(s) in the myoplasm of ascidian eggs, one of the typical examples of maternal determinants discovered by E.G. Conklin. Takuya was my first PhD student at Kyoto University and his research objective was to confirm the existence of maternal factor(s). As originally reported by Richard Whittaker (see below), when division of the ascidian 8-cell embryo was arrested by cytochalasin-B during a time when control normal embryos give rise to the mid-tailbud stage, they eventually develop acetylcholine esterase (AChE) activity, a marker of muscle differentiation, in a pair of B4.1 blastomeres, but not in the other three pairs. Using the ascidian, *Halocynthia roretzi*, Takuya transplanted cytoplasm of B4.1 into A4.1 of another embryo to examine the determinants. Research on *Halocynthia* is limited to only a couple of months in the winter. After repetition of microinjection of myoplasm for two seasons, Takuya finally got several injected embryos that showed AChE activity in A4.1.

The evening at Asamushi was a celebration of his research paper publication (Deno & Satoh, 1984). Our enthusiasm about the success led us to drink too much. We drank to discuss, and discussed to drink. However, we both completely agreed that the molecular nature of muscle determinants would never be discovered during our lifetimes. That was a time, when there were no tools or techniques to characterize them. A few decades after our deaths, someone might possibly succeed, we concluded.

Although a preferential accumulation of RNAs in myoplasm was shown in the 1980s by William Jeffery, the molecular nature of muscle determinants had long proven elusive, far from any hope of reaching the final peak. Yusuke Marikawa and Shoko Yoshida of my lab developed a method to fragment *Ciona savignyi* eggs. Only the black-colored, nonnuclear fragments contained the determinant, suggesting its maternal mRNA nature. Taking advantage of this system, they succeeded in isolating a gene, named *pem* (*posterior end mark*), and determined that maternal *pem* mRNA is localized in eggs and early embryos, just as the muscle determinant is (Yoshida, Marikawa, & Satoh, 1996, cover picture of *Development* 122(7)). Although *pem* plays a role in development, including germ cell formation, this maternal factor was not the muscle determinant.

Finally in 2001, so rapidly, or so it seemed, after several years of subtractive hybridization between blastomeres with and without *Halocynthia* embryonic myoplasm, Nishida and Sawada (2001) identified a gene, named *macho-1*, which encodes a zinc finger-type transcription factor that acts as a muscle determinant. Publication of the manuscript in *Nature* greatly impressed developmental biologists since it confirmed Conklin's insight. Later, a *macho-1* ortholog was isolated from *Ciona intestinalis* (*Ci-macho1*), and since then, a genetic cascade of muscle differentiation has been well characterized, beginning with maternal *macho1*, through zygotic activity of *Tbx2b* and *MyoD*, and later, various muscle structural genes leading to the formation of muscle cells (e.g., Satoh, 2014).

In science, generally every finding further stimulates the imagination of other researchers, especially of young scientists. Ascidian embryology illustrates this pattern.

2. TIMING MECHANISMS

I started my own ascidian research around 1976. My PhD thesis at the University of Tokyo was about sex differentiation mechanisms of medaka fish, specifically whether sex steroid hormones are natural inducers in this process. A year-long electron microscopic search of steroid-producing cells in the gonad that had already commenced differentiation, forced me to conclude that sex steroids are powerful artificial inducers, but not natural inducers. When I got a job at the Department of Zoology, at Kyoto University, and began to participate in research on early embryogenesis of animals, I gradually became interested in temporal control or timing mechanisms of animal development. I initiated my studies of this complex problem by questioning how the embryo determines the timing of gastrulation. It was possible that the embryo might determine timing by the number of cell division (cytokinesis), nuclear division, or DNA replication. With this aim, I selected ascidians as an experimental system, since it was known that the ascidian embryo initiates gastrulation at the 110-cell stage after seven rounds of cleavage.

Soon after I started this line of research, I read about some very stimulating experiments done by Richard Whittaker in 1973, in which he showed that even if cleavage of *Ciona* eggs is arrested, manipulated embryos are able to produce enzymes known as differentiation markers of muscle or endoderm (Whittaker, 1973). This meant that cytokinesis is not required for the initiation of marker enzyme production. In addition, blockade of nuclear division by colchicine and other reagents did not affect the appearance of marker enzyme production, suggesting that nuclear division itself is not the timing mechanism. I was so impressed with this study and repeated his experiment using *Halocynthia* embryos and confirmed his results. Then, unfortunately, Whittaker and I fell into a competition to block DNA replication. Reagents known to block DNA replication did not work in the ascidian system. I had an opportunity to meet Susumu Ikegami who found that aphidicolin acts as an inhibitor of DNA replication in starfish eggs. Aphidicolin blocked DNA replication of ascidian embryos too, and this blockage resulted in failure of marker enzyme development (Satoh & Ikegami, 1981). In addition, the number of DNA replications appeared intimately associated with muscle cell differentiation. A detailed mechanism underlying the requirement of DNA replication for initiation of transcription factor gene activation still remains to be elucidated.

3. MARINE BIOLOGICAL LABORATORY AT WOODS HOLE

One day in May 1992, the telephone rang in my lab at Kyoto University. When I answered, I heard the voice of Eric Davidson. He invited me to participate in the Embryology Course at Woods Hole as a member of the teaching staff to introduce the ascidian system. My dream came true. To every student and scholar like me who works in marine invertebrate embryogenesis, the Marine Biological Laboratory (MBL) is the place they wish to be. It is so stimulating to learn in a scientific atmosphere in which ideas are freely discussed and exchanged, inspiring future research.

My first chance to visit MBL was in 1984, the year of the Los Angeles Olympics, when I was selected as a MEXT fellow to be a visiting researcher in the United States, I asked Whittaker to host my stay in his lab at MBL. Unfortunately, our relationship was so bad as a result of our earlier competition over temporal control mechanisms of ascidians that he refused to accept my appointment at MBL. Instead, I stayed 10 months at Vanderbilt University. In late August, my family and I drove from Nashville to Woods Hole, but I could only see MBL from the outside. During this stay however, I had the opportunity to visit Bill Jeffery at the University of Texas, Austin for the first time. Later, Bill gave me the chance to stay at Woods Hole when he organized a course there in 1987. At that time, I met Billie Swalla and others, who have become long-standing friends and competitors.

A 5-year Embryology Course at MBL from 1992 to 1996, coorganized by Eric Davidson, Dave McClay, and Michael Levine was such a nice experience for me, one that still brings back warm memories. First, I was able to learn about annual progress in various disciplines within developmental biology, surely more than students of the course learned: gene regulatory systems in sea urchins, the function of maternally and zygotically expressed genes in *Drosophila*, lineage-dependent expression and function of genes in

Caenorhabditis, gastrulation movement in *Xenopus*, and so on. Eric's inexhaustible energy for pursuing research was far beyond my scientific scope, but I really wished that the ascidian system would one day catch up with these other systems. Every year, on the final day when I left Woods Hole, I asked Eric for his appraisal of my lecture and lab exercises regarding ascidian embryos. How delighted I was by his answer, "Yes, you did well this year, so please come again next summer." With satisfaction, I smiled as I watched the changing scenery as I took the bus from Cape Cod to Boston.

Second, I became acquainted with many top researchers in developmental biology, including some predestined meet with Michael Levine. His lecture about *Drosophila* embryos was always brilliant and very impressive. His lab exhibition of whole-mount *in situ* hybridization and antibody staining of *Drosophila* embryos was so beautiful. I was impressed by his interest in the ascidian system. Soon after, his work using a *Ciona* system greatly advanced molecular developmental biology of ascidians. Reflecting the most advanced and/or sophisticated techniques used in the *Drosophila* system, *in situ* hybridization pictures from Mike's lab were always very beautiful, including the "smile" expression of *Brachyury* (*Developmental Biology 244*(2) cover picture).

4. FROM *HALOCYNTHIA* TO *CIONA*

One day in April 1997, I told my lab members that our research material should change from *H. roretzi* to *C. intestinalis*; there would be no further use of *H. roretzi* unless dictated by special circumstances. To us, this decision was difficult to make, since we had accumulated molecular data only in *H. roretzi*. This meant that we had to reinitiate our painstaking studies in *C. intestinalis*. The major reasons that convinced me to make this change were internationality and the future introduction of a genetic approach into this system. First, *H. roretzi* is local to Japan and Korea, and the cosmopolitan *C. intestinalis* was more popular among European and American researchers. Second, in contrast to the limited availability of *H. roretzi* gametes for only a few winter months, and their longer generation time (~3 years), *C. intestinalis* provides gametes year-round, with a shorter generation time (~3 months). This might promise future introduction of genetic approaches into this system.

We worked diligently to establish the modern *Ciona* system. In Japan, *C. intestinalis* has been approved as a bioresource organism, in which the maintenance and distribution are supported by National Bio-Resource Project (NBRP) grant (Kazuko Hirayama had contributed enormously in the establishment of this project). Through the NBRP system, users can obtain ripe *C. intestinalis* anytime they need to use them. In addition, a system of *Minos* transposon-based transgenic lines has been established by the efforts of Sasakura, Awazu, Chiba, Kano, and Satoh (2003). Not only are there many transgenic lines expressing GFP or RFP in specific cells and/or tissues, but also those with mutations of selected genes have recently been used to explore molecular mechanisms underlying processes of tissue and organ formation. One example is a mutation named "*swimming juveniles*," in which larvae proceed into juvenile organ formation without resorption of the larval tail (Sasakura et al., 2005). Later, it was revealed that a transposon was inserted into the 5′ regulatory region of the *CesA* gene, which encodes cellulose synthase, a gene highly likely to have originated by horizontal transfer of bacterial *CesA* into a tunicate ancestor. Today, *C. intestinalis* is a standard organism used in various fields of biology. Various techniques including transgenic line formation, and gene manipulation with TALEN and the CRISPER/Cas9 system has been successfully introduced into the *Ciona* system. These have greatly advanced our understanding of molecular mechanisms of development.

5. *CIONA* GENOME PROJECT

One day in February 2002, Michael Levine's Toyota "Spider" ran fast along California freeway R24 toward Walnut Creek, hometown of the Joint Genome Institute, DOE. Mike and I had an appointment to meet Daniel Rokhsar to discuss the *C. intestinalis* genome project. Two months ago, Mike sent me an e-mail to suggest a possible decoding of *Ciona* genome at JGI since sequencing machines that had been used for the *Fugu* genome project were now free for the next target. I soon decided to visit JGI with Mike to meet Dan. At that time, Dan was managing sequencing and assembly of genomes at JGI, where the main focus was the human genome. We agreed that *C. intestinalis* was the best ascidian target for a genome-decoding project, by virtue of the large number of researchers worldwide and because it had a large set of EST data (see below).

At the beginning of the twenty-first century, the Japanese Government decided to promote "millennium projects" in the life sciences including brain, cancer, and genome research. The genome project targeted not only medical fields but also basic biology. Yuji Kohara, a pioneer of clone-based genome sequencing, led this project. At the first meeting to discuss the project after the big grant had been awarded, I asked Yuji how I could pursue the *Ciona* genome project. His answer was disappointing. "At the moment, in Japan, it is impossible to undertake metazoan genome sequencing projects, so you should concentrate on EST analyses, which will benefit the real genome project; genome sequencing project may be possible in several years." In about 2001, the human genome project was completed. Sequencing power in Japan was not always high compared to USA and European countries.

According to Yuji's suggestion, we collected a large quantity of EST data from *C. intestinalis*. With a great help of Yutaka Satou and other members of my lab, we produced the "*C. intestinalis* gene collection," a package of 36 384-well plates containing ~13,500 independent cDNA clones that covered almost all transcription factors, signaling molecules, and other developmentally relevant molecules (Satou et al., 2002). We distributed the collection to every laboratory of *Ciona* researchers in the world. I personally took the collection to several labs when I visited those cities. From that time onward, *Ciona* developmental biologists no longer needed to isolate cDNA clones of interest if the clone was included in the collection. For 2 or 3 years since, many *Ciona* research papers acknowledged this collection. After a while, however, use of the collection became so common that acknowledgments became progressively more rare in *Ciona* manuscripts.

The sequencing power of JGI, one of the biggest sequencing centers in the world, was astonishing. The sequenced genome of a single wild-type specimen collected at Half Moon Bay, CA was ready to be annotated in late March 2002. Dan Rokhsar organized a *C. intestinalis* genome annotation jamboree in April 2012, inviting ascidian developmental biologists from worldwide to JGI. An exiting week had passed very quickly. Everyone who attended the jamboree was so impressed when gene models of interest appeared on laptop computers. Of course, one week was too little to annotate all the genes needed for future analyses. We left Walnut Creek with promises to continue further annotation in own laboratories.

Late August of 2002, I visited JGI once again to initiate manuscript preparation with Dan and Mike. Mike was not able to join the discussion for the first day. Dan wanted the *Ciona* genome manuscript to be like that of *Caenorhabditis elegans*, which was published in *Science* in 1999, comprising several articles, each of which discussed different topics, such as genes involved in development or neuronal functions. Accordingly, Dan and

I discussed research topics and researchers who were involved in each topic. Late on the second day, Mike came to join the discussion. He felt strongly that an entirely different approach was necessary. He argued that four key issues should guide the preparation of the *Ciona* genome manuscript: (1) it should focus on the origin of chordates and vertebrates so that it would attract many readers; (2) it should be published as one article, considering the breadth of ascidian research field; (3) it should be published that year in *Nature* or *Science*, since in near future many metazoan genome-decoding articles might appear, which might compete with the publication of the *Ciona* genome manuscript. He had already decided on an approximate submission deadline, which was very close to the end of 2002, leaving only a few months; and (4) we should get *Ciona* as a cover photo of the issue of the journal to signal this triumph.

Soon after, Mike and I concentrated on preparation of the manuscript by asking all the jamboree members to send us results. However, Dan did not work on it at all, because his first baby was soon to be born. The deadline to submit the manuscript to *Nature* had passed, and that for *Science* was close. Mike lost contact with Dan and asked me to contact him, but he did not reply. Finally Mike gave up on the project. In late October, Dan suddenly returned and began to work with amazing speed. He collected all data, prepared a very rough manuscript, and submitted it to *Science* in early November. One reviewer criticized this is far from complete manuscript. Nonetheless, a paper entitled "The draft genome of *Ciona intestinalis*: insights into chordate and vertebrate origins" was published in the December 13th issue of *Science* with a beautiful cover picture of many *Ciona* larvae with *Brachyury* expression in the tail notochord (Dehal, Satou, Campbell, et al., 2002).

I did not feel any physical changes in my body during the project, but soon after the paper appeared I became a very sever sick with autoimmune hemolytic anemia. Stress during the project may have caused my immune system to destroy my own red blood cells. Starting in late December 2002, I remained hospitalized for one and a half months until my red blood cell count returned to normal. More than 2 years medical treatment was required for complete recovery. One finding of the *Ciona* genome project is that ascidians have no genes for biosynthesis of corticosteroids, which are likely a vertebrate innovation. On my bed in the hospital, with daily injections of corticosteroids, I came to a profound understanding that I am a vertebrate.

6. THE *CIONA* SYSTEM WITH A MASS OF GENE EXPRESSION DATA

Kyoto is a city surrounded by mountains on all but the south side. Summer is hot and humid, not conducive to experiments with *Ciona* embryos that develop at ~18 °C. Following the *Ciona* genome jamboree, in the summer of 2002, all members of my lab concentrated on annotating genes associated with *Ciona* development. We chose 10 topics for gene annotation, including all categories of transcription factors, signal transduction molecules, cell cycle, cell junction, muscle, and others. Adding two articles of introduction and cDNA counts for inferring temporal and spatial expression of specific genes, a total of 12 articles were published as the May/June issue of Development, Genes, and Evolution in 2003 (Satou & Satoh, 2003). This special issue with cover photo of *Brachyury* sequences spread out from siphons of an adult *Ciona* is one of my fondest memories. I am still very proud of harmonious and vigorous research activity of my lab at that time.

Thereafter, we collected a large number of gene expression profiles of *Ciona* embryos. Expression profiles of most transcription factors and signaling molecules, in addition to randomly selected genes at five different stages of *Ciona* embryogenesis, provide a basis for researching genes involved in various aspects of organ formation.

7. ON THE ORIGIN OF CHORDATES

In April 1991, my laboratory accepted two new PhD students, Hiroshi Wada and Hitoyoshi Yasuo. At that time, molecular phylogenic approaches had just been introduced to infer metazoan relationships. I became gradually interested not only in ascidian embryogenesis but also in chordate evolution. For example, since the notochord is one of the most characteristic features of chordates and since ascidians develop a notochord consisting of exactly 40 cells, aligned at the midline of the larval tail, I reasoned that discovery of molecular mechanisms involved in ascidian notochord development might lead to an understanding of the evolutionary origin of the notochord. I suggested that Hiroshi should tackle molecular phylogeny of deuterostomes and Hitoyoshi should address molecular mechanisms underlying notochord formation in ascidian embryos.

Hiroshi very quickly became familiar with molecular phylogeny methods based on comparisons of 18S rRNA sequences. His manuscript, published in PNAS in 1994, failed to resolve the phylogenic position of ascidians, due to the so-called "long branch attraction problem," caused by comparatively rapid molecular evolutionary, but clearly demonstrated the phylogenetic affinities of echinoderms and hemichordates (Wada & Satoh, 1994). This result was evaluated by Halayanch next year to join these two groups of deuterostomes as the Ambulacraria. At that time, rRNA sequencing was performed by reading ladders of ^{35}S-labeled nucleotides, first electrophoresed on agarose gels and then visualized on X-ray films. At that time, Shigeki Fujiwara of Kochi University was in my lab for a 1-year sabbatical to pursue research to get a PhD degree. He was interested in localizing maternal mRNAs from *Halocynthia* 8-cell embryos that have possible developmental functions. One day an excited Shigeki came into a common room where all my lab members had desks. Showing his X-ray film, he said, "Yay, I finally got sequences of localized factor(s)." "Really, that's great!" said Hiroshi, picking up Shigeki's film and gazing at it briefly. Then Hiroshi said, "Fujiwara-san, this is an rRNA sequence." An expert can recognize an 18S rRNA nucleotide sequence ladder at a glance.

On the other hand, following the cloning of the mouse *Brachyu*ry gene, Hitoyoshi found that ascidian *Brachyu*ry is expressed specifically in the notochord (Yasuo & Satoh, 1993). Later, it was shown that *Brachyury* plays a key role in ascidian notochord formation. This finding promoted extensive work on molecular mechanisms involved in ascidian embryonic notochord formation. Maternally expressed *β-catenin*, zygotically expressed *FoxD*, *FoxA*, *ZicL*, and *Fgf9/16/19* have been characterized as essential components of an upstream cascade leading to transcriptional initiation of *Brachyury*. In collaboration with Mike Levine's laboratory, Hiroki Takahashi and Kohji Hotta succeeded in isolating and characterizing nearly 40 genes that are expressed in the developing notochord as downstream components of *Brachyury* (Takahashi et al., 1999). The work was published in *Genes and Development* with a beautiful cover photo of an *in situ* hybridization of a certain notochord-specific gene.

At the same time, several PhD students including Yoshito Harada and Kuni Tagawa attempted to isolate *Brachyury* from sea urchins, acorn worms, and amphioxus. *Brachyury* of nonchordate deuterostomes is expressed in the archenteron invagination and stomodaeum invagination regions. Injection of mRNA encoding urchin or acorn worm *Brachyury* into *Ciona* eggs resulted in the formation of extra notochord cells in manipulated embryos (Satoh, Harada, & Satoh, 2000). This suggests that *Brachyury* is originally expressed and functions in morphogenetic movement during gastrulation, and that later during chordate evolution, the gene was recruited into notochord formation (Satoh, Tagawa, & Takahashi, 2012). Our efforts to explore molecular mechanisms involved in notochord formation are still continuing. I have proposed the "aboral-dorsalization hypothesis" to explain how the notochord developed in lineages leading to chordate evolution (Satoh, 2008). We also proposed that three groups of chordates, namely cephalochordates, urochordates, and vertebrates, so far recognized as subphyla of the phylum Chordata, should be ranked as separate phyla, with chordates being a superphylum (Satoh, Rokhsar, & Nishikawa, 2014).

In summary, I am very happy to have had many opportunities to contribute to the establishment of the ascidian system as one of the best experimental models for exploring genome-wide mechanisms involved in transcriptional control of development and evolution. Looking back over the last two decades, I always feel that I was not a researcher who accumulated data to make new discoveries, as many PhD students in my Kyoto lab did, but I am pleased that I was one of the pioneers who opened the gate of the ascidian system.

ACKNOWLEDGMENT

Steven Aird is acknowledged to help in improvement of this manuscript.

REFERENCES

Dehal, P., Satou, Y., Campbell, R. K., et al. (2002). The draft genome of *Ciona intestinalis*: Insights into chordate and vertebrate origins. *Science*, *298*, 2157–2167.

Deno, T., & Satoh, N. (1984). Studies on the cytoplasmic determinant for muscle cell differentiation in ascidian embryos: An attempt at transplantation of the myoplasm. *Development, Growth & Differentiation*, *26*, 43–48.

Nishida, H., & Sawada, K. (2001). Macho-1 encodes a localized mRNA in ascidian eggs that specifies muscle fate during embryogenesis. *Nature*, *409*, 724–729.

Sasakura, Y., Awazu, S., Chiba, S., Kano, S., & Satoh, N. (2003). Application of *Minos*, one of the Tc1/*mariner* superfamily transposable elements, to ascidian embryos as a tool for insertional mutagenesis. *Gene*, *308*, 11–20.

Sasakura, Y., Nakashima, K., Awazu, S., Matsuoka, T., Nakayama, A., Azuma, J., et al. (2005). Transposon-mediated insertional mutagenesis revealed the functions of animal cellulose synthase in the ascidian *Ciona intestinalis*. *Proceedings of the National Academy of Sciences of the United States of America*, *102*, 15134–15139.

Satoh, N. (2008). An aboral-dorsalization hypothesis for chordate origin. *Genesis*, *46*, 614–622.

Satoh, N. (2014). *Developmental genomics of ascidians*. Hoboken, NJ, USA: Wiley/Blackwell.

Satoh, G., Harada, Y., & Satoh, N. (2000). The expression of nonchordate deuterostome *Brachyury* genes in the ascidian *Ciona* embryo can promote the differentiation of extra notochord cells. *Mechanisms of Development, 96*, 155–163.

Satoh, N., & Ikegami, S. (1981). A definite number of aphidicolin-sensitive cell-cycle events are required for acetylcholinesterase development in the presumptive muscle cells of the ascidian embryos. *Journal of Embryology and Experimental Morphology, 61*, 1–13.

Satoh, N., Rokhsar, D., & Nishikawa, T. (2014). Chordate evolution and the three-phylum system. *Proceeding of the Royal Society B, 281*, 2014179.

Satoh, N., Tagawa, K., & Takahashi, H. (2012). How was the notochord born? *Evolution & Development, 14*, 56–75.

Satou, Y., & Satoh, N. (2003). Genomewide surveys of developmentally relevant genes in *Ciona intestinalis. Development Genes and Evolution, 213*, 211–212.

Satou, Y., et al. (2002). A cDNA resource from the basal chordate *Ciona intestinalis. Genesis, 33*, 153–154.

Takahashi, H., Hotta, K., Erives, A., Di Gregorio, A., Zeller, R. W., Levine, M., et al. (1999). *Brachyury* downstream notochord differentiation in the ascidian embryo. *Genes & Development, 13*, 1519–1523.

Wada, H., & Satoh, N. (1994). Details of the evolutionary history from invertebrates to vertebrates, as deduced from the sequences of 18S rDNA. *Proceedings of the National Academy of Sciences of the United States of America, 91*, 1801–1804.

Whittaker, J. R. (1973). Segregation during ascidian embryogenesis of egg cytoplasmic information for tissue-specific enzyme development. *Proceedings of the National Academy of Sciences of the United States of America, 70*, 2096–2100.

Yasuo, H., & Satoh, N. (1993). Function of vertebrate *T* gene. *Nature, 364*, 582–583.

Yoshida, S., Marikawa, Y., & Satoh, N. (1996). Posterior end mark, a novel maternal gene encoding a localized factor in the ascidian embryo. *Development, 122*, 2005–2012.

CHAPTER SEVENTEEN

Lineage Segregation in the Totipotent Embryo

Guangming Wu*, Hans R. Schöler*,[1]
*Department of Cell and Developmental Biology, Max Planck Institute for Molecular Biomedicine, Münster, Germany
[1]Corresponding author: e-mail address: office@mpi-muenster.mpg.de

Contents

Abstract

After a spermatozoon enters an oocyte, maternal factors accumulated in the oocyte reprogram the genomes of the terminally differentiated oocyte and spermatozoon epigenetically and turn the zygote into a totipotent cell, with the capacity to differentiate into all types of somatic cells in a highly organized manner and generate the entire organism, a feature referred to as totipotency. Differentiation of the first lineage begins after three cleavages, when the early embryo compacts and becomes polarized, followed by segregation of the first lineages—the inner cell mass (ICM) and the trophectoderm (TE). To date, a full understanding of the molecular mechanisms that underlie the establishment of totipotency and the ICM/TE lineage segregation remains unclear. In this review, we discuss recent findings in the mechanism of transcriptional regulation networks and signaling pathways in the first lineage separation in the totipotent mouse embryo.

1. INTRODUCTION

A new circle of life starts with the fusion of two haploid gametes, oocyte and spermatozoon, forming a totipotent diploid zygote. All biological processes occurring during the first cell cycle of the embryo rely on maternal

Current Topics in Developmental Biology, Volume 117
ISSN 0070-2153
http://dx.doi.org/10.1016/bs.ctdb.2015.10.014

factors, which have accumulated during the long growth phase of the germinal vesicle oocyte (Bachvarova & De Leon, 1980). From the late 1-cell stage to 2-cell stage, there is a burst of transcription—known as zygotic genome activation (ZGA; Schultz, 1993), following extensive epigenetic modification mediated by maternal regulatory factors. ZGA is characterized by more efficient utilization of TATA-less promoters (Davis & Schultz, 2000); activation of repetitive elements (Peaston et al., 2004), particularly endogenous retrotransposons, e.g., MERVL at the 2-cell stage as a marker for totipotent cells (Macfarlan et al., 2012); uncoupling of transcription and translation in 1-cell zygotes (Nothias, Miranda, & DePamphilis, 1996); and activation of enhancers for transcription in 2-cell embryos (Wiekowski, Miranda, & DePamphilis, 1991). ZGA provides the first step in the establishment of totipotency. Currently, the term "totipotency" is used in two related but different levels: i.e., (1) the ability of a single cell to contribute to all cell lineages, including the TE, of an organism (Ishiuchi & Torres-Padilla, 2013); and (2) more stringently, the ability of a single cell to develop into a complete organism (Edwards & Beard, 1997). The zygote is the ultimate totipotent cell, as it is capable, by itself, of developing into a full organism, with all embryonic and extraembryonic lineages. In embryos, totipotency is lost gradually during the first few divisions (Tarkowski, Suwinska, Czolowska, & Ozdzenski, 2010). Blastomeres from the 2-cell-stage embryos fulfill the more stringent definition for totipotency in mice (Papaioannou, Mkandawire, & Biggers, 1989; Tarkowski, 1959; Tsunoda & McLaren, 1983). Some blastomeres from 8-cell-stage embryos contribute to the development of all lineages in chimeric mice (Balakier & Pedersen, 1982; Kelly, 1977; Rossant, 1976) and thus provide evidence for totipotency based on the less stringent definition.

By the 8-cell stage, mouse embryos compact and polarize as tight junctions appear, followed by asymmetric cell division to give rise to inner cells and outer cells—a positional difference that eventually leads to the formation of the first two embryonic lineages, the ICM and TE. Our current understanding of the transcription factor networks and signaling pathway involved in the emergence of totipotency and the first lineage separation will be discussed in this review.

2. ESTABLISHMENT OF TOTIPOTENCY AND PLURIPOTENCY

Unlike pluripotency, the molecular basis of totipotency remains largely unknown. In mice, besides the zygote, 2-cell-stage blastomeres can also generate a whole organism on their own and therefore regarded to be totipotent cells (Tarkowski, 1959). Morphologically, 2-cell embryos are characterized by lacking chromocenters (Probst, Santos, Reik, Almouzni, & Dean, 2007), which are 4,6-diamidino-2-phenylindole stained foci in nuclei. Transcriptionally, 2-cell embryos are characterized by activation of major satellites, MERVL and 2-cell-specific genes such as Eif1a-like genes (which include Gm5662, Gm2022, Gm4027, BB287469, Gm2016, Gm21319, Gm8300, and Gm10264), Zscan4 genes (Zscan4b–Zscan4f), Zfp352, and Tdpoz genes (Tdpoz1–Tdpoz5) (Evsikov et al., 2004). Recently, it is demonstrated that depletion of either the p150 or the p60 subunits of chromatin assembly factor-1 (CAF-1) in embryonic stem (ES) cells led to the formation of 2-cell-like cells with a morphology feature and a transcriptional profile similar to that of 2-cell-stage embryos (Ishiuchi et al., 2015). Because CAF-1 performs the first step of the chromatin assembly process by bringing H3 and H4 to the daughter DNA strands (Smith & Stillman, 1991) and the absence of functional CAF-1 delays nucleosome assembly (Takami, Ono, Fukagawa, Shibahara, & Nakayama, 2007), the authors proposed that inefficiency in chromatin assembly in 2-cell embryos may be a key mechanism in the establishment of totipotency (Ishiuchi et al., 2015). Notably, it is shown that the absence of nuclear receptor subfamily 0, group B, member 1 (Nr0b1, also known as Dax1), an important component of the transcription factor network that governs pluripotency in mouse ES cells, also led to overexpression of 2-cell-embryo-specific transcripts including Zscan4c and prevented normal self-renewal by inducing arrest at G2 phase followed by cell death (Fujii et al., 2015).

As one of the 27 maternal-effect genes that is regarded to be functionally important for ZGA (Li, Zheng, & Dean, 2010), maternal Oct4 (octamer-binding transcription factor 4, encoded by gene *Pou5f1*) is considered to be the genetic "master switch" in igniting the establishment of totipotency (Pesce, Anastassiadis, & Scholer, 1999), and it is presumed to be the most upstream gene in the molecular circuitry of pluripotency (Jaenisch & Young, 2008).

However, several recent studies have found that after genetic removal of maternal Oct4 in oocytes by crossing $Oct4^{flox/flox}/ZP3^{Cre/+}$ female mice with wild-type male mice, $Oct4^{flox/flox}/ZP3^{Cre/+}$ female mice were fully fertile, indicating that Oct4 is not required for initiating totipotency or pluripotency in embryos (Frum et al., 2013; Le Bin et al., 2014; Wu et al., 2013).

Still, Oct4 is at the top of the pluripotency regulatory hierarchy in pluripotent cells (Chen et al., 2008; Rodda et al., 2005). It forms a positive feedback loop (Okumura-Nakanishi, Saito, Niwa, & Ishikawa, 2005) and is essential to maintaining pluripotency (Nichols et al., 1998). Therefore, identifying maternal factors critical in *Oct4* transcriptional activation is key in understanding how embryos regain its potency after fertilization. Recent studies have found that certain maternal factors are involved in the regulation of *Oct4* expression. Cancer-associated factor Tpt1 has been reported to activate the transcription of *Oct4* and *Nanog* in transplanted somatic nuclei in the Xenopus oocyte (Koziol, Garrett, & Gurdon, 2007), but another study failed to confirm this finding in mouse embryos upon knockdown of *Tpt1* by small interfering RNA (siRNA; Wu et al., 2013). Promyelocytic leukemia (Pml) protein was found to be required for *Oct4* gene expression and the maintenance of its open chromatin conformation in stem cells. In proliferating stem cells, Pml-nuclear body, along with the transcription factors TR2, SF1, and Sp1, and the Brg1-dependent chromatin remodeling complex, associates with the *Oct4* promoter to maintain a nucleosome-free region for gene activity (Chuang et al., 2011). Studies in search for master genes in the oocyte have revealed a novel oocyte-specific eukaryotic translation initiation factor 4E (*Eif4eloo*) (Evsikov et al., 2006) and a large number of oocyte-specific genes with yet unknown functions, such as those belonging to the homeodomain transcription factor Obox family (Rajkovic, Yan, Yan, Klysik, & Matzuk, 2002). The maternal transcription factor Sall4 (spalt-like transcription factor 4) binds to the *Oct4* distal enhancer, and it is regarded to be a transcriptional activator of *Oct4* expression based on evidence that reduction in *Sall4* mRNA level in blastocysts at merely 50% knockdown efficiency by *Sall4* siRNA injection into zygotes led to a 70% reduction in *Oct4* expression (Zhang et al., 2006). However, the efficient knockdown of *Sall4* by injecting *Sall4* siRNA into maternal *Oct4*-deficient zygotes—to avoid any possible effect of maternal Oct4 as a positive autoregulator—did not lead to any changes in *Oct4* expression at the blastocyst stage (Wu et al., 2013). Nr5a2 (nuclear receptor subfamily 5, group A, member 2, also known as LRH-1, liver receptor homolog-1) was found to maintain Oct4 expression at the epiblast (EPI) stage of embryonic development by binding to the proximal enhancer and proximal promoter regions of *Oct4*, but to play no evident role in the self-renewal of embryonic stem cells (ESCs; Gu et al., 2005). However, Nr5a2 can induce EPI stem cells into ground-state pluripotency, a basal proliferative state that is free of epigenetic restriction (Guo & Smith, 2010), and replace Oct4 in the reprogramming of somatic cells into pluripotent cells (Heng et al., 2010).

Zscan4 (zinc finger and SCAN domain containing 4), showed to be regulated by PI3K kinase and directly interact with transcriptional repressors the lysine-specific demethylase 1 (LSD1) and C-terminal binding protein-2 (CtBP2) (Storm et al., 2014) and act as an activator of spontaneous telomere sister chromatid exchange (T-SCE) and telomere elongation in mouse ES cells (Zalzman et al., 2010), is activated during ZGA, and its reduction in transcript levels by siRNAs delays the progression from the 2-cell to the 4-cell stage, leading to the formation of blastocysts that fail to implant or proliferate in blastocyst outgrowth culture (Falco et al., 2007). Zscan4 is essential for induction of induced pluripotent stem cells (iPSCs), and its ectopic expression can activate early embryonic genes and improve the efficiency of iPSC generation (Hirata et al., 2012). However, knockdown of *Zscan4* in preimplantation embryos by siRNA against all six isoforms of *Zscan4* (*a–f*) had no impact on *Oct4* expression (Wu et al., 2013). As a component of an active DNA demethylase, *AID* (activation-induced cytidine deaminase) was also shown to be required for reprogramming (Bhutani et al., 2010). A genome-scale RNA interference (RNAi) screen in ESCs identified components of the Paf1 complex with strong effects on *Oct4* expression and showed that Paf1C overexpression blocks the differentiation of ESCs, while Paf1C knockdown causes expression changes in ESCs that are similar to those observed with Oct4 or Nanog depletion (Ding et al., 2009). To this day, it is unclear how *Oct4* is activated in the embryo.

3. THE FIRST LINEAGE SEGREGATION

Upon fertilization, the unicellular totipotent zygote undergoes three equal cell divisions, followed by compaction of 8 cell embryos, along with junction formation and cellular polarization. After another two rounds of division, the embryo cavitates and forms the blastocyst, in which the first embryonic lineages—TE and ICM—can be clearly distinguished. The initiation of these two lineages has been the subject of active investigations for the past 40 years, but it remains controversial. Key transcriptional regulators of these early lineages identified in the mammalian preimplantation embryo include Oct4, Nanog, and Sox2 for the ICM, and Cdx2 (caudal-type homeobox protein 2), Eomes, and Gata3 (Cockburn & Rossant, 2010) and Tead4 (Nishioka et al., 2008; Yagi et al., 2007) for the TE. Several models of lineage segregation have been proposed.

3.1 The Inside–Outside Model

After compaction at the 8-cell stage, cell polarization and asymmetric division are observed. Due to the orientation of the cleavage plane, incubated 8-cell embryos that divided once showed 62% outer/outer divisions (symmetric) and 38% outer/inner divisions (asymmetric). At the fifth cleavage (16- to 32-cell stage), 70% were found to be outer/outer divisions and 30% were outer/inner divisions. Two to four inner cells are allocated at the 16-cell stage and 2–4 inner cells are allocated at the 32-cell stage, giving a total number of 10–12 inner cells after the fifth cleavage division (Pedersen, Wu, & Balakier, 1986). It was proposed that lineage segregation is directed by the position of the cell (Tarkowski & Wroblewska, 1967): outside cells develop into TE and inside cells develop into ICM. According to this model, cells on the inside and cells on the outside are exposed to distinct environments and to different amounts of cell contact, resulting in distinct fates. Apical localization of members of the Par complex (Par3, Par6, and aPKC) in the polarized outer cells may play an important role in promoting TE formation (Alarcon, 2010; Plusa et al., 2005). The maternally derived FILIA–MATER complex is restricted to the apical cytocortex of 2-cell embryos, due to its absence in regions of cell–cell contact, and was later found in the outer cells, but its role in cell fate determination is unclear (Ohsugi, Zheng, Baibakov, Li, & Dean, 2008).

Individual blastomeres devoid of cell–cell contact by continuing separation at each division after 2-cell stage in preimplantation embryos do not show a lineage-specific gene expression pattern, but rather develop a unique pattern that is similar to the TE (Lorthongpanich, Doris, Limviphuvadh, Knowles, & Solter, 2012). It indicates that the correct patterning of lineage-specific gene expression requires positional signals and cell–cell interaction, excluding the presence of a dominant TE-specific determinant.

3.2 Prepattern Model

Unfertilized oocytes from many invertebrates and vertebrates, such as ascidians and amphibians, store maternal materials unevenly in the cytoplasm to form "the animal–vegetal (A–V) axis." The basic body pattern of the embryo is built by placing morphogenetic determinants in the form of protein or mRNA in certain regions of the oocyte cytoplasm. As the embryo divides, the determinants are apportioned to the different cells to specify the cell type (Gilbert, 2006). However, whether this nonmammalian "prepatterning model" system is conserved in mammalian preimplantation development has long been subjected to debate, against the "regulative model," in which the fate of embryonic cells is determined by cellular interactions and the surrounding niche. Numerous studies claim that the axes of the mouse blastocyst are already primed in the egg (Gardner, 1997, 2001; Piotrowska, Wianny, Pedersen, & Zernicka-Goetz, 2001; Zernicka-Goetz, 2005). It was also reported that the descendants of the 2-cell blastomeres tend to distribute in either the embryonic or abembryonic territory (Gardner, 2001; Piotrowska et al., 2001; Plusa et al., 2005), but other reports found no such preference (Alarcon & Marikawa, 2003; Chroscicka, Komorowski, & Maleszewski, 2004; Motosugi, Bauer, Polanski, Solter, & Hiiragi, 2005). Cdx2 is a TE-specific transcription factor that plays a critical role in TE formation, and is proposed to be a determinant of TE formation, providing the long-sought proof for the prepattern model. This evidence came from a report which showed that maternal Cdx2 mRNA become exclusively segregated into one of the 2-cell-stage blastomeres (Deb, Sivaguru, Yong, & Roberts, 2006) and to contribute solely to the TE of the blastocyst, while the other blastomere contributed to the ICM. This finding contradicts not only the observations that go against the prepatterning model (Alarcon & Marikawa, 2003, 2005; Chroscicka et al., 2004; Motosugi et al., 2005), but also those that support the prepatterning model (Gardner, 2001; Piotrowska et al., 2001; Plusa et al., 2005). In any case, this report was subject to investigation shortly after its publication and retracted from the literature due to scientific misconduct and image manipulation. Indeed, upon comparison of transcription profiles of individual twin blastomeres from the same 2-cell embryo, the same research group could not find any determinant candidates that were asymmetrically distributed between the blastomeres in 2-cell embryos (Roberts, Katayama, Magnuson, Falduto, & Torres, 2011). However, the Cdx2 story did not end after the retraction. Some researchers insist that the conclusion might still be valid, and showed that *Cdx2* mRNA becomes polarized at the apical cortex of polarized cells (Jedrusik et al., 2008) and that depletion of maternal and zygotic *Cdx2* mRNA results in delayed embryo development with increased cell cycle length and problems in initiating compaction, therefore failing to develop the TE. The authors therefore supported the maternal Cdx2 as a TE determinant, as this phenotype is more severe than the zygotic knockout of *Cdx2* (Jedrusik et al., 2010). This observation contradicts two studies in which maternal/zygotic *Cdx2*-null embryos were found to reach the blastocyst stage and collapse around the time of implantation, like the zygotic *Cdx2*-null embryos (Blij, Frum, Akyol, Fearon, &

Ralston, 2012; Wu et al., 2010). The discrepancy between these reports could be due to the injection manipulation and selective observation (Wu & Scholer, 2011).

To date, "determinants" in the oocyte which play a key role in lineage differentiation and body planning, comparable to the nonmammalian system, have not been found.

3.3 Stochastic Model

It was reported that, during the onset of the first lineage separation, the expression of Nanog, Oct4, and Cdx2 was stochastic at the protein level among the blastomeres of the mouse embryo, followed by a phase of positional change (sorting) depending on the global differences in gene expression (Dietrich & Hiiragi, 2007). Using time-lapse video, it was shown that the blastomeres move extensively at each cleavage stage (Kurotaki, Hatta, Nakao, Nabeshima, & Fujimori, 2007), supporting the model of cell sorting and consistent with the highly regulative capacity of the embryo. Such sorting model was also found in the second lineage segregation. Initially, the EPI and primitive endoderm-specific transcription factors Nanog and Gata6, respectively, are expressed in a random "salt and pepper" pattern in the ICM, followed by segregation into the appropriate cell lineages (Chazaud, Yamanaka, Pawson, & Rossant, 2006). This model fits well with the new "Seesaw" model, proposed by Shu et al. (2013), which stipulated that the pluripotent state has a precarious balancing equilibrium that results from continuous mutual competition between rival lineage specification forces. That model integrates two coupled modules—the canonical pluripotency module and the lineage-antagonism module. The former module is represented by the mutual activation of Oct4 and Sox2, whereas the latter module is represented by the mutual inhibition and balance of the mesoendoderm (ME) and ectoderm (ECT) genes. The self-activating pluripotency module gets activated when all of the lineage-specifying forces are counteracted at the dynamic balance point of the "seesaw," then Sox2 and Oct4 block the ME and ECT lineage fates, respectively. As a result, the pluripotent state is maintained.

How important is the compaction to the first lineage specification? The compaction process initiates cell positioning, cell polarity, and tight junction formation and is probably an essential step in TE lineage commitment. However, interruption of compaction by incubating embryos with ECCD-1 antibody did not prevent TE formation as indicated by *Cdx2* activation (Wu et al., 2010). Moreover, we observed dynamic movement of blastomeres of the 8-cell embryo. When each of the 144 blastomeres from 18 8-cell embryos was placed at the outside of a compacted 8-cell embryos, only 54 remained outside and became TE (symmetric division, 37.5%), while the remaining 90 became both TE and ICM (asymmetric division, 62.5%; unpublished data; Fig. 1). This observation suggests that we might overestimate the role of events at around the 8-cell compaction stage and that ignore the importance of cell movement during the first lineage specification. Indeed, developmental bias in cleavage-stage mouse blastomeres was reported using newly developed Rainbow transgenic mice (Tabansky et al., 2013), and Oct4 kinetics in 4- and 8-cell-stage embryos was found to predict cell lineage (Plachta, Bollenbach, Pease, Fraser, & Pantazis, 2011), indicating that cell lineage specification may start before the compaction process and that the TE/ICM lineage segregation may be not totally stochastic.

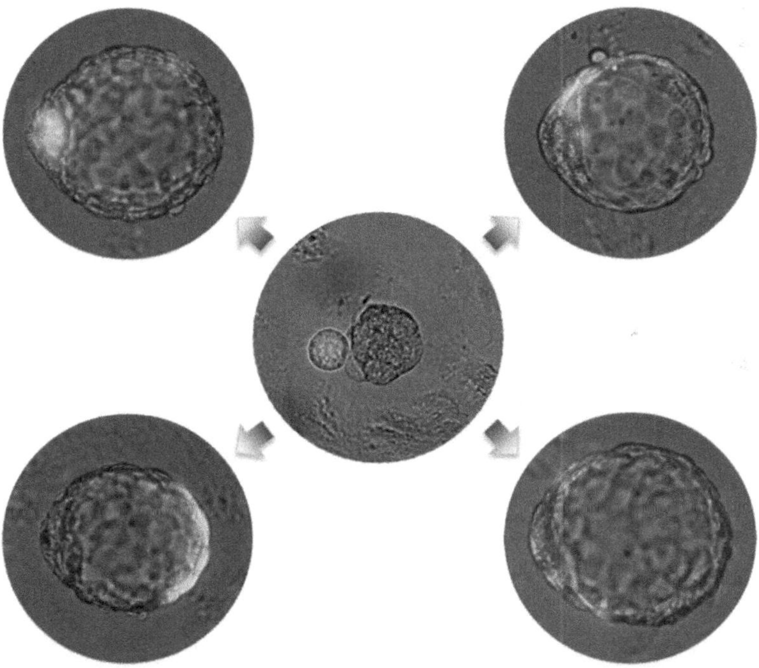

Figure 1 When a blastomere of 8-cell embryo from GOF18-GFP transgenic mouse was placed at the outside of a compacted 8-cell embryos, it could become mural or polar TE (symmetric division), and more likely, it could be differentiated into both TE and ICM (asymmetric division), demonstrating that the initial position of blastomeres in a 8-cell embryo is not the cause of its fate specification, but rather, it is the result of the blastomere with different migration property. Green: GFP. (A color version of this figure is located at the end of this chapter.)

3.4 Reciprocal Interaction of Oct4/Cdx2 Model

The establishment of the ICM and TE lineages in mice begins with the upregulation of *Cdx2* in the outside cells followed by downregulation of Oct4, Sox2, and Nanog in the same cells. Initially, Cdx2 and Oct4 are coexpressed in the morula; the reciprocal repression occurs in blastocysts to determine the first lineage specification (Niwa et al., 2005). However, depletion of maternal and zygotic *Cdx2* mRNA was reported to have no effect on the initial specification of the TE (Blij et al., 2012; Wu et al., 2010). On the other hand, after elimination of maternal and zygotic Oct4 expression (Wu et al., 2013), Oct4-null embryos were still found to cavitate and form a distinct ICM, which did not express Cdx2 but expressed Nanog. Later on, Nanog and Cdx2 are colocalized in some EPI nuclei. Thus, the reciprocal Oct4/Cdx2 interaction does not initiate the specification of the first lineage differentiation, but rather maintains the ICM/TE lineages after segregation.

3.5 Hippo Signaling Pathway Regulates TE Lineage Specification

Tead4 (TEA domain family member 4) is ubiquitously expressed from the 2-cell stage onward in the preimplantation embryo and has been proposed to act upstream of Cdx2 to regulate TE formation (Nishioka et al., 2008; Yagi et al., 2007). *Tead4* mutant embryos exhibit defects in the specification and development of the TE and consequently do not form a blastocoel. Tead4 plays a crucial role in the initiation of TE formation—expression of Oct4 up to the blastocyst stage could suppress *Cdx2* expression and thus TE initiation, until Tead4 starts to promote *Cdx2* expression by overcoming this negative input (Nishioka et al., 2008; Yagi et al., 2007). Moreover, *Gata3* expression becomes restricted to TE cells, and this expression is also dependent on Tead4, suggesting that Gata3 and Cdx2 act in parallel pathways downstream of Tead4 and function to activate target genes within the TE lineage (Ralston et al., 2010).

So how does Tead4 only activate Cdx2 in the outer cells? Recently, the Hippo signaling pathway—a major regulator of cell growth, proliferation, and apoptosis, and that is conserved from Drosophila to mammals—has been described in the TE/ICM lineage specification in the mouse embryo (Nishioka et al., 2009). This signaling pathway regulates activity of Tead4 through its coactivator Yap (Yes-associated protein)/Taz (transcriptional coactivator with PDZ-binding motif). Cell density signals from cell contact and/or position may activate/derepress the Hippo signaling, resulting in Yap phosphorylation by the Ser/Thr kinases Lats1/2 (large tumor suppressor homologue 1/2) and subsequently nuclear exclusion in the inner cells, leading to inactive Tead4 in the inner cells, which are fated to form ICM cells (Cockburn & Rossant, 2010). In the outside cells, the Hippo pathway is switched off, and unphosphorylated Yap is shuttled into the nuclei where it can, in cooperation with Tead4, induce *Cdx2* expression, leading to acquisition of the TE fate (Fig. 2).

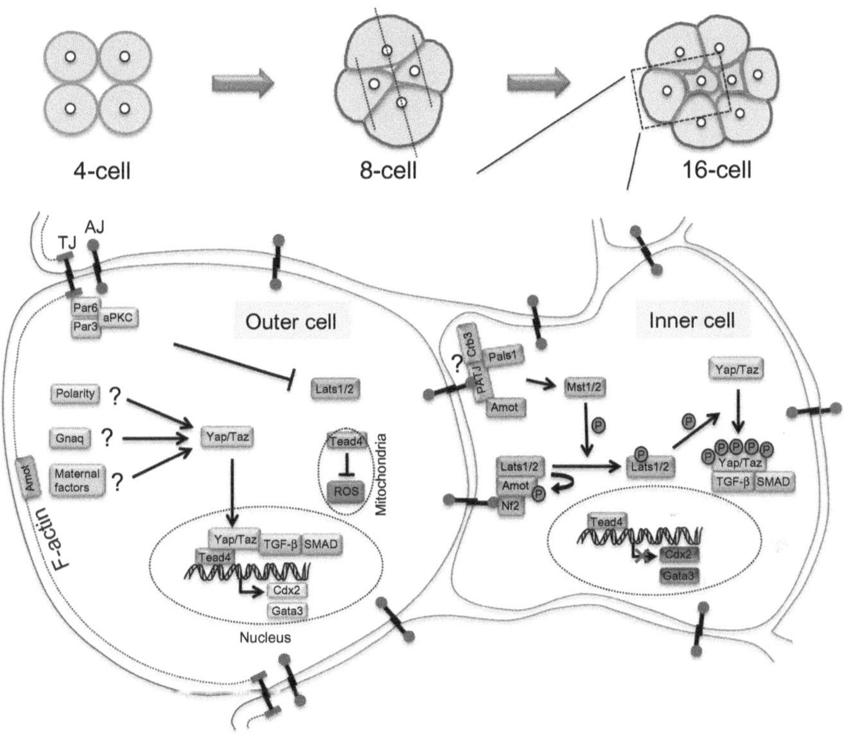

Figure 2 A model of TE/ICM fate specification in the early embryo involves the Hippo pathway. Mouse embryo compacts at the 8-cell stage. At division to the 16-cell stage, dependent on division planes (dashed lines), some cells divide conservatively to generate two outer polar cells at the 16-cell stage, whereas others divide to give one outer polar and one inner nonpolar cell. In the outer cells, Yap and Taz translocate into the nucleus, bind Tead4, and then transcriptionally activate TE genes, such as *Cdx2* and *Gata3*, to specify TE fate. In the inner cells, the Hippo pathway is active, and Mst kinases phosphorylate Lats proteins, which then phosphorylate Yap/Taz. Phosphorylated Yap and Taz are excluded from the nucleus and accumulate in the cytoplasm, where they are degraded by the proteasome, and these cells are specified as ICM. TJ, tight junction; AJ, adherens junction; ROS, reactive oxygen species. (A color version of this figure is located at the end of this chapter.)

Embryos lacking *Tead4* fail to develop the TE and do not maintain *Cdx2* expression (Nishioka et al., 2008; Yagi et al., 2007). Correspondingly, overexpression of Lats2 kinase (an activator of Hippo) reduces expression of Cdx2 in the outer cells, whereas embryos lacking *Lats1/2* express Cdx2 in the inner cells (Lorthongpanich et al., 2013; Nishioka et al., 2009). Similar results are obtained when other components of the pathway, such as *Nf2* and *Amot*, are disrupted (Cockburn, Biechele, Garner, & Rossant, 2013; Hirate et al., 2013). How these components are integrated to fully define lineage restriction in the blastocyst is largely unknown, and additional inputs remain to be identified (Wennekamp, Mesecke, Nedelec, & Hiiragi, 2013). For example, the Crumbs proteins located at tight junctions can activate the Hippo pathway depending on the cell density (Varelas et al., 2010), but in embryos, tight junctions are only formed in the outside cells with the inactivated Hippo pathway. It is reported that it is the cell polarity, rather than cell position, that is linked to the Hippo pathway activity. One study found that outer apolar cells were eventually internalized by the surrounding polar cells. The apolar cells activated the Hippo signaling prior to assuming their inner positions (Anani, Bhat, Honma-Yamanaka, Krawchuk, & Yamanaka, 2014).

However, the Hippo pathway may not be the only pathway involved in the first lineage segregation in the mouse embryo. It is reported that Notch is active specifically in the outer cells of the blastocyst. Analysis of double mutants for *Tead4* and the Notch effector *Rbpj* shows that the Notch and Hippo pathways converge on *Cdx2* to activate its expression in the TE, and that forced activation of Notch directs blastomeres to the TE (Rayon et al., 2014). It was also reported that Tead4 actually is not essential for specification, maintenance, nor differentiation of TE but it plays crucial role in maintaining energy homeostasis during TE specification (Kaneko & DePamphilis, 2013), suggesting the involvement of other Yap-interacting transcription factors, or even other pathways critical in TE specification. Culturing mouse embryos with BMP4 blocks their development at the compaction stage and results in the unusual colocalization of Cdx2 and Tead4 in the nuclei of the inner cells (Home et al., 2012, Nishioka et al., 2009). Moreover, YAP can be activated by GNAQ (guanine nucleotide binding protein, alpha q polypeptide) in uveal melanoma cells through a Trio-regulated Rho GTPase signaling circuitry independent of the Hippo pathway (Feng et al., 2014).

The last few years have seen quite some exciting breakthroughs in the study of lineage specification in the early mouse embryo. Nevertheless, many questions still remain unanswered. To date, the upstream "master" signals that reside on top of the Hippo pathway to switch it off in the outer cells and on in the inner cells have not been identified. It would be interesting to see how the upstream signals relate to cell position, cell junctions, cell polarity, and maternal factors, particularly those distributed in the cortex of the oocyte. The finding of the Hippo pathway brings us one step closer to ending the long-lasting debate over the first lineage specification and to integrating the different models to solve this fundamental question in developmental biology.

REFERENCES

Alarcon, V. B. (2010). Cell polarity regulator PARD6B is essential for trophectoderm formation in the preimplantation mouse embryo. *Biology of Reproduction*, *83*, 347–358.

Alarcon, V. B., & Marikawa, Y. (2003). Deviation of the blastocyst axis from the first cleavage plane does not affect the quality of mouse postimplantation development. *Biology of Reproduction*, *69*, 1208–1212.

Alarcon, V. B., & Marikawa, Y. (2005). Unbiased contribution of the first two blastomeres to mouse blastocyst development. *Molecular Reproduction and Development*, *72*, 354–361.

Anani, S., Bhat, S., Honma-Yamanaka, N., Krawchuk, D., & Yamanaka, Y. (2014). Initiation of Hippo signaling is linked to polarity rather than to cell position in the pre-implantation mouse embryo. *Development*, *141*, 2813–2824.

Bachvarova, R., & De Leon, V. (1980). Polyadenylated RNA of mouse ova and loss of maternal RNA in early development. *Developmental Biology*, *74*, 1–8.

Balakier, H., & Pedersen, R. A. (1982). Allocation of cells to inner cell mass and trophectoderm lineages in preimplantation mouse embryos. *Developmental Biology*, *90*, 352–362.

Bhutani, N., Brady, J. J., Damian, M., Sacco, A., Corbel, S. Y., & Blau, H. M. (2010). Reprogramming towards pluripotency requires AID-dependent DNA demethylation. *Nature*, *463*, 1042–1047.

Blij, S., Frum, T., Akyol, A., Fearon, E., & Ralston, A. (2012). Maternal Cdx2 is dispensable for mouse development. *Development*, *139*, 3969–3972.

Chazaud, C., Yamanaka, Y., Pawson, T., & Rossant, J. (2006). Early lineage segregation between epiblast and primitive endoderm in mouse blastocysts through the Grb2-MAPK pathway. *Developmental Cell*, *10*, 615–624.

Chen, X., Xu, H., Yuan, P., Fang, F., Huss, M., Vega, V. B., et al. (2008). Integration of external signaling pathways with the core transcriptional network in embryonic stem cells. *Cell*, *133*, 1106–1117.

Chroscicka, A., Komorowski, S., & Maleszewski, M. (2004). Both blastomeres of the mouse 2-cell embryo contribute to the embryonic portion of the blastocyst. *Molecular Reproduction and Development*, *68*, 308–312.

Chuang, Y. S., Huang, W. H., Park, S. W., Persaud, S. D., Hung, C. H., Ho, P. C., et al. (2011). Promyelocytic leukemia protein in retinoic acid-induced chromatin remodeling of Oct4 gene promoter. *Stem Cells*, *29*, 660–669.

Cockburn, K., Biechele, S., Garner, J., & Rossant, J. (2013). The Hippo pathway member Nf2 is required for inner cell mass specification. *Current Biology*, *23*, 1195–1201.

Cockburn, K., & Rossant, J. (2010). Making the blastocyst: Lessons from the mouse. *The Journal of Clinical Investigation*, *120*, 995–1003.

Davis, W., Jr., & Schultz, R. M. (2000). Developmental change in TATA-box utilization during preimplantation mouse development. *Developmental Biology*, *218*, 275–283.

Deb, K., Sivaguru, M., Yong, H. Y., & Roberts, R. M. (2006). Cdx2 gene expression and trophectoderm lineage specification in mouse embryos. *Science*, *311*, 992–996.

Dietrich, J. E., & Hiiragi, T. (2007). Stochastic patterning in the mouse pre-implantation embryo. *Development*, *134*, 4219–4231.

Ding, L., Paszkowski-Rogacz, M., Nitzsche, A., Slabicki, M. M., Heninger, A. K., de Vries, I., et al. (2009). A genome-scale RNAi screen for Oct4 modulators defines a role of the Paf1 complex for embryonic stem cell identity. *Cell Stem Cell*, *4*, 403–415.

Edwards, R. G., & Beard, H. K. (1997). Oocyte polarity and cell determination in early mammalian embryos. *Molecular Human Reproduction*, *3*, 863–905.

Evsikov, A. V., de Vries, W. N., Peaston, A. E., Radford, E. E., Fancher, K. S., Chen, F. H., et al. (2004). Systems biology of the 2-cell mouse embryo. *Cytogenetic and Genome Research*, *105*, 240–250.

Evsikov, A. V., Graber, J. H., Brockman, J. M., Hampl, A., Holbrook, A. E., Singh, P., et al. (2006). Cracking the egg: Molecular dynamics and evolutionary aspects of the transition from the fully grown oocyte to embryo. *Genes & Development*, *20*, 2713–2727.

Falco, G., Lee, S. L., Stanghellini, I., Bassey, U. C., Hamatani, T., & Ko, M. S. (2007). Zscan4: A novel gene expressed exclusively in late 2-cell embryos and embryonic stem cells. *Developmental Biology*, *307*, 539–550.

Feng, X., Degese, M. S., Iglesias-Bartolome, R., Vaque, J. P., Molinolo, A. A., Rodrigues, M., et al. (2014). Hippo-independent activation of YAP by the GNAQ uveal melanoma oncogene through a trio-regulated rho GTPase signaling circuitry. *Cancer Cell*, *25*, 831–845.

Frum, T., Halbisen, M. A., Wang, C., Amiri, H., Robson, P., & Ralston, A. (2013). Oct4 cell-autonomously promotes primitive endoderm development in the mouse blastocyst. *Developmental Cell*, *25*, 610–622.

Fujii, S., Nishikawa-Torikai, S., Futatsugi, Y., Toyooka, Y., Yamane, M., Ohtsuka, S., et al. (2015). Nr0b1 is a negative regulator of Zscan4c in mouse embryonic stem cells. *Scientific Reports*, *5*, 9146.

Gardner, R. L. (1997). The early blastocyst is bilaterally symmetrical and its axis of symmetry is aligned with the animal-vegetal axis of the zygote in the mouse. *Development*, *124*, 289–301.

Gardner, R. L. (2001). Specification of embryonic axes begins before cleavage in normal mouse development. *Development*, *128*, 839–847.

Gilbert, S. F. (2006). *Developmental biology*. Sunderland, MA: Sinauer Associates Inc.

Gu, P., Goodwin, B., Chung, A. C., Xu, X., Wheeler, D. A., Price, R. R., et al. (2005). Orphan nuclear receptor LRH-1 is required to maintain Oct4 expression at the epiblast stage of embryonic development. *Molecular and Cellular Biology*, *25*, 3492–3505.

Guo, G., & Smith, A. (2010). A genome-wide screen in EpiSCs identifies Nr5a nuclear receptors as potent inducers of ground state pluripotency. *Development*, *137*, 3185–3192.

Heng, J. C., Feng, B., Han, J., Jiang, J., Kraus, P., Ng, J. H., et al. (2010). The nuclear receptor Nr5a2 can replace Oct4 in the reprogramming of murine somatic cells to pluripotent cells. *Cell Stem Cell*, *6*, 167–174.

Hirata, T., Amano, T., Nakatake, Y., Amano, M., Piao, Y., Hoang, H. G., et al. (2012). Zscan4 transiently reactivates early embryonic genes during the generation of induced pluripotent stem cells. *Scientific Reports*, *2*, 208.

Hirate, Y., Hirahara, S., Inoue, K., Suzuki, A., Alarcon, V. B., Akimoto, K., et al. (2013). Polarity-dependent distribution of angiomotin localizes Hippo signaling in preimplantation embryos. *Current Biology*, *23*, 1181–1194.

Home, P., Saha, B., Ray, S., Dutta, D., Gunewardena, S., Yoo, B., et al. (2012). Altered subcellular localization of transcription factor TEAD4 regulates first mammalian cell lineage commitment. *Proceedings of the National Academy of Sciences of the United States of America, 109*, 7362–7367.

Ishiuchi, T., Enriquez-Gasca, R., Mizutani, E., Boskovic, A., Ziegler-Birling, C., Rodriguez-Terrones, D., et al. (2015). Early embryonic-like cells are induced by downregulating replication-dependent chromatin assembly. *Nature Structural & Molecular Biology, 22*, 662–671.

Ishiuchi, T., & Torres-Padilla, M. E. (2013). Towards an understanding of the regulatory mechanisms of totipotency. *Current Opinion in Genetics & Development, 23*, 512–518.

Jaenisch, R., & Young, R. (2008). Stem cells, the molecular circuitry of pluripotency and nuclear reprogramming. *Cell, 132*, 567–582.

Jedrusik, A., Bruce, A. W., Tan, M. H., Leong, D. E., Skamagki, M., Yao, M., et al. (2010). Maternally and zygotically provided Cdx2 have novel and critical roles for early development of the mouse embryo. *Developmental Biology, 344*, 66–78.

Jedrusik, A., Parfitt, D. E., Guo, G., Skamagki, M., Grabarek, J. B., Johnson, M. H., et al. (2008). Role of Cdx2 and cell polarity in cell allocation and specification of trophectoderm and inner cell mass in the mouse embryo. *Genes & Development, 22*, 2692–2706.

Kaneko, K. J., & DePamphilis, M. L. (2013). TEAD4 establishes the energy homeostasis essential for blastocoel formation. *Development, 140*, 3680–3690.

Kelly, S. J. (1977). Studies of the developmental potential of 4- and 8-cell stage mouse blastomeres. *The Journal of Experimental Zoology, 200*, 365–376.

Koziol, M. J., Garrett, N., & Gurdon, J. B. (2007). Tpt1 activates transcription of oct4 and nanog in transplanted somatic nuclei. *Current Biology, 17*, 801–807.

Kurotaki, Y., Hatta, K., Nakao, K., Nabeshima, Y., & Fujimori, T. (2007). Blastocyst axis is specified independently of early cell lineage but aligns with the ZP shape. *Science, 316*, 719–723.

Le Bin, G. C., Muñoz-Descalzo, S., Kurowski, A., Leitch, H., Lou, X., Mansfield, W., et al. (2014). Oct4 is required for lineage priming in the developing inner cell mass of the mouse blastocyst. *Development, 141*, 1001–1010.

Li, L., Zheng, P., & Dean, J. (2010). Maternal control of early mouse development. *Development, 137*, 859–870.

Lorthongpanich, C., Doris, T. P., Limviphuvadh, V., Knowles, B. B., & Solter, D. (2012). Developmental fate and lineage commitment of singled mouse blastomeres. *Development, 139*, 3722–3731.

Lorthongpanich, C., Messerschmidt, D. M., Chan, S. W., Hong, W., Knowles, B. B., & Solter, D. (2013). Temporal reduction of LATS kinases in the early preimplantation embryo prevents ICM lineage differentiation. *Genes & Development, 27*, 1441–1446.

Macfarlan, T. S., Gifford, W. D., Driscoll, S., Lettieri, K., Rowe, H. M., Bonanomi, D., et al. (2012). Embryonic stem cell potency fluctuates with endogenous retrovirus activity. *Nature, 487*, 57–63.

Motosugi, N., Bauer, T., Polanski, Z., Solter, D., & Hiiragi, T. (2005). Polarity of the mouse embryo is established at blastocyst and is not prepatterned. *Genes & Development, 19*, 1081–1092.

Nichols, J., Zevnik, B., Anastassiadis, K., Niwa, H., Klewe-Nebenius, D., Chambers, I., et al. (1998). Formation of pluripotent stem cells in the mammalian embryo depends on the POU transcription factor Oct4. *Cell, 95*, 379–391.

Nishioka, N., Inoue, K., Adachi, K., Kiyonari, H., Ota, M., Ralston, A., et al. (2009). The Hippo signaling pathway components Lats and Yap pattern Tead4 activity to distinguish mouse trophectoderm from inner cell mass. *Developmental Cell, 16*, 398–410.

Nishioka, N., Yamamoto, S., Kiyonari, H., Sato, H., Sawada, A., Ota, M., et al. (2008). Tead4 is required for specification of trophectoderm in pre-implantation mouse embryos. *Mechanisms of Development, 125*, 270–283.

Niwa, H., Toyooka, Y., Shimosato, D., Strumpf, D., Takahashi, K., Yagi, R., et al. (2005). Interaction between Oct3/4 and Cdx2 determines trophectoderm differentiation. *Cell, 123*, 917–929.

Nothias, J. Y., Miranda, M., & DePamphilis, M. L. (1996). Uncoupling of transcription and translation during zygotic gene activation in the mouse. *The EMBO Journal, 15*, 5715–5725.

Ohsugi, M., Zheng, P., Baibakov, B., Li, L., & Dean, J. (2008). Maternally derived FILIA-MATER complex localizes asymmetrically in cleavage-stage mouse embryos. *Development, 135*, 259–269.

Okumura-Nakanishi, S., Saito, M., Niwa, H., & Ishikawa, F. (2005). Oct-3/4 and Sox2 regulate Oct-3/4 gene in embryonic stem cells. *The Journal of Biological Chemistry, 280*, 5307–5317.

Papaioannou, V. E., Mkandawire, J., & Biggers, J. D. (1989). Development and phenotypic variability of genetically identical half mouse embryos. *Development, 106*, 817–827.

Peaston, A. E., Evsikov, A. V., Graber, J. H., de Vries, W. N., Holbrook, A. E., Solter, D., et al. (2004). Retrotransposons regulate host genes in mouse oocytes and preimplantation embryos. *Developmental Cell, 7*, 597–606.

Pedersen, R. A., Wu, K., & Balakier, H. (1986). Origin of the inner cell mass in mouse embryos: Cell lineage analysis by microinjection. *Developmental Biology, 117*, 581–595.

Pesce, M., Anastassiadis, K., & Scholer, H. R. (1999). Oct-4: Lessons of totipotency from embryonic stem cells. *Cells, Tissues, Organs, 165*, 144–152.

Piotrowska, K., Wianny, F., Pedersen, R. A., & Zernicka-Goetz, M. (2001). Blastomeres arising from the first cleavage division have distinguishable fates in normal mouse development. *Development, 128*, 3739–3748.

Plachta, N., Bollenbach, T., Pease, S., Fraser, S. E., & Pantazis, P. (2011). Oct4 kinetics predict cell lineage patterning in the early mammalian embryo. *Nature Cell Biology, 13*, 117–123.

Plusa, B., Frankenberg, S., Chalmers, A., Hadjantonakis, A. K., Moore, C. A., Papalopulu, N., et al. (2005). Downregulation of Par3 and aPKC function directs cells towards the ICM in the preimplantation mouse embryo. *Journal of Cell Science, 118*, 505–515.

Probst, A. V., Santos, F., Reik, W., Almouzni, G., & Dean, W. (2007). Structural differences in centromeric heterochromatin are spatially reconciled on fertilisation in the mouse zygote. *Chromosoma, 116*, 403–415.

Rajkovic, A., Yan, C., Yan, W., Klysik, M., & Matzuk, M. M. (2002). Obox, a family of homeobox genes preferentially expressed in germ cells. *Genomics, 79*, 711–717.

Ralston, A., Cox, B. J., Nishioka, N., Sasaki, H., Chea, E., Rugg-Gunn, P., et al. (2010). Gata3 regulates trophoblast development downstream of Tead4 and in parallel to Cdx2. *Development, 137*, 395–403.

Rayon, T., Menchero, S., Nieto, A., Xenopoulos, P., Crespo, M., Cockburn, K., et al. (2014). Notch and hippo converge on Cdx2 to specify the trophectoderm lineage in the mouse blastocyst. *Developmental Cell, 30*, 410–422.

Roberts, R. M., Katayama, M., Magnuson, S. R., Falduto, M. T., & Torres, K. E. (2011). Transcript profiling of individual twin blastomeres derived by splitting two-cell stage murine embryos. *Biology of Reproduction, 84*, 487–494.

Rodda, D. J., Chew, J. L., Lim, L. H., Loh, Y. H., Wang, B., Ng, H. H., et al. (2005). Transcriptional regulation of nanog by OCT4 and SOX2. *The Journal of Biological Chemistry, 280*, 24731–24737.

Rossant, J. (1976). Postimplantation development of blastomeres isolated from 4- and 8-cell mouse eggs. *Journal of Embryology and Experimental Morphology, 36*, 283–290.

Schultz, R. M. (1993). Regulation of zygotic gene activation in the mouse. *Bioessays, 15*, 531–538.

Shu, J., Wu, C., Wu, Y., Li, Z., Shao, S., Zhao, W., et al. (2013). Induction of pluripotency in mouse somatic cells with lineage specifiers. *Cell, 153*, 963–975.

Smith, S., & Stillman, B. (1991). Stepwise assembly of chromatin during DNA replication in vitro. *The EMBO Journal, 10*, 971–980.

Storm, M. P., Kumpfmueller, B., Bone, H. K., Buchholz, M., Sanchez Ripoll, Y., Chaudhuri, J. B., et al. (2014). Zscan4 is regulated by PI3-kinase and DNA-damaging agents and directly interacts with the transcriptional repressors LSD1 and CtBP2 in mouse embryonic stem cells. *PLoS One, 9*, e89821.

Tabansky, I., Lenarcic, A., Draft, R. W., Loulier, K., Keskin, D. B., Rosains, J., et al. (2013). Developmental bias in cleavage-stage mouse blastomeres. *Current Biology, 23*, 21–31.

Takami, Y., Ono, T., Fukagawa, T., Shibahara, K., & Nakayama, T. (2007). Essential role of chromatin assembly factor-1-mediated rapid nucleosome assembly for DNA replication and cell division in vertebrate cells. *Molecular Biology of the Cell, 18*, 129–141.

Tarkowski, A. K. (1959). Experiments on the development of isolated blastomeres of mouse eggs. *Nature, 184*, 1286–1287.

Tarkowski, A. K., Suwinska, A., Czolowska, R., & Ozdzenski, W. (2010). Individual blastomeres of 16- and 32-cell mouse embryos are able to develop into foetuses and mice. *Developmental Biology, 348*, 190–198.

Tarkowski, A. K., & Wroblewska, J. (1967). Development of blastomeres of mouse eggs isolated at the 4- and 8-cell stage. *Journal of Embryology and Experimental Morphology, 18*, 155–180.

Tsunoda, Y., & McLaren, A. (1983). Effect of various procedures on the viability of mouse embryos containing half the normal number of blastomeres. *Journal of Reproduction and Fertility, 69*, 315–322.

Varelas, X., Samavarchi-Tehrani, P., Narimatsu, M., Weiss, A., Cockburn, K., Larsen, B. G., et al. (2010). The Crumbs complex couples cell density sensing to Hippo-dependent control of the TGF-beta-SMAD pathway. *Developmental Cell, 19*, 831–844.

Wennekamp, S., Mesecke, S., Nedelec, F., & Hiiragi, T. (2013). A self-organization framework for symmetry breaking in the mammalian embryo. *Nature Reviews. Molecular Cell Biology, 14*, 452–459.

Wiekowski, M., Miranda, M., & DePamphilis, M. L. (1991). Regulation of gene expression in preimplantation mouse embryos: Effects of the zygotic clock and the first mitosis on promoter and enhancer activities. *Developmental Biology, 147*, 403–414.

Wu, G., Gentile, L., Fuchikami, T., Sutter, J., Psathaki, K., Esteves, T. C., et al. (2010). Initiation of trophectoderm lineage specification in mouse embryos is independent of Cdx2. *Development, 137*, 4159–4169.

Wu, G., Han, D., Gong, Y., Sebastiano, V., Gentile, L., Singhal, N., et al. (2013). Establishment of totipotency does not depend on Oct4A. *Nature Cell Biology, 15*, 1089–1097.

Wu, G., & Scholer, H. R. (2011). Role of mouse maternal Cdx2: What's the debate all about? *Reproductive Biomedicine Online, 22*, 516–518. discussion 519–520.

Yagi, R., Kohn, M. J., Karavanova, I., Kaneko, K. J., Vullhorst, D., DePamphilis, M. L., et al. (2007). Transcription factor TEAD4 specifies the trophectoderm lineage at the beginning of mammalian development. *Development, 134*, 3827–3836.

Zalzman, M., Falco, G., Sharova, L. V., Nishiyama, A., Thomas, M., Lee, S. L., et al. (2010). Zscan4 regulates telomere elongation and genomic stability in ES cells. *Nature, 464*, 858–863.

Zernicka-Goetz, M. (2005). Cleavage pattern and emerging asymmetry of the mouse embryo. *Nature Reviews. Molecular Cell Biology, 6*, 919–928.

Zhang, J., Tam, W. L., Tong, G. Q., Wu, Q., Chan, H. Y., Soh, B. S., et al. (2006). Sall4 modulates embryonic stem cell pluripotency and early embryonic development by the transcriptional regulation of Pou5f1. *Nature Cell Biology, 8*, 1114–1123.

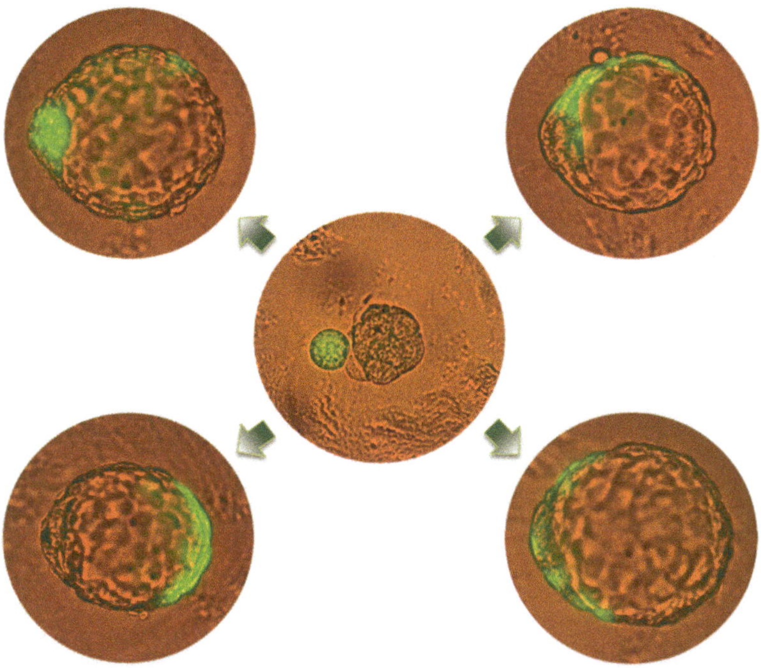

Guangming Wu and Hans R. Schöler, Figure 1 When a blastomere of 8-cell embryo from GOF18-GFP transgenic mouse was placed at the outside of a compacted 8-cell embryos, it could become mural or polar TE (symmetric division), and more likely, it could be differentiated into both TE and ICM (asymmetric division), demonstrating that the initial position of blastomeres in a 8-cell embryo is not the cause of its fate specification, but rather, it is the result of the blastomere with different migration property. Green: GFP.

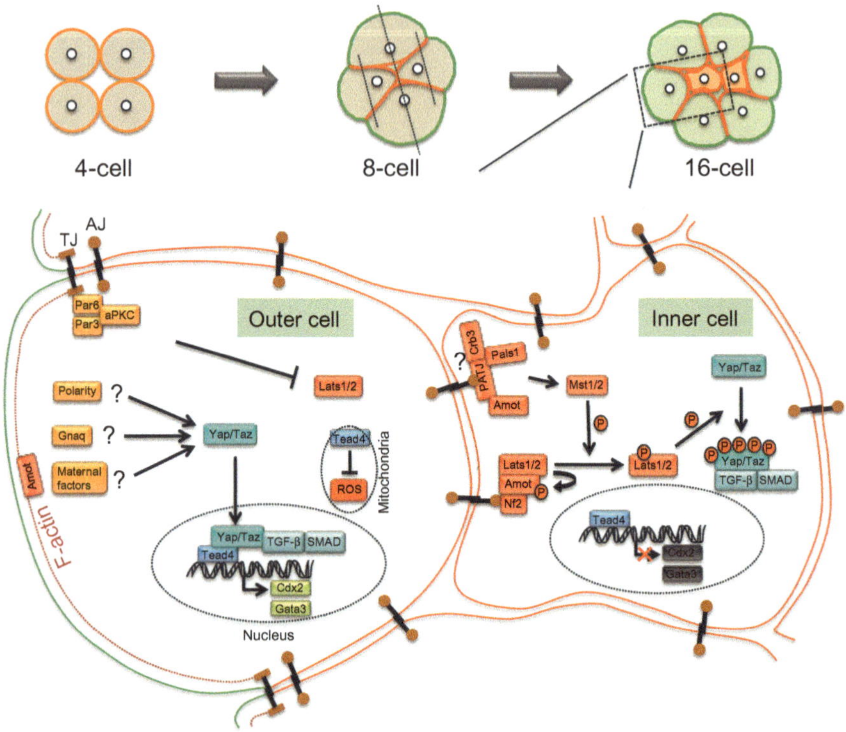

Guangming Wu and Hans R. Schöler, Figure 2 A model of TE/ICM fate specification in the early embryo involves the Hippo pathway. Mouse embryo compacts at the 8-cell stage. At division to the 16-cell stage, dependent on division planes (dashed lines), some cells divide conservatively to generate two outer polar cells at the 16-cell stage, whereas others divide to give one outer polar and one inner nonpolar cell. In the outer cells, Yap and Taz translocate into the nucleus, bind Tead4, and then transcriptionally activate TE genes, such as *Cdx2* and *Gata3*, to specify TE fate. In the inner cells, the Hippo pathway is active, and Mst kinases phosphorylate Lats proteins, which then phosphorylate Yap/Taz. Phosphorylated Yap and Taz are excluded from the nucleus and accumulate in the cytoplasm, where they are degraded by the proteasome, and these cells are specified as ICM. TJ, tight junction; AJ, adherens junction; ROS, reactive oxygen species.

CHAPTER EIGHTEEN

The Complexities and Unexpected Insights of Developmental Genetic Analysis

Trudi Schupbach[1]
Department of Molecular Biology, Princeton University, Princeton, New Jersey, USA
[1]Corresponding author: e-mail address: schupbac@princeton.edu

Abstract

The study of development involves many important techniques. Here I am trying to reflect on the strength of genetic analysis and its ability to uncover unexpected relationships and regulatory inputs from seemingly unrelated pathways.

As we have concentrated on studying the EGF receptor pathway in *Drosophila* oogenesis for many years, this essay allows me to look back and marvel at the wealth of information we have gained on this model pathway, and how much there is still to explore. What became clear over this time period was the understanding that analyzing a single developmental pathway as a linear structure in isolation does not provide an adequate picture. Investigations into such pathways that initially seem linear invariably will lead to an exploration of other processes and pathways that provide input or that may interact in various ways with core factors in the pathway. Such interactors may differ between cell types and developmental stages and their discovery therefore depends on the specific tissue and cell type that is studied. This reflects the reality of development where cells have to make choices in their differentiation pathways based on a variety of signals and intrinsic factors. One might therefore come to the conclusion that genetic analysis focused initially on single genes, using maximally double and triple mutants, will not be able to dissect a developmental decision in its entirety. In recent years, this has led to the notion that only genomic

Current Topics in Developmental Biology, Volume 117
ISSN 0070-2153
http://dx.doi.org/10.1016/bs.ctdb.2015.10.015

approaches will be able to give us true pictures of biological phenomena. On the other hand, looking at networks that are generated by such genomic techniques, such as mapping protein–protein interactions, or measuring transcriptome responses to environmental changes, one is easily overwhelmed by the wealth of potential interactions suggested by these experiments and finds it difficult to devise approaches that will efficiently tease out major versus minor interactions. For this reason, genetic analysis starting with mutations that have a clear phenotypic effect on a developmental process is still a very valuable path to take. Genetics can often indicate the core members of a pathway, which are usually the ones giving strong and reliable phenotypes, and then over time and with further analysis help to reveal other regulators and modulators that fine tune the outcome—often in a tissue- and time-specific manner. This is particularly insightful in the study of development, which has to describe activities that are tightly regulated in time and space. I will use our investigations of the EGF receptor pathway here as an example of how much can be learned and how unexpected results gained from modifying mutations that may often have a variable or subtle phenotype by themselves can lead to novel insights, making the pursuit of developmental processes with the help of classical and molecular genetics exciting and rewarding.

In screens for mutations that affect pattern formation of egg and embryo in *Drosophila*, we isolated a large number of lines in which the mutant females produced eggs with dorso-ventral patterning defects (Schupbach & Wieschaus, 1991). This was usually manifest in the morphology of the dorsal appendages, which are very sensitive markers of changes in dorso-ventral patterning. These appendages are normally formed by two anterior dorso-lateral groups of follicle cells that migrate from their original position in the epithelium anteriorward giving rise to two tubes of characteristic shape and length (Berg, 2005; Osterfield, Du, Schüpbach, Wieschaus, & Shvartsman, 2013). In the screens, we isolated mutations that either increased the populations of these dorsal appendage forming follicle cells (dorsalizing mutations) or reduced the populations, fusing them into one single population or failing to establish them altogether (ventralizing mutations) (Schupbach, 1987). Since the global pattern of the follicle cell epithelium is transmitted to the future embryo, these mutations not only affected the egg shell and the dorsal appendage morphology but also the dorso-ventral pattern of the embryo developing inside such eggs. Some of the lines showed a very reproducible, strong phenotype, whereas in others, the phenotype was rather variable and did not become notably stronger even when the mutant was tested in *trans* to a deficiency for the locus. The logical procedure was then to first concentrate on genes where the mutants led to a more uniform phenotype, in particular, we began with the characterization of mutations where all the eggs were strongly ventralized. Analysis of such mutations in mosaic egg chambers indicated that the genes participated in signaling processes that involve interactions of the germline and the surrounding somatic follicle cells (Schupbach, 1987). Eventual molecular cloning revealed that we had isolated mutations in the *Drosophila* EGF receptor and one of its ligands, a TGF-alpha like molecule, which we called Gurken (which means cucumber in German) (Neuman-Silberberg & Schüpbach, 1993; Price, Clifford, & Schupbach, 1989; Schejter & Shilo, 1989). We also found that Gurken (*grk*) was produced in a localized fashion during oogenesis and this local signal then activated the EGF receptor in the neighboring follicle cells, but not in follicle cells far from the source, thus setting up the dorso-ventral axis of the egg and future embryo. Later, we and the laboratory of St Johnston showed that this signaling process in fact already acts early during oogenesis and is also required for the proper establishment of the anterior–posterior axis of egg and embryo (Gonzalez-Reyes, Elliott, & St Johnston, 1995; Roth, Neuman-Silberberg, Barcelo, & Schüpbach, 1995).

What then was the function of all these other genes for which we had isolated mutations with variable effects on dorso-ventral axis formation? As the *grk* RNA has to be localized during oogenesis to the future dorsal side of the oocyte, in close proximity to the oocyte nucleus (Fig. 1A), we expected a number of genes to be involved in RNA transport and translational control of *grk* mRNA. Interestingly, the majority of genes we and others isolated and analyzed that affected *grk* localization and translational control showed a dorsalized phenotype of eggs and embryos, indicating that specific factors are involved in the negative regulation of *grk* protein expression, and that localization factors, when missing can lead to the mislocalized RNA being translated in ectopic places (e.g., *capu*, *spire*, *fs(1)K10*, *squid*, *cup*, *Hrp48*, *otu*) (Clouse, Ferguson, & Schüpbach, 2008; Goodrich, Clouse, & Schüpbach, 2004; Manseau & Schüpbach, 1989; Norvell, Kelley, Wehr, & Schupbach, 1999; Wieschaus, Marsh, & Gehring, 1978). What about factors that would be positively involved in the upregulation of translation after localization of the RNA, which should have resulted in a ventralized mutant phenotype? Even though there were a considerable number of ventralizing mutants, there were fewer of these shown to be involved in directly regulating translation of the localized RNA, presumably

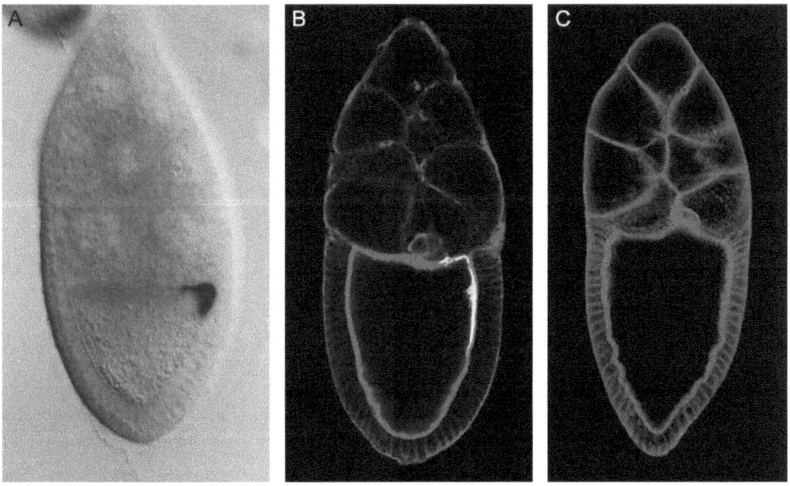

Figure 1 Gurken accumulation in developing egg chambers at stages 9 and 10. (A) A wild-type egg chamber at stage 9 with *grk* mRNA accumulation shown by *in situ* hybridization (blue) on the dorsal anterior corner in close proximity to the oocyte nucleus. (B and C) Egg chambers at stage 10, stained with an antibody to Grk protein (yellow) and with rhodamine-labeled phalloidin to outline the actin cytoskeleton (red). The protein is readily detectable in the wild-type egg chamber (B) but is absent from the egg chamber mutant for *spnB* (C). (A color version of this figure is located at the end of this chapter.)

because this process involves more general factors, that will interfere with early egg chamber formation, when mutant, and would therefore not be isolated in a screen aimed at isolating dorso-ventral patterning mutants. Weaker alleles still allowed the identification of the genes *vasa* (Styhler, Nakamura, Swan, Suter, & Lasko, 1998; Tomancak, Guichet, Zavorszky, & Ephrussi, 1998), *Orb* (Christerson & McKearin, 1994; Lantz, Chang, Horabin, Bopp, & Schedl, 1994), and *encore* (Hawkins, Van Buskirk, Grossniklaus, & Schupbach, 1997). While this was an obvious conclusion from the screens, it is often hard to predict which aspects of a process involve specific factors and which ones rely on more general cellular machineries and will require specific screening techniques to capture some of the major core factors of the process. In our case, the dorsalizing mutations provided important insights into the RNA localization and translational control process of Grk and served as a starting point for biochemical isolation of further factors and interactors (Clouse et al., 2008; Delanoue, Herpers, Soetaert, Davis, & Rabouille, 2007; Goodrich et al., 2004).

The bigger surprises, however, arose from the analysis of the more variable ventralizing mutations. This class of mutations is often collectively called "spindle mutants" because many of the eggs produced by the mutant females do not possess any dorsal appendages, are elongated and rather pointy at both ends of the egg, whereas a wild-type egg has a rounded posterior end. However, the phenotypes of these lines were very variable and the same female could lay eggs ranging from wild-type-like morphology to strongly ventralized phenotypes. The Nusslein-Volhard research group had isolated such mutations on the third chromosome and called them *spindle-A*, *B*, *C*, etc., (González-Reyes, Elliott, & St Johnston, 1997; Tearle & Nüsslein-Volhard, 1987), whereas we kept with the "*gurken*," i.e., vegetable theme for the mutations on the second chromosome and named them after other vegetables with elongated shapes such as *okra*, *aubergine*, *zucchini*, *squash* (Schupbach & Wieschaus, 1991). When we molecularly identified the first two members of the group, *okra* and *spnB*, we were surprised to find that they encoded homologs of DNA repair enzymes, known from yeast and mammalian cell culture to be involved in homologous DNA repair (Ghabrial, Ray, & Schüpbach, 1998). Judging from antibody staining and Western analysis, it appeared that the mutations somehow downregulated or blocked the normal upregulation of *grk* mRNA translation and resulted in egg chambers with severely reduced levels of Grk protein (Fig. 1B and C). Furthermore, immunostaining showed that there was a variability in the reduction of Grk protein, correlated with the variability in egg shell phenotypes. It was only after consideration of the literature on DNA repair that a possible interpretation arose and we realized that the effects on Grk protein accumulation and dorsoventral patterning might be caused by a DNA damage checkpoint, given that such checkpoints had been shown to interfere with a variety of processes in cells with sustained DNA damage. We tested this possibility by constructing double mutants with mutations in various checkpoint kinases and found that indeed, mutations in the *Drosophila* ATR homolog *mei-41* and the *Chk2* homolog *mnk*, both could restore Grk protein levels and egg shell patterning to wild-type phenotypes (Abdu, Brodsky, & Schupbach, 2002; Ghabrial & Schüpbach, 1999). In addition, using mutations in the topoisomerase *mei-W68* we could also show that the defects were suppressed, when the initiation of meiotic double-strand breaks in the DNA was prevented (Ghabrial & Schüpbach, 1999). These experiments demonstrated again the power of genetics, as the suppression now clearly indicated that there is a pathway from meiotic recombination and failure of DNA repair to the activation of the checkpoint kinases and a block in *grk* mRNA translation, a very unexpected result that could only be obtained through genetic analysis.

A few years later we, and simultaneously the laboratory of Dr. William Theurkauf, found another unpredicted input into the activation of the meiotic checkpoint. In our laboratory, we had continued the analysis of other spindle-like mutations, notably *cut-off*, *zucchini*, and *squash*. We had tested these mutations for suppression by the meiotic checkpoint and found that *cut-off* and *squash* were suppressed in combination with mutations in Chk2. The mutant egg chambers also showed a persistence of DNA breaks, similar to the DNA repair mutations, which, in combination with the Chk2 suppression, suggested that the phenotype was linked to some type of DNA damage. To our perplexion, our molecular identification seemed to point to a possible involvement of the genes in RNA regulation, possibly having RNA nuclease activity, and did not give us a clear indication of what type of DNA damage process these genes might be involved in. However, several groups had reported that *aubergine* and *spnE*, two genes that produced the mutant spindle phenotype, were involved in the newly discovered piRNA pathway, a pathway that is required to produce small RNAs that are directed against retrotransposons in the germline and in some somatic cells during oogenesis (Aravin et al., 2003, 2001; Cook, Koppetsch, Wu, & Theurkauf, 2004; Kennerdell, Yamaguchi, & Carthew, 2002; Vagin et al., 2006). We therefore checked the production of retrotransposons in the mutants and found that the level of retrotransposon RNA was greatly increased. We also tested *aubergine* alleles and were able to show that the dorso-ventral defects of these mutations were also suppressed by mutations in Chk2 (Chen, Pane, & Schupbach, 2007). Similarly, the Theurkauf group showed that loss of Chk2 and Mei-41/ATR could suppress the dorso-ventral defects of *aubergine* and *armitage*, another factor involved in the piRNA pathway (Klattenhoff et al., 2007). It therefore became clear that the meiotic checkpoint was also activated to some extent when the piRNA pathway was not functioning to suppress retrotransposon transposition. While initially the fact that Mei-41, the *Drosophila* ATR homolog, was not able to completely suppress the dorso-ventral defect of the piRNA mutants was puzzling, as was the observation that *zucchini* and *piwi* mutants were insignificantly suppressed by mutants in Chk2, it is likely that full suppression might require that both upstream DNA damage kinases, ATR and ATM be blocked simultaneously, and in addition, that some of the *piRNA* genes might have additional oogenesis roles, not suppressable by checkpoint mutations. This may point to a more diverse range of DNA damage effected in the absence of the piRNA pathway, compared to the homologous recombination repair pathway that specifically affects double-strand breaks. But both pathways ultimately signal through Chk2 to limit *grk* expression.

Why would such a link from meiosis to dorso-ventral patterning exist? In *Drosophila*, meiosis is initiated in the germarium, and under normal circumstances, DSBs are repaired by about stage 2 of oogenesis, just when the egg chamber leaves the germarium, the region of the ovary where it is assembled (Mehrotra & McKim, 2006). Subsequently, it begins a more rapid descent toward the lower end of the ovarioles where it then enters the oviduct as a mature stage 14 egg. During stages 2–14, the oocyte nucleus proceeds through later parts of prophase, the chromosomes condense and finally enter metaphase of meiosis I where they arrest until egg activation (Von Stetina & Orr-Weaver, 2011). During the same developmental stages, Gurken protein is continuously produced first at the posterior end of the egg chamber and after stage 7 at the dorsal side of the oocyte (Neuman-Silberberg & Schüpbach, 1996). It would seem reasonable to expect a close coordination between the progression of the meiotic cell cycle and the developmental signal (Grk) that establishes first the anterior-posterior and then the dorso-ventral pattern of the egg. With a block in the cell cycle due to unrepaired DNA damage, one might therefore expect a concominant block in egg chamber development. However, when the meiotic checkpoint is activated, the egg chambers do not halt in their progression through the ovarioles, nor do they remain at stage 2/3. They grow in size and undergo characteristic morphological changes, such as elongation and initiate border cell migration, events that are characteristic of wild-type egg chambers, even though their nuclei are delayed in progression such that the chromosomes never condense properly into the so called karyosome structure. Very specifically, Grk protein is not accumulating properly, and we also found that accumulation of another developmental important protein, *fs(1)K10*, is affected by the meiotic checkpoint (Ghabrial & Schüpbach, 1999). It appears that checkpoint activity is in particular directed against certain developmental regulators, but does not arrest development of the egg chambers overall. Possibly in normal development, this checkpoint activity just has to serve to slow down developmental signals for a short period of time in early oogenesis, until all the DNA breaks are successfully repaired. But in the DNA repair mutants, when these breaks remain unrepaired, the situation seems to diverge to such an extent that in spite of checkpoint activation, some processes proceed and others do not. This results in eggs that are abnormal and infertile. While it may seem a waste of energy and material to produce such eggs, given that *Drosophila* females can lay many eggs every day, it may not be sufficiently detrimental to produce an occasional abnormal egg—compared to a complete arrest of egg production and possibly a block inside an ovariole. *Drosophila* ovaries do have other checkpoints where egg chambers can be induced to undergo apoptosis, both in the germarium and around stage 8 in starved females (Bakhrat, Pritchett, Peretz, McCall, & Abdu, 2010; Drummond-Barbosa & Spradling, 2001), but those conditions may be more drastically impinging on the energy balance of the female and therefore cause wide spread apoptosis, whereas a very occasional egg chamber with an unrepaired DNA break may not have a sufficiently negative impact to require elimination before the egg is laid. In any case, there are still a number of significant questions that remain unresolved as to the mechanism and biological function of the meiotic checkpoint.

One of the intriguing findings in trying to uncover the mechanism of checkpoint effects on *grk* mRNA translation was the observation that Vasa, a translational regulator, was phosphorylated when the checkpoint was active (Abdu et al., 2002; Chen et al., 2007; Ghabrial & Schüpbach, 1999; Klattenhoff et al., 2007). We performed mass spec analysis of Vasa in wild-type and spnB mutant ovaries and located at least one consistent site of phosphorylation to Serine 43 when the checkpoint was active (A. Pane, W. Li, T. Schupbach, unpublished). However, when we mutated this serine to alanine, we could not measure any effect on the meiotic checkpoint process. Overexpression of both the wild-type and the mutant Vasa protein could suppress the dorso-ventral phenotype of spindle mutations to a large extent; however, since both wild-type and mutant Vasa were able to do so, the phosphorylation of this residue does not appear to make a critical difference in the process. We also constructed a mutant gene containing a glutamic acid instead of the serine 43, and again did not measure any difference in Grk production between wild-type controls and the mutant expressing this form of Vasa. We now believe that the phosphorylation may represent a downstream effect of stalled translation rather than a decisive regulatory event. While these results were disappointing, they nevertheless show again the strength of the genetic analysis that allows us to distinguish between protein modifications with important regulatory function and others that are acquired as a consequence of a developmental change.

Another genetic result derived from a suppressor screen that we performed isolating mutations that could restore Grk protein production even in the presence of a mutation in the DNA repair enzyme encoded by *spnB*. Again unexpectedly, we found that loss of function mutations in a highly conserved translational regulator, eIF1A, leads to a significant upregulation of Grk protein in heterozygous condition in the *spnB* mutant background (Li, Klovstad, & Schüpbach, 2014). It is presently mechanistically not clear why loss of one copy of this translational regulator can enable more efficient translation of *grk* mRNA under the checkpoint mediated translational stalling conditions. Nevertheless, the results were clear and clearly point to a more complex role of eIF1A in the translational regulation process. To resolve this question a more biochemical approach will certainly be helpful in the future, but it was again the genetic analysis that made us aware of this interesting and unexpected regulatory event. Just using protein–protein interactions and biochemistry, one would have expected Vasa to be "connected" in a network to eIF1A, but the prediction would have been that of course, if translation is affected in a checkpoint-dependent manner, eIF1A might play a role in the process. Indeed, complete loss of eIF1A leads to a block in translation and cell death. However, the unexpected observation that the loss of one copy of eIF1A serves to "restore" a significant amount of *grk* mRNA translation under checkpoint conditions was only achievable by genetic analysis.

While genetics has certainly provided the major insights into the patterning process so far, several additional techniques have dramatically augmented the value of mutant analysis. In particular, the refinement of microscopy in the past 20 years has been instrumental in providing insight into the location and accumulation of the important proteins within the oocyte and follicle cells. In the very recent past, microscopy has made further enormous advances and is now able to provide novel insights into developmental processes at an even higher level of accuracy and quantitative information. For instance, the laboratories of Dr. Ilan Davis and Dr. Catherine Rabouille have been able to make considerable progress on the question of *grk* mRNA localization using sophisticated microscopic methods and immunogold EM analysis of proteins and RNA (Delanoue et al., 2007; Weil et al., 2012). It is therefore to be expected that the question of *grk* mRNA regulation by the meiotic checkpoint can be further addressed in the near future through a combination of imaging, genetics, and biochemistry.

Considering the wealth of information on patterning in oogenesis that has been obtained by many laboratories in the past 30 years, it is remarkable how much of the fundamental insights have come from genetic analysis. While the way screens are conducted these days is often different, involving RNAi or generation of mosaic tissues, i.e., using cell type targeted analysis, genetic screens do allow the identification of key factors with clear phenotypes that then can serve to build and link pathways further and test members of different molecular pathways to understand which proteins and complexes perform the major task of directing development, and which ones might provide an input that is more peripheral, or depends on other, parallel developmental processes. Without question, once the genetic backbone has been uncovered, to resolve all the issues in time and space methods of imaging possibly coupled to computational modeling and eventually some genomic approaches will be necessary. And as those approaches become more sophisticated, they will build on the genetic foundation and will certainly uncover new questions and surprises and keep developmental biology exciting for many years to come.

ACKNOWLEDGMENTS

First and foremost I would like to thank all the present and former members of my laboratory who have made the analysis of patterning in *Drosophila* oogenesis so exciting and rewarding. I would also like to thank Eric Wieschaus for intellectual support and discussion. I am also grateful to acknowledge the Howard Hughes Medical Institute and US Public Health Service Grant NIH-RO1 GM077620 for supporting our work.

REFERENCES

Abdu, U., Brodsky, M., & Schupbach, T. (2002). Activation of a meiotic checkpoint during *Drosophila* oogenesis regulates the translation of Gurken through Chk2/Mnk. *Current Biology*, *12*, 1645–1651.

Aravin, A. A., Lagos-Quintana, M., Yalcin, A., Zavolan, M., Marks, D., Snyder, B., et al. (2003). The small RNA profile during *Drosophila melanogaster* development. *Developmental Cell*, *5*, 337–350.

Aravin, A. A., Naumova, N. M., Tulin, A. V., Vagin, V. V., Rozovsky, Y. M., & Gvozdev, V. A. (2001). Double-stranded RNA-mediated silencing of genomic tandem repeats and transposable elements in the *D. melanogaster* germline. *Current Biology*, *11*, 1017–1027.

Bakhrat, A., Pritchett, T., Peretz, G., McCall, K., & Abdu, U. (2010). Drosophila Chk2 and p53 proteins induce stage-specific cell death independently during oogenesis. *Apoptosis*, *15*, 1425–1434.

Berg, C. A. (2005). The Drosophila shell game: Patterning genes and morphological change. *Trends in Genetics*, *21*, 346–355.

Chen, Y., Pane, A., & Schupbach, T. (2007). Cutoff and aubergine mutations result in retrotransposon upregulation and checkpoint activation in Drosophila. *Current Biology*, *17*, 637–642.

Christerson, L. B., & McKearin, D. M. (1994). orb is required for anteroposterior and dorsoventral patterning during *Drosophila* oogenesis. *Genes & Development*, *8*, 614–628.

Clouse, K. N., Ferguson, S. B., & Schüpbach, T. (2008). Squid, Cup, and PABP55B function together to regulate *gurken* translation in Drosophila. *Developmental Biology*, *313*, 713–724.

Cook, H. A., Koppetsch, B. S., Wu, J., & Theurkauf, W. E. (2004). The *Drosophila* SDE3 homolog *armitage* is required for *oskar* mRNA silencing and embryonic axis specification. *Cell*, *116*, 817–829.

Delanoue, R., Herpers, B., Soetaert, J., Davis, I., & Rabouille, C. (2007). Drosophila Squid/hnRNP helps Dynein switch from a *gurken* mRNA transport motor to an ultrastructural static anchor in sponge bodies. *Developmental Cell*, *13*, 523–538.

Drummond-Barbosa, D., & Spradling, A. C. (2001). Stem cells and their progeny respond to nutritional changes during Drosophila oogenesis. *Developmental Biology*, *231*, 265–278.

Ghabrial, A., Ray, R. P., & Schüpbach, T. (1998). okra and spindle-B encode components of the RAD52 DNA repair pathway and affect meiosis and patterning in Drosophila oogenesis. *Genes & Development*, *12*, 2711–2723.

Ghabrial, A., & Schüpbach, T. (1999). Activation of a meiotic checkpoint regulates translation of Gurken during *Drosophila* oogenesis. *Nature Cell Biology*, *1*, 354–357.

Gonzalez-Reyes, A., Elliott, H., & St Johnston, D. (1995). Polarization of both major body axes in Drosophila by gurken-torpedo signalling. *Nature*, *375*, 654–658.

González-Reyes, A., Elliott, H., & St Johnston, D. (1997). Oocyte determination and the origin of polarity in Drosophila: The role of the spindle genes. *Development*, *124*, 4927–4937.

Goodrich, J. S., Clouse, K. N., & Schüpbach, T. (2004). Hrb27C, Sqd and Otu cooperatively regulate *gurken* RNA localization and mediate nurse cell chromosome dispersion in Drosophila oogenesis. *Development*, *131*, 1949–1958.

Hawkins, N. C., Van Buskirk, C., Grossniklaus, U., & Schupbach, T. (1997). Post-transcriptional regulation of *gurken* by *encore* is required for axis determination in Drosophila. *Development*, *124*, 4801–4810.

Kennerdell, J. R., Yamaguchi, S., & Carthew, R. W. (2002). RNAi is activated during *Drosophila* oocyte maturation in a manner dependent on aubergine and spindle-E. *Genes & Development*, *16*, 1884–1889.

Klattenhoff, C., Bratu, D. P., McGinnis-Schultz, N., Koppetsch, B. S., Cook, H. A., & Theurkauf, W. E. (2007). *Drosophila* rasiRNA pathway mutations disrupt embryonic axis specification through activation of an ATR/Chk2 DNA damage response. *Developmental Cell*, *12*, 45–55.

Lantz, V., Chang, J. S., Horabin, J. I., Bopp, D., & Schedl, P. (1994). The Drosophila orb RNA binding protein is required for the formation of the egg chamber and establishment of polarity. *Genes & Development*, *8*, 598–613.

Li, W., Klovstad, M., & Schüpbach, T. (2014). Repression of Gurken translation by a meiotic checkpoint in Drosophila oogenesis is suppressed by a reduction in the dose of eIF1A. *Development*, *141*, 3910–3921.

Manseau, L. J., & Schüpbach, T. (1989). *cappucino* and *spire*: Two unique maternal effect loci that are required for both the anterior-posterior and dorso-ventral patterns of the Drosophila embryo. *Genes & Development*, *3*, 1437–1452.

Mehrotra, S., & McKim, K. S. (2006). Temporal analysis of meiotic DNA Double-strand break formation and repair in Drosophila females. *PLoS Genetics*, *2*.

Neuman-Silberberg, F. S., & Schüpbach, T. (1993). The Drosophila dorso-ventral patterning gene *gurken* produces a dorsally localized RNA and encodes a TGF-alpha like protein. *Cell*, *75*, 165–174.

Neuman-Silberberg, F. S., & Schüpbach, T. (1996). The *Drosophila* TGF-alpha-like protein Gurken: Expression and cellular localization during *Drosophila* oogenesis. *Mechanisms of Development*, *59*, 105–113.

Norvell, A., Kelley, R. L., Wehr, K., & Schupbach, T. (1999). Specific isoforms of Squid, a Drosophila hnRNP, perform distinct roles in Gurken localization during oogenesis. *Genes & Development*, *13*, 864–876.

Osterfield, M., Du, X., Schüpbach, T., Wieschaus, E., & Shvartsman, S. Y. (2013). Three-dimensional epithelial morphogenesis in the developing Drosophila egg. *Developmental Cell*, *24*, 400–410.

Price, J. V., Clifford, R. J., & Schupbach, T. (1989). The maternal ventralizing locus torpedo is allelic to faint little ball, an embryonic lethal, and encodes the Drosophila EGF receptor homolog. *Cell*, *56*, 1085–1092.

Roth, S., Neuman-Silberberg, F. S., Barcelo, G., & Schüpbach, T. (1995). *cornichon* and the EGF receptor signaling process are necessary for both anterior-posterior and dorsal-ventral Pattern formation in Drosophila. *Cell*, *81*, 967–978.
Schejter, E. D., & Shilo, B. Z. (1989). The Drosophila EGF receptor homolog (DER) gene is allelic to *faint little ball*, a locus essential for embryonic development. *Cell*, *56*, 1093–1114.
Schupbach, T. (1987). Germ line and soma cooperate during oogenesis to establish the dorsoventral pattern of egg shell and embryo in *Drosophila melanogaster*. *Cell*, *49*, 699–707.
Schupbach, T., & Wieschaus, E. (1991). Female sterile mutations on the second chromosome of *Drosophila melanogaster*. II. Mutations blocking oogenesis or altering egg morphology. *Genetics*, *129*, 1119–1136.
Styhler, S., Nakamura, A., Swan, A., Suter, B., & Lasko, P. (1998). Vasa is required for Gurken accumulation in the oocyte, and is involved in oocyte differentiation and germline cyst development. *Development*, *125*, 1569–1578.
Tearle, R., & Nüsslein-Volhard, C. (1987). Tübingen mutants stocklist. *Drosophila Information Service*, *66*, 209–226.
Tomancak, P., Guichet, A., Zavorszky, P., & Ephrussi, A. (1998). Oocyte polarity depends on regulation of *gurken* by Vasa. *Development*, *125*, 1723–1732.
Vagin, V. V., Sigova, A., Li, C., Seitz, H., Gvozdev, V., & Zamore, P. D. (2006). A distinct small RNA pathway silences selfish genetic elements in the germline. *Science*, *313*, 320–324.
Von Stetina, J. R., & Orr-Weaver, T. L. (2011). Developmental control of oocyte maturation and egg activation in metazoan models. *Cold Spring Harbor Perspectives in Biology*, *3*, a005553 (Review).
Weil, T. T., Parton, R. M., Herpers, B., Soetaert, J., Veenendaal, T., Xanthakis, D., et al. (2012). Drosophila patterning is established by differential association of mRNAs with P bodies. *Nature Cell Biology*, *14*, 1305–1313.
Wieschaus, E., Marsh, J. L., & Gehring, W. (1978). *fs(1)K10*, a germline-dependent female sterile mutation causing abnormal chorion morphology in *Drosophila melanogaster*. *Roux's Archives of Developmental Biology*, *184*, 75–82.

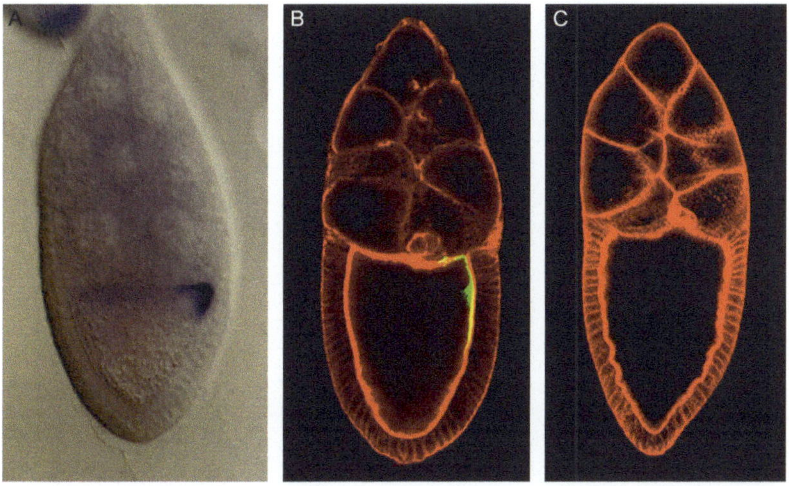

Trudi Schupbach, Figure 1 Gurken accumulation in developing egg chambers at stages 9 and 10. (A) A wild-type egg chamber at stage 9 with *grk* mRNA accumulation shown by *in situ* hybridization (blue) on the dorsal anterior corner in close proximity to the oocyte nucleus. (B and C) Egg chambers at stage 10, stained with an antibody to Grk protein (yellow) and with rhodamine-labeled phalloidin to outline the actin cytoskeleton (red). The protein is readily detectable in the wild-type egg chamber (B) but is absent from the egg chamber mutant for *spnB* (C).

CHAPTER NINETEEN

Homeodomains, Hedgehogs, and Happiness

Matthew P. Scott[1]
Carnegie Institution for Science, Washington, D.C., USA; Stanford University School of Medicine, Stanford, California, USA
[1]Corresponding author: e-mail address: mscott@carnegiescience.edu

Abstract

Developmental biologists have had a spectacular quarter century of discoveries, building on many decades of work earlier, discovering molecular, cellular, and genetic mechanisms that underlie the magical process by which an egg becomes a plant or animal. Among the discoveries were homeodomains, DNA-binding domains that allow transcription factors to recognize their target genes, and the Hedgehog signaling pathway, which is used in many organs and tissues for communication among cells. The experience of unveiling the mechanisms and molecules connected to both of these findings has been remarkable, joyful, difficult, and a time of great teamwork and collaboration within and between laboratory groups. More than ever it is possible to discern the evolutionary processes, and their mechanisms, that led to the diversity of life on earth. A huge amount of work remains to be done to obtain a broad understanding of what happened and how development works.

It was a dark and stormy night, well, dark, and late anyway, when I developed my first DNA sequencing gel. The $14 \times 17''$ film, still dripping, showed clear defects, with bands of varying clarity, lanes running askew, and uneven sample amounts. The radioactive gel was safely put away and I could begin wondering what the results would show, if anything. I gleaned what I could, writing down the sequences from each lane. I had a specific question in mind: why were small pieces of DNA from two different *Drosophila* genes, a segmentation gene and a homeotic gene, weakly cross-hybridizing? I quickly found that they both had As, Ts, Cs, and Gs. Not much help. Maybe I should translate the sequence into protein? The building was largely silent, there were no computers, I was alone, and I didn't know the genetic code by heart. The library had limited choices, but they did have the *Cold Spring Harbor Symposium* books and one of those had an early publication of the genetic code. I went and found it. It was incomplete, because at the time the stop codons were not fully characterized, but it was enough. I translated my bits

Current Topics in Developmental Biology, Volume 117
ISSN 0070-2153
http://dx.doi.org/10.1016/bs.ctdb.2015.11.034

of DNA sequence, caught some errors, and found that there were short bits of protein sequence that were similar. Years and years of work by my advisor, geneticist Thomas Kaufman, had led me to isolate those genes, and guidance from my other advisor, Barry Polisky, a molecular biologist, had made it possible. Here was the beginning of the homeodomain, a region of similarity in protein-coding potential between *Antennapedia* and *fushi tarazu*. Meanwhile in Walter Gehring's lab in Switzerland, a similar discovery was being made by Bill McGinnis and others. It was Ed Lewis's original conclusion that homeotic genes were serial repeats of some kind that instantly made sense of the homeodomain as a common feature of *Hox* proteins.

The sequences on the film that day brought us abruptly from thinking at the level of DNA to thinking at the level of proteins...and proteins with suggestive properties. The cross-hybridizing sequences could have been transposons, or related *cis*-acting control elements, but they were instead protein-coding exons. When Dr. Allen Laughon found that the homeodomain sequence resembled the helix-turn-helix sequence of bacterial and yeast DNA-binding proteins (a result also found in Walter's lab), it all came together. We were studying gene-controlling proteins. All our classes on the lac operon and yeast mating type were suddenly directly applicable to thinking about pattern formation in developing animals. What a time.

Later when we saw the first stripes of *fushi tarazu* protein in early embryos, and Dr. Sean Carroll's champagne cork made a lasting indent on our lab ceiling, the location of the protein in nuclei provided another hint...maybe these developmental genes are so powerful because they coordinate activities of other genes.

Our voyage back out of the nucleus began with the arrival of Joan Hooper as a postdoc. She had been scrutinizing, as many had, the stunning discoveries made by Christiane Nüsslein-Volhard, Eric Wieschaus, and Gerd Jürgens in their genetic screen for segmentation mutants. Joan had noted that mutants for *patched* were close to full-size, yet repatterned, suggesting that the mirror-image polarity defects were due to mispatterning rather than to cell death followed by some sort of adaptation. The polarity defects—cells confusing north for south—suggested an altered signaling process. Her decision to pursue *patched* opened up a new world for us outside the nucleus, since indeed as we and Philip Ingham found, *patched* encodes a receptor for a signal and not a transcription factor. Now we had both ends of the problem of development: signals that coordinate cells and gene control that directs cells to differentiate. Much was missing, particularly, how the cell's machinery controls shape, localization of molecules, and movement...but it was a start.

Phil's genetic studies of double mutants revealed a fascinating relationship between *patched* and the *Hedgehog* gene, a set of findings and puzzles that led to understanding the Hedgehog signal as inducing production of its own antagonist, the Patched receptor. The relationships determined in flies, and rest of the machinery that fell into place from fly genetics and molecular biology in many labs, held true in mice. As with *Hox* genes, far more was evolutionarily conserved than anyone had imagined.

It's worth reflecting on that last point. Usually in scientific history you can go back and find someone somewhere who said the crucial idea, even if they were at the time dismissed as crazy. In the case of developmental genes, in retrospect, it is hard to imagine the moment in earth's history when the referee's whistle would blow and all the genes would change in concert to make, say, a mammal instead of a worm. It makes much more sense to think that regulatory properties of a system of genes and proteins that make patterned tissues would change over time to give rise to different animals. Yet there were at best, to my knowledge, only the most indirect suggestions that the main "hardware" of developmental regulatory genes would be evolutionarily conserved. Instead, great emphasis was often put on the expected major *differences* between animal species' developmental genes that *must* exist—because mammals and insects and other animals look so different. At some level, it was like saying cars of different colors are fundamentally different, but of course it was more subtle and we are still seeing new surprises about conservation of all sorts of processes we'd thought were areas of evolutionary novelty. Somewhere along the line, though, it became more high-profile-journal-worthy to find something that was *not* conserved rather than something that was.

That transition in thinking to regarding most development regulators as ancient and shared, still incomplete, went a long way to unifying a fragmented field of science. The molecular biology of animal development dramatically demonstrates common evolutionary origins. Everyone celebrated, and still celebrates, discovering that some highly specialized aspect of their findings suddenly has direct implications for most tissues and organs, or most animals, or even most organisms. Sometimes the implications connect to medicine or agriculture. At that point, you can finally explain to your extended family that what you are doing leads to useful outcomes and maybe why you still do not have a real job!

For us the medical connection was cancer, and it came from hard work by Ron Johnson, a postdoc, Lisa Goodrich, a grad student, and later Ljiljana Milenkovic, a Ph.D. Research Specialist. They set out on a quest on which others had already failed to find mammalian *patched*. They did it by an evolutionary stepping-stone approach, learning what parts of the *patched* gene are most conserved by looking at genes from organisms closer to *Drosophila* before trying the leap to mice. This worked, but their bravery was substantial—it was an all or none project where if they did not get the mammalian gene they would have nothing to show for long effort. The mouse gene showed us the conservation of the relationship with Hedgehog and allowed the isolation of the human gene. The mapping of the human *PATCHED* gene, with human geneticist Prof. Richard Myers, gave Ron a candidate disease, Basal Cell Nevus (Gorlin's) Syndrome, which involved birth defects and cancer. With Prof. Ervin Epstein of UCSF, we were able to prove the association in families, while Lisa and Ljiljana's knockout *patched* mice showed that the disease, like the genes, is conserved.

Suddenly, seemingly overnight, we were cancer researchers...something neither intended nor anticipated. However, we had in fact asserted that something like this would happen. When? Way back when we began working on isolating fly developmental genes and argued to the American Cancer Society, in grant applications they funded, that genes that control development control growth. Such genes should eventually shed light on uncontrolled growth. Great thanks to the ACS for allowing this leap of faith and supporting a young faculty member.

Our entry into medical research took good advantage of my lab's location at Stanford School of Medicine. My stem cell-studying wife Margaret Fuller and I had moved from the University of Colorado, where we had wonderful colleagues and many scientific successes, to Stanford. One reason was the expectation that our work might connect to medicine and that being within a medical school would help us exploit such opportunities. That proved to be true, and helped compensate for missing our Colorado friends, mountains, and rivers. We did start diving in Monterey, which also helped.

At Stanford, people with deep experience in medicine joined our labs, such as Anthony Oro, a dermatologist, who pursued Hedgehog/Patched signaling in the context of skin development and cancer. Damage to the human *PATCHED* gene is responsible for basal cell carcinoma, the most common human cancer. Robert Wechsler-Reya joined to learn the normal role of Hedgehog/Patched signaling in the cerebellum, from which tumors arise that are characteristic of Gorlin's Syndrome. He found that Hedgehog is a powerful mitogenic signal in the cerebellum, so loss of the restraining influence of *patched*, as Lisa had shown, causes excessive growth and medulloblastoma—the most common childhood malignant brain tumor. In recent years, successors to these researchers, in my lab and many others, have used knowledge of the pathway to identify candidate drug targets and candidate drugs, and some of these look promising as future therapies for devastating childhood brain tumors and other cancers.

With the perspective of our Hedgehog/Patched experience, we can tell how long it took to move from basic science, aimed at the mysteries of development, to directly apply medical science and a FDA-approved drug. It took 32 years from the report of *Drosophila* segmentation genes to approval of Vismodegib. Half those years were occupied, by research in a plethora of labs, in identifying the relevant genes and proteins and figuring out something of their molecular functions and relationships. Then came the discovery of the cancer connection, in our lab and Allan Bale's lab at Yale. It took another 16 years for the studies of targets, drugs, and cancer-afflicted people to get to a useful treatment. Like most cancer treatments, a single drug is eventually evaded by resistant tumor cells but it is still useful. Ongoing work is yielding other drugs. Combinations of drugs and other therapies hold promise for substantial impacts on a number of kinds of human cancer.

Thirty-two years! Couldn't we have moved faster? Perhaps, and with other pathways and diseases maybe some of the Hedgehog/Patched experiences will be instructive. To a family with a sick child, that is a staggering period of time. The experience does show that long-term funding, first of untargeted basic science and then of applied science, is necessary if truly new therapies are to be found. This is not monkeys at typewriters, or trying everything on the shelf, but instead a logical process of discovery, investigation, and deduction. Like many others, I am concerned that our national funding systems have somehow "unlearned" the lessons of their own history. The push to do "translational research" seems to be premature in many cases. Some obvious incremental test is done, and probably will not work, instead of a deep investigation into mechanisms that should precede the attempt at translation. If the original novel is short, derivative, and incoherent, there's no value in translating it. In developmental biology as in other fields of discovery science, immense value to medicine has emerged from exploration and risk taking, whereas much less progress has been made going headlong at a "developmental disease." Nothing has changed about how to discover important ideas and treatments except a misconceived funding urgency that subverts truly important discoveries in favor of obvious short-term steps.

The underpinnings of the excitement we experienced in our lab were interdisciplinary, a usually delightful and sometimes tense coming-together of geneticists and molecular biologists (and, in Sean, an immunologist), with developmental biology as the centerpiece mystery. These days we also depend on mass spectrometry, crystallography, advanced microscopy, and computer science—for most developmental biology challenges interdisciplinary work is not optional. We are privileged to be biologists at a time when the code for the structure of organisms is being unveiled. The power of combining new technology with old mysteries became, for me, the basis for career steps such as organizing the Bio-X program at Stanford, which involved more than 400 faculty in 50 departments, and now at Carnegie Science a broad span from cosmology to physics, chemistry, geology, biology, and ecology.

In risking new approaches and dubious projects, we owe much to our mentors, in my case Mary Lou Pardue at MIT, Thom and Barry at Indiana, Tom Cech as a fellow Pardue lab member and later colleague at U. Colorado, and Allan Spradling and Joe Gall at various places but now, like me, at Carnegie. It is a slight oddity of science that you can never fully repay your own mentors, so instead you try to do for your students and postdocs what your mentors did for you. In the process you again acquire debt, because faculty constantly learn from their more junior coworkers. You try, as a faculty member, to encourage exploration and risk and the pursuit of excellent controls, and you hope to support people during the inevitable rough times. This is mutual, of course; the personal strengths, the humor and cheer, of people in my lab have carried me through challenging periods.

Which brings me to happiness. What in these homeodomain and Hedgehog experiences brought moments of joy? Primarily it was teamwork, usually small teams with everyone contributing in their different ways. Research is a refuge for a bunch of characters, often people who care more about seeking adventures and surprises than about money and stability. The shared roller coaster of successes and agonizing failures, the eventual insights, the feeling of having been present at historic moments, the laughter and rebelliousness of good science done by good people—all sources of joy. Another source of joy is seeing people new to science learn and prosper, then take their own independent leadership positions. I have mentioned some of the people who played a role in two lines of research in my lab, all of whom continue to make important contributions to science. Many others contributed along the way to these and other projects but cannot be included here—too many stories to tell. But deep deep thanks…you know who you are. I do have to mention Kaye Suyama and Diane Bush, who along with Ljiljana Milenkovic gave the lab decades of service. Their efforts led to many discoveries, rooted in the strong society they made in the lab.

I was fortunate in coming up through the *Drosophila* community, where scientific leaders much senior to me set standards of sharing, open communication, and good behavior that rippled though the entire community. These traditions greatly accelerated progress in the field, shamed some people who were being secretive or selfish, and made gatherings of the clan times of fun and excitement. I have been "burned" by being "too open," but those events are overshadowed by the many times when open discussions or exchanges of materials with competitors helped all of us.

Happiness for times past...but what about now? After the many insights of modern developmental biology, are we pretty much done? That's not a happy prospect. Fortunately, we are not even close to done. We know quite a lot about gene control, and we know quite a lot about signaling. We are learning how to make isolated cells, such as cultured stem cells, dance to various tunes. We know a lot of ways to break genes, and a lot of things that go wrong when we do. We have a lot still to learn about how development works, and how evolution of animal forms works. Can we prescribe the shapes and connections of muscles? Do we understand how a person's face is shaped to resemble her parents? Do we understand why some tissues regenerate well and others do not? What is the total information contained in an egg? Progress toward these questions and many others of developmental biology is substantial, but we are not there yet. We need many more years of teamwork—including across disciplines, many more years of sharing ideas and tools and reagents, many more years of strong lab societies that support scientists-in-training, many more free-wheeling debates. During this past quarter century or so we have identified much of the developmental machinery by taking it apart and now we have the difficult challenge of putting it all together. *That* is a happy prospect.

CHAPTER TWENTY

Turning One Cell Type into Another

Jonathan M.W. Slack[*,†,1]
[*]Department of Biology and Biochemistry, University of Bath, Bath, United Kingdom
[†]Department of Genetics, Cell Biology and Development, University of Minnesota, Minneapolis, Minnesota, USA
[1]Corresponding author: e-mail address: slack017@umn.edu

Contents

Abstract

The nature of cells in early embryos may be respecified simply by exposure to inducing factors. In later stage embryos, determined cell populations do not respond to inducing factors but may be respecified by other stimuli, especially the introduction of specific transcription factors. Fully differentiated cell types are hard to respecify by any method, but some degree of success can be achieved using selected combinations of transcription factors, and this may have clinical significance in the future.

1. INTRODUCTION

I came into developmental biology as a postdoc in 1974, some time before the molecular biology revolution of the 1980s turned it into the textbook subject it is today. At that time, the majority view was that development was all about the control of gene expression, so the way to study development was to investigate, for example, the mechanism of globin gene expression. This rather narrow focus seemed to me to ignore most of the interesting developmental problems. At the time, I was inspired by the analysis of Wolpert (1969) which drew attention to the problem of spatial

Current Topics in Developmental Biology, Volume 117
ISSN 0070-2153
http://dx.doi.org/10.1016/bs.ctdb.2015.11.017

patterning, also called regional specification, as being the real core of developmental biology. This led me to read voraciously in the literature of experimental embryology and, despite never having been actually taught the subject, all this reading eventually made me quite an expert. At the time, Wolpert's own experimental work was on the limb, an organ with a quasi-periodic pattern of cartilage and muscle. But it seemed to me that it would be hard to produce any biochemical assay to study this problem since all parts of the limb had the same cell types in them. To be able to distinguish anterior from posterior or dorsal from ventral in an *in vitro* test situation, we should need another system where different cell types were formed at different positions. This is why I chose initially to work on the early amphibian embryo. As in other vertebrates, each body level of the mesoderm from dorsal to ventral generates different cell types, so regional character could potentially be identified in a small tissue explant by identifying the predominant cell type.

My work developed through various stages: changing cell types in early embryos, which can be done by manipulating the inductive signals; changing cell types in regeneration, which occurs less often than is generally supposed; and changing mature differentiated cell types by overexpression of transcription factors. I also make an excursion into the study of naturally occurring metaplasias, which give some insight into the mechanisms of cell type interconversion. Here, I shall briefly recap each of these phases and the conclusions that they produced.

2. CHANGING CELL TYPE IN EARLY EMBRYOS

As indicated above, I was attracted to work on the problem of dorsoventral pattern in the early amphibian embryo. The classic work of Yamada (1937) had indicated that some sort of gradient controlled the formation of different cell types at different dorsoventral levels of the gastrula. Using combinations of dorsal and ventral tissue explants, I succeeded in showing that there really was a dorsalizing signal operating, distinct from the mesoderm-inducing signal previously discovered by Nieuwkoop (Nieuwkoop, 1969; Slack & Forman, 1980; Fig. 1A and B). Furthermore, this signal appeared very similar to the famous "organizer" of Spemann, but operating throughout gastrulation rather than just at the beginning. With the help of Jim Smith, my second postdoc, I reanalyzed using *Xenopus* the organizer graft, which had originally been studied in newt embryos. We found that the phenomenon was very similar to what Spemann had described in newts and that

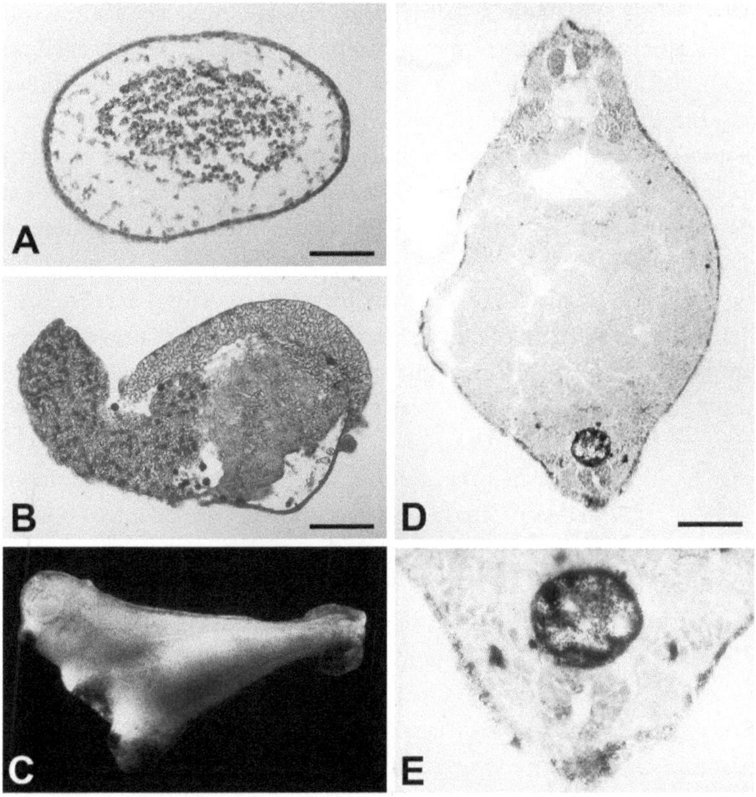

Figure 1 Respecifying cell type in early embryos. (A) An explant from the ventral marginal zone of a *Xenopus* gastrula, which has generated blood islands and mesenchyme. Scale bar: 100 μm. (B) A combination of a similar ventral-derived explant with the dorsal marginal zone from an axolotl gastrula (note the large red nuclei of the axolotl tissue). Here, the ventral-derived tissue has formed a large mass of muscle. Scale bar: 100 μm. (C) An organizer graft conducted with *Xenopus*. (D) Section through a similar specimen. Scale bar: 200 μm. The donor tissue was labeled with horseradish peroxidase. (E) Higher power view, showing the induction of both a neural tube and somites from the ventral tissue of the host. (A color version of this figure is located at the end of this chapter.)

there was certainly an element of dorsalization of the ventral tissues by the organizer (Smith & Slack, 1983). In particular, when the dorsal marginal zone was joined to the ventral region, the surrounding ventral mesoderm formed large blocks of muscle instead of the usual mesenchyme and blood islands (Fig. 1C–E). Up until this point, "the organizer" had been seen in most literature as a miraculous "instructive" agent, in some mysterious way conjuring up a complete secondary embryo from the ventral side of the host. Such an entity could not be explained in terms of any plausible mechanism. But explanation did become possible once we had established that it could be viewed as a number of separate simple systems, each resembling a gradient with threshold responses, and occurring in a sequence to generate the early body plan (Slack, Dale, & Smith, 1984).

During the early 1980s, my lab did a lot of fate mapping, explant, and combination experiments focusing on the formation of the various mesodermal tissue types: notochord, somitic muscle, kidney, lateral plate, and blood islands. We arrived at a "three signal model" in which the mesodermal-inducing signal came in dorsal and ventral flavors and was joined by the dorsalizing signal emanating from the organizer region (Dale & Slack, 1987a, 1987b). This simple model gave an outline explanation of the formation of pattern in early vertebrate development and focused attention on the identity of the presumed inductive signals. There had been a previous "gold rush" for the organizer, or at least the neural-inducing component of it, back in the 1930s; which had come to the eventual conclusion that the process was very unspecific and many substances could elicit a similar response (Needham, 1942). With the benefit of hindsight, we can see that workers of the 1930s could not possibly have isolated the growth factors that are responsible for embryonic induction because of their protein nature and very low concentrations. But they were also wrong about the lack of specificity. At least in *Xenopus*, it is not the case that many substances will induce neural tissue from gastrula ectoderm, still less have the full dorsalizing effect of an organizer graft. In fact rather few substances do so. After some trial and error, we found that the fibroblast growth factors (FGFs) had mesoderm-inducing activity and identified this growth factor family with the "ventral vegetal signal" of our model (Slack, Darlington, Heath, & Godsave, 1987; Fig. 2). Shortly afterward, a similar report came from David Kimelman at UCSF (Kimelman & Kirschner, 1987). A couple of years later, Jim Smith, now in his own lab at Mill Hill, purified a mesoderm-inducing factor from a *Xenopus* cell line and identified at as activin (Smith, Price, Van Nimmen, & Huylebroek, 1990). This had the properties of our presumed "dorsal vegetal signal."

The dorsalizing, or organizer, factor was identified by two lines of work. First, an expression cloning exercise for neural-inducing activity by Richard Harland produced a novel gene product called "noggin" (Smith & Harland, 1992). Second, work with BMP inhibitors indicated that dorsalization could result from inhibition of BMP normally produced by the ectoderm (Hemmati Brivanlou & Melton, 1994). Noggin was, in fact, a BMP inhibitor. So developmental biologists now seemed to have all three of the

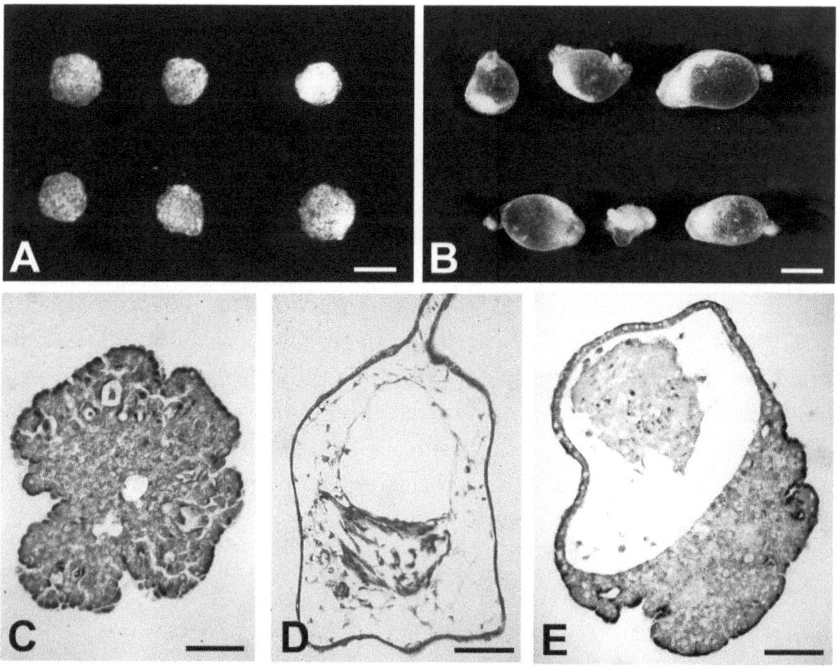

Figure 2 Effects of fibroblast growth factor (FGF) on *Xenopus* embryo explants. (A) Animal pole explants from *Xenopus* blastulae, allowed to differentiate *in vitro* for 3 days. (B) The same, but treated with FGF. (C and D) Sections through untreated and treated explants. (E) A muscle mass arising from a higher dose of FGF. Scale bars: A and B, 200 μm; C, 50 μm; D and E, 100 μm. (A color version of this figure is located at the end of this chapter.)

components of our three signal model in their hands: activin, FGF, and noggin. However, by this time, molecular biology had progressed enough to allow the study of actual gene expression in the early embryo and the situation started to become more complex.

It turns out that the mesoderm-inducing signal is not activin, but a mixture of activin-like proteins called "nodal-relateds." The difference in dorsal- and ventral-type signals arises from the activation of the Wnt signaling pathway on the dorsal side, resulting from cytoplasmic movements postfertilization (Weaver & Kimelman, 2004). FGF signaling is necessary for mesoderm induction, but it has only a permissive role at this early stage. The dorsalizing signal is a mixture of BMP inhibitors, particularly noggin and chordin, secreted from the organizer region where both Wnt and nodal signaling was high earlier on (Kimelman, 2006).

When we started studying FGFs in the embryo itself, we found that several of them are expressed in the early mesoderm and one of their roles is stabilizing the newly formed mesoderm through a feedback loop with the transcription factor brachyury (Isaacs, Pownall, & Slack, 1994; Fig. 3A and B). But the most important role for FGFs in early development actually turned out to be their role as "posteriorizing factors" (earlier called "transforming signals" in a prescient but hard-to-read paper of Nieuwkoop, 1952). They are secreted throughout gastrulation in the blastopore region, which becomes the posterior, and the nested upregulation of *Hox* genes which controls the patter of the posterior part of the body is regulated by these FGF signals, via upregulation of Cdx transcription factors (Isaacs, Pownall, & Slack, 1998; Pownall, Tucker, Slack, & Isaacs, 1996; Fig. 3C and D). It later turned out that there is a parallel system involving Wnt signals (Kiecker & Niehrs, 2001), so both FGF and Wnt are important for posterior patterning.

Thirty years after the work of the 1980s, the succession of signals and responses involved in early amphibian development is now well-known

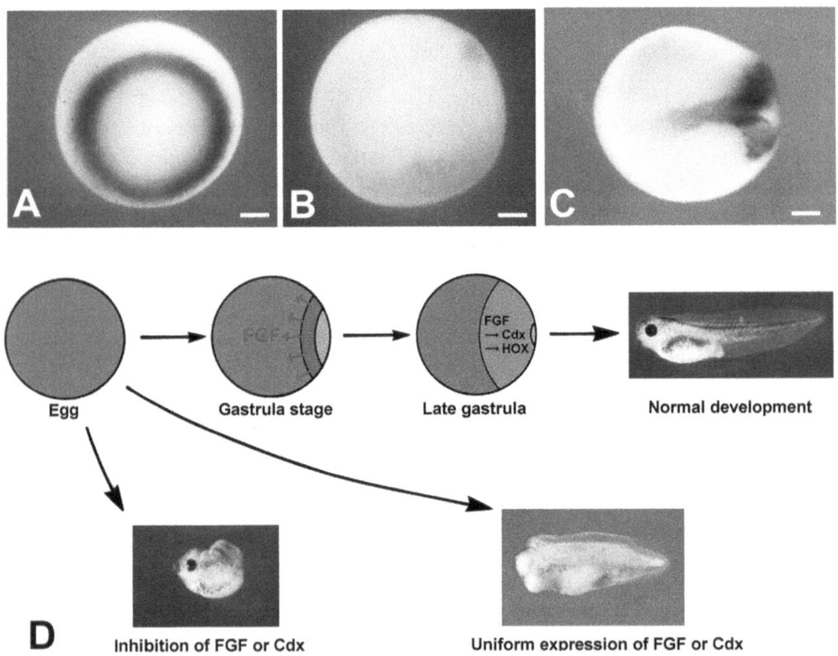

Figure 3 Endogenous expression and function of FGFs in *Xenopus* embryos. (A and B) Expression of *brachyury* in early gastrula without (A) and with (B) inhibition of FGF signaling by a dominant-negative receptor. (C) Normal expression of *eFGF* (*FGF4*) in a closed gastrula. (D) Role of FGFs in controlling the anteroposterior pattern: inhibition deletes the posterior while overexpression deletes the anterior. Scale bars: 230 μm. (A color version of this figure is located at the end of this chapter.)

textbook material (Gilbert, 2013; Slack, 2013; Wolpert & Tickle, 2011). The course of events in the zebrafish has turned out to be almost exactly the same (Schier & Talbot, 2005). That in the mouse and chick share many features, particularly the posteriorizing role of Wnts and FGFs, although there are also some differences (Skromne & Stern, 2001; Stern, 2005; Takaoka & Hamada, 2012). The subject of early regional specification and the inducing factors that control it is not now as exciting and fashionable as it once was. However, it left an important legacy because the substances now used to control the differentiation of human pluripotent stem cells are precisely those inducing factors discovered by developmental biologists in the 1980s: FGFs, activins, Wnts, and hedgehogs, together with their low-molecular-weight agonists and antagonists (Murry & Keller, 2008). So although understanding of the process of vertebrate early pattern formation had no practical consequences at the time of discovery, it did help feed into a technology that can at some stage be expected to deliver effective transplantation therapies for diabetes, heart failure, and various neural pathologies.

3. CHANGING AND NOT CHANGING CELL TYPES IN REGENERATION

My initial forays into regeneration arose because I had early on found a method of generating beautifully symmetrical double-posterior limbs by various grafts carried out on the prelimb bud stages of axolotl embryos (Slack, 1976; Fig. 4A). It was an easy experiment to amputate these and I found that they also regenerated as double-posterior duplications (Fig. 4B). This was of some interest as there was at the time an influential theory around that predicted that such structures should not regenerate at all (French, Bryant, & Bryant, 1976). Not only did my duplications regenerate very nicely, they also tended to become "narrower," for example, a limb with a 4-3-2-2-3-4 digit pattern might regenerate as a 4-3-3-4 pattern (Slack & Savage, 1978). This made the situation highly reminiscent of the posterior to anterior gradient controlling limb pattern during development (Tickle, Summerbell, & Wolpert, 1975) (now known to be composed of sonic hedgehog; Johnson & Tabin, 1997). While it seemed very likely that there would be a similar gradient in the regeneration blastema, the posterior signaling property could evidently also be "remembered" in the mature tissues of the limb, to become reactivated when a blastema forms after amputation. Doing grafts with immunologically tolerant tissue (Fig. 4C and D), I found that the posterior property of skin could last for at least 1 year in an

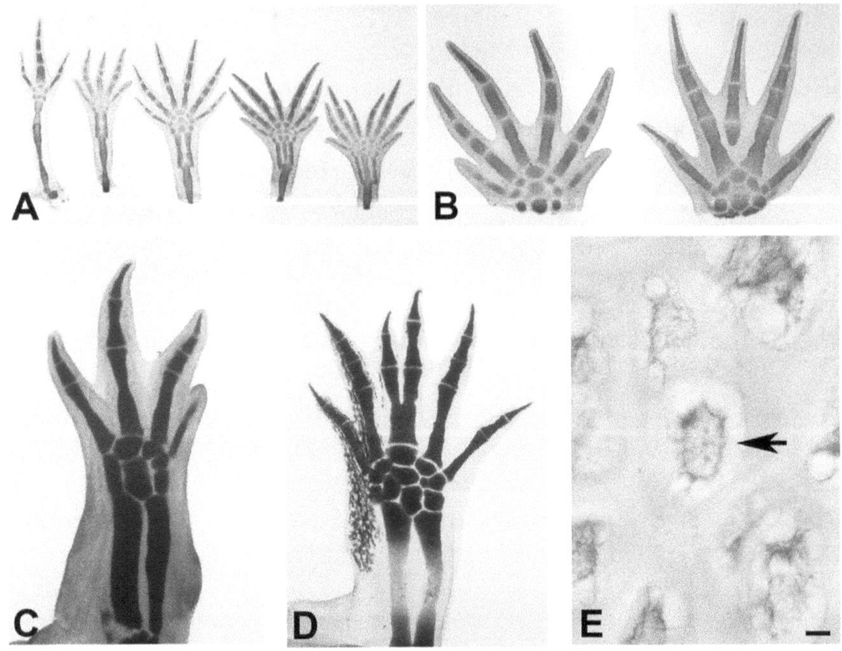

Figure 4 Limb regeneration. (A) Set of double-posterior duplicated limbs made by operations on the limb rudiment of axolotl embryos. (B) Amputation of one such through the wrist (left) gives rise to a regenerated limb (right) which is also double-posterior but of narrower pattern. (C and D) Immunocompatible skin grafts onto fully developed limbs followed by amputation through the graft. In (C), the graft is anterior skin to posterior limb and in (D), it is posterior skin to anterior limb. (E) After a similar skin graft, done using a triploid donor, some triploid cells can later been seen in regenerated cartilage (arrow). Scale bar: 15 μm. (A color version of this figure is located at the end of this chapter.)

anterior position, indicating that it was encoded by something much more stable than the diffusion gradient presumed to be active during development (Slack, 1980). Exactly what this property is remains unclear today, and it is significant that the term "positional information" is still used in regeneration research but not elsewhere in developmental biology. These early regeneration studies really concerned pattern rather than cell differentiation although I did in the course of the skin grafting work confirm previous reports that cells from the dermis could become cartilage during regeneration, indicating that there is real metaplasia in limb regeneration at least between the various connective tissue cell types (Slack, 1980; Fig. 4E).

When I returned to regeneration research 20 years later, I used *Xenopus* rather than axolotl because I felt that the range of techniques available for working with an established model organism like *Xenopus* were so much better. I focused on regeneration of the tadpole tail, as this followed naturally from some previous investigations we had made on normal tail development (Beck & Slack, 1999; Tucker & Slack, 1995a, 1995b). By then transgenic methods with *Xenopus* had become reliable (Amaya & Kroll, 1999) and so we could make uniformly GFP-transgenic embryos as donors for grafts. We made grafts from neurula stage embryos of pure pieces of neural plate, notochord, or somite, into unlabelled hosts. When these were grown to tadpole stage, the relevant part of the tail had just a single tissue type labeled with GFP. We amputated such tadpoles, allowing them to regenerate, and were surprised to discover a complete lack of metaplasia between these tissue types. Notochord and spinal cord regenerated from the same tissue of the stump, skeletal muscle arose from satellite cells in the stump, and pigment cells of the skin of the regenerated tail came from neural crest-derived precursors in the stump (Chen, Lin, & Slack, 2006; Gargioli & Slack, 2004; Lin, Chen, & Slack, 2007; Fig. 5A–E). Subsequently, the same technique was used by the lab of Elly Tanaka to show that this was also true of limb regeneration in the axolotl (Kragl et al., 2009). There is no metaplasia here either, except the previously demonstrated interconversion of dermis, cartilage, and tendons. Using dominant-negative inhibitors in transgenics, we showed that the FGF and Wnt pathways were needed for regeneration of both spinal cord and muscle (Fig. 5F). In common with other regeneration systems, we found that the spinal cord is required for regeneration of the muscle but not vice versa (Beck, Christen, & Slack, 2003; Lin, Chen, & Slack, 2012; Fig. 5G and H).

4. NATURALLY OCCURRING CHANGES OF CELL TYPE

Back in the early 1980s, I had been intrigued by the fact that arthropods, most notably *Drosophila*, but other arthropods as well, exhibited homeotic mutants, in which there was a total and qualitative change of one body part into another, whereas vertebrates did not. If developmental mechanisms were broadly similar in all animals, then one might expect to find mutations that caused changes of body part in all animals. When I looked into it more carefully, it turned out that there are such changes in vertebrates after all (Fig. 6). In human pathology, they are called metaplasias if they occur during adult life, or heterotopias if they occur during embryonic development. They had not previously been recognized as homeotic transformations because superficially they do not much resemble the homeotic mutations of *Drosophila*. Metaplasias tend to be small foci of a

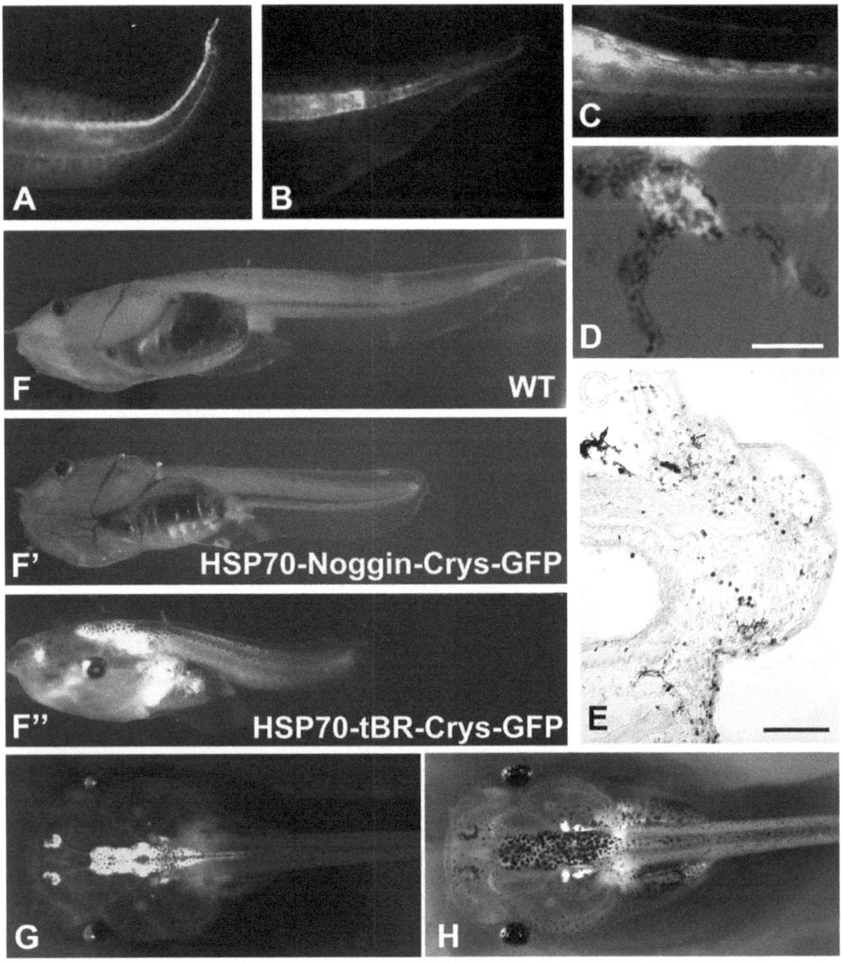

Figure 5 Lack of metaplasia in tail regeneration. (A–D) Regenerated *Xenopus* tadpole tails in which one tissue has been labeled by grafting of a *GFP*-transgenic tissue to an unlabelled host embryo. (A) Spinal cord. (B) Notochord. (C) Skeletal muscle. (D) Melanocyte. Scale bar: 10 μm. (E) Pax7+ muscle satellite cells in the tail regeneration bud. Scale bar: 100 μm. (F–F″) Inhibition of tail regeneration by blockage of BMP signaling using transgenics with inducible dominant-negative inhibitors. (G and H) N-tubulin (G) and cardiac actin (H) promoters used to drive dominant-negative inhibitor expression in experiments on regeneration of specific tissue types. (A color version of this figure is located at the end of this chapter.)

certain tissue type in the wrong position, whereas homeoses are classically transformation of whole organs, like legs or wings, into each other. However, the differences are more apparent than real. The homeotic mutations of *Drosophila* have been selectively bred in the lab to become much more

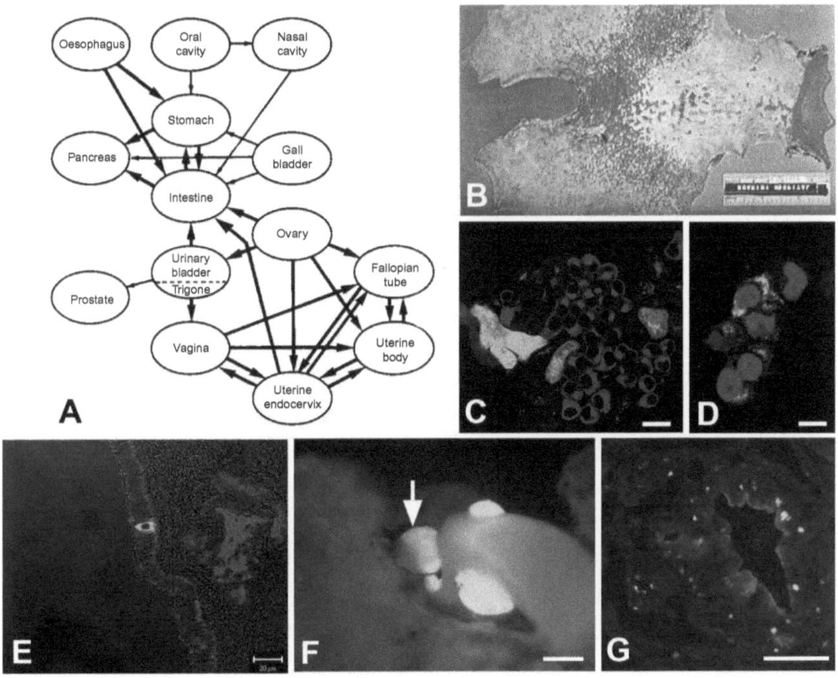

Figure 6 Metaplasia. (A) "Kinematic graph" summarizing metaplastic transformations of epithelia described in the human pathology literature (Slack, 1985). (B) Intestinal metaplasia in a human stomach, visualized by histochemical stain for alkaline phosphatase, also visible in the duodenum fragment to the right (Stemmermann & Hayashi, 1968). (C) Transformation of AR42j cells from a pancreatic to hepatic phenotype using dexamethasone. Green cells express glucose-6-phosphatase, normally found in hepatocytes but not pancreatic cells. Scale bar: 10 μm. (D) A similar transformation brought about by the transcription factor C/EBPβ. The green cells express transferrin. Scale bar: 10 μm. (E) An isolated insulin-positive endocrine cell (green) in the epithelium of a mouse bile duct. Scale bar: 20 μm. (F) Formation of ectopic pancreas (arrow) in a *Xenopus* tadpole by overexpression of *Pdx1–VP16* under control of a liver-specific promoter. The two right hand blobs of green are the normal pancreas. Scale bar: 200 μm. (G) Mixed phenotype cells of a small bile duct in the mouse liver overexpressing *Pdx1* + *Ngn3* + *MafA*. Red is insulin, and green is EdU indicating dividing cells. Scale bar: 100 μm. (A color version of this figure is located at the end of this chapter.)

penetrant than they originally were in nature. Also they are manifest in a single tissue layer, the epidermis, which forms all the hairs and bristles that enable identification of body parts in insects. An archetypal metaplasia commonly found in humans is intestinal metaplasia of the stomach (Fig. 6), where patches of intestinal epithelium are found in the stomach. These patches may be quite large and where several foci have joined together can cover most of the stomach (Stemmermann & Hayashi, 1968). There are different histological subtypes, but in the so-called complete metaplasia the ectopic tissue contains all the normal cell types of the intestine in the normal arrangement. Since both stomach and intestinal epithelia are composed of several cell types arising from their respective types of stem cell, and they are normally found at different body levels, this ectopic formation of intestine seems very similar to a homeotic event.

Because of this I proposed in 1985 that metaplasias arose from a switch in the character of tissue-specific stem cells caused by misexpression of one or more of the developmental control genes that normally distinguished the tissue types in question (Slack, 1985). This general explanation has subsequently been confirmed. For example, intestinal metaplasia of the stomach is associated with misexpression of *Cdx* genes (Silberg et al., 1997), which normally distinguishes intestine from stomach, and squamous metaplasia of the bronchus with misexpression of *p63* (Sheikh, Fuhrer, Cieply, & Yousem, 2004) which is a master control gene for squamous epithelia. Furthermore, I predicted that individual foci of metaplasia would be monoclonal, as the initiating event is likely to occur in a single cell and to be heritable on cell division. This was recently confirmed for intestinal metaplasia using a method for human cell lineage tracing based on somatic mutation leading to loss of function of cytochrome oxidase in mitochondria (Gutierrez-Gonzalez et al., 2011).

Back in 1985, it was hard to think of feasible experiments to pursue this question further. But by the 2000s, molecular biology had progressed sufficiently to make it relatively easy to overexpress specific genes or dominant-negative inhibitors in cells or whole animals, and at this point my interest in metaplasias and their mechanism led to experiments on what is now called "direct reprogramming."

5. CHANGING CELL TYPE BY BRUTE FORCE

Nowadays, the term "direct reprogramming" means converting one cell type to another by introduction of transcription factors. My introduction to this issue came partly from the consideration of metaplasias mentioned above, and also from a chance observation by my colleague David Tosh that a rat pancreatic exocrine cell line, AR42j-B13, could be converted to hepatocyte phenotype by treatment with the synthetic glucocorticoid dexamethasone. An *in vitro* system like this seemed easier to experiment with than whole human beings. On further investigation, we showed that the transformation could also be achieved by introduction of a single transcription factor gene, *C/EBPβ*, and that once the transformation had occurred it was stable in the absence of the original stimulus (Shen, Slack, & Tosh, 2000; Fig. 6). This was the first case of direct reprogramming apart from the discovery of MyoD by Harold Weintraub some years previously (Davis, Weintraub, & Lassar, 1987). David Tosh went on to characterize the hepatocytes in detail and showed that they were of good quality, with characteristic features of mature hepatocytes such as the presence of bile canaliculi, secretion of urea, and the presence of class 2 detoxifying enzymes (Tosh, Shen, & Slack, 2002).

This experience gave us the idea of trying to go the opposite way and using transcription factors to convert hepatocytes to pancreatic cells. We achieved an early success with *Xenopus*, where we could use our transgenic expertise to show that an activated form of the pancreatic transcription factor Pdx1 (Pdx1–VP16) would respecify the immature liver to produce patches of ectopic pancreas (Horb, Shen, Tosh, & Slack, 2003; Fig. 6F). However, this effect is lost by about 7 days of development, as the liver differentiates. This suggests that differentiated but immature cell types can be more readily reprogrammed than fully differentiated ones. At the same time, they are less readily respecified than early embryo cells. For example, the 4-day liver of a *Xenopus* tadpole is already "determined" in the embryological sense of being irreversibly committed and unable to be respecified by grafting to different positions or exposure to inducing factors.

Reprogramming mammalian liver to pancreas seemed to be a more difficult task. The real holy grail was the production of pancreatic beta cells because these could potentially be used in the treatment of diabetes (Bonner-Weir & Weir, 2005). There is a clinical procedure known as islet transplantation which is used especially for diabetic patients who have "unawareness of hypoglycemia." This condition is very dangerous as it can lead to unconsciousness without warning or to death during sleep. In islet transplantation, the islets from a donor pancreas are infused into the portal vein of the patient and lodge in the liver. The grafted islets remain active and are very successful at restoring awareness of hypoglycemia, although somewhat less successful at obviating the requirement for regular insulin injections (Robertson, 2010). The main problem with islet transplants is the lack of donors, and so many stem cell and developmental biology labs have become interested in novel ways of making beta cells or beta cell substitutes.

Early reports by the lab of Sarah Ferber in Israel showed that introduction of *Pdx1* into the liver upregulated various pancreatic and beta cell genes (Ber et al., 2003; Ferber et al., 2000). However, these experiments were hard to reproduce, perhaps because just the right degree of liver damage is required to get an effect. One by-product of our efforts to do so was the discovery that the liver actually does normally contain a few pancreatic endocrine cells, associated with large bile ducts in the hilar region, something which had never been observed before (Dutton et al., 2007; Fig. 6E). We found that these cells develop from the bile duct epithelium during late embryonic development (Eberhard, Tosh, & Slack, 2008).

In 2008, Doug Melton's lab published a paper claiming complete *in vivo* transformation of mouse pancreatic exocrine cells to beta cells using a three gene cocktail: *Pdx1*, *Ngn3*, and *MafA* (Zhou, Brown, Kanarek, Rajagopal, & Melton, 2008). We then studied the effects of this gene combination in some detail. We confirmed that it did induce insulin-positive cells in the pancreas, although an *in vitro* study of the effects on the exocrine-like AR42j-B13 cells suggested that the reprogramming was partial and that the key property of glucose-stimulated insulin secretion was missing (Akinci, Banga, Greder, Dutton, & Slack, 2012). We did find that fetal hepatoblasts gave a much higher percentage of insulin-positive cells than mature liver (Yang, Akinci, Dutton, Banga, & Slack, 2013), rather similar to the transgenic *Xenopus* model studied previously. The most interesting result was the discovery that it was possible to induce glucose-responsive cells from the small bile ducts of the adult liver (Banga, Akinci, Greder, Dutton, & Slack, 2012; Banga, Greder, Dutton, & Slack, 2014; Fig. 6G). These also are not fully reprogrammed to beta cells as they retain various morphological and biochemical characteristics of ducts. However, they are able to relieve experimental diabetes in mice and a therapy based on this phenomenon might be of use in a veterinary context, especially for dogs. As far as the quality of other examples of direct reprogramming is concerned, a recent study of genetic regulatory networks has indicated that none of the well-publicized examples, such as transformation of fibroblasts to neurons or cardiomyocytes, are really complete, all yield mixed phenotypes to some degree (Cahan et al., 2014). This indicates that it is not easy to reprogram mature differentiated cell types completely and that many partial and incomplete phenotypes can be stable, although from a clinical point of view this may not matter so long as they are functionally competent for the purpose required.

6. CONCLUSIONS

It has been known for a long time that regions of early embryos could often be respecified simply by grafting from one place to another. This is because the environment in terms of inducing factors is altered, and nowadays similar effects can be obtained by treating with purified protein factors or by suitable gene overexpression experiments. We can understand the process in terms of activation or repression of combination of genes required to specify early body regions (for example, *brachyury/T* for mesoderm or *cdx* genes for posterior body levels). The target genes are "open for business" and so the stimulation of appropriate signal transduction pathways by inducing factors is sufficient to bring about a new configuration of gene activity and a change in state of commitment. The effects of FGFs on early *Xenopus* ectoderm are a typical example of this (Slack et al., 1987).

The classical embryological concept of "determination" indicated an invisible, but stable and irreversible, commitment to a specific body part or cell type (Slack, 1983). Such states are no longer invisible due to the current techniques of *in situ* hybridization and RNAseq. By definition, such states are stable to changes of inducing factors in the environment. But they are evidently less stable to perturbation than fully differentiated cell types. This is indicated by many observations indicating the relative instability of determined cell types, including for example, the fact that embryonic hepatoblasts are more labile to overexpression of the gene combination *Pdx1* + *Ngn3* + *MafA* than are mature hepatocytes (Yang et al., 2013). It is probable that the nature of determination is similar to the state of tissue-specific stem cells. We know that these can occasionally alter their change state of commitment, as in cases of metaplasia, or the very rare but occasionally well documented "transdifferentiations" of grafted hematopoietic stem cells (Tosh & Slack, 2002). The nature of this state is still unclear. The simple introduction of a transcription factor is not necessarily sufficient to activate its target genes, so there must be some chromatin structure stabilizing the current configuration of gene activation and repression, which needs to be opened up to enable the transcription factor to see its targets in the DNA sequence.

Fully differentiated cell types have long been known to have a number of specialized gene products expressed at very high levels which are those proteins required to do the function of the cell: such as muscle contraction, nervous conduction, or secretion. Inactive regions of the genome may by this stage be condensed into heterochromatin and be very difficult to reactivate, although the environment of the oocyte nucleus does seem able to open up even the most tightly condensed chromatin (Tamada et al., 2006). It has recently been found that some transcription factors, called "pioneer factors" have the ability to find their target sequences in DNA even when they are located in closed chromatin (Zaret & Carroll, 2011). The cocktails used for direct reprogramming usually incorporate at least one pioneer factor which explains how they can work at all.

Another issue highlighted by this topic is the difference between cell types, tissue types, organs, and body parts. One of the key insights of experimental embryology was that the early inductive signals did not control the formation of specific differentiated cell types as such but instead controlled the formation of whole body regions, such as germ layers or anteroposterior body levels. In the early days, phenomena such as the partial loss-of-function allele of *engrailed* in *Drosophila* indicated that activity of this gene encoded "posterior" status in developing imaginal discs rather than any specific cell type (Morata & Lawrence, 1975). Once molecular biology became able to clone genes of developmental interest, many were found which did encode body position, the most famous being the *Hox* genes which control, among other things, the anteroposterior body pattern of all animals (McGinnis & Krumlauf, 1992). In the hierarchy of things, tissue types and organs evidently lie somewhere between body parts and cell types. Despite the vast literature of histology and pathology, mostly concerned with study of tissues, there seems to be no generally accepted definition of a tissue. But to me, a tissue is the ensemble of cell types formed from a single sort of stem cell or progenitor cell. This is pretty clear when considering tissue-specific stem cells. For example, the intestinal epithelium contains stem cells which generate all of the component cell types: absorptive, goblet, enteroendocrine, and Paneth cells (Barker, van de Wetering, & Clevers, 2008). So it makes sense to consider the intestinal epithelium as one tissue. It is combined with various other tissues: muscle layers, loose connective tissues, lymphatic tissue, nerves, and blood vessels, to form the intestine, which may be regarded as an organ. This viewpoint has some merit in giving an objective foundation to the terms "tissue" and "organ." Future research can certainly be expected to give a much fuller understanding of both determined and differentiated states, and a much better repertoire of techniques for turning one of these states into another.

REFERENCES

Akinci, E., Banga, A., Greder, L. V., Dutton, J. R., & Slack, J. M. W. (2012). Reprogramming of pancreatic exocrine cells towards a beta cell character using Pdx1, Ngn3 and MafA. *Biochemical Journal, 442*, 539–550.

Amaya, E., & Kroll, K. L. (1999). A method for generating transgenic frog embryos. *Methods in Molecular Biology, 97*, 393–414.

Banga, A., Akinci, E., Greder, L. V., Dutton, J. R., & Slack, J. M. W. (2012). In vivo reprogramming of Sox9 + cells in the liver to insulin-secreting ducts. *Proceedings of the National Academy of Sciences of the United States of America, 109*, 15336–15341.

Banga, A., Greder, L. V., Dutton, J. R., & Slack, J. M. W. (2014). Reprogramming of progenitor cells in the liver to a pancreatic endocrine phenotype using a three gene cocktail and a PPAR agonist. *Gene Therapy, 21*, 19–27.

Barker, N., van de Wetering, M., & Clevers, H. (2008). The intestinal stem cell. *Genes & Development, 22*, 1856–1864.

Beck, C. W., Christen, B., & Slack, J. M. W. (2003). Molecular pathways needed for regeneration of spinal cord and muscle in a vertebrate. *Developmental Cell, 5*, 429–439.

Beck, C., & Slack, J. M. W. (1999). A developmental pathway controlling outgrowth of the Xenopus tail bud. *Development, 126*, 1611–1620.

Ber, I., Shternhall, K., Perl, S., Ohanuna, Z., Goldberg, I., Barshack, I., et al. (2003). Functional, persistent, and extended liver to pancreas transdifferentiation. *The Journal of Biological Chemistry, 278*, 31950–31957.

Bonner-Weir, S., & Weir, G. C. (2005). New sources of pancreatic beta-cells. *Nature Biotechnology, 23*, 857–861.

Cahan, P., Li, H., Morris, S. A., Lummertz da Rocha, E., Daley, G. Q., & Collins, J. J. (2014). CellNet: Network biology applied to stem cell engineering. *Cell, 158*, 903–915.

Chen, Y., Lin, G. F., & Slack, J. M. W. (2006). Control of muscle regeneration in the Xenopus tadpole tail by Pax7. *Development, 133*, 2303–2313.

Dale, L., & Slack, J. M. W. (1987a). Fate map for the 32 cell stage of *Xenopus laevis*. *Development, 99*, 527–551.

Dale, L., & Slack, J. M. W. (1987b). Regional specification within the mesoderm of early embryos of *Xenopus laevis*. *Development, 100*, 279–295.

Davis, R. L., Weintraub, H., & Lassar, A. B. (1987). Expression of a single transfected cDNA converts fibroblasts to myoblasts. *Cell, 51*, 987–1000.

Dutton, J. R., Chillingworth, N. L., Eberhard, D., Brannon, C. R., Hornsey, M. A., Tosh, D., et al. (2007). Beta cells occur naturally in extrahepatic bile ducts of mice. *Journal of Cell Science, 120*, 239–245.

Eberhard, D., Tosh, D., & Slack, J. M. W. (2008). Origin of pancreatic endocrine cells from biliary duct epithelium. *Cellular and Molecular Life Sciences, 65*, 3467–3480.

Ferber, S., Halkin, A., Cohen, H., Ber, I., Einav, Y., Goldberg, I., et al. (2000). Pancreatic and duodenal homeobox gene 1 induces expression of insulin genes in liver and ameliorates streptozotocin-induced hyperglycemia. *Nature Medicine, 6*, 568–572.

French, V., Bryant, P. J., & Bryant, S. V. (1976). Pattern regulation in epimorphic fields. *Science, 193*, 969–981.

Gargioli, C., & Slack, J. M. W. (2004). Cell lineage tracing during Xenopus tail regeneration. *Development, 131*, 2669–2679.

Gilbert, S. F. (2013). *Developmental biology*. Sunderland, MA: Sinauer Associates Inc.

Gutierrez-Gonzalez, L., Graham, T. A., Rodriguez-Justo, M., Leedham, S. J., Novelli, M. R., Gay, L. J., et al. (2011). The clonal origins of dysplasia from intestinal metaplasia in the human stomach. *Gastroenterology, 140*, 1251–1260. e6.

Hemmati Brivanlou, A., & Melton, D. A. (1994). Inhibition of activin receptor signalling promotes neuralization in Xenopus. *Cell, 77*, 273–281.

Horb, M. E., Shen, C. N., Tosh, D., & Slack, J. M. W. (2003). Experimental conversion of liver to pancreas. *Current Biology, 13*, 105–115.

Isaacs, H. V., Pownall, M. E., & Slack, J. M. W. (1994). eFGF regulates Xbra expression during Xenopus gastrulation. *The EMBO Journal, 13*, 4469–4481.

Isaacs, H. V., Pownall, M. E., & Slack, J. M. W. (1998). Regulation of Hox gene expression and posterior development by the Xenopus caudal homologue Xcad3. *EMBO Journal, 17*, 3413–3427.

Johnson, R. L., & Tabin, C. J. (1997). Molecular models for vertebrate limb development. *Cell, 90*, 979–990.

Kiecker, C., & Niehrs, C. (2001). A morphogen gradient of Wnt/beta-catenin signalling regulates anteroposterior neural patterning in Xenopus. *Development, 128*, 4189–4201.

Kimelman, D. (2006). Mesoderm induction: From caps to chips. *Nature Reviews. Genetics, 7*, 360–372.

Kimelman, D., & Kirschner, M. (1987). Synergistic induction of mesoderm by FGF and TGF-b and the identification of an mRNA coding for FGF in the early Xenopus embryo. *Cell, 51*, 869–877.

Kragl, M., Knapp, D., Nacu, E., Khattak, S., Maden, M., Epperlein, H. H., et al. (2009). Cells keep a memory of their tissue origin during axolotl limb regeneration. *Nature, 460*, 60–65.

Lin, G., Chen, Y., & Slack, J. M. W. (2007). Regeneration of melanophores and other neural crest derivatives in the Xenopus tadpole tail. *BMC Developmental Biology, 7*, 56.

Lin, G., Chen, Y., & Slack, J. M. W. (2012). Transgenic analysis of signaling pathways required for Xenopus tadpole spinal cord and muscle regeneration. *Anatomical Record, 295*, 1532–1540.

McGinnis, W., & Krumlauf, R. (1992). Homeobox genes and axial patterning. *Cell, 68*, 283–302.

Morata, G., & Lawrence, P. A. (1975). Control of compartment development by the engrailed gene in Drosophila. *Nature, 255*, 614–617.

Murry, C. E., & Keller, G. (2008). Differentiation of embryonic stem cells to clinically relevant populations: Lessons from embryonic development. *Cell, 132*, 661–680.

Needham, J. (1942). *Biochemistry and morphogenesis*. Cambridge: Cambridge University Press.

Nieuwkoop, P. D. (1952). Activation and organization of the central nervous system in amphibians III. Synthesis of a new working hypothesis. *The Journal of Experimental Zoology, 120*, 83–108.

Nieuwkoop, P. D. (1969). The formation of the mesoderm in urodelean amphibians I. Induction by the endoderm. *Wilhelm Roux' Archiv für Entwicklungsmechanik der Organismen, 162*, 341–373.

Pownall, M. E., Tucker, A. S., Slack, J. M., & Isaacs, H. V. (1996). eFGF, Xcad3 and Hox genes form a molecular pathway that establishes the anteroposterior axis in Xenopus. *Development, 122*, 3881–3892.

Robertson, R. P. (2010). Islet transplantation a decade later and strategies for filling a half-full glass. *Diabetes, 59*, 1285–1291.

Schier, A. F., & Talbot, W. S. (2005). Molecular genetics of axis formation in zebrafish. *Annual Review of Genetics, 39*, 561–613.

Sheikh, H. A., Fuhrer, K., Cieply, K., & Yousem, S. (2004). p63 expression in assessment of bronchioloalveolar proliferations of the lung. *Modern Pathology, 17*, 1134–1140.

Shen, C. N., Slack, J. M. W., & Tosh, D. (2000). Molecular basis of transdifferentiation of pancreas to liver. *Nature Cell Biology, 2*, 879–887.

Silberg, D. G., Furth, E. E., Taylor, J. K., Schuck, T., Chiou, T., & Traber, P. G. (1997). CDX1 protein expression in normal, metaplastic, and neoplastic human alimentary tract epithelium. *Gastroenterology, 113*, 478–486.

Skromne, I., & Stern, C. D. (2001). Interactions between Wnt and Vg1 signalling pathways initiate primitive streak formation in the chick embryo. *Development, 128*, 2915–2927.

Slack, J. M. W. (1976). Determination of polarity in the amphibian limb. *Nature, 261*, 44–46.

Slack, J. M. W. (1980). Morphogenetic properties of the skin axolotl limb regeneration. *Journal of Embryology and Experimental Morphology, 58*, 265–288.

Slack, J. M. W. (1983). *From egg to embryo. Determinative events in early development.* Cambridge: Cambridge University Press.

Slack, J. M. W. (1985). Homeotic transformations in Man: Implications for the mechanism of embryonic development and for the organization of epithelia. *Journal of Theoretical Biology, 114*, 463–490.

Slack, J. M. W. (2013). *Essential developmental biology.* Oxford: Wiley-Blackwell.

Slack, J. M. W., Dale, L., & Smith, J. C. (1984). Analysis of embryonic induction by using cell lineage markers. *Philosophical Transactions of the Royal Society of London. Series B, 307*, 331–336.

Slack, J. M. W., Darlington, B. G., Heath, J. K., & Godsave, S. F. (1987). Mesoderm induction in early Xenopus embryos by heparin-binding growth factors. *Nature, 326*, 197–200.

Slack, J. M. W., & Forman, D. (1980). An interaction between dorsal and ventral regions of the marginal zone in early amphibian embryos. *Journal of Embryology and Experimental Morphology, 56*, 283–299.

Slack, J. M. W., & Savage, S. (1978). Regeneration of mirror symmetrical limbs in the axolotl. *Cell, 14*, 1–8.

Smith, W. C., & Harland, R. M. (1992). Expression cloning of noggin, a new dorsalizing factor localised to the Spemann organizer in Xenopus embryos. *Cell, 70*, 829–840.

Smith, J. C., Price, B. M. J., Van Nimmen, K., & Huylebroek, D. (1990). Identification of a potent Xenopus mesoderm inducing factor as a homologue of active A. *Nature, 345*, 729–731.

Smith, J. C., & Slack, J. M. W. (1983). Dorsalization and neural induction: Properties of the organizer in Xenopus laevis. *Journal of Embryology and Experimental Morphology, 78*, 299–317.

Stemmermann, G. N., & Hayashi, T. (1968). Intestinal metaplasia of the gastric mucosa. A gross and microscopic study of its distribution in various disease states. *Journal of the National Cancer Institute, 41*, 627–634.

Stern, C. D. (2005). Neural induction: Old problem, new findings, yet more questions. *Development, 132*, 2007–2021.

Takaoka, K., & Hamada, H. (2012). Cell fate decisions and axis determination in the early mouse embryo. *Development, 139*, 3–14.

Tamada, H., Van Thuan, N., Reed, P., Nelson, D., Katoku-Kikyo, N., Wudel, J., et al. (2006). Chromatin decondensation and nuclear reprogramming by nucleoplasmin. *Molecular and Cellular Biology, 26*, 1259–1271.

Tickle, C., Summerbell, D., & Wolpert, L. (1975). Positional signalling and specification of digits in chick limb morphogenesis. *Nature, 254*, 199–202.

Tosh, D., Shen, C. N., & Slack, J. M. W. (2002). Differentiated properties of hepatocytes induced from pancreatic cells. *Hepatology, 36*, 534–543.

Tosh, D., & Slack, J. M. W. (2002). How cells change their phenotype. *Nature Reviews. Molecular Cell Biology, 3*, 187–194.

Tucker, A. S., & Slack, J. M. W. (1995a). Tail bud determination in the vertebrate embryo. *Current Biology, 5*, 807–813.

Tucker, A. S., & Slack, J. M. W. (1995b). The Xenopus laevis tail-forming region. *Development, 121*, 249–262.

Weaver, C., & Kimelman, D. (2004). Move it or lose it: Axis specification in Xenopus. *Development, 131*, 3491–3499.

Wolpert, L. (1969). Positional information and the spatial pattern of cellular differentiation. *Journal of Theoretical Biology, 25*, 1–47.

Wolpert, L., & Tickle, C. (2011). *Principles of development.* Oxford and New York: Oxford University Press.

Yamada, T. (1937). Das Determinationzustand des Rumpfmesoderms im Molchkeim nach der Gastrulation. *Wilhelm Roux' Archiv für Entwicklungsmechanik der Organismen, 137*, 151–270.

Yang, Y., Akinci, E., Dutton, J. R., Banga, A., & Slack, J. M. W. (2013). Stage specific reprogramming of mouse embryo liver cells to a beta cell-like phenotype. *Mechanisms of Development, 130*, 601–612.

Zaret, K. S., & Carroll, J. S. (2011). Pioneer transcription factors: Establishing competence for gene expression. *Genes & Development, 25*, 2227–2241.

Zhou, Q., Brown, J., Kanarek, A., Rajagopal, J., & Melton, D. A. (2008). In vivo reprogramming of adult pancreatic exocrine cells to beta-cells. *Nature, 455*, 627–632.

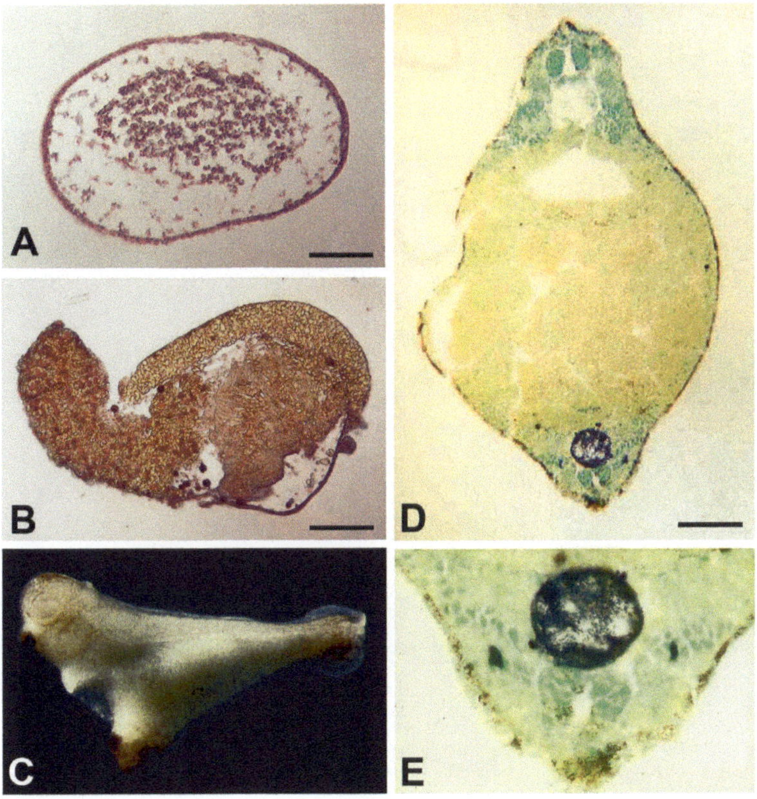

Jonathan M.W. Slack, Figure 1 Respecifying cell type in early embryos. (A) An explant from the ventral marginal zone of a *Xenopus* gastrula, which has generated blood islands and mesenchyme. Scale bar: 100 μm. (B) A combination of a similar ventral-derived explant with the dorsal marginal zone from an axolotl gastrula (note the large red nuclei of the axolotl tissue). Here, the ventral-derived tissue has formed a large mass of muscle. Scale bar: 100 μm. (C) An organizer graft conducted with *Xenopus*. (D) Section through a similar specimen. Scale bar: 200 μm. The donor tissue was labeled with horse-radish peroxidase. (E) Higher power view, showing the induction of both a neural tube and somites from the ventral tissue of the host.

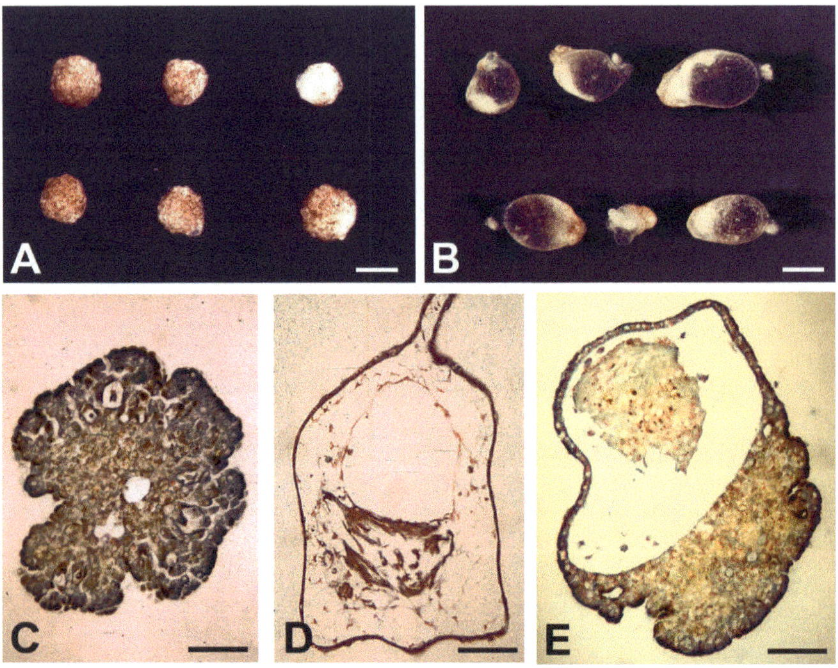

Jonathan M.W. Slack, Figure 2 Effects of fibroblast growth factor (FGF) on *Xenopus* embryo explants. (A) Animal pole explants from *Xenopus* blastulae, allowed to differentiate *in vitro* for 3 days. (B) The same, but treated with FGF. (C and D) Sections through untreated and treated explants. (E) A muscle mass arising from a higher dose of FGF. Scale bars: A and B, 200 μm; C, 50 μm; D and E, 100 μm.

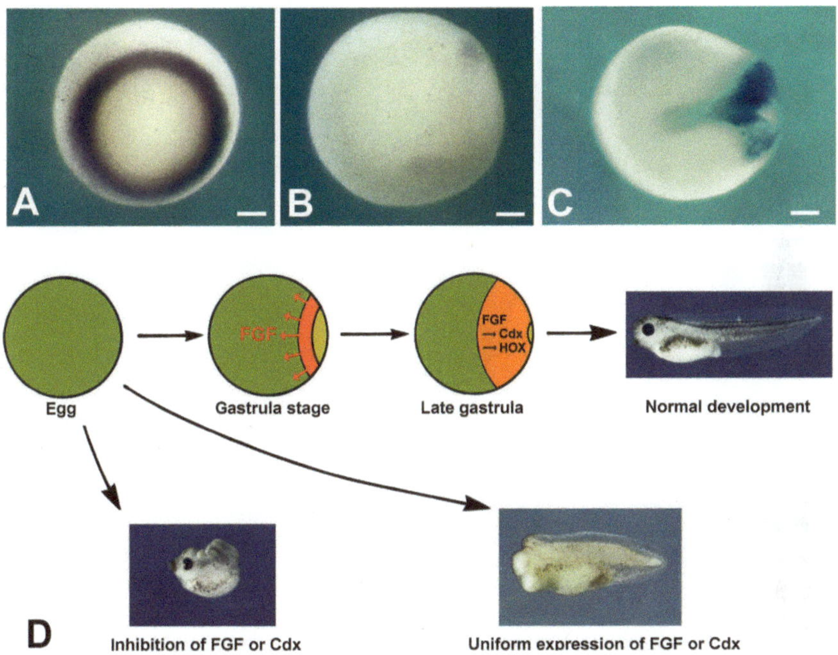

Jonathan M.W. Slack, Figure 3 Endogenous expression and function of FGFs in *Xenopus* embryos. (A and B) Expression of *brachyury* in early gastrula without (A) and with (B) inhibition of FGF signaling by a dominant-negative receptor. (C) Normal expression of *eFGF* (*FGF4*) in a closed gastrula. (D) Role of FGFs in controlling the anteroposterior pattern: inhibition deletes the posterior while overexpression deletes the anterior. Scale bars: 230 μm.

Jonathan M.W. Slack, Figure 4 Limb regeneration. (A) Set of double-posterior duplicated limbs made by operations on the limb rudiment of axolotl embryos. (B) Amputation of one such through the wrist (left) gives rise to a regenerated limb (right) which is also double-posterior but of narrower pattern. (C and D) Immunocompatible skin grafts onto fully developed limbs followed by amputation through the graft. In (C), the graft is anterior skin to posterior limb and in (D), it is posterior skin to anterior limb. (E) After a similar skin graft, done using a triploid donor, some triploid cells can later been seen in regenerated cartilage (arrow). Scale bar: 15 μm.

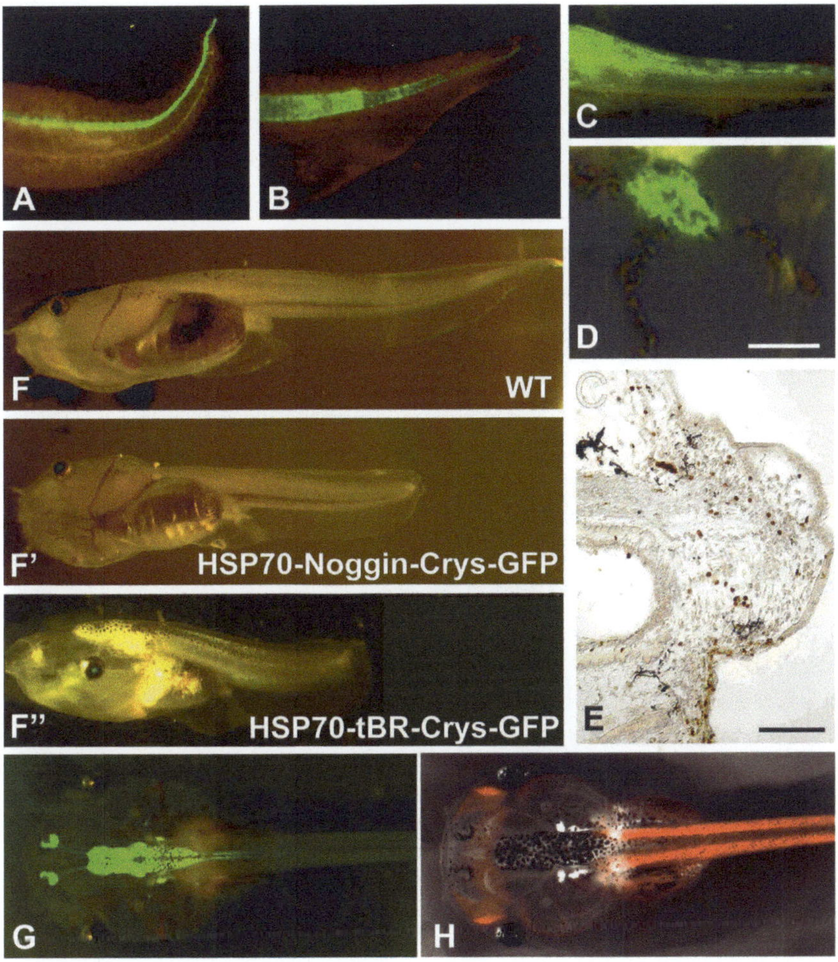

Jonathan M.W. Slack, Figure 5 Lack of metaplasia in tail regeneration. (A–D) Regenerated *Xenopus* tadpole tails in which one tissue has been labeled by grafting of a *GFP*-transgenic tissue to an unlabelled host embryo. (A) Spinal cord. (B) Notochord. (C) Skeletal muscle. (D) Melanocyte. Scale bar: 10 μm. (E) Pax7+ muscle satellite cells in the tail regeneration bud. Scale bar: 100 μm. (F–F″) Inhibition of tail regeneration by blockage of BMP signaling using transgenics with inducible dominant-negative inhibitors. (G and H) N-tubulin (G) and cardiac actin (H) promoters used to drive dominant-negative inhibitor expression in experiments on regeneration of specific tissue types.

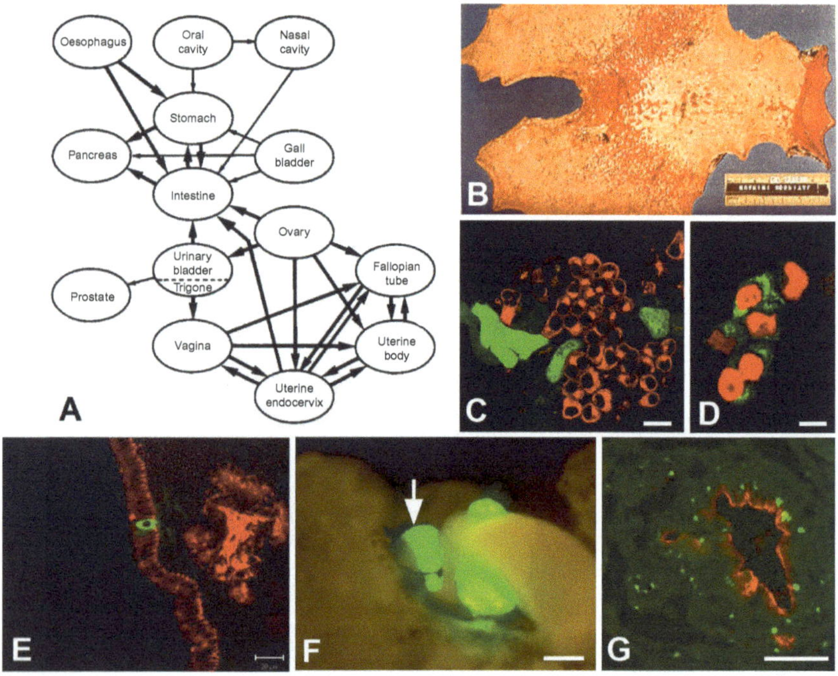

Jonathan M.W. Slack, Figure 6 Metaplasia. (A) "Kinematic graph" summarizing metaplastic transformations of epithelia described in the human pathology literature (Slack, 1985). (B) Intestinal metaplasia in a human stomach, visualized by histochemical stain for alkaline phosphatase, also visible in the duodenum fragment to the right (Stemmermann & Hayashi, 1968). (C) Transformation of AR42j cells from a pancreatic to hepatic phenotype using dexamethasone. Green cells express glucose-6-phosphatase, normally found in hepatocytes but not pancreatic cells. Scale bar: 10 μm. (D) A similar transformation brought about by the transcription factor C/EBPβ. The green cells express transferrin. Scale bar: 10 μm. (E) An isolated insulin-positive endocrine cell (green) in the epithelium of a mouse bile duct. Scale bar: 20 μm. (F) Formation of ectopic pancreas (arrow) in a *Xenopus* tadpole by overexpression of *Pdx1–VP16* under control of a liver-specific promoter. The two right hand blobs of green are the normal pancreas. Scale bar: 200 μm. (G) Mixed phenotype cells of a small bile duct in the mouse liver over-expressing *Pdx1* + *Ngn3* + *MafA*. Red is insulin, and green is EdU indicating dividing cells. Scale bar: 100 μm.

CHAPTER TWENTY-ONE

Mechanotransduction During Vertebrate Neurulation

Sergei Y. Sokol[1]
Department of Developmental and Regenerative Biology, Icahn School of Medicine at Mount Sinai, New York, USA
[1]Corresponding author: e-mail address: sergei.sokol@mssm.edu

Contents

Abstract

Vertebrate neural tube formation is a complex morphogenetic process, which involves hundreds of genes dynamically coordinating various behaviors in different cell populations of neural tissue. The challenge remains to determine the relative contributions of physical forces and biochemical signaling events to neural tube closure and accompanying cell fate specification. Planar cell polarity (PCP) molecules are prime candidate factors for the production of actomyosin-dependent mechanical signals necessary for morphogenesis. Conversely, physical forces may contribute to the polarized distribution of PCP proteins. Understanding mechanosensory and mechanotransducing properties of diverse molecules should help define the direction and amplitude of physical stresses that are critical for neurulation.

1. INTRODUCTION

When compared to chemical and biochemical stimuli, mechanical signals represent equally important means of modulating eukaryotic cell behaviors during embryonic development and adult physiology. Among well-known examples of mechanosensation are the response of skin mechanoreceptors to touch and the ability of inner ear hair cells to mediate

Current Topics in Developmental Biology, Volume 117
ISSN 0070-2153
http://dx.doi.org/10.1016/bs.ctdb.2015.11.036

hearing and control balance. The idea that physical stresses can shape early embryos attracted the attention of early developmental biologists, including D'Arcy Thomson and Wilhelm Roux, more than a century ago (Karfunkel, 1974; Keller, 2012). However, in the past 30–40 years, early studies of *Entwicklungsmechanik* (i.e., developmental mechanics, the term coined by Roux) were overshadowed by technological success of molecular biology emphasizing the biochemical basis of signaling in cells. During the last decade, advances in modern biophysical and imaging approaches and accumulation of data on the structure and function of cytoskeletal and cell adhesion molecules revived interest to physical stresses as an underlying cause of embryogenesis.

Morphogenetic events during early development are prominent among many processes regulated by physical forces (reviewed by Orr, Helmke, Blackman, & Schwartz, 2006; Wozniak & Chen, 2009). These include *Drosophila* ventral closure and border cell migration, *C. elegans* endoderm ingression and vertebrate gastrulation and neurulation. Although tissue deformations can be described by principles of mechanics, the main challenge is to understand *mechanotransduction*, i.e., the ability of embryonic cells to sense mechanical stimuli and transform them into biochemical reactions in the nucleus or other cellular compartments. Another challenge is to distinguish whether molecular changes that are physiologically relevant to morphogenesis are caused by mechanical or biochemical signals. There are several excellent recent reviews on the topic (Fernandez-Sanchez, Brunet, Roper, & Farge, 2015; Heisenberg & Bellaiche, 2013; Lecuit & Yap, 2015; Orr et al., 2006; Wozniak & Chen, 2009). I do not intend to provide comprehensive coverage of the available literature, but will focus on vertebrate neurulation as one example of a morphogenetic process that likely involves both biochemical and mechanical modes of cell–cell communication. This essay also considers possible mechanisms underlying transmission of mechanical signals to the nucleus, where they control cell fate specification in addition to cell shape and behavior.

2. FORCES OPERATING DURING VERTEBRATE NEURAL TUBE CLOSURE

Vertebrate neurulation, using *Xenopus laevis* as a model (Davidson & Keller, 1999; Schroeder, 1971; Suzuki, Morita, & Ueno, 2012), is initiated with formation of the neural plate consisting of superficial and inner cell layers (Fig. 1). In the beginning of neural fold formation epithelial cells of

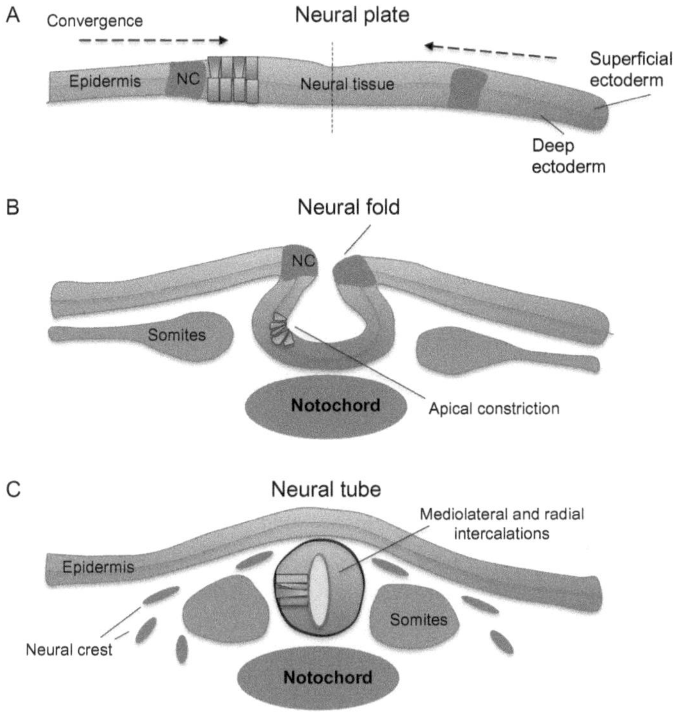

Figure 1 *Xenopus* neural tube as a model for morphogenetic movements. Scheme illustrating main stages of *Xenopus* neural tube closure. (A) Neural plate consisting of two cell layers. (B) Apical constriction leads to neural plate folding. (C) Neural folds fuse into the neural tube, and the neural tube elongates due to radial and mediolateral cell intercalations. (A color version of this figure is located at the end of this chapter.)

the superficial layer reduce their apical surface to become bottle-shaped. This process, known as *apical constriction*, promotes epithelial bending in many morphogenetic events (Martin & Goldstein, 2014). Additionally, cells of the deep layer elongate basolaterally, thereby promoting neural fold elevation. Neural folds are brought into apposition and fuse to complete neural tube closure. At the same time, midline cells intercalate mediolaterally to undergo *convergent extension* (Edlund, Davidson, & Keller, 2013; Keller, Shih, & Sater, 1992). Furthermore, *radial intercalations* convert the 2–3-cell-layered neural plate into a single-layered pseudo-stratified neural tube. Besides cells that constitute the neural tube, other tissues such as neural crest, nonneural ectoderm and axial and paraxial mesoderm participate in neurulation by undergoing *individual or collective migration.*

Neural tube closure may be affected by forces coming from neuroepithelial cells themselves (*internal forces*) and those originating outside of the neural plate (*external forces*). Roux (1885) showed that an isolated chick neural plate is able to fold properly (cited in Karfunkel, 1974), suggesting that neural tube closure is driven by internally derived mechanical signals. These forces may synergize with the force of converging dorsal epithelium as pointed out by Schroeder (1971). In contrast, Lewis found that neural tube closes faster when neural folds are separated from nonneural ectoderm by incision and proposed that dorsal epidermis antagonizes neural plate folding (Karfunkel, 1974; Lewis, 1947). These observations illustrate a debate over cellular mechanisms of neurulation that has been ongoing for more than a hundred years. Besides the epidermis, underlying mesoderm regulates neural tube formation by secreting a number of inducing factors and provides mechanical coupling by adhering tightly to the neural plate (Elul & Keller, 2000; Keller, 2002; Schroeder, 1971).

To gain insight into force-dependent morphogenetic mechanisms, one must understand how mechanical signals are *generated* in embryonic cells and tissues. Many studies have elucidated candidate molecules that are responsible for production of mechanical forces during various morphogenetic events including neural tube closure. One such molecule is nonmuscle Myosin II, a motor protein regulated by Rho GTPase-dependent phosphorylation (Rolo, Skoglund, & Keller, 2009; Vicente-Manzanares, Ma, Adelstein, & Horwitz, 2009). Myosin II functions together with actin microfilaments to control protrusive and contractile cell behavior. Different modes of actin polymerization are required *in vivo* for Myosin II accumulation and activation in different models (Lecuit, Lenne, & Munro, 2011). F-actin assembly is controlled by many molecular players, such as the Arp2/3 complex, formins, WASP, and WAVE family members, ADF/cofilin and actin capping proteins (Campellone & Welch, 2010; Krause & Gautreau, 2014). Specific functions of these molecules have been recognized by *in vitro* studies. For example, formin proteins regulate linear elongation of microfilaments in filopodia; whereas, the Arp2/3 complex is largely involved in actin filament branching in lamellipodia. Despite knowledge accumulated from tissue culture models, how actomyosin networks control diverse cell behaviors *in vivo* remains largely unclear.

Once a mechanical signal has been produced, it needs to be *sensed and transmitted* through embryonic tissues. Early studies proposed that neural tube closure depends on collective behavior of cells, in which physical stresses are propagated from one cell to another via cell junctions (Beloussov, 1980; Odell, Oster, Alberch, & Burnside, 1981). Physical stresses generated by actomyosin complexes during apical constriction can be transmitted across the tissue by surface adhesion molecules, such as integrins and cadherins. Integrin-based focal adhesion clusters and cadherin-based cell junctional complexes couple mechanical tension and adhesion to allow cell shape changes and tissue restructuring. The transmembrane protein complexes are connected to actin cytoskeleton via diverse actin-binding proteins, including α-catenin, filamins, vinculin, and others (del Rio et al., 2009; le Duc et al., 2010; Lecuit & Yap, 2015; Razinia, Makela, Ylanne, & Calderwood, 2012; Yonemura, Wada, Watanabe, Nagafuchi, & Shibata, 2010). Tensile forces acting upon these molecules can be measured using molecular mechanosensors, in which protein conformational changes are visualized by FRET, such as those based on vinculin and E-cadherin (Borghi et al., 2012; Grashoff et al., 2010). Force-dependent changes in conformation cause the appearance of new adaptor protein binding sites that are needed for efficient mechanical coupling or biochemical signaling (Cai et al., 2014; del Rio et al., 2009; Rauskolb, Sun, Sun, Pan, & Irvine, 2014; Yonemura et al., 2010). Whereas many proteins are likely to undergo conformational changes upon physical strain, only some are expected to trigger physiologically relevant responses. The idea that mechanical stresses are propagated in the tissue exclusively through cell junctions has been recently challenged by a study using *Drosophila* mutants that are deficient in cellularization and membrane formation (He, Doubrovinski, Polyakov, & Wieschaus, 2014). These mutants exhibit similar changes in embryo shape that normally accompany ventral furrow formation, suggesting that junctions are not absolutely required to coordinate trans-epithelial cytoskeletal networks. Multicellular tissue may, therefore, behave as one large "super-cell" as has been proposed by cell biologists studying motility of epithelial cell clusters *in vitro* (Khalil & Friedl, 2010).

3. MORPHOGENETIC FORCES ARE GENERATED BY COORDINATED CELL POLARIZATION

Tissue folding is a universal morphogenetic event that occurs at different stages in the majority of metazoan embryos. During this process, the orientation of cell behaviors and the relevant proteins along the axis of folding *must be coordinated with tissue polarization.* Consistent with this hypothesis, force-associated molecules, such as Myosin II, Rho-associated kinase (ROCK) and the actin-binding protein Shroom3, were found polarized in the plane of the tissue during germband elongation in *Drosophila* embryos (Bertet, Sulak, & Lecuit, 2004; Fernandez-Gonzalez, Simoes

Sde, Roper, Eaton, & Zallen, 2009; Simoes Sde, Mainieri, & Zallen, 2014; Zallen & Wieschaus, 2004). Also, planar cell polarity (PCP) proteins become segregated to different cell edges in many epithelial tissues. This group includes the core PCP proteins Frizzled, Vang/Stbm, Prickle, Flamingo/CELSR, Diego/Diversin, and their effectors (reviewed by Devenport, 2014; Peng & Axelrod, 2012; Simons & Mlodzik, 2008). PCP proteins are prime candidates for activating ROCK and Myosin II at specific cellular locations, but the biochemical support for this conclusion has been limited (Bosveld et al., 2012; Nishimura, Honda, & Takeichi, 2012; Shindo & Wallingford, 2014; Winter et al., 2001). The localization of PCP protein complexes to distinct sides of epithelial cells appears critical for morphogenesis, because epithelial polarity is disrupted in several fly tissues in PCP mutants (Peng & Axelrod, 2012; Strutt & Strutt, 2009). Although PCP has not been extensively studied in *C. elegans*, mutations in worm Wnt/PCP gene homologues affect cell ingression during gastrulation (Lee et al., 2006). The essential function of PCP proteins in morphogenesis extends to ascidians based on the properties of *Ciona* Prickle (Jiang, Munro, & Smith, 2005; Newman-Smith, Kourakis, Reeves, Veeman, & Smith, 2015) and vertebrate PCP homologues (reviewed by Gray, Roszko, & Solnica-Krezel, 2011; Sokol, 2015).

Given that the vertebrate neural plate folds anisotropically, in the direction perpendicular to the anteroposterior body axis, it may be expected to exhibit PCP. Indeed, neuroectodermal cells are characterized by medially directed protrusions (Elul & Keller, 2000) and contain several polarized PCP molecules (Ciruna, Jenny, Lee, Mlodzik, & Schier, 2006; Nishimura et al., 2012; Ossipova, Kim, & Sokol, 2015; Yin, Kiskowski, Pouille, Farge, & Solnica-Krezel, 2008). Moreover, interference with the function of vertebrate PCP proteins prevents neural tube folding in *Xenopus*, zebrafish, chick, and mice (Gray et al., 2011; Montcouquiol et al., 2003; Nishimura et al., 2012; Sokol, 1996; Wallingford, Niswander, Shaw, & Finnell, 2013). Notably, the neural plate reveals two orthogonal axes of planar polarity represented by the polarization of Prickle, Vangl2, and Dishevelled along the anteroposterior axis (Ciruna et al., 2006; Ossipova et al., 2015; Yin et al., 2008) and Rab11 and Diversin along the mediolateral axis (Ossipova et al., 2014). Loss-of-function studies indicate the relevance of Rab11 and Diversin localization for neural tube closure (Ossipova et al., 2014). Flamingo/Celsr1, ROCK, Myosin II, and the formin DAAM1 are enriched at anteroposterior cell junctions that are perpendicular to the anteroposterior axis, but whether they are concentrated to one or both sides of cells is unknown (Mahaffey, Grego-Bessa, Liem, & Anderson, 2013;

McGreevy, Vijayraghavan, Davidson, & Hildebrand, 2015; Nishimura et al., 2012). Although the significance of these observations is not entirely clear, the existence of two PCP axes in the neural plate suggests that different physical forces underlying diverse cell behaviors may operate during neural tube closure (Sokol, 2015).

A putative Wnt/PCP signaling pathway driving morphogenesis has been proposed based on the ability of Wnt ligands, ROR, and Frizzled receptors to trigger phosphorylation of Dishevelled and Vangl2 and activate small Rho GTPases RhoA and Rac1 (Gao et al., 2011; Habas, Dawid, & He, 2003; Habas, Kato, & He, 2001; Sun et al., 2001; Yanagawa, van Leeuwen, Wodarz, Klingensmith, & Nusse, 1995). These proximal events may stimulate downstream PCP effectors, such as ROCK and Myosin II (Marlow, Topczewski, Sepich, & Solnica-Krezel, 2002; Nishimura et al., 2012; Winter et al., 2001), and locally modulate actomyosin dynamics at anteroposterior or mediolateral cell junctions. Indeed, in ascidians, GFP-MLC (regulatory Myosin II light chain) colocalizes with Prickle to the anterior edge of each cell (Newman-Smith et al., 2015), but whether the same result holds in vertebrate embryos remains to be demonstrated. A different molecular mechanism has been proposed for the PCP effector Fuzzy which affects cell polarity via Rab-dependent vesicular trafficking (Gray et al., 2009). Rab proteins likely control morphogenetic events by recycling cadherins and PCP proteins such as Celsr1 and Vangl2 (Classen, Anderson, Marois, & Eaton, 2005; Devenport, Oristian, Heller, & Fuchs, 2011; Mahaffey et al., 2013; Ossipova et al., 2014).

Based on the above evidence, PCP signaling can be mechanistically linked to neural tube closure *in vivo*. Nevertheless, it remains unclear how the same molecules execute a variety of cellular responses. One possibility is that participation in diverse intracellular events is regulated posttranslationally by different signaling pathways. For example, different roles of Myosin II in apical or lateral constriction of endoderm cells during ascidian gastrulation have been attributed to differential phosphorylation (Sherrard, Robin, Lemaire, & Munro, 2010). Alternatively, different myosin classes or isoforms may be involved in cell shape changes during neural tube closure, similar to what was shown for mitotic spindle positioning during *Xenopus* epiboly (Woolner & Papalopulu, 2012) or mammalian cell migration *in vitro* (Juanes-Garcia et al., 2015). In-depth future analysis of temporal and spatial actomyosin dynamics and its regulation by PCP signaling will help understand molecular differences between major cell behaviors, such as apical constriction and cell intercalations, during neural tube closure.

4. MECHANICAL STRESSES CAN SHAPE CELLS AND CREATE TISSUE POLARITY

Long before PCP molecules were discovered, mechanical forces were suggested to regulate morphogenesis by triggering contact-dependent cell polarization (Beloussov, 1980; Odell et al., 1981; Tickle & Trinkaus, 1976). More recently, this idea has been supported by findings that planar polarity of ciliated cells in *Xenopus* epidermis and the mouse node can be controlled by flow and stretch (Chien, Keller, Kintner, & Shook, 2015; Hashimoto et al., 2010; Mitchell, Jacobs, Li, Chien, & Kintner, 2007; Nonaka, Shiratori, Saijoh, & Hamada, 2002). Also, PCP components are reoriented by cell flow along the axis of tissue elongation during *Drosophila* wing growth (Aigouy et al., 2010). In chordate embryos, inhibition of Myosin II contractility disrupted PCP in *Ciona* notochord (Newman-Smith et al., 2015) and *Xenopus* neural plate (Ossipova et al., 2015). Similar disruption of PCP is observed in the mouse skin lacking WDR1, a protein required for cortical tension (Luxenburg et al., 2015). These observations reiterate the existence of feedback regulation between physical forces and PCP.

How does one determine the location and direction of physical forces acting in the neural plate? Some studies compare embryonic tissues to viscoelastic materials (Beloussov, 1980; Zhou, Kim, & Davidson, 2009) that "obey" the Poisson rule. According to this rule, the stretched material is compressed perpendicular to the direction of stretching. Consistent with this notion, measurement of tensile forces in epithelial cells during wound repair indicated increased tension at the lateral sides of the protruding clump of cells that is perpendicular to the direction of migration (Reffay et al., 2014). Of note, Rab11 and Diversin, two proteins needed for neural tube closure, are strikingly polarized in the direction perpendicular to the anteroposterior axis of the folding *Xenopus* neural plate (Ossipova et al., 2014), possibly reflecting the physical stresses that are responsible for axis elongation.

Supporting this hypothesis, the subcellular distribution of Rab11 is sensitive to contractile forces in embryonic ectoderm (Fig. 2). Rab11 becomes highly polarized in cells that undergo apical constriction in response to Shroom3 (Ossipova et al., 2014). Remarkably, this polarization is observed not only in cells expressing Shroom3, but also in cells adjacent to the Shroom3-expressing tissue (Ossipova et al., 2014). This suggests that

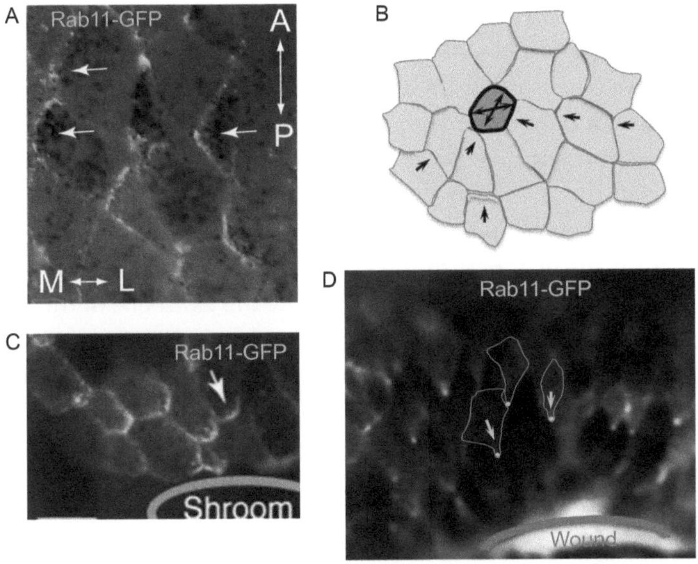

Figure 2 Rab11 as a mechanosensor. (A) Endogenous Rab11 is polarized to medioapical corners of neuroepithelial cells in the neural plate. Body axes are indicated. Costaining was done with anti-ZO1 antibody. (B) Rab11 planar polarization (arrows) induced by a constricting Shroom3-expressing cell (orange) in adjacent ectodermal cells. (C) Polarization of Rab11-GFP near a Shroom3-expressing cell clone. (D) Rab11 planar polarization in embryonic ectoderm during wound healing. Locations of Shroom3-expressing cells (C) and the position of the wound (D) are indicated. *Panels (A) and (D) modified from Ossipova et al. (2014).* (A color version of this figure is located at the end of this chapter.)

Shroom3 triggers a signal that is directly or indirectly sensed by polarized Rab11 endosomes and transmitted through the responding tissue. This is consistent with the model in which apical constriction is relayed from one contracting cell to the other (Odell et al., 1981). Moreover, Rab11 endosomes are similarly polarized in the plane of the wounded embryonic epidermis, being enriched at cell corners that are closest to the wound (Ossipova et al., 2014). In both cases, planar polarization results from contractile forces generated during apical constriction of Shroom3-containing cells or contraction of the supracellular actomyosin belt surrounding the wound. These experiments suggest that Rab11 is a mechanosensor, which reveals effects of physical forces on PCP and might contribute to propagation of these forces through the tissue. Consistent with the roles for Rab11 in force propagation and planar polarity, interference with Rab11 function inhibits Shroom3-mediated constriction and delays wound healing (Ossipova et al., 2014).

Rab11 enrichment at medioapical corners of cells in the neural plate suggests that physical forces act on neuroectoderm along the anteroposterior axis (Fig. 2A). Since other PCP proteins, including Vangl2 and Celsr1, exhibit a distinct polarity pattern (anteroposterior vs. mediolateral enrichment), these proteins may sense differently oriented forces. In summary, although mechanical stresses can arise as a consequence of PCP signaling through activation of actomyosin dynamics and junction remodeling, mechanical signals may ensure the robustness of morphogenetic processes through feedback regulation of PCP. Despite the existence of this feedback, an open question remains whether oriented cell behaviors in the neural plate result from mechanotransduction or intercellular biochemical signaling.

5. REGULATION OF TRANSCRIPTION BY MECHANOTRANSDUCTION DURING NEURULATION

Besides the direct control of cell shape and behavior, mechanical forces may regulate gene transcription and cell fate specification. Regulation of gene transcription by physical forces has been proposed for the activation of *Twist* in early *Drosophila* embryos (Desprat, Supatto, Pouille, Beaurepaire, & Farge, 2008), *Brachyury* in zebrafish (Brunet et al., 2013), and *Sox3* and *muscle-specific actin* in *Xenopus* embryos (Kornikova, Troshina, Kremnyov, & Beloussov, 2010). In-depth analysis is needed to confirm putative molecular mechanisms and determine whether these observations can be extended to other models. Whether mechanical signals control embryonic cell fates remains a challenging question for developmental biologists since the location of forces in early embryos is largely unknown (Zhou, Pal, Maiti, & Davidson, 2015). A more simple question is *how progenitor fates are affected by molecules associated with physical forces, cell shape, and cell polarity*. Among these are Rho GTPases, PCP proteins and actomyosin-remodeling factors. Modulation of Rho GTPase and Myosin II activity and substrate stiffness has been shown to influence mesenchymal stem cell decisions to undergo osteogenic and adipogenic differentiation *in vitro* (Dupont et al., 2011; McBeath, Pirone, Nelson, Bhadriraju, & Chen, 2004; Sordella, Jiang, Chen, Curto, & Settleman, 2003). Similarly, epidermal stem cell commitment is affected by actomyosin dynamics (Connelly et al., 2010). Embryonic stem cell lineages also depend on substrate stiffness and actomyosin contractility (Engler, Sen, Sweeney, & Discher, 2006; Kim, Ossipova, & Sokol, 2015). *In vivo*, mechanotransduction likely occurs in embryonic progenitor cells that experience physical forces or undergo asymmetric division during development. Changes in polarized protein distribution are known to regulate daughter cell fates during asymmetric cell divisions in *Drosophila* embryos (Bellaiche et al., 2001; Emery et al., 2005; Knoblich, 2010). Moreover, vertebrate polarity proteins were shown to affect lineage decisions of asymmetrically dividing progenitors in epidermis and the central nervous system in *Xenopus*, zebrafish, and mouse embryos (Lake & Sokol, 2009; Ossipova, Ezan, & Sokol, 2009; Ossipova, Tabler, Green, & Sokol, 2007; Plusa et al., 2005; Segalen et al., 2010).

Although orientation of physical stresses during neural tube closure is unknown, one may speculate that these forces act *upon cell populations that normally undergo changes in shape and polarity*. One such population is the apically constricting cells of the neural plate; however, the relationship between apical constriction and cell fates in the neural plate has not been studied. Another population is the neural crest, a population of multipotent progenitor cells that originate at the neural plate border, undergo epithelial–mesenchymal transition (EMT) and migrate to diverse locations to produce multiple cell types. Notably, modeling of neural plate closure suggested that the neural plate border experiences high shear stress (Jacobson & Gordon, 1976). A recent study used different antagonists of ROCK and Myosin II to understand how physical forces might affect neural crest development (Kim et al., 2015). Interference with ROCK and Myosin II activities stimulated neural crest specification in *Xenopus* embryos and differentiating human embryonic stem cells (Kim et al., 2015). In the neural plate, these reagents delayed neural tube closure, consistent with the inhibition of actomyosin contractility. These observations suggest that forces operating during neural tube closure act to inhibit neural crest formation.

The mechanism for neural crest gene activation by ROCK and Myosin II inhibitors is currently unclear. A force-dependent cytoskeletal change would cause a messenger protein to translocate between the cortex or cyto plasm and the nucleus, leading to altered gene transcription. Such a protein could associate with a *bona fide* resident transcription factor and function as a transcriptional coactivator or corepressor (Fig. 3). One example is β-catenin that is present at cadherin-based adherens junctions, but may associate with T-cell factors to control Wnt-dependent gene expression in the nucleus. Indeed, β-catenin was proposed to transduce a mechanical signal leading to *Twist* activation in *Drosophila* (Desprat et al., 2008).

Another possible transducer of actomyosin regulation is megakaryocytic acute leukemia (MAL), also known as MRTF or MKL (Miralles, Posern, Zaromytidou, & Treisman, 2003; Olson & Nordheim, 2010). MAL binds

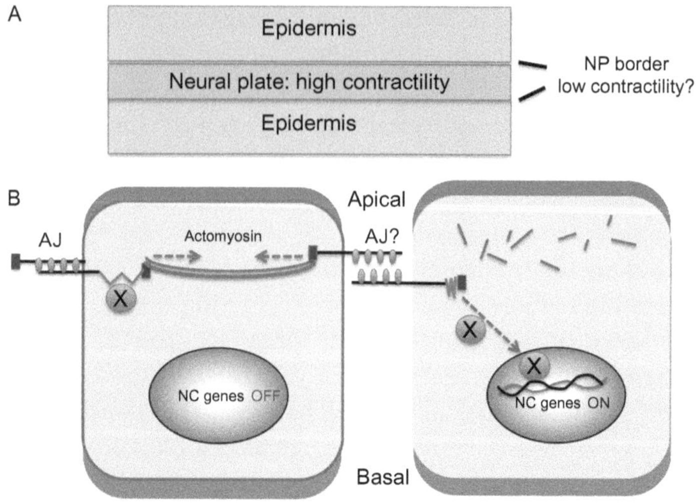

Figure 3 Force-dependent transcriptional regulation of the neural crest. (A) Hypothetical variations in actomyosin-generated forces that affect neural crest specification at the neural plate border. (B) A model for neural crest gene activation. Mechanical stress triggers cytoskeletal reorganization causing messenger translocation from cell junctions to the nucleus and leading to altered gene transcription. (A color version of this figure is located at the end of this chapter.)

to the serum response factor (SRF) to stimulate a number of transcriptional targets, but this binding is inhibited by monomeric globular actin (G-actin). Actin polymerization reduces the amount of G-actin, leading to SRF target activation (Olson & Nordheim, 2010). It has been proposed that this mechanism underlies *Drosophila* border cell migration (Somogyi & Rorth, 2004) and differentiation of mouse epidermal stem cells (Connelly et al., 2010). SRF also contributes to craniofacial development (Vasudevan & Soriano, 2014) suggesting that it may participate in neural crest specification.

Actomyosin complexes also regulate the localization of Yes-associated protein (YAP) and the related protein TAZ. YAP/TAZ are transcriptional factors of the Hippo signaling pathway (Yu & Guan, 2013), which has been shown to respond to mechanical stresses, Myosin II and Wnt signaling in cultured cells (Benham-Pyle, Pruitt, & Nelson, 2015; Dupont et al., 2011; Park et al., 2015). Cytoskeletal tension inhibits Hippo signaling and promotes *Drosophila* wing growth through Ajuba Lim family proteins, revealing *in vivo* significance of this signaling pathway (Rauskolb et al., 2014). Of note, both YAP and Ajuba LIM proteins have been implicated in neural crest development (Gee, Milgram, Kramer, Conlon, & Moody, 2011; Kim et al., 2015; Langer et al., 2008).

In addition to these candidate signal transducers, other junctional proteins and actomyosin modulators can shuttle to the nucleus and may be responsible for transcriptional control in response to mechanical signals. These include nuclear myosins and actin-remodeling factors, p120 catenin and the Y-box protein ZONAB (Balda & Matter, 2000; Lee, Ji, Furuta, Park, & McCrea, 2014; Lundquist et al., 2014; Miyamoto et al., 2013; Vartiainen, Guettler, Larijani, & Treisman, 2007). Further studies are needed to establish specific roles for these proteins in neural crest regulation and neural tube closure.

ACKNOWLEDGMENTS

I am grateful to Lance Davidson, Chi Pak, and Paul Wassarman for critical reading of this essay, Olga Ossipova for providing images shown in Fig. 2A and C, and members of the Sokol laboratory for discussions. I apologize to those authors whose work has not been cited here due to limited space. The work in the Sokol laboratory is supported by the National Institutes of Health.

REFERENCES

Aigouy, B., Farhadifar, R., Staple, D. B., Sagner, A., Roper, J. C., Julicher, F., et al. (2010). Cell flow reorients the axis of planar polarity in the wing epithelium of Drosophila. *Cell*, *142*, 773–786.

Balda, M. S., & Matter, K. (2000). The tight junction protein ZO-1 and an interacting transcription factor regulate ErbB-2 expression. *EMBO Journal*, *19*, 2024–2033.

Bellaiche, Y., Radovic, A., Woods, D. F., Hough, C. D., Parmentier, M. L., O'Kane, C. J., et al. (2001). The partner of inscuteable/discs-large complex is required to establish planar polarity during asymmetric cell division in Drosophila. *Cell*, *106*, 355–366.

Beloussov, L. V. (1980). The role of tensile fields and contact cell polarization in the morphogenesis of amphibian axial rudiments. *Roux's Archives of Developmental Biology*, *188*, 1–7.

Benham-Pyle, B. W., Pruitt, B. L., & Nelson, W. J. (2015). Cell adhesion. Mechanical strain induces E-cadherin-dependent Yap1 and beta-catenin activation to drive cell cycle entry. *Science*, *348*, 1024–1027.

Bertet, C., Sulak, L., & Lecuit, T. (2004). Myosin-dependent junction remodelling controls planar cell intercalation and axis elongation. *Nature*, *429*, 667–671.

Borghi, N., Sorokina, M., Shcherbakova, O. G., Weis, W. I., Pruitt, B. L., Nelson, W. J., et al. (2012). E-cadherin is under constitutive actomyosin-generated tension that is increased at cell-cell contacts upon externally applied stretch. *Proceedings of the National Academy of Sciences of the United States of America*, *109*, 12568–12573.

Bosveld, F., Bonnet, I., Guirao, B., Tlili, S., Wang, Z., Petitalot, A., et al. (2012). Mechanical control of morphogenesis by fat/dachsous/four-jointed planar cell polarity pathway. *Science*, *336*, 724–727.

Brunet, T., Bouclet, A., Ahmadi, P., Mitrossilis, D., Driquez, B., Brunet, A. C., et al. (2013). Evolutionary conservation of early mesoderm specification by mechanotransduction in Bilateria. *Nature Communications*, *4*, 2821.

Cai, D., Chen, S. C., Prasad, M., He, L., Wang, X., Choesmel-Cadamuro, V., et al. (2014). Mechanical feedback through E-cadherin promotes direction sensing during collective cell migration. *Cell*, *157*, 1146–1159.

Campellone, K. G., & Welch, M. D. (2010). A nucleator arms race: Cellular control of actin assembly. *Nature Reviews. Molecular Cell Biology*, *11*, 237–251.
Chien, Y. H., Keller, R., Kintner, C., & Shook, D. R. (2015). Mechanical strain determines the axis of planar polarity in ciliated epithelia. *Current Biology*, *25*, 2774–2784.
Ciruna, B., Jenny, A., Lee, D., Mlodzik, M., & Schier, A. F. (2006). Planar cell polarity signalling couples cell division and morphogenesis during neurulation. *Nature*, *439*, 220–224.
Classen, A. K., Anderson, K. I., Marois, E., & Eaton, S. (2005). Hexagonal packing of Drosophila wing epithelial cells by the planar cell polarity pathway. *Developmental Cell*, *9*, 805–817.
Connelly, J. T., Gautrot, J. E., Trappmann, B., Tan, D. W., Donati, G., Huck, W. T., et al. (2010). Actin and serum response factor transduce physical cues from the microenvironment to regulate epidermal stem cell fate decisions. *Nature Cell Biology*, *12*, 711–718.
Davidson, L. A., & Keller, R. E. (1999). Neural tube closure in Xenopus laevis involves medial migration, directed protrusive activity, cell intercalation and convergent extension. *Development*, *126*, 4547–4556.
del Rio, A., Perez-Jimenez, R., Liu, R., Roca-Cusachs, P., Fernandez, J. M., & Sheetz, M. P. (2009). Stretching single talin rod molecules activates vinculin binding. *Science*, *323*, 638–641.
Desprat, N., Supatto, W., Pouille, P. A., Beaurepaire, E., & Farge, E. (2008). Tissue deformation modulates twist expression to determine anterior midgut differentiation in Drosophila embryos. *Developmental Cell*, *15*, 470–477.
Devenport, D. (2014). The cell biology of planar cell polarity. *Journal of Cell Biology*, *207*, 171–179.
Devenport, D., Oristian, D., Heller, E., & Fuchs, E. (2011). Mitotic internalization of planar cell polarity proteins preserves tissue polarity. *Nature Cell Biology*, *13*, 893–902.
Dupont, S., Morsut, L., Aragona, M., Enzo, E., Giulitti, S., Cordenonsi, M., et al. (2011). Role of YAP/TAZ in mechanotransduction. *Nature*, *474*, 179–183.
Edlund, A. F., Davidson, L. A., & Keller, R. E. (2013). Cell segregation, mixing, and tissue pattern in the spinal cord of the Xenopus laevis neurula. *Developmental Dynamics*, *242*, 1134–1146.
Elul, T., & Keller, R. (2000). Monopolar protrusive activity: A new morphogenic cell behavior in the neural plate dependent on vertical interactions with the mesoderm in Xenopus. *Developmental Biology*, *224*, 3–19.
Emery, G., Hutterer, A., Berdnik, D., Mayer, B., Wirtz-Peitz, F., Gaitan, M. G., et al. (2005). Asymmetric Rab 11 endosomes regulate delta recycling and specify cell fate in the Drosophila nervous system. *Cell*, *122*, 763–773.
Engler, A. J., Sen, S., Sweeney, H. L., & Discher, D. E. (2006). Matrix elasticity directs stem cell lineage specification. *Cell*, *126*, 677–689.
Fernandez-Gonzalez, R., Simoes Sde, M., Roper, J. C., Eaton, S., & Zallen, J. A. (2009). Myosin II dynamics are regulated by tension in intercalating cells. *Developmental Cell*, *17*, 736–743.
Fernandez-Sanchez, M. E., Brunet, T., Roper, J. C., & Farge, E. (2015). Mechanotransduction's impact in animal development, evolution, and tumorigenesis. *Annual Review of Cell and Developmental Biology*, *31*, 373–397.
Gao, B., Song, H., Bishop, K., Elliot, G., Garrett, L., English, M. A., et al. (2011). Wnt signaling gradients establish planar cell polarity by inducing Vangl2 phosphorylation through Ror2. *Developmental Cell*, *20*, 163–176.
Gee, S. T., Milgram, S. L., Kramer, K. L., Conlon, F. L., & Moody, S. A. (2011). Yes-associated protein 65 (YAP) expands neural progenitors and regulates Pax3 expression in the neural plate border zone. *PLoS One*, *6*, e20309.

Grashoff, C., Hoffman, B. D., Brenner, M. D., Zhou, R., Parsons, M., Yang, M. T., et al. (2010). Measuring mechanical tension across vinculin reveals regulation of focal adhesion dynamics. *Nature*, *466*, 263–266.

Gray, R. S., Abitua, P. B., Wlodarczyk, B. J., Szabo-Rogers, H. L., Blanchard, O., Lee, I., et al. (2009). The planar cell polarity effector Fuz is essential for targeted membrane trafficking, ciliogenesis and mouse embryonic development. *Nature Cell Biology*, *11*, 1225–1232.

Gray, R. S., Roszko, I., & Solnica-Krezel, L. (2011). Planar cell polarity: Coordinating morphogenetic cell behaviors with embryonic polarity. *Developmental Cell*, *21*, 120–133.

Habas, R., Dawid, I. B., & He, X. (2003). Coactivation of Rac and Rho by Wnt/frizzled signaling is required for vertebrate gastrulation. *Genes & Development*, *17*, 295–309.

Habas, R., Kato, Y., & He, X. (2001). Wnt/Frizzled activation of Rho regulates vertebrate gastrulation and requires a novel Formin homology protein Daam1. *Cell*, *107*, 843–854.

Hashimoto, M., Shinohara, K., Wang, J., Ikeuchi, S., Yoshiba, S., Meno, C., et al. (2010). Planar polarization of node cells determines the rotational axis of node cilia. *Nature Cell Biology*, *12*, 170–176.

He, B., Doubrovinski, K., Polyakov, O., & Wieschaus, E. (2014). Apical constriction drives tissue-scale hydrodynamic flow to mediate cell elongation. *Nature*, *508*, 392–396.

Heisenberg, C. P., & Bellaiche, Y. (2013). Forces in tissue morphogenesis and patterning. *Cell*, *153*, 948–962.

Jacobson, A. G., & Gordon, R. (1976). Changes in the shape of the developing vertebrate nervous system analyzed experimentally, mathematically and by computer simulation. *Journal of Experimental Zoology*, *197*, 191–246.

Jiang, D., Munro, E. M., & Smith, W. C. (2005). Ascidian prickle regulates both mediolateral and anterior-posterior cell polarity of notochord cells. *Current Biology*, *15*, 79–85.

Juanes-Garcia, A., Chapman, J. R., Aguilar-Cuenca, R., Delgado-Arevalo, C., Hodges, J., Whitmore, L. A., et al. (2015). A regulatory motif in nonmuscle myosin II-B regulates its role in migratory front-back polarity. *Journal of Cell Biology*, *209*, 23–32.

Karfunkel, P. (1974). The mechanisms of neural tube formation. *International Review of Cytology*, *38*, 245–271.

Keller, R. (2002). Shaping the vertebrate body plan by polarized embryonic cell movements. *Science*, *298*, 1950–1954.

Keller, R. (2012). Developmental biology. Physical biology returns to morphogenesis. *Science*, *338*, 201–203.

Keller, R., Shih, J., & Sater, A. (1992). The cellular basis of the convergence and extension of the Xenopus neural plate. *Developmental Dynamics*, *193*, 199–217.

Khalil, A. A., & Friedl, P. (2010). Determinants of leader cells in collective cell migration. *Integrative Biology: Quantitative Biosciences from Nano to Macro*, *2*, 568–574.

Kim, K., Ossipova, O., & Sokol, S. Y. (2015). Neural crest specification by inhibition of the ROCK/myosin II pathway. *Stem Cells*, *33*, 674–685.

Knoblich, J. A. (2010). Asymmetric cell division: Recent developments and their implications for tumour biology. *Nature Reviews. Molecular Cell Biology*, *11*, 849–860.

Kornikova, E. S., Troshina, T. G., Kremnyov, S. V., & Beloussov, L. V. (2010). Neuro-mesodermal patterns in artificially deformed embryonic explants: A role for mechano-geometry in tissue differentiation. *Developmental Dynamics*, *239*, 885–896.

Krause, M., & Gautreau, A. (2014). Steering cell migration: Lamellipodium dynamics and the regulation of directional persistence. *Nature Reviews. Molecular Cell Biology*, *15*, 577–590.

Lake, B. B., & Sokol, S. Y. (2009). Strabismus regulates asymmetric cell divisions and cell fate determination in the mouse brain. *Journal of Cell Biology*, *185*, 59–66.

Langer, E. M., Feng, Y., Zhaoyuan, H., Rauscher, F. J., 3rd., Kroll, K. L., & Longmore, G. D. (2008). Ajuba LIM proteins are snail/slug corepressors required for neural crest development in Xenopus. *Developmental Cell*, *14*, 424–436.

le Duc, Q., Shi, Q., Blonk, I., Sonnenberg, A., Wang, N., Leckband, D., et al. (2010). Vinculin potentiates E-cadherin mechanosensing and is recruited to actin-anchored sites within adherens junctions in a myosin II-dependent manner. *Journal of Cell Biology*, *189*, 1107–1115.

Lecuit, T., Lenne, P. F., & Munro, E. (2011). Force generation, transmission, and integration during cell and tissue morphogenesis. *Annual Review of Cell and Developmental Biology*, *27*, 157–184.

Lecuit, T., & Yap, A. S. (2015). E-cadherin junctions as active mechanical integrators in tissue dynamics. *Nature Cell Biology*, *17*, 533–539.

Lee, M., Ji, H., Furuta, Y., Park, J. I., & McCrea, P. D. (2014). p120-catenin regulates REST and CoREST, and modulates mouse embryonic stem cell differentiation. *Journal of Cell Science*, *127*, 4037–4051.

Lee, J. Y., Marston, D. J., Walston, T., Hardin, J., Halberstadt, A., & Goldstein, B. (2006). Wnt/frizzled signaling controls C. elegans gastrulation by activating actomyosin contractility. *Current Biology*, *16*, 1986–1997.

Lewis, W. H. (1947). Mechanics of invagination. *Anatomical Record*, *97*, 139–156.

Lundquist, M. R., Storaska, A. J., Liu, T. C., Larsen, S. D., Evans, T., Neubig, R. R., et al. (2014). Redox modification of nuclear actin by MICAL-2 regulates SRF signaling. *Cell*, *156*, 563–576.

Luxenburg, C., Heller, E., Pasolli, H. A., Chai, S., Nikolova, M., Stokes, N., et al. (2015). Wdr1-mediated cell shape dynamics and cortical tension are essential for epidermal planar cell polarity. *Nature Cell Biology*, *17*, 592–604.

Mahaffey, J. P., Grego-Bessa, J., Liem, K. F., Jr., & Anderson, K. V. (2013). Cofilin and Vangl2 cooperate in the initiation of planar cell polarity in the mouse embryo. *Development*, *140*, 1262–1271.

Marlow, F., Topczewski, J., Sepich, D., & Solnica-Krezel, L. (2002). Zebrafish Rho kinase 2 acts downstream of Wnt11 to mediate cell polarity and effective convergence and extension movements. *Current Biology*, *12*, 876–884.

Martin, A. C., & Goldstein, B. (2014). Apical constriction: Themes and variations on a cellular mechanism driving morphogenesis. *Development*, *141*, 1987–1998.

McBeath, R., Pirone, D. M., Nelson, C. M., Bhadriraju, K., & Chen, C. S. (2004). Cell shape, cytoskeletal tension, and RhoA regulate stem cell lineage commitment. *Developmental Cell*, *6*, 483–495.

McGreevy, E. M., Vijayraghavan, D., Davidson, L. A., & Hildebrand, J. D. (2015). Shroom3 functions downstream of planar cell polarity to regulate myosin II distribution and cellular organization during neural tube closure. *Biology Open*, *4*, 186–196.

Miralles, F., Posern, G., Zaromytidou, A. I., & Treisman, R. (2003). Actin dynamics control SRF activity by regulation of its coactivator MAL. *Cell*, *113*, 329–342.

Mitchell, B., Jacobs, R., Li, J., Chien, S., & Kintner, C. (2007). A positive feedback mechanism governs the polarity and motion of motile cilia. *Nature*, *447*, 97–101.

Miyamoto, K., Teperek, M., Yusa, K., Allen, G. E., Bradshaw, C. R., & Gurdon, J. B. (2013). Nuclear wave1 is required for reprogramming transcription in oocytes and for normal development. *Science*, *341*, 1002–1005.

Montcouquiol, M., Rachel, R. A., Lanford, P. J., Copeland, N. G., Jenkins, N. A., & Kelley, M. W. (2003). Identification of Vangl2 and Scrb1 as planar polarity genes in mammals. *Nature*, *423*, 173–177.

Newman-Smith, E., Kourakis, M. J., Reeves, W., Veeman, M., & Smith, W. C. (2015). Reciprocal and dynamic polarization of planar cell polarity core components and myosin. *Elife*, *4*, e05361.

Nishimura, T., Honda, H., & Takeichi, M. (2012). Planar cell polarity links axes of spatial dynamics in neural-tube closure. *Cell*, *149*, 1084–1097.

Nonaka, S., Shiratori, H., Saijoh, Y., & Hamada, H. (2002). Determination of left-right patterning of the mouse embryo by artificial nodal flow. *Nature*, *418*, 96–99.
Odell, G. M., Oster, G., Alberch, P., & Burnside, B. (1981). The mechanical basis of morphogenesis. I. Epithelial folding and invagination. *Developmental Biology*, *85*, 446–462.
Olson, E. N., & Nordheim, A. (2010). Linking actin dynamics and gene transcription to drive cellular motile functions. *Nature Reviews. Molecular Cell Biology*, *11*, 353–365.
Orr, A. W., Helmke, B. P., Blackman, B. R., & Schwartz, M. A. (2006). Mechanisms of mechanotransduction. *Developmental Cell*, *10*, 11–20.
Ossipova, O., Ezan, J., & Sokol, S. Y. (2009). PAR-1 phosphorylates mind bomb to promote vertebrate neurogenesis. *Developmental Cell*, *17*, 222–233.
Ossipova, O., Kim, K., Lake, B. B., Itoh, K., Ioannou, A., & Sokol, S. Y. (2014). Role of Rab11 in planar cell polarity and apical constriction during vertebrate neural tube closure. *Nature Communications*, *5*, 3734.
Ossipova, O., Kim, K., & Sokol, S. Y. (2015). Planar polarization of Vangl2 in the vertebrate neural plate is controlled by Wnt and Myosin II signaling. *Biology Open*, *4*, 722–730.
Ossipova, O., Tabler, J., Green, J. B., & Sokol, S. Y. (2007). PAR1 specifies ciliated cells in vertebrate ectoderm downstream of aPKC. *Development*, *134*, 4297–4306.
Park, H. W., Kim, Y. C., Yu, B., Moroishi, T., Mo, J. S., Plouffe, S. W., et al. (2015). Alternative Wnt signaling activates YAP/TAZ. *Cell*, *162*, 780–794.
Peng, Y., & Axelrod, J. D. (2012). Asymmetric protein localization in planar cell polarity: Mechanisms, puzzles, and challenges. *Current Topics in Developmental Biology*, *101*, 33–53.
Plusa, B., Frankenberg, S., Chalmers, A., Hadjantonakis, A. K., Moore, C. A., Papalopulu, N., et al. (2005). Downregulation of Par3 and aPKC function directs cells towards the ICM in the preimplantation mouse embryo. *Journal of Cell Science*, *118*, 505–515.
Rauskolb, C., Sun, S., Sun, G., Pan, Y., & Irvine, K. D. (2014). Cytoskeletal tension inhibits Hippo signaling through an Ajuba-Warts complex. *Cell*, *158*, 143–156.
Razinia, Z., Makela, T., Ylanne, J., & Calderwood, D. A. (2012). Filamins in mechanosensing and signaling. *Annual Review of Biophysics*, *41*, 227–246.
Reffay, M., Parrini, M. C., Cochet-Escartin, O., Ladoux, B., Buguin, A., Coscoy, S., et al. (2014). Interplay of RhoA and mechanical forces in collective cell migration driven by leader cells. *Nature Cell Biology*, *16*, 217–223.
Rolo, A., Skoglund, P., & Keller, R. (2009). Morphogenetic movements driving neural tube closure in Xenopus require myosin IIB. *Developmental Biology*, *327*, 327–338.
Schroeder, T. E. (1971). Mechanisms of morphogenesis: The embryonic neural tube. *International Journal of Neuroscience*, *2*, 183–197.
Segalen, M., Johnston, C. A., Martin, C. A., Dumortier, J. G., Prehoda, K. E., David, N. B., et al. (2010). The Fz-Dsh planar cell polarity pathway induces oriented cell division via Mud/NuMA in Drosophila and zebrafish. *Developmental Cell*, *19*, 740–752.
Sherrard, K., Robin, F., Lemaire, P., & Munro, E. (2010). Sequential activation of apical and basolateral contractility drives ascidian endoderm invagination. *Current Biology*, *20*, 1499–1510.
Shindo, A., & Wallingford, J. B. (2014). PCP and septins compartmentalize cortical actomyosin to direct collective cell movement. *Science*, *343*, 649–652.
Simoes Sde, M., Mainieri, A., & Zallen, J. A. (2014). Rho GTPase and shroom direct planar polarized actomyosin contractility during convergent extension. *Journal of Cell Biology*, *204*, 575–589.
Simons, M., & Mlodzik, M. (2008). Planar cell polarity signaling: From fly development to human disease. *Annual Review of Genetics*, *42*, 517–540.
Sokol, S. Y. (1996). Analysis of dishevelled signalling pathways during Xenopus development. *Current Biology*, *6*, 1456–1467.

Sokol, S. Y. (2015). Spatial and temporal aspects of Wnt signaling and planar cell polarity during vertebrate embryonic development. *Seminars in Cell & Developmental Biology*, *42*, 78–85.
Somogyi, K., & Rorth, P. (2004). Evidence for tension-based regulation of Drosophila MAL and SRF during invasive cell migration. *Developmental Cell*, *7*, 85–93.
Sordella, R., Jiang, W., Chen, G. C., Curto, M., & Settleman, J. (2003). Modulation of Rho GTPase signaling regulates a switch between adipogenesis and myogenesis. *Cell*, *113*, 147–158.
Strutt, H., & Strutt, D. (2009). Asymmetric localisation of planar polarity proteins: Mechanisms and consequences. *Seminars in Cell & Developmental Biology*, *20*, 957–963.
Sun, T. Q., Lu, B., Feng, J. J., Reinhard, C., Jan, Y. N., Fantl, W. J., et al. (2001). PAR-1 is a dishevelled-associated kinase and a positive regulator of Wnt signalling. *Nature Cell Biology*, *3*, 628–636.
Suzuki, M., Morita, H., & Ueno, N. (2012). Molecular mechanisms of cell shape changes that contribute to vertebrate neural tube closure. *Development, Growth & Differentiation*, *54*, 266–276.
Tickle, C., & Trinkaus, J. P. (1976). Observations on nudging cells in culture. *Nature*, *261*, 413.
Vartiainen, M. K., Guettler, S., Larijani, B., & Treisman, R. (2007). Nuclear actin regulates dynamic subcellular localization and activity of the SRF cofactor MAL. *Science*, *316*, 1749–1752.
Vasudevan, H. N., & Soriano, P. (2014). SRF regulates craniofacial development through selective recruitment of MRTF cofactors by PDGF signaling. *Developmental Cell*, *31*, 332–344.
Vicente-Manzanares, M., Ma, X., Adelstein, R. S., & Horwitz, A. R. (2009). Non-muscle myosin II takes centre stage in cell adhesion and migration. *Nature Reviews. Molecular Cell Biology*, *10*, 778–790.
Wallingford, J. B., Niswander, L. A., Shaw, G. M., & Finnell, R. H. (2013). The continuing challenge of understanding, preventing, and treating neural tube defects. *Science*, *339*, 1222002.
Winter, C. G., Wang, B., Ballew, A., Royou, A., Karess, R., Axelrod, J. D., et al. (2001). Drosophila Rho-associated kinase (Drok) links frizzled-mediated planar cell polarity signaling to the actin cytoskeleton. *Cell*, *105*, 81–91.
Woolner, S., & Papalopulu, N. (2012). Spindle position in symmetric cell divisions during epiboly is controlled by opposing and dynamic apicobasal forces. *Developmental Cell*, *22*, 775–787.
Wozniak, M. A., & Chen, C. S. (2009). Mechanotransduction in development: A growing role for contractility. *Nature Reviews. Molecular Cell Biology*, *10*, 34–43.
Yanagawa, S., van Leeuwen, F., Wodarz, A., Klingensmith, J., & Nusse, R. (1995). The dishevelled protein is modified by wingless signaling in Drosophila. *Genes & Development*, *9*, 1087–1097.
Yin, C., Kiskowski, M., Pouille, P. A., Farge, E., & Solnica-Krezel, L. (2008). Cooperation of polarized cell intercalations drives convergence and extension of presomitic mesoderm during zebrafish gastrulation. *Journal of Cell Biology*, *180*, 221–232.
Yonemura, S., Wada, Y., Watanabe, T., Nagafuchi, A., & Shibata, M. (2010). alpha-Catenin as a tension transducer that induces adherens junction development. *Nature Cell Biology*, *12*, 533–542.
Yu, F. X., & Guan, K. L. (2013). The hippo pathway: Regulators and regulations. *Genes & Development*, *27*, 355–371.
Zallen, J. A., & Wieschaus, E. (2004). Patterned gene expression directs bipolar planar polarity in Drosophila. *Developmental Cell*, *6*, 343–355.
Zhou, J., Kim, H. Y., & Davidson, L. A. (2009). Actomyosin stiffens the vertebrate embryo during crucial stages of elongation and neural tube closure. *Development*, *136*, 677–688.
Zhou, J., Pal, S., Maiti, S., & Davidson, L. A. (2015). Force production and mechanical accommodation during convergent extension. *Development*, *142*, 692–701.

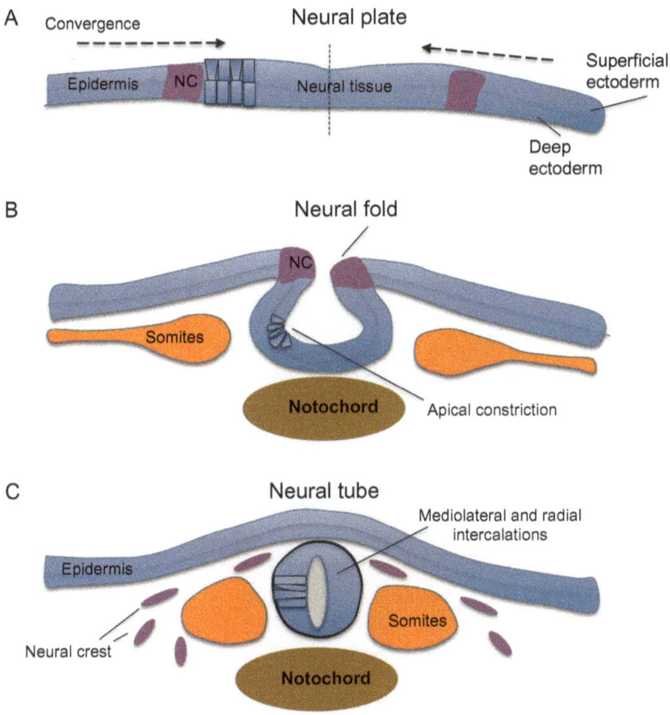

Sergei Y. Sokol, Figure 1 *Xenopus* neural tube as a model for morphogenetic movements. Scheme illustrating main stages of *Xenopus* neural tube closure. (A) Neural plate consisting of two cell layers. (B) Apical constriction leads to neural plate folding. (C) Neural folds fuse into the neural tube, and the neural tube elongates due to radial and mediolateral cell intercalations.

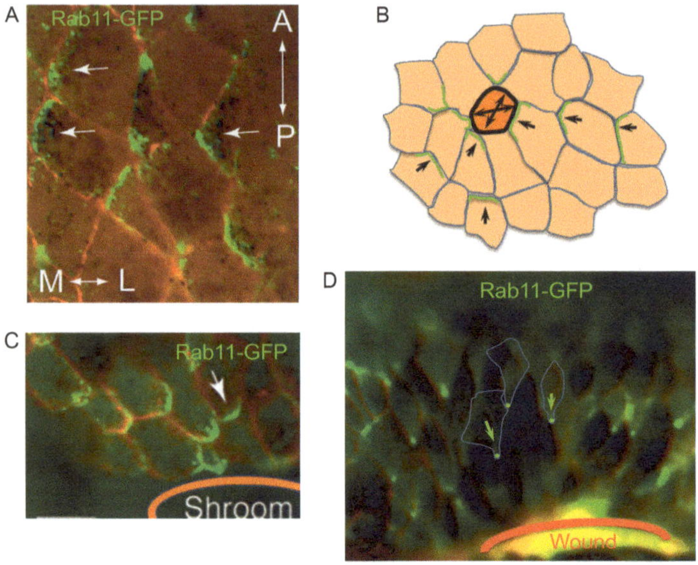

Sergei Y. Sokol, Figure 2 Rab11 as a mechanosensor. (A) Endogenous Rab11 is polarized to medioapical corners of neuroepithelial cells in the neural plate. Body axes are indicated. Costaining was done with anti-ZO1 antibody. (B) Rab11 planar polarization (arrows) induced by a constricting Shroom3-expressing cell (orange) in adjacent ectodermal cells. (C) Polarization of Rab11-GFP near a Shroom3-expressing cell clone. (D) Rab11 planar polarization in embryonic ectoderm during wound healing. Locations of Shroom3-expressing cells (C) and the position of the wound (D) are indicated. *Panels (A) and (D) modified from Ossipova et al. (2014).*

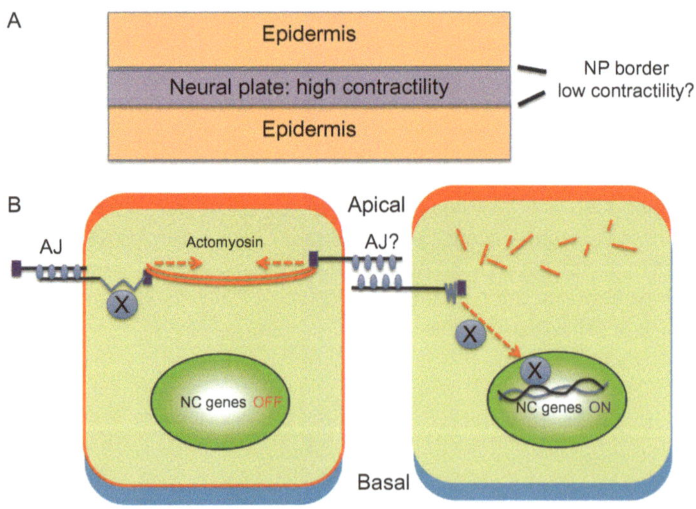

Sergei Y. Sokol, Figure 3 Force-dependent transcriptional regulation of the neural crest. (A) Hypothetical variations in actomyosin-generated forces that affect neural crest specification at the neural plate border. (B) A model for neural crest gene activation. Mechanical stress triggers cytoskeletal reorganization causing messenger translocation from cell junctions to the nucleus and leading to altered gene transcription.

CHAPTER TWENTY-TWO

Preformation Versus Epigenesis in Early Mammalian Development

Davor Solter[1]
Siriraj Center of Excellence in Stem Cell Research, Mahidol University, Bangkok, Thailand
Max Planck Institute of Immunobiology and Epigenetics, Freiburg, Germany
[1]Corresponding author: e-mail address: davorsolter@mac.com

Contents

Abstract

Whether or not early mammalian development results from preformation or epigenesis remains an unresolved issue. Evidence for or against either is weak, inconclusive, and often misinterpreted. Yet, one can parsimoniously conceptualize formation of the mouse blastocyst as a series of random, stochastic events stemming from initial and sequential small asymmetries in egg, zygote, and cleavage stages. Differential compartmentalized gene expression does not lead but follows the morphogenesis and cell fate allocation in the mammalian blastocyst.

Everyone is entitled to his own opinion, but not his own facts.
Attributed to Daniel Patrick Moynihan, James R. Schlesinger and several others
The harmony of the world is made manifest in form and number.
Sir D'Arcy Wentworth Thompson

Current Topics in Developmental Biology, Volume 117
ISSN 0070-2153
http://dx.doi.org/10.1016/bs.ctdb.2015.11.006

1. INTRODUCTION

Forty years ago if asked what is the most interesting question in mammalian development, I would most likely have started by drawing a blastocyst and then say: "here it is and how do you make it?" The blastocyst would have had somewhere between 50 and 100 cells comprised of four distinct parts: mural and polar trophectoderm, primitive endoderm, and epiblast (see Fig. 1). I knew that each blastocyst is composed of cells with unique characteristics and with a precisely defined fate: mural trophectoderm would develop into trophoblast giant cells, polar trophectoderm into fetal placenta, primitive endoderm into yolk sac endoderm, and epiblast into all fetal tissues, yolk sac mesoderm, amnion, and allantois (Albert & Peters, 2009; Cockburn & Rossant, 2010; Dard, Breuer, Maro, & Louvet-Vallée, 2008; Rossant & Tam, 2009; Takaoka & Hamada, 2012; Zernicka-Goetz, Morris, & Bruce, 2009). When asked what determines the character and fate of cells in each different blastocyst compartment, I would have answered: their position and contacts with their surrounding cells and with the environment and also signals they receive from adjacent and distant compartments. Both interactions and signals are unique and specific for each blastocyst compartment. With such an answer, I would have invoked two leading theoretical mechanisms that regulate everything in developmental biology: positional information and reaction–diffusion (Green & Sharpe, 2015).

The next obvious question would be how did the blastocyst develop from a fertilized egg and now the answers become less assured. At that time, we had and still have two hypothetical answers: the so-called inside–outside hypothesis

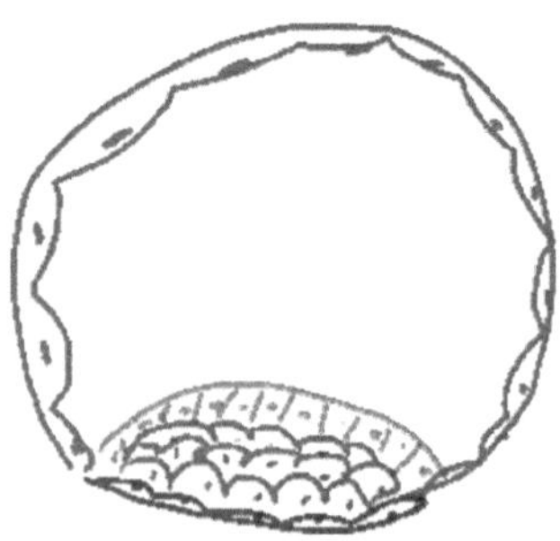

Figure 1 Idealized mouse blastocyst with four compartments each with a different developmental fate. Red—mural trophectoderm, blue—polar trophectoderm, green—epiblast, and yellow—primitive endoderm. (A color version of this figure is located at the end of this chapter.)

(Tarkowski & Wróblewska, 1967) and the polarization hypothesis (Johnson & Ziomek, 1981). The inside–outside hypothesis basically stated that following the fourth division of the newly formed embryo (from the 8- to 16-cell stage), some blastomeres are located inside while others are outside. The inside blastomeres will eventually give rise to the inner cell mass (ICM) composed of primitive endoderm and epiblast, while the outside blastomeres will become the trophectoderm. The polarization hypothesis is actually a refinement of the inside–outside hypothesis, stating that at the 8-cell stage the blastomeres become polarized, each with a distinct apical and basal side. Those that divide symmetrically (with the plane of division parallel with the axis of polarity) will remain outside to become the trophectoderm. Those that divide asymmetrically (with a plane of division perpendicular to the axis of polarity) will give rise to a cell remaining outside and one that will localize inside and become part of ICM. It is clear that neither of these hypotheses explains how the blastocyst develops but merely states what is happening in morphological terms. The questions as to how some blastomeres move inside or why and how blastomeres become polarized remain unanswered.

When trying to understand the emergence of blastocyst in mechanistic terms, we have to remember that there are two basic conceptual models of development: mosaic and regulative (prepatterning and epigenesis). However, these are abstract creations and there is probably no development that follows exclusively either one or the other of these two models. The mosaic development model proposes that early on, in the unfertilized or fertilized egg (zygote), specific molecules are already distributed in a precisely defined pattern and their presence or absence determines the developmental fate of the area in which they reside. In essence, if we knew not only the molecules present in these stages but also their distribution, we should be able to superimpose on the zygote the precise image of the future organism. In contrast, pure regulative development would proceed from a zygote that is completely homogeneous, the form of the future embryo and adult cannot be superimposed on it, but rather the organism gradually develops through environmental signals and intercellular interactions. Many examples of predominantly mosaic development have been described and almost, if not all, nonmammalian species develop in accord with this developmental pattern (e.g., the common model organisms *Caenorhabditis elegans*, *Drosophila melanogaster*, *Xenopus laevis*). The situation with mammalian development has never been absolutely clear and both types of development have been suggested. In the mid-twentieth century, the cytoplasmic localization of several molecules in the egg was described with the suggestion that they are later portioned into specific blastocyst components, thus suggesting that mammals also follow mosaic development paradigm (Dalcq, 1966; Mulnard, 1955). [However, we were not able to reproduce these earlier results using better histochemical techniques (Solter, Damjanov, & Škreb, 1973)].

When experimental manipulations *in vitro* of (predominantly) mouse embryos became possible 50 years ago, the results of several experiments suggested that there are no cytoplasmic localizations in the mouse zygote and, in distinction to the mosaic development hypothesis, all blastomeres resulting from the first two or three cleavage divisions are identical and totipotent. The community of mammalian developmental biologists passionately discussed this issue, and though we all believed that mouse development is in principle regulative, most of us had some reservations about the strength of the experimental evidence. It would require too much space to go into detail here, so one example of the controversy will have to suffice (for more details, see Gardner, 1996). A single mouse blastomere from a 2-cell stage embryo can develop into an entire animal and a single blastomere from either 4- or 8-cell stage embryos can participate in all embryonic and extraembryonic tissues in chimeras, suggesting their identity and totipotency (Kelly, 1975, 1977). These experiments weigh against developmentally relevant cytoplasmic localizations in the zygote and indicate that all components necessary for development are equally distributed throughout the egg. However, one could argue (since not every blastomere from a 2-cell stage embryo develops into a mouse nor every blastomere from 4- and 8-cell stage embryos participates in every tissue in a chimera) that there is cytoplasmic localization of essential components in all zygotes and when these components are allocated to one blastomere of the 2-cell stage embryo, this one will develop into an adult and its sister blastomere will not. Thus, proponents of regulative mammalian development would argue that all experimental failures are due to technical difficulties (which are substantial), while champions of mosaic development would argue that developmental failures are the obvious consequence of an unequal distribution of localized molecules.

Late in the last century, Gardner (1996) presented a very detailed and scholarly criticism (not of the experimental results *per se* but rather of the meaning ascribed to them) of experiments interpreted as evidence of regulative development in mammals. He concluded that all of the evidence was weak and did not disprove the existence of mosaic development and prepatterning. He then took a step further, forgetting the age-old saying that the absence of evidence is not the evidence of absence, and concluded that prepatterning probably did exist and that it should be experimentally demonstrable. The rest of this brief exposition will discuss experiments that were, and are, presented as demonstrating the prepatterning of early mammalian development and I will evaluate their evidentiary strength. I will also attempt to present a simple, form-driven view of early mammalian development, which is in my view stochastic and does not require prepatterning, but is rather designed to complete successfully in the absence of any preformed pattern.

2. EGG ASYMMETRY AND HOW IT AFFECTS THE FIRST CLEAVAGE DIVISION

There is an obvious morphological asymmetry visible in the mouse oocyte and zygote. The first and/or second polar body is located to one side so that the egg and zygote are not perfectly round but flattened on the side where the polar body is located. In response to this absence of radial symmetry, it is generally agreed to call the part of the egg abutting the polar bodies the animal pole and the opposite side of the egg the vegetal pole. These terms have functional implications in nonmammalian, especially amphibian, embryos, and one must be careful not to automatically assume similar functional implications in the mammalian egg and zygote. Nevertheless, as the search for prepatterning in mammalian development began, the apparent asymmetry of the egg represented a convenient starting point.

By observing a series of developmental stages, starting with the asymmetric oocyte and zygote (Fig. 2), one is tempted to imagine a specific developmental pattern so that the animal–vegetal (A–V) axis is constant and successive developmental stages are predicated on the morphology of the previous ones, thus invoking prepatterning mechanism of development. A series of papers proposed that the plane of the first division is predicated on and contains the A–V axis (Bedzhov, Graham, Leung, & Zernicka-Goetz, 2014; Bischoff,

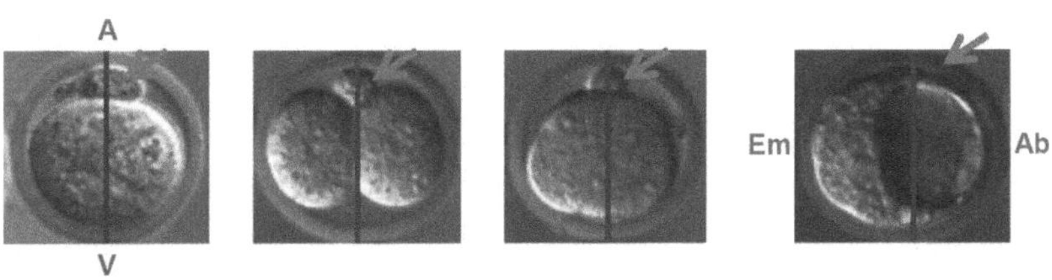

Figure 2 Zygote and cleavage stages of mouse preimplantation development: zygote, 2-cell stage, compacted 8-cell stage and blastocyst, all in transparent zonae pellucidae. Blue lines denote putative animal–vegetal (A–V) axis and red arrows show the position of the second polar body. Em and Ab indicate embryonic and abembryonic part of the blastocyst, respectively. (A color version of this figure is located at the end of this chapter.)

Parfitt, & Zernicka-Goetz, 2008; Gardner, 1997, 2001, 2007; Gardner & Davies, 2006; Piotrowska, Wianny, Pedersen, & Zernicka-Goetz, 2001; Plusa, Grabarek, Piotrowska, Glover, & Zernicka-Goetz, 2002; Plusa et al., 2005; Zernicka-Goetz, 2005) and, in addition, is determined by the sperm entry point (Piotrowska & Zernicka-Goetz, 2001). Following this mindset, the position of the A–V axis of the zygote will be maintained until blastocyst formation and becomes perpendicular to the embryonic–abembryonic (Em–Ab) axis of the blastocyst. This view implies that in the 2-cell stage embryo, each blastomere contains both animal and vegetal cytoplasm but the blastomeres are not necessarily identical, since at the time of blastocyst formation the progeny of one blastomere will contribute to the embryonic part (containing ICM and polar trophectoderm) and the progeny of the other will contribute to the abembryonic part (containing the mural trophectoderm). However, it is not possible to predict which of the two blastomere will develop into embryonic and which into abembryonic part of the blastocyst.

This scenario depends on several assumptions, the most notable being that the position of the second polar body in the zygote, particularly in the 2-cell stage (where it is always located in an intercellular furrow—Fig. 2) and subsequently until blastocyst formation, does not change. However, both we and others observed (Bauer, Motosugi, Miura, Sabe, & Hiiragi, 2008; Hiiragi & Solter, 2004; Kurotaki, Hatta, Nakao, Nabeshima, & Fujimori, 2007) that the second polar body moves freely. Indeed, the movement starts before the beginning of cleavage and is always directed toward the cleavage plane. The result is that, regardless of the initial position of the second polar body, it always ends up in the furrow between the two blastomeres. In subsequent divisions, the polar body usually lies in a cleavage furrow, and until at the eight-cell stage, its association with individual blastomeres dynamically changes (Kurotaki et al., 2007). The final position of the polar body in the blastocyst does not correspond with its initial position and association although, probably for mechanical reasons, it is usually found along the circumference of the short diameter of the ellipsoid blastocyst (Fig. 2). Initially, we thought this was due to spatial constraints, with the polar body pushed into any available free space within the zona pellucida. However, this is not the case; the polar body moves into the cleavage furrow even in the absence of the space-constraining zona pellucida. The movement of the polar body also does not depend on the existence of a tether that normally attaches the polar body to the zygote, since an additional polar body can be placed on the zygote surface where it adheres and then it also moves into the cleavage furrow at the next cleavage (Bauer et al., 2008).

Polar body movement is passive, caused by the streaming of many, if not all, plasma membrane proteins and anything attached to them (i.e., antibodies or beads), toward the cleavage furrow at the beginning of each cell division. This membrane mobility is not unique to the zygote but is also observed in other dividing cells (Bauer et al., 2008).

Considering this cleavage- and division-associated movement of cell surface proteins and the structures used to label them toward the cleavage furrow, the significance of the sperm entry point (marked by fluorescent beads) in determining cleavage plane and developmental fate of blastomere containing it (Piotrowska & Zernicka-Goetz, 2001) should be reexamined. In deciding what comes first, there are two possibilities: (1) the second polar body and the sperm entry point determine the plane of the first cleavage, or (2) cleavage happens and then the second polar body and any marker for the sperm entry point are passively carried into the cleavage furrow. The second option is equally if not more probable based on the experimental data (Bauer et al., 2008).

If the second polar body and sperm entry point do not determine the plane of the first division, what does, and is the position of the plane of the first division entirely random? We suggested (Hiiragi & Solter, 2004) the final plane of the apposition of the two pronuclei either marks or determines the plane of the first division. Both pronuclei form at the periphery of the zygote, the female pronucleus close to the position of the second polar body and male pronucleus in the vicinity of the sperm entry point and fertilization cone. Subsequently, they both move toward the center of the zygote, compact against each other, their nuclear membranes dissolve, and the two sets of chromosomes are brought into a single metaphase plate (Hiiragi & Solter, 2004). A barrel-shaped acentriolar spindle, derived from multiple microtubule foci in the zygote cytoplasm, organizes the metaphase plate. These unique features of the first mitosis make it possible to predict the plane of the first division, i.e., from the plane of the pronuclear apposition. If the second polar body (Bedzhov et al., 2014; Bischoff et al., 2008; Gardner, 1997, 2001, 2007; Gardner & Davies, 2006; Piotrowska et al., 2001; Plusa et al., 2002, 2005; Zernicka-Goetz, 2005) and sperm entry point (Piotrowska & Zernicka-Goetz, 2001) indeed determine the plane of the first division, the plane would be always meridional, dividing the zygote into two halves containing equally animal and vegetal cytoplasm. However, if the plane is determined by pronuclear apposition, it could be meridional, equatorial, or anything in between (Hiiragi & Solter, 2004). Obviously, if the plane of cleavage is equatorial or close to it, the resulting blastomeres will be different, one containing the animal and the other the vegetal half of the zygote. Thus, distinguishing between the two determinative modes for the plane of the first division is crucial to conceptualization of early mammalian development. If meridional cleavage is predetermined and essential, prepatterning of the mammalian zygote can be assumed, while if development is possible after any random stochastic positioning of the first cleavage plane, absence of prepatterning in the zygote is strongly indicated.

Although we suggest that the second polar body and the sperm entry point cannot define the plane of the first division, the position of this plane is not entirely random. In a high number of embryos (60–70%), the plane of the first division is indeed within 30° of both the second polar body and the sperm entry point (Hiiragi & Solter, 2004; Plusa et al., 2005; Zernicka-Goetz, 2005). The reasons for this are the unique morphological and functional features of the mammalian egg. In the process of fertilization, the mouse sperm penetrates the zona pellucida at random; however, the site of sperm entry into the oocyte does not coincide with the site of its zona penetration (Motosugi, Dietrich, Polanski, Solter, & Hiiragi, 2006). Fertilization occurs significantly more often in the animal half of the embryo, despite the fact that the relatively large, microvilli-free region overlaying the metaphase chromosomes is excluded from sperm penetration. Preferential sperm penetration in the animal half of the embryo is caused by the presence of the large space created by the first polar body (Fig. 3). By removing a large amount of cytoplasm, or by inserting another polar body, the frequency of sperm penetration changes and approximates the available surface area of the egg (Fig. 3) (Motosugi et al., 2006). Thus, in the normal situation, fertilization takes place most often close to the second polar body (Fig. 4), where the male and female pronuclei form close to each other and move in tandem to the center of the zygote. In this case, the apposition plane (future plane of the first division) is formed close to the virtual plane that passes close to the second polar body and sperm entry point (Fig. 4) (Hiiragi & Solter, 2004).

In summary, we suggest that the position of the plane of the first division cannot be predicted with certainty. Moreover, successful development does not require a predetermined position of the first cleavage division, and thus each of the two blastomeres does not receive a specific portion of the zygote cytoplasm. Yet, based strictly on the form of the egg and zygote, and the effects of these forms on the initial morphogenetic events, one can bet that the plane of the first division will be close to meridional and close to the second polar body. One could say that the position of the first polar body, the space created by this position, and the movement resulting in apposition

A

B

C

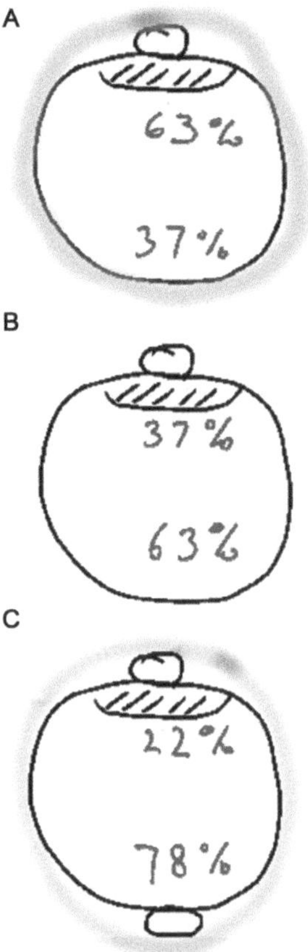

Figure 3 The role of the space between the oocyte and the zona pellucida in determining the frequency of fertilization in the animal and vegetal half of the egg. (A) Zona intact normal egg; (B) egg with zona removed; and (C) zona intact egg with an additional polar body inserted at the vegetal pole. Hatched areas represent the microvilli-free surface overlaying the metaphase chromosomes that excludes sperm penetration. Numbers represent the experimentally derived frequency of sperm penetration in the animal and vegetal half in percentages. *Adapted from Motosugi et al. (2006).* (A color version of this figure is located at the end of this chapter.)

of pronuclei are "prepatterning," but this is certainly not prepatterning in the classical sense that is observed in nonmammalian embryos.

3. IS THERE EVIDENCE FOR PREPATTERNING IN DEVELOPMENT FROM THE 2-CELL STAGE TO BLASTOCYST?

It has been repeatedly suggested that the second division and subsequent formation of the 4-cell stage embryo is also predominantly

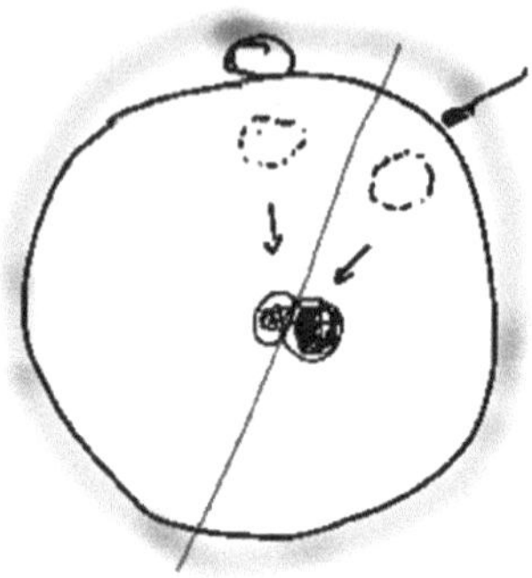

Figure 4 Following fertilization, the male and female pronuclei form under the egg plasma membrane near the sperm entry point and first polar body, respectively (dotted circles). They move to the center of the egg (solid filled circles) and create a plane of apposition that predicts the plane of the first cleavage (red line). *Adapted from Hiiragi and Solter (2004).* (A color version of this figure is located at the end of this chapter.)

prepatterned. If the two divisions (going from 2- to 4-cell stage) follow a specific cleavage pattern, the blastomeres of the 4-cell stage embryo are not identical and will have different developmental fates (Gardner, 2010; Piotrowska-Nitsche, Perea-Gomez, Haraguchi, & Zernicka-Goetz, 2004; Piotrowska-Nitsche & Zernicka-Goetz, 2005). However, others did not observe any differences between the blastomeres of the 4-cell stage embryos using similar approaches (Alarcón & Marikawa, 2008; Louvet-Vallée, Vinot, & Maro, 2005; Marikawa & Alarcón, 2009). There are several possible reasons for these discrepant results: chiefly, the reproducibility in identification of the cleavage patterns, and the use, or not, of embryos within the zona pellucida when studying reconstructed embryos (as discussed by Gardner, 2010; Marikawa & Alarcón, 2009).

In all of these experiments, the location of the second polar body in the cleavage furrow of the 2-cell stage embryo was used as a marker of the animal pole. Thus, it was assumed that equatorial division of a blastomere of the 2-cell stage would result in one blastomere with "animal" and other with "vegetal" pole cytoplasm. However, the assumption that the second polar body identifies the "animal" pole in 2-cell stage embryos is not correct for a certain percentage of embryos. Thus, reliance on the position of the second polar body to identify blastomeres containing vegetal cytoplasm could contribute to the reported discrepancies.

Several investigators have proposed that the cleavage plane of the 2-cell stage embryo is perpendicular to the embryonic–abembryonic axis of the blastocyst (see Fig. 2), thus predicting and then finding that one blastomere contributes predominantly to the embryonic half, including the ICM, and the other to the abembryonic half and the mural trophectoderm of the embryo (Bedzhov et al., 2014; Bischoff et al., 2008; Gardner, 1997, 2001, 2007; Gardner & Davies, 2006; Piotrowska et al., 2001; Plusa et al., 2002, 2005; Zernicka-Goetz, 2005). In other words, these investigators suggested that the plane dividing the clonal progeny of each of two blastomeres is identical, or at least parallel, to the plane that divides the embryonic from the abembryonic part of the blastocyst and both planes are perpendicular to the Em–Ab axis. However, several other groups suggested that the position of Em–Ab axis is determined by the ellipsoid (not round) shape of the zona pellucida so that the ICM and blastocoel always lie on opposite ends of the zona (the mechanical restraint hypothesis) (Alarcón & Marikawa, 2008; Kurotaki et al., 2007; Marikawa & Alarcón, 2009; Motosugi, Bauer, Polanski, Solter, & Hiiragi, 2005; Watanabe, Biggins, Tannan, & Srinivas, 2014). These investigators also observed that the plane dividing the clonal progenies of the 2-cell stage blastomeres and the plane dividing the embryonic and abembryonic part of the blastocyst are not parallel but can cross at any angle between 0° and 90° (Motosugi et al., 2005).

The prepatterning model of blastocyst formation assumes a differential developmental fate of each blastomere from a 2-cell stage embryo and that there is either no or minimal movement of blastomeres and embryos within the zona pellucida. A regulative model assumes the opposite. Both sides presented movies as supportive evidence for their point of view, though only those favoring a regulative model stated that embryos shown in the movies were able to implant and develop to term.

Thus, one can visualize early mammalian development progressing and being directed through a series of small morphological irregularities such as flattening of the egg at the animal pole, the bulk of the second polar body creating a space within the zona pellucida that traps mobile sperm, which in turn favors the prevalent site of fertilization leading to the preferred plane of the first division. At the 8-cell stage, increased surface tension and contractility of the blastomeres (Maître, Niwayama, Turlier, Nédélec, & Hiiragi, 2015), possibly caused by the increased cell–cell contact at this stage, lead to compaction and polarization as the blastomeres transform into an epithelium. Space constraints then force some of the blastomeres inside, and the space constraints exerted by ellipsoid zona pellucida shape the position of the ICM and blastocoel in the newly formed blastocyst. There appears to be no need to invoke prepatterning in the zygote to explain any of these events and, like trying to balance a pencil on its sharpened tip, and trying to guess which way is going to fall, none of it is exactly predictable.

I think of preimplantation development unfolding through the changing forms of the embryos in the zona pellucida driven by the series of small, sequential asymmetries. The differences in contacts and positions of the blastomeres as the embryo develops would then lead to differential gene expression, e.g., activation of Hippo pathway at the time of compaction (Lorthongpanich, Doris, Limviphuvadh, Knowles, & Solter, 2012; Lorthongpanich & Issaragrisil, 2015; Lorthongpanich et al., 2013; Sasaki, 2015). Differential gene expression between blastomeres proceeds gradually and unique gene expression patterns are finally established, together with the formation of distinct components of the blastocyst (Guo et al., 2010). One can argue that in early mammalian development, differential gene expression does not lead but rather follows the morphological patterning. Early mammalian development is in this way unique and different from development of nonmammalian species. In early nonmammalian development, cells are portioned into different future body compartments following cues established in the egg. However, since the first goal of mammalian development is to assure nutrition of the embryo in the uterus, trophectoderm and primitive endoderm are necessarily the first products of cellular differentiation. Prepatterning the mammalian egg with cues for a future body plan makes no sense when it is not possible to determine which part of the egg will actually develop into the somatic compartments. We can thus view mammalian development as precocious and, in terms of developmental information and organization, the early gastrula is the mammalian equivalent of *Drosophila* egg or *Xenopus* zygote.

4. CODA

Is the mouse zygote prepatterned or is the development of mammals predominantly or completely regulative and can this issue be experimentally addressed? We would agree that evidence "proving" regulative development does not convince (Gardner, 1996), but the evidence for prepatterning is equally weak and often misinterpreted. It is difficult to imagine experiments that would convincingly eliminate one or the other mode of development (*evidence against*). It is equally difficult to imagine experiments that would prove regulative development (*evidence for*) although evidence for prepatterning (finding a zygote-localized gene product with clear developmental function) can be easily imagined but such evidence has thus far been elusive. The simplest question, and the one most easily addressed at this time, is: are the two blastomeres of the 2-cell stage embryo identical or different? Two reports using transcriptome analysis and bioinformatics comparisons have been carried out and each oppositely concluded: there are reproducible differences in the transcriptomes of two paired blastomeres from 2-cell stage embryo (Biase, Cao, & Zhong, 2014), and there are not reproducible differences between them (VerMilyea et al., 2011).

Considering how simple the mouse blastocyst is, merely 50–100 cells, considering how easy it is to experimentally manipulate the early mammalian embryo, and considering the sophisticated visual and molecular techniques at our disposal today, it should be possible to understand the zygote-to-blastocyst transition in totality. Yet, despite the significant amount of data acquired in the last 50 years, unanswered questions and controversies still predominate.

ACKNOWLEDGMENTS

Many relevant references were not included due to space constraints for which I apologize. My thanks to Barbara B. Knowles for the critical reading of the manuscript and to my long-term collaborator Takashi Hiiragi whose experiments significantly influenced my thinking about the development of early mouse embryo.

REFERENCES

Alarcón, V. B., & Marikawa, Y. (2008). Spatial alignment of the mouse blastocyst axis across the first cleavage plane is caused by mechanical constraint rather than developmental bias among blastomeres. *Molecular Reproduction and Development, 75*, 1143–1153.

Albert, M., & Peters, A. H. F. M. (2009). Genetic and epigenetic control of early mouse development. *Current Opinion in Genetics & Development, 19*, 1–9.

Bauer, T., Motosugi, N., Miura, K., Sabe, H., & Hiiragi, T. (2008). Dynamic rearrangement of surface proteins is essential for cytokinesis. *Genesis, 46*, 152–162.

Bedzhov, I., Graham, S. J. L., Leung, C. Y., & Zernicka-Goetz, M. (2014). Developmental plasticity, cell fate specification and morphogenesis in the early mouse embryo. *Philosophical Transactions of the Royal Society B, 369*, 20130538.

Biase, F. H., Cao, X., & Zhong, S. (2014). Cell fate inclination within 2-cell and 4-cell mouse embryo revealed by single-cell RNA sequencing. *Genome Research, 24*, 1787–1796.

Bischoff, M., Parfitt, D.-E., & Zernicka-Goetz, M. (2008). Formation of the embryonic-abembryonic axis of the mouse blastocyst: Relationships between orientation of early cleavage divisions and pattern of symmetric/asymmetric divisions. *Development, 135*, 953–962.

Cockburn, K., & Rossant, J. (2010). Making the blastocyst: Lessons from the mouse. *The Journal of Clinical Investigation, 140*, 995–1003.

Dalcq, A. M. (1966). Détection des enzymes de déphosphorylation dans les œufs de rat et de souris fixes au formol and traits *in toto*. *Archives de Biologie (Liège)*, 77, 205–344.

Dard, N., Breuer, M., Maro, B., & Louvet-Vallée, S. (2008). Morphogenesis of the mammalian blastocyst. *Molecular and Cellular Endocrinology, 282*, 70–77.

Gardner, R. (1996). Can developmentally significant spatial patterning of the egg be discounted in mammals? *Human Reproduction Update*, *2*, 3–27.
Gardner, R. (1997). The early blastocyst is bilaterally symmetrical and its axis of symmetry is aligned with the animal-vegetal axis of the zygote in the mouse. *Development*, *124*, 289–301.
Gardner, R. (2001). Specification of embryonic axes begins before cleavage in normal mouse development. *Development*, *128*, 839–847.
Gardner, R. (2007). The axis of polarity of the mouse blastocyst is specified before blastulation and independently of the zona pellucida. *Human Reproduction*, *22*, 798–806.
Gardner, R. (2010). Normal bias in the direction of fetal rotation depends on blastomere composition during early cleavage in the mouse. *PLoS One*, *5*, e9610.
Gardner, R., & Davies, T. J. (2006). An investigation of the origin and significance of bilateral symmetry of the pronuclear zygote in the mouse. *Human Reproduction*, *21*, 492–502.
Green, J. B. A., & Sharpe, J. (2015). Positional information and reaction–diffusion: Two big ideas in developmental biology combine. *Development*, *142*, 1203–1211.
Guo, G., Huss, M., Tong, G. Q., Nang, C., Sun, L. L., Clarke, N. D., et al. (2010). Resolution of cell fate decisions revealed by single-cell gene expression analysis from zygote to blastocyst. *Developmental Cell*, *18*, 675–685.
Hiiragi, T., & Solter, D. (2004). First cleavage plane of the mouse egg is not predetermined but defined by the topology of the two apposing pronuclei. *Nature*, *430*, 360–364.
Johnson, M. H., & Ziomek, C. A. (1981). The foundation of the two distinct cell lineages within the mouse morula. *Cell*, *24*, 71–80.
Kelly, S. J. (1975). Studies of the potency of early cleavage blastomeres of the mouse. In M. Balls & A. E. Wild (Eds.), *The early development in mammals* (pp. 97–105). Cambridge: Cambridge University Press.
Kelly, S. J. (1977). Studies of the developmental potential of 4- and 8-cell stage mouse blastomeres. *Journal of Experimental Zoology*, *200*, 365–376.
Kurotaki, Y., Hatta, K., Nakao, K., Nabeshima, Y., & Fujimori, T. (2007). Blastocyst axis is specified independently of early cell lineage but aligns with the ZP shape. *Science*, *316*, 1081–1092.
Lorthongpanich, C., Doris, T. P. Y., Limviphuvadh, V., Knowles, B. B., & Solter, D. (2012). Developmental fate and lineage commitment of singled mouse blastomeres. *Development*, *139*, 3722–3731.
Lorthongpanich, C., & Issaragrisil, S. (2015). Emerging role of the Hippo signaling pathway in position sensing and lineage specification in mammalian preimplantation embryos. *Biology of Reproduction*, *92*, 1–10.
Lorthongpanich, C., Messerschmidt, D. M., Chan, S. W., Hong, W., Knowles, B. B., & Solter, D. (2013). Temporal reduction of LATS1/2 kinases in early preimplantation embryos prevents ICM lineage differentiation. *Genes & Development*, *27*, 1441–1446.
Louvet-Vallée, S., Vinot, S., & Maro, B. (2005). Mitotic spindles and cleavage planes are oriented randomly in the two-cell mouse embryo. *Current Biology*, *15*, 464–469.
Maître, J.-L., Niwayama, R., Turlier, H., Nédélec, F., & Hiiragi, T. (2015). Pulsatile cell-autonomous contractility drives compaction in the mouse embryo. *Nature Cell Biology*, *17*, 845–855.
Marikawa, Y., & Alarcón, V. B. (2009). Establishment of trophectoderm and inner cell mass lineages in the mouse embryo. *Molecular Reproduction and Development*, *76*, 1019–1032.
Motosugi, N., Bauer, T., Polanski, Z., Solter, D., & Hiiragi, T. (2005). Polarity of the mouse embryo is established at blastocyst and is not prepatterned. *Genes & Development*, *19*, 1081–1092.
Motosugi, N., Dietrich, J.-E., Polanski, Z., Solter, D., & Hiiragi, T. (2006). Space asymmetry directs preferential sperm entry in the absence of polarity in the mouse oocyte. *PLoS Biology*, *4*, e135.

Mulnard, J. (1955). Contribution à la connaissance des enzymes dans l'ontogénèse. Les phosphomonoestérases acide et alkaline dans le development du Rat et de la Souris. *Archives de Biologie (Liège), 66*, 525–685.
Piotrowska, K., Wianny, F., Pedersen, R. A., & Zernicka-Goetz, M. (2001). Blastomeres arising from the first cleavage division have distinguishable fates in normal mouse development. *Development, 128*, 3739–3748.
Piotrowska-Nitsche, K., Perea-Gomez, A., Haraguchi, S., & Zernicka-Goetz, M. (2004). Four-cell stage mouse blastomeres have different developmental properties. *Development, 132*, 479–490.
Piotrowska, K., & Zernicka-Goetz, M. (2001). Role for sperm in spatial patterning of the early mouse embryo. *Nature, 409*, 517–521.
Piotrowska-Nitsche, K., & Zernicka-Goetz, M. (2005). Spatial arrangement of individual 4-cell stage blastomeres and the order in which they are generated correlate with blastocyst pattern in the mouse embryo. *Mechanisms of Development, 122*, 487–500.
Plusa, B., Grabarek, J. B., Piotrowska, K., Glover, D. M., & Zernicka-Goetz, M. (2002). Site of the previous meiotic division defines cleavage orientation in the mouse embryo. *Nature Cell Biology, 4*, 811–815.
Plusa, B., Hadjantonakis, A.-K., Gray, D., Piotrowska-Nitsche, K., Jedrusik, A., Papaioannou, V. E., et al. (2005). The first cleavage of the mouse zygote predicts the blastocyst axis. *Nature, 434*, 391–395.
Rossant, J., & Tam, P. P. L. (2009). Blastocyst lineage formation, early embryonic asymmetries and axis patterning in the mouse. *Development, 136*, 701–713.
Sasaki, H. (2015). Position- and polarity-dependent Hippo signaling regulated cell fates in preimplantation mouse embryos. *Seminars in Cell & Developmental Biology*. http://dx.doi.org/10.1016/j.semcdb.2015.05.003.
Solter, D., Damjanov, I., & Škreb, N. (1973). Distribution of hydrolytic enzymes in early rat and mouse embryos. A reappraisal. *Zeitschrift für Anatomie und Entwicklungsgeschichte, 139*, 119–126.
Takaoka, K., & Hamada, H. (2012). Cell fate decisions and axis determination in the early mouse embryo. *Development, 139*, 3–14.
Tarkowski, A. K., & Wróblewska, J. (1967). Development of blastomeres of mouse eggs isolated at the 4- and 8-cell stage. *Journal of Embryology and Experimental Morphology, 18*, 155–180.
VerMilyea, M. D., Maneck, M., Yoshida, N., Blochberger, I., Suzuki, E., Suzuki, T., et al. (2011). Transcriptome asymmetry within mouse zygotes but not between early embryonic sister blastomeres. *The EMBO Journal, 30*, 1841–1851.
Watanabe, T., Biggins, J. S., Tannan, N. B., & Srinivas, S. (2014). Limited predictive value of blastomere angle of division in trophectoderm and inner cell mass specification. *Development, 141*, 2279–2287.
Zernicka-Goetz, M. (2005). Cleavage pattern and emerging asymmetry of the mouse embryo. *Nature Reviews. Molecular Cell Biology, 6*, 919–928.
Zernicka-Goetz, M., Morris, S. A., & Bruce, A. W. (2009). Making a firm decision: Multifaceted regulation of cell fate in the early mouse embryo. *Nature Reviews. Genetics, 10*, 467–477.

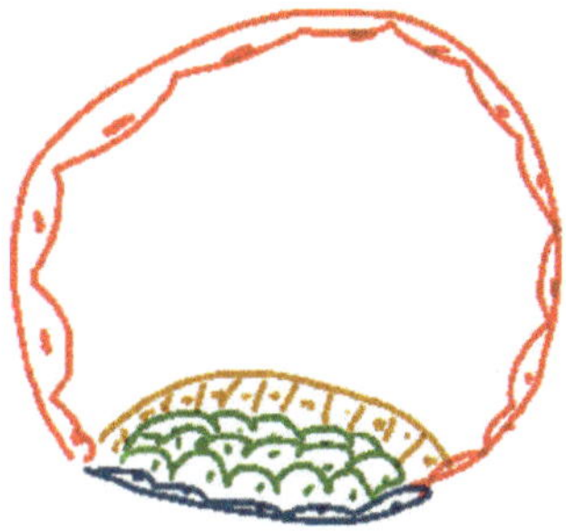

Davor Solter, Figure 1 Idealized mouse blastocyst with four compartments each with a different developmental fate. Red—mural trophectoderm, blue—polar trophectoderm, green—epiblast, and yellow—primitive endoderm.

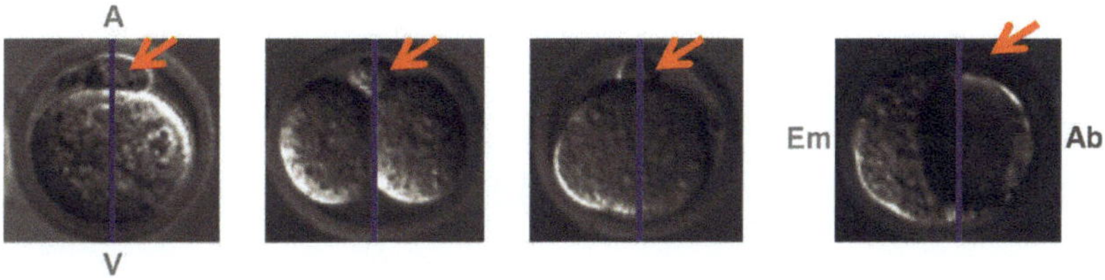

Davor Solter, Figure 2 Zygote and cleavage stages of mouse preimplantation development: zygote, 2-cell stage, compacted 8-cell stage and blastocyst, all in transparent zonae pellucidae. Blue lines denote putative animal–vegetal (A–V) axis and red arrows show the position of the second polar body. Em and Ab indicate embryonic and abembryonic part of the blastocyst, respectively.

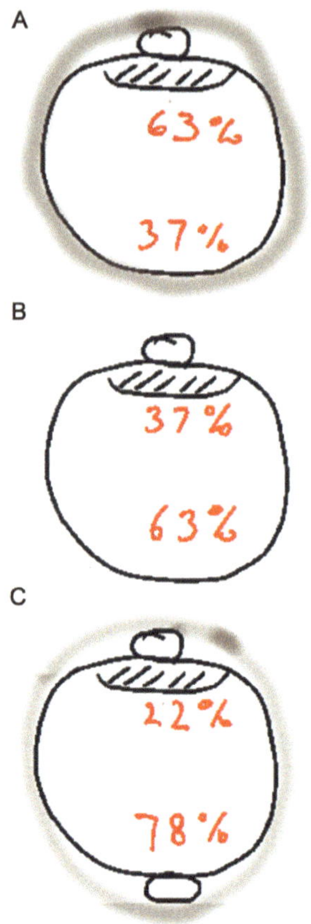

Davor Solter, Figure 3 The role of the space between the oocyte and the zona pellucida in determining the frequency of fertilization in the animal and vegetal half of the egg. (A) Zona intact normal egg; (B) egg with zona removed; and (C) zona intact egg with an additional polar body inserted at the vegetal pole. Hatched areas represent the microvilli-free surface overlaying the metaphase chromosomes that excludes sperm penetration. Numbers represent the experimentally derived frequency of sperm penetration in the animal and vegetal half in percentages. *Adapted from Motosugi et al. (2006).*

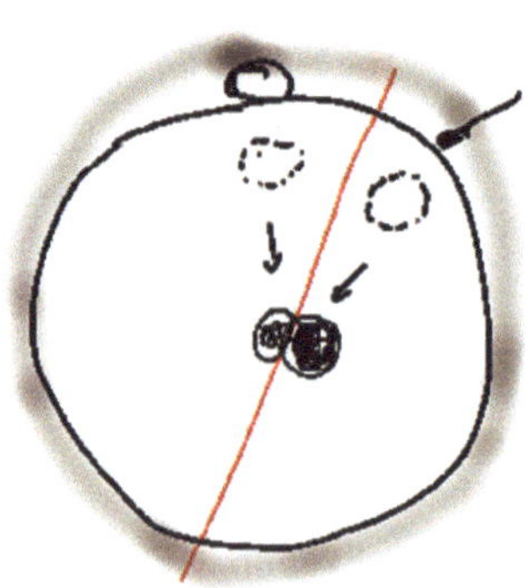

Davor Solter, Figure 4 Following fertilization, the male and female pronuclei form under the egg plasma membrane near the sperm entry point and first polar body, respectively (dotted circles). They move to the center of the egg (solid filled circles) and create a plane of apposition that predicts the plane of the first cleavage (red line). *Adapted from Hiiragi and Solter (2004).*

CHAPTER TWENTY-THREE

A Thousand and One Receptor Tyrosine Kinases: Wherein the Specificity?

Harish N. Vasudevan, Philippe Soriano[1]
Department of Developmental and Regenerative Biology, The Tisch Cancer Institute, Icahn School of Medicine at Mount Sinai, New York, USA
[1]Corresponding author: e-mail address: philippe.soriano@mssm.edu

Contents

Abstract

Over 20 years ago, a series of seminal studies in PC12 neurons provided a framework for how receptor tyrosine kinases generate many different outcomes despite activating a set of shared intracellular pathways. In this essay, we revisit the question of receptor tyrosine kinase specificity. We first examine the relationship between receptor phosphorylation and intracellular pathway activation. We then consider the mechanisms through which signaling dynamics encode distinct cellular outcomes and finally discuss how two different receptors drive divergent transcriptional responses within the same developmental context. Establishing the key parameters that dictate the response to growth factor stimulation is critical for determining how receptor tyrosine kinases orchestrate development, an essential prerequisite for understanding the pathological consequences when such signaling processes go awry.

1. INTRODUCTION

Receptor tyrosine kinases (RTKs) are a family of cell surface receptors important across a range of physiological and disease processes. Upon ligand treatment, RTKs undergo phosphorylation at intracellular tyrosine residues,

Current Topics in Developmental Biology, Volume 117
ISSN 0070-2153
http://dx.doi.org/10.1016/bs.ctdb.2015.10.016

leading to the binding of signaling proteins at these phosphotyrosine sites (Lemmon & Schlessinger, 2010). Many RTKs bind overlapping sets of signaling proteins, ultimately leading to the activation of many shared intracellular pathways, but the outcomes dictated by different RTKs can be quite distinct. The resulting puzzle of how distinct cellular outputs arise from a common set of signaling inputs has intrigued biologists for decades. Many models to explain signal specificity have been proposed (Hunter, 2000; Marshall, 1995; Pawson, 2004; Simon, 2000; Volinsky & Kholodenko, 2013). In theory, each signaling event could determine a particular cellular output; under this linear model, perturbation of any single input would impair a specific outcome. For instance, epidermal growth factor (EGF) promotes osteoblast differentiation in human mesenchymal stem cells while platelet-derived growth factor (PDGF) represses this process; this effect is PI3K dependent and explained at least in part by selective activation of PI3K in response to PDGF, but not EGF, stimulation (Kratchmarova, Blagoev, Haack-Sorensen, Kassem, & Mann, 2005). However, the number of RTK-mediated outcomes is greater than the number of intracellular pathways, making a strictly linear model insufficient to fully explain the range of behaviors controlled by RTKs. Alternatively, differences in the magnitude or duration of a signal could determine cellular output, and a common input would then specify multiple outputs based on distinct signaling dynamics. A classic example of such dynamic encoding is in PC12 neurons, where a transient phospho-ERK pulse in response to EGF treatment results in cell proliferation while a sustained phospho-ERK wave in response to NGF stimulation drives cell differentiation (Marshall, 1995). Further, the additive integration of multiple linear or dynamic inputs can generate an even greater range of outputs. Since RTK activation generally results in the concerted induction of many intracellular pathways, such combinatorial encoding likely contributes toward the outcomes mediated by RTK signaling. Given the limited number of intracellular pathways in comparison to the diversity of cell types and outcomes regulated by RTKs, both dynamic and combinatorial encoding are likely to play important roles in many contexts.

The framework and associated examples described above predominantly describe intracellular pathway activation in response to growth factor stimulation. However, it is important to distinguish between signaling events at the receptor (i.e., direct binding of proteins to phosphorylated tyrosine residues on the receptor) and intracellular pathway level (i.e., the consequent pattern of intracellular pathway activation in response to RTK signaling), as they constitute distinct steps at which information is encoded. Given that effector binding at the receptor drives the pattern of intracellular pathway activation, these two parameters are closely related; indeed, some effectors are themselves enzymes (e.g., RasGAP, PI3K, and PLCϒ) that directly modulate downstream signaling, but others function as scaffolds (e.g., FRS2 and IRS1), serving to recruit more signaling proteins. Nonetheless, events at the receptor and intracellular pathway level are mechanistically separate. For example, perturbation of PI3K binding at the receptor and chemical inhibition of the PI3K pathway directly can have different effects on signal transduction (Francavilla et al., 2013), likely due to crosstalk between pathways and the ability of multiple effectors at the receptor to coordinately regulate a single downstream intracellular pathway.

In this essay, we explore the mechanisms underlying RTK specificity. We first discuss experiments examining the importance of individual phosphotyrosine residues on the receptor in dictating intracellular pathway activation and developmental outcomes. Focusing on phospho-ERK, we next consider a pair of recent studies elucidating the mechanisms through which differences in intracellular signaling dynamics are converted into distinct outcomes. Finally, we summarize our recent work examining how two different receptors elicit divergent transcriptional responses in a common context. We conclude by outlining two avenues for future investigation we find particularly intriguing: the systematic mapping of phosphotyrosine residues to intracellular pathway activation and the functional relevance of single-cell heterogeneities within signaling responses.

2. STARTING AT THE RECEPTOR: MULTIFUNCTIONAL PHOSPHOTYROSINES

The first step upon ligand binding is phosphorylation of intracellular tyrosine residues on the receptor itself; therefore, a critical question is how effector binding at these phosphotyrosines shapes signal specificity. One can imagine two scenarios: individual binding events could modulate distinct signaling outcomes or alternatively, the combined pattern of many binding events may function additively to drive the desired response. Using embryonic phenotypes as a readout for PDGF receptor α (PDGFRα) and PDGF receptor β (PDGFRβ) signaling, analysis of mice harboring point mutations at tyrosine residues that abrogate binding of specific effector proteins suggest both models apply in vivo. An allelic series at the *Pdgfra* locus identified specific phenotypes associated with loss of particular binding events, suggesting each effector transmits a distinct biological signal with a central role for PI3K

(Klinghoffer, Hamilton, Hoch, & Soriano, 2002). In contrast, an allelic series at the *Pdgfrb* locus did not identify distinct phenotypes when the binding of individual effectors was perturbed (Tallquist, French, & Soriano, 2003); instead, combinatorial loss of multiple effectors exacerbates a shared vascular smooth muscle phenotype, suggesting additive output across multiple binding events drives cellular outcomes downstream of PDGFRβ. Interestingly, mutation of multiple PDGFRβ intracellular tyrosines reduces the magnitude of phospho-ERK and phospo-Akt induction but does not completely abolish activation of these pathways, underscoring the phenotypic effect of quantitative differences in intracellular pathway activation.

While powerful in revealing the *in vivo* importance of specific effectors, these genetic experiments come with some important caveats. As with many loss-of-function approaches, these results do not distinguish between changes in signaling dynamics and total attenuation of a signaling event. Indeed, abrogation of PI3K binding to PDGFRα abolishes phospho-Akt induction entirely (Klinghoffer et al., 2002), making it difficult to ascertain whether phospho-Akt induction or dynamics is critical. Alternatively, crosstalk with other signal transduction cascades may underlie the observed phenotypes, as disruption of one signaling event often leads to compensatory induction of others. Again taking the case of the *Pdgfra* allelic series, loss of PI3K binding leads to increased SHP-2 recruitment, which may be functionally relevant. Further, the temporal dynamics of tyrosine phosphorylation itself may be important for specifying different signals (Zheng et al., 2013), and simple loss-of-function mutations do not provide insight regarding the function of phosphorylation kinetics. As discussed below, modern proteomic approaches show promise as a method to distinguish between these various possibilities, allowing systems-level measurements at high temporal resolution. Finally, the issue of cellular context is confounding: cell type-specific requirements for an effector may occur independently of the receptor itself, perhaps based on the relative expression of a key signaling molecule (e.g., the receptor itself) or the local chromatin architecture permitting modulation of only a particular pathway's target genes in a given cell type. For example, the same receptor can induce different outcomes based on many context-dependent parameters such as expression level, reflected by the observation that overexpressing the EGF receptor in PC12 neurons switches the effect of EGF stimulation from proliferation to differentiation (Traverse et al., 1994). Nonetheless, these genetic experiments reveal that individual tyrosine residues (and the effectors that bind them) can function either independently or additively to specify developmental outcomes, hinting at the degree of mechanistic diversity at the very first step of signal transduction.

More recent phosphoproteomic approaches, in which the phosphorylation status of multiple residues can be simultaneously measured, have shed light on how different tyrosine phosphorylation patterns at the receptor encode distinct cellular outcomes. In one elegant study, the authors use proteomics to investigate the mechanisms underlying the specificity of fibroblast growth factor (FGF) signaling (Francavilla et al., 2013). They demonstrate how two ligands, FGF7 and FGF10, drive different phosphorylation patterns at their shared receptor, FGFR2, thus modulating distinct intracellular pathway activation patterns and cellular outcomes within the same cell type. Harkening back to the importance of phospho-ERK duration established in the PC12 paradigm, FGF7 induces a sustained pERK response and drives cell proliferation while FGF10 results in a transient pERK response associated with cell migration in this system. This distinction originates from differential phosphorylation on a single FGFR2 intracellular tyrosine residue that binds PI3K subunits and acts as a switch between the two modes of signaling dynamics by regulating receptor degradation and recycling. Mutation of this tyrosine eliminates the differential response to FGF7 and FGF10, as FGFR2 is now "locked" into the FGF7-proliferation axis and exhibits limited migration in response to FGF10. In addition, this single tyrosine mutation affects multiple signaling pathways in addition to PI3K, underscoring the high degree of crosstalk and complexity in linking individual phosphotyrosines to downstream pathway activation. This study provides an example in which a qualitative switch in tyrosine phosphorylation of a particular residue at the receptor results in altered signaling dynamics and cellular outcome, thus highlighting the distinction between information encoding at the receptor and intracellular pathway level. Such unbiased approaches to measure phosphorylation on a systems level will provide valuable insight into how patterns of intracellular tyrosine phosphorylation at the receptor encode downstream signaling.

With regard to the framework presented earlier, these results highlight the importance of dynamic and combinatorial encoding in the response to RTK activation, as evidenced by the requirement of multiple phosphotyrosines to coordinately regulate a shared set of pathways downstream of PDGFRβ and the ability of FGF7 and FGF10 to induce distinct signaling kinetics through differential phosphorylation status of a single FGFR2 intracellular tyrosine residue. Further, when this FGFR2 phosphotyrosine residue is mutated, the signaling dynamics of multiple pathways are affected, arguing against a simple linear model in which each signaling event maps one-to-one to intracellular pathway activation and cellular outcome. In the case of PDGFRα, it is plausible that a similar mechanism may occur in mouse mutants where PI3K binding is disrupted at PDGFRα, and more generally, such mutations of a single tyrosine residue may have multiple effects on signal transduction that are difficult to predict *a priori*.

3. INTRACELLULAR SIGNALING DYNAMICS: ALL ROADS LEAD TO ERK

The importance of quantitative differences in the kinetics and magnitude of intracellular pathway activation naturally leads to the question of how cells interpret such differences in signaling dynamics. Particular attention has been given to ERK, which can be activated in response to a diverse range of extracellular stimuli and result in many cellular outcomes, including proliferation, differentiation, and migration. The importance of differences in phospho-ERK duration for specifying distinct cellular outputs has been long appreciated in PC12 neurons; modern work investigating the basis for this phenomenon have illuminated the underlying mechanisms through which phospho-ERK dynamics are decoded to specify distinct outcomes. A recent study utilized optogenetics to selectively control ERK activation itself, allowing direct modulation of ERK signaling dynamics through control of upstream Ras activity (Toettcher, Weiner, & Lim, 2013); in contrast to loss-of-function studies (e.g., mutating individual tyrosine residues on the receptor or treating with pharmacological inhibitors), this type of approach defines signals for which ERK activation is sufficient and more generally, permits uncoupling of pathway activation from pathway dynamics. By combining their optogenetic tool with a reverse phase protein array, the authors delineate a proteomic "signature" for distinct frequencies of ERK activity. Notably, not all proteins regulated by PDGF stimulation are ERK-responsive, indicating ERK alone is not sufficient to drive the full RTK response and highlighting the utility of this approach in delineating sufficiency from necessity. Thus, these other responses are either ERK-independent or more likely, require multiple intracellular pathway inputs, again underscoring the importance of combinatorial encoding.

The advent of computational models describing the RTK signaling network has advanced our understanding of how dynamic changes in an input encode a desired output. The iterative refinement of such models based on experimental results designed to test model fidelity is a powerful method for understanding mechanisms of RTK specificity, as evidenced by a study focusing on how different phospho-ERK durations regulate downstream transcription (Nakakuki et al., 2010). The authors initially build a model for how the transcription factor *Fos* distinguishes between transient and sustained phospho-ERK, with the primary parameters being (1) the combination of nuclear phospho-ERK and nuclear phospho-RSK (to drive *Fos* transcription) and (2) cytoplasmic phospho-ERK and Fos protein (to generate phosphorylated Fos protein). However, upon finding experimental results are not in full concordance with this model's predictions, an updated model incorporating a transcriptional repressor is developed that does fit the observed data, making the system more robust to changes in input parameters and minimizing noise. Thus, the decoding of differences in ERK dynamics requires an intricate network integrating changes at the transcriptional and translational levels. Although the transcriptional repressor itself was not identified in this study, one can imagine the biological importance of such a gene that acts as a superintendent of the core transcriptional response to RTK signaling. Of note, both the optogenetics and computational modeling studies discussed above extrapolated their results to PC12 cells, highlighting the relevance of the PC12 paradigm over two decades later.

While elegant, a critical drawback to these studies is they have been primarily conducted *in vitro*, and it is not obvious how to extrapolate these parameters to an *in vivo* system. One potentially promising technology to interrogate these processes is live imaging using biosensors of ERK activity, which has revealed intriguing properties of phospho-ERK signaling patterns both *in vitro* (Aoki et al., 2013) and *in vivo* (Hiratsuka et al., 2015). In the first study (Aoki et al., 2013), the authors use a FRET biosensor to demonstrate that the frequency of ERK pulsing is correlated with cell proliferation in a density-dependent manner *in vitro*. By utilizing a light dependent ERK activation system, they then confirm that synthetically producing ERK oscillations drives a greater proliferative response than continuous ERK induction, demonstrating the functional outcome of alterations in ERK signaling dynamics. In a second study (Hiratsuka et al., 2015), the same group extended their FRET-based analysis of ERK dynamics to the skin, recording changes in ERK dynamics *in vivo*. At steady-state, the authors observed occasional "ripples" of ERK activation, with spontaneous ERK pulses beginning at foci and then migrating radially outward. This radial propagation of ERK activity was associated with cell proliferation, and the authors use a combination of pharmacological agents to induce or inhibit mitotic activity to confirm this relationship. Such approaches to both visualize and control ERK dynamics *in vivo* will facilitate the assignment of functional roles for different modes of ERK signaling in developmental processes.

4. ONE CONTEXT, TWO RECEPTORS, TWO RESPONSES: PDGF AND FGF IN THE MIDFACE

The above examples center on the mechanisms through which a single receptor or intracellular pathway can mediate multiple cellular outcomes. However, an equally critical question is how two different RTKs elicit divergent responses. To address this issue, we recently compared the transcriptional responses to PDGF and FGF signaling in the palate (Vasudevan, Mazot, He, & Soriano, 2015), a context of particular interest given the palatal clefting phenotypes observed upon loss of either RTK. Interestingly, we found divergent gene expression programs and cellular outcomes in response to PDGF and FGF. These differences seem to stem from differences in intracellular pathway usage, with PDGF regulating a PI3K-dependent differentiation axis and FGF driving an ERK-dependent proliferation circuit. In addition to this qualitative delineation, the FGF-mediated phospho-ERK response displayed a greater magnitude and longer duration compared to the PDGF-mediated phospho-ERK response. Building upon the differential gene expression profiles observed upon PDGF and FGF stimulation in the palate, we further explored the function of a single shared target gene, serum response factor (SRF) (Vasudevan & Soriano, 2014). Although both PDGF and FGF induce SRF mRNA and protein, the two RTKs promote different cofactor interactions and consequently, distinct gene expression programs, adding yet another layer of regulatory complexity.

These studies illustrate two important lessons. First, intracellular signaling dynamics are a critical parameter when comparing the effects of two different RTKs. For example, the PDGF and FGF-mediated phospho-ERK response in the palate is nearly identical after 5 min; only upon measuring phospho-ERK at multiple timepoints is a difference appreciated, and this likely contributes to the ERK-dependence of FGF-regulated proliferation but not PDGF-mediated differentiation. Second, the integration of many parameters encodes the resultant gene expression program; in the case of SRF activity downstream of PDGF and FGF, intracellular pathway dynamics, mRNA induction, and protein-protein interactions are all key inputs delineating the ultimate transcriptional output. Further experimental and computational work is needed to fully define the full range and relative weights of the many input parameters that define the transcriptional response to RTK signaling.

5. CONCLUDING REMARKS

We began this essay by putting forward two models for RTK signal specificity: a linear model where distinct intracellular pathways lead to particular cellular outcomes and a dynamic model in which the magnitude and duration of pathway activation encode distinct outcomes. As research into RTKs and cell signaling has progressed over the last two decades, the linear model suggested by early genetic and pharmacologic experiments has given way to a more complex system in which RTKs convert growth factor inputs into a myriad of intracellular signals with specific magnitudes and frequencies, which are further fine-tuned by a complex network of feedback regulation. In many respects, seminal studies in PC12 neurons were a harbinger of this shift, providing an early example of dynamic encoding that recent studies often return to for experimental validation. Despite the considerable progress toward unraveling the mechanisms underlying RTK signal specificity, our understanding remains far from complete. Two points of particular interest are how patterns of tyrosine phosphorylation at the receptor map to intracellular pathway activation and how signaling dynamics on a single-cell level translate into behaviors within a population.

It is well appreciated that either loss-of-function or gain-of-function mutations in phosphotyrosine residues can have detrimental effects, but a systematic understanding of how these mutations alter intracellular pathway activation and dynamics is lacking. Can each loss-of-function mutation be linked to a particular set of intracellular pathways, and are any such dependencies "hard-wired" into the signaling network or context dependent? Conversely, are all gain-of-function mutations created equal, or do different mutations at the receptor lead to distinct patterns of intracellular pathway activation? By applying phosphoproteomic approaches to obtain an unbiased picture of the signaling events mediated by each mutation, empirical answers to these questions are beginning to emerge. We believe combining this signaling data with genetic approaches to interrogate the developmental function of individual phosphotyrosine residues *in vivo* will provide a detailed understanding of how RTK-mediated outcomes are generated.

Both the optogenetic and FRET-based studies discussed in this essay observed a degree of cellular heterogeneity, with the responses of individual cells within a population exhibiting high variability. Similarly, in the classic PC12 system, not all cells treated with NGF undergo cell differentiation. Thus, it is naturally of interest to examine how much variability in signaling dynamics is present during development and if is there an underlying organization to the heterogeneity of responses within a population. One could imagine the multiple lineages that comprise a given tissue to be differentially primed for RTK responsiveness, providing an intuitive basis for such heterogeneity *in vivo*. Recently, a single-cell RNA-seq study revealed marked variation within populations of dendritic cells identically treated with various stimuli *in vitro* (Shalek et al., 2014), and we suspect the complexity of RTK signaling coupled with the high diversity of cell types within the embryo will only amplify this effect for growth factors utilized across multiple developmental processes. It will be enlightening to determine if particular transcriptional programs predispose a cell to respond in a specific manner and further, how the expression of various components within the transcriptional feedback/feedforward loops vary at the single-cell level, both at baseline and in response to RTK signaling. To our knowledge, no such study comparing the response to different RTKs or simultaneously interrogating multiple intracellular pathways in addition to phospho-ERK has been carried out. Taken together, such information may contribute toward a predictive model utilizing gene expression, proteomic, and other data types as an input to identify those cells and tissues most likely to execute a particular RTK response.

Modern technologies allow elegant manipulation of specific nodes within an intracellular pathway, perturbation of pathway dynamics without affecting pathway activation, and the measurement of signaling parameters on a systems level, in many cases at single-cell resolution. Applying these techniques to the developing embryo will elucidate the answer to a number of outstanding issues involving RTK specificity. The consequent knowledge will pay great dividends in understanding how growth factors dictate cellular behaviors, laying the requisite foundation for studying how such processes can be pathologically corrupted. More broadly, the conceptual frameworks and emerging technologies outlined above can likely be expanded beyond RTKs to any number of signaling events during development (Hoch & Soriano, 2015), and insights gleaned across the full range of signaling processes in the embryo will further our understanding of the critical parameters underlying cellular decision making.

ACKNOWLEDGMENTS

Work in the author's laboratory is supported by grants to P.S. from NIH/NIDCR (RO1DE022363 and RO1 DE022778) and NYSTEM (IIRP N11G-131). H.N.V. was supported by NIH/NIDCR fellowship F31 DE023456.

REFERENCES

Aoki, K., Kumagai, Y., Sakurai, A., Komatsu, N., Fujita, Y., et al. (2013). Stochastic ERK activation induced by noise and cell-to-cell propagation regulates cell density-dependent proliferation. *Molecular Cell, 52*, 1–12.

Francavilla, C., Rigbolt, K., Emdal, K., Carraro, G., Vernet, E., et al. (2013). Functional proteomics defines the molecular switch underlying FGF receptor trafficking and cellular outputs. *Molecular Cell, 51*, 707–722.

Hiratsuka, T., Fujita, Y., Naoki, H., Aoki, K., Kamioka, Y., et al. (2015). Intercellular propagation of extracellular signal-regulated kinase activation revealed by in vivo imaging of mouse skin. *eLife, 4*, 1–18.

Hoch, R. V., & Soriano, P. (2015). Generating diversity and specificity through developmental cell signaling. *Principles of Developmental Genetics, 1*, 3–36.

Hunter, T. (2000). Signaling—2000 and beyond. *Cell, 100*, 113–127.

Klinghoffer, R. A., Hamilton, T. G., Hoch, R., & Soriano, P. (2002). An allelic series at the PDGFαR locus indicates unequal contributions of distinct signaling pathways during development. *Developmental Cell, 2*, 103–113.

Kratchmarova, I., Blagoev, B., Haack-Sorensen, M., Kassem, M., & Mann, M. (2005). Mechanism of divergent growth factor effects in mesenchymal stem cell differentiation. *Science, 308*, 1472–1477.

Lemmon, M. A., & Schlessinger, J. (2010). Cell signaling by receptor tyrosine kinases. *Cell, 141*, 1117–1134.

Marshall, C. J. (1995). Specificity of receptor tyrosine kinase signaling: Transient versus sustained extracellular signal-regulated kinase activation. *Cell, 80*, 179–185.

Nakakuki, T., Birtwistle, M., Saeki, Y., Yumoto, N., Ide, K., et al. (2010). Ligand-specific c-fos expression emerges from the spatiotemporal control of ErbB network dynamics. *Cell, 141*, 884–896.

Pawson, T. (2004). Specificity in signal transduction: From phosphotyrosine-SH2 domain interactions to complex cellular systems. *Cell, 116*, 191–203.

Shalek, A. K., Satija, R., Shuga, J., Trombetta, J., Gennert, D., et al. (2014). Single-cell RNA-seq reveals dynamic paracrine control of cellular variation. *Nature, 509*, 363–369.

Simon, M. A. (2000). Receptor tyrosine kinases: Specific outcomes from general signals. *Cell, 103*(1), 13–15.

Tallquist, M. D., French, W., & Soriano, P. (2003). Additive effects of PDGF receptor β signaling pathways in vascular smooth muscle cell development. *PLoS Biology, 1*, 288–299.

Toettcher, J. E., Weiner, O. D., & Lim, W. A. (2013). Using optogenetics to interrogate the dynamic control of signal transmission by the Ras/Erk module. *Cell, 155*, 1422–1434.

Traverse, S., Seedorf, K., Paterson, H., Marshall, C. J., Cohen, P., et al. (1994). EGF triggers neuronal differentiation of PC12 cells that overexpress the EGF receptor. *Current Biology, 4*, 694–701.

Vasudevan, H. N., Mazot, P., He, F., & Soriano, P. (2015). Receptor tyrosine kinases modulate distinct transcriptional programs by differential usage of intracellular pathways. *eLife, 4*(e07186), 1–22.

Vasudevan, H. N., & Soriano, P. (2014). SRF regulates craniofacial development through selective recruitment of MRTF cofactors by PDGF signaling. *Developmental Cell*, *31*, 332–344.

Volinsky, N., & Kholodenko, B. N. (2013). Complexity of receptor tyrosine kinase signal processing. *Cold Spring Harbor perspectives in biology*, *5*, a009043.

Zheng, Y., Zhang, C., Croucher, D., Soliman, M., St-Denis, N., et al. (2013). Temporal regulation of EGF signalling networks by the scaffold protein Shc1. *Nature*, *499*, 166–171.

CHAPTER TWENTY-FOUR

The Carnegie Department of Embryology at 100: Looking Forward

Allan C. Spradling[1]
Department of Embryology, Carnegie Institution for Science, Baltimore, Maryland, USA
[1]Corresponding author: e-mail address: spradling@ciwemb.edu

Abstract

Biological research has a realistic chance within the next 50 years of discovering the basic mechanisms by which metazoan genomes encode the complex morphological structures and capabilities that characterize life as we know it. However, achieving those goals is now threatened by researchers who advocate an end to basic research on nonmammalian organisms. For the sake of society, medicine, and the science of biology, the focus of biomedical research should place more emphasis on basic studies guided by the underlying evolutionary commonality of all major animals, as manifested in their genes, pathways, cells, and organs.

These remarks were originally delivered on September 26, 2013 to celebrate the 100th anniversary of the Department of Embryology of the Carnegie Institution for Science. They were revised for inclusion in this volume. The Department of Embryology (Brown, 1987; Maienschein, Glitz, & Allen, 2004; Spradling, 1997) is a small research institute that was founded in 1913 on the campus of Johns Hopkins University to study human embryology. The interests of its 8–12 faculty members subsequently evolved to encompass reproductive physiology, cellular differentiation, molecular genetics, and other topics. The department has nurtured creative research on the role of genes in development for the last 50 years, initially under the Directorship of Don Brown. Highlights include Don Brown's anatomical analyses of purified genes and functional mapping of eukaryotic gene regulatory sequences by in vitro transcription, Steve McKnight's characterization of eukaryotic promoter structure and transcription factors, Gerry Rubin and Allan Spradling's development of transposon-mediated transformation in Drosophila, Sue Wessler and Nina

Current Topics in Developmental Biology, Volume 117
ISSN 0070-2153
http://dx.doi.org/10.1016/bs.ctdb.2015.11.038

Fedoroff's cloning and study of a maize transposable element, Doug Koshland's genetic analyses of yeast chromosome behavior and of SMC proteins, Joe Gall's studies of amphibian lampbrush chromosomes, and Andy Fire's discovery of RNAi (with Craig Mello).

I would like to start this talk by extending my sincere thanks to the Carnegie Institution for supporting my laboratory for the last 33 years. For a budding young biologist, getting a job in the Embryology Department felt like winning the lottery. And that was before I could appreciate the full value of what Carnegie offers, or how much the Institution would shape my career by providing consistent, crucial support at every level. What the Department means to all of us who work here has been beautifully expressed during this anniversary Symposium, and in videotaped recollections (see www.ciwemb.edu) from former Carnegie developmental biologists.

Carnegie's biggest impact over the years has always been through the people who train here and then migrate to locations throughout the research and education enterprise. One could think of the Department as a stem cell, its graduates as transit-amplifying cells, and the graduates' own trainees as downstream cells. It's extremely gratifying to see how successful Carnegie's progeny have been, to read their papers, and to communicate, visit, and interact with many on an almost daily basis. Notice that I did not describe Carnegie's offspring as "differentiated cells," because it is now clear that the Embryology Department is not just any stem cell, but a germline stem cell, so there need be no limit to the process we are celebrating. Because of this, and because the last 100 years have been described so eloquently throughout the day, I propose to talk this evening about the future of developmental biology.

Speaking about the future of science is a hazardous business. Whether it has ever been done successfully is doubtful. About the only thing you can be sure of is that the future will scarcely resemble whatever is predicted. In the spirit of Carnegie, however, I will nonetheless proceed. Mostly, I wish to focus on a more limited question: Is the scientific style that has characterized research here still relevant for advancing our field? Can a department with 10 laboratories make a significant impact in an era of worldwide mega-biology? Or have we become an institution that is starting to look back too much on its glorious past? In fact, I believe that Embryology's scientific style is critically needed as biology goes forward, perhaps more than at any time during the last 100 years.

The biggest insight from developmental research so far gets less attention than it deserves. It has come from comparing development across different animal groups, or "evo-devo," a field that Nipam Patel helped get started while he was an Embryology Staff Associate in the mid-1980s. Comparative studies have now revealed a remarkable and largely unexpected unity of metazoan life. All major multicellular animal groups on Earth today appear to have developed from a single early metazoan, a creature that already displayed most of the organs, tissues, cells, pathways, and genes that constitute the basic ground plan of animal life. This realization represents one of the greatest insights in the history of science and makes the task facing the field of developmental biology seem, for the first time, truly feasible. It is simply amazing and wonderful to know that humans share a basic body plan, organ systems, cell types, and genes with so many of Earth's other creatures—that we are literally variants of a common design.

When I was growing up, I heard that Albert Schweitzer valued all animal life so greatly that he refused even to slap a mosquito. It seemed impossible to me then that mosquitoes could be comparable to ourselves. Now we know that mammals and insects do share core developmental mechanisms. But I still slap mosquitoes. How can I justify that, you might ask? I do it to help them—as a species—evolve to evade humans even better than they already do.

If we can understand this core developmental program, we can understand the development of any particular animal group and species in terms of how they have diverged from that foundation during evolution. Understanding developmental strategies this way should become a unifying paradigm for thinking about animal groups and their characteristics. It means that the insights gained from studying any organism's development can contribute to our knowledge of both its special features and the core. Thus, evo-devo reveals a sort of natural biological coordinate system for thinking about the great diversity of life around us.

The powerful model genetic organisms that have dominated our attention during recent decades are defining major aspects of the core developmental program. Studying species with unusual properties or phylogenetic positions is becoming increasingly valuable as the core becomes better known. The Embryology Department has done much to advance technology in the major model systems, and also to pioneer new systems, such as the work by Alejandro Sanchez-Alvarado and Phil Newmark on *Planaria*. Curiously, during my time at Carnegie, most faculty members have worked on different organisms. But we have been studying a common program of metazoan development.

Despite the importance of understanding metazoan development and life's evolutionary history, the science of developmental biology today is at a turning point. Some of the most important gaps in our knowledge can now be addressed as a result of recently established experimental methods. These are largely the "high-throughput genomic" techniques that have generated so much excitement since the days of the genome projects. The Embryology Department's own technical contributions, such as Andy Fire's discovery of RNAi, rank along with sequencing among the most valuable such technologies. With the approaches now available, we can realistically contemplate closing the gap between genotype and phenotype, the holy grail of developmental genetics. If basic developmental genetic research continues to be supported and pursued by talented researchers, we will map the core program of development at single-cell resolution in several model organisms and determine the effects of altering every important gene and control element. Carrying out such a program in conjunction with fundamental biochemistry and cell biology has the potential to reveal the basic physiological processes underlying metazoan life and to change the way we think about many of them. The undertaking might require 25–50 years, but would ultimately be successful. With such knowledge, DNA sequences would become catalogs of detailed information about phenotype and disease susceptibility, transforming biology and medicine.

However, there are many people today, mostly outside developmental biology, who argue against a research program focused on basic studies of embryos and tissues. Like the old classical geneticists, they think it is possible to deduce one's way across the genotype–phenotype divide simply by looking at the nature of mutations and their final outcomes. These people believe that adding a few single-cell assays and building some models using high-throughput data will reveal how the genome controls phenotype without ever confronting the complexity of actual developing embryos and tissues. Given the appeal of a shortcut to politicians and to many leaders of biomedicine, it is not clear whether our society still possesses the wisdom and long-term vision to complete the genetic analysis of metazoan development and physiology, despite the incalculable benefit that it would provide. Choosing the path of shortcuts and hype would likely surrender our current leadership in biological science and medicine.

Why do we need the Embryology Department and places like it to advance the systematic analysis of metazoan development? Even if our will holds and our technology proves adequate for analyzing development bit by genomic bit, our current intellectual picture of how the genome encodes developmental information may fail us. Biology today resembles the world of classical physics at the end of the nineteenth century. As the story goes, this was a time when physicists thought all the laws of the universe were known and the only remaining challenge would be to calculate fundamental constants to greater and greater precision. Biological enthusiasts today believe that every important biological law has already been discovered. Information is being collected on how proteins interact and where they bind to chromatin, in databases of greater and greater precision. This information is valuable, especially if one analyzes data within a single-cell type (which enthusiasts rarely do). But data analysis cannot begin to address the fact that critical laws of developmental genetics have not yet been discovered. There is a lot we still don't understand, even in principle, about how genomes encode metazoan form and function. We need places like the Department of Embryology in order to develop the biological equivalent of quantum mechanics.

But can't one find missing laws by taking vast biological databases and subjecting them to mathematical analysis? Hasn't biology always been driven more by techniques than by ideas? Well, as long as one is guided by a good set of ideas, experiments are often rate limiting. But ideas eventually have their own limits that no amount of data can overcome. To Charles Darwin, doing experiments in geology without reference to ideas was like counting the stones in a gravel pit. In fact, the world is full of data, everywhere we turn. But most data we shouldn't even want to know, and knowing those data is likely only to slow productivity. Data can be accumulated wisely and take on revolutionary meaning only if collected and examined in the context of the right ideas. Our brains are devices that filter out the vast majority of sensory input, and then in fiendishly clever ways construct something sensible from a highly processed fraction. Our brains evolved to protect us from big data. The proponents of data mining will mostly discover how frequently statistical significance does not correspond to biological or medical significance. Abandoning the search for new ideas represents the perfect corollary to abandoning the study of the actual organisms we seek to understand.

How can new ideas and experimental directions that have the potential to change developmental genetics be encouraged? There is no known way to guarantee such research, but the chance of discovering new principles and mechanisms can be maximized by a department's scientific style and research atmosphere. This atmosphere, like a niche, plays a big role in controlling what the departmental stem cells produce. Of primary importance is ensuring that true stem cells are in residence, cells that are flexible and multipotent, rather than committed to a single product. The Embryology Department has never had a specific mission or scientific program. It feels safe to doubt anything here—if you can defend your position. However, stem cells can proliferate rapidly only in a supportive and nutrient-rich microenvironment. The Department is most fortunate to reside in the new Maxine Singer Building, one of the best niches ever for developmental research. We have access to state-of-the-art equipment and are assisted by highly talented specialists. Finally, we recognize the influence of niche signals on stem cell behavior. We place the highest value on discovery and believe that working in new areas and using new techniques should be encouraged. In 1981, neither Gerry Rubin nor I had ever injected a Drosophila embryo. We planned most of our experiments on transformation of Drosophila during a Gordon Conference coffee break.

There is nothing so stimulating of stem cell activity as what I think of as "creative disbelief"—doubting an experiment, a paper, a fad, and especially questioning some tenet of our current science. Disbelief suggests many new experiments, whereas belief suggests only experiments that are already being done by others. Overturning an old idea frequently leads to something new; sticking with existing ideas means you will always work in the mainstream. Fields that have accepted a fundamentally wrong idea cannot progress further until they are rescued by creative disbelief. Developmental biology probably has some intellectual roadblocks that need to be cleared away in the coming years at places like Carnegie. In contrast, at many venues, creative skepticism has become less welcome, and it is too often being replaced with hype.

So what might be developmental biology's roadblocks? For starters, the field should beware of relying too heavily on gain-of-function experiments to define fundamental aspects of biology. What cells can do in a situation for which evolution never prepared them is a very shaky foundation on which to build biological understanding. Another looming danger is trying to use simple cell cultures to address developmental rather than cell biological problems. The study of stem cells reveals just how profoundly cells depend on their microenvironment to maintain both their immediate activities, such as cell cycling, and their state of differentiation. It would be fabulous to create an artificial niche that could cause a Drosophila stem cell to respond normally to a wide range of challenges. However, today the order in which such experiments can be done successfully is usually reversed. One must thoroughly understand the biology *in vivo* first, in order to guide *in vitro* system development, rather than the opposite.

However, the biggest roadblock may be the "standard model" of developmental genetics itself. This model is based on the notion that cells of various "cell types" have well-defined programs of differential gene expression. Understanding development would mean understanding how cytoplasmic and environmental asymmetries, along with the starting pattern of gene expression, induce an altered program of expression in daughter cells. If a closed cycle of such changes can be understood throughout the life cycle, and if the changes at each step are predicted by the functions of a cell's expressed genes, then the entire program of gene expression will be explained.

Now that the dominant role of chromatin in controlling gene expression has become clear, hopes have declined for predicting gene activity from lists of expressed regulators. Every cell's chromatin state, by determining how gene regulatory sequences are organized and how they interact with binding factors, specifies both its constitutive and inducible expression program. Acquiring a stable chromatin state must largely correspond to the process of cell differentiation, even though some changes in chromatin state can still take place throughout life. Unfortunately, how chromatin states are specified and how they regulate gene expression remain very poorly understood. It was much easier to imagine predicting gene activity from local combinatorial transcription factor binding than from chromatin-modifying factors. The idea of enhancer looping remains too simplistic to bridge this gap. How chromatin states are duplicated during S phase and accurately inherited by daughter cells remains equally mystifying.

Our limited understanding of chromatin is not the only problem with the standard model. In a strange parallel to classical physics, the model has been highly successful in explaining cell behavior at a "high" level, where the properties of many similar cells are averaged together. But gene expression in individual tissue cells can be harder to reconcile with the concept of a few uniform cell types. Some discrepancies are probably caused by undescribed, minor cell types with their own expression patterns, something that was found in the Drosophila ovary. However, there is increasing evidence of epigenetic and genetic differences among tissue cells of the same type.

Cellular individuality implies that individual chromatin states are not exactly the same, even among cells of the same type. Differences may arise during differentiation because of imperfections in chromatin inheritance or result from local external influences affecting only some of the cells. Regulation does not go only from chromatin to cellular phenotype; differences in metabolism, mechanical forces, and microenvironment may feed back and alter the chromatin state. As a result, how individual tissue cells will behave is subject to a small amount of uncertainty. Developmental programs seem to finesse cellular differences by using patterning mechanisms that are nonautonomous. The exact number of cells and their lineage relationships in a developing tissue rarely matters within broad limits. When cells synthesize a product, signals and feedback systems ensure that the overall product distribution is more accurate than the autonomous responses of the component cells. The weaker or more distracted cells either synthesize longer or their portion of the structure is supplemented by material from neighbors.

The deterministic standard model encounters further problems with progenitor cells, i.e., cells that proliferate prior to the onset of differentiation. Gene expression in progenitor cells varies much more than in differentiated cells, and these changes are stochastic but transiently heritable. The chromatin states of progenitor cells must be subject to organizational shifts that influence expression levels. Progenitor cells may be epigenetically unstable to facilitate their ability to differentiate into multiple cell states. But the molecular nature of these instabilities and how they are resolved during differentiation remain virtually unknown. Current ideas include changes in histone modification or in the levels of higher-order compaction.

Another instability likely derives from the genome itself. It is clear that differentiated cells retain most of the germline genome. But that is very different from assuming, as we do today, that the entire genome is precisely replicated in all somatic cells except for immune cells. Additional diversity outside that coded in the genome appears to be valuable in multiple situations. Site-specific recombination and somatic mutation diversify immunoglobulin proteins, RNA editing alters the structure of membrane-spanning regions of channel proteins, and somatic transposition is proposed to diversify neurons in developing brains. Replication errors provide a significant source of genome variation, especially if a region fails to complete duplication prior to the next cell cycle. We lack the technical means to detect sequence differences between individual tissue cells, so we know very little about their nature and frequency. However, it's not necessary to have an army of clones to run a tissue any more than it is to run a research lab. Tissues cells may resemble cooperating individuals with different strengths and weaknesses. How cell-to-cell genetic variation is produced and integrated with epigenetic differences represents another challenge in moving beyond the standard model of development.

So what is the new quantum mechanics of development going to be? It's too early to know. But this brief summary makes clear how much more we need to learn about chromatin states, how they arise, how they regulate gene expression, and how they're inherited and stabilized. The most exciting and important discoveries in developmental biology during the coming years will reveal the molecular events underlying these processes. Many of these solutions will be unlike what we're currently thinking about, and will not come from churning more data according to current thinking. These solutions will come from resourceful individual scientists following their own ideas in the spirit of creative disbelief.

The field of developmental biology needs to put more emphasis on testing new ideas and new alternatives to the standard model. What if the core events of development are controlled by conserved intercellular signals and regulatory proteins, but the fine tuning that distinguishes phylogenetic groups and individual species uses species-specific genes and chromatin feedback mechanisms from developing tissues? It might mean that one could differentiate generic cell types *in vitro* from a pluripotent stem cell, but they would lack detailed features requiring signals that cannot be recapitulated outside of an organized tissue. What if nuclear RNAs that currently lack known functions—such as lncRNAs, pseudogene transcripts, transposon transcripts, and read-through transcripts—can contribute to chromatin states? A subset of chromosome-bound RNAs might undergo duplication during S phase and associate with both sister chromatids to mediate epigenetic inheritance.

One thing seems certain. How long it takes to decipher the laws of chromatin action and epigenetic inheritance will ultimately determine the rate at which we learn to understand our genomes. We should be embracing the great challenge of solving how metazoans develop and how genomes really work, not celebrating data collection and hype. We should be training individuals who are ready to develop the quantum mechanisms of development. The mechanisms will be simple, beautiful, robust, and, very likely, unexpected. They will be discovered by scientists with the freedom and resources to pursue their own ideas.

Thus, the future of developmental biology is promising but not assured. If conditions for progress remain favorable, within 50 years, and perhaps sooner, we'll know in detail how genomes control development, allowing genotype and phenotype to be connected at the levels of the cell and the organism. However, this will happen only if society continues to support basic developmental biology research and if the molecular genetic analysis of embryos and tissues remains central to these studies. Career opportunities in developmental biology must continue to attract their share of the brightest young scientists. Finally, the right ideas must be found to explain how genomes control development. Consequently, places like the Carnegie Department of Embryology, where new ideas can be proposed, developed, and criticized, remain essential for developmental biology to prosper.

It's a privilege to work in developmental biology—so much fun, so fascinating, so rich, so much to discover. But analyzing development also carries a responsibility to society. Nearly all of us have been aided in our careers by public support through taxes, or from charitable foundations. A scientists' first responsibility to society is to produce research that is true. It's very difficult to know in basic research what is important and what is not. But work that is untrue is not going to repay society for its trust and investment.

Working on development also carries a particular responsibility because society is rightly concerned about understanding the connection between genotype and phenotype. There is widespread fear of genetic manipulation, especially human genetic manipulation. Having engineered thousands of Drosophila genes, I am probably the last person to worry that genetic manipulation will lead to superhumans and superdiseases. This is not going to happen, because so far only the evolutionary process has shown the capacity to improve fitness in wild populations. But gene therapy will eventually work efficiently in humans, just as it does in other diverse organisms. Developmental geneticists are in a unique position to provide insight and guidance to the public on the ethical issues surrounding advances in gene therapy and genome editing. We should be prepared to provide sensible advice and to oppose those who play upon fears of the unknown for their own aggrandizement and to advance their personal agendas.

Having the opportunity to do science for a living must already be considered great good fortune. To search for truth, to immerse oneself in the beauty of biological forms, to look for new knowledge that is potentially of great benefit to mankind, all while working with and mentoring exceptionally smart, creative, and fascinating people is to live without truly working at all. Indeed, it is profoundly satisfying to join the line of humans throughout the ages who searched to understand the universe by applying skepticism and reason rather than remaining satisfied with myth. Working at Carnegie has provided these satisfactions in a highly enriched form. I'd like to close this talk in the same way that I began: by expressing my profoundest gratitude to Carnegie and wishing the Department of Embryology another prosperous century of science.

REFERENCES

Brown, D. D. (1987). The Department of Embryology of the Carnegie Institution of Washington. *Bioessays*, *6*, 92–96.

Maienschein, J., Glitz, M., & Allen, G. E. (Eds.), (2004). *Centennial History of the Carnegie Institution of Washington. V. The Department of Embryology*. Cambridge, UK: Cambridge University Press. 227 pp.

Spradling, A. C. (1997). The Carnegie Institution of Washington, Department of Embryology. *Molecular Medicine*, *3*, 417–419.

CHAPTER TWENTY-FIVE

Organ Function as a Modulator of Organ Formation: Lessons from Zebrafish

Michelle M. Collins, Didier Y.R. Stainier[1]
Department of Developmental Genetics, Max Planck Institute for Heart and Lung Research, Bad Nauheim, Germany
[1]Corresponding author: e-mail address: didier.stainier@mpi-bn.mpg.de

Contents

Abstract

Organogenesis requires an intricate balance between cell differentiation and tissue growth to generate a complex and fully functional organ. However, organogenesis is not solely driven by genetic inputs, as the development of several organ systems requires their own functionality. This theme is particularly evident in the developing heart as progression of cardiac development is accompanied by increased and altered hemodynamic forces. In the absence or disruption of these forces, heart development

Current Topics in Developmental Biology, Volume 117
ISSN 0070-2153
http://dx.doi.org/10.1016/bs.ctdb.2015.10.017

is abnormal, suggesting that the heart must sense these changes and respond appropriately. Here, we discuss concepts of how embryonic heart function contributes to heart development using lessons learned mostly from studies in zebrafish.

1. INTRODUCTION

Several different types of mechanical force are required during embryonic development. Such forces may be extrinsic to a cell, such as the frictional force from fluid flow across the cell surface (termed shear stress), fluid pressure within a closed lumen exerted against the wall (termed transmural pressure), and stretch force resulting from contraction of a tissue. Intrinsic to a cell are forces that function to stabilize cell shape, including tensional forces generated by intracellular microtubules that balance compression forces, or mechanical interactions between the cell and the extracellular matrix (ECM) (Mammoto & Ingber, 2010).

The concept that forces generated from blood flow affect the morphogenesis of the cardiovascular system was first proposed more than a century ago from observations of blood vessel remodeling in response to changes in blood flow in the chick embryo (Chapman, 1918; Thoma, 1893). The large size of the developing chick embryo and accessibility of its vasculature made the system highly attractive to developmental biologists who showed that manipulation of flow led to severe consequences in cardiovascular development, including valvular, septal, and vascular defects (Clark & Rosenquist, 1978; Gessner, 1966; Jaffee, 1965).

These observations have also been reported in other vertebrate models. Large-scale forward genetic screens in zebrafish have led to the identification and subsequent analysis of several heart mutants that revealed that cardiac formation, and especially maturation, was perturbed when cardiac function was affected (Chen et al., 1996; Stainier et al., 1996). Because of its *ex utero* development, optical clarity, and ability to survive much longer than other developmental models in the absence of a functional heart and circulatory system (Sehnert et al., 2002; Stainier, 2001), the zebrafish embryo has been instrumental in unraveling how cardiac function impacts its formation. Exploiting these advantages, Hove et al. (2003) showed that physically blocking blood influx or efflux resulted in severe cardiac phenotypes, including heart looping defects and a failure to form the bulbus arteriosus, and at later stages, defective cardiac valve formation and chamber morphogenesis, further indicating that hemodynamic forces are critical for many aspects of cardiac development. In this review, we summarize the major findings in zebrafish that have led to the current understanding of how hemodynamic forces affect heart development.

2. OVERVIEW OF HEART FORMATION IN ZEBRAFISH

2.1 Early Specification of Cardiac Progenitors and Formation of the Linear Heart Tube

Development of the heart begins with the specification of the myocardial and endocardial lineages during early embryonic development. In the late blastula (40% epiboly stage), myocardial progenitors are located in the lateral marginal zone (Stainier, Lee, & Fishman, 1993), and chamber progenitors are organized such that ventricular myocardial progenitors are located more marginally and dorsally compared to atrial myocardial progenitors (Keegan, Meyer, & Yelon, 2004). Endocardial progenitors are also distributed within the marginal zone (Lee, Stainier, Weinstein, & Fishman, 1994), although at this stage, unlike myocardial progenitors, they do not exhibit chamber-specific spatial organization (Keegan et al., 2004).

The bilateral pools of cardiac progenitors move dorsally and anteriorly during gastrulation and early somitogenesis stages, eventually becoming positioned within the anterior lateral plate mesoderm, while maintaining regionalized populations of atrial and ventricular progenitors (Berdougo, Coleman, Lee, Stainier, & Yelon, 2003). These bilateral populations move toward the midline, where endocardial precursors migrate slightly in advance of myocardial precursors, and subsequently fuse at the midline to form the cardiac cone whereby the atrial precursors are located at the outer circumference, adjacent to a ring of ventricular precursors, and containing endocardial precursors at the center (Stainier et al., 1993). Following formation of the cardiac cone, myocardial cells involute asymmetrically and extend, causing the linear heart tube to telescope out in an anterior and leftward direction (Chen et al., 1997). Consequently, endocardial cells become positioned in the interior of the linear heart tube, the venous pole becomes located on the left side of the embryo, and the arterial pole remains at the midline.

At the onset of contractions, around 22 hours postfertilization (hpf), the linear heart tube in zebrafish has been reported to function as a dynamic suction pump (Forouhar et al., 2006). Pumping appears to result from propagation of an elastic wave, resulting in nonlinear flow and mismatched impedance that allow for suction via the creation of high- and low-pressure zones. These properties are hypothesized to be necessary for the subsequent morphological changes that occur during cardiac development.

2.2 Cardiac Looping and Chamber Emergence

The linear heart tube undergoes a rightward looping event to position the ventricle to the right of the midline, while the atrium remains positioned on the left. At this point, bidirectional blood flow concentrated at the atrioventricular canal (AVC) becomes stronger. Coincident with the onset of vigorous blood flow, the atrium and ventricle become morphologically distinct through the process of chamber ballooning. This process is driven by regional changes in cardiomyocyte cell shape in the ventricle that cause the expansion of the outer curvature and compaction of the inner curvature (Auman et al., 2007). Endocardial chamber ballooning is driven by similar changes in cell shape and surface area (Dietrich, Lombardo, Veerkamp, Priller, & Abdelilah-Seyfried, 2014), and both myocardial and endocardial ballooning are affected by altered blood flow (Auman et al., 2007; Dietrich et al., 2014).

2.3 Atrioventricular Valve Formation

Cardiac valves have a crucial role in maintaining unidirectional blood flow through the sequentially contracting cardiac chambers. At 48 hpf, a subset of endocardial cells become cuboidal and coalesce to form a ring at the AVC. By 60 hpf, neighboring endocardial cells invaginate to help form the valve leaflets, and by 72 hpf, the leaflets are more evident and completely block retrograde blood flow (Beis et al., 2005; Scherz, Huisken, Sahai-Hernandez, & Stainier, 2008). Maturation into more developed leaflets is apparent already by 120 hpf.

Imaging of developing hearts has shown the presence of bidirectional blood flow, and quantification of the fraction of the cardiac cycle when there is retrograde flow (termed the retrograde flow fraction (RFF)) revealed that the highest RFF at the AVC is observed during stages just prior to valve formation. Further, these oscillatory flows become concentrated at the AVC during cardiac looping, as the AVC diameter becomes smaller and surrounded by the endothelial ring (Vermot et al., 2009). Importantly, early steps of endothelial ring formation prior to 48 hpf do not seem to be dependent on flow (Bartman et al., 2004; Vermot et al., 2009), but the later steps of valvulogenesis appear to require oscillatory flows at the AVC.

2.4 Trabeculation

The process of trabeculation begins at the end of the cardiac looping stage, as some cardiomyocytes in the outer curvature of the ventricle delaminate from the basal side of the compact myocardial layer, eventually forming transverse muscular ridges (Liu et al., 2010). By 96 hpf, these ridges have thickened and by 5 days postfertilization (dpf), the trabecular network has become more complex and extensive, spanning the outer curvature of the ventricle. Trabeculae fail to form in the absence of blood flow (Chi et al., 2008; Staudt et al., 2014), and the process of trabeculation likely requires the endocardium, as *cloche* mutant hearts, which lack an endocardial layer (Stainier, Weinstein, Detrich, Zon, & Fishman, 1995), fail to trabeculate (Peshkovsky, Totong, & Yelon, 2011).

2.5 Emergence of the Cardiac Conduction System

The cardiac conduction system (CCS) cooperates with the valves to ensure that blood flow is unidirectional by controlling electrical impulses from the apex of the ventricle to its base, allowing efficient blood ejection into the outflow tract. Development of the CCS has been described in four phases that correlate to morphological changes occurring throughout early cardiac development (Chi et al., 2008). The first phase corresponds to the linear conduction across the linear heart tube. During looping phases, the second phase leads to a significant delay across the forming AVC. The third phase occurs during the formation of trabeculation whereby an immature fast conduction network emerges in the ventricle. The final phase is completed by 2–3 weeks postfertilization, when the fully mature fast conduction network shows an apex-to-base activation pattern (Chi et al., 2008).

3. INTRACARDIAC FORCES THAT ARE REQUIRED TO SHAPE THE HEART

3.1 Chamber Emergence and Ballooning

Cardiac chamber emergence is driven by a process called ballooning, in which local bulges create a bean-shaped morphology characterized by an outer convex and inner concave curvature. In addition to morphological differences, the inner and outer curvatures may also be distinguished by different gene and protein expression profiles (Stainier & Fishman, 1992; Yelon, Horne, & Stainier, 1999). Both the endocardium and myocardium require vigorous blood flow for correct emergence and ballooning of the atrial and ventricular chambers, and crosstalk between these two cell layers is critical for these phases of cardiac development (Auman et al., 2007; Dietrich et al., 2014; Yang, Hartjes, Nelson, & Xu, 2014).

Zebrafish mutants identified in forward genetics screens were instrumental to show that heart function is required for correct chamber emergence and ballooning. *Weak atrium* (*wea*) mutants, which carry a loss-of-function mutation in the atrial-specific myosin gene *myh6*, lack atrial sarcomeres and thus, contractility is impaired only within the atrium (Berdougo et al., 2003). However, *myh6/wea* mutant hearts display dysmorphic ventricles and less pronounced inner and outer curvatures. During chamber emergence, cardiomyocytes in the outer curvature of *myh6/wea* mutant ventricles are smaller and less elongated than in wild type, and an excess of cuboidal cells is observed and not restricted to the inner curvature (Auman et al., 2007). Similarly, modulating heart contractility by treatment with 2,3-butanedione 2-monoxime (BDM), a myosin ATPase inhibitor, resulted in similar cell shape defects (Auman et al., 2007). Furthermore, a complete absence of cardiac contractility, such as that observed in *tnnt2a/silent heart* (*sih*) mutants, leads to defects in cardiomyocyte cell shape changes, including the failure of the inner curvature atrial cardiomyocytes to reach, or maintain, their characteristic size, as well as the failure of AVC cardiomyocytes to undergo apical constriction (B. Jungblut, J. Huisken, and D.Y.R. Stainier, unpublished observations).

The functional complement to the *myh6/wea* mutant is *half-hearted* (*haf*), which carries a mutation in the ventricular myosin gene *vmhc* and lacks ventricular contractility (Auman et al., 2007). *vmhc/haf* mutant ventricular cardiomyocytes exhibit an increased surface area, and elongated cells are found throughout the chamber rather than restricted to the outer curvature, resulting in a large and distended ventricle. Data from cell transplantation experiments revealed a cell autonomous role for *vmhc/haf* in maintaining cardiomyocyte morphology, suggesting that *vmhc/haf* functions intrinsically within the cardiomyocytes. Altogether, these data indicate that contractility is important for regionalized cell shape changes, including cardiomyocyte enlargement and elongation. Further dissection of the role of blood flow and contractility in cardiomyocyte cell shape changes will benefit from the use of experimental conditions where contractility is unaffected but blood flow is manipulated in controlled and reproducible ways.

Shear stress from blood flow is required for regionalized cell shape changes within the endocardium of the ventricle during cardiac ballooning (Dietrich et al., 2014). Additionally, hemodynamic forces also appear to be required for endocardial cell proliferation, as the ventricle in *tnnt2a/sih* mutants and *gata2* morpholino-injected (morphant) embryos shows a significant decrease in endocardial cell numbers. Data from morpholino-based approaches to modulate the hematocrit (the proportion of blood volume composed of red blood cells) or reduce retrograde flow at the AVC further suggest that retrograde flow is more important for ventricular endocardial cell proliferation than blood viscosity or shear stress.

Taken together, these data suggest that there is extensive crosstalk between the endocardium and myocardium. Correct establishment of the actomyosin network within the myocardium allows contractility, and hence, hemodynamic forces, which are consequently sensed by the endocardium. Hemodynamic forces promote ventricular chamber emergence in both the endocardium and myocardium, leading to regionalized changes in cell shape that contribute to the morphology of the inner and outer curvatures. Intrinsic contractile forces within cardiomyocytes constrain cell elongation to balance extrinsic mechanical forces. Myocardial chamber defects observed in embryos that have absent or aberrant endocardial formation, such as *cloche* (Stainier et al., 1995) or *scl/tal1* mutants (Bussmann, Bakkers, & Schulte-Merker, 2007; Schumacher, Bloomekatz, Garavito-Aguilar, & Yelon, 2013), suggest that endocardial cells likely participate in sensing flow and have critical roles in transmitting mechanical information to the myocardium. Communication between endocardial and myocardial cells could occur not only via the secretion of signaling molecules but also via the modulation of the cardiac jelly, the specialized ECM separating the endocardium and myocardium (Mellman, Huisken, Dinsmore, Hoppe, & Stainier, 2012).

3.2 The AVC and Valve Formation

Hemodynamic forces play a key role in the formation of the atrioventricular valves throughout their development. Surgical implantation of beads in the early heart suggested that blood flow is required for endocardial cushion formation and valve remodeling (Hove et al., 2003). Observations from both genetic and pharmacologic approaches indicate that myocardial function, and not shear stress, is required for endocardial ring formation at the AVC. The *acta1b/cardiofunk (cfk)* and *tnnt2a/sih* mutants fail to drive blood out of the heart and subsequently fail to accumulate endocardial cells at the AV boundary by 48 hpf (Bartman et al., 2004). However, BDM treatment also abolishes blood flow, but endocardial ring formation appears to occur in more than half of the treated embryos, suggesting that it is reduced myocardial function and not the lack of blood flow that is responsible for endocardial ring formation defects (Bartman et al., 2004). These data are supported by observations from Vermot et al. (2009) showing that blood flow is not necessary for endocardial ring formation, but rather for the cell behaviors driving leaflet formation.

The importance of reversing flows at the AVC for the subsequent maturation of the valves was investigated using a combination of approaches. Reducing blood viscosity (and thus shear stress) by injecting the *gata1* morpholino did not affect valve formation, despite an increased RFF at the AVC. In contrast, injection of the *gata2* morpholino strongly reduced the RFF and resulted in abnormal valve development, suggesting that the reversing flow is more important than shear stress for valve maturation (Vermot et al., 2009). Further modulation of the RFF through the use of lidocaine or temperature led to the conclusions that the valves are most sensitive to decreased RFF between 36 and 48 hpf (Vermot et al., 2009). More recently, Kalogirou et al. (2014) analyzed two zebrafish mutants with altered intracardiac flow patterns and concluded that intracardiac flow dynamics regulate valve morphogenesis independently of myocardial contractility. For example, in *myh6/wea* mutant hearts, where flow patterns are disrupted due to the inactivity of the atrium, the AV valves fail to form four cuspids. Thus, while the tools used to study the role of myocardial contractility and shear stress in cardiac valve formation has become somewhat more sophisticated, it is clear that much work remains to be done to achieve a detailed understanding.

3.3 Trabeculation

Observations of *tnnt2a/sih* mutants showed that trabeculation does not occur in the absence of heartbeat (Chi et al., 2008; Samsa et al., 2015; Staudt et al., 2014). This observation was confirmed using the *myh6/wea* atrial-contractility mutant (Peshkovsky et al., 2011). Although luminal protrusions could be observed in *myh6/wea* mutant ventricles, they formed aberrantly and failed to progress to form ridges by 4 dpf. These data suggest that blood flow is important for trabeculation, although initiating steps may still occur in the presence of aberrant flow patterns. Given that the endocardium is required for trabeculae to form (Peshkovsky et al., 2011; Staudt & Stainier, 2012), it may potentially be involved with sensing flow and participate in signaling to the myocardium. One consideration, however, is that the ventricular wall is already dysmorphic as a result of aberrant atrial function in *myh6/wea* mutants (Auman et al., 2007), thereby potentially disrupting trabeculation indirectly. It remains to be determined exactly how altering blood flow impacts this process; future experiments may be able to tease apart the specific role of flow for initiation versus maintenance and maturation of trabeculae (Staudt et al., 2014).

3.4 The Cardiac Conduction System

The role for hemodynamic forces during the development of the CCS has nicely been demonstrated by experiments using the chick embryo; the resulting data suggest that a stretch and pressure-sensitive mechanism regulate the early events of Purkinje fiber differentiation (Bressan, Louie, & Mikawa, 2014; Hall et al., 2004). Furthermore, modulating mechanical loading by left atrial ligation delayed maturation of the CCS, whereas increased mechanical loading accelerated maturation of the Purkinje system (Reckova et al., 2003). These data indicate that mechanical stretching of cardiomyocytes is required during early stages of CCS differentiation in the chick.

In zebrafish, optical mapping of calcium activation in 48 hpf *tnnt2a/sih* mutants showed an AV conduction delay similar to that observed in wild-type embryos, suggesting that the early stages of CCS differentiation are not affected by the absence of heartbeat (Chi et al., 2008). However, at 100 hpf, the ventricular outer curvature to base calcium activation pattern was disrupted in *tnnt2a/sih* mutants; rather than an apex-to-base activation, lateral calcium activation across the myocardium was observed. These larvae also lack trabeculae, which could potentially influence CCS development. Therefore, in zebrafish as in chick, hemodynamic forces modulate the development of the CCS, although possibly at different steps of the process. In addition to hemodynamic forces, electrical forces within the heart also appear to be able to modulate cardiac chamber morphology (Chi et al., 2010), although we know little about the underlying mechanisms.

The generation and use of an optogenetic zebrafish line in combination with light-sheet microscopy allowed the identification of cardiac pacemaker cells during development (Arrenberg, Stainier, Baier, & Huisken, 2010). This system was also exploited to precisely and reversibly control heart rate and simulate pathological states such as bradycardia, tachycardia, and AV blocks (Arrenberg et al., 2010). As the majority of the methods used to modulate hemodynamic forces are generally in action from the onset of cardiac development, this method constitutes a powerful tool whereby hemodynamic forces may be controlled more precisely during developmental time. Using such an approach should allow for finer resolution and limit potential confounding early developmental defects in order to better understand later cardiac maturation processes.

4. EXTRACARDIAC FORCES AND THEIR FUNCTION DURING HEART DEVELOPMENT

4.1 Epicardium Formation

In addition to intracardiac forces, forces generated extracardiacally also play important roles during heart development. The epicardium is derived from an extracardiac cluster of mesothelial cells. Following formation of the proepicardial cluster, these cells migrate within the pericardial cavity toward the surface of the myocardium where they adhere and progressively form the mature epicardium. Observations of different animal models have proposed two mechanisms for how proepicardial cells reach the myocardium: (1) via transient ECM bridge structures and (2) via flow-directed release and progressive attachment to the myocardial surface. Whether both of these processes occur in zebrafish remains a topic of discussion (Peralta et al., 2013; Plavicki et al., 2014), as does the specific role of hemodynamic forces in epicardial development.

Flow generated by heartbeat is critical for epicardium formation (Peralta et al., 2013; Plavicki et al., 2014), particularly for migration toward and adherence to the myocardium, as pharmacological manipulations of the heartbeat at various points of epicardial development abrogated this process (Peralta et al., 2013; Plavicki et al., 2014). Additionally, not only is the presence of flow important, but also differential flow appears to facilitate both the release and adhesion of proepicardial cells. High flow forces are observed near the AVC and venous pole, areas from which proepicardial cells are released, whereas the distal ventricle is in an area of weak flow forces, allowing for adhesion to the myocardium to occur (Peralta et al., 2013).

However, Peralta et al. (2013) reported that proepicardial clustering did not occur in the absence of heartbeat, whereas Plavicki et al. (2014) reported that it did. Further, there is no consensus whether the proepicardial cells reach the myocardium via ECM bridges. With rather limited data investigating the role of hemodynamic forces during epicardial development in zebrafish, it is difficult to understand this process. Further experiments are thus required to clarify the presence and/or function of ECM bridges in zebrafish, and whether fluid flow or simply mechanical tension during contraction is required within the pericardial cavity.

5. HOW DOES FORCE AFFECT FORMATION?

Several hypotheses have been put forth to describe how hemodynamic forces are sensed by cells, and how these forces are subsequently able to elicit biochemical and morphological effects.

In vitro data have clearly shown that cultured endothelial cells are able to respond to artificial flow, thereby changing gene expression profiles and rearranging their cytoskeleton (Nagel, Resnick, Dewey, & Gimbrone, 1999; Ostrowski et al., 2014; Wang, Baker, Chen, & Schwartz, 2013). Shear stress forces less than 1 dyn cm^{-2} are detectable by endothelial cells in culture, and shear stress forces in the range of 8–15 dyn cm^{-2} cause cytoskeletal rearrangements (Malek & Izumo, 1996; Olesen, Claphamt, & Davies, 1988). These values are well below the predicted wall shear stress observed in the zebrafish heart at 4.5 dpf (76 dyn cm^{-2}) and within the range of detection at 37 hpf (2.5 dyn cm^{-2}) (Hove et al., 2003). Endothelial cells in culture are also sensitive to changes in flow direction, to gradients within flow, and turbulent or oscillatory flows (Ostrowski et al., 2014; Sorescu et al., 2003), suggesting that mechanisms to detect very subtle flow alterations must be at play as organ function impacts its formation.

Mechanotransducers convert physical stresses into biochemical responses at the cellular level (Hoffman, Grashoff, & Schwartz, 2011). Several putative mechanotransducers have been identified in endothelial cells, including ion channels (Folgering et al., 2008), the apical glycocalyx (Tarbell, Simon, & Curry, 2014), heterotrimeric G proteins (Chachisvilis, Zhang, & Frangos, 2006), and primary cilia (Espinha, Hoey, Fernandes, Rodrigues, & Jacobs, 2014; Goetz et al., 2014). Many of these components can also discriminate between variations in flow regimes *in vivo* or *in vitro*; therefore, it is likely that these mechanotransduction pathways function similarly within the endocardium to transduce subtle changes in hemodynamic forces during development. However, it remains to be determined how these and other components coordinate responses to shear stress, contractility, and stretching.

Biochemical outputs downstream of hemodynamic forces include changes in gene expression, induction of paracrine signaling pathways, and cytoskeletal rearrangements. Several flow-responsive genes have been identified and play critical roles during heart development, including *klf2a*

(Vermot et al., 2009) and *miR-21* (Banjo et al., 2013). Importantly, flow-responsive genes do not behave equally; unique temporal profiles and different responses to various types of flow have been reported (Banjo et al., 2013; Fisher, Chien, Barakat, & Nerem, 2001). But ultimately, the question remains regarding how flow is linked to intracellular changes, including transcriptional regulation. It was recently reported that the membrane-bound mechanosensitive TRP calcium channels activate an intracellular signaling cascade upstream of *klf2a* expression in response to oscillatory flows at the AVC (Heckel et al., 2015), providing a framework for understanding the downstream effects of flow.

It is also likely that fluid forces within the developing heart elicit morphological changes through the stretching of cardiomyocytes. As described, changes in ventricular cardiomyocyte shape during chamber emergence are thought to be balanced by contractile forces that provide resistance to such changes (Auman et al., 2007). As cellular tension is due to balanced forces across the cytoskeleton, stretching may directly influence cell morphology by reorganization of actomyosin cytoskeletal structures to transmit forces across a single cell or even between neighboring cells via contact with the ECM (Mammoto & Ingber, 2010; Wang, Tytell, & Ingber, 2009).

Changes in flow also cause ECM remodeling and may therefore regulate cell or tissue behavior. To test the role of fluid flow on the developing AV valve ECM, Tan et al. (2013) generated a fluid flow bioreactor culture system. To mimic the increased hemodynamic forces during development, the flow rate was artificially ramped up during the culture of AV cushions from HH stage 25 chick embryos and AV explants responded by increased deposition of fibrous ECM components and altered morphological development of the AV cushions. As oscillatory flow at the AVC seems to be required for endothelial cells to invaginate to form the valve cushions (Vermot et al., 2009), these data suggest that flow may also direct ECM remodeling during valve formation.

6. IMPLICATIONS FOR HUMAN HEART DEVELOPMENT AND DISEASE

Congenital heart disease is one of the most common types of birth defects, occurring in roughly 8 per 1000 live births (Go et al., 2013). Given the major heart defects observed in animal models in which hemodynamic forces are disrupted, it is clear that the role of biomechanical forces, in addition to genetic and environmental exposures, should be considered when evaluating causes and treatments of congenital and adult heart disease. As described here, it is clear that even subtle changes in blood flow can impact cardiac development in multiple ways.

Both cell autonomous and cell nonautonomous mechanisms should be considered as potential etiologies of cardiac disease. This hypothesis may be especially pertinent in cases of human cardiomyopathies in which mutations in cardiac contractile genes are significant drivers of heart failure and death (Aksel, Choe Yu, Sutton, Ruppel, & Spudich, 2015; Schmitt et al., 2006), and data from zebrafish demonstrate both cell autonomous and cell non-autonomous roles for myosin genes during chamber ballooning (Auman et al., 2007). Recently, investigation of a mouse model of hypertrophic cardiomyopathy in which animals carried an Arg403Gln α-MHC transgene that could be temporally inhibited revealed that the time window in which mutation-silencing therapies could be effective is limited to embryonic life (Cannon et al., 2015). These data indicate that cardiomyocyte defects during embryonic development act as a trigger for disease manifestation in postnatal life.

Understanding how hemodynamic forces affect development is also important for therapeutic purposes. For example, congenital valvular disease detected during the embryonic period is often treated by surgical replacement of diseased valve tissue. These treatments typically use mechanical prostheses that fail to grow and remodel during postnatal development, thereby requiring multiple replacements during childhood and adolescent growth. Tissue-engineered heart valves are thought to circumvent many of these issues (Sacks, Schoen, & Mayer, 2009). Thus, many of the lessons learned about the role of forces during formation of the valves will be critical for the successful engineering of heart valve tissue. Furthermore, it is likely that hemodynamic as well as electrical forces are key epigenetic factors during cardiac remodeling following injury or disease, and better models to study these events at the cellular and biochemical levels would be most rewarding.

7. SUMMARY

Formation of the heart is a complex process, requiring precise integration of genetic and mechanical inputs to shape a functional organ. The zebrafish model has provided ample mechanistic insights into how cardiac developmental processes are influenced by organ function. Many of the lessons learned from these studies will influence the way we think about cardiac disease in humans, as it is quite clear that subtle alterations in hemodynamic forces during embryonic development can lead to drastic morphogenetic changes. With the recent advent of efficient genome engineering techniques, studies in zebrafish will continue to uncover mechanisms of how the function of the heart impacts its development.

ACKNOWLEDGMENTS

We apologize to those authors whose work we were unable to include due to length constraints. We thank Hyouk-Bum Kwon for critical reading of the manuscript and thoughtful discussions.

REFERENCES

Aksel, T., Choe Yu, E., Sutton, S., Ruppel, K. M., & Spudich, J. A. (2015). Ensemble force changes that result from human cardiac myosin mutations and a small-molecule effector. *Cell Reports*, *11*(6), 910–920.

Arrenberg, A. B., Stainier, D. Y., Baier, H., & Huisken, J. (2010). Optogenetic control of cardiac function. *Science*, *330*(6006), 971–974.

Auman, H. J., Coleman, H., Riley, H. E., Olale, F., Tsai, H. J., & Yelon, D. (2007). Functional modulation of cardiac form through regionally confined cell shape changes. *PLoS Biology*, *5*(3), e53.

Banjo, T., Grajcarek, J., Yoshino, D., Osada, H., Miyasaka, K. Y., Kida, Y. S., et al. (2013). Haemodynamically dependent valvulogenesis of zebrafish heart is mediated by flow-dependent expression of miR-21. *Nature Communications*, *4*, 1978.

Bartman, T., Walsh, E. C., Wen, K. K., McKane, M., Ren, J., Alexander, J., et al. (2004). Early myocardial function affects endocardial cushion development in zebrafish. *PLoS Biology*, *2*(5), E129.

Beis, D., Bartman, T., Jin, S. W., Scott, I. C., D'Amico, L. A., Ober, E. A., et al. (2005). Genetic and cellular analyses of zebrafish atrioventricular cushion and valve development. *Development*, *132*(18), 4193–4204.

Berdougo, E., Coleman, H., Lee, D. H., Stainier, D. Y., & Yelon, D. (2003). Mutation of weak atrium/atrial myosin heavy chain disrupts atrial function and influences ventricular morphogenesis in zebrafish. *Development*, *130*(24), 6121–6129.

Bressan, M. C., Louie, J. D., & Mikawa, T. (2014). Hemodynamic forces regulate developmental patterning of atrial conduction. *PLoS One*, *9*(12), e115207.

Bussmann, J., Bakkers, J., & Schulte-Merker, S. (2007). Early endocardial morphogenesis requires Scl/Tal1. *PLoS Genetics*, *3*(8), e140.

Cannon, L., Yu, Z. Y., Marciniec, T., Waardenberg, A. J., Iismaa, S. E., Nikolova-Krstevski, V., et al. (2015). Irreversible triggers for hypertrophic cardiomyopathy are established in the early postnatal period. *Journal of the American College of Cardiology*, *65*(6), 560–569.

Chachisvilis, M., Zhang, Y. L., & Frangos, J. A. (2006). G protein-coupled receptors sense fluid shear stress in endothelial cells. *Proceedings of the National Academy of Sciences of the United States of America*, *103*(42), 15463–15468.

Chapman, W. B. (1918). The effect of the heart-beat upon the development of the vascular system in the chick. *American Journal of Anatomy*, *23*(1), 175–203.

Chen, J. N., Haffter, P., Odenthal, J., Vogelsang, E., Brand, M., van Eeden, F. J., et al. (1996). Mutations affecting the cardiovascular system and other internal organs in zebrafish. *Development*, *123*, 293–302.

Chen, J. N., van Eeden, F. J., Warren, K. S., Chin, A., Nusslein-Volhard, C., Haffter, P., et al. (1997). Left-right pattern of cardiac BMP4 may drive asymmetry of the heart in zebrafish. *Development, 124*(21), 4373–4382.

Chi, N. C., Bussen, M., Brand-Arzamendi, K., Ding, C., Olgin, J. E., Shaw, R. M., et al. (2010). Cardiac conduction is required to preserve cardiac chamber morphology. *Proceedings of the National Academy of Sciences of the United States of America, 107*(33), 14662–14667.

Chi, N. C., Shaw, R. M., Jungblut, B., Huisken, J., Ferrer, T., Arnaout, R., et al. (2008). Genetic and physiologic dissection of the vertebrate cardiac conduction system. *PLoS Biology, 6*(5), e109.

Clark, E. B., & Rosenquist, G. C. (1978). Spectrum of cardiovascular anomalies following cardiac loop constriction in the chick embryo. *Birth Defects Original Article Series, 14*(7), 431–442.

Dietrich, A. C., Lombardo, V. A., Veerkamp, J., Priller, F., & Abdelilah-Seyfried, S. (2014). Blood flow and Bmp signaling control endocardial chamber morphogenesis. *Developmental Cell, 30*(4), 367–377.

Espinha, L. C., Hoey, D. A., Fernandes, P. R., Rodrigues, H. C., & Jacobs, C. R. (2014). Oscillatory fluid flow influences primary cilia and microtubule mechanics. *Cytoskeleton (Hoboken), 71*(7), 435–445.

Fisher, A. B., Chien, S., Barakat, A. I., & Nerem, R. M. (2001). Endothelial cellular response to altered shear stress. *American Journal of Physiology. Lung Cellular and Molecular Physiology, 281*(3), L529–L533.

Folgering, J. H., Sharif-Naeini, R., Dedman, A., Patel, A., Delmas, P., & Honore, E. (2008). Molecular basis of the mammalian pressure-sensitive ion channels: Focus on vascular mechanotransduction. *Progress in Biophysics and Molecular Biology, 97*(2–3), 180–195.

Forouhar, A. S., Liebling, M., Hickerson, A., Nasiraei-Moghaddam, A., Tsai, H. J., Hove, J. R., et al. (2006). The embryonic vertebrate heart tube is a dynamic suction pump. *Science, 312*(5774), 751–753.

Gessner, I. H. (1966). Spectrum of congenital cardiac anomalies produced in chick embryos by mechanical interference with cardiogenesis. *Circulation Research, 18*(6), 625–633.

Go, A. S., Mozaffarian, D., Roger, V. L., Benjamin, E. J., Berry, J. D., Borden, W. B., et al. (2013). Heart disease and stroke statistics—2013 update: A report from the American Heart Association. *Circulation, 127*(1), e6–e245.

Goetz, J. G., Steed, E., Ferreira, R. R., Roth, S., Ramspacher, C., Boselli, F., et al. (2014). Endothelial cilia mediate low flow sensing during zebrafish vascular development. *Cell Reports, 6*(5), 799–808.

Hall, C. E., Hurtado, R., Hewett, K. W., Shulimovich, M., Poma, C. P., Reckova, M., et al. (2004). Hemodynamic-dependent patterning of endothelin converting enzyme 1 expression and differentiation of impulse-conducting Purkinje fibers in the embryonic heart. *Development, 131*(3), 581–592.

Heckel, E., Boselli, F., Roth, S., Krudewig, A., Belting, H. G., Charvin, G., et al. (2015). Oscillatory Flow Modulates Mechanosensitive klf2a Expression through trpv4 and trpp2 during Heart Valve Development. *Current Biology, 25*(10), 1354–1361.

Hoffman, B. D., Grashoff, C., & Schwartz, M. A. (2011). Dynamic molecular processes mediate cellular mechanotransduction. *Nature, 475*(7356), 316–323. http://dx.doi.org/10.1038/nature10316.

Hove, J. R., Koster, R. W., Forouhar, A. S., Acevedo-Bolton, G., Fraser, S. E., & Gharib, M. (2003). Intracardiac fluid forces are an essential epigenetic factor for embryonic cardiogenesis. *Nature, 421*(6919), 172–177.

Jaffee, O. C. (1965). Hemodynamic factors in the development of the chick embryo heart. *The Anatomical Record, 151*, 69–75.

Kalogirou, S., Malissovas, N., Moro, E., Argenton, F., Stainier, D. Y., & Beis, D. (2014). Intracardiac flow dynamics regulate atrioventricular valve morphogenesis. *Cardiovascular Research*, *104*(1), 49–60. http://dx.doi.org/10.1093/cvr/cvu186.

Keegan, B. R., Meyer, D., & Yelon, D. (2004). Organization of cardiac chamber progenitors in the zebrafish blastula. *Development*, *131*(13), 3081–3091.

Lee, R. K., Stainier, D. Y., Weinstein, B. M., & Fishman, M. C. (1994). Cardiovascular development in the zebrafish. II. Endocardial progenitors are sequestered within the heart field. *Development*, *120*(12), 3361–3366.

Liu, J., Bressan, M., Hassel, D., Huisken, J., Staudt, D., Kikuchi, K., et al. (2010). A dual role for ErbB2 signaling in cardiac trabeculation. *Development*, *137*(22), 3867–3875.

Malek, A. M., & Izumo, S. (1996). Mechanism of endothelial cell shape change and cytoskeletal remodeling in response to fluid shear stress. *Journal of Cell Science*, *109*(Pt 4), 713–726.

Mammoto, T., & Ingber, D. E. (2010). Mechanical control of tissue and organ development. *Development*, *137*(9), 1407–1420.

Mellman, K., Huisken, J., Dinsmore, C., Hoppe, C., & Stainier, D. Y. (2012). Fibrillin-2b regulates endocardial morphogenesis in zebrafish. *Developmental Biology*, *372*(1), 111–119.

Nagel, T., Resnick, N., Dewey, C. F., Jr., & Gimbrone, M. A., Jr. (1999). Vascular endothelial cells respond to spatial gradients in fluid shear stress by enhanced activation of transcription factors. *Arteriosclerosis, Thrombosis, and Vascular Biology*, *19*(8), 1825–1834.

Olesen, S.-P., Claphamt, D., & Davies, P. (1988). Haemodynamic shear stress activates a K+ current in vascular endothelial cells. *Nature*, *331*(6152), 168–170.

Ostrowski, M. A., Huang, N. F., Walker, T. W., Verwijlen, T., Poplawski, C., Khoo, A. S., et al. (2014). Microvascular endothelial cells migrate upstream and align against the shear stress field created by impinging flow. *Biophysical Journal*, *106*(2), 366–374.

Peralta, M., Steed, E., Harlepp, S., Gonzalez-Rosa, J. M., Monduc, F., Ariza-Cosano, A., et al. (2013). Heartbeat-driven pericardiac fluid forces contribute to epicardium morphogenesis. *Current Biology*, *23*(18), 1726–1735.

Peshkovsky, C., Totong, R., & Yelon, D. (2011). Dependence of cardiac trabeculation on neuregulin signaling and blood flow in zebrafish. *Developmental Dynamics*, *240*(2), 446–456.

Plavicki, J. S., Hofsteen, P., Yue, M. S., Lanham, K. A., Peterson, R. E., & Heideman, W. (2014). Multiple modes of proepicardial cell migration require heartbeat. *BMC Developmental Biology*, *14*, 18.

Reckova, M., Rosengarten, C., deAlmeida, A., Stanley, C. P., Wessels, A., Gourdie, R. G., et al. (2003). Hemodynamics is a key epigenetic factor in development of the cardiac conduction system. *Circulation Research*, *93*(1), 77–85.

Sacks, M. S., Schoen, F. J., & Mayer, J. E. (2009). Bioengineering challenges for heart valve tissue engineering. *Annual Review of Biomedical Engineering*, *11*, 289–313.

Samsa, L. A., Givens, C., Tzima, E., Stainier, D. Y., Qian, L., & Liu, J. (2015). Cardiac contraction activates endocardial Notch signaling to modulate chamber maturation in zebrafish. *Development*, *142*(23), 4080–4091. http://dx.doi.org/10.1242/dev.125724.

Scherz, P. J., Huisken, J., Sahai-Hernandez, P., & Stainier, D. Y. (2008). High-speed imaging of developing heart valves reveals interplay of morphogenesis and function. *Development*, *135*(6), 1179–1187.

Schmitt, J. P., Debold, E. P., Ahmad, F., Armstrong, A., Frederico, A., Conner, D. A., et al. (2006). Cardiac myosin missense mutations cause dilated cardiomyopathy in mouse models and depress molecular motor function. *Proceedings of the National Academy of Sciences of the United States of America*, *103*(39), 14525–14530.

Schumacher, J. A., Bloomekatz, J., Garavito-Aguilar, Z. V., & Yelon, D. (2013). tal1 Regulates the formation of intercellular junctions and the maintenance of identity in the endocardium. *Developmental Biology*, *383*(2), 214–226.

Sehnert, A. J., Huq, A., Weinstein, B. M., Walker, C., Fishman, M., & Stainier, D. Y. (2002). Cardiac troponin T is essential in sarcomere assembly and cardiac contractility. *Nature Genetics*, *31*(1), 106–110.

Sorescu, G. P., Sykes, M., Weiss, D., Platt, M. O., Saha, A., Hwang, J., et al. (2003). Bone morphogenic protein 4 produced in endothelial cells by oscillatory shear stress stimulates an inflammatory response. *The Journal of Biological Chemistry*, *278*(33), 31128–31135.

Stainier, D. Y. (2001). Zebrafish genetics and vertebrate heart formation. *Nature Reviews. Genetics*, *2*(1), 39–48.

Stainier, D. Y., & Fishman, M. C. (1992). Patterning the zebrafish heart tube: Acquisition of anteroposterior polarity. *Developmental Biology*, *153*(1), 91–101.

Stainier, D. Y., Fouquet, B., Chen, J. N., Warren, K. S., Weinstein, B. M., Meiler, S. E., et al. (1996). Mutations affecting the formation and function of the cardiovascular system in the zebrafish embryo. *Development*, *123*, 285–292.

Stainier, D. Y., Lee, R. K., & Fishman, M. C. (1993). Cardiovascular development in the zebrafish. I. Myocardial fate map and heart tube formation. *Development*, *119*(1), 31–40.

Stainier, D. Y., Weinstein, B. M., Detrich, H. W., 3rd., Zon, L. I., & Fishman, M. C. (1995). Cloche, an early acting zebrafish gene, is required by both the endothelial and hematopoietic lineages. *Development*, *121*(10), 3141–3150.

Staudt, D. W., Liu, J., Thorn, K. S., Stuurman, N., Liebling, M., & Stainier, D. Y. (2014). High-resolution imaging of cardiomyocyte behavior reveals two distinct steps in ventricular trabeculation. *Development*, *141*(3), 585–593.

Staudt, D., & Stainier, D. (2012). Uncovering the molecular and cellular mechanisms of heart development using the zebrafish. *Annual Review of Genetics*, *46*, 397–418.

Tan, H., Biechler, S., Junor, L., Yost, M. J., Dean, D., Li, J., et al. (2013). Fluid flow forces and rhoA regulate fibrous development of the atrioventricular valves. *Developmental Biology*, *374*(2), 345–356.

Tarbell, J. M., Simon, S. I., & Curry, F. R. (2014). Mechanosensing at the vascular interface. *Annual Review of Biomedical Engineering*, *16*, 505–532.

Thoma, R. (1893). *Untersuchungen über die Histogenes und Histomechanic des Gefasssystems*. Stuttgart: Ferdinand Enke.

Vermot, J., Forouhar, A. S., Liebling, M., Wu, D., Plummer, D., Gharib, M., et al. (2009). Reversing blood flows act through klf2a to ensure normal valvulogenesis in the developing heart. *PLoS Biology*, 7(11), e1000246.

Wang, C., Baker, B. M., Chen, C. S., & Schwartz, M. A. (2013). Endothelial cell sensing of flow direction. *Arteriosclerosis, Thrombosis, and Vascular Biology*, *33*(9), 2130–2136.

Wang, N., Tytell, J. D., & Ingber, D. E. (2009). Mechanotransduction at a distance: Mechanically coupling the extracellular matrix with the nucleus. *Nature Reviews. Molecular Cell Biology*, *10*(1), 75–82.

Yang, J., Hartjes, K. A., Nelson, T. J., & Xu, X. (2014). Cessation of contraction induces cardiomyocyte remodeling during zebrafish cardiogenesis. *American Journal of Physiology. Heart and Circulatory Physiology*, *306*(3), H382–H395.

Yelon, D., Horne, S. A., & Stainier, D. Y. (1999). Restricted expression of cardiac myosin genes reveals regulated aspects of heart tube assembly in zebrafish. *Developmental Biology*, *214*(1), 23–37.

CHAPTER TWENTY-SIX

Organizers in Development

Claire Anderson, Claudio D. Stern[1]
Department of Cell & Developmental Biology, University College London, London, United Kingdom
[1]Corresponding author: e-mail address: c.stern@ucl.ac.uk

Contents

Abstract

An "organizer" is formally defined as a region, or group of cells in an embryo that can both induce (change the fate) and pattern (generate an organized set of structures) adjacent embryonic cells. To date, about four such regions have been demonstrated: the primary or Spemann organizer (Hensen's node in amniotes), the notochord, the zone of polarizing activity of the limb bud, and the mid-hindbrain boundary. Here we review the evidence for these and compare them with a few other regions which have been proposed to represent other organizers and we speculate on why so few such regions have been discovered.

More than 100 years ago, Ethel Browne carried out experiments in Hydra to understand how this simple organism regenerates its head after cutting. Using two differently colored species, she showed that grafting a region just adjacent to the head (the hypostome) could induce a complete new body axis derived from the host, rather than the graft (Browne, 1909). Based on this result, she proposed that tissues could influence each others' fates, resulting in a coherent, organized body plan. Intercellular signals that can divert the fate of a responding tissue in a new direction are what is known in developmental biology as "induction" (Gurdon, 1987), and the

Current Topics in Developmental Biology, Volume 117
ISSN 0070-2153
http://dx.doi.org/10.1016/bs.ctdb.2015.11.023

combination of this with the formation of a coherent, organized set of structures is commonly referred to as "patterning."

1. THE "SPEMANN ORGANIZER"

Initially, there was some debate concerning whether one particular structure in the early amphibian embryo (the dorsal lip of the blastopore, the site of ingression of cells during gastrulation) itself gives rise to the embryonic axis or whether it also acts by influencing neighboring cells (Hamburger, 1988; Lewis, 1907; Oppenheimer, 1991; Sander & Faessler, 2001). The critical experiments that finally resolved the controversy were performed by Hilde Mangold and Hans Spemann in 1921–1924 (Oppenheimer, 1964; Spemann, 1921; Spemann & Mangold, 1924). Probably inspired by Browne's earlier Hydra experiments (Lenhoff, 1991), Spemann and Mangold exchanged pieces of tissue from gastrula stage embryos between three differently pigmented species of newt (*Triton taeniatus*, *T. cristatus*, and *T. alpestris*). The key finding was that a graft of dorsal blastopore lip could induce a well-organized axis from cells of the host embryo. The graft itself contributed mainly to notochord, some somite, and floor plate tissue, whereas the host gave rise to all other structures including the rest of the central nervous system. Spemann wrote: "a piece of the upper blastopore lip of a gastrulating amphibian embryo exerts an organizing effect on its environment in such a way that, following its transplantation to an indifferent region of another embryo, it there causes the formation of a secondary embryo. Such a piece can therefore be designated as an organizer" (Oppenheimer, 1964; Spemann & Mangold, 1924).

Soon after Spemann and Mangold's discovery of the dorsal lip of the blastopore as an amphibian organizer, Waddington showed that the tip of the primitive streak in avian (Waddington, 1930, 1932, 1933; Waddington & Schmidt, 1933) and mammalian (Waddington, 1934, 1937) embryos has similar properties to Spemann's organizer: it can induce the formation of a neural plate including head and more caudal structures when grafted onto epiblast outside the prospective neural plate of a host embryo. The tip of the primitive streak, called Hensen's node, is the amniote equivalent of the amphibian dorsal lip as it is also the site of ingression of the most dorsal/axial mesendoderm during gastrulation. In Teleost fishes, the equivalent structure is the center of the embryonic shield (Oppenheimer, 1934a, 1934b, 1939). Many studies have shown that grafts of the organizer can act when grafted into a host of a different species, even across different vertebrate classes: chick to frog (Kintner & Dodd, 1991), mouse to frog (Blum et al., 1992), chick to zebrafish (Hatta & Takahashi, 1996), fish to frog (Oppenheimer, 1936), chick to rabbit (Waddington, 1934), rabbit to chick (Waddington, 1937), and mouse to chick (Zhu, Belo, De Robertis, & Stern, 1999).

In his 1921 paper where a footnote announced the discovery of the organizer based on a single example that had then been obtained by Hilde Mangold, Spemann wrote: "such a fragment of an organizing center can briefly be called an 'organizer'; it creates for itself in the indifferent material in which it lies, or into which it is transplanted, a 'field of organization' of definite orientation and extent" (Oppenheimer, 1970; Spemann, 1921). This statement arises because Spemann had observed that the blastopore spreads from the dorsal lip gradually in a ventral direction during gastrulation. The early dorsal lip is the only part that can induce a full axis upon transplantation, accounting for his use of "briefly." The idea of a "field of organization" comes from a combination of the spreading of the blastopore and the spreading of the inductive influence of the dorsal lip upon grafting onto a responding host region (Hamburger, 1988).

But the idea of an organizer being only a part of a more complex "organizing center" quickly evolved into the notion that several separate organizers might be located within such a center. This could account for the distinct inducing and regionalizing (patterning) properties of the organizing center. Evidence apparently supporting this notion was first obtained by Spemann himself (Spemann, 1929, 1931a, 1931b): blastopore lips of different stages, or different parts of the blastopore, were transplanted into hosts. Spemann's idea was that the early blastopore lip contains all the organizers required to induce a complete axis, and that organizers for increasingly caudal (tail) structures are gradually lost as they migrate out into the archenteron roof. This was then further refined in experiments performed by Otto Mangold (Spemann's colleague and widower of Hilde Mangold, who had died in 1924). He lifted the ectodermal covering of an older embryo (late gastrula/early neurula stage) to reveal the underlying archenteron roof (future axial and paraxial mesoderm), which was then cut into pieces and each grafted into the blastocoel of a younger (early gastrula stage) host ("Einsteck" assay). The most anterior mesoderm induced head structures only, middle sections of the mesoderm induced trunk, and the most posterior part induced a tail protruding from the host embryo (Mangold, 1933). Interestingly, Spemann and O. Mangold also found that (neuro)ectoderm from equivalent regions had similar regional inducing ability—it would normally generate neuroectoderm of the same region as that from which it had been excised. This was called "homeogenetic induction" (Mangold, 1929, 1933; Mangold & Spemann, 1927; Nieuwkoop, 1952). Together, Spemann and O. Mangold's results led to the idea that a larger "organization center" might contain individual head, trunk, and tail organizers, and perhaps more. Although this conclusion still has some followers (and the terms "head organizer," "tail organizer," etc. are still in occasional use), it should be noted that in the Einsteck assay, the graft lies adjacent to at least some prospective neural plate of the host, which does not allow pure inductive properties to be distinguished from patterning an already induced neural plate.

Waddington's experiments in several species (chick, duck, rabbit, and occasional experiments in amphibians during his visits to Naples with Dorothy and Joseph Needham) led him to a different view of how induction and patterning might relate to each other. Waddington proposed a two-step, "evocation-individuation" model. Pure induction (which Waddington initially believed to be mainly dependent on the graft and not the host, "non-assimilative induction") was called "evocation" (Needham, Waddington, & Needham, 1934) and the regional character of the induced structure which is influenced by the host (assimilative induction) is called "individuation" (Stern, 2000; Waddington, 1940; Waddington & Schmidt, 1933). Later, as he further clarified the concept of "competence" of the responding tissue (Waddington, 1932), he no longer placed so much importance in graft or host for defining evocation (induction) and individuation (patterning) (Stern, 2000). Therefore, unlike Spemann and O. Mangold, Waddington considered the organizer as a single region rather than a population of separate organizers with different functions.

A rather different model to account for induction and patterning was proposed by Nieuwkoop and is known as the "Activation–Transformation model" (Nieuwkoop, 1952; Nieuwkoop & Nigtevecht, 1954). This came partly from experiments in which prospective neuroectoderm grafted into the dorsal midline of a host embryo would end up in different positions within the neural axis. They observed that the most rostral part of the ectopic nervous system that arose was coherent with the same level of the host neural plate, and that the secondary structures then extended distally into a protruding tail. In a variant of this experiment, prospective notochord from a neurula stage was sandwiched between two pieces of presumptive neuroectoderm from two gastrula stage embryos. The results led Nieuwkoop to propose a two-step model. The first step, "activation," involved the induction of anterior structures, whereas later influences emanating from the archenteron roof (or from the organizer as it retreats into the tail bud in normal embryos) gradually "transform" (i.e., caudalize or posteriorize) the initially induced structures (Nieuwkoop, 1952). This model was subsequently modified further into a three-step model (Stern, 2001), which introduces a stabilization step between activation and transformation.

Recent work on neural induction and patterning of the nervous system has not yet succeeded in providing clear results to distinguish between these three models, and many molecular and cellular studies suggest that the ability of the organizer to induce a complete, coherently patterned nervous system could involve elements of all three mechanisms. Whatever the answer, the idea that there may be as many "organizers" as distinct regions to be induced is particularly unattractive, because the complexity of the nervous system that needs to be generated by the embryo is huge, implying a vast number of such signaling regions (Stern, 2001).

Spemann's and O. Mangold's idea that an "organization center" contains (and gradually releases) individual organizers at successive stages in development led to the idea that the whole of development may be orchestrated by a sequence of organizers, each responsible for inducing and patterning a particular region or organ system. The "primary organizer," the dorsal blastopore lip, would be succeeded by a series of other "secondary organizers," followed by "tertiary organizers," and so on: a sequence of inductions generating the complexity and pattern of the organism only made possible by induction of the tissue before it. Although now inaccurate, the idea inspired Holtfreter to draw a chain reaction diagram of newt development based on sequential induction of tissues (Fig. 1). In the following sections, we will briefly review the main embryonic cell populations that have been considered to act in this way.

2. THE NOTOCHORD AND FLOOR PLATE

Following gastrulation, the dorsal blastopore (or shield in fish, Hensen's node in amniotes) gradually gives rise to a rod of axial mesoderm underling the neural plate: the notochord. A graft of notochord adjacent to the lateral neural tube of the trunk changes the fate of the latter and transforms them into ventral cell types: a floor plate and motor neurons, arranged in their normal relation to each other, recapitulating the pattern of the ventral neural tube (Placzek, Tessier-Lavigne, Yamada, Jessell, & Dodd, 1990b; van Straaten, Hekking, Thors, Wiertz-Hoessels, & Drukker, 1985; Yamada,

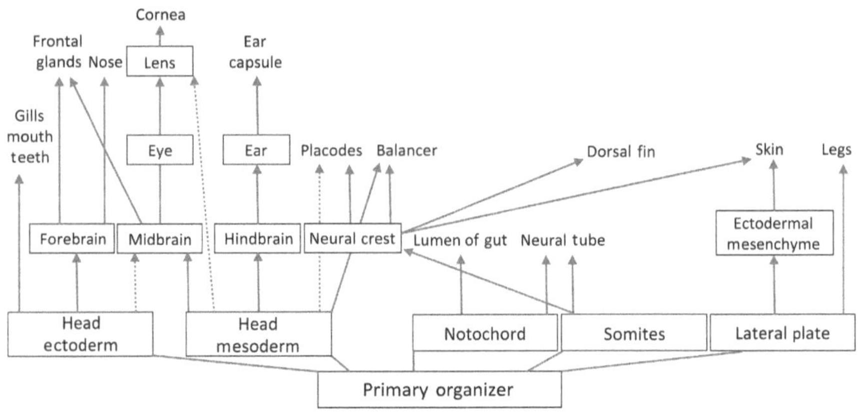

Figure 1 A now outdated chain reaction diagram of amphibian development based on sequential induction of tissues, illustrating the idea of secondary and tertiary organizers; after Waddington (1940), in turn based on Holtfreter (1938). Waddington's legend reads: "The main organizer regions are enclosed in rectangles. The lines join organs which are developed one from the other, while the arrows show inducing relations, the dotted ones being of less importance." The significance of sequential inductions in development was described thus by Holtfreter: "… not only the entire neural system, but also all the other tissue derivatives of the ectoderm, are likewise products of induction … I, therefore, maintain that embryonic induction is the most important device by which differentiation comes about in amphibian development" *Holtfreter (1991)*. (A color version of this figure is located at the end of this chapter.)

Placzek, Tanaka, Dodd, & Jessell, 1991). Further away, other progressively less ventral neuronal populations (interneurons) arise, in the same dorsoventral arrangement as in the normal spinal cord. A notochord graft is also able to pattern the somite mesoderm, which lies adjacent to the neural tube and notochord, by inducing ventral (sclerotome) fate and specifying medial (epaxial) muscle as opposed to lateral (hypaxial) muscle derivatives from the dermomyotome (Brand-Saberi, Ebensperger, Wilting, Balling, & Christ, 1993; Marcelle, Ahlgren, & Bronner-Fraser, 1999; Marcelle, Stark, & Bronner-Fraser, 1997; Pourquie, Coltey, Teillet, Ordahl, & Le Douarin, 1993; Pourquie et al., 1996; Pownall, Strunk, & Emerson, 1996; Vasiliauskas, Hancock, & Stern, 1999). In zebrafish, where there is virtually no sclerotome, the notochord induces slow muscle fate from myotome cells, defining an "adaxial" population (Blagden, Currie, Ingham, & Hughes, 1997). The notochord can also specify different regions of the underlying endoderm including pancreatic fate (Hebrok, Kim, & Melton, 1998; Kim, Hebrok, & Melton, 1997). Therefore, as the notochord possesses both inducing and patterning properties on more than one germ layer, it should be considered as a true organizer. Moreover like Spemann's organizer, its patterning ability influences all three germ layers (neuroectoderm, mesoderm [somites], and endoderm).

A notochord graft adjacent to the lateral neural tube of older embryos induces motor neurons without prior induction of a floor plate, indicating that the notochord alone is able to induce and pattern ventral cell types in the neural tube (Placzek, Tessier-Lavigne, Jessell, & Dodd, 1990a; van Straaten, Hekking, Wiertz-Hoessels, Thors, & Drukker, 1988; Yamada et al., 1991). The floor plate can also induce organized ventral cell types in the lateral neural tube of the trunk and ventral somitic derivatives in the dorsal somite in a similar manner to a graft of notochord (Pourquie et al., 1993; Yamada et al., 1991). Therefore, the floor plate can also act as an organizer.

There is some evidence suggesting that the dorsal midline of the neural tube, the roof plate, also has some organizing properties. In co-culture, the roof plate can induce an organized pattern of dorsal interneurons in chick lateral neural plate explants (Liem, Tremml, & Jessell, 1997) and can also lead to dorsal axon repulsion in rat spinal cord explants (Augsburger, Schuchardt, Hoskins, Dodd, & Butler, 1999). Targeted genetic ablation of the roof plate with diphtheria toxin leads to disorganization of dorsal interneurons (Lee, Dietrich, & Jessell, 2000), supporting a role for the roof plate in the induction and patterning of dorsal cell types in the trunk neural tube. Therefore, the roof plate can change the fate of intermediate neural plate by inducing dorsal cell types, including a normal arrangement of dorsal interneurons.

3. ORGANIZERS AND BRAIN PATTERNING

The notochord and floor plate end anteriorly in the prospective pituitary/Rathke's pouch area, underlying the confluence between midbrain and forebrain. A floor plate-like region, expressing several common markers including Sonic hedgehog (*Shh*), extends rostrally in the ventral brain: the hypothalamus and basal plate. From the middle of its trajectory, a dorsally projecting extension of the *Shh*-expressing region, the zona limitans intrathalamica (ZLI) marks the border between the first two subdivisions of the diencephalon (ventral and dorsal thalamus, now referred to as prethalamus and thalamus, respectively). It has been suggested that these *Shh*-expressing regions, along with other regions, expressing FGFs (like the anterior neural ridge (ANR) and the mid-hindbrain boundary) are important signaling centers, and we shall briefly review the evidence below.

The ZLI is a transverse morphological furrow bisecting the posterior forebrain (diencephalon) and is the only part of the central nervous system where floor plate/basal plate markers are expressed dorsally in the alar plate (Echelard et al., 1993). A graft of ZLI anteriorly into the prospective telencephalon induces ectopic expression of the prethalamic and telencephalic marker, *DLX2*, but not the thalamic marker *GBX2* (Vieira & Martinez, 2006). Conversely, a graft of ZLI into the prospective diencephalon induces ectopic expression of *GBX2* in the posterior diencephalon (pretectum) but is unable to induce *DLX2*. Thus, the ZLI is only able to induce anterior markers anteriorly and posterior markers posteriorly within the forebrain; moreover, with the markers available, this study could not distinguish between prethalamus and telencephalon. Preventing the dorsal extension of the ZLI with a physical barrier does not prevent the induction of prethalamus and thalamus adjacent to the basal plate but does reduce the dorsal extension of *DLX2* and *GBX2*, resulting in loss of differentiated diencephalic nuclei that are normally close to the ZLI (perirotundic area and interstitial nucleus of the posterior optic tract and nucleus intercalatus anterior) (Vieira & Martinez, 2006). Therefore at best, the ZLI possesses only modest patterning functions. To demonstrate that the ZLI is a true organizer, it will be necessary to show that a graft can generate a coherent pattern of structures (for example, specific functional nuclei and/or a set of several markers specific for distinct regions) around it. Interestingly, basal plate from either anterior or posterior to the ZLI is able to induce thalamic marker expression in the pretectum in a similar manner to the ZLI (Vieira & Martinez, 2006), raising the possibility that induction by the ZLI is an extension of floor plate-like induction and patterning to the dorsal diencephalon (i.e., vertical signaling from the floor plate becomes planar signaling with the dorsal extension of the ZLI). Indeed, both the basal plate and the ZLI secrete SHH, which was proposed to mediate the function of the ZLI (Kiecker & Lumsden, 2004). However, genetic ablation of the basal plate is followed by normal expression of diencephalic markers in zebrafish (Scholpp, Wolf, Brand, & Lumsden, 2006).

Perhaps in a similar fashion to how floor plate properties may extend rostrally into the basal plate and possibly the ZLI, the roof plate also retains some signaling properties in the brain, at least in hindbrain regions. Co-culture of chick hindbrain roof plate with hindbrain neuroepithelium generates an ectopic roof plate and *ATOH1*-expressing dorsal-most interneurons adjacent to this, but the study did not distinguish host and graft cells (Broom, Gilthorpe, Butts, Campo-Paysaa, & Wingate, 2012). Stronger evidence for a role of the roof plate in induction of hindbrain dorsal interneurons comes from mouse models: *Lmx1a* is expressed in the hindbrain and gain and loss of *Lmx1a* function results in either an increase or a decrease in the dorsal-most interneurons in the anterior hindbrain (rhombomere 1) (Chizhikov et al., 2006) and the cerebellum, a derivative of rhombomere 1, is reduced in *Dreher/Lmx1a* mutant embryos (Millonig, Millen, & Hatten, 2000). Therefore, the hindbrain roof plate may play a role in the induction of dorsal-most interneurons and their organization, analogous to the function of the roof plate of the trunk.

Further rostrally, in the forebrain, the roof plate embeds between the cerebral hemispheres forming the cortical hem and choroid plexus, adjacent to which are the hippocampi. So far, there is no evidence for hippocampus induction by the cortical hem. Mutant mouse models implicate the roof plate of the forebrain in proliferation and expansion of the hippocampus and cortex rather than induction (Lee, Dietrich, et al., 2000; Lee, Tole, Grove, & McMahon, 2000). In conclusion, as for the ZLI, evidence that the roof plate of the brain is a true organizer is limited.

The ANR in amniotes (also known as the anterior neural border or ANB in zebrafish) is a morphologically defined structure where the anterior neural plate abuts the non-neural ectoderm, eventually forming the commissural plate in the anterior forebrain (telencephalon). Exclusion of the ANR from mouse neural plate explants or ablation of the ANB in zebrafish results in loss of telencephalic markers (Houart, Westerfield, & Wilson, 1998; Shimamura & Rubenstein, 1997). A graft of ANB into more posterior locations induces ectopic expression of specific telencephalic genes *emx1* and *foxg1* (*bf1*) and can suppress mesencephalic (midbrain) gene expression (Houart et al., 2002, 1998). So far, a role for the ANR/ANB in patterning the telencephalon is not as clear, but FGF8, which is expressed in the ANR, can rescue *Foxg1 (Bf1)* expression in neural plate lacking the ANR (Shimamura & Rubenstein, 1997). Furthermore, mis-expression of *Fgf8* in the posterior telencephalon of mouse embryos produces partial mirror-image duplication of the somatosensory barrel cortex (Fukuchi-Shimogori & Grove, 2001), indicating that signaling from the ANR/commissural plate may be responsible for patterning anterior telencephalic structures. FGF8 is not the only signal involved: *sfrp/tlc*, a Frizzled-related Wnt antagonist, also expressed in the ANB, can both rescue ANB ablation and mimic a graft of ANB (Houart et al., 2002). An interesting possibility is that the ANR/ANB (perhaps through Wnt inhibition) acts by antagonizing caudalizing signals like those suggested by the Nieuwkoop model (Wilson & Houart, 2004). In conclusion, like the

ZLI, there are suggestive indications that the ANR/ANB could be an organizer, but direct evidence for distinct inducing and patterning functions remains to be demonstrated.

The midbrain–hindbrain boundary (MHB), also known as the isthmus, is a physical groove between the anterior hindbrain (rhombomere 1) and the posterior mesencephalon. The MHB and immediately adjacent hindbrain and posterior mesencephalic regions are characterized by the expression of Engrailed 1/2 (EN1/2). When this region is transplanted to the posterior forebrain (diencephalon), the graft differentiates into cerebellum as it would normally, but also induces expression of EN1/2 in the host (Gardner & Barald, 1991) and an ectopic tectum/posterior mesencephalon (Martinez, Wassef, & Alvarado-Mallart, 1991). A similar graft of anterior hindbrain including the MHB from rat or mouse embryos into the chick diencephalon induces ectopic expression of EN1/2 (Martinez et al., 1991). Furthermore, inversion of the mesencephalic alar plate so that the anterior midbrain is now contacting the MHB results in respecification of the anterior midbrain to a posterior fate (Martinez & Alvarado-Mallart, 1990), and inversion of the mesencephalon together with the MHB results in duplication of posterior midbrain structures as a mirror image (Marin & Puelles, 1994).

Grafting either quail or mouse MHB into the posterior hindbrain of a chick induces ectopic EN1/2 expression and differentiation of cerebellum (an anterior hindbrain structure) in the host (Martinez, Marin, Nieto, & Puelles, 1995). The MHB rapidly regenerates following ablation (Irving & Mason, 1999), but the lack of a MHB in *Wnt1−/−* knockout mouse embryos results in loss of both posterior mesencephalon and anterior hindbrain derivatives (McMahon & Bradley, 1990; McMahon, Joyner, Bradley, & McMahon, 1992). These studies indicate that the MHB has induction and patterning roles both anteriorly in the mesencephalon and posteriorly in the hindbrain. The MHB can therefore be considered as a true organizer.

4. PATTERNING THE LIMB

Two major signaling regions have been implicated in limb patterning: the zone of polarizing activity (ZPA) situated in the posterior limb bud mesenchyme, and the apical ectodermal ridge (AER), running along the edge of the ectodermal covering of the bud. A ZPA graft into the anterior limb mesenchyme generates anteroposterior mirror-image duplication of limb skeletal structures and associated soft tissues; the earlier this is done, the more complete the duplication (Saunders & Gasseling, 1968; Tickle, Shellswell, Crawley, & Wolpert, 1976). An AER graft induces outgrowth of ectopic, patterned limb structures distal to the structures that had already been specified at the time of grafting (Saunders & Gasseling, 1968; Zwilling, 1956). The AER requires the presence of the ZPA to be maintained and vice versa; therefore, generation of the anterior-posterior and proximo-distal axes is intimately linked (Saunders & Gasseling, 1968; Todt & Fallon, 1987; Zwilling, 1956). Therefore, the ZPA and AER together are required for normal limb patterning. Cross-species graft experiments show that a mammalian (Maccabe & Parker, 1976; Tickle et al., 1976) or reptilian (Fallon & Crosby, 1977) ZPA can induce extra wing digits in the chick.

SHH can mimic a ZPA graft in the anterior limb and induce mirror-image duplication (Riddle, Johnson, Laufer, & Tabin, 1993), while FGFs can rescue limb outgrowth after removal of the AER (Fallon et al., 1994; Niswander, Tickle, Vogel, Booth, & Martin, 1993). Genetic studies in mouse have revealed that *Shh* from the posterior ZPA is required to generate a gradient of *Gli3* repressor (highest anteriorly) that limits the number of digits formed and determines the anterior-posterior identity of each digit (Litingtung, Dahn, Li, Fallon, & Chiang, 2002; te Welscher et al., 2002). SHH-dependent BMP signaling (Drossopoulou et al., 2000) and AER-derived FGF8 (Sanz-Ezquerro & Tickle, 2003) are also proposed to play roles in determining digit identity in chick along both axes. Although in this case it is more difficult to define separate inducing and patterning functions, it is clear that the ZPA, perhaps acting in conjunction with the AER, can generate new structures that form a coherent pattern along all three axes (proximo-distal, anterior-posterior, and dorso-ventral), and it can therefore probably be considered as a bona fide organizer.

5. SPEMANN'S EXPERIMENTS ON THE LENS

Spemann had approached the famous organizer experiment after having studied the cell interactions that lead to formation of the lens of the eye, two decades earlier. There he had already thought of the ideas of inductive and patterning influences between tissues. The lens develops from the non-neural ectoderm overlying the neural optic vesicle in the forebrain. Spemann suggested that the optic vesicle induces the lens and showed that injury to eye rudiment in the frog (*Rana fusca*) results in optic cup formation from the optic vesicle, but no lens developed (Spemann, 1901). This was supported by optic vesicle graft experiments to the head or belly ectoderm, which is not fated to form eye, resulted in an ectopic lens (Lewis, 1904). However, this simple explanation for lens induction received strong criticism (Mencl, 1903) and similar injury experiments in *Rana palustris* demonstrated "free-lenses" when the optic vesicle was ablated (King, 1905). It later became evident that it is very difficult to remove all lens precursor cells from the optic vesicle prior to grafting experiments (Spemann, 1908) and *R. palustris* represents one of many examples of species-specific differences in lens induction (Saha, 1991). Later studies in amphibians showed that lens formation is a multistep induction process beginning around gastrulation (Fisher & Grainger, 2004). Grafting experiments in Xenopus showed that the optic vesicle is not able to induce lens in nonlens ectoderm, but it is able to induce lens if the ectoderm has been exposed to early specification steps during gastrulation and neurulation (Henry & Grainger, 1987)—in other words, it only serves a permissive role, rather than being an instructive induction. More recently, further light has been thrown onto this: the optic vesicle seems to act mainly as a barrier for neural crest migration. In turn, the neural crest inhibits lens fate. Therefore, a key role of the optic vesicle is to prevent neural crest cells from inhibiting lens fate in the appropriate region and to align the two structures (Bailey, Bhattacharyya, Bronner-Fraser, & Streit, 2006). Therefore, the optic vesicle is neither a lens inducer nor is there evidence for specific patterning functions, and so cannot be considered an organizer.

6. COMMON FEATURES BETWEEN ORGANIZERS?

When it was discovered as a marker for the dorsal lip/shield/Hensen's node of vertebrate embryos at the time when this structure has organizing ability, the homeobox gene *Goosecoid* (*GSC*) was hailed as a universal marker of the organizer (Blum et al., 1992; Cho, Blumberg, Steinbeisser, & De Robertis, 1991; Izpisua-Belmonte, De Robertis, Storey, & Stern, 1993; Zhu et al., 1999). Although it is not expressed in most of the "secondary" organizers listed above, it does seem to play a role in regulating signaling in the limb (Heanue et al., 1997).

GSC orthologs are also found in some invertebrates. In Hydra, *Cngsc* is expressed in the "bud," which represents the future hypostome (the region originally found by Ethel Browne to act as an organizer) (Browne, 1909). mRNA transcribed from the Hydra gene can induce secondary axes in

Xenopus embryos when injected in one of the early blastomeres (Broun, Sokol, & Bode, 1999). Drosophila also has an orthologous gene, *D-gsc*, expressed in cells fated to become part of the nervous system (central and stomatogastric nervous system) and in invaginating stomodeum/foregut. *D-gsc* injection into UV irradiated Xenopus embryos can rescue axis polarity (Goriely et al., 1996). Whether or not the cells expressing *D-gsc* in fly represent a true organizer, with the ability to induce and pattern neighboring tissues, remains to be tested. This will require transplantation of cells, which is of course very difficult in the fly embryo.

Are the key signals emitted from different organizers the same? This notion was first raised from a few heterochronic and heterotopic transplantation experiments: chick Hensen's node, notochord, and floor plate can mimic an anterior ZPA graft and induce mirror-image duplication of digits in the limb (Hornbruch & Wolpert, 1986; Wagner, Thaller, Jessell, & Eichele, 1990). Moreover, several of the regions that can be considered as having organizer properties share expression of members of particular families of secreted factors, mainly Hedgehog, FGFs, and/or Wnt antagonists. This makes it tempting to speculate, as already done by Waddington, that signals from organizers may be universal but that the nature of the induction and patterning events are determined by the specific competence of the responding tissue.

7. CONCLUSIONS

We have briefly presented evidence (see Table 1) that suggests that at least four regions of the vertebrate embryo can act as true organizers in that they possess both inducing and patterning functions: the classical Spemann organizer (the shield in fish, Hensen's node in amniotes), the notochord, the midbrain–hindbrain boundary, and the ZPA (perhaps together with the AER) of the limb bud. There are other regions that have been, or could be, considered to possess organizing properties (for example, the prechordal mesendoderm or the Nieuwkoop center at very early stages of frog development) in vertebrates but the evidence is not yet compelling enough. One might expect, as the early pioneers in this field did, that organizers should exist throughout animal development, including in invertebrates and at later stages of development. The challenge remains to arrive at an efficient strategy to point to good candidate regions, whose inducing and patterning functions can then be tested directly by transplantation.

Table 1 Regions Considered to Have Some Organizer Function, Listed Along with the Main Evidence for Their Ability to Induce (Column 2) and Pattern (Column 3) Responding Tissues

Signaling Region	Induction (Transforms the Fate of Responding Tissue)	Patterning (Coherent Spatial Arrangement of Several Markers in Responding Tissue)
Spemann organizer	Induces neural plate in tissue not normally fated to give rise to neural tissue	Induced nervous system patterned along all three axes. Also gives dorso-ventral (axial/lateral) identity to adjacent mesoderm
Notochord	Changes the fate of the lateral neural tube by inducing floor plate and motor neurons. Induces sclerotome (amniotes) and slow muscle (fish) in somite. Induces pancreatic fate in endoderm	Induction of a coherent arrangement of neuronal cell types (motor neurons, V1 and V2 interneurons, etc.) in neural tube. Patterns somite along dorso-ventral and medio-lateral axes. Also patterns the endoderm, specifying dorsal/midline identity.
Floor plate	Shares some of the notochord's properties	Shares some of the notochord's properties
Roof plate	Induces dorsal interneurons in trunk neural plate explants and dorsal interneurons in hindbrain neuroepithelium	Dorsal D1 and D2 interneurons generated in neural plate explants in an organized manner. Genetic ablation of roof plate in mouse embryos causes disorganization of dorsal interneurons in the spinal cord
Zona limitans intrathalamica (ZLI)	ZLI grafts induce thalamic *GBX2* in the pretectum	A physical barrier causes disorganization of structures adjacent to the ZLI
Anterior neural ridge/border (ANR/ANB)	ANB grafts induce telencephalic *emx1* and *foxg1* in more posterior locations	No direct evidence
Midbrain–hindbrain boundary (MHB)/isthmus	Induces ectopic Engrailed 1/2 (EN1/2) expression and tectum in the diencephalon (would normally form thalamus and pretectum)	Can repolarize an inverted mesencephalon and induce mirror-image duplication of midbrain structures
ZPA and AER	ZPA grafts to the anterior limb induces mirror-image duplication of skeletal and soft tissue elements (requires AER activity. Grafts of AER alone induce distal growth of limb bud)	Structures duplicated by ZPA grafts to the anterior limb have correct A–P, P–D, and D–V pattern

ACKNOWLEDGMENT

Our research on this topic was funded by the BBSRC.

REFERENCES

Augsburger, A., Schuchardt, A., Hoskins, S., Dodd, J., & Butler, S. (1999). BMPs as mediators of roof plate repulsion of commissural neurons. *Neuron, 24*, 127–141.

Bailey, A. P., Bhattacharyya, S., Bronner-Fraser, M., & Streit, A. (2006). Lens specification is the ground state of all sensory placodes, from which FGF promotes olfactory identity. *Developmental Cell, 11*, 505–517.

Blagden, C. S., Currie, P. D., Ingham, P. W., & Hughes, S. M. (1997). Notochord induction of zebrafish slow muscle mediated by Sonic hedgehog. *Genes and Development, 11*, 2163–2175.

Blum, M., Gaunt, S. J., Cho, K. W., Steinbeisser, H., Blumberg, B., Bittner, D., et al. (1992). Gastrulation in the mouse: The role of the homeobox gene goosecoid. *Cell, 69*, 1097–1106.

Brand-Saberi, B., Ebensperger, C., Wilting, J., Balling, R., & Christ, B. (1993). The ventralizing effect of the notochord on somite differentiation in chick embryos. *Anatomy and Embryology (Berlin), 188*, 239–245.

Broom, E. R., Gilthorpe, J. D., Butts, T., Campo-Paysaa, F., & Wingate, R. J. (2012). The roof plate boundary is a bi-directional organiser of dorsal neural tube and choroid plexus development. *Development, 139*, 4261–4270.

Broun, M., Sokol, S., & Bode, H. R. (1999). Cngsc, a homologue of goosecoid, participates in the patterning of the head, and is expressed in the organizer region of Hydra. *Development, 126*, 5245–5254.

Browne, E. N. (1909). The production of new hydranths in hydra by the insertion of small grafts. *Journal of Experimental Zoology, 7*, 1–37.

Chizhikov, V. V., Lindgren, A. G., Currle, D. S., Rose, M. F., Monuki, E. S., & Millen, K. J. (2006). The roof plate regulates cerebellar cell-type specification and proliferation. *Development, 133*, 2793–2804.

Cho, K. W., Blumberg, B., Steinbeisser, H., & De Robertis, E. M. (1991). Molecular nature of Spemann's organizer: The role of the Xenopus homeobox gene goosecoid. *Cell, 67*, 1111–1120.

Drossopoulou, G., Lewis, K. E., Sanz-Ezquerro, J. J., Nikbakht, N., McMahon, A. P., Hofmann, C., et al. (2000). A model for anteroposterior patterning of the vertebrate limb based on sequential long- and short-range Shh signalling and Bmp signalling. *Development, 127*, 1337–1348.

Echelard, Y., Epstein, D. J., St-Jacques, B., Shen, L., Mohler, J., McMahon, J. A., et al. (1993). Sonic hedgehog, a member of a family of putative signaling molecules, is implicated in the regulation of CNS polarity. *Cell, 75*, 1417–1430.

Fallon, J. F., & Crosby, G. M. (1977). Polarising zone activity in limb buds of amniotes. In D. A. Ede, J. R. Hinchliffe, & M. Balls (Eds.), *Vertebrate limb and somite morphogenesis* (pp. 55–69). New York: Cambridge University Press.

Fallon, J. F., Lopez, A., Ros, M. A., Savage, M. P., Olwin, B. B., & Simandl, B. K. (1994). FGF-2: Apical ectodermal ridge growth signal for chick limb development. *Science, 264*, 104–107.

Fisher, M., & Grainger, R. M. (2004). Lens induction and determination. In F. J. Lovicu & M. L. Robinson (Eds.), *Development of the ocular lens* (pp. 27–47). New York: Cambridge University Press.

Fukuchi-Shimogori, T., & Grove, E. A. (2001). Neocortex patterning by the secreted signaling molecule FGF8. *Science, 294*, 1071–1074.

Gardner, C. A., & Barald, K. F. (1991). The cellular environment controls the expression of engrailed-like protein in the cranial neuroepithelium of quail-chick chimeric embryos. *Development*, *113*, 1037–1048.

Goriely, A., Stella, M., Coffinier, C., Kessler, D., Mailhos, C., Dessain, S., et al. (1996). A functional homologue of goosecoid in Drosophila. *Development*, *122*, 1641–1650.

Gurdon, J. B. (1987). Embryonic induction—Molecular prospects. *Development*, *99*, 285–306.

Hamburger, V. (1988). *The heritage of experimental embryology: Hans Spemann and the organizer*. New York: Oxford University Press.

Hatta, K., & Takahashi, Y. (1996). Secondary axis induction by heterospecific organizers in zebrafish. *Developmental Dynamics*, *205*, 183–195.

Heanue, T. A., Johnson, R. L., Izpisua-Belmonte, J. C., Stern, C. D., De Robertis, E. M., & Tabin, C. J. (1997). Goosecoid misexpression alters the morphology and Hox gene expression of the developing chick limb bud. *Mechanisms of Development*, *69*, 31–37.

Hebrok, M., Kim, S. K., & Melton, D. A. (1998). Notochord repression of endodermal Sonic hedgehog permits pancreas development. *Genes and Development*, *12*, 1705–1713.

Henry, J. J., & Grainger, R. M. (1987). Inductive interactions in the spatial and temporal restriction of lens-forming potential in embryonic ectoderm of *Xenopus laevis*. *Developmental Biology*, *124*, 200–214.

Holtfreter, J. (1938). Differenzierungspotenzen isolierter Teile der Urodelengastrula. *Wilhelm Roux' Archiv für Entwicklungsmechanik der Organismen*, *138*, 522–652.

Holtfreter, J. (1991). Reminiscences on the life and work of Johannes Holtfreter. In S. F. Gilbert (Ed.), *A conceptual history of modern embryology* (pp. 109–126). New York: Plenum Press.

Hornbruch, A., & Wolpert, L. (1986). Positional signalling by Hensen's node when grafted to the chick limb bud. *Journal of Embryology and Experimental Morphology*, *94*, 257–265.

Houart, C., Caneparo, L., Heisenberg, C., Barth, K., Take-Uchi, M., & Wilson, S. (2002). Establishment of the telencephalon during gastrulation by local antagonism of Wnt signaling. *Neuron*, *35*, 255–265.

Houart, C., Westerfield, M., & Wilson, S. W. (1998). A small population of anterior cells patterns the forebrain during zebrafish gastrulation. *Nature*, *391*, 788–792.

Irving, C., & Mason, I. (1999). Regeneration of isthmic tissue is the result of a specific and direct interaction between rhombomere 1 and midbrain. *Development*, *126*, 3981–3989.

Izpisua-Belmonte, J. C., De Robertis, E. M., Storey, K. G., & Stern, C. D. (1993). The homeobox gene goosecoid and the origin of organizer cells in the early chick blastoderm. *Cell*, *74*, 645–659.

Kiecker, C., & Lumsden, A. (2004). Hedgehog signaling from the ZLI regulates diencephalic regional identity. *Nature Neuroscience*, 7, 1242–1249.

Kim, S. K., Hebrok, M., & Melton, D. A. (1997). Notochord to endoderm signaling is required for pancreas development. *Development*, *124*, 4243–4252.

King, H. D. (1905). Experimental studies on the eye of the frog embryo. *Archiv für Entwicklungsmechanik der Organismen*, *19*, 85–107.

Kintner, C. R., & Dodd, J. (1991). Hensen's node induces neural tissue in Xenopus ectoderm. Implications for the action of the organizer in neural induction. *Development*, *113*, 1495–1505.

Lee, K. J., Dietrich, P., & Jessell, T. M. (2000). Genetic ablation reveals that the roof plate is essential for dorsal interneuron specification. *Nature*, *403*, 734–740.

Lee, S. M., Tole, S., Grove, E., & McMahon, A. P. (2000). A local Wnt-3a signal is required for development of the mammalian hippocampus. *Development*, *127*, 457–467.

Lenhoff, H. M. (1991). Ethel Browne, Hans Spemann, and the discovery of the organizer phenomenon. *Biology Bulletin*, *181*, 72–80.

Lewis, W. H. (1904). Experimental studies on the development of the eye in amphibia. I. On the origin of the lens. *The American Journal of Anatomy*, *3*, 505–536.
Lewis, W. H. (1907). Transplantation of the lips of the blastopore in *Rana palustris*. *The American Journal of Anatomy*, 7, 137–141.
Liem, K. F., Jr., Tremml, G., & Jessell, T. M. (1997). A role for the roof plate and its resident TGFbeta-related proteins in neuronal patterning in the dorsal spinal cord. *Cell*, *91*, 127–138.
Litingtung, Y., Dahn, R. D., Li, Y., Fallon, J. F., & Chiang, C. (2002). Shh and Gli3 are dispensable for limb skeleton formation but regulate digit number and identity. *Nature*, *418*, 979–983.
Maccabe, J. A., & Parker, B. W. (1976). Polarizing activity in the developing limb of the Syrian hamster. *Journal of Experimental Zoology*, *195*, 311–317.
Mangold, O. (1929). Experimente zur Analyse der determination und induktion der Medullarplatte. *Wilhelm Roux' Archiv für Entwicklungsmechanik der Organismen*, *117*, 586–696.
Mangold, O. (1933). Über die induktionsfähighkeit der verschiedenen bezirke der neurula von Urodelen. *Naturwissenschaften*, *21*, 761–766.
Mangold, O., & Spemann, H. (1927). Über Induktion von Medullarplatte durch Medullarplatte im jüngeren Keim, ein Beispiel homöogenetischer oder assimilatorischer Induktion. *Wilhelm Roux' Archiv für Entwicklungsmechanik der Organismen*, *111*, 341–422.
Marcelle, C., Ahlgren, S., & Bronner-Fraser, M. (1999). In vivo regulation of somite differentiation and proliferation by Sonic Hedgehog. *Developmental Biology*, *214*, 277–287.
Marcelle, C., Stark, M. R., & Bronner-Fraser, M. (1997). Coordinate actions of BMPs, Wnts, Shh and noggin mediate patterning of the dorsal somite. *Development*, *124*, 3955–3963.
Marin, F., & Puelles, L. (1994). Patterning of the embryonic avian midbrain after experimental inversions: A polarizing activity from the isthmus. *Developmental Biology*, *163*, 19–37.
Martinez, S., & Alvarado-Mallart, R. M. (1990). Expression of the homeobox Chick-en gene in chick/quail chimeras with inverted mes-metencephalic grafts. *Developmental Biology*, *139*, 432–436.
Martinez, S., Marin, F., Nieto, M. A., & Puelles, L. (1995). Induction of ectopic engrailed expression and fate change in avian rhombomeres: Intersegmental boundaries as barriers. *Mechanisms of Development*, *51*, 289–303.
Martinez, S., Wassef, M., & Alvarado-Mallart, R. M. (1991). Induction of a mesencephalic phenotype in the 2-day-old chick prosencephalon is preceded by the early expression of the homeobox gene en. *Neuron*, *6*, 971–981.
McMahon, A. P., & Bradley, A. (1990). The Wnt-1 (int-1) proto-oncogene is required for development of a large region of the mouse brain. *Cell*, *62*, 1073–1085.
McMahon, A. P., Joyner, A. L., Bradley, A., & McMahon, J. A. (1992). The midbrain-hindbrain phenotype of Wnt-1-/Wnt-1-mice results from stepwise deletion of engrailed-expressing cells by 9.5 days postcoitum. *Cell*, *69*, 581–595.
Mencl, E. (1903). Ein Fall von beiderseitiger Augenlinsenausbildung während der Abwesenheit von Augenblasen. *Archiv für Entwicklungsmechanik der Organismen*, *16*, 328–339.
Millonig, J. H., Millen, K. J., & Hatten, M. E. (2000). The mouse Dreher gene Lmx1a controls formation of the roof plate in the vertebrate CNS. *Nature*, *403*, 764–769.
Needham, J., Waddington, C. H., & Needham, D. M. (1934). Physico-chemical experiments on the amphibian organiser. *Proceedings of the Royal Society of London B*, *114*, 393–422.
Nieuwkoop, P. D. (1952). Activation and organization of the central nervous system in amphibians. Parts I-III. I. Induction and activation. II. Differentiation and organization. III. Synthesis of a new working hypothesis. *Journal of Experimental Zoology*, *120*, 1–108.

Nieuwkoop, P. D., & Nigtevecht, G. V. (1954). Neural activation and transformation in explants of competent ectoderm under the influence of fragments of anterior notochord in Urodeles. *Journal of Embryology & Experimental Morphology*, *2*, 175–193.
Niswander, L., Tickle, C., Vogel, A., Booth, I., & Martin, G. R. (1993). FGF-4 replaces the apical ectodermal ridge and directs outgrowth and patterning of the limb. *Cell*, *75*, 579–587.
Oppenheimer, J. M. (1934a). Experimental studies on the developing perch (*Perca flavescens* Mitchill). *Proceedings of the Society for Experimental Biology and Medicine*, *31*, 1123–1124.
Oppenheimer, J. M. (1934b). Experiments on early developing stages of Fundulus. *Proceedings of the National Academy of Sciences of the United States of America*, *20*, 536–538.
Oppenheimer, J. M. (1936). Structures developed in amphibians by implantation of living fish organizers. *Proceedings of the Society for Experimental Biology and Medicine*, *34*, 461–463.
Oppenheimer, J. M. (1939). Transplantation experiments on developing teleosts (*Fundulus* and *Perca*). *Journal of Experimental Zoology*, *72*, 409–437.
Oppenheimer, J. M. (1964). Induction of embryonic primordia by implantation of organizers from a different species. In B. H. Willier & J. M. Oppenheimer (Eds.), *Foundations of experimental embryology* (pp. 145–184). New Jersey, USA: Prentice Hall, Inc.
Oppenheimer, J. M. (1970). Cells and organizers. *American Zoologist*, *10*, 75–88.
Oppenheimer, J. M. (1991). Curt Herbst's contributions to the concept of embryonic induction. In S. F. Gilbert (Ed.), *A conceptual history of modern embryology* (pp. 63–89). New York: Plenum Press.
Placzek, M., Tessier-Lavigne, M., Jessell, T., & Dodd, J. (1990a). Orientation of commissural axons in vitro in response to a floor plate-derived chemoattractant. *Development*, *110*, 19–30.
Placzek, M., Tessier-Lavigne, M., Yamada, T., Jessell, T., & Dodd, J. (1990b). Mesodermal control of neural cell identity: Floor plate induction by the notochord. *Science*, *250*, 985–988.
Pourquie, O., Coltey, M., Teillet, M. A., Ordahl, C., & Le Douarin, N. M. (1993). Control of dorsoventral patterning of somitic derivatives by notochord and floor plate. *Proceedings of the National Academy of Sciences of the United States of America*, *90*, 5242–5246.
Pourquie, O., Fan, C. M., Coltey, M., Hirsinger, E., Watanabe, Y., Breant, C., et al. (1996). Lateral and axial signals involved in avian somite patterning: A role for BMP4. *Cell*, *84*, 461–471.
Pownall, M. E., Strunk, K. E., & Emerson, C. P., Jr. (1996). Notochord signals control the transcriptional cascade of myogenic bHLH genes in somites of quail embryos. *Development*, *122*, 1475–1488.
Riddle, R. D., Johnson, R. L., Laufer, E., & Tabin, C. (1993). Sonic hedgehog mediates the polarizing activity of the ZPA. *Cell*, *75*, 1401–1416.
Saha, M. (1991). Spemann seen through a lens. In S. F. Glibert (Ed.), *A conceptual history of modern embryology* (pp. 91–108). New York: Plenum Press.
Sander, K., & Faessler, P. E. (2001). Introducing the Spemann-Mangold organizer: Experiments and insights that generated a key concept in developmental biology. *International Journal of Developmental Biology*, *45*, 1–11.
Sanz-Ezquerro, J. J., & Tickle, C. (2003). Fgf signaling controls the number of phalanges and tip formation in developing digits. *Current Biology*, *13*, 1830–1836.
Saunders, J. W., Jr., & Gasseling, M. T. (1968). Ectoderm–mesenchymal interaction in the origins of wing symmetry. In R. Fleischmajer & R. E. Billingham (Eds.), *Epithelial–mesenchymal interactions* (pp. 78–97). Baltimore: Williams Wilkins.
Scholpp, S., Wolf, O., Brand, M., & Lumsden, A. (2006). Hedgehog signalling from the zona limitans intrathalamica orchestrates patterning of the zebrafish diencephalon. *Development*, *133*, 855–864.
Shimamura, K., & Rubenstein, J. L. (1997). Inductive interactions direct early regionalization of the mouse forebrain. *Development*, *124*, 2709–2718.

Spemann, H. (1901). Über Korrelationen in der Enlwickelung des Auges. *Verhandlungen der Anatomischen Gesellschaft, 15*, 61–79.
Spemann, H. (1908). Neue Versuche zur Entwicklung des Wirbeltierauges. *Verhandlungen der Deutschen Zoologischen Gesellschaft, 18*, 101–110.
Spemann, H. (1921). Die Erzeugung tierischer Chimären durch heteroplastiche embryonale Transplantation zwischen *Triton cristatus* und *taeniatus*. *Archiv für Entwicklungsmechanik der Organismen, 48*, 533–570.
Spemann, H. (1929). Über den Anteil von Organisator u. Wirtskeim am Zustandekommen der Induktion. *Naturwissenschaften, 17*, 287–289.
Spemann, H. (1931a). Das Verhalten von Organisatoren nach Zerstörung ihrer Struktur. *Verhandlungen der Deutschen Zoologischen Gesellschaft, 1931*, 129–132.
Spemann, H. (1931b). Über den Anteil von Implantat und Wirskeim an der Orientierung und Beschaffenheit der induzierten Embryonalanlage. *Wilhelm Roux' Archiv für Entwicklungsmechanik der Organismen, 123*, 390–517.
Spemann, H., & Mangold, H. (1924). Über Induktion von Embryonalanlagen durch Implantation artfremder Organisatoren. *Wilhelm Roux' Archiv für Entwicklungsmechanik der Organismen, 100*, 599–638.
Stern, C. D. (2000). Conrad H. Waddington's contributions to avian and mammalian development, 1930-1940. *International Journal of Developmental Biology, 44*, 15–22.
Stern, C. D. (2001). Initial patterning of the central nervous system: How many organizers? *Nature Reviews. Neuroscience, 2*, 92–98.
te Welscher, P., Zuniga, A., Kuijper, S., Drenth, T., Goedemans, H. J., Meijlink, F., et al. (2002). Progression of vertebrate limb development through SHH-mediated counteraction of GLI3. *Science, 298*, 827–830.
Tickle, C., Shellswell, G., Crawley, A., & Wolpert, L. (1976). Positional signalling by mouse limb polarising region in the chick wing bud. *Nature, 259*, 396–397.
Todt, W. L., & Fallon, J. F. (1987). Posterior apical ectodermal ridge removal in the chick wing bud triggers a series of events resulting in defective anterior pattern formation. *Development, 101*, 501–515.
van Straaten, H. W., Hekking, J. W., Thors, F., Wiertz-Hoessels, E. L., & Drukker, J. (1985). Induction of an additional floor plate in the neural tube. *Acta Morphologica Neerlando-Scandinavica, 23*, 91–97.
van Straaten, H. W., Hekking, J. W., Wiertz-Hoessels, E. J., Thors, F., & Drukker, J. (1988). Effect of the notochord on the differentiation of a floor plate area in the neural tube of the chick embryo. *Anatomy and Embryology (Berlin), 177*, 317–324.
Vasiliauskas, D., Hancock, S., & Stern, C. D. (1999). SWiP-1: Novel SOCS box containing WD-protein regulated by signalling centres and by Shh during development. *Mechanisms of Development, 82*, 79–94.
Vieira, C., & Martinez, S. (2006). Sonic hedgehog from the basal plate and the zona limitans intrathalamica exhibits differential activity on diencephalic molecular regionalization and nuclear structure. *Neuroscience, 143*, 129–140.
Waddington, C. H. (1930). Developmental mechanics of chick and duck embryos. *Nature, 125*, 924–925.
Waddington, C. H. (1932). Experiments on the development of chick and duck embryos cultivated *in vitro*. *Philosophical Transactions of the Royal Society of London (Biology), 221*, 179–230.
Waddington, C. H. (1933). Induction by the primitive streak and its derivatives in the chick. *Journal of Experimental Biology, 10*, 38–46.
Waddington, C. H. (1934). Experiments on embryonic induction. III. A note on inductions by the chick primitive streak transplanted to the rabbit embryo. *Journal of Experimental Biology, 11*, 224–226.

Waddington, C. H. (1937). Experiments on determination in the rabbit embryo. *Archives of Biology*, *48*, 273–290.
Waddington, C. H. (1940). *Organisers and genes*. London: Cambridge University Press.
Waddington, C. H., & Schmidt, G. A. (1933). Induction by heteroplastic grafts of the primitive streak in birds. *Wilhelm Roux' Archiv für Entwicklungsmechanik der Organismen*, *128*, 522–563.
Wagner, M., Thaller, C., Jessell, T., & Eichele, G. (1990). Polarizing activity and retinoid synthesis in the floor plate of the neural tube. *Nature*, *345*, 819–822.
Wilson, S. W., & Houart, C. (2004). Early steps in the development of the forebrain. *Developmental Cell*, *6*, 167–181.
Yamada, T., Placzek, M., Tanaka, H., Dodd, J., & Jessell, T. M. (1991). Control of cell pattern in the developing nervous system: Polarizing activity of the floor plate and notochord. *Cell*, *64*, 635–647.
Zhu, L., Belo, J. A., De Robertis, E. M., & Stern, C. D. (1999). Goosecoid regulates the neural inducing strength of the mouse node. *Developmental Biology*, *216*, 276–281.
Zwilling, E. (1956). Interaction between limb bud ectoderm and mesoderm in the chick embryo. II. Experimental limb duplication. *Journal of Experimental Zoology*, *132*, 173–188.

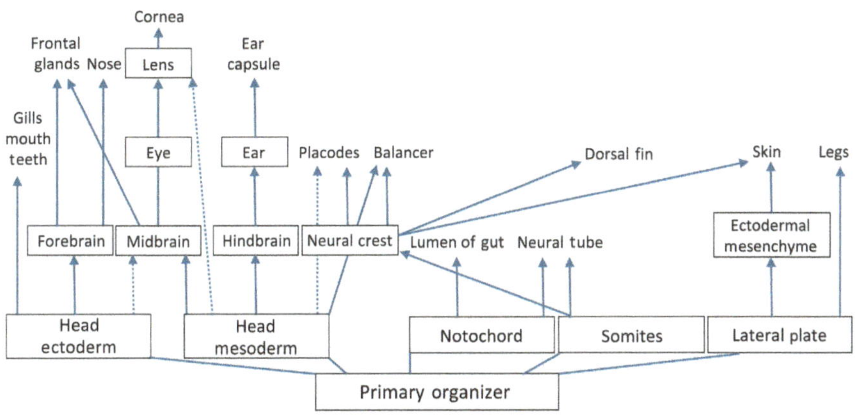

Claire Anderson and Claudio D. Stern, Figure 1 A now outdated chain reaction diagram of amphibian development based on sequential induction of tissues, illustrating the idea of secondary and tertiary organizers; after Waddington (1940), in turn based on Holtfreter (1938). Waddington's legend reads: "The main organizer regions are enclosed in rectangles. The lines join organs which are developed one from the other, while the arrows show inducing relations, the dotted ones being of less importance." The significance of sequential inductions in development was described thus by Holtfreter: "… not only the entire neural system, but also all the other tissue derivatives of the ectoderm, are likewise products of induction … I, therefore, maintain that embryonic induction is the most important device by which differentiation comes about in amphibian development" *Holtfreter (1991).*

CHAPTER TWENTY-SEVEN

The Soft Touch: Low-Affinity Transcription Factor Binding Sites in Development and Evolution

Justin Crocker[1], Ella Preger-Ben Noon[1], David L. Stern[2]
Janelia Research Campus, Howard Hughes Medical Institute, Ashburn, Virginia, USA
[2]Corresponding author: e-mail address: sternd@janelia.hhmi.org

Contents

Abstract

Transcription factor proteins regulate gene expression by binding to specific DNA regions. Most studies of transcription factor binding sites have focused on the highest affinity sites for each factor. There is abundant evidence, however, that binding sites with a range of affinities, including very low affinities, are critical to gene regulation. Here, we present the theoretical and experimental evidence for the importance of low-affinity sites in gene regulation and development. We also discuss the implications of the widespread use of low-affinity sites in eukaryotic genomes for robustness, precision, specificity, and evolution of gene regulation.

Get low, get low. Get low, get low. Get low, get low.

Lil Jon & The East Side Boyz (2002)

[1] These authors contributed equally to this work.

Current Topics in Developmental Biology, Volume 117
ISSN 0070-2153
http://dx.doi.org/10.1016/bs.ctdb.2015.11.018

1. INTRODUCTION

Gene expression patterns in development are determined by the binding of transcription factors to accessible DNA near genes. Transcription factor proteins are thought to regulate genes primarily by binding to high-affinity, "cognate," binding sites that confer high specificity for the protein–DNA interaction. This widespread assumption is readily apparent in genomic studies that attempt to predict transcription factor binding sites; these studies almost always use cognate sites or position weight matrices and employ high thresholds for calling binding sites. However, both theoretical and empirical studies over the past several decades challenge the assumption that high-affinity binding sites dominate transcriptional regulation. Here, we briefly review the theory and empirical evidence for the role of low-affinity binding sites in regulating gene expression. We further discuss theoretical ideas for the importance of low-affinity binding sites and some evolutionary implications of these observations.

2. THEORY FOR THE ROLE OF LOW-AFFINITY BINDING SITES

The binding of transcription factors to DNA depends primarily on hydrogen bonding between specific amino acid residues in the protein and specific bases in the DNA sequence. For each transcription factor, there will thus be one DNA sequence with the lowest (most favorable) free energy of binding and a large collection of DNA sequences with less favorable binding energies. The importance of considering a spectrum of binding site affinities, relative to the highest affinity binding sites, was put into a theoretical framework by Berg and von Hippel (1987). Since each base-pair position contributes mostly additively to the overall binding free energy of a binding site, there will be many possible replacements of individual nucleotide sites that result in similar reductions in the relative affinity of binding (Berg & von Hippel, 1987). While there are few binding sites that confer the highest affinities, there are a vast number of binding sites that can confer equivalent lower levels of affinity. Experimental measures of transcription factor–DNA affinity for collections of random sequences support this theoretical conclusion (Badis et al., 2009; Slattery et al., 2011).

For a variety of reasons, enhancers may not encode the highest affinity binding sites. In fact, the highest affinity binding sites are rarely used in nature. Instead, individual transcription factors tend to bind to a variety of binding sites exhibiting a distribution of affinities.

3. EARLY EVIDENCE FOR THE ROLE OF LOW-AFFINITY BINDING SITES

Foundational work on the phage λ operator and the yeast *Gal1* promoter identified binding sites with a range of affinities that are crucial for gene regulation (Giniger & Ptashne, 1988; Hochschild, Douhan, & Ptashne, 1986). In the case of phage λ, three binding sites (O_R1–O_R3) for λ repressor and λ cro comprise an operator for the *cl* gene, which encodes the repressor protein, and for the *cro* gene. Although both λ repressor and λ Cro can bind to these DNA sites, the specificities of the two proteins differ; λ repressor binds more tightly to O_R1 than to O_R3, allowing the transcription of *cl* and the repression of *cro*. Cro, on the other hand, binds more strongly to O_R3 than to O_R1, resulting in the repression of *cl* (Hochschild et al., 1986). This difference allows the genetic switch between lysogenic and lytic phage growth to function properly. Similarly, the yeast *Gal1* promoter contains four low-affinity binding sites for the activator protein Gal4. Binding is cooperative for at least two of these sites and promoter activity is roughly proportional to the number of Gal4 molecules bound (Giniger & Ptashne, 1988).

In animal systems, the importance of low-affinity binding sites was demonstrated in the SV40 enhancer, the first DNA sequence found to "enhance" the expression of a transgene in HeLa cells (Banerji, Rusconi, & Schaffner, 1981). Multiple regions within the 72 base pair SV40 enhancer contribute to cell-type specific expression (Davidson et al., 1986). Activation of these regions is achieved by at least two factors, Sp1 and AP1, which bind to multiple sites of varying affinity (Lee, Haslinger, Karin, & Tjian, 1987). Indeed, a hallmark of enhancers is that they are bound by multiple transcription factors, often containing clusters of "degenerate" binding sites for the same factor—that is, sites that differ from the cognate site.

Therefore, early evidence demonstrated that low-affinity binding sites are biologically important.

4. FUNCTIONAL EVIDENCE FOR THE IMPORTANCE OF LOW-AFFINITY BINDING SITES FROM DEVELOPMENT

In developmental systems, transcriptional enhancers encode information specifying the locations, levels, and timing of gene expression (Arnone & Davidson, 1997). This spatio-temporal information can be modulated through variation in binding site affinities. For example, a low-affinity Pax6-binding site in an enhancer of the mouse *Neurogenenin2* gene is required to restrict expression to domains of high Pax6 expression. Replacing this conserved, low-affinity site with a high-affinity site causes ectopic expression (Scardigli, 2003) (Fig. 1A). Similarly, low-affinity sites facilitate the restricted expression of *Drosophila* genes responsive to the Dorsal protein gradient (Jiang & Levine, 1993). Changing these low-affinity sites to high-affinity sites generates ectopic expression and higher levels of expression in the normal expression domain.

Variation in binding site affinity also plays critical roles in determining the timing of gene expression. For example, different target genes of the *Caenorhabditis elegans* forkhead box A transcription factor PHA-4-containing binding sites of different affinities, which are important to regulate the timing of gene expression (Gaudet & Mango, 2002). Manipulation of PHA-4 binding sites revealed that high-affinity sites drive target gene expression early in development, when PHA-4 levels are low, and low-affinity sites drive expression later, after PHA-4 levels rise (Fig. 1B).

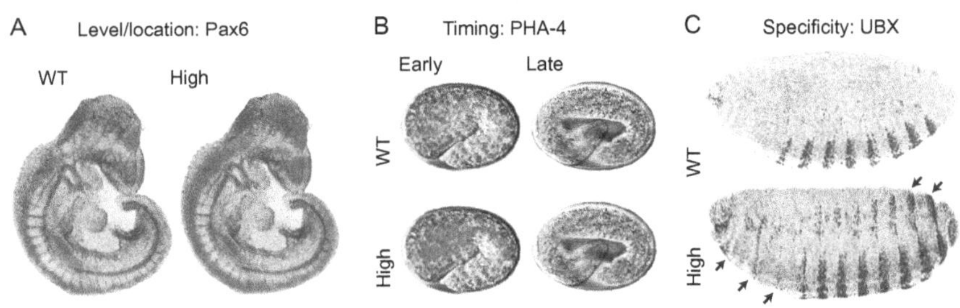

Figure 1 Functional evidence for the importance of low-affinity binding sites from development. (A) Replacing low-affinity PAX6 sites with a high-affinity site causes ectopic expression in a *Neurogenenin2* enhancer (Scardigli, 2003). (B) Replacing low-affinity PHA-4 sites with a high-affinity sites leads to precocious expression (Gaudet & Mango, 2002). (C) Replacing a low-affinity Ultrabithorax site with a high-affinity site in a *Drosophila shavenbaby* enhancer leads to ectopic expression in anterior regions (arrows) (Crocker et al., 2015). (A color version of this figure is located at the end of this chapter.)

Similarly, a mouse enhancer of the *Pax6* gene contains two evolutionarily conserved binding sites with relatively low affinity for the transcription factor Prep1 (Rowan et al., 2010). Replacement of these sites with higher affinity sites alters the timing of transcriptional activation and causes ectopic expression.

Finally, low-affinity binding sites may play critical roles in facilitating precise and specific interactions with transcription factors. For example, in enhancers of several Hedgehog-responsive *Drosophila* genes, low-affinity sites for Cubitus interruptus (a member of the Gli family of transcription factors) mediate activation of expression in response to high and low levels of Hedgehog signaling. Increasing the affinity of these Cubitus interruptus sites switched the enhancer output from activation to repression at low levels of Hedgehog (Ramos & Barolo, 2013). In addition, we recently showed that low-affinity binding sites for the Hox transcription factor Ultrabithorax in *Drosophila shavenbaby* enhancers confer specificity for Ultrabithorax binding *in vivo* (Crocker et al., 2015). Replacing these sites with high-affinity sites resulted in ectopic expression in embryonic tissues that do not express Ultrabithorax (Fig. 1C, arrow), most likely through binding of other homeodomain-containing transcription factors. Therefore, low-affinity binding sites are required to specify precise gene expression patterns during development.

5. GENOMIC EVIDENCE FOR THE IMPORTANCE OF LOW-AFFINITY BINDING SITES

Genome-wide assays of transcription factor occupancy, such as ChIP-on-chip and ChIP-seq, show a continuum of binding strengths. Unfortunately, widely used analysis techniques ignore the quantitative information that is provided by these assays and transform the data into a binary readout, thus treating genomic regions as either bound or not bound. Arbitrarily setting thresholds for peak calling and for motif detection in called peaks—and even calling "peaks" rather than considering the full distribution of apparent binding—may fail to identify weak binding with potential functionality. Yet, the few genomic studies to date that used quantitative approaches to analyze ChIP data generated compelling genomic evidence both that transcription factors bind to low-affinity binding sites *in vivo* and that this binding might influence gene transcription (Li et al., 2008; Tanay, 2006; Wang, Malecka, Trøen, & Delabie, 2015).

In a groundbreaking study, Tanay showed that low-affinity interactions are abundant in the yeast genome and comprise a substantial fraction of the interactions between transcription factors and DNA (Tanay, 2006). Evolutionary analysis revealed that the predicted binding energies of these interactions are conserved in many promoters, further suggesting that low-affinity binding sites are functionally important. Indeed, promoters with low-affinity sites drive significant expression at levels that correlate with the strength of binding. This study demonstrated that interactions with a range of affinities should be considered, rather than focusing on interactions that pass an arbitrary threshold.

In a recent study, Wang et al. developed a biophysical model that allows the detection of both high- and low-affinity sites in ChIP-seq data from human cell lines. Using this model, they showed that the transcription factor SPIB predominantly binds and regulates genes that contain *both* high- and low-affinity sites in their regulatory regions (Wang et al., 2015).

In contrast, a study of transcription in the fly embryo blastoderm suggests that, at least in this setting, genomic regions displaying low levels of transcription factor occupancy may not be critically important (Li et al., 2008). Li et al. found that individual transcription factors bound to thousands of genomic targets with a range of occupancy levels. They found that highly and moderately bound regions were preferably found in the regulatory regions of blastoderm transcribed genes, whereas regions with low occupancy were found preferentially distant from these genes and in coding regions. The poorly bound regions were less well conserved than moderately and highly bound regions. Transgenic reporter gene assays showed that the majority of weakly bound regions drove no expression in the early embryo (Fisher et al., 2012). These observations suggest that weak binding is nonfunctional in this context. Nonetheless, functional enhancers contain sites with a range of binding affinities (Berger et al., 2008; Li et al., 2008) and it is not yet clear how low-affinity sites contribute to function of these regions at a genome-wide scale.

Therefore, genomic studies provide evidence that binding of transcription factors to low-affinity binding sites is functionally important *in vivo*, although there may be some developmental settings in which stronger affinity sites play a dominant role in regulating transcription.

One caveat regarding the genome-wide studies is that high binding can result either from a small number of sites with high affinity or a large number of clustered sites with low affinity, or any combination in between. Thus, while the studies discussed above draw conclusions about binding site affinity from levels of occupancy, it is not clear that this conclusion is warranted for regions that show high binding. Wider use of biophysical models in binding site prediction, such as the one developed by Wang et al. (2015), may help resolve this question.

6. POTENTIAL REASONS FOR ABUNDANCE OF LOW-AFFINITY TRANSCRIPTION FACTOR BINDING SITES

There are several nonexclusive hypotheses for the role of low-affinity binding sites. Most of these models are currently untested, mainly because few studies focus on the contribution of low-affinity sites.

First, it is possible that enhancer specificity is increased by decreasing binding site affinities while simultaneously increasing the diversity of factors that bind to a regulatory region (Fig. 2A). Consistent with this hypothesis, prokaryotes and eukaryotes display strikingly different modes of gene regulation (Wunderlich & Mirny, 2009). The length of transcription factor binding motifs in prokaryotes—and the information content within these sites—is, on average, nearly double that of eukaryotes. In eukaryotes, the use of multiple sites for multiple factors in a single enhancer compensates for the relatively lower affinity of individual transcription factor binding sites (Wunderlich & Mirny, 2009). The use of multiple, relatively low-affinity transcription factors in eukaryotes may have allowed the evolution of transcriptional mechanisms that can support the regulatory logic required to build multicellular eukaryotes.

Second, lower affinity binding sites can be exploited to construct "digital" ON/OFF gene regulation through cooperative binding across multiple sites (Giniger & Ptashne, 1988) (Fig. 2B). For example, in *Drosophila melanogaster*, clustered binding sites at an enhancer for the *hunchback* gene mediate a sharp spatial boundary of expression in response to a gradient of Bicoid expression (Burz, Rivera-Pomar, Jäckle, & Hanes, 1998; Lebrecht et al., 2005). Low-affinity binding sites are also important for heterotypic cooperative binding. Replacing a conserved low-affinity Pax6 site with a high-affinity site prevents formation of an activation-competent Sox2–Pax6 complex (Kamachi, Uchikawa, Tanouchi, Sekido, & Kondoh, 2001). Alternatively, multiple low-affinity binding sites could mediate noncooperative, graded responses to variable inputs (Fig. 2C). While this has not been explicitly demonstrated yet for low-affinity sites, a cluster of high-affinity binding sites for the NF-κB transcription factor in the human *NFKBIA* promoter region confers quantitative changes in transcription in

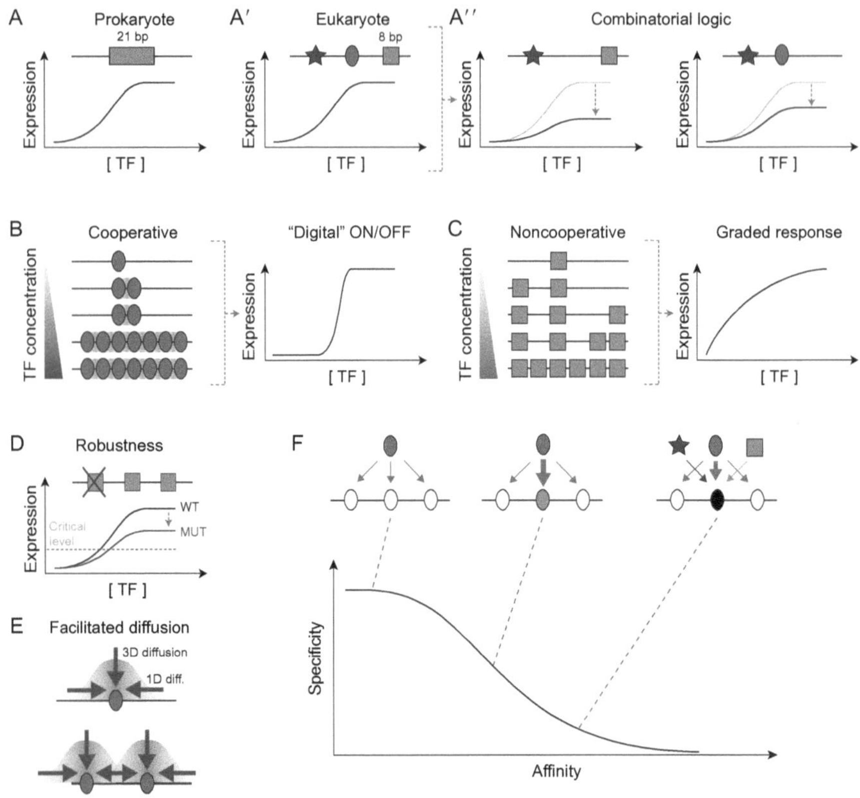

Figure 2 Potential reasons for abundance of low-affinity transcription factor binding sites. (A) The size of transcription factor binding motifs in prokaryotes is, on average, nearly double that of eukaryotes. (A′) In eukaryotes, the low affinity of individual transcription factors is compensated for by the use of sites for multiple factors in a single enhancer. (A″) The use of multiple binding sites facilitates combinatorial logic and refined patterns of expression. (B) Clusters of low-affinity sites can be organized so as to mediate cooperative interactions between the transcription factors that bind to enhancers, leading to ON/OFF responses to variable inputs. (C) Clusters of low-affinity binding sites can act in a noncooperative manner to confer graded responses to variable inputs. (D) Clusters of low-affinity binding sites may serve to increase robustness of transcription enhancers, such that the loss of one binding site still allows the enhancer to provide a critical level of expression from the remaining sites. (E) Clusters of binding sites can decrease the search time for transcription factor binding sites through local one-dimensional diffusion between binding sites—providing a larger binding site "target." (F) Clustering of low-affinity homotypic binding sites can result from a fundamental tradeoff between affinity and specificity. Low-affinity binding sites provide specificity for unique Hox factors, whereas high-affinity binding sites are bound by multiple *Hox* genes (Crocker et al., 2015). (A color version of this figure is located at the end of this chapter.)

response to varied levels of NF-κB (Giorgetti et al., 2010). This architecture allows the *NFKBIA* gene to respond to graded levels of environmental variables. Similar mechanisms act during development as well to determine transcription levels in response to specific transcription factor concentrations (Driever, Thoma, & Nüsslein-Volhard, 1989; Gaudet & Mango, 2002).

Third, clusters of low-affinity binding sites may serve to increase robustness of transcriptional enhancers (Fig. 2D). This was suggested by theoretical studies (Giorgetti et al., 2010; Spivakov, 2014) and hinted at by studies of evolutionarily chimeric enhancers (Ludwig, Bergman, Patel, & Kreitman, 2000). We recently demonstrated that clusters of low-affinity Ultrabithorax binding sites increase robustness to environmental and genetic perturbations for *shavenbaby* enhancers in *Drosophila* (Crocker et al., 2015).

Fourth, when low-affinity binding sites cluster near a high-affinity binding site they may, at least theoretically, increase the local concentration of transcription factors to maximize the probability that some transcription factors occupy the cognate site (Fig. 2E) (Zhang et al., 2006). Clusters of binding sites may decrease the search time for transcription factor binding sites, because localization of transcription factors to binding sites may be faster through one-dimensional diffusion along DNA than by three-dimensional diffusion through the nucleoplasm (Elf, Li, & Xie, 2007; Kabata et al., 1993; Leith et al., 2012; Ruusala & Crothers, 1992).

Finally, the use of low-affinity binding sites can result from a fundamental tradeoff between affinity and specificity for some transcription factor families (Fig. 2F) (Crocker et al., 2015). High-affinity Hox binding sites can be bound by multiple *Hox* genes and potentially by many homeodomain-containing genes (Berger et al., 2008; Noyes et al., 2008). In contrast, low-affinity binding sites provide specificity for unique Hox factors (Crocker et al., 2015). Similarly, the *Drosophila* homeodomain transcription factor Slouch regulates several enhancers through Slouch specific sequences. Swapping these sequences for higher affinity sites that can bind multiple homeodomain family members alters normal expression (Busser et al., 2012). Binding through low-affinity sites might be a general feature of members of the homeodomain family. In fact, several studies have found that homeodomain transcription factors bind *in vivo* to low-affinity sites (Berger et al., 2008; Busser et al., 2012; Jaeger et al., 2010; Tanay, 2006). For example, the majority of sequences bound *in vivo* by Tcf1/Hnf1 do not contain the best *in vitro* binding sequences, suggesting that these transcription factors bind to low-affinity sites (Berger et al., 2008). A similar tradeoff between affinity and specificity may apply to other paralogous families of transcription factors that share similar DNA-binding motifs (Gordân et al., 2013; Park & Hwang, 2012). However, it is unlikely that this affinity-specificity tradeoff model will apply to all transcription factors. Many transcription factors bind to high-affinity sites *in vivo*, including some homeodomain-containing transcription factors (Berger et al., 2008). For example, many target genes of the *Drosophila* transcription factor Dorsal bind to well defined, high-affinity binding sites (Markstein, Markstein, Markstein, & Levine, 2002). Similarly, many human transcription factors preferentially associate *in vivo* with high-affinity, cognate binding sites (Bernstein et al., 2012; Kheradpour & Kellis, 2014). Some transcription factor families, such as zinc-finger containing transcription factors may bind to sequences with sufficiently diverse DNA sequences that low affinity is not required to confer specificity (Bulyk, Huang, Choo, & Church, 2001). However, the possibility that some classes of DNA sequences are disfavored evolutionarily because of competitive binding either within or between classes of transcription factors may shape the distribution of transcription factor binding sites used in genomes.

7. CLUSTERS OF LOW-AFFINITY BINDING SITES CAN LEAD TO RAPID EVOLUTION OF ENHANCERS

Evolution may help explain the widespread use of low-affinity sites in enhancers. Low-affinity sites may be a common feature of most enhancers because low-affinity sites can evolve through fewer mutational steps than the less common high-affinity sites (Ramos & Barolo, 2013). Thus, evolutionary models predict that enhancers containing clusters of low-affinity binding sites will be abundant (He, Duque, & Sinha, 2012) (Fig. 3). Clusters of

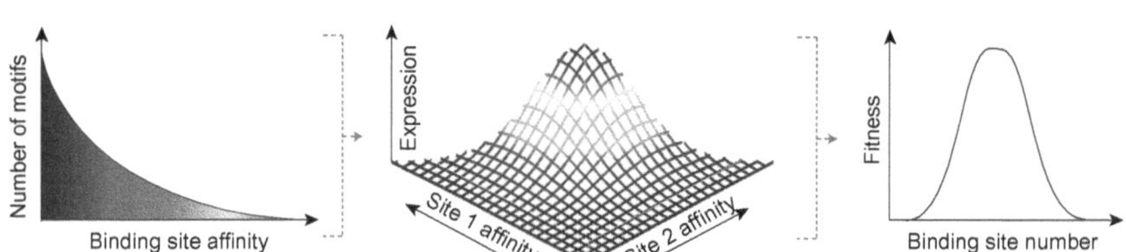

Figure 3 More DNA sequences can confer weak affinity than can confer strong affinity. Considering a two binding site model, varying the affinity of each individual binding site leads to a range of possible expression levels from the enhancer. Because more DNA sequences can confer weak affinity than can confer strong affinity, and because clusters of weak sites can provide high expression levels, evolution will tend to evolve enhancers containing multiple low-affinity binding sites. *Adapted from He et al. (2012).* (A color version of this figure is located at the end of this chapter.)

low-affinity binding sites can thus be considered the evolutionary null expectation for enhancer architecture.

The widespread use of low-affinity sites in enhancers may, in turn, explain the rapid evolutionary turnover in binding sites that is observed for most enhancers. Because low-affinity sites are abundant in sequence space, one low-affinity site can easily be replaced by a different low-affinity site of equal function (He et al., 2012). Rapid binding site turnover has been demonstrated for multiple enhancers (Bradley et al., 2010; Cotney et al., 2013; Crocker, Potter, & Erives, 2010; Ludwig et al., 2000; Miller et al., 2007; Moses et al., 2006; Stefflova et al., 2013) and theoretical investigation suggests that this turnover allows the rapid evolution of novel transcription factor binding sites (Lusk & Eisen, 2010; MacArthur & Brookfield, 2004; Stone & Wray, 2001). Thus, evolution may drive the widespread use of low-affinity sites and these sites may contribute to the most common patterns of enhancer evolution.

8. CONCLUSIONS

Eukaryotic genomes contain many low-affinity, poorly conserved binding sites that are likely to play an important and general role in *cis*-regulatory control and evolution (Jaeger et al., 2010; Ramos & Barolo, 2013; Tanay, 2006). These low-affinity sites will sometimes be clustered to increase enhancer robustness or for other mechanistic reasons, and clusters of low-affinity sites are likely to lead to rapid evolutionary turnover of binding sites. For transcription factor families with similar DNA-binding domains, the use of low-affinity sites may result from a fundamental tradeoff between affinity and specificity.

The prevalence of low-affinity, poorly conserved binding sites (Dermitzakis & Clark, 2002; Doniger, Huh, & Fay, 2005) complicates the prediction of enhancer regions using bioinformatics tools that focus on high-affinity and/or highly conserved binding sites. Models of *Drosophila* segmentation gene regulation that include clusters of homotypic weak binding sites have higher predictive power and are more robust to simulated mutation than models without these sites (Segal, Raveh-Sadka, Schroeder, Unnerstall, & Gaul, 2008). Therefore, considering a spectrum of binding affinities can, in some cases, improve bioinformatic predictions of enhancer function.

Several lines of research are required to understand how low-affinity interactions are choreographed on transcriptional enhancers. First, we require a complete description of the information required for an enhancer to regulate expression during development. Currently, we have a complete understanding of the factors required for expression for only a few enhancers (Merika & Thanos, 2001). Notably, these well-characterized enhancers regulate environmental responses, which may have fundamentally different regulatory requirements than developmental enhancers (Giorgetti et al., 2010). Second, we need to understand the dynamics of binding to low-affinity sites within a cellular context. Single molecule experiments are currently providing the resolution required to understand weak interactions *in vivo* (Liu et al., 2015). In principle, these assays could be used to track single transcription factors binding to individual enhancers. Finally, in order to fully understand the function of low-affinity binding sites, it would be useful to understand how binding to these sites occurs in the context of the 3D architecture of the genome—including interactions with a single promoter, with multiple promoters, and with additional enhancers. A prime example of these interactions includes the recent comparison of strong versus weak "shadow" enhancers, where it was shown that weak enhancers function additively, whereas strong enhancers behave subadditively due to competition with the target promoter (Bothma et al., 2015).

Genomes contain DNA sequences displaying a range of affinities for transcription factors, usually including both the highest affinity site and a wide spectrum of lower affinity sites. A comprehensive understanding of transcriptional regulation will account for the full affinity range of functional binding sites.

ACKNOWLEDGMENTS

We thank Albert Tsai and Jessica Cande for critical comments on drafts of this chapter. E.P.-B.N. was supported by postdoctoral fellowships from the Human Frontier Science Program and EMBO.

REFERENCES

Arnone, M. I., & Davidson, E. H. (1997). The hardwiring of development: Organization and function of genomic regulatory systems. *Development*, *124*, 1851–1864.

Badis, G., Berger, M. F., Philippakis, A. A., Talukder, S., Gehrke, A. R., Jaeger, S. A., et al. (2009). Diversity and complexity in DNA recognition by transcription factors. *Science*, *324*, 1720–1723.

Banerji, J., Rusconi, S., & Schaffner, W. (1981). Expression of a beta-globin gene is enhanced by remote SV40 DNA sequences. *Cell*, *27*, 299–308.

Berg, O. G., & von Hippel, P. H. (1987). Selection of DNA binding sites by regulatory proteins. Statistical–mechanical theory and application to operators and promoters. *Journal of Molecular Biology*, *193*, 723–750.

Berger, M. F., Badis, G., Gehrke, A. R., Talukder, S., Philippakis, A. A., Peña-Castillo, L., et al. (2008). Variation in homeodomain DNA binding revealed by high-resolution analysis of sequence preferences. *Cell, 133*, 1266–1276.

Bernstein, B. E., Birney, E., Dunham, I., Green, E. D., Gunter, C., & Snyder, M. (2012). An integrated encyclopedia of DNA elements in the human genome. *Nature, 489*, 57–74.

Bothma, J. P., Garcia, H. G., Ng, S., Perry, M. W., Gregor, T., & Levine, M. (2015). Enhancer additivity and non-additivity are determined by enhancer strength in the Drosophila embryo. *eLife, 4*, e07956.

Bradley, R. K., Li, X.-Y., Trapnell, C., Davidson, S., Pachter, L., Chu, H. C., et al. (2010). Binding site turnover produces pervasive quantitative changes in transcription factor binding between closely related Drosophila species. *PLoS Biology, 8*, e1000343.

Bulyk, M. L., Huang, X., Choo, Y., & Church, G. M. (2001). Exploring the DNA-binding specificities of zinc fingers with DNA microarrays. *Proceedings of the National Academy of Sciences of the United States of America, 98*, 7158–7163.

Burz, D. S., Rivera-Pomar, R., Jäckle, H., & Hanes, S. D. (1998). Cooperative DNA-binding by Bicoid provides a mechanism for threshold-dependent gene activation in the Drosophila embryo. *The EMBO Journal, 17*, 5998–6009.

Busser, B. W., Shokri, L., Jaeger, S. A., Gisselbrecht, S. S., Singhania, A., Berger, M. F., et al. (2012). Molecular mechanism underlying the regulatory specificity of a Drosophila homeodomain protein that specifies myoblast identity. *Development, 139*, 1164–1174.

Cotney, J., Leng, J., Yin, J., Reilly, S. K., DeMare, L. E., Emera, D., et al. (2013). The evolution of lineage-specific regulatory activities in the human embryonic limb. *Cell, 154*, 185–196.

Crocker, J., Abe, N., Rinaldi, L., McGregor, A. P., Frankel, N., Wang, S., et al. (2015). Low affinity binding site clusters confer hox specificity and regulatory robustness. *Cell, 160*, 191–203.

Crocker, J., Potter, N., & Erives, A. (2010). Dynamic evolution of precise regulatory encodings creates the clustered site signature of enhancers. *Nature Communications, 1*, 99.

Davidson, I., Fromental, C., Augereau, P., Wildeman, A., Zenke, M., & Chambon, P. (1986). Cell-type specific protein binding to the enhancer of simian virus 40 in nuclear extracts. *Nature, 323*, 544–548.

Dermitzakis, E. T., & Clark, A. G. (2002). Evolution of transcription factor binding sites in Mammalian gene regulatory regions: Conservation and turnover. *Molecular Biology and Evolution, 19*, 1114–1121.

Doniger, S. W., Huh, J., & Fay, J. C. (2005). Identification of functional transcription factor binding sites using closely related Saccharomyces species. *Genome Research, 15*, 701–709.

Driever, W., Thoma, G., & Nüsslein-Volhard, C. (1989). Determination of spatial domains of zygotic gene expression in the Drosophila embryo by the affinity of binding sites for the Bicoid morphogen. *Nature, 340*, 363–367.

Elf, J., Li, G.-W., & Xie, X. S. (2007). Probing transcription factor dynamics at the single-molecule level in a living cell. *Science, 316*, 1191–1194.

Fisher, W. W., Li, J. J., Hammonds, A. S., Brown, J. B., Pfeiffer, B. D., Weiszmann, R., et al. (2012). DNA regions bound at low occupancy by transcription factors do not drive patterned reporter gene expression in Drosophila. *Proceedings of the National Academy of Sciences of the United States of America, 109*, 21330–21335.

Gaudet, J., & Mango, S. E. (2002). Regulation of organogenesis by the Caenorhabditis elegans FoxA protein PHA-4. *Science, 295*, 821–825.

Giniger, E., & Ptashne, M. (1988). Cooperative DNA binding of the yeast transcriptional activator GAL4. *Proceedings of the National Academy of Sciences of the United States of America, 85*, 382–386.

Giorgetti, L., Siggers, T., Tiana, G., Caprara, G., Notarbartolo, S., Corona, T., et al. (2010). Noncooperative interactions between transcription factors and clustered DNA binding

sites enable graded transcriptional responses to environmental inputs. *Molecular Cell, 37*, 418–428.
Gordân, R., Shen, N., Dror, I., Zhou, T., Horton, J., Rohs, R., et al. (2013). Genomic regions flanking E-box binding sites influence DNA binding specificity of bHLH transcription factors through DNA shape. *Cell Reports, 3*, 1093–1104.
He, X., Duque, T. S. P. C., & Sinha, S. (2012). Evolutionary origins of transcription factor binding site clusters. *Molecular Biology and Evolution, 29*, 1059–1070.
Hochschild, A., Douhan, J., & Ptashne, M. (1986). How λ repressor and λ Cro distinguish between OR1 and OR3. *Cell, 47*, 807–816.
Jaeger, S. A., Chan, E. T., Berger, M. F., Stottmann, R., Hughes, T. R., & Bulyk, M. L. (2010). Conservation and regulatory associations of a wide affinity range of mouse transcription factor binding sites. *Genomics, 95*, 185–195.
Jiang, J., & Levine, M. (1993). Binding affinities and cooperative interactions with bHLH activators delimit threshold responses to the dorsal gradient morphogen. *Cell, 72*, 741–752.
Jon, L., & the East Side Boyz. (2002). Get low. Lil Jon, 2002.
Kabata, H., Kurosawa, O., Arai, I., Washizu, M., Margarson, S., Glass, R., et al. (1993). Visualization of single molecules of RNA polymerase sliding along DNA. *Science, 262*, 1561–1563.
Kamachi, Y., Uchikawa, M., Tanouchi, A., Sekido, R., & Kondoh, H. (2001). Pax6 and SOX2 form a co-DNA-binding partner complex that regulates initiation of lens development. *Genes & Development, 15*, 1272–1286.
Kheradpour, P., & Kellis, M. (2014). Systematic discovery and characterization of regulatory motifs in ENCODE TF binding experiments. *Nucleic Acids Research, 42*, 2976–2987.
Lebrecht, D., Foehr, M., Smith, E., Lopes, F. J. P., Vanario-Alonso, C. E., Reinitz, J., et al. (2005). Bicoid cooperative DNA binding is critical for embryonic patterning in Drosophila. *Proceedings of the National Academy of Sciences of the United States of America, 102*, 13176–13181.
Lee, W., Haslinger, A., Karin, M., & Tjian, R. (1987). Activation of transcription by two factors that bind promoter and enhancer sequences of the human metallothionein gene and SV40. *Nature, 325*, 368–372.
Leith, J. S., Tafvizi, A., Huang, F., Uspal, W. E., Doyle, P. S., Fersht, A. R., et al. (2012). Sequence-dependent sliding kinetics of p53. *Proceedings of the National Academy of Sciences of the United States of America, 109*, 16552–16557.
Li, X. Y., MacArthur, S., Bourgon, R., Nix, D., Pollard, D. a., Iyer, V. N., et al. (2008). Transcription factors bind thousands of active and inactive regions in the Drosophila blastoderm. *PLoS Biology, 6*, 0365–0388.
Liu, Z., Legant, W. R., Chen, B.-C., Li, L., Grimm, J. B., Lavis, L. D., et al. (2015). 3D imaging of Sox2 enhancer clusters in embryonic stem cells. *eLife, 4*, e07956.
Ludwig, M. Z., Bergman, C., Patel, N. H., & Kreitman, M. (2000). Evidence for stabilizing selection in a eukaryotic enhancer element. *Nature, 403*, 564–567.
Lusk, R. W., & Eisen, M. B. (2010). Evolutionary mirages: Selection on binding site composition creates the illusion of conserved grammars in Drosophila enhancers. *PLoS Genetics, 6*, e1000829.
MacArthur, S., & Brookfield, J. F. (2004). Expected rates and modes of evolution of enhancer sequences. *Molecular Biology and Evolution, 21*, 1064–1073.
Markstein, M., Markstein, P., Markstein, V., & Levine, M. S. (2002). Genome-wide analysis of clustered Dorsal binding sites identifies putative target genes in the Drosophila embryo. *Proceedings of the National Academy of Sciences of the United States of America, 99*, 763–768.
Merika, M., & Thanos, D. (2001). Enhanceosomes. *Current Opinion in Genetics & Development, 11*, 205–208.

Miller, C. T., Beleza, S., Pollen, A. A., Schluter, D., Kittles, R. A., Shriver, M. D., et al. (2007). cis-Regulatory changes in Kit ligand expression and parallel evolution of pigmentation in sticklebacks and humans. *Cell, 131*, 1179–1189.

Moses, A. M., Pollard, D. A., Nix, D. A., Iyer, V. N., Li, X.-Y., Biggin, M. D., et al. (2006). Large-scale turnover of functional transcription factor binding sites in Drosophila. *PLoS Computational Biology, 2*, e130.

Noyes, M. B., Christensen, R. G., Wakabayashi, A., Stormo, G. D., Brodsky, M. H., & Wolfe, S. A. (2008). Analysis of homeodomain specificities allows the family-wide prediction of preferred recognition sites. *Cell, 133*, 1277–1289.

Park, S. K., & Hwang, B. J. (2012). Regulatory regions of the C elegans genome contain more low-affinity REF-1 transcription-factor binding sites than high-affinity sites. *Genes & Genomics, 34*, 703–708.

Ramos, A. I., & Barolo, S. (2013). Low-affinity transcription factor binding sites shape morphogen responses and enhancer evolution. *Philosophical Transactions of the Royal Society of London. Series B, Biological Sciences, 368*, 20130018.

Rowan, S., Siggers, T., Lachke, S. A., Yue, Y., Bulyk, M. L., & Maas, R. L. (2010). Precise temporal control of the eye regulatory gene Pax6 via enhancer-binding site affinity. *Genes & Development, 24*, 980–985.

Ruusala, T., & Crothers, D. M. (1992). Sliding and intermolecular transfer of the lac repressor: Kinetic perturbation of a reaction intermediate by a distant DNA sequence. *Proceedings of the National Academy of Sciences of the United States of America, 89*, 4903–4907.

Scardigli, R. (2003). Direct and concentration-dependent regulation of the proneural gene Neurogenin2 by Pax6. *Development, 130*, 3269–3281.

Segal, E., Raveh-Sadka, T., Schroeder, M., Unnerstall, U., & Gaul, U. (2008). Predicting expression patterns from regulatory sequence in Drosophila segmentation. *Nature, 451*, 535–540.

Slattery, M., Riley, T., Liu, P., Abe, N., Gomez-Alcala, P., Dror, I., et al. (2011). Cofactor binding evokes latent differences in DNA binding specificity between Hox proteins. *Cell, 147*, 1270–1282.

Spivakov, M. (2014). Spurious transcription factor binding: Non-functional or genetically redundant? *Bioessays, 36*, 798–806.

Stefflova, K., Thybert, D., Wilson, M. D., Streeter, I., Aleksic, J., Karagianni, P., et al. (2013). Cooperativity and rapid evolution of cobound transcription factors in closely related mammals. *Cell, 154*, 530–540.

Stone, J. R., & Wray, G. A. (2001). Rapid evolution of cis-regulatory sequences via local point mutations. *Molecular Biology and Evolution, 18*, 1764–1770.

Tanay, A. (2006). Extensive low-affinity transcriptional interactions in the yeast genome. *Genome Research, 16*, 962–972.

Wang, J., Malecka, A., Trøen, G., & Delabie, J. (2015). Comprehensive genome-wide transcription factor analysis reveals that a combination of high affinity and low affinity DNA binding is needed for human gene regulation. *BMC Genomics, 16*(Suppl. 7), S12.

Wunderlich, Z., & Mirny, L. A. (2009). Different gene regulation strategies revealed by analysis of binding motifs. *Trends in Genetics, 25*, 434–440.

Zhang, C., Xuan, Z., Otto, S., Hover, J. R., McCorkle, S. R., Mandel, G., et al. (2006). A clustering property of highly-degenerate transcription factor binding sites in the mammalian genome. *Nucleic Acids Research, 34*, 2238–2246.

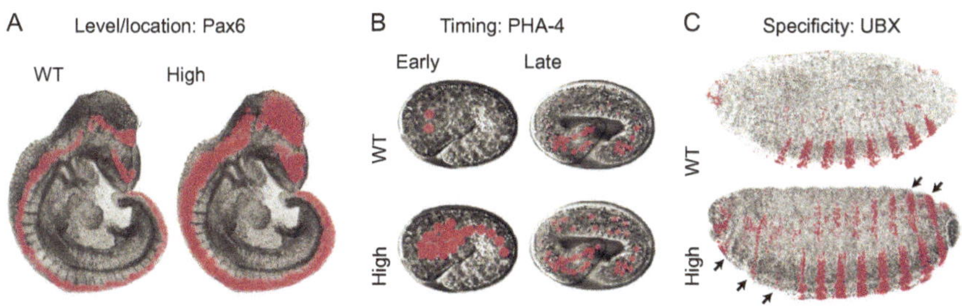

Justin Crocker *et al.*, Figure 1 Functional evidence for the importance of low-affinity binding sites from development. (A) Replacing low-affinity PAX6 sites with a high-affinity site causes ectopic expression in a *Neurogenenin2* enhancer (Scardigli, 2003). (B) Replacing low-affinity PHA-4 sites with a high-affinity sites leads to precocious expression (Gaudet & Mango, 2002). (C) Replacing a low-affinity Ultrabithorax site with a high-affinity site in a *Drosophila shavenbaby* enhancer leads to ectopic expression in anterior regions (arrows) (Crocker et al., 2015).

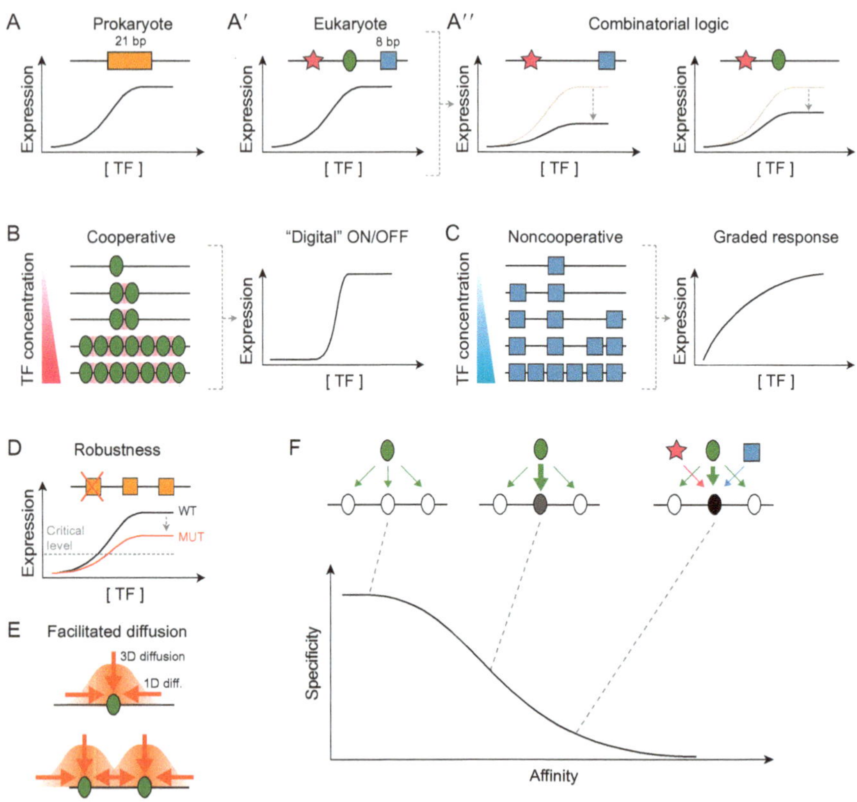

Justin Crocker *et al.*, Figure 2 Potential reasons for abundance of low-affinity transcription factor binding sites. (A) The size of transcription factor binding motifs in prokaryotes is, on average, nearly double that of eukaryotes. (A′) In eukaryotes, the low affinity of individual transcription factors is compensated for by the use of sites for multiple factors in a single enhancer. (A″) The use of multiple binding sites facilitates combinatorial logic and refined patterns of expression. (B) Clusters of low-affinity sites can be organized so as to mediate cooperative interactions between the transcription factors that bind to enhancers, leading to ON/OFF responses to variable inputs. (C) Clusters of low-affinity binding sites can act in a noncooperative manner to confer graded responses to variable inputs. (D) Clusters of low-affinity binding sites may serve to increase robustness of transcription enhancers, such that the loss of one binding site still allows the enhancer to provide a critical level of expression from the remaining sites. (E) Clusters of binding sites can decrease the search time for transcription factor binding sites through local one-dimensional diffusion between binding sites—providing a larger binding site "target." (F) Clustering of low-affinity homotypic binding sites can result from a fundamental tradeoff between affinity and specificity. Low-affinity binding sites provide specificity for unique Hox factors, whereas high-affinity binding sites are bound by multiple *Hox* genes (Crocker et al., 2015).

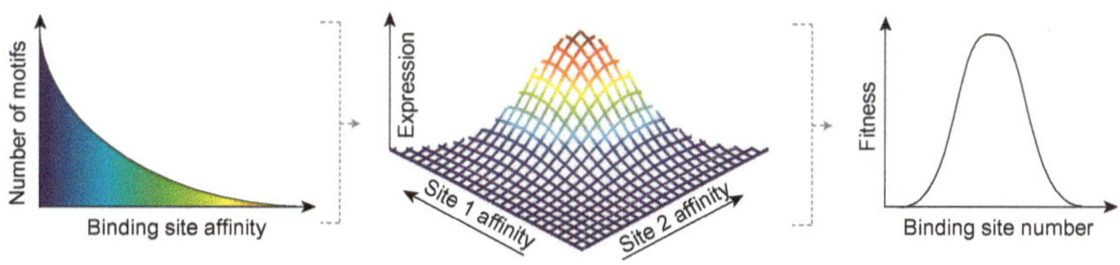

Justin Crocker *et al.*, Figure 3 More DNA sequences can confer weak affinity than can confer strong affinity. Considering a two binding site model, varying the affinity of each individual binding site leads to a range of possible expression levels from the enhancer. Because more DNA sequences can confer weak affinity than can confer strong affinity, and because clusters of weak sites can provide high expression levels, evolution will tend to evolve enhancers containing multiple low-affinity binding sites. *Adapted from He et al. (2012).*

CHAPTER TWENTY-EIGHT

Developmental Competence for Primordial Germ Cell Fate

Ufuk Günesdogan*,†,1, M. Azim Surani*,†,1

*Wellcome Trust/Cancer Research UK Gurdon Institute, University of Cambridge, Cambridge, United Kingdom

†Department of Physiology, Development and Neuroscience, University of Cambridge, Cambridge, United Kingdom

[1]Corresponding authors: e-mail address: u.gunesdogan@gurdon.cam.ac.uk; a.surani@gurdon.cam.ac.uk

Contents

Abstract

During mammalian embryonic development, the trophectoderm and primitive endoderm give rise to extraembryonic tissues, while the epiblast differentiates into all

ISSN 0070-2153
http://dx.doi.org/10.1016/bs.ctdb.2015.11.007

somatic lineages and the germline. Remarkably, only a few classes of signaling pathways induce the differentiation of these progenitor cells into diverse lineages. Accordingly, the functional outcome of a particular signal depends on the developmental competence of the target cells. Thus, developmental competence can be defined as the ability of a cell to integrate intrinsic and extrinsic cues to execute a specific developmental program toward a specific cell fate. Downstream of signaling, there is the combinatorial activity of transcription factors and their cofactors, which is modulated by the chromatin state of the target cells. Here, we discuss the concept of developmental competence, and the factors that regulate this state with reference to the specification of mammalian primordial germ cells.

1. INTRODUCTION

Although every cell in the organism harbors the same genetic information, each cell fate decision requires only a specific set of genes, which defines its form and function. Thus, a fundamental question in biology is how the genetic network in subsets of cells is differentially controlled to orchestrate the development of a whole organism. This involves the context-dependent role of key transcription factors, which bind to *cis*-regulatory DNA sequences to regulate cell fates.

Signaling pathways directly or indirectly control the activity and/or expression of transcription factors. Given the complexity of a developing embryo, it is intriguing that only a few classes of signaling pathways are required to drive the differentiation of all cell types of an organism. In the mammalian postimplantation epiblast, bone morphogenetic protein (BMP) and WNT signaling are not only required for the induction of mesoderm fate (Winnier, Blessing, Labosky, & Hogan, 1995), but also for the specification of primordial germ cells (PGCs) (Aramaki et al., 2013; Lawson et al., 1999; Ohinata et al., 2009), which are the precursors of the gametes. In prospective PGCs, the signals induce the expression of *Prdm1* (encoding BLIMP1), *Prdm14* and *Tfap2c* (encoding AP2γ). Each of these factors has distinct context-dependent roles in development, but in appropriately competent cells, the combination of them drives a PGC-specific transcriptional program.

The specification of PGCs is one example of the intricate interplay of intrinsic and extrinsic signals that results in the activation of signal-dependent transcription factors, which in turn can modulate the chromatin state of the target cells and activate the expression of lineage-specific transcription factors. Their combinatorial action is modified by cofactors and the chromatin state of their binding sites, suggesting that signaling pathways, the proteome, and the chromatin state facilitate developmental competence.

Here, we provide an overview of the factors that influence the acquisition of developmental competence during the specification of PGCs from the posterior proximal epiblast in embryos. We also discuss *in vitro* model systems, which recapitulate these events starting with naïve pluripotent stem cells. We refer to mouse development unless stated otherwise. We first consider the acquisition of developmental competence of postimplantation epiblast cells toward both the mesoderm and PGC fate, and summarize in this context, the complex interplay of signaling pathways. We then discuss the combinatorial role of the key transcription factors of the PGC fate, BLIMP1, PRDM14, AP2γ, and their context-dependent roles in embryonic development. Finally, we discuss one target element of transcription factors, the transcriptional enhancer, and its role in refining the transcriptional output downstream of signaling.

2. ACQUISITION OF DEVELOPMENTAL COMPETENCE

2.1 Early Embryonic Development

Developmental competence follows after fertilization and establishment of the totipotent zygote. After a few cell divisions, the zygote gives rise to the morula, which contains outer and smaller inner cells that will become trophectoderm (TE) and the inner cell mass (ICM), respectively, during the formation of the blastocyst (Arnold & Robertson, 2009). The ICM subsequently segregates into the primitive endoderm (PrE) and the epiblast. After blastocyst implantation, TE and PrE serve as the precursors of extraembryonic lineages, while the epiblast becomes the founder population of the embryo proper including the germline. Hence, there is a sequential commitment of developmental competence to spatially separated cell populations during early embryonic development. *In vivo* the competent state is transient, but can be captured in cells *in vitro* in embryonic stem cells (ESCs) from the preimplantation epiblast (Evans & Kaufman, 1981; Martin, 1981), extraembryonic endoderm cells from the PrE (Kunath et al., 2005), and trophoblast stem cells (TSCs) from the TE (Tanaka, Kunath, Hadjantonakis, Nagy, & Rossant, 1998).

2.2 Primordial Germ Cells

Signaling from extraembryonic tissues induces the differentiation of mesoderm, ectoderm, definitive endoderm, and PGCs from the postimplantation epiblast (Fig. 1A). The TE lineage forms the extraembryonic ectoderm (ExE), which is proximal to the epiblast. The PrE lineage gives rise to the visceral endoderm, which surrounds the epiblast and ExE. BMP4 signaling from the ExE induces PGCs in a few cells of the proximal posterior epiblast (Fig. 1A; Lawson et al., 1999). Subsequently, PGCs migrate toward the prospective gonads, while they undergo extensive transcriptional and epigenetic changes before they enter meiosis to form the gametes, sperm, and oocytes, respectively (Hackett et al., 2013; Hajkova et al., 2008; Kurimoto et al., 2008, 2015; Magnúsdóttir et al., 2013; Seki et al., 2007). PGCs are unipotent as they differentiate either into eggs or sperm depending on the sex of the embryo. Specification of PGCs is also accompanied by reexpression of a number of pluripotency genes (Kurimoto et al., 2008; Magnúsdóttir et al., 2013). Accordingly, PGCs from embryos at embryonic day (E) 7.5–E12.5 can dedifferentiate to pluripotent embryonic germ cells (EGCs) in culture and form testicular teratomas (Matsui, Zsebo, & Hogan, 1992; Resnick, Bixler, Cheng, & Donovan, 1992; Stevens, 1967). EGCs are very similar to ESCs (Leitch et al., 2010), except for their DNA methylation status at imprinted loci (Durcova-Hills, Ainscough, & McLaren, 2001; Shovlin, Durcova-Hills, Surani, & McLaren, 2008; Tada et al., 1998). Nonetheless, PGCs are highly specialized cells, which do not contribute to chimeras after injection into embryos (Leitch et al., 2014), and from ~E12.5 onward, they undergo differentiation only toward gametes. The precise role of pluripotency factors in early PGCs is not entirely clear except that they might help to promote epigenetic reprogramming (Papp & Plath, 2013).

The developmental competence for PGC specification is established in the postimplantation epiblast, but only a subset of cells in the posterior of the embryo acquires PGC fate (Fig. 1A). However, distal epiblast cells transplanted to the posterior can also respond to BMP4 signaling and become PGCs (Tam & Zhou, 1996). Indeed, a large number of epiblast cells when separated from the extraembryonic tissues have the potential to become PGCs in response to high levels of BMP4 in *ex vivo* culture (Fig. 1B, Ohinata et al., 2009). The developmental competence for the PGC fate is restricted to a short developmental time window from ~E5.5 to 6.5 (Ohinata et al., 2009), prior to the formation of the primitive streak and the onset of gastrulation (Arnold & Robertson, 2009).

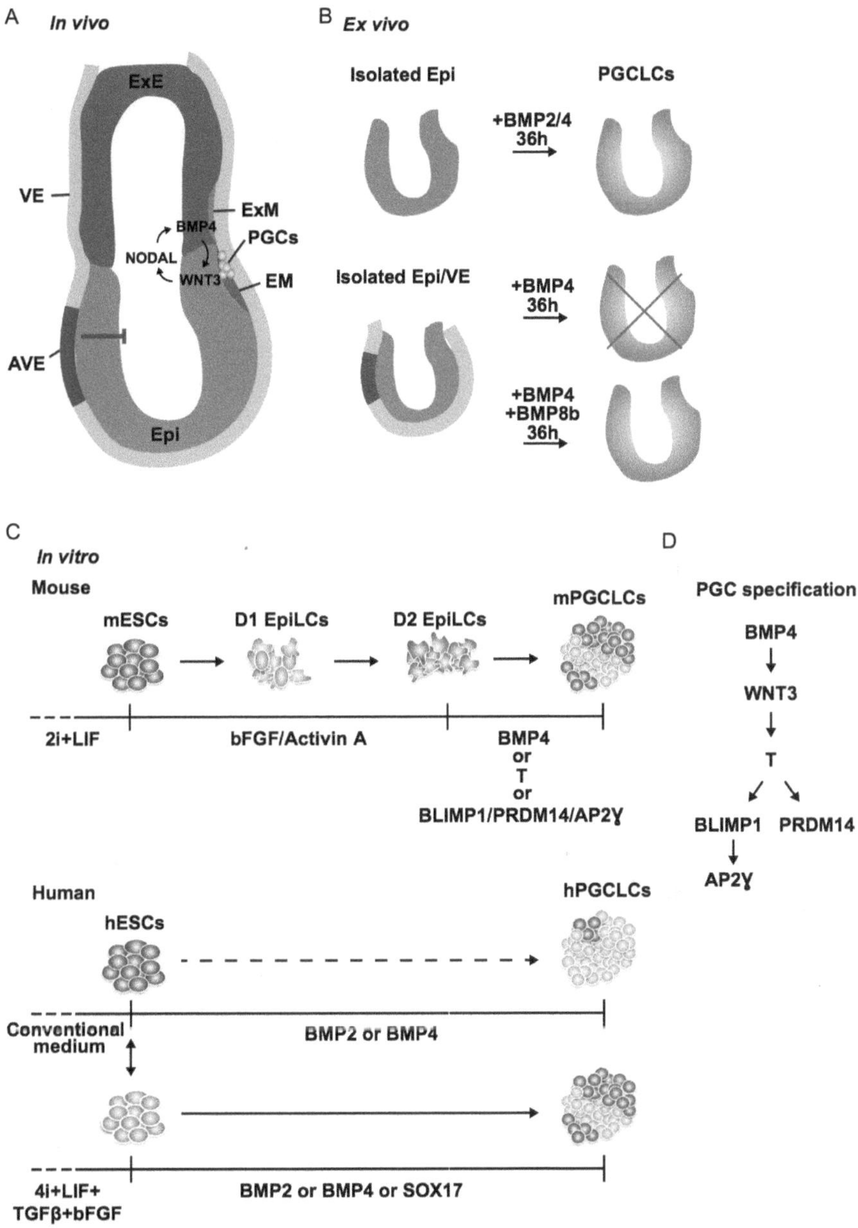

Figure 1 PGC specification *in vivo* and *in vitro*. (A) BMP4 signaling from the ExE specifies PGCs in the posterior epiblast of postimplantation embryos at ~E6.5. Nodal signaling induces the expression of *Bmp4* in the ExE, which in turn induces the expression of *Wnt3* in the epiblast. WNT3 maintains *Nodal* expression. It was shown that BMP and WNT signaling are required for the induction of PGCs. The AVE is source for inhibitory (*Continued*)

2.3 Epiblast Stem Cells

The postimplantation epiblast can give rise to pluripotent epiblast stem cells (EpiSCs) *in vitro* in the presence of bFGF and Activin A (Brons et al., 2007; Tesar et al., 2007). EpiSCs can differentiate into derivates of the three germ layers, but the induction of PGCs is very inefficient (Hayashi & Surani, 2009). Also, they contribute very poorly, if at all, to chimeras after blastocyst injections (Brons et al., 2007; Han et al., 2010; Tesar et al., 2007). EpiSCs appear to be transcriptionally close to the ectoderm of the late gastrula (Kojima et al., 2014), suggesting that EpiSCs become heterogeneous in culture (Han et al., 2010; Tsakiridis et al., 2014), where the majority of cells progress to a later developmental stage and consequently loose the developmental competence to become PGCs. This is also reflected by their epigenetic state, since putative enhancer elements and promoters of genes important for somatic differentiation show enrichment for active histone marks (Factor et al., 2014; Kurimoto et al., 2015). The inhibition of WNT signaling results in a more homogeneous population of EpiSC, which can contribute to the germline, when transplanted into the postimplantation embryo (Sugimoto et al., 2015; Sumi, Oki, Kitajima, & Meno, 2013). Further studies are required to address the question whether EpiSCs maintained or derived under these culture conditions, can respond to BMP4 and give rise to PGCLCs efficiently as observed with epiblast-like cells (EpiLCs; see below).

Figure 1—Cont'd signals against posteriorization of the anterior embryo. ExE, extraembryonic ectoderm, (A) VE, (anterior) visceral endoderm; Epi, epiblast; EM, embryonic mesoderm; PGCs, primordial germ cells; ExM, extraembryonic mesoderm. (B) Postimplantation epiblasts separated from the extraembryonic tissues respond to BMP2 or BMP4 and large numbers of cells acquire PGC-like fate in *ex vivo* culture. In the presence of the AVE, the addition of both BMP4 and BMP8b but not of BMP4 alone is sufficient for the induction of PGCLCs. PGCLCs, PGC-like cells. (C) In mouse, ESCs (mESCs) are differentiated into EpiLCs in the presence of bFGF and Activin A. Day 2 EpiLCs respond to BMP4 and give rise to mouse PGCLCs (mPGCLCs). Similarly, the induced expression of BLIMP1/PRDM14/AP2γ or T results in mPGCLC induction. The direct induction of human PGCLCs (hPGCLCs) with BMP2 or BMP4 from human ESCs (hESCs) cultured in conventional medium is very inefficient. In contrast, hESCs in 4i conditions reversibly gain and retain competence for hPGCLCs and acquire germ cell fate in response to BMP2 or BMP4. Unlike in the mouse, the induced expression of SOX17 results in the upregulation of germ cell markers. (D) The specification of PGCs in mice relies on BMP4, which induces the expression of WNT3. WNT3 induces the expression of T. T upregulates the expression of BLIMP1 and PRDM14. BLIMP1 induces AP2γ. Together, these transcription factors drive the PGC fate. (A color version of this figure is located at the end of this chapter.)

2.4 Epiblast-Like Cells

ESCs cultured with a GSK3 inhibitor (CHIR99021), a MEK/ERK inhibitor (PD0325901), and leukemia inhibitory factor (LIF), represent a relative homogeneous population of naïve pluripotent cells (Ying et al., 2008), which are transcriptionally close to the preimplantation epiblast (Boroviak, Loos, Bertone, Smith, & Nichols, 2014). These ESCs give rise to EpiLCs following culture in bFGF and Activin A (Fig. 1C; Hayashi, Ohta, Kurimoto, Aramaki, & Saitou, 2011). EpiLCs undergo extensive cell death after 3 days of culture and thus appear to represent a transient cell state. They are to some extent different from EpiSCs on transcriptional level, but similar to the pregastrulating epiblast (Hayashi et al., 2011). Accordingly, a large proportion of EpiLCs at day 2 respond to BMP4 and become PGC-like cells (PGCLCs) very efficiently, which can give rise to sperm or oocytes after transplantation experiments (Hayashi et al., 2011, 2012). However, this response was not observed with EpiLCs at day 1, suggesting that the transit to the competent state for the PGC fate is not completed at this stage. Future studies might reveal the key differences between day 1 and day 2 EpiLCs and what restricts or promotes the induction of the PGC fate as they progress toward the competence for germ cell and somatic fates.

2.5 Human PGCLCs

Recent studies have explored, whether human PGCLCs (hPGCLCs) can be derived from human ESCs (hESCs) *in vitro* (Irie et al., 2015; Sasaki et al., 2015; Sugawa et al., 2015). Irie et al. cultured hESCs with four inhibitors (for GSK3, MEK, p38, JNK; "4i") together with LIF, bFGF, and TGFβ (Gafni et al., 2013). hESCs in 4i can be directly induced into hPGCLCs upon addition of BMP2 or BMP4 (Fig. 1C), which are transcriptionally close to human PGCs from week 7 embryos 4 days after induction. Thus, hESCs in 4i gain and retain competence for germ cell fate while undergoing self-renewal. This is unlike mouse EpiLCs where the competence for PGCs occurs only transiently. Intriguingly, hESCs in 4i culture exhibit a gene expression profile that includes expression of mesoderm and primitive streak markers. Hence, the expression of lineage-specific markers such as T (also known as BRACHYURY) could render hESCs competent for the hPGCLC fate, since T plays an important role in the specification of mice PGCLCs (see below). Consistent with this, hESCs cultured in conventional medium show a poor response to hPGCLC induction (Irie et al., 2015; Sugawa et al., 2015). Notably, the two hESC states, in 4i and conventional culture, respectively, are reversible and depend on the appropriate environmental signal (Fig. 1C; Irie et al., 2015).

3. INTERPLAY OF SIGNALING PATHWAYS

3.1 BMP Pathway

Signaling molecules pattern the embryo, and they act during specific time windows to induce developmental processes, which results in the differentiation of groups of cells. These pathways do not act individually but show a highly complex interplay with other signals. Thus, the balance of activating and inhibitory signals refines the temporal and spatial output during cell fate specification.

BMPs induce the receptor-mediated phosphorylation of the transcription factors SMAD1/5/8, which can form a complex with SMAD4 and subsequently translocate into the nucleus, and induce or repress transcription of target genes. The output of SMAD activation is modulated on different levels, since SMADs interact with many coactivators or repressors. For example, the transcriptional coactivator p300/CBP interacts with SMAD1 to SMAD4 (Pouponnot, Jayaraman, & Massagué, 1998). p300/CBP in turn not only interacts with transcription factors of all major families, but also has acetyltransferase activity (Holmqvist & Mannervik, 2014). It was shown to bind *cis*-regulatory sequences genome wide, including putative enhancer elements (Visel et al., 2009), often resulting in the acetylation of histone H3 lysine 27 (H3K27ac), a hallmark of active enhancers (Creyghton et al., 2010). However, not all p300/CBP-bound enhancers are enriched for H3K27ac, suggesting that p300/CBP-bound enhancer elements become "poised" in the first instance before activation. Therefore, the activity of the BMP pathway not only results in a direct activation or repression of target genes, but could also modulate the enhancer landscape and consequently the developmental competence of the target cells.

BMP4 is required for the induction of mesoderm as well as the PGC fate in the postimplantation epiblast (Lawson et al., 1999; Mishina, Suzuki, Ueno, & Behringer, 1995; Winnier et al., 1995). Accordingly, a mutation in the *Bmp4* gene, results in embryonic lethality between E6.5 and E9.5 due to early patterning defects (Mishina et al., 1995; Winnier et al., 1995). BMP4 has diverse roles during development and was recently also implicated in preimplantation embryos for the development of extraembryonic lineages as well as for cell division (de Mochel et al., 2015; Graham et al., 2014).

3.2 Nodal/BMP/WNT Signaling Network in the Posterior Epiblast

The initiation of gastrulation is marked by the appearance of the primitive streak in the postimplantation epiblast, the site where cells ingress to form the mesoderm. Not only BMP but also Nodal, FGF, and WNT signaling are required to control this complex developmental process (Ciruna, Schwartz, Harpal, Yamaguchi, & Rossant, 1997; Conlon et al., 1994; Dunn, Vincent, Oxburgh, Robertson, & Bikoff, 2004; Johansson & Wiles, 1995; Kelly, Pinson, & Skarnes, 2004; Liu et al., 1999; Mishina et al., 1995; Winnier et al., 1995; Yamaguchi, Harpal, Henkemeyer, & Rossant, 1994). Together, these pathways regulate each other in feedback loops to mediate the activation of the mesoderm-specific transcriptional network (Fig. 1A). It should be noted that Nodal, like the BMPs, is a member of the TGFβ family and can act through SMAD2/3 (Arnold & Robertson, 2009). However, the SMAD2-dependent function controls the patterning of the anterior VE, which is a source of inhibitory signals such as *Cer1* against posteriorization of the anterior epiblast (Brennan et al., 2001). The SMAD2-independent role of Nodal is based on the secretion of its precursor at high levels from the posterior epiblast to the ExE resulting in the expression of Nodal convertases and *Bmp4* (Ben-Haim et al., 2006). BMP4 is secreted from the ExE and activates the expression of *Wnt3* in the posterior epiblast, which in turn maintains *Nodal* expression. Accordingly, *Bmp4* and *Wnt3* expression is absent in *Nodal* mutant embryos (Brennan et al., 2001).

3.3 PGC Induction

The BMP and WNT pathways also induce PGCs in the posterior epiblast proximal to the ExE (Fig. 1A; Aramaki et al., 2013; Lawson et al., 1999; Ohinata et al., 2009). BMP4 and BMP2, which are structurally very similar, are sufficient to induce the PGC fate in *ex vivo* cultures of isolated whole epiblasts (Fig. 1B; Ohinata et al., 2009). BMP8b, also secreted from the ExE, has an indirect role in PGC specification by restricting the development of the anterior VE. *Bmp8b* mutant embryos do not induce PGCs and exhibit an abnormal VE morphology with an expanded expression domain of inhibitory signals including *Cer1*, which are important for the anterior identity of the epiblast. Also, in *ex vivo* cultures of isolated epiblasts with the VE still attached, the addition of exogenous BMP4 without BMP8b does not result in the induction of PGCs (Fig. 1B), presumably due to the expansion of the inhibitory signals emanating from the VE. Thus, the primary signal for the induction of PGC fate is BMP4, which is supported by BMP2, while BMP8b indirectly affects this process.

WNT3 appears to be the mediator of BMP signaling to induce the PGC fate (Aramaki et al., 2013). *Wnt3a* alone does not induce PGCs in *ex vivo* epiblast culture (Ohinata et al., 2009). However, *Wnt3* mutant embryos fail to induce PGCs, although *Bmp4* expression in the ExE is not affected. A direct target of WNT3 is the transcription factor T, which is known to play an integral part in mesoderm development. In this developmental context, T binds to the enhancers of the genes encoding the key transcription factors of the germline *Prdm1* and *Prdm14* to activate their expression (Fig. 1D; Aramaki et al., 2013). It is important to note, however, that *T* mutant embryos show initial expression of BLIMP1 in the posterior epiblast, but not of PRDM14. This suggests that additional factors might be responsible for the initial upregulation of BLIMP1 in prospective PGCs, which then might be augmented by T. As suggested by Aramaki et al., one possible candidate could be the transcription factor EOMES, since forced expression of *eomes* results in a moderate upregulation of BLIMP1 after PGCLC induction. T but not WNT3a is sufficient to induce PGCLCs. The pretreatment of EpiLCs with exogenous WNT3a results in a decrease of PGCLC induction efficiency via T, suggesting that WNT activates other targets, which could drive the EpiLCs toward a mesodermal fate rather than PGCLC fate providing a possible explanation for the distinct effects of T and WNT3a in the *in vitro* experiments.

The role of Nodal signaling in PGC specification has not been extensively explored. It is likely that Nodal has at least an indirect role in PGC specification, since it affects the expression of *Bmp4* and consequently *Wnt3* in the posterior epiblast (Brennan et al., 2001). However, after E6.0, the addition of exogenous Nodal or its inhibition in the epiblast did not affect PGC induction via BMP4 (Ohinata et al., 2009), suggesting that in this context Nodal does not play an important role just prior PGC specification.

Taken together, these findings demonstrate the interplay between different signaling pathways during postimplantation development inducing distinct cell lineages from the same population of epiblast cells. These cells do show at least similar developmental competence in the early postimplantation epiblast, but the signaling can potentially not only directly induce a particular cell fate, but also change the developmental competence of the cells. This is exemplified by BMP signaling, which is apparently required to create a permissive environment in epiblast cells, which can then respond to WNT signaling to induce PGCs (Aramaki et al., 2013).

However, the exact mechanistic function of BMP–SMADs during PGC specification requires further investigation. It is important to note that subtle intrinsic, maybe stochastic, differences in the postimplantation epiblast can prime subgroup of cells to respond differentially to the same signal.

4. COMBINATORIAL ACTION OF TRANSCRIPTION FACTORS

4.1 Context Dependency

Transcription factors contain a DNA-binding domain and bind to short DNA sequences (8–21 bp), referred to as transcription factor binding motif. However, the occurrence of binding motifs in the genome is not necessarily predictive for transcription factor occupancy, as usually the number of motifs vastly outnumbers actual binding sites of a transcription factor. Accordingly, many transcription factors show a highly cell type-specific binding pattern, which is due to the combinatorial action of transcription factors with cofactors, dose-dependent effects, and chromatin state (including accessibility) of DNA-binding sites (Spitz & Furlong, 2012). For example, TGFβ signaling is mediated via the SMAD2 and SMAD3 transcription factors and is required for diverse developmental processes. Accordingly, SMAD3 co-occupies different sets of regulatory sequences together with lineage-specific transcription factors; for example, in ESCs with OCT4, in myotubes with MYOD1, and in pro-B cells with PU.1 (Mullen et al., 2011). In addition, the level of phosphorylated SMAD2 induced dictates its binding pattern to distinct sets of regulatory elements in ESCs, which is reflected in the differentiation into distinct lineages (Lee et al., 2011). Further, a comparison of multiple datasets for human transcription factors showed a correlation of transcription factor binding with nucleosome-depleted and DNase I-accessible regions (Wang et al., 2012). However, the accessibility of a DNA-binding site can not only be the cause but also a direct consequence of transcription factor binding as it is the case upon binding of pioneer factors (Zaret & Carroll, 2011).

4.2 BLIMP1/PRDM14/AP2γ in PGCs

Upon specification of PGCs via the BMP and WNT pathway, the key transcription factors BLIMP1 and PRDM14 are upregulated (Fig. 1D). BLIMP1 mainly represses genes of the somatic lineages but can also act as an activator, as it induces the expression of *Tfap2c*. Similarly, BLIMP1 also induces the expression of AP2α during neural crest formation in zebrafish (Powell, Hernandez-Lagunas, LaMonica, & Artinger, 2013). The unique combinatorial action of BLIMP1/PRDM14/AP2γ in PGCs is critical in directing the early development of the germline. Accordingly, a mutation in any one of the three genes results in the loss of PGCs (Ohinata et al., 2005; Weber et al., 2010; Yamaji et al., 2008). Similar to the response of EpiLCs to BMP4, the induced expression of the three factors specifies the PGC fate. In contrast, the expression of the three factors in ESCs does not result in PGCLC induction; instead, these cells remain transcriptionally similar to ESCs (Nakaki et al., 2013). This clearly demonstrates the distinct developmental competence of ESCs and EpiLCs.

4.3 BLIMP1

BLIMP1, PRDM14, and AP2γ are not only expressed in the germline but also have different roles during development. BLIMP1 was initially discovered as a transcriptional repressor of the human IFNβ promoter (Keller & Maniatis, 1991). It has a critical role in B and T lymphocytes as well as in multiple other lineages during development (Bikoff, Morgan, & Robertson, 2009). For example, it was recently shown that BLIMP1 is involved in the differentiation of trophoblast subtypes (Mould, Morgan, Li, Bikoff, & Robertson, 2012). Accordingly, *Prdm1* mutant embryos arrest at E10.5 of embryonic development due to placental defects (Vincent et al., 2005), which demonstrates that the same transcription factor is not only critical for the development of embryonic but also extraembryonic lineages. BLIMP1 has a zinc finger DNA-binding domain as well as a PR domain, which has similarity to the SET domains of histone methyltransferases, but appears to be catalytically inactive (Ancelin et al., 2006). Rather, BLIMP1 associates with distinct chromatin modifiers depending on the developmental context including histone deacetylases (Yu, Angelin-Duclos, Greenwood, Liao, & Calame, 2000), G9a H3 methyl transferase (Gyory, Wu, Fejér, Seto, & Wright, 2004), a lysine-specific demethylase (LSD1) (Su et al., 2009), and the arginine methyltransferase PRMT5 (Ancelin et al., 2006). In PGCLCs as well as in P19 embryonal carcinoma (P19EC) cells, which were used as a model for PGCs, BLIMP1 binds mainly promoters of cell cycle and somatic genes resulting in their repression, which implies that BLIMP1 suppresses somatic differentiation, paving the way for epigenetic programming and expression of pluripotency genes (Kurimoto et al., 2015; Magnúsdóttir et al., 2013). Interestingly, BLIMP1-bound regions at day 2 of PGCLC differentiation, which are still present in day 6 PGCLCs, accumulate the repressive polycomb repressive complex (PRC2)-catalyzed modification H3K27me3, suggesting that BLIMP1 might directly or indirectly recruit PRC2 (Kurimoto et al., 2015). This is consistent with the data on P19EC cells, which shows that BLIMP1 has a high overlap with PRC-binding regions in ESCs (Magnúsdóttir et al., 2013). There might also be a link to the regulation of *Mad2l2*, since in *Mad2l2* mutants, the process of global enrichment of H3K27me3 in PGCs is abrogated and PGCs are lost by E9.5 (Pirouz, Pilarski, & Kessel, 2013).

4.4 PRDM14

PRDM14 also belongs to the family of PR domain-containing proteins with zinc finger domains. But as for BLIMP1, any histone methyltransferase activity for PRDM14 has not so far been demonstrated. PRDM14 interacts with TETs (Okashita et al., 2013), which mediate the conversion of 5-methylcytosine to 5-hydroxymethylcytosine to promote DNA demethylation (Hackett & Surani, 2013). PRDM14 also activates the expression of pluripotency factors, represses FGF signaling and components of the DNA methylation machinery to maintain naïve pluripotency (Grabole et al., 2013; Ma, Swigut, Valouev, Rada-Iglesias, & Wysocka, 2011; Nakaki & Saitou, 2014; Yamaji et al., 2013). PRDM14 exerts its repressive function at least in part by interacting with components of the PRC2 complex, which is also required for X chromosome reactivation in female embryos (Chan et al., 2013; Payer et al., 2013; Yamaji et al., 2013). Concomitant with the role of PRDM14 in pluripotency, it is expressed in preimplantation embryos including the ICM at the blastocyst stage, before its expression ceases upon implantation. It then becomes specifically upregulated in PGCs (Burton et al., 2013; Kurimoto et al., 2008; Yamaji et al., 2008). Nevertheless, the role of PRDM14 during early embryonic development requires further investigation, since a null mutation in the *Prdm14* gene results in viable, but sterile mice, which however depends on the genetic background (Payer et al., 2013). In PGCs, it appears that while BLIMP1 represses somatic differentiation, PRDM14 mainly binds putative enhancer elements to induce reexpression of pluripotency factors as well as to promote epigenetic programming including global DNA demethylation (Grabole et al., 2013; Magnúsdóttir et al., 2013). However, to characterize the function of PRDM14 in detail, it will be interesting to map the binding sites of PRDM14 directly in PGCs or PGCLCs. It is important to note that while *Prdm14* alone is sufficient to induce mouse PGCLC fate *in vitro* (Nakaki et al., 2013), it appears that its function is not conserved. At the onset of hPGCLC induction, the expression of *Prdm14* is initially repressed, suggesting that it might not play a prominent role at early stages of human PGC development (Irie et al., 2015; Sasaki et al., 2015; Sugawa et al., 2015). Interestingly, the transcription factor SOX17, which is required for endoderm development, is involved in hPGCLC induction, since loss of *SOX17* abrogates hPGCLC specification and its induced expression is sufficient to upregulate germ cell markers (Fig. 1C; Irie et al., 2015). This suggests that in humans a different combination of transcription factors including SOX17 and BLIMP1 drive the PGC fate.

4.5 AP2γ

AP2γ, a member of the activator protein-2 (AP-2) family, with a DNA-binding, dimerization, and a transactivation domain, has like BLIMP1 and PRDM14 multiple context-dependent roles during embryonic development. A mutation in *Tfap2c* results in embryonic lethality at E7.0–9.0 due to placental defects (Auman et al., 2002; Werling & Schorle, 2002), since AP2γ is involved in the maintenance of the TE lineage (Kuckenberg, Kubaczka, & Schorle, 2012). Interestingly, in the developmental context of this extraembryonic lineage, AP2γ interacts with SOX2; together they mediate self-renewal and the control of the transcriptional network of TSCs *in vitro* (Adachi et al., 2013), which are derived from TE cells (Tanaka et al., 1998). This also provides an example for the context-dependent role of the pluripotency factor SOX2, which integrates LIF–STAT signaling in ESCs and FGF–ERK signaling in TSCs (Adachi et al., 2013). In PGCs, the expression of *Tfap2c* is induced by BLIMP1 shortly after specification, where AP2γ binds promoters and enhancers, acting cooperatively with BLIMP1 and PRDM14 to induce genes essential for the PGC lineage (Magnúsdóttir et al., 2013; Schemmer et al., 2013). Of note, SOX2 is also induced in PGCs and the conditional loss of *Sox2* results in the loss of PGCs (Campolo et al., 2013; Kurimoto et al., 2008; Magnúsdóttir et al., 2013). Thus, it could be that AP2γ and SOX2 also act together in PGCs as in TSCs.

In summary, the function of many transcription factors is not restricted to one particular lineage, but there is a lineage-specific combination of proteins that cooperatively define a unique functional outcome. Thus, the developmental competence of a cell is also defined by its proteome, because the presence or absence of cofactors, e.g., chromatin modifiers, defines the functional output of transcription factors that were initially induced by signaling. To gain insight into the context-dependent roles of transcription factors in future studies, it is key to perform biochemical studies to identify cofactors of critical transcription factors.

5. TRANSCRIPTIONAL ENHANCER ELEMENTS

5.1 Enhancer Setup

Transcription factors bind *cis*-regulatory DNA sequences including enhancer elements to control transcription. It appears that enhancers show a highly cell type-specific epigenetic setup, which could facilitate the binding of transcription factors and thus potentially contribute to the developmental competence of a cell. It is important to note that the presence of histone modifications at enhancers could also be a consequence of transcription factor binding (Calo & Wysocka, 2013). The generation of specific histone mutants would be the most stringent way to test for the functional role of histone modifications. Since canonical histone genes are present in many copies in the genome of higher eukaryotes, the generation of defined histone mutants is technically challenging and was achieved, for example, in *Drosophila* but not in mammals so far (Günesdogan, Jäckle, & Herzig, 2010). However, recent advances in genome editing tools including the CRISPR–Cas9 technology have enabled the locus-specific editing of histone modifications in mammalian cells. For example, the guide RNA-mediated recruitment of a fusion protein, which consists of the catalytically inactive Cas9 (dCas9) protein and the catalytic domain of the acetyltransferase p300, results in enrichment of H3K27ac at selected enhancers and transcriptional activation of target genes (Hilton et al., 2015). Further, the fusion of dCas9 to the LSD1 results in a decrease of H3K4me2 concomitantly with a decrease in H3K27ac at target enhancers as well as reduced target gene expression (Kearns et al., 2015). The apparent interdependence of H3K4me2 and H3K27ac suggests recruitment of histone deacetylases via LSD1 (Mendenhall et al., 2013). Taken together, these and other studies demonstrate elegantly that the combinatorial presence of histone modifications is required for the activity of enhancer elements and might thus be involved in the acquirement of developmental competence. Nevertheless, there are some limitations of such experiments that need to be considered, since histone modifiers usually have many nonhistone targets. Also, the binding of dCas9 to an enhancer could spatially interfere with the binding of transcription factors, which could also effect target gene expression.

5.2 Enhancer Setup and Developmental Competence

Recent studies characterized the dynamic histone modification enrichment at enhancers during the differentiation of multipotent progenitor cells, e.g., during haematopoiesis (Lara-Astiaso et al., 2014) or hESC differentiation (Wang et al., 2015). Wang et al. profiled the enrichment of H3K4me1 and H3K27ac at enhancers during the sequential differentiation of hESCs toward pancreatic endoderm cells (Fig. 2). Enhancers enriched only for H3K4me1 were categorized as "poised," while enhancers enriched for both H3K4me1 and H3K27ac were categorized as "active" (Creyghton et al., 2010; Heintzman et al., 2007). A subset of enhancers became poised at the gut tube (GT) stage, which upon differentiation into different lineages became activated. This suggests that the poised state of enhancers can contribute or define the developmental competence of GT cells. In addition,

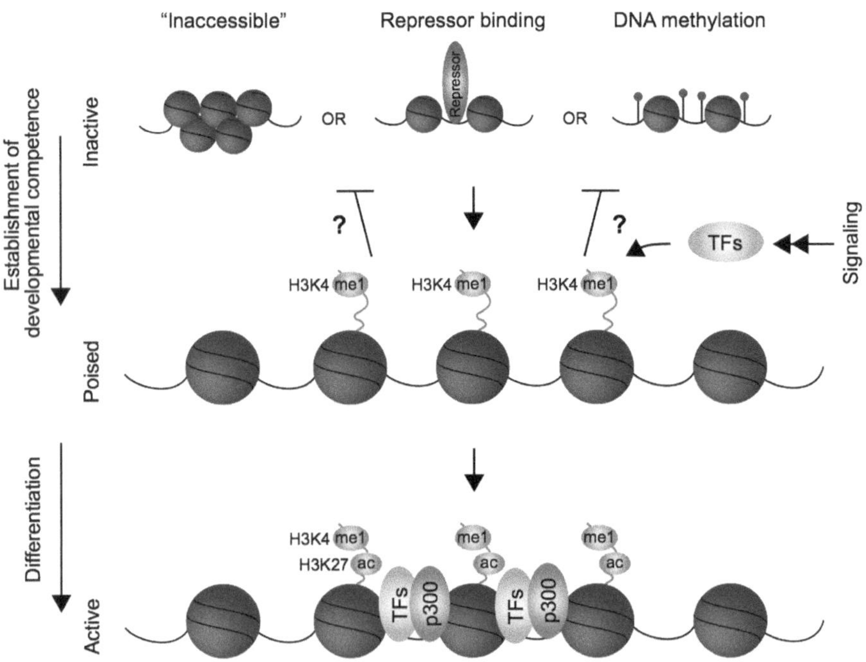

Figure 2 Enhancer setup. Transcriptional enhancers show a combinatorial enrichment of histone modifications. The enrichment of H3K4me1 at enhancers could result in an open chromatin structure, avoid repressor binding or DNA methylation and thus could act as a "placeholder" for transcription factor binding. Upon signaling, transcription factors can engage these enhancer elements to activate transcription of target genes. The activation of an enhancer element is marked by the enrichment of H3K27ac, which could be catalyzed by p300, which is also bound by many enhancers. (A color version of this figure is located at the end of this chapter.)

poised enhancers are bound by the pioneer transcription factor FOXA1, which can modulate local chromatin structure to mediate enhancer function (Zaret & Carroll, 2011). While knockdown of *FOXA1* did not affect H3K4me1 enrichment at tested enhancers, further studies are required to determine the precise role of FOXA1 in setting up the poised state, considering reported redundancy between FOXA proteins (Gao et al., 2008; Li et al., 2009; Wan et al., 2005).

5.3 Enhancer–Promoter Contacts

Enhancers do not always target the nearest promoter, but also act in a long-range fashion. Thereby, the intervening sequence between enhancer and promoter loops out, bringing these elements in close spatial proximity. This "interaction" could allow the delivery of enhancer-bound transcription factors and RNA polymerase II to the promoter to activate or enhance transcription. The regulation of promoter activity appears to be very complex, since promoters often interact with multiple enhancers, and promoters and enhancers also interact with each other (Ghavi-Helm et al., 2014; Schoenfelder, Furlan-Magaril, et al., 2015).

The PRC1 component RING1B binds promoters in ESCs, which are predominantly in contact with poised enhancers that are enriched for H3K4me1, H3K27me3, and p300 (Fig. 3A; Schoenfelder, Sugar, et al., 2015). In *Ring1A/B* knockout cells, these enhancers loose H3K27me3, but gain the active H3K27ac mark concomitantly with the transcriptional induction of associated genes (Fig. 3B). Similarly, inactive promoters also interact with poised enhancers, which then become selectively activated upon differentiation in *Drosophila* (Ghavi-Helm et al., 2014). Thus, a feature of poised enhancers could be the "preformation" of enhancer–promoter contacts, which could facilitate the quick response to signal-dependent and/or lineage-specific transcription factors.

5.4 Enhancer Setup for PGC Fate

In mice, the gain of developmental competence for the PGC fate is established during the differentiation of ESCs to EpiLCs (Hayashi et al., 2011). This transition is accompanied by the dynamic reorganization of the enhancer landscape displaying distinct sets of histone modifications accompanied by differential binding of transcription factors (Buecker et al., 2014). Notably, OCT4 exhibits a marked cell type-specific genomic occupancy, with enrichment peaks close to genes associated with naïve pluripotency in ESCs, close to

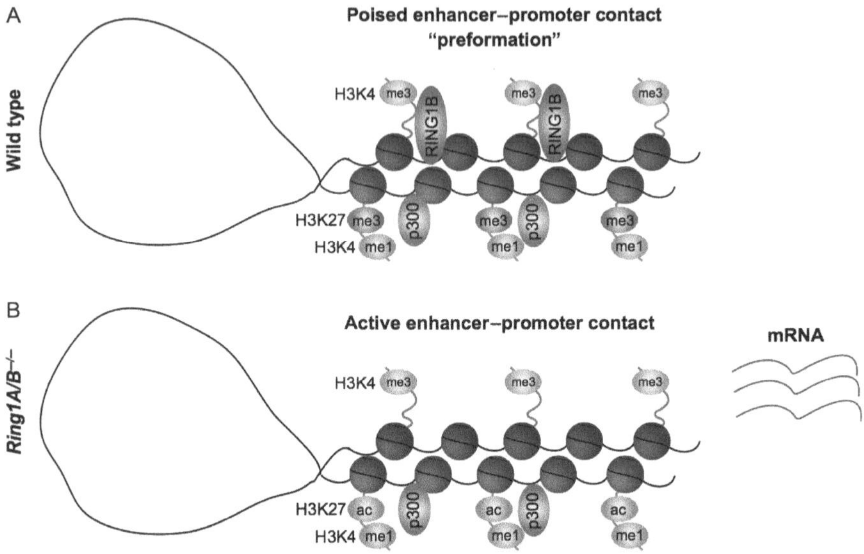

Figure 3 Preformation of enhancer–promoter contacts. (A) In ESCs, RING1B-bound promoters are in contact with poised enhancers that are enriched for H3K4me1, H3K27me3, and p300. (B) In *Ring1A/B* knockout cells, these enhancers loose H3K27me3 but gain H3K27ac, which results in the upregulation of associated target gene expression. Of note, p300, which can catalyze the acetylation of H3K27, appears to be inhibited at poised enhancers. Since the loss of *Ring1A/B* results in enrichment of H3K27ac, perhaps the PRC1 complex is involved in the inhibition of p300 histone acetyltransferase activity. (A color version of this figure is located at the end of this chapter.)

genes associated with the postimplantation epiblast in EpiLCs, and shared between both cell types. This redistribution of OCT4 correlates with H3K27ac and p300 enrichment. ESC-specific OCT4-binding sites are also enriched for H3K27ac in PGCLCs, suggesting that OCT4 controls similar subsets of genes in both cell types (Kurimoto et al., 2015). The dynamic behavior of OCT4 can be explained by a recurring concept for gene expression control. While OCT4 is expressed at similar levels in ESCs and EpiLCs, its binding is modulated by other cell type-specific transcription factors such as OTX2 in EpiLCs (Buecker et al., 2014; Yang et al., 2014). Interestingly, the comparison of BLIMP1-binding targets to the transcriptional consequences of loss of *Blimp1* during the reversion of PGCs to EGCs revealed that BLIMP1 binds to and restricts the expression of a subset of OCT4 and other pluripotency transcription factor targets. This suggests that BLIMP1 limits the pluripotency network and maintains the lineage restriction of PGCs

(Nagamatsu, Saito, Takubo, & Suda, 2015). Accordingly, BLIMP1 is repressed early during EGC derivation, prior to the acquisition of functional pluripotency (Durcova-Hills, Tang, Doody, Tooze, & Surani, 2008; Leitch et al., 2010).

In sum, histone modifications at enhancer elements can predict their activity and in some examples also the developmental competence of a cell. However, the functional implications of these marks are less clear; e.g., how does the presence of H3K4me1 results in the poised state of an enhancer? The underlying mechanism could be that the presence of H3K4me1 results in an open chromatin structure at putative enhancer elements, prevents the binding of transcriptional repressor proteins, or inhibits DNA methylation indirectly (Fig. 2; Calo & Wysocka, 2013), enabling the binding of activating transcription factors upon signaling. In agreement with this, recent studies show that the induced loss or gain of these modifications can influence enhancer activity.

6. CONCLUDING REMARKS

Developmental competence for cell fate decisions involves multiple interconnected aspects including intrinsic and extrinsic signaling, transcription factors, chromatin modifiers, enhancer landscape among others. The latter might include DNA methylation, cell cycle progression, and cell–cell contacts. In PGCs, the downstream effectors SMADs and T of the BMP and WNT pathway, respectively, activate the expression of BLIMP1/PRDM14, which drive the transcriptional program of the PGC fate. There are many more binding sites for these factors, which could activate the expression of essential cofactors and/or change the epigenetic state of enhancer elements in prospective PGCs. Future studies will benefit from the recent development of new techniques based on CRISPR/Cas9 for genome and epigenome editing, which might shed light on the underlying control mechanisms for a robust and refined transcriptional output to determine cell fates.

ACKNOWLEDGMENTS

We thank Erna Magnúsdóttir, Wolfram Gruhn, and Toshihiro Kobayashi for helpful comments on the chapter. The Gurdon Institute is funded by a Wellcome Trust Core Grant (092096) and a Cancer Research UK Grant (C6946/A14492). U.G. is supported by the Isaac Newton Trust and the Leverhulme Trust Early Career Fellowship. M.A.S. is supported by a Wellcome Trust Investigator Award.

REFERENCES

Adachi, K., Nikaido, I., Ohta, H., Ohtsuka, S., Ura, H., Kadota, M., et al. (2013). Context-dependent wiring of Sox2 regulatory networks for self-renewal of embryonic and trophoblast stem cells. *Molecular Cell, 52*, 380–392.

Ancelin, K., Lange, U. C., Hajkova, P., Schneider, R., Bannister, A. J., Kouzarides, T., et al. (2006). Blimp1 associates with Prmt5 and directs histone arginine methylation in mouse germ cells. *Nature Cell Biology, 8*, 623–630.

Aramaki, S., Hayashi, K., Kurimoto, K., Ohta, H., Yabuta, Y., Iwanari, H., et al. (2013). A mesodermal factor, T, specifies mouse germ cell fate by directly activating germline determinants. *Developmental Cell, 27*, 516–529.

Arnold, S. J., & Robertson, E. J. (2009). Making a commitment: Cell lineage allocation and axis patterning in the early mouse embryo. *Nature Reviews. Molecular Cell Biology, 10*, 91–103.

Auman, H. J., Nottoli, T., Lakiza, O., Winger, Q., Donaldson, S., & Williams, T. (2002). Transcription factor AP-2gamma is essential in the extra-embryonic lineages for early postimplantation development. *Development, 129*, 2733–2747.

Ben-Haim, N., Lu, C., Guzman-Ayala, M., Pescatore, L., Mesnard, D., Bischofberger, M., et al. (2006). The nodal precursor acting via activin receptors induces mesoderm by maintaining a source of its convertases and BMP4. *Developmental Cell, 11*, 313–323.

Bikoff, E. K., Morgan, M. A., & Robertson, E. J. (2009). An expanding job description for Blimp-1/PRDM1. *Current Opinion in Genetics & Development, 19*, 379–385.

Boroviak, T., Loos, R., Bertone, P., Smith, A., & Nichols, J. (2014). The ability of inner-cell-mass cells to self-renew as embryonic stem cells is acquired following epiblast specification. *Nature Cell Biology, 16*, 516–528.

Brennan, J., Lu, C. C., Norris, D. P., Rodriguez, T. A., Beddington, R. S., & Robertson, E. J. (2001). Nodal signalling in the epiblast patterns the early mouse embryo. *Nature, 411*, 965–969.

Brons, I. G. M., Smithers, L. E., Trotter, M. W. B., Rugg-Gunn, P., Sun, B., Chuva de Sousa Lopes, S. M., et al. (2007). Derivation of pluripotent epiblast stem cells from mammalian embryos. *Nature, 448*, 191–195.

Buecker, C., Srinivasan, R., Wu, Z., Calo, E., Acampora, D., Faial, T., et al. (2014). Reorganization of enhancer patterns in transition from naive to primed pluripotency. *Cell Stem Cell, 14*, 838–853.

Burton, A., Muller, J., Tu, S., Padilla-Longoria, P., Guccione, E., & Torres-Padilla, M.-E. (2013). Single-cell profiling of epigenetic modifiers identifies PRDM14 as an inducer of cell fate in the mammalian embryo. *Cell Reports, 5*, 687–701.

Calo, E., & Wysocka, J. (2013). Modification of enhancer chromatin: What, how, and why? *Molecular Cell, 49*, 825–837.

Campolo, F., Gori, M., Favaro, R., Nicolis, S., Pellegrini, M., Botti, F., et al. (2013). Essential role of Sox2 for the establishment and maintenance of the germ cell line. *Stem Cells, 31*, 1408–1421.

Chan, Y.-S., Göke, J., Lu, X., Venkatesan, N., Feng, B., Su, I.-H., et al. (2013). A PRC2-dependent repressive role of PRDM14 in human embryonic stem cells and induced pluripotent stem cell reprogramming. *Stem Cells, 31*, 682–692.

Ciruna, B. G., Schwartz, L., Harpal, K., Yamaguchi, T. P., & Rossant, J. (1997). Chimeric analysis of fibroblast growth factor receptor-1 (Fgfr1) function: A role for FGFR1 in morphogenetic movement through the primitive streak. *Development, 124*, 2829–2841.

Conlon, F. L., Lyons, K. M., Takaesu, N., Barth, K. S., Kispert, A., Herrmann, B., et al. (1994). A primary requirement for nodal in the formation and maintenance of the primitive streak in the mouse. *Development, 120*, 1919–1928.

Creyghton, M. P., Cheng, A. W., Welstead, G. G., Kooistra, T., Carey, B. W., Steine, E. J., et al. (2010). Histone H3K27ac separates active from poised enhancers and predicts developmental state. *Proceedings of the National Academy of Sciences of the United States of America, 107*, 21931–21936.

de Mochel, N. S. R., Luong, M., Chiang, M., Javier, A. L., Luu, E., Toshihiko, F., et al. (2015). BMP signaling regulates both cell proliferation and ICM lineage commitment in preimplantation-stage mouse embryos. *Developmental Biology, 397*, 45–55.

Dunn, N. R., Vincent, S. D., Oxburgh, L., Robertson, E. J., & Bikoff, E. K. (2004). Combinatorial activities of Smad2 and Smad3 regulate mesoderm formation and patterning in the mouse embryo. *Development, 131*, 1717–1728.

Durcova-Hills, G., Ainscough, J., & McLaren, A. (2001). Pluripotential stem cells derived from migrating primordial germ cells. *Differentiation, 68*, 220–226.

Durcova-Hills, G., Tang, F., Doody, G., Tooze, R., & Surani, M. A. (2008). Reprogramming primordial germ cells into pluripotent stem cells. *PLoS One, 3*, e3531.

Evans, M. J., & Kaufman, M. H. (1981). Establishment in culture of pluripotential cells from mouse embryos. *Nature, 292*, 154–156.

Factor, D. C., Corradin, O., Zentner, G. E., Saiakhova, A., Song, L., Chenoweth, J. G., et al. (2014). Epigenomic comparison reveals activation of "seed" enhancers during transition from naive to primed pluripotency. *Stem Cell, 14*, 854–863.

Gafni, O., Weinberger, L., Mansour, A. A., Manor, Y. S., Chomsky, E., Ben-Yosef, D., et al. (2013). Derivation of novel human ground state naive pluripotent stem cells. *Nature, 504*, 282–286.

Gao, N., LeLay, J., Vatamaniuk, M. Z., Rieck, S., Friedman, J. R., & Kaestner, K. H. (2008). Dynamic regulation of Pdx1 enhancers by Foxa1 and Foxa2 is essential for pancreas development. *Genes & Development, 22*, 3435–3448.

Ghavi-Helm, Y., Klein, F. A., Pakozdi, T., Ciglar, L., Noordermeer, D., Huber, W., et al. (2014). Enhancer loops appear stable during development and are associated with paused polymerase. *Nature, 512*, 96–100.

Grabole, N., Tischler, J., Hackett, J. A., Kim, S., Tang, F., Leitch, H. G., et al. (2013). Prdm14 promotes germline fate and naive pluripotency by repressing FGF signalling and DNA methylation. *EMBO Reports, 14*, 629–637.

Graham, S. J. L., Wicher, K. B., Jedrusik, A., Guo, G., Herath, W., Robson, P., et al. (2014). BMP signalling regulates the pre-implantation development of extra-embryonic cell lineages in the mouse embryo. *Nature Communications, 5*, 5667.

Günesdogan, U., Jäckle, H., & Herzig, A. (2010). A genetic system to assess *in vivo* the functions of histones and histone modifications in higher eukaryotes. *EMBO Reports, 11*, 772–776.

Gyory, I., Wu, J., Fejér, G., Seto, E., & Wright, K. L. (2004). PRDI-BF1 recruits the histone H3 methyltransferase G9a in transcriptional silencing. *Nature Immunology, 5*, 299–308.

Hackett, J. A., Sengupta, R., Zylicz, J. J., Murakami, K., Lee, C., Down, T. A., et al. (2013). Germline DNA demethylation dynamics and imprint erasure through 5-hydroxymethylcytosine. *Science, 339*, 448–452.

Hackett, J. A., & Surani, M. A. (2013). DNA methylation dynamics during the mammalian life cycle. *Philosophical Transactions of the Royal Society of London Series B, Biological Sciences, 368*. 20110328.

Hajkova, P., Ancelin, K., Waldmann, T., Lacoste, N., Lange, U. C., Cesari, F., et al. (2008). Chromatin dynamics during epigenetic reprogramming in the mouse germ line. *Nature, 452*, 877–881.

Han, D. W., Tapia, N., Joo, J. Y., Greber, B., Araúzo-Bravo, M. J., Bernemann, C., et al. (2010). Epiblast stem cell subpopulations represent mouse embryosof distinct pregastrulation stages. *Cell, 143*, 617–627.

Hayashi, K., Ogushi, S., Kurimoto, K., Shimamoto, S., Ohta, H., & Saitou, M. (2012). Offspring from oocytes derived from *in vitro* primordial germ cell-like cells in mice. *Science, 338*, 971–975.

Hayashi, K., Ohta, H., Kurimoto, K., Aramaki, S., & Saitou, M. (2011). Reconstitution of the mouse germ cell specification pathway in culture by pluripotent stem cells. *Cell, 146*, 519–532.

Hayashi, K., & Surani, M. A. (2009). Self-renewing epiblast stem cells exhibit continual delineation of germ cells with epigenetic reprogramming *in vitro*. *Development, 136*, 3549–3556.

Heintzman, N. D., Stuart, R. K., Hon, G., Fu, Y., Ching, C. W., Hawkins, R. D., et al. (2007). Distinct and predictive chromatin signatures of transcriptional promoters and enhancers in the human genome. *Nature Genetics, 39*, 311–318.

Hilton, I. B., D'Ippolito, A. M., Vockley, C. M., Thakore, P. I., Crawford, G. E., Reddy, T. E., et al. (2015). Epigenome editing by a CRISPR-Cas9-based acetyltransferase activates genes from promoters and enhancers. *Nature Biotechnology, 33*, 510–517.

Holmqvist, P.-H., & Mannervik, M. (2014). Genomic occupancy of the transcriptional co-activators p300 and CBP. *Transcription, 4*, 18–23.

Irie, N., Weinberger, L., Tang, W. W. C., Kobayashi, T., Viukov, S., Manor, Y. S., et al. (2015). SOX17 is a critical specifier of human primordial germ cell fate. *Cell, 160*, 253–268.

Johansson, B. M., & Wiles, M. V. (1995). Evidence for involvement of activin A and bone morphogenetic protein 4 in mammalian mesoderm and hematopoietic development. *Molecular and Cellular Biology, 15*, 141–151.

Kearns, N. A., Pham, H., Tabak, B., Genga, R. M., Silverstein, N. J., Garber, M., et al. (2015). Functional annotation of native enhancers with a Cas9-histone demethylase fusion. *Nature Methods, 12*, 401–403.

Keller, A. D., & Maniatis, T. (1991). Identification and characterization of a novel repressor of beta-interferon gene expression. *Genes & Development, 5*, 868–879.

Kelly, O. G., Pinson, K. I., & Skarnes, W. C. (2004). The Wnt co-receptors Lrp5 and Lrp6 are essential for gastrulation in mice. *Development, 131*, 2803–2815.

Kojima, Y., Kaufman-Francis, K., Studdert, J. B., Steiner, K. A., Power, M. D., Loebel, D. A. F., et al. (2014). The transcriptional and functional properties of mouse epiblast stem cells resemble the anterior primitive streak. *Cell Stem Cell, 14*, 107–120.

Kuckenberg, P., Kubaczka, C., & Schorle, H. (2012). The role of transcription factor Tcfap2c/TFAP2C in trophectoderm development. *Reproductive BioMedicine Online, 25*, 12–20.

Kunath, T., Arnaud, D., Uy, G. D., Okamoto, I., Chureau, C., Yamanaka, Y., et al. (2005). Imprinted X-inactivation in extra-embryonic endoderm cell lines from mouse blastocysts. *Development, 132*, 1649–1661.

Kurimoto, K., Yabuta, Y., Hayashi, K., Ohta, H., Kiyonari, H., Mitani, T., et al. (2015). Quantitative dynamics of chromatin remodeling during germ cell specification from mouse embryonic stem cells. *Cell Stem Cell, 16*, 517–532.

Kurimoto, K., Yabuta, Y., Ohinata, Y., Shigeta, M., Yamanaka, K., & Saitou, M. (2008). Complex genome-wide transcription dynamics orchestrated by Blimp1 for the specification of the germ cell lineage in mice. *Genes & Development, 22*, 1617–1635.

Lara-Astiaso, D., Weiner, A., Lorenzo-Vivas, E., Zaretsky, I., Jaitin, D. A., David, E., et al. (2014). Chromatin state dynamics during blood formation. *Science, 345*, 943–949.

Lawson, K. A., Dunn, N. R., Roelen, B. A., Zeinstra, L. M., Davis, A. M., Wright, C. V., et al. (1999). Bmp4 is required for the generation of primordial germ cells in the mouse embryo. *Genes & Development, 13*, 424–436.

Lee, K. L., Lim, S. K., Orlov, Y. L., Yit, L. Y., Yang, H., Ang, L. T., et al. (2011). Graded nodal/activin signaling titrates conversion of quantitative phospho-Smad2 levels into qualitative embryonic stem cell fate decisions. *PLoS Genetics*, 7, e1002130.

Leitch, H. G., Blair, K., Mansfield, W., Ayetey, H., Humphreys, P., Nichols, J., et al. (2010). Embryonic germ cells from mice and rats exhibit properties consistent with a generic pluripotent ground state. *Development, 137*, 2279–2287.

Leitch, H. G., Okamura, D., Durcova-Hills, G., Stewart, C. L., Gardner, R. L., Matsui, Y., et al. (2014). On the fate of primordial germ cells injected into early mouse embryos. *Developmental Biology, 385*, 155–159.

Li, Z., White, P., Tuteja, G., Rubins, N., Sackett, S., & Kaestner, K. H. (2009). Foxa1 and Foxa2 regulate bile duct development in mice. *The Journal of Clinical Investigation, 119*, 1537–1545.

Liu, P., Wakamiya, M., Shea, M. J., Albrecht, U., Behringer, R. R., & Bradley, A. (1999). Requirement for Wnt3 in vertebrate axis formation. *Nature Genetics, 22*, 361–365.

Ma, Z., Swigut, T., Valouev, A., Rada-Iglesias, A., & Wysocka, J. (2011). Sequence-specific regulator Prdm14 safeguards mouse ESCs from entering extraembryonic endoderm fates. *Nature Publishing Group, 18*, 120–127.

Magnúsdóttir, E., Dietmann, S., Murakami, K., Günesdogan, U., Tang, F., Bao, S., et al. (2013). A tripartite transcription factor network regulates primordial germ cell specification in mice. *Nature Cell Biology, 15*, 905–915.

Martin, G. R. (1981). Isolation of a pluripotent cell line from early mouse embryos cultured in medium conditioned by teratocarcinoma stem cells. *Proceedings of the National Academy of Sciences of the United States of America, 78*, 7634–7638.

Matsui, Y., Zsebo, K., & Hogan, B. L. (1992). Derivation of pluripotential embryonic stem cells from murine primordial germ cells in culture. *Cell, 70*, 841–847.

Mendenhall, E. M., Williamson, K. E., Reyon, D., Zou, J. Y., Ram, O., Joung, J. K., et al. (2013). Locus-specific editing of histone modifications at endogenous enhancers. *Nature Biotechnology, 31*, 1133–1136.

Mishina, Y., Suzuki, A., Ueno, N., & Behringer, R. R. (1995). Bmpr encodes a type I bone morphogenetic protein receptor that is essential for gastrulation during mouse embryogenesis. *Genes & Development, 9*, 3027–3037.

Mould, A., Morgan, M. A. J., Li, L., Bikoff, E. K., & Robertson, E. J. (2012). Blimp1/Prdm1 governs terminal differentiation of endovascular trophoblast giant cells and defines multipotent progenitors in the developing placenta. *Genes & Development, 26*, 2063–2074.

Mullen, A. C., Orlando, D. A., Newman, J. J., Lovén, J., Kumar, R. M., Bilodeau, S., et al. (2011). Master transcription factors determine cell-type-specific responses to TGF-b signaling. *Cell, 147*, 565–576.

Nagamatsu, G., Saito, S., Takubo, K., & Suda, T. (2015). Integrative analysis of the acquisition of pluripotency in PGCs reveals the mutually exclusive roles of Blimp-1 and AKT signaling. *Stem Cell Reports, 5*, 111–124.

Nakaki, F., Hayashi, K., Ohta, H., Kurimoto, K., Yabuta, Y., & Saitou, M. (2013). Induction of mouse germ-cell fate by transcription factors *in vitro*. *Nature, 501*, 222–226.

Nakaki, F., & Saitou, M. (2014). PRDM14: A unique regulator for pluripotency and epigenetic reprogramming. *Trends in Biochemical Sciences, 39*, 289–298.

Ohinata, Y., Ohta, H., Shigeta, M., Yamanaka, K., Wakayama, T., & Saitou, M. (2009). A signaling principle for the specification of the germ cell lineage in mice. *Cell, 137*, 571–584.

Ohinata, Y., Payer, B., O'Carroll, D., Ancelin, K., Ono, Y., Sano, M., et al. (2005). Blimp1 is a critical determinant of the germ cell lineage in mice. *Nature*, *436*, 207–213.

Okashita, N., Kumaki, Y., Ebi, K., Nishi, M., Okamoto, Y., Nakayama, M., et al. (2013). PRDM14 promotes active DNA demethylation through the Ten-eleven translocation (TET)-mediated base excision repair pathway in embryonic stem cells. *Development*, *141*, 269–280.

Papp, B., & Plath, K. (2013). Epigenetics of reprogramming to induced pluripotency. *Cell*, *152*, 1324–1343.

Payer, B., Rosenberg, M., Yamaji, M., Yabuta, Y., Koyanagi-Aoi, M., Hayashi, K., et al. (2013). Tsix RNA and the germline factor, PRDM14, link X reactivation and stem cell reprogramming. *Molecular Cell*, *52*, 805–818.

Pirouz, M., Pilarski, S., & Kessel, M. (2013). A critical function of Mad2l2 in primordial germ cell development of mice. *PLoS Genetics*, *9*, e1003712.

Pouponnot, C., Jayaraman, L., & Massagué, J. (1998). Physical and functional interaction of SMADs and p300/CBP. *The Journal of Biological Chemistry*, *273*, 22865–22868.

Powell, D. R., Hernandez-Lagunas, L., LaMonica, K., & Artinger, K. B. (2013). Prdm1a directly activates foxd3 and tfap2a during zebrafish neural crest specification. *Development*, *140*, 3445–3455.

Resnick, J. L., Bixler, L. S., Cheng, L., & Donovan, P. J. (1992). Long-term proliferation of mouse primordial germ cells in culture. *Nature*, *359*, 550–551.

Sasaki, K., Yokobayashi, S., Nakamura, T., Okamoto, I., Yabuta, Y., Kurimoto, K., et al. (2015). Robust *in vitro* induction of human germ cell fate from pluripotent stem cells. *Cell Stem Cell*, *17*, 178–194.

Schemmer, J., Araúzo-Bravo, M. J., Haas, N., Schäfer, S., Weber, S. N., Becker, A., et al. (2013). Transcription factor TFAP2C regulates major programs required for murine fetal germ cell maintenance and haploinsufficiency predisposes to teratomas in male mice. *PLoS One*, *8*, e71113.

Schoenfelder, S., Furlan-Magaril, M., Mifsud, B., Tavares-Cadete, F., Sugar, R., Javierre, B.-M., et al. (2015). The pluripotent regulatory circuitry connecting promoters to their long-range interacting elements. *Genome Research*, *25*, 582–597.

Schoenfelder, S., Sugar, R., Dimond, A., Javierre, B.-M., Armstrong, H., Mifsud, B., et al. (2015). Polycomb repressive complex PRC1 spatially constrains the mouse embryonic stem cell genome. *Nature Genetics*, *47*, 1179–1186.

Seki, Y., Yamaji, M., Yabuta, Y., Sano, M., Shigeta, M., Matsui, Y., et al. (2007). Cellular dynamics associated with the genome-wide epigenetic reprogramming in migrating primordial germ cells in mice. *Development*, *134*, 2627–2638.

Shovlin, T. C., Durcova-Hills, G., Surani, A., & McLaren, A. (2008). Heterogeneity in imprinted methylation patterns of pluripotent embryonic germ cells derived from pre-migratory mouse germ cells. *Developmental Biology*, *313*, 674–681.

Spitz, F., & Furlong, E. E. M. (2012). Transcription factors: From enhancer binding to developmental control. *Nature Reviews. Genetics*, *13*, 613–626.

Stevens, L. C. (1967). Origin of testicular teratomas from primordial germ cells in mice. *Journal of the National Cancer Institute*, *38*, 549–552.

Su, S.-T., Ying, H.-Y., Chiu, Y.-K., Lin, F.-R., Chen, M.-Y., & Lin, K.-I. (2009). Involvement of histone demethylase LSD1 in Blimp-1-mediated gene repression during plasma cell differentiation. *Molecular and Cellular Biology*, *29*, 1421–1431.

Sugawa, F., Araúzo-Bravo, M. J., Yoon, J., Kim, K.-P., Aramaki, S., Wu, G., et al. (2015). Human primordial germ cell commitment *in vitro* associates with a unique PRDM14 expression profile. *The EMBO Journal*, *34*, 1009–1024.

Sugimoto, M., Kondo, M., Koga, Y., Shiura, H., Ikeda, R., Hirose, M., et al. (2015). A simple and robust method for establishing homogeneous mouse epiblast stem cell lines by wnt inhibition. *Stem Cell Reports*, *4*, 744–757.

Sumi, T., Oki, S., Kitajima, K., & Meno, C. (2013). Epiblast ground state is controlled by canonical Wnt/β-catenin signaling in the postimplantation mouse embryo and epiblast stem cells. *PLoS One*, *8*, e63378.

Tada, T., Tada, M., Hilton, K., Barton, S. C., Sado, T., Takagi, N., et al. (1998). Epigenotype switching of imprintable loci in embryonic germ cells. *Development Genes and Evolution*, *207*, 551–561.

Tam, P. P., & Zhou, S. X. (1996). The allocation of epiblast cells to ectodermal and germ-line lineages is influenced by the position of the cells in the gastrulating mouse embryo. *Developmental Biology*, *178*, 124–132.

Tanaka, S., Kunath, T., Hadjantonakis, A. K., Nagy, A., & Rossant, J. (1998). Promotion of trophoblast stem cell proliferation by FGF4. *Science*, *282*, 2072–2075.

Tesar, P. J., Chenoweth, J. G., Brook, F. A., Davies, T. J., Evans, E. P., Mack, D. L., et al. (2007). New cell lines from mouse epiblast share defining features with human embryonic stem cells. *Nature*, *448*, 196–199.

Tsakiridis, A., Huang, Y., Blin, G., Skylaki, S., Wymeersch, F., Osorno, R., et al. (2014). Distinct Wnt-driven primitive streak-like populations reflect *in vivo* lineage precursors. *Development*, *141*, 1209–1221.

Vincent, S. D., Dunn, N. R., Sciammas, R., Shapiro-Shalef, M., Davis, M. M., Calame, K., et al. (2005). The zinc finger transcriptional repressor Blimp1/Prdm1 is dispensable for early axis formation but is required for specification of primordial germ cells in the mouse. *Development*, *132*, 1315–1325.

Visel, A., Blow, M. J., Li, Z., Zhang, T., Akiyama, J. A., Holt, A., et al. (2009). ChIP-seq accurately predicts tissue-specific activity of enhancers. *Nature*, *457*, 854–858.

Wan, H., Dingle, S., Xu, Y., Besnard, V., Kaestner, K. H., Ang, S.-L., et al. (2005). Compensatory roles of Foxa1 and Foxa2 during lung morphogenesis. *The Journal of Biological Chemistry*, *280*, 13809–13816.

Wang, A., Yue, F., Li, Y., Xie, R., Harper, T., Patel, N. A., et al. (2015). Epigenetic priming of enhancers predicts developmental competence of hESC-derived endodermal lineage intermediates. *Cell Stem Cell*, *16*, 386–399.

Wang, J., Zhuang, J., Iyer, S., Lin, X., Whitfield, T. W., Greven, M. C., et al. (2012). Sequence features and chromatin structure around the genomic regions bound by 119 human transcription factors. *Genome Research*, *22*, 1798–1812.

Weber, S., Eckert, D., Nettersheim, D., Gillis, A. J. M., Schäfer, S., Kuckenberg, P., et al. (2010). Critical function of AP-2 gamma/TCFAP2C in mouse embryonic germ cell maintenance. *Biology of Reproduction*, *82*, 214–223.

Werling, U., & Schorle, H. (2002). Transcription factor gene AP-2 gamma essential for early murine development. *Molecular and Cellular Biology*, *22*, 3149–3156.

Winnier, G., Blessing, M., Labosky, P. A., & Hogan, B. L. (1995). Bone morphogenetic protein 4 is required for mesoderm formation and patterning in the mouse. *Genes & Development*, *9*, 2105–2116.

Yamaguchi, T. P., Harpal, K., Henkemeyer, M., & Rossant, J. (1994). fgfr-1 is required for embryonic growth and mesodermal patterning during mouse gastrulation. *Genes & Development*, *8*, 3032–3044.

Yamaji, M., Seki, Y., Kurimoto, K., Yabuta, Y., Yuasa, M., Shigeta, M., et al. (2008). Critical function of Prdm14 for the establishment of the germ cell lineage in mice. *Nature Genetics*, *40*, 1016–1022.

Yamaji, M., Ueda, J., Hayashi, K., Ohta, H., Yabuta, Y., Kurimoto, K., et al. (2013). PRDM14 ensures naive pluripotency through dual regulation of signaling and epigenetic pathways in mouse embryonic stem cells. *Cell Stem Cell*, *12*, 368–382.

Yang, S.-H., Kalkan, T., Morissroe, C., Marks, H., Stunnenberg, H., Smith, A., et al. (2014). Otx2 and Oct4 drive early enhancer activation during embryonic stem cell transition from naive pluripotency. *Cell Reports*, *7*, 1968–1981.

Ying, Q.-L., Wray, J., Nichols, J., Batlle-Morera, L., Doble, B., Woodgett, J., et al. (2008). The ground state of embryonic stem cell self-renewal. *Nature*, *453*, 519–523.
Yu, J., Angelin-Duclos, C., Greenwood, J., Liao, J., & Calame, K. (2000). Transcriptional repression by blimp-1 (PRDI-BF1) involves recruitment of histone deacetylase. *Molecular and Cellular Biology*, *20*, 2592–2603.
Zaret, K. S., & Carroll, J. S. (2011). Pioneer transcription factors: Establishing competence for gene expression. *Genes & Development*, *25*, 2227–2241.

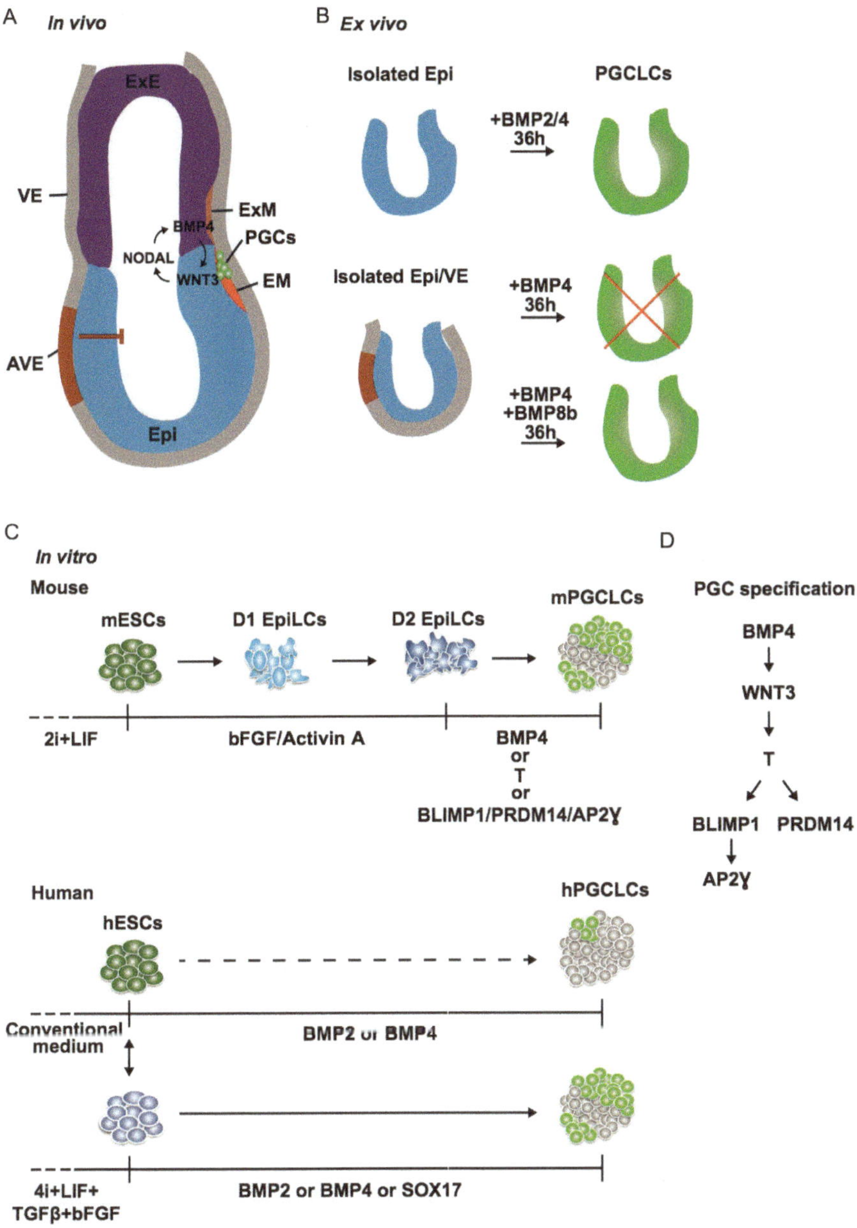

Ufuk Günesdogan and M. Azim Surani, Figure 1 PGC specification *in vivo* and *in vitro*. (A) BMP4 signaling from the ExE specifies PGCs in the posterior epiblast of postimplantation embryos at ~E6.5. Nodal signaling induces the expression of *Bmp4* in the ExE, which in turn induces the expression of *Wnt3* in the epiblast. WNT3 maintains

(Continued)

Ufuk Günesdogan and M. Azim Surani, Figure 1—Cont'd *Nodal* expression. It was shown that BMP and WNT signaling are required for the induction of PGCs. The AVE is source for inhibitory signals against posteriorization of the anterior embryo. ExE, extraembryonic ectoderm, (A) VE, (anterior) visceral endoderm; Epi, epiblast; EM, embryonic mesoderm; PGCs, primordial germ cells; ExM, extraembryonic mesoderm. (B) Postimplantation epiblasts separated from the extraembryonic tissues respond to BMP2 or BMP4 and large numbers of cells acquire PGC-like fate in *ex vivo* culture. In the presence of the AVE, the addition of both BMP4 and BMP8b but not of BMP4 alone is sufficient for the induction of PGCLCs. PGCLCs, PGC-like cells. (C) In mouse, ESCs (mESCs) are differentiated into EpiLCs in the presence of bFGF and Activin A. Day 2 EpiLCs respond to BMP4 and give rise to mouse PGCLCs (mPGCLCs). Similarly, the induced expression of BLIMP1/PRDM14/AP2γ or T results in mPGCLC induction. The direct induction of human PGCLCs (hPGCLCs) with BMP2 or BMP4 from human ESCs (hESCs) cultured in conventional medium is very inefficient. In contrast, hESCs in 4i conditions reversibly gain and retain competence for hPGCLCs and acquire germ cell fate in response to BMP2 or BMP4. Unlike in the mouse, the induced expression of SOX17 results in the upregulation of germ cell markers. (D) The specification of PGCs in mice relies on BMP4, which induces the expression of WNT3. WNT3 induces the expression of T. T upregulates the expression of BLIMP1 and PRDM14. BLIMP1 induces AP2γ. Together, these transcription factors drive the PGC fate.

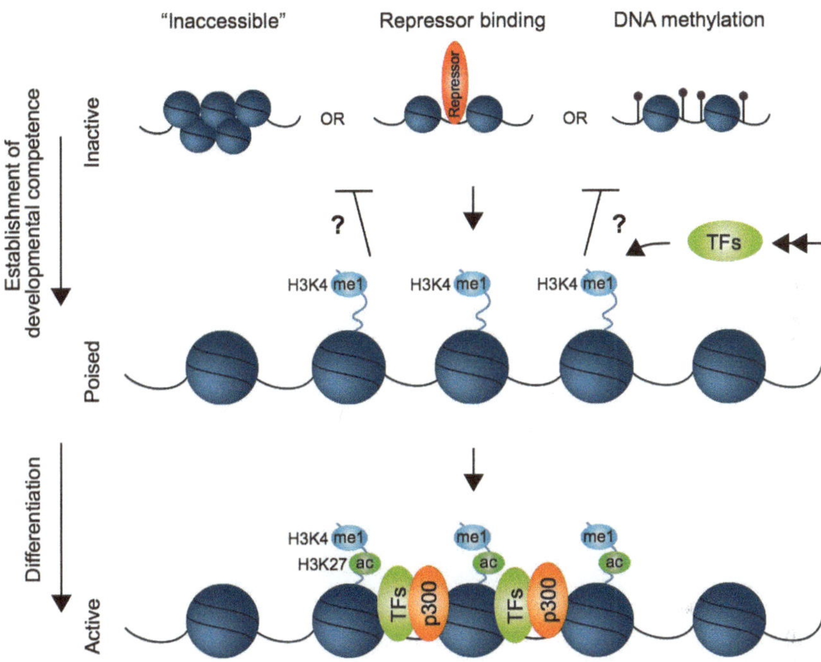

Ufuk Günesdogan and M. Azim Surani, Figure 2 Enhancer setup. Transcriptional enhancers show a combinatorial enrichment of histone modifications. The enrichment of H3K4me1 at enhancers could result in an open chromatin structure, avoid repressor binding or DNA methylation and thus could act as a "placeholder" for transcription factor binding. Upon signaling, transcription factors can engage these enhancer elements to activate transcription of target genes. The activation of an enhancer element is marked by the enrichment of H3K27ac, which could be catalyzed by p300, which is also bound by many enhancers.

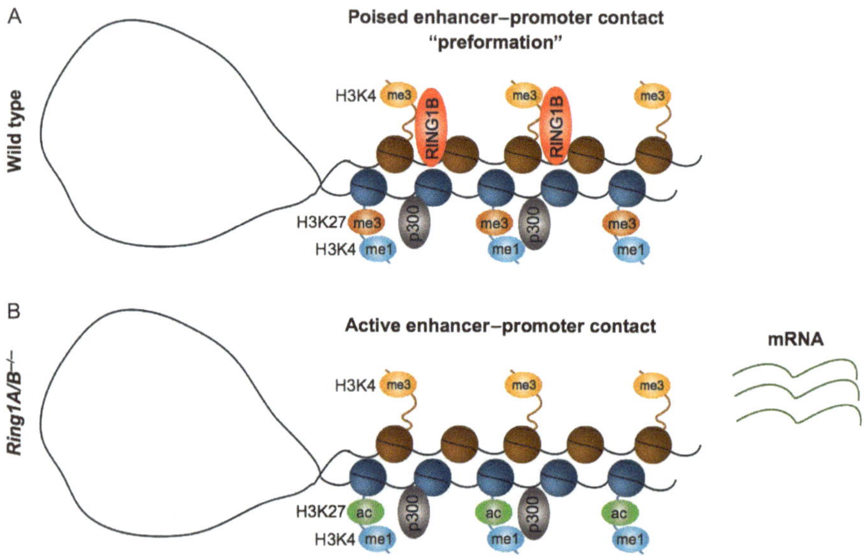

Ufuk Günesdogan and M. Azim Surani, Figure 3 Preformation of enhancer–promoter contacts. (A) In ESCs, RING1B-bound promoters are in contact with poised enhancers that are enriched for H3K4me1, H3K27me3, and p300. (B) In *Ring1A/B* knockout cells, these enhancers loose H3K27me3 but gain H3K27ac, which results in the upregulation of associated target gene expression. Of note, p300, which can catalyze the acetylation of H3K27, appears to be inhibited at poised enhancers. Since the loss of *Ring1A/B* results in enrichment of H3K27ac, perhaps the PRC1 complex is involved in the inhibition of p300 histone acetyltransferase activity.

CHAPTER TWENTY-NINE

Formation of the Embryonic Head in the Mouse: Attributes of a Gene Regulatory Network

Patrick P.L. Tam*,†,1, Nicolas Fossat*,†, Emilie Wilkie*,‡, David A.F. Loebel*,†, Chi Kin Ip*,†, Mirana Ramialison§,¶

*Embryology Unit, Children's Medical Research Institute, Westmead, New South Wales, Australia
†Discipline of Medicine, Sydney Medical School, University of Sydney, Sydney, New South Wales, Australia
‡Bioinformatics Unit, Children's Medical Research Institute, Westmead, New South Wales, Australia
§Australian Regenerative Medicine Institute, Monash University, Clayton, Victoria, Australia
¶Systems Biology Institute Australia, Monash University, Clayton, Victoria, Australia
[1]Corresponding author: e-mail address: ptam@cmri.org.au

We would like to dedicate this publication to Eric Davidson (1937–2015), who pioneered and established the building blocks of gene regulatory networks during embryogenesis. From establishing the core biological principles underlying developmental GRNs, to describing their mathematical properties and establishing software for network visualization, he has paved the grounds for the holistic investigation of complex embryonic systems. His work has inspired developmental and computational biologists (authors included) and will continue to inspire future generations of systems developmental biologists.

Contents

Abstract

The embryonic head is the first major body part to be constructed during embryogenesis. The allocation and the assembly of the progenitor tissues, which start at gastrulation, are accompanied by the spatiotemporal activity of transcription factors and signaling pathways that drives lineage specification, germ layer formation, and cell/tissue movement. The morphogenesis, regionalization, and patterning of the brain

Current Topics in Developmental Biology, Volume 117
ISSN 0070-2153
http://dx.doi.org/10.1016/bs.ctdb.2015.11.019

and craniofacial structures rely on the function of LIM-domain, homeodomain, and basic helix-loop-helix transcription factors. These factors constitute the central nodes of a gene regulatory network (GRN) which encompasses and intersects with signaling pathways involved with head formation. It is predicted that the functional output of this "head GRN" impacts on cellular function and cell–cell interactions that are essential for lineage differentiation and tissue modeling, which are key processes underpinning the formation of the head.

1. ASSEMBLING THE PROGENITOR TISSUES OF THE EMBRYONIC HEAD

Gastrulation is a momentous milestone of early embryogenesis in mammals that marks the formation of the primary germ layers from which all the tissues of the embryo will be generated. A significant outcome of this process is the allocation of the descendants of the pluripotent embryonic cells (the epiblast cells) to different tissue lineages in the body. Lineage allocation is accomplished by the specification of germ layer precursors, followed by progressive restriction of differentiation potency, culminating in the formation of lineage-restricted progenitors for specific tissue types. Fate-mapping studies in mouse embryos at successive developmental stages have generated detailed information on the timing of emergence and the whereabouts of the progenitors of major body parts. Shortly after the establishment of this basic body plan, the primordium of the embryonic head emerges, making it the first major body part to form. Studying the formation of the head is therefore of particular interest to the understanding of the early processes necessary to build the embryo.

The embryonic head is comprised of derivatives of all three primary germ layers. The ectoderm contributes to the neuroectoderm of the brain, surface ectoderm (the skin), and the cranial neural crest, which contributes to peripheral nerves, melanocytes, and parts of the craniofacial skeleton. The cranial mesoderm consists of the precursors of the muscles, connective tissues, and skeletal tissues. The definitive endoderm forms the lining of the pharynx and the foregut. A distinct cell population that emerges from the anterior primitive streak (the organizer region) constitutes the axial mesendoderm (AME) that underlies the midline of the brain and consists of the prechordal plate and the anterior notochord (Arkell & Tam, 2012; Fossat, Jones, Garcia-Garcia, & Tam, 2012; Robb & Tam, 2004; Tam & Behringer, 1997). All the germ layer precursors are initially contained within the epiblast—the tissue descending from the pluripotent inner cell mass cells of the preimplantation embryo. The onset of gastrulation heralds the formation of the mesoderm, the definitive endoderm, and the AME through the morphogenetic activity at the primitive streak whereby cells undergo ingression and transition between epithelial and mesenchymal states. Following the allocation to these germ layers, the remaining epiblast cell population acquires an ectodermal fate. Cell fate analysis showed that precursors of the cranial mesoderm, the anterior (foregut) endoderm, and the anterior AME emerge simultaneously during gastrulation (Robb & Tam, 2004). These tissue precursors are then displaced, by coordinated cell migration, toward the prospective anterior region of the embryo, suggesting that different germ layer derivatives contributing to the embryonic head maintain a consistent neighboring relationship from their inception through the subsequent morphogenetic journey. On reaching their anterior destination these tissue precursors become juxtaposed with the anterior ectoderm that is fated for the brain, and together they constitute the building blocks of the embryonic head.

Morphogenesis of the embryonic head ensues with the formation of the head folds, shaped by the differential growth of specific regions of the anterior neuroectoderm and the physical forces generated by the expanding cranial mesenchyme. The neuroectoderm tissue of the head folds is patterned in the dorsal–ventral dimension to form the roof, lateral walls, and floor plate of the cephalic neural tube. The head folds are also regionalized in the anterior–posterior axis into fore-, mid- and hindbrain, and are further modeled into neuromeres that are segmented domains in the three major brain parts: prosomeres (forebrain), mesomeres (midbrain), and rhombomeres (hindbrain) (Puelles & Rubenstein, 2003). A subtle meristic pattern of somitomeres, which is spatially aligned with the neuromeres, has also been described in the cranial mesoderm (Meier & Tam, 1982; Trainor, Tan, & Tam, 1994). Cranial neural crest cells, a nascent population of ectomesenchymal cells, emerge from the lateral–dorsal margin of the neuroectoderm of the midbrain and the hindbrain and migrate into the cranial mesenchyme. Neural crest cells derived from specific neuromeres contribute to the same branchial arch structures as the neighboring mesoderm-derived mesenchyme (Trainor & Tam, 1995). These two progenitor tissues, together with the endoderm from the spatially aligned segment of the foregut and the overlying surface ectoderm (the "ectomere": Brugmann, Tapadia, & Helms, 2006; Couly & Le Douarin, 1990), give rise to specific facial and pharyngeal primordia. The spatial/segmental alignment of the germ layer- and neural crest-derivatives in the embryonic head resonates with the neighboring relationship of the anterior population of endoderm and mesoderm and anterior AME during gastrulation. It is likely, therefore, that the regionalization and patterning of the embryonic head is driven by a continual morphogenetic interaction that is established during the generation of the progenitor tissues.

2. SIGNALING ACTIVITY AND TISSUE PATTERNING

A requisite step in the establishment of embryonic patterning is the acquisition of asymmetry in the body, which initially occurs in the anterior–posterior embryonic axis. An immediate outcome of this is the patterning of the early embryo into domains of tissues that are allocated to the anterior (head) and posterior (trunk and tail) body parts. Studies in zebrafish and *Xenopus* show that the delineation of tissue domains could be accomplished by modulating the global pattern of signaling activities, such as WNT, BMP, and retinoic acid (RA) in the embryo (Aramaki, Sasai, Yakura, & Sasai, 2010; Dosch & Niehrs, 2000; Inomata, Shibata, Haraguchi, & Sasai, 2013; Lee & Skromne, 2014; Niehrs, 2004). An appealing conjecture is that a gradient of signaling activity is created through the combinatorial action of signaling factors, receptors, agonists, and antagonists of the BMP and WNT pathways, such that varying levels of signaling activity are set up along the anterior–posterior axis of the embryo. In this regard, germ layer tissues in the anterior domain of the embryonic field, where there is diminished signaling activity, are allocated to the head structures, whereas tissues on the opposite side with higher level of signaling activity acquire the posterior fates. While there is no compelling evidence showing the presence of an anterior–posterior gradient of signal or morphogen activity, the findings of progressive loss of anterior structures and/or the posteriorization of the presumptive anterior tissues that accompany a reduced antagonistic activity or the gain in signaling function are consistent with such a paradigm.

In the mouse embryo at the pregastrulation stage, germ layer progenitors are broadly regionalized in the epiblast: ectoderm in the prospective anterior and distal domains and endoderm and mesoderm in the prospective posterior domain, with a predominantly mesoderm domain intercalated between these two regions (Arkell & Tam, 2012; Fossat et al., 2012). Analysis of the expression pattern of signaling pathway genes shows that transcripts encoding signaling factors (e.g., Wnt3, Bmp4, and Nodal) are found predominantly in the posterior epiblast. In contrast, genes encoding the secreted antagonists (e.g., Dkk1, Cer1, and Lefty1) are expressed in the anterior tissues (Perea-Gomez et al., 2002; Pfister, Steiner, & Tam, 2007; Tam & Loebel, 2007). This is consistent with the concept that the anterior tissues secrete antagonists that negatively regulate the level of signaling activity perceived by the germ layer tissues that are allocated to the formation of the embryonic head. The spatial pattern of the Lef/Tcf-LacZ reporter activity and the nuclear localization of β-catenin reveals that germ layer tissues in the anterior domain of the gastrula stage embryo perceive little or no WNT signals where tissues in the posterior domain respond robustly to the WNT signals. This regionalized pattern of signaling activity is maintained across the anterior–posterior axis of the embryo when the building blocks of the embryonic head are being assembled (Fossat et al., 2012, 2011; Lewis et al., 2007, 2008). By the head-fold stage, the "WNT-silent" anterior domain of the embryo encompasses the neuroectoderm of the fore- and midbrain, the cranial mesoderm associated with these brain regions and the foregut endoderm, whereas the more posterior brain parts are responding actively to WNT signals (Fossat et al., 2012, 2011; Lewis et al., 2007, 2008).

The critical role of the signaling antagonists is revealed by the impact of the loss of function of DKK1 (a WNT antagonist), Noggin and Chordin (two BMP antagonists) on head formation. Mutant embryos that are homozygous-null for *Dkk1* (Mukhopadhyay et al., 2001), or heterozygous-null for *Dkk1* and *Nog* (del Barco Barrantes, Davidson, Grone, Westphal, & Niehrs, 2003) or *Nog* and *Chrd* (Bachiller et al., 2000), display a truncation of the anterior (head) structures. It is unclear, however, how the WNT and BMP antagonists interact to modulate the output of the signaling pathways. It has been postulated that the DKK1 and BMP antagonists interact with the intracellular signal transducers (e.g., SMADs), with DKK1 acting via WNT signaling to regulate the expression of SMADs (Singh et al., 2012; Terada, Misao, Katase, Nishimatsu, & Nohno, 2013) and the BMP antagonists act upstream at the ligand–receptor step to regulate transducer activity (Bachiller et al., 2000). Together these antagonists act synergistically to control the level of signaling activity, which points to the co-requirement for regulating both signaling pathways to facilitate head morphogenesis. A study focusing on the WNT pathway has revealed that a stringent control of the signaling activity is critical for the development of the embryonic head (Fossat et al., 2011). Gain-of-function point mutations of *Lrp6* and *Ctnnb1*, encoding the coreceptor and the transducer (β-catenin) of the WNT pathway, respectively, produce a partial phenocopy of the head truncation of the *Dkk1*-null mutant. Compound mutant embryos harboring different combinations of the mutant *Dkk1*, *Lrp6*, and *Ctnnb1* alleles display a range of head truncation phenotypes with varying degrees of loss of anterior structures, from minimal reduction of the forebrain (*Ctnnb1* heterozygous mutant) to complete loss of head structures anterior to the mid–hindbrain junction (triple heterozygous *Dkk1*; *Lrp6*;*Ctnnb1* mutants). The severity of the truncation phenotype is correlated with the gain in WNT signaling activity corresponding to the combination of mutant alleles carried by the embryo (Fossat et al., 2011). This genetic interaction model is one of the first *in vivo* demonstrations of the dependence of the execution of a morphogenetic process on the level of signaling activity perceived by the cells. An immediate implication is that the formation of the head structure at specific anterior–posterior positions is permissible within an intransigent threshold of WNT activity, and this threshold increases progressively for more posterior structures. The role of the antagonists is therefore to fine-tune the signaling activity in order to facilitate the morphogenesis of head structures at specific positions along the anterior–posterior axis (Fossat et al., 2012).

3. LHX1: A TRANSCRIPTIONAL REGULATOR OF HEAD MORPHOGENESIS

A multitude of transcription factors, such as Hesx1, Lhx1, Otx2, Hhex, Foxg1, Tcf7l1, Emx2, Gsc, and Twist1, are known to play critical roles in head development (Andoniadou et al., 2011; Bildsoe et al., 2013; Fossat, Chatelain, Brun, & Lamonerie, 2006; Ip, Fossat, Jones, Lamonerie, & Tam, 2014; Lewis et al., 2007; Martinez Barbera et al., 2000; Soo et al., 2002; Xuan et al., 1995; Yoshida et al., 1997). Mutations of the genes encoding these transcription factors lead to abnormal development of craniofacial structures, which can be attributed to the disruption of the process that coordinates the assembly and the differentiation of the tissue precursors and the morphogenesis of the embryonic head. As such, they constitute the head "kernel," defined by E. Davidson as a subnetwork of genes critical for the building of the head, which are embedded at the core of a more complex developmental gene regulatory network (GRN) (Davidson & Erwin, 2006).

Lhx1 encodes a LIM-domain homeobox transcription factor that is expressed in the epiblast-derived mesoderm, anterior endoderm, and AME that all contribute to the progenitor tissues of the embryonic head (Barnes, Crosby, Jones, Wright, & Hogan, 1994; Fossat et al., 2015; Shimono & Behringer, 1999). Null *Lhx1* mutant embryos develop poorly during gastrulation and the mesoderm and the endoderm fail to form properly (Shawlot & Behringer, 1995; Tam, Khoo, Wong, Tsang, & Behringer, 2004). Mutant embryos that develop past the gastrulation stage display a truncated head, which phenocopies the *Dkk1*-null mutant (Shawlot & Behringer, 1995; Fig. 1A). This complex and variable phenotype indicates that Lhx1 may act in more than one type of tissue that contributes to the head structures before and during gastrulation. Previous studies have demonstrated that head formation requires the activity of Lhx1 in the visceral endoderm (an extraembryonic structure) before gastrulation (Shawlot et al., 1999; Shimono & Behringer, 2003).

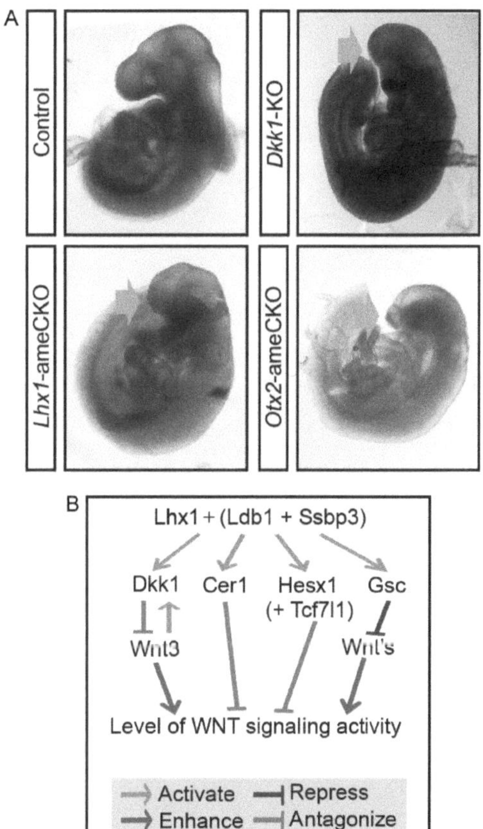

Figure 1 (A) Phenotype of *Dkk1* knockout (KO), *Lhx1* anterior mesendoderm (ame) conditional knockout (CKO), and *Otx2*-ameCKO mutant embryos. Control = wild-type embryo; embryonic age = E9.5; arrow, head truncation. (B) Lhx1 downstream molecular activities in the anterior mesendoderm impacting on the level of WNT signaling activity. (A color version of this figure is located at the end of this chapter.)

Our recent study, conducted using a conditional gene knockout (CKO) approach, has shown that *Lhx1* activity is required at two sequential stages during gastrulation (Fossat et al., 2015). Ablation of *Lhx1* activity in the epiblast early in gastrulation results in head truncation and impacts on the germ layer tissues that are derived from the epiblast (mesoderm, endoderm, and AME) and in which *Lhx1* is expressed. *Lhx1*-deficient mesoderm loses the expression of *Pcdh8*, which encodes the paraxial mesoderm protocadherin. As a consequence, the movement of the mesoderm cells is impeded (Hukriede et al., 2003). However, the ablation of *Lhx1* specifically in the mesoderm (Fossat et al., 2015) and the inactivation of *Pcdh8* in the embryo (Yamamoto, Kemp, Bachiller, Geissert, & De Robertis, 2000) are inconsequential for head formation, suggesting that *Lhx1* function is not essential in the mesoderm for this process. In the *Lhx1* epiblast-CKO mutant, the AME is absent as revealed by the loss of expression of tissue-specific genes (e.g., *Shh*, *Dkk1*, *Cer1*, *Hesx1*, and *Gsc*) (Fossat et al., 2015). The failure to form the AME is associated with the downregulation of another protocadherin, encoded by *Pcdh7*, which is a transcriptional target of Lhx1 (Fossat et al., 2015). Pcdh7 is involved with cell movement, cell shape, and cell adhesion (Li et al., 2013; Rashid, Newell, Shama, & Bradley, 2006; Yoshida, 2003). Downregulation of *Pcdh7* in the *Lhx1* epiblast-CKO mutant is associated with the impeded translocation of the AME precursors (expressing *Foxa2*, *Nog*, and *Chrd*) from the epiblast to the endoderm and the anterior extension of the AME (Fossat et al., 2015).

Conditional knockout of *Lhx1* in the AME (AME-CKO) results in the loss of head structures (Fig. 1A). Whereas the AME is formed (and expresses *Shh*), the expression of *Dkk1*, *Hesx1*, *Gsc*, and *Cer1* is reduced or abolished (Fossat et al., 2015). These genes possess functional Lhx1-binding motifs in their regulatory regions, can be activated by Lhx1 (Fossat et al., 2015) and are involved with the negative modulation of WNT signaling activity: *Dkk1* (acting on Lrp6 to curtail receptor activity) (Zorn, 2001), *Hesx1* (enabling the repression of WNT target by interacting with the inhibitory TCF, Tcf7l1) (Andoniadou et al., 2011), *Gsc* (repressor of *Wnt* transcription) (Lewis et al., 2007; Yao & Kessler, 2001), and *Cer1* (secreted antagonist to BMP and WNT signaling) (Piccolo et al., 1999; Fig. 1B). Loss of the antagonistic activity against WNT signaling in both AME- and epiblast-CKO mutants could lead to excessive WNT signaling activity (Fossat et al., 2015). The head defects in embryos that are heterozygous for a *Lhx1*-null allele is exacerbated on a *Dkk1*-null heterozygous or a *Ctnnb1* gain-of-function heterozygous background (Fossat et al., 2015), suggesting

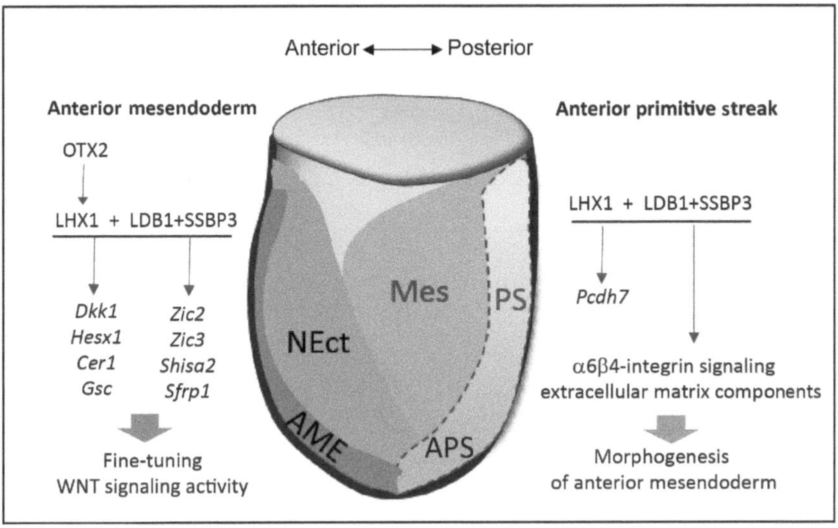

Figure 2 Functional context-specific requirement of Lhx1 activity for the formation of anterior mesoderm via the influence on the transit of progenitor from the anterior primitive streak (APS) to the midline and the anterior extension of the tissue, and for the regulation of the transcription of genes in the anterior mesendoderm (AME) that encode factors for modulating the level of WNT signaling activity in the anterior germ layer tissues. Mes, mesoderm; NEct, neuroectoderm; PS, primitive streak. (A color version of this figure is located at the end of this chapter.)

that gain in WNT signaling activity may be the factor underpinning the truncation of head structures in *Lhx1*-deficient embryos. In the context of silencing WNT signaling activity, Lhx1 acts at three points in the WNT signaling cascade: the antagonist against the ligand–receptor activity, the inhibitory transcription cofactor and the transcriptional repressor of signaling factors. Lhx1 is required for context-related functions: first for the formation of the AME that is the source of WNT modulating signals during gastrulation and, later at the onset of head-fold formation, for the activation of genes that encode the modulating factors for WNT signaling (Fig. 2; Fossat et al., 2015, unpublished data).

4. THE OTX2–LHX1–WNT CASCADE

Transcriptome analysis of *Lhx1* epiblast-CKO and *Dkk1*-null mutant embryos at the late gastrulation stage (unpublished data) offers an opportunity to address the molecular consequence of an excess of WNT signaling. Comparison of the gene expression profiles of these two mutant embryos

Table 1 Common Downstream Activity in the Head Progenitor Tissues of the Late Gastrula Stage *Lhx1* Epiblast Conditional Knockout and *Dkk1* Knockout Mutant Embryos

Functional Category	Downregulated	Upregulated
WNT antagonist	*Shisa2, Sfrp1, Dkk1*	
WNT agonist	*Fzd5*	*Fzd10, Rspo3, Wnt8a*
WNT target		*Fst, Tnfrsf19, Axin2*
Neural and neural crest differentiation	*Sox2, Sox9, Pou3f1, Otx2*	*Gbx2, Arg1*
Endoderm differentiation	*Hhex*	*Cdx2, Nkx2.1, Zfp703*
Expression in anterior mesendoderm/notochord	*Parm1*	
Posterior tissue differentiation		*Hoxa1, Hoxb1, Tbx6, Fgf10, Meox1, Tshz1, Dll1, Meis2*
Biochemical pathways	*Slc27a2*	*Gna14, Aldh1a2, Rnf213*

(Table 1) revealed the concurrent downregulation of antagonists and upregulation of agonists and signaling targets. This is accompanied by the downregulation of genes associated with the differentiation of neural crest cells and neural progenitors, the anterior endoderm and the mesendoderm, which is consistent with the loss of anterior neural tissues and the anterior mesendoderm. In contrast, there is increased expression of genes associated with differentiation of posterior embryonic tissues: hindbrain, posterior foregut, and trunk mesoderm (Table 1). Raising the level of WNT signaling activity above a threshold may therefore not only block the differentiation of the head progenitors but divert them to acquire the posterior tissue fates. The head truncation phenotype may therefore be the result of both loss of anterior attributes and the posteriorization of the progenitor tissues.

In both the *Lhx1* epiblast-CKO and the *Dkk1* mutant embryos, differentiation of anterior neural tissues is disrupted, and the progenitor tissues lose their neural identity as revealed by the loss of expression of lineage-specific genes including the orthodenticle homeobox transcription factor, *Otx2* (Fossat et al., 2015 and unpublished data). Loss-of-function mutation of *Otx2* results in head truncation, which phenocopies that of the *Lhx1* and *Dkk1* mutant embryos (Fig. 1A), implying that Otx2 may interface with the Lhx1–WNT molecular cascade for head formation. Otx2 is known to regulate *Dkk1*, *Cer1*, *Gsc*, and *Hesx1* (Ip et al., 2014; Kimura-Yoshida et al., 2005; Spieler et al., 2004; Sudou, Yamamoto, Ogino, & Taira,

2012; Yasuoka et al., 2014), all transcriptional targets of Lhx1. Otx2 also regulates the expression of *Lhx1*, therefore placing it upstream of the Lhx1–WNT cascade (Ip et al., 2014; Fig. 2). Otx2 and Lhx1 can also directly interact to control gene expression (Nakano, Murata, Matsuo, & Aizawa, 2000; Yasuoka et al., 2014). While Otx2 may function via other downstream molecular pathways (particularly in the anterior neurectoderm where *Lhx1* is not expressed), it is likely that the role of Otx2 for head formation in the AME overlaps with that of Lhx1 and is mediated principally by Lhx1 activity through the regulation of morphogenetic cell movement and WNT signaling activity.

5. THE HEAD GENE REGULATORY NETWORK

The findings so far strongly indicate that *Lhx1* is a cornerstone gene in the cascade of molecular activity that drives head development. Lhx1 regulates the transcription of genes that are required for the fine-tuning of signaling, in particular by controlling WNT signaling activity. Lhx1 also controls the activity of genes, such as *Pcdh7*, that drive morphogenetic cells movements (Fossat et al., 2015). Mutations in many Lhx1 downstream genes result in abnormal development of the head and face structures (Table 2), indicating that while they may collectively account for the severe head truncation phenotype, the activity of individual genes may contribute to specific components of the developmental process.

While studies undertaking a gene-by-gene approach have provided useful insight into the functional role of individual genes, a systems-based approach is required to elucidate the collective functional output of the array of genes that are necessary for head formation. Identifying the ensemble of Lhx1-dependent genes that interact in a GRN constitutes a starting point to comprehend the molecular mechanisms that control head formation. Elucidation of the GRN must also take into account other regulators that act in

Table 2 Functional Correlates of Lhx1 Downstream Genes in Embryonic Development

Genes associated with abnormal head development/brain morphology	*Hesx1*, *Sfrp1*, *Shisa2 Taku*, *Zic2*, *Fyn*, *Ccnd2*, *Flnb*, *Lamc1 Sama3f*, *Epha7*, *Ephb1*, *Robo1*, *Gbx2*, *Rxrg*, *Tead1*, *Il17rd*, *Dll3*, *Sall4*, *Zic3*, *Col2a1*, *Ipmk*, *Cyp26a1*, *Fgfr2*
Genes associated with abnormal patterning	*Lama1*, *Lamb1*, *Eomes*, *Tjp2*, *Ext1*, *Map4k4*

concert with Lhx1. Otx2 is an example of a transcriptional regulator acting in conjunction with Lhx1 in the GRN (Ip et al., 2014; Nakano et al., 2000; Yasuoka et al., 2014; Zakin et al., 2000). Genome-wide screen of binding motifs for Lhx1 and Otx2 in the *Xenopus* embryo at the stage of embryonic head formation has revealed binding sites for these two transcription factors in the putative regulatory element of many genes, pointing to the likelihood that Lhx1 and Otx2 may cooperatively regulate the transcription of a significant subset of genes in the genome for head formation (Yasuoka et al., 2014).

As the information of the functional interaction among the Lhx1 and Otx2 downstream genes in the mouse embryo is not yet available, a hypothetical GRN associated with head formation was assembled based on the analysis of the Lhx1 and Otx2 occupancy data available in *Xenopus* (Yasuoka et al., 2014). We then identified the genes that are associated with conserved binding sites in the *Xenopus* and the mouse and that are expressed in the anterior region of the late gastrula stage mouse embryo. Initial functional analysis of these genes, conducted by testing for enriched Gene Ontology Biological Pathways (Table 3), places the Lhx1 targets in the canonical pathway of Erk-MAPK/planar cell polarity, cell motility, WNT-calcium signaling, and glucose/glycogen metabolism, and the Otx2 targets in TGFβ/Wnt signaling, steroid signaling, cell cycle regulation, and stem cell potency. Genes that may be cooperatively regulated by Lhx1 and Otx2 are primarily involved with signaling pathways, such as those for RA, WNT/β-catenin, Notch, and Ras GTPase.

Through incorporating protein–protein interaction data, a model GRN was constructed, revealing the interactions between the putative target genes and potential downstream effectors (Cowley et al., 2012; Newman, 2006). The resulting third neighbor network of 3507 nodes and 7944 interactions has a high level of interconnectivity, and of the 38 communities identified, 31 where functionally enriched for biological pathways (Fig. 3; Gene Ontology Consortium, 2015; Mi et al., 2013). These pathways include processes relating to transcription, RNA processing, regulation of signaling, activation of MAPK activity, WNT signaling pathway, BMP signaling pathway, G protein-coupled receptor signaling pathway, fusion and transport, cell–cell adhesion, cell communication, and cell migration. In essence, the collective functional output of the Lhx1/Otx2 target genes is predominantly for regulating the signaling activity that influences cell–cell interaction, differentiation, and morphogenetic movement.

The landscape of WNT signaling activity in the anterior (head) tissue changes during the development of the embryonic head (Fossat et al., 2012).

Table 3 Lhx1–Otx2 Downstream Molecular Pathways

	Molecular Pathways	Genes	Upstream Regulators
Lhx1	Planar cell polarity/ Erk-MAPK	*Celsr1, Adra2b, Adora2b, Gnai2, Dusp1, Dusp4*	Commd3-Bmi1, Ctnnb1, Pitx2, Ptbp1, E2f1
	Regulation of actin-based motility	*Rhob, Rac3, Pip4k2c*	
	WNT-calcium signaling	*Fzd4, Fzd10, Id4*	
	Glucose/glycogen metabolism	*Pgm1, Pgm5*	
Otx2	Androgen signaling	*Hspa4, Jun, Kat2b, Ercc3, Crebbp, Prkch, Gng3, Gtf2a1, Taf2, Gnao1, Gna11, Gna13*	Sox2, Ctnnb1, Foxc1, Foxc2, Histone H3
	Gene regulatory network: Stem cells	*Gbx2, Isl1, Zic3, Cdx2, Skil, Otx1, Hoxb1*	
	Molecular mechanisms of cancer/cell cycle regulation	*Rap2b, Rhot2, Nfkbia, Rhou, E2f3, E2f5, Rras, Ptch1, Ptch2, Ccne1, Ccnd2, Rabif, Aph1a, Prkch, Cdk2, Cdk5*	
	TGFβ/Wnt signaling	*Hoxb5, Nkx2-5, Sp4, Jun, Ube2b, Hsp90ab1, C5nk1d, Pten, Dkk1, Tcf7, Tcf7l2, Znf423, Samd1, Smad6, Tgfb2*	
Lhx1 and Otx2	Retinoic acid receptor activation/ EGF/ PDGF signaling	*Sra1, Ccnh, Rdh10, Cyp26a1, Arid2m Map3k1, Mapk11, Nr2f1, Pik3r3, Csnk2a2, Brd7, Prkar2b, Aldh1a2, Carm1, Prkcb, Srf, Rasa1*	Ctnnb1, Sox2, Tcf7l1, Gdf11/ Bmp11, Sf3b2
	Wnt/β-catenin signaling	*Sox3, Sox4, Sox11, Gnaq, Tle1, Tel4, Csnk2a2, Fzd2, Fzd5, Fzd8, Ccnd1, Sfrp5, Lef1*	
	Notch signaling	*Dll1, Hes5, Lfng, Hes1, Jag1, Hey1*	
	Pluripotency	*Sox2, Pik3r3, Dzd8, Id2, T, Left1, Fzd5, Mapk11, Dzd2, Mesp1, Hesx1, Gata6, Eomes, Foxc1, Gsc*	
	Ras family of small GTPase signaling	*Kpanb1, Cseil, Xpo1, Ipo5*	

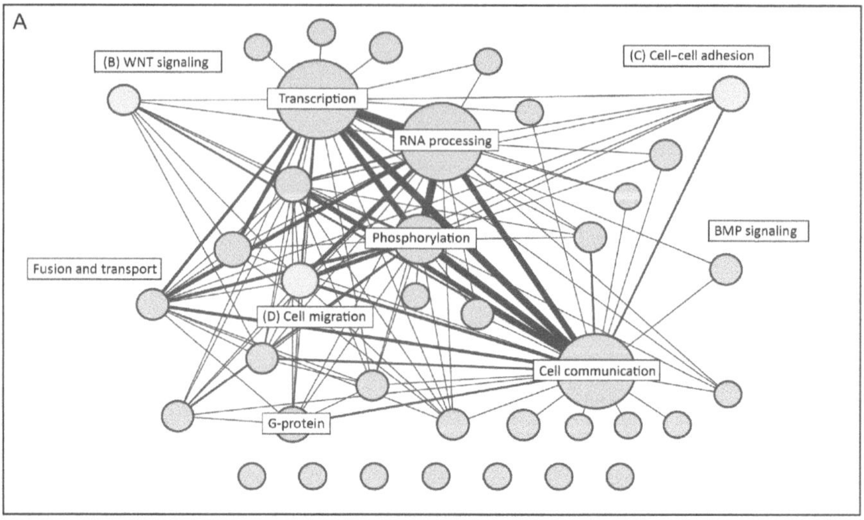

Figure 3 The head gene regulatory network (GRN). This network was constructed using data from a ChIP-Seq study on *Xenopus* embryos (Yasuoka et al., 2014). Conserved binding sites were identified in the mouse and the associated genes were identified based on binding by Otx2 and/or Lhx1 within a 5000 bp upstream and 1000 bp downstream of the transcription start site. This list of candidate target genes was compared to the transcriptome of the anterior region of the late gastrula stage mouse embryo obtained by microarray analysis. This analysis identified 102 Lhx1, 595 Otx2, and 386 Lhx1 and Otx2 putative target genes that are expressed in the anterior region of the mouse embryo. By incorporating protein–protein interaction data from the PINA database (Cowley et al., 2012) a GRN to the depth of third neighbors from the Lhx1 and Otx2 putative gene targets was constructed. The resulting network contains 3507 nodes and 7944 interactions. To gain functional insight into the network, community analysis was performed through modularity clustering (Newman, 2006), revealing 38 communities. Functional Analysis of each community was performed by hypergeometric testing (p-value < 0.001) using the Biological Pathways from the Gene Ontology Consortium (2015) and GOSlim Biological Pathways from the Panther database (Mi, Muruganujan, Casagrande, & Thomas, 2013), consistency between the 10 most significant terms for each module was used for assessment of functional relevance. (A) Network of communities. Edge thickness is weighted according to number of interactions traversing communities. The size of each node is weighted by the number of genes assigned to a particular community. The network structure of interacting genes of three communities (highlighted in yellow) identified by the functional analysis are shown: (B) WNT signaling pathway, (C) cell–cell adhesion, and (D) cell migration. Nodes colored in red are binding targets of Lhx1 and Otx2, diamond node shape denotes transcription factors. The genes corresponding to each of these three communities are listed in Table 4. (A color version of this figure is located at the end of this chapter.)

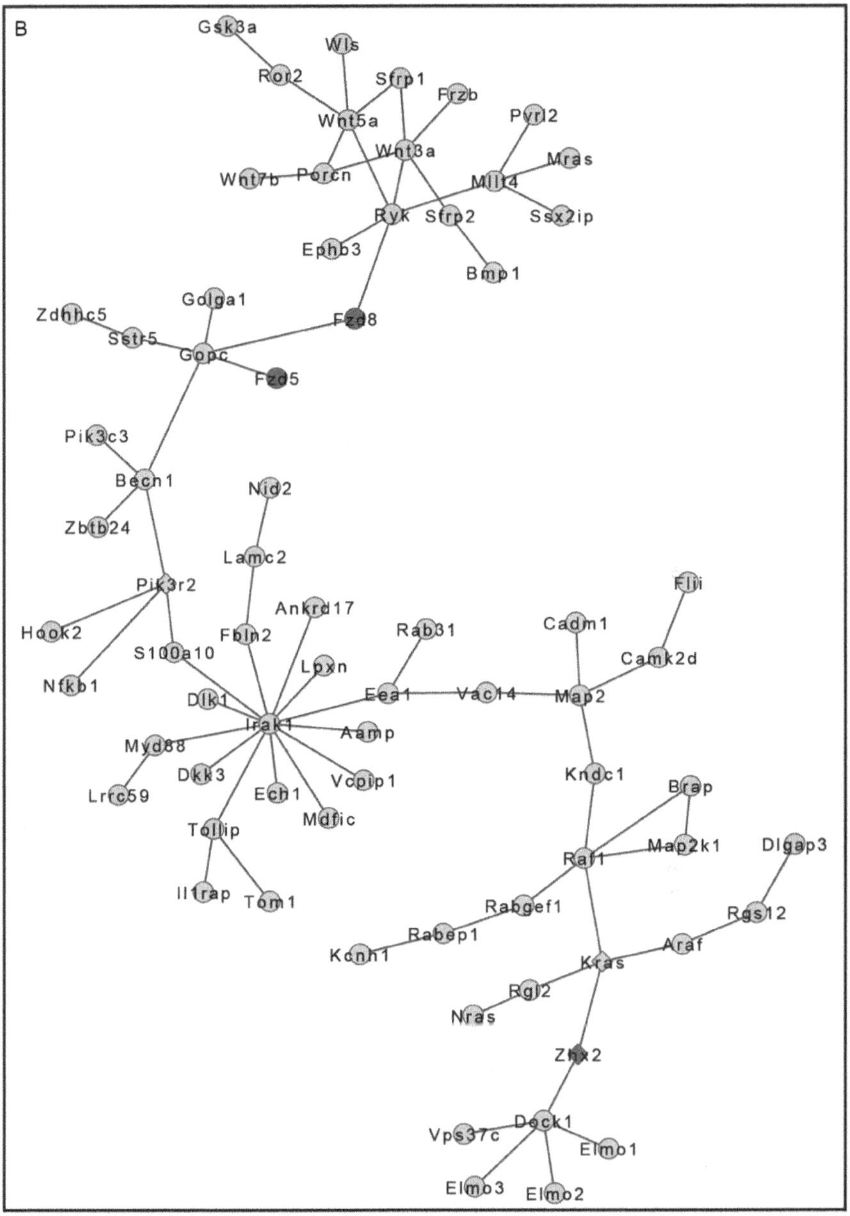

Figure 3—Cont'd

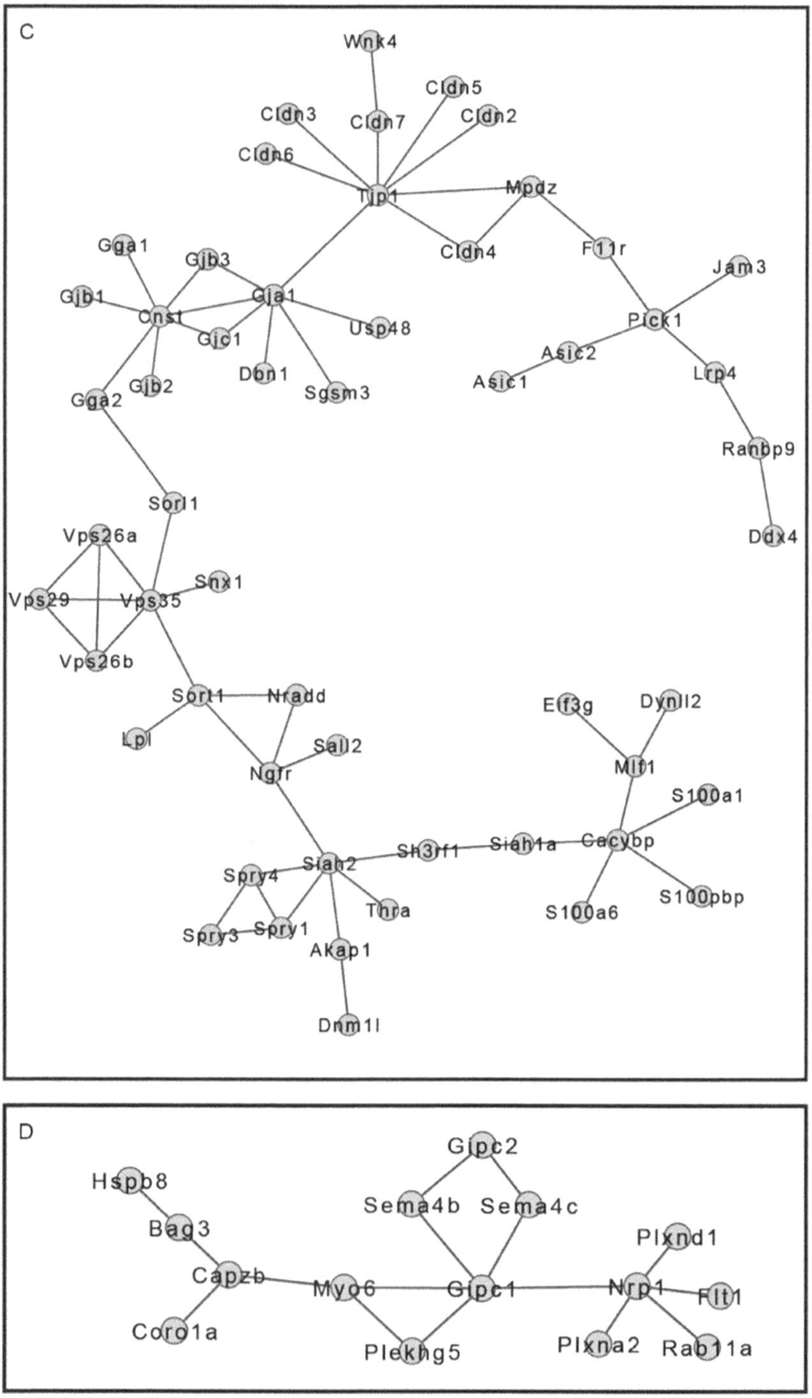

Figure 3—Cont'd

Table 4 Genes in the Communities of the Head Gene Regulatory Network

Community	Functionally Enriched Genes
WNT signaling	*Aamp, Ankrd17, Araf, Becn1, Bmp1, Brap, Cadm1, Camk2d, Dkk3, Dlgap3, Dlk1, Dock1, Ech1, Eea1, Elmo1, Elmo2, Elmo3, Ephb3, Fbln2, Flii, Frzb, Fzd5, Fzd8, Golga1, Gopc, Gsk3a, Hook2, Il1rap, Irak1, Kcnh1, Kndc1, Kras, Lamc2, Lpxn, Lrrc59, Map2, Map2k1, Mdfic, Mllt4, Mras, Myd88, Nfkb1, Nid2, Nras, Pik3c3, Pik3r2, Porcn, Pvrl2, Rab31, Rabep1, Rabgef1, Raf1, Rgl2, Rgs12, Ror2, Ryk, S100a10, Sfrp1, Sfrp2, Sstr5, Ssx2ip, Tollip, Tom1, Vac14, Vcpip1, Vps37c, Wls, Wnt3a, Wnt5a, Wnt7b, Zbtb24, Zdhhc5, Zhx2*
Cell–cell adhesion	*Akap1, Asic1, Asic2, Cacybp, Cldn2, Cldn3, Cldn4, Cldn5, Cldn6, Cldn7, Cnst, Dbn1, Ddx4, Dnm1l, Dynll2, Eif3g, F11r, Gga1, Gga2, Gja1, Gjb1, Gjb2, Gjb3, Gjc1, Jam3, Lpl, Lrp4, Mlf1, Mpdz, Ngfr, Nradd, Pick1, Ranbp9, S100a1, S100a6, S100pbp, Sall2, Sgsm3, Sh3rf1, Siah1a, Siah2, Snx1, Sorl1, Sort1, Spry1, Spry3, Spry4, Thra, Tjp1, Usp48, Vps26a, Vps26b, Vps29, Vps35, Wnk4*
Cell migration	*Bag3, Capzb, Coro1a, Flt1, Gipc1, Gipc2, Hspb8, Myo6, Nrp1, Plekhg5, Plxna2, Plxnd1, Rab11a, Sema4b, Sema4c*

Initially, the entire head tissue progenitors are shrouded within a WNT signaling-silent space. As the head folds develop, WNT signaling activity remains low in the neuroectoderm of the prospective forebrain, in which *Dkk1* is also expressed. In contrast, tissue progenitors for the posterior brain regions actively respond to WNT signaling ligands (e.g., Wnt1, Wnt3, and Wnt3a) that are expressed in the dorsal neuroectoderm and neighboring germ layer tissues. The active WNT signaling environment is therefore conducive for eliciting the transcriptional program that drives the morphogenesis of craniofacial structures derived from the head progenitors in the WNT active domain.

The basic helix-loop-helix transcription factor Twist1 can be activated by Wnt1, Wnt3, and Wnt3a (Howe, Watanabe, Leonard, & Brown, 2003; Reinhold, Kapadia, Liao, & Naski, 2006; Wu et al., 2012), and is potentially a direct target of Ctnnb1 (Hamburg-Shields, DiNuoscio, Mullin, Lafayatis, & Atit, 2015). *Twist1*, which is expressed in the cranial neural crest cells originating from the mid- and hindbrain neuroectoderm and in the cranial mesoderm, plays a key role in the development of the facial and pharyngeal structures, being required for cell survival and differentiation in the cranial neural crest and for maintaining the mesenchymal cell phenotype of the cranial mesoderm (Bildsoe et al., 2009, 2013).

As a WNT target gene and key player in craniofacial morphogenesis, Twist1 integrates the activities of multiple transcription factors and signaling pathways that act on craniofacial development and differentiation (Fig. 4). Cranial neural crest cells destined for a skeletogenic fate normally downregulate *Sox10* in a Twist1-dependent manner to facilitate osteogenic or chondrogenic differentiation (Bildsoe et al., 2009; Ota et al., 2004; Soo et al., 2002). Other Twist1-dependent genes include members of the Aristaless-related family of transcription factors (*Alx1*, *Alx3*, *Alx4*), that are required for normal skeletal development (Bildsoe et al., 2009; Loebel et al., 2012; Loebel, O'Rourke, Steiner, Banyer, & Tam, 2002; O'Rourke, Soo, Behringer, Hui, & Tam, 2002; Soo et al., 2002). *Alx3*, is directly activated by Twist1 (Garcia-Sanz, Fernandez-Perez, & Vallejo, 2013) and *Twist1* and *Alx4* are implicated in restricting signaling by the Shh pathway (Loebel et al., 2012; Takahashi et al., 1998). Twist1 appears to interact with the Shh pathway in a complex manner, most likely via regulation of expression of Gli transcription factors. In mouse embryos, expression of *Gli3* requires Twist1 activity (Loebel et al., 2012; Soo et al., 2002),

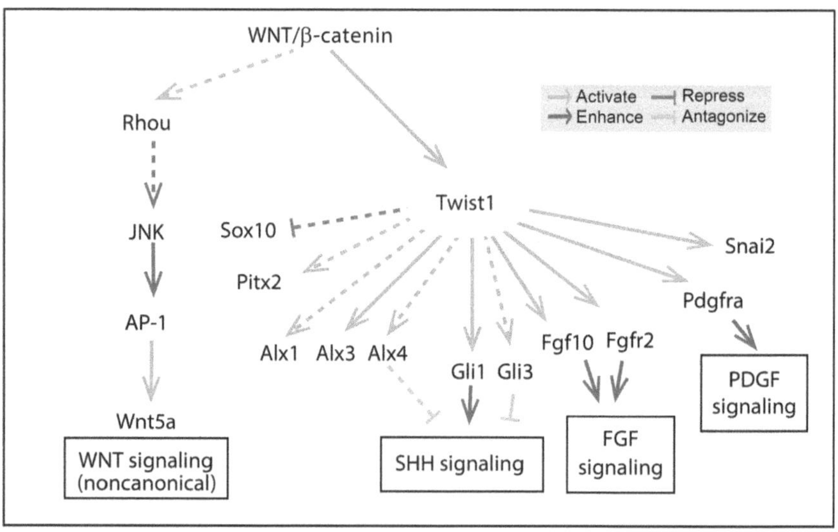

Figure 4 The WNT/β-catenin signaling community of the head gene regulatory network, putatively downstream of Wnt3a, Wnt5a, and Wnt7b, may drive the molecular cascades of the bHLH factor, Twist1 and the Rho-related small GTPase, Rhou. The functional output of these two cascades is channeled to the regulation of the activity of several signaling pathways that are involved with the morphogenesis of the craniofacial structures. Solid connector represents known interaction and dashed connector for putative interaction. (A color version of this figure is located at the end of this chapter.)

but the *Gli1* expression domain is expanded in the absence of Twist1 (Soo et al., 2002). Twist1 directly activates the expression of *Pdgfra* (in tumor cells, Eckert et al., 2011), which is required for normal development of the craniofacial skeleton (Bildsoe et al., 2013; Soriano, 1997). Twist1 function also impacts on FGF signaling. Fgf10 expression is reduced when Twist1 is lost from the cranial neural crest (Bildsoe et al., 2009). Twist1 could directly activate the expression of *Fgf10* and the gene encoding the FGF receptor, *Fgfr2* (Connerney et al., 2006, 2008).

In the cranial mesoderm, Twist1 functions for enhancing the mesenchymal cell state (Bildsoe et al., 2013). Twist1 is an inducer of epithelial mesenchymal transition (EMT) (Khan, Chen, Zhang, & Fu, 2013) and upregulates genes associated with EMT, including a direct target, *Snai2* (Casas et al., 2011), and interacts functionally with *Snai1* or *Snai2* in development (Oram & Gridley, 2005). Twist1 interacts with myogenic bHLH factors (Hamamori, Wu, Sartorelli, & Kedes, 1997) is required for *Pitx2* expression in the extraocular muscles (Bildsoe et al., 2013). Whether this is due to a direct influence of Twist1 in *Pitx2* expression, or the loss of muscle progenitors is not known.

WNT signaling also impacts on the development of the foregut of mouse embryos. In response to WNT/β-catenin signaling (Loebel et al., 2011; Tao, Pennica, Xu, Kalejta, & Levine, 2001), the expression of a member of the Rho-related small GTPases family, *Rhou* (Loebel et al., 2011) is upregulated in the foregut endoderm. Analysis of the consequences of knockdown of *Rhou* indicates that it is required for the proper epithelial morphogenesis and differentiation of the foregut endoderm. Rhou may act by promoting JNK activity and the upregulation of c-Jun/AP-1 transcriptional targets including Wnt5a, thereby providing a potential link between β-catenin-dependent and -independent signaling in the foregut.

6. FUTURE PERSPECTIVES

While several transcription factors have been identified as the key regulators for head specification and morphogenesis, the nature of their interactions and the genes that they control are yet to be fully characterized. Of particular interest is to determine how transcriptional regulation intersects with activities of signaling pathways that are vital for head formation. Our preliminary analysis demonstrated that a working model of a putative mammalian GRN could be constructed based on the assumption of genome conservation of regulatory input and with prior knowledge of functional interaction in an alternative model organism. The task ahead is to build the tissue- and stage-specific GRN for head development for a model mammalian organism such as the mouse. This is likely to be technically challenging in view of the limited quantity of material for -omics analysis that can be extracted from early mouse embryos. The next step in the quest for an understanding of the functional attributes of the GRN is to elucidate the dynamics of gene regulation and the output of the network that is meaningful and relevant to developmental mechanisms *in vivo*. To gain a deeper insight into the functioning of the GRN, a novel experimental paradigm of *in vivo* systems biology analysis is required: one that could measure the interactions of the nodes and edges of the GRN and the functional output through the validation of the dynamics of molecular activity and the phenotypic outcomes. To this end, an innovative strategy could be developed by harnessing the capacity of pluripotent stem cells (PSC) to be scaled-up for precise genome editing to engineer modifications of multiple GRN components. The impacts of manipulating the dynamics of the GRN on development could be assessed by exploiting the avenues availed by the generation of completely PSC-derived mouse chimeras. These experiments, in combination with cell-based systems biology investigations, should provide significant insights into how the genome functions in the construction of a major body part of the embryo.

ACKNOWLEDGMENTS

Our work is supported by the National Health and Medical Research Council of Australia (NHMRC Project Grants 632777, 1003100, 1066832, and Research Fellowship 1049980), Australian Research Council (Discovery Project grants 0985952, 1094008, 120100099, 140101067), and Mr James Fairfax.

REFERENCES

Andoniadou, C. L., Signore, M., Young, R. M., Gaston-Massuet, C., Wilson, S. W., Fuchs, E., et al. (2011). HESX1- and TCF3-mediated repression of Wnt/beta-catenin targets is required for normal development of the anterior forebrain. *Development*, *138*(22), 4931–4942.

Aramaki, T., Sasai, N., Yakura, R., & Sasai, Y. (2010). Jiraiya attenuates BMP signaling by interfering with type II BMP receptors in neuroectodermal patterning. *Developmental Cell*, *19*(4), 547–561.

Arkell, R. M., & Tam, P. P. L. (2012). Initiating head development in mouse embryos: Integrating signalling and transcriptional activity. *Open Biology*, *2*(3), 120030.

Bachiller, D., Klingensmith, J., Kemp, C., Belo, J. A., Anderson, R. M., May, S. R., et al. (2000). The organizer factors chordin and noggin are required for mouse forebrain development. *Nature*, *403*(6770), 658–661.

Barnes, J. D., Crosby, J. L., Jones, C. M., Wright, C. V., & Hogan, B. L. (1994). Embryonic expression of Lim-1, the mouse homolog of xenopus xlim-1, suggests a role in lateral mesoderm differentiation and neurogenesis. *Developmental Biology*, *161*(1), 168–178.

Bildsoe, H., Loebel, D. A. F., Jones, V. J., Chen, Y. T., Behringer, R. R., & Tam, P. P. L. (2009). Requirement for Twist1 in frontonasal and skull vault development in the mouse embryo. *Developmental Biology*, *331*(2), 176–188.

Bildsoe, H., Loebel, D. A., Jones, V. J., Hor, A. C., Braithwaite, A. W., Chen, Y. T., et al. (2013). The mesenchymal architecture of the cranial mesoderm of mouse embryos is disrupted by the loss of Twist1 function. *Developmental Biology*, *374*(2), 295–307.

Brugmann, S. A., Tapadia, M. D., & Helms, J. A. (2006). The molecular origins of species-specific facial pattern. *Current Topics in Developmental Biology*, *73*, 1–42.

Casas, E., Kim, J., Bendesky, A., Ohno-Machado, L., Wolfe, C. J., & Yang, J. (2011). Snail2 is an essential mediator of Twist1-induced epithelial mesenchymal transition and metastasis. *Cancer Research*, *71*(1), 245–254.

Connerney, J., Andreeva, V., Leshem, Y., Mercado, M. A., Dowell, K., Yang, X. H., et al. (2008). Twist1 homodimers enhance FGF responsiveness of the cranial sutures and promote suture closure. *Developmental Biology*, *318*(2), 323–334.

Connerney, J., Andreeva, V., Leshem, Y., Muentener, C., Mercado, M. A., & Spicer, D. B. (2006). Twist1 dimer selection regulates cranial suture patterning and fusion. *Developmental Dynamics*, *235*(5), 1345–1357.

Couly, G., & Le Douarin, N. M. (1990). Head morphogenesis in embryonic avian chimeras: Evidence for a segmental pattern in the ectoderm corresponding to the neuromeres. *Development*, *108*(4), 543–558.

Cowley, M. J., Pinese, M., Kassahn, K. S., Waddell, N., Pearson, J. V., Grimmond, S. M., et al. (2012). PINA v2.0: Mining interactome modules. *Nucleic Acids Research*, *40*(Database issue), D862–D865.

Davidson, E. H., & Erwin, D. H. (2006). Gene regulatory networks and the evolution of animal body plans. *Science*, *311*(5762), 796–800.

del Barco Barrantes, I., Davidson, G., Grone, H. J., Westphal, H., & Niehrs, C. (2003). Dkk1 and noggin cooperate in mammalian head induction. *Genes & Development*, *17*(18), 2239–2244.

Dosch, R., & Niehrs, C. (2000). Requirement for anti-dorsalizing morphogenetic protein in organizer patterning. *Mechanisms of Development*, *90*(2), 195–203.

Eckert, M. A., Lwin, T. M., Chang, A. T., Kim, J., Danis, E., Ohno-Machado, L., et al. (2011). Twist1-induced invadopodia formation promotes tumor metastasis. *Cancer Cell*, *19*(3), 372–386.

Fossat, N., Chatelain, G., Brun, G., & Lamonerie, T. (2006). Temporal and spatial delineation of mouse Otx2 functions by conditional self-knockout. *EMBO Reports*, 7(8), 824–830.

Fossat, N., Ip, C. K., Jones, V. J., Studdert, J. B., Khoo, P. L., Lewis, S. L., et al. (2015). Context-specific function of the LIM homeobox 1 transcription factor in head formation of the mouse embryo. *Development*, *142*(11), 2069–2079.

Fossat, N., Jones, V., Garcia-Garcia, M. J., & Tam, P. P. L. (2012). Modulation of WNT signaling activity is key to the formation of the embryonic head. *Cell Cycle*, *11*(1), 26–32.

Fossat, N., Jones, V., Khoo, P. L., Bogani, D., Hardy, A., Steiner, K., et al. (2011). Stringent requirement of a proper level of canonical WNT signalling activity for head formation in mouse embryo. *Development*, *138*(4), 667–676.

Garcia-Sanz, P., Fernandez-Perez, A., & Vallejo, M. (2013). Differential configurations involving binding of USF transcription factors and Twist1 regulate Alx3 promoter activity in mesenchymal and pancreatic cells. *Biochemical Journal*, *450*, 199–208.

Gene Ontology Consortium. (2015). Gene ontology consortium: Going forward. *Nucleic Acids Research*, *43*(Database issue), D1049–D1056.

Hamamori, Y., Wu, H. Y., Sartorelli, V., & Kedes, L. (1997). The basic domain of myogenic basic helix-loop-helix (bHLH) proteins is the novel target for direct inhibition by another bHLH protein, twist. *Molecular and Cellular Biology*, *17*(11), 6563–6573.

Hamburg-Shields, E., DiNuoscio, G. J., Mullin, N. K., Lafayatis, R., & Atit, R. P. (2015). Sustained beta-catenin activity in dermal fibroblasts promotes fibrosis by up-regulating expression of extracellular matrix protein-coding genes. *The Journal of Pathology*, *235*(5), 686–697.

Howe, L. R., Watanabe, O., Leonard, J., & Brown, A. M. C. (2003). Twist is up-regulated in response to Wnt1 and inhibits mouse mammary cell differentiation. *Cancer Research*, *63*(8), 1906–1913.

Hukriede, N. A., Tsang, T. E., Habas, R., Khoo, P. L., Steiner, K., Weeks, D. L., et al. (2003). Conserved requirement of Lim1 function for cell movements during gastrulation. *Developmental Cell*, *4*(1), 83–94.

Inomata, H., Shibata, T., Haraguchi, T., & Sasai, Y. (2013). Scaling of dorsal-ventral patterning by embryo size-dependent degradation of Spemann's organizer signals. *Cell*, *153*(6), 1296–1311.

Ip, C. K., Fossat, N., Jones, V., Lamonerie, T., & Tam, P. P. L. (2014). Head formation: OTX2 regulates Dkk1 and Lhx1 activity in the anterior mesendoderm. *Development*, *141*(20), 3859–3867.

Khan, M. A., Chen, H. C., Zhang, D. Z., & Fu, J. J. (2013). Twist: A molecular target in cancer therapeutics. *Tumor Biology*, *34*(5), 2497–2506.

Kimura-Yoshida, C., Nakano, H., Okamura, D., Nakao, K., Yonemura, S., Belo, J. A., et al. (2005). Canonical Wnt signaling and its antagonist regulate anterior-posterior axis polarization by guiding cell migration in mouse visceral endoderm. *Developmental Cell*, *9*(5), 639–650.

Lee, K., & Skromne, I. (2014). Retinoic acid regulates size, pattern and alignment of tissues at the head-trunk transition. *Development*, *141*(22), 4375–4384.

Lewis, S. L., Khoo, P. L., Andrea De Young, R., Bildsoe, H., Wakamiya, M., Behringer, R. R., et al. (2007). Genetic interaction of Gsc and Dkk1 in head morphogenesis of the mouse. *Mechanisms of Development*, *124*(2), 157–165.

Lewis, S. L., Khoo, P. L., De Young, R. A., Steiner, K., Wilcock, C., Mukhopadhyay, M., et al. (2008). Dkk1 and Wnt3 interact to control head morphogenesis in the mouse. *Development*, *135*(10), 1791–1801.

Li, A. M., Tian, A. X., Zhang, R. X., Ge, J., Sun, X., & Cao, X. C. (2013). Protocadherin-7 induces bone metastasis of breast cancer. *Biochemical and Biophysical Research Communications*, *436*(3), 486–490.

Loebel, D. A., Hor, A. C., Bildsoe, H., Jones, V., Chen, Y. T., Behringer, R. R., et al. (2012). Regionalized Twist1 activity in the forelimb bud drives the morphogenesis of the proximal and preaxial skeleton. *Developmental Biology*, *362*(2), 132–140.

Loebel, D. A. F., O'Rourke, M. P., Steiner, K. A., Banyer, J., & Tam, P. P. L. (2002). Isolation of differentially expressed genes from wild-type and twist mutant mouse limb buds. *Genesis*, *33*(3), 103–113.

Loebel, D. A. F., Studdert, J. B., Power, M., Radziewic, T., Jones, V., Coultas, L., et al. (2011). Rhou maintains the epithelial architecture and facilitates differentiation of the foregut endoderm. *Development*, *138*(20), 4511–4522.

Martinez Barbera, J. P., Clements, M., Thomas, P., Rodriguez, T., Meloy, D., Kioussis, D., et al. (2000). The homeobox gene Hex is required in definitive endodermal tissues for normal forebrain, liver and thyroid formation. *Development*, *127*(11), 2433–2445.

Meier, S., & Tam, P. P. L. (1982). Metameric pattern development in the embryonic axis of the mouse. I. Differentiation of the cranial segments. *Differentiation; Research in Biological Diversity*, *21*(2), 95–108.

Mi, H., Muruganujan, A., Casagrande, J. T., & Thomas, P. D. (2013). Large-scale gene function analysis with the PANTHER classification system. *Nature Protocols*, *8*(8), 1551–1566.

Mukhopadhyay, M., Shtrom, S., Rodriguez-Esteban, C., Chen, L., Tsukui, T., Gomer, L., et al. (2001). Dickkopf1 is required for embryonic head induction and limb morphogenesis in the mouse. *Developmental Cell, 1*(3), 423–434.

Nakano, T., Murata, T., Matsuo, I., & Aizawa, S. (2000). OTX2 directly interacts with LIM1 and HNF-3beta. *Biochemical and Biophysical Research Communications, 267*(1), 64–70.

Newman, M. E. (2006). Modularity and community structure in networks. *Proceedings of the National Academy of Sciences of the United States of America, 103*(23), 8577–8582.

Niehrs, C. (2004). Regionally specific induction by the Spemann-Mangold organizer. Nature reviews *Genetics, 5*(6), 425–434.

Oram, K. F., & Gridley, T. (2005). Mutations in snail family genes enhance craniosynostosis of Twist1 haplo-insufficient mice: Implications for Saethre-Chotzen syndrome. *Genetics, 170*(2), 971–974.

O'Rourke, M. P., Soo, K., Behringer, R. R., Hui, C. C., & Tam, P. P. L. (2002). Twist plays an essential role in FGF and SHH signal transduction during mouse limb development. *Developmental Biology, 248*(1), 143–156.

Ota, M. S., Loebel, D. A. F., O'Rourke, M. P., Wong, N., Tsoi, B., & Tam, P. P. L. (2004). Twist is required for patterning the cranial nerves and maintaining the viability of mesodermal cells. *Developmental Dynamics, 230*(2), 216–228.

Perea-Gomez, A., Vella, F. D., Shawlot, W., Oulad-Abdelghani, M., Chazaud, C., Meno, C., et al. (2002). Nodal antagonists in the anterior visceral endoderm prevent the formation of multiple primitive streaks. *Developmental Cell, 3*(5), 745–756.

Pfister, S., Steiner, K. A., & Tam, P. P. L. (2007). Gene expression pattern and progression of embryogenesis in the immediate post-implantation period of mouse development. *Gene expression patterns: GEP*, 7(5), 558–573.

Piccolo, S., Agius, E., Leyns, L., Bhattacharyya, S., Grunz, H., Bouwmeester, T., et al. (1999). The head inducer Cerberus is a multifunctional antagonist of Nodal, BMP and Wnt signals. *Nature, 397*(6721), 707–710.

Puelles, L., & Rubenstein, J. L. (2003). Forebrain gene expression domains and the evolving prosomeric model. *Trends in Neurosciences, 26*(9), 469–476.

Rashid, D., Newell, K., Shama, L., & Bradley, R. (2006). A requirement for NF-protocadherin and TAF1/Set in cell adhesion and neural tube formation. *Developmental Biology, 291*(1), 170–181.

Reinhold, M. I., Kapadia, R. M., Liao, Z. X., & Naski, M. C. (2006). The Wnt-inducible transcription factor Twist1 inhibits chondrogenesis. *Journal of Biological Chemistry, 281*(3), 1381–1388.

Robb, L., & Tam, P. P. L. (2004). Gastrula organiser and embryonic patterning in the mouse. *Seminars in Cell & Developmental Biology, 15*(5), 543–554.

Shawlot, W., & Behringer, R. R. (1995). Requirement for Lim1 in head-organizer function. *Nature, 374*(6521), 425–430.

Shawlot, W., Wakamiya, M., Kwan, K. M., Kania, A., Jessell, T. M., & Behringer, R. R. (1999). Lim1 is required in both primitive streak-derived tissues and visceral endoderm for head formation in the mouse. *Development, 126*(22), 4925–4932.

Shimono, A., & Behringer, R. R. (1999). Isolation of novel cDNAs by subtractions between the anterior mesendoderm of single mouse gastrula stage embryos. *Developmental Biology, 209*(2), 369–380.

Shimono, A., & Behringer, R. R. (2003). Angiomotin regulates visceral endoderm movements during mouse embryogenesis. *Current Biology, 13*(7), 613–617.

Singh, A. M., Reynolds, D., Cliff, T., Ohtsuka, S., Mattheyses, A. L., Sun, Y., et al. (2012). Signaling network crosstalk in human pluripotent cells: A Smad2/3-regulated switch that controls the balance between self-renewal and differentiation. *Cell Stem Cell, 10*(3), 312–326.

Soo, K., O'Rourke, M. P., Khoo, P. L., Steiner, K. A., Wong, N., Behringer, R. R., et al. (2002). Twist function is required for the morphogenesis of the cephalic neural tube and the differentiation of the cranial neural crest cells in the mouse embryo. *Developmental Biology*, *247*(2), 251–270.

Soriano, P. (1997). The PDGF alpha receptor is required for neural crest cell development and for normal patterning of the somites. *Development*, *124*(14), 2691–2700.

Spieler, D., Baumer, N., Stebler, J., Koprunner, M., Reichman-Fried, M., Teichmann, U., et al. (2004). Involvement of Pax6 and Otx2 in the forebrain-specific regulation of the vertebrate homeobox gene ANF/Hesx1. *Developmental Biology*, *269*(2), 567–579.

Sudou, N., Yamamoto, S., Ogino, H., & Taira, M. (2012). Dynamic *in vivo* binding of transcription factors to cis-regulatory modules of cer and gsc in the stepwise formation of the Spemann-Mangold organizer. *Development*, *139*(9), 1651–1661.

Takahashi, M., Tamura, K., Buscher, D., Masuya, H., Yonei-Tamura, S., Matsumoto, K., et al. (1998). The role of Alx-4 in the establishment of anteroposterior polarity during vertebrate limb development. *Development*, *125*(22), 4417–4425.

Tam, P. P. L., & Behringer, R. R. (1997). Mouse gastrulation: The formation of a mammalian body plan. *Mechanisms of Development*, *68*(1–2), 3–25.

Tam, P. P. L., Khoo, P. L., Wong, N., Tsang, T. E., & Behringer, R. R. (2004). Regionalization of cell fates and cell movement in the endoderm of the mouse gastrula and the impact of loss of Lhx1(Lim1) function. *Developmental Biology*, *274*(1), 171–187.

Tam, P. P. L., & Loebel, D. A. (2007). Gene function in mouse embryogenesis: Get set for gastrulation. Nature reviews *Genetics*, *8*(5), 368–381.

Tao, W. K., Pennica, D., Xu, L. F., Kalejta, R. F., & Levine, A. J. (2001). Wrch-1, a novel member of the Rho gene family that is regulated by Wnt-1. *Genes & Development*, *15*(14), 1796–1807.

Terada, K., Misao, S., Katase, N., Nishimatsu, S., & Nohno, T. (2013). Interaction of Wnt signaling with BMP/smad signaling during the transition from cell proliferation to myogenic differentiation in mouse myoblast-derived cells. *International Journal of Cell Biology*, *2013*. 616294.

Trainor, P. A., & Tam, P. P. L. (1995). Cranial paraxial mesoderm and neural crest cells of the mouse embryo: Co-distribution in the craniofacial mesenchyme but distinct segregation in branchial arches. *Development*, *121*(8), 2569–2582.

Trainor, P. A., Tan, S. S., & Tam, P. P. L. (1994). Cranial paraxial mesoderm: Regionalisation of cell fate and impact on craniofacial development in mouse embryos. *Development*, *120*(9), 2397–2408.

Wu, Y. Y., Ginther, C., Kim, J., Mosher, N., Chung, S. Y., Slamon, D., et al. (2012). Expression of Wnt3 activates Wnt/beta-catenin pathway and promotes EMT-like phenotype in Trastuzumab-resistant HER2-overexpressing breast cancer cells. *Molecular Cancer Research*, *10*(12), 1597–1606.

Xuan, S. H., Baptista, C. A., Balas, G., Tao, W. F., Soares, V. C., & Lai, E. (1995). Winged helix transcription factor Bf-1 is essential for the development of the cerebral hemispheres. *Neuron*, *14*(6), 1141–1152.

Yamamoto, A., Kemp, C., Bachiller, D., Geissert, D., & De Robertis, E. M. (2000). Mouse paraxial protocadherin is expressed in trunk mesoderm and is not essential for mouse development. *Genesis*, *27*(2), 49–57.

Yao, J., & Kessler, D. S. (2001). Goosecoid promotes head organizer activity by direct repression of Xwnt8 in Spemann's organizer. *Development*, *128*(15), 2975–2987.

Yasuoka, Y., Suzuki, Y., Takahashi, S., Someya, H., Sudou, N., Haramoto, Y., et al. (2014). Occupancy of tissue-specific cis-regulatory modules by Otx2 and TLE/Groucho for embryonic head specification. *Nature Communications*, *5*, 4322.

Yoshida, K. (2003). Fibroblast cell shape and adhesion *in vitro* is altered by overexpression of the 7a and 7b isoforms of protocadherin 7, but not the 7c isoform. *Cellular & Molecular Biology Letters*, *8*(3), 735–741.

Yoshida, M., Suda, Y., Matsuo, I., Miyamoto, N., Takeda, N., Kuratani, S., et al. (1997). Emx1 and Emx2 functions in development of dorsal telencephalon. *Development*, *124*(1), 101–111.

Zakin, L., Reversade, B., Virlon, B., Rusniok, C., Glaser, P., Elalouf, J. M., et al. (2000). Gene expression profiles in normal and Otx2-/- early gastrulating mouse embryos. *Proceedings of the National Academy of Sciences of the United States of America*, *97*(26), 14388–14393.

Zorn, A. M. (2001). Wnt signalling: Antagonistic Dickkopfs. *Current biology*, *11*(15), R592–R595.

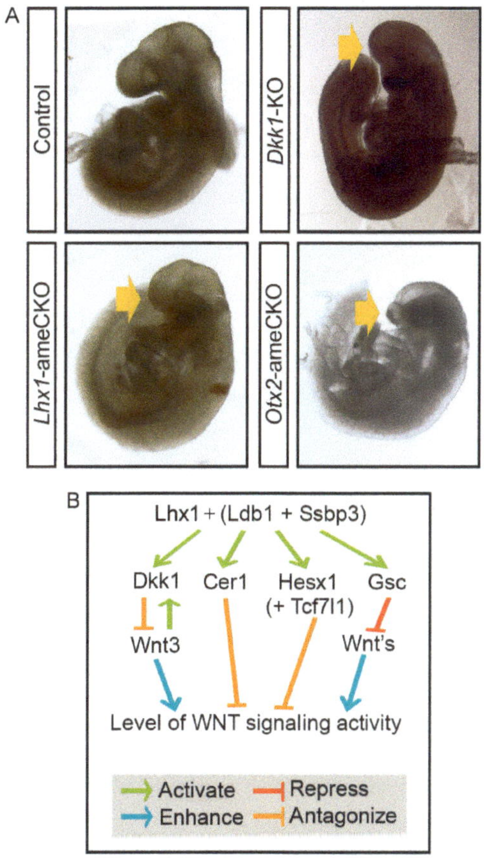

Patrick P.L. Tam *et al.*, Figure 1 (A) Phenotype of *Dkk1* knockout (KO), *Lhx1* anterior mesendoderm (ame) conditional knockout (CKO), and *Otx2*-ameCKO mutant embryos. Control = wild-type embryo; embryonic age = E9.5; arrow, head truncation. (B) Lhx1 downstream molecular activities in the anterior mesendoderm impacting on the level of WNT signaling activity.

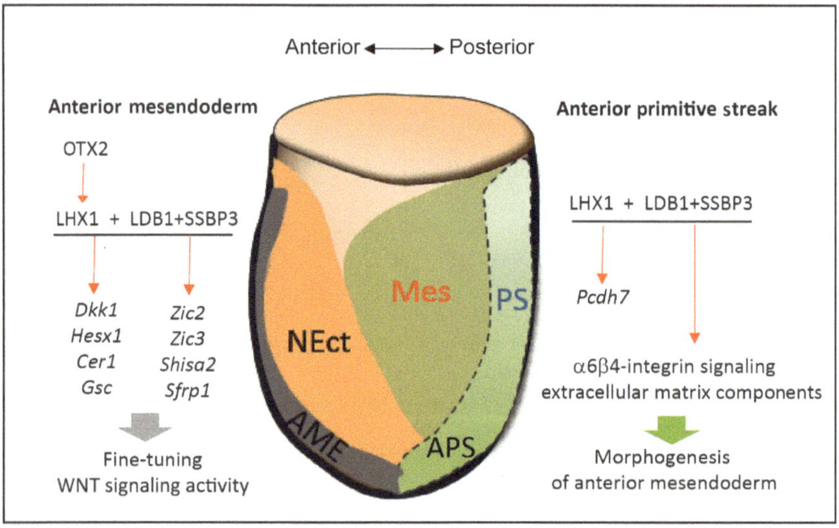

Patrick P.L. Tam *et al.*, Figure 2 Functional context-specific requirement of Lhx1 activity for the formation of anterior mesoderm via the influence on the transit of progenitor from the anterior primitive streak (APS) to the midline and the anterior extension of the tissue, and for the regulation of the transcription of genes in the anterior mesendoderm (AME) that encode factors for modulating the level of WNT signaling activity in the anterior germ layer tissues. Mes, mesoderm; NEct, neuroectoderm; PS, primitive streak.

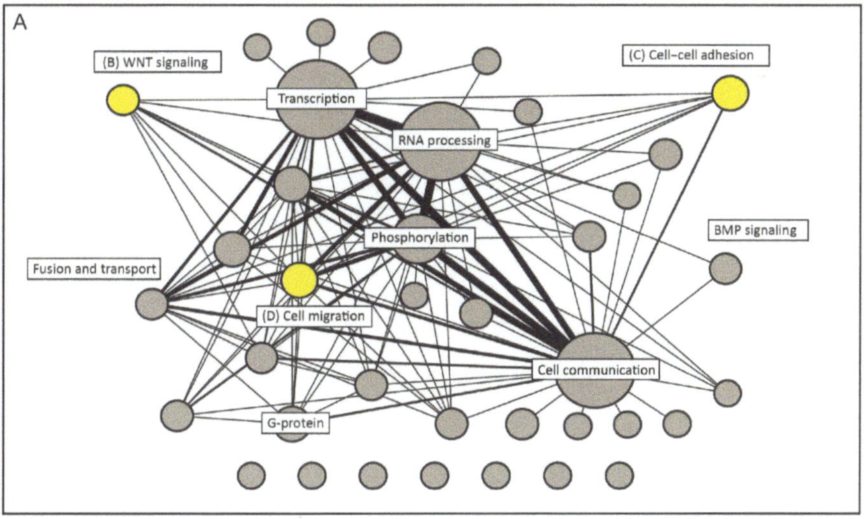

Patrick P.L. Tam *et al.*, Figure 3 The head gene regulatory network (GRN). This network was constructed using data from a ChIP-Seq study on *Xenopus* embryos (Yasuoka et al., 2014). Conserved binding sites were identified in the mouse and the associated genes were identified based on binding by Otx2 and/or Lhx1 within a 5000 bp upstream and 1000 bp downstream of the transcription start site. This list of candidate target genes was compared to the transcriptome of the anterior region of the late gastrula stage mouse embryo obtained by microarray analysis. This analysis identified 102 Lhx1, 595 Otx2, and 386 Lhx1 and Otx2 putative target genes that are expressed in the anterior region of the mouse embryo. By incorporating protein–protein interaction data from the PINA database (Cowley et al., 2012) a GRN to the depth of third neighbors from the Lhx1 and Otx2 putative gene targets was constructed. The resulting network contains 3507 nodes and 7944 interactions. To gain functional insight into the network, community analysis was performed through modularity clustering (Newman, 2006), revealing 38 communities. Functional Analysis of each community was performed by hypergeometric testing (p-value < 0.001) using the Biological Pathways from the Gene Ontology Consortium (2015) and GOSlim Biological Pathways from the Panther database (Mi, Muruganujan, Casagrande, & Thomas, 2013), consistency between the 10 most significant terms for each module was used for assessment of functional relevance. (A) Network of communities. Edge thickness is weighted according to number of interactions traversing communities. The size of each node is weighted by the number of genes assigned to a particular community. The network structure of interacting genes of three communities (highlighted in yellow) identified by the functional analysis are shown: (B) WNT signaling pathway, (C) cell–cell adhesion, and (D) cell migration. Nodes colored in red are binding targets of Lhx1 and Otx2, diamond node shape denotes transcription factors. The genes corresponding to each of these three communities are listed in Table 4.

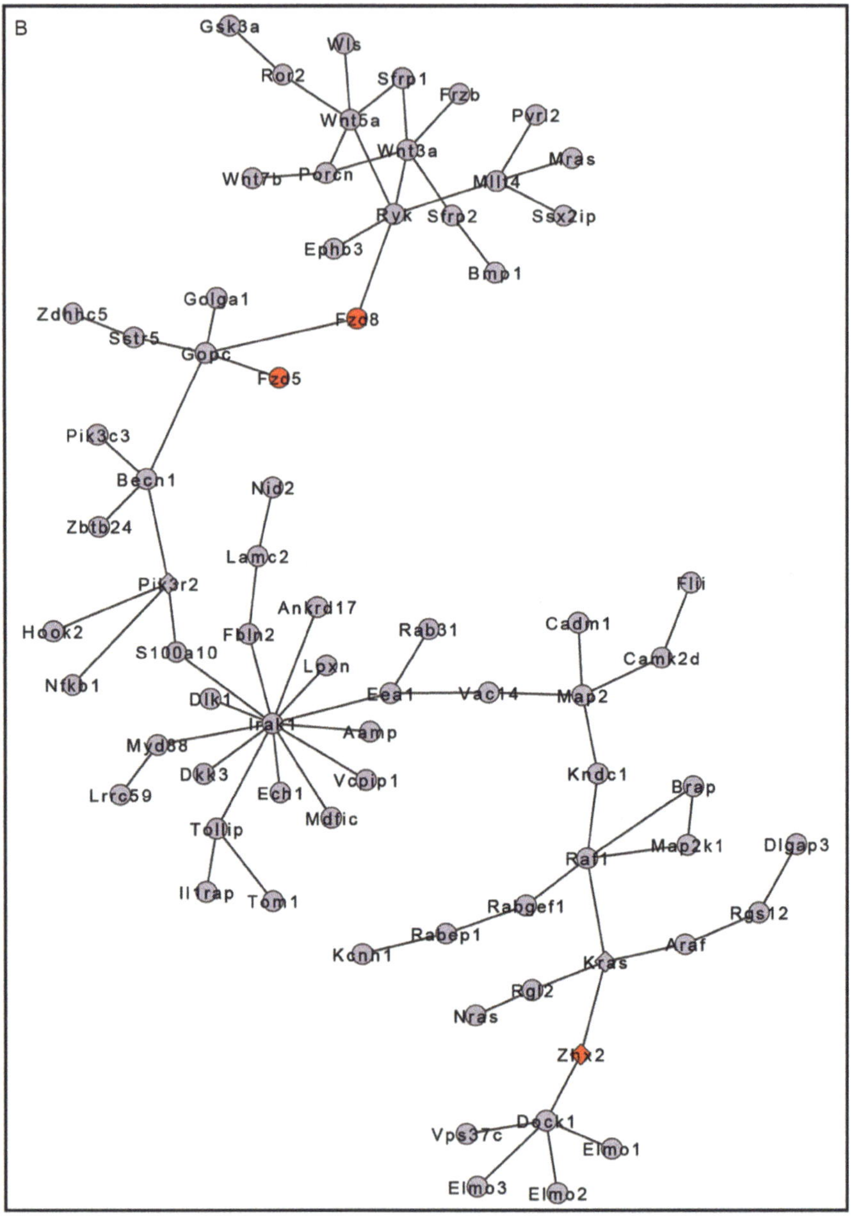

Patrick P.L. Tam *et al.*, Figure 3—Cont'd

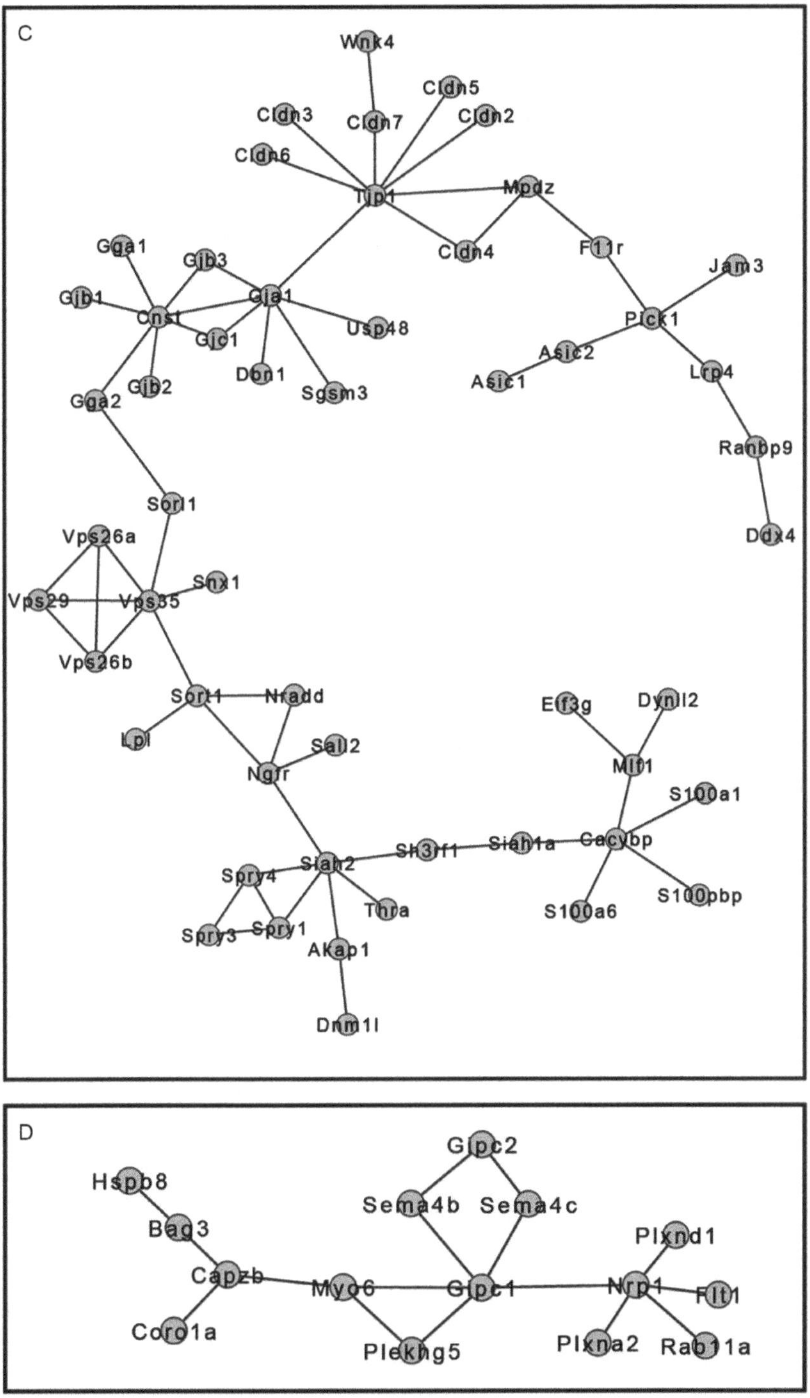

Patrick P.L. Tam *et al.*, Figure 3—Cont'd

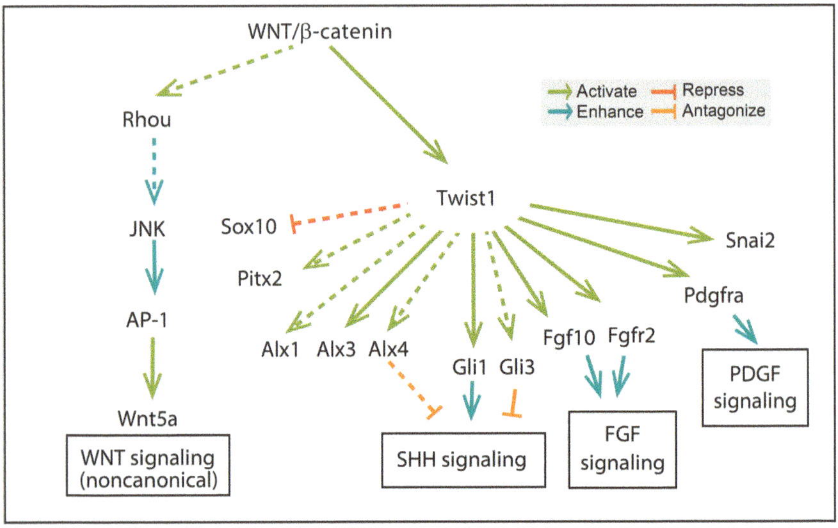

Patrick P.L. Tam *et al.*, Figure 4 The WNT/β-catenin signaling community of the head gene regulatory network, putatively downstream of Wnt3a, Wnt5a, and Wnt7b, may drive the molecular cascades of the bHLH factor, Twist1 and the Rho-related small GTPase, Rhou. The functional output of these two cascades is channeled to the regulation of the activity of several signaling pathways that are involved with the morphogenesis of the craniofacial structures. Solid connector represents known interaction and dashed connector for putative interaction.

CHAPTER THIRTY

Developmental Biology: We Are All Walking Mutants

Paul A. Trainor[1]
Stowers Institute for Medical Research, Kansas City, Missouri, USA
Department of Anatomy and Cell Biology, University of Kansas Medical Center, Kansas City, Kansas, USA
[1]Corresponding author: e-mail address: pat@stowers.org

Abstract

What is Developmental Biology? Developmental Biology is a discipline that evolved from the collective fields of embryology, morphology, and anatomy, which firmly established that structure underpins function. In its simplest terms, Developmental Biology has come to describe how a single cell becomes a completely formed organism. However, this definition of Developmental Biology is too narrow. Developmental Biology describes the properties of individual cells; their organization into tissues, organs, and organisms; their homeostasis, regeneration, aging, and ultimately death. Developmental Biology provides a context for cellular reprogramming, stem cell biology, regeneration, tissue engineering, evolutionary development and ecology, and involves the reiterated use of the same cellular mechanisms and signaling pathways throughout the lifespan of an organism. Using neural crest cells as an example, this review explores the contribution of Developmental Biology to our understanding of development, evolution, and disease.

Humpty Dumpty sat on a wall
Humpty Dumpty had a great fall
All the King's horses and all the king's men
Couldn't put Humpty together again

Opie and Opie (1951)

If ever there was a nursery rhyme that captured the essence and perhaps ultimate translational goal of Developmental Biology, it is the story of Humpty Dumpty. Mending or rebuilding Humpty Dumpty would signify that our mechanistic understanding of the development of a complex living organism was so complete that we could recapitulate it at will. This idea, however, is

Current Topics in Developmental Biology, Volume 117
ISSN 0070-2153
http://dx.doi.org/10.1016/bs.ctdb.2015.11.029

currently still a dream and very far from reality. Nonetheless, the tremendous progress that has been made during the past 50 years in the field of Developmental Biology offers enormous hope and promise for the future.

But what is Developmental Biology? This is a question often naively asked by students enamored with the wonders of stem cells, tissue engineering, and regenerative medicine. In simplistic terms, Developmental Biology evolved from the collective fields of embryology, morphology, and anatomy. While these fields in their heyday may have been predominantly descriptive, they firmly established the principle that structure underpins function. Developmental Biology has since come to describe in its purest form, how a single cell becomes a completely formed organism and implicit within that definition is the concept of time, which is one of the Developmental Biology's founding and underlying principles. However, one could easily argue that this definition of Developmental Biology is too narrow.

There are a limited number of possible behaviors for any given cell or tissue in an organism. Therefore, one of the characteristic features of development is the reiterated use of the same cellular mechanisms and signaling pathways in different tissues at different times. For example, epithelial-to-mesenchymal transition (EMT) is a process in which epithelial cells alter their polarity and cell adhesion characteristics to become migratory and invasive mesenchymal cells. EMT was initially described as an embryonic feature of gastrulation (Hay, 1963, 1995; Trelstad, Hay, & Revel, 1967), but has since been shown to be an underlying feature of placenta development, neural crest cell formation, neural tube closure, heart valve formation, organogenesis, and palatogenesis (Thiery, Acloque, Huang, & Nieto, 2009). EMT is, however, also an essential regulator of tissue homeostasis and repair in mature or adult organisms, particularly during fibrosis and wound healing (Kalluri & Weinberg, 2009; Nieto, 2009). Furthermore, EMT is a critical driver of cancer metastasis (Barrallo-Gimeno & Nieto, 2005; Vega et al., 2004; Ye et al., 2015).

The loss of E-cadherin is considered a fundamental component of EMT, and numerous molecular pathways are reiteratively used to regulate EMT during embryogenesis and adult tissue homeostasis including TGF-b, Ras-MAPK, FGF, Wnt, and Notch signaling (Hay, 1995). Taken together with the fact that certain childhood solid tumors and metastatic cancers such as melanoma and neuroblastoma derive from specific embryonic progenitor cells, this lends support to the idea that cancer in some instances is a disorder of development.

Thus, the established definition of Developmental Biology is too narrow, as the process of development continues long after embryogenesis, throughout maturation, and even during aging. In fact, it could be argued that even when an organism ceases to function, developmental mechanisms still kick in to try and preserve cells, organelles, and other machinery despite the ultimate futility of doing so. Therefore, one could consider Developmental Biology to be a process that begins with a single cell and ends only with the death of the organism it gave rise to.

Developmental Biology, which defines how a progenitor cell becomes a tissue and then an organ and an organism, has spawned new fields such as cellular reprogramming, stem cell biology, regeneration, tissue engineering, evolutionary development (Evo-Devo), while also contributing to newly emerging fields such as evolutionary developmental ecology (Evo-Devo-Eco). It is important to realize that without Developmental Biology these new fields would have no context. The fact that *in vitro* fertilization is possible and that we can grow organs and also model human diseases *in vitro* (Lee & Studer, 2010) is largely dependent upon our understanding of development. This is where our knowledge of progenitor cells, lineage, fate, tissue interactions, signaling, survival, and differentiation has come from.

In the 1960s, the administration of thalidomide to reduce nausea in women during their first trimester of pregnancy resulted in numerous babies being born with major limb and other developmental abnormalities (McBride, 1963, 1976, 1977, 1978, 1981). This phenomenon unequivocally illustrated that maternal environmental exposures could cross the placenta to the detriment of the embryo or fetus. These observations also became a lightning rod for studies investigating the etiology and pathogenesis of birth defects. However, before one can understand the etiology and pathogenesis of abnormal development, it is essential to understand the etiology and pathogenesis of normal development.

But what is normal versus abnormal when there is such an incredibly broad phenotypic spectrum of variation within and between species? Developmental Biology helps us to understand we are all walking mutants, and that morphological traits which may be advantageous to one organism could be disadvantageous to another. For example, the reduced or absent limb morphology of certain species of snakes, lizards, and skinks has evolved numerous times and is thought to facilitate burrowing and movement through dense vegetation (Gans & Wever, 1975; Wiens, Brandley, & Reeder, 2006). However, in contrast, the equivalent reduced (phocomelia, micromelia, meromelia) and absent (amelia) limb conditions in humans are recognized as birth defects that impair grasping, mobility, and other functions. Similarly, ectrodactyly (split/cleft hand or foot) and syndactyly (fused digits) are features of a congenital syndrome in humans known as split-hand/split-foot malformation (SHSFM). While disadvantageous in humans, ectrodactyly and syndactyly are characteristic features of Chamaeleonidae (chameleons) and have facilitated their adaptive radiation throughout arboreal niches (Diaz & Trainor, 2015).

It would be remiss to think the 1960s represent the beginning of the study of Developmental Biology. The German embryologist Hans Driesch, for example, demonstrated as far back as 1892 that after splitting two-cell and four-cell sea urchin embryos into single cells, each cell had the potential of a zygote and could develop normally into a mature sea urchin (Driesch, 1892). Perhaps more remarkably, Hans Spemann and Hilde Mangold discovered in 1923 that the dorsal lip of the blastopore of a tailed amphibian (*Triturus*) could when transplanted into another embryo induce host tissue to form a complete secondary axis (Spemann & Mangold, 2001). This region became known as the embryonic organizer, and its true nature was subsequently revealed through the discovery of growth factors (Cohen, Levi-Montalcini, & Hamburger, 1954; Levi-Montalcini, Meyer, & Hamburger, 1954).

One particular cell type that has always fascinated Developmental Biologists is the neural crest cell, which is considered a uniquely vertebrate cell type. Neural crest cells comprise a migratory stem and progenitor cell population and are synonymous with vertebrate development and evolution. Neural crest cells are thought to have been first recognized and described by William His (His, 1868). However, the term neural crest cell is attributed to Arthur Milnes Marshall (Hall, 2009) and it reflects the anatomical origin of the cells in the closing or closed folds of the neural plate or neural tube, respectively. Over the past 145 years, neural crest cells have provided a unique paradigm with which to study various developmental processes such as morphogenetic induction, multipotency, EMT, migratory chemotaxis, and contact inhibition of locomotion, lineage, and fate determination (Trainor, 2014).

During the first half of the twentieth century, the majority of neural crest cell research was undertaken in amphibian embryos and often involved interspecies transplantations (Horstadius, 1950). The 1960s saw the introduction of tritiated thymidine cell-labeling techniques to visualize and trace the migration of neural crest cells throughout developing amphibian and avian embryos (Chibon, 1964, 1965, 1967; Weston, 1963, 1970). The end of the decade culminated with the seminal introduction of the quail-chick marking system (Le Douarin & Barq, 1969; Le Douarin, 1973). While similar in principle to the earlier generation of amphibian chimeras by German embryologists (Andres, 1946, 1949; Wagner, 1949, 1959), the quail nuclear stain and subsequent quail-specific antibody enabled embryologists (Developmental Biologists) to easily distinguish individual neural crest cells of quail origin from the surrounding tissue of another host avian species. With this technique, generations of scientists reliably marked and studied the properties of neural crest cells (Le Douarin & Kalcheim, 1999). These approaches were instrumental in Drew Noden's demonstrations of the remarkable properties and contributions of cranial neural crest cells particularly to the head (Noden, 1975, 1983, 1984, 1986) and remain a fundamental staple of neural crest cell research today. In the late 1980s, new vital dye cell-labeling techniques opened the door to visualizing neural crest cells in any species, but particularly fish, avians, and mice (Bronner-Fraser & Fraser, 1988, 1989; Schilling & Kimmel, 1994; Trainor & Tam, 1995; Trainor, Tan, & Tam, 1994). These experimental techniques in combination with advances in imaging and transgenic and knockout animal models afforded for the first time opportunities to follow in real time, the dynamics of neural crest cells, as well as their developmental plasticity and fates. Collectively, these techniques facilitated comparative analyses of neural crest cell development, fate, and evolution, highlighting the interplay between plasticity and commitment, and species-specific patterning properties, which helped spur the quest to identify the evolutionary origins of neural crest cells.

The importance of neural crest cells and the central role they played in the evolution of diverse phenotypic variation in vertebrates, particularly with respect to craniofacial structures, was captured in the "New Head" hypothesis (Gans & Northcutt, 1983). This model essentially suggests that most of the morphological and functional differences between vertebrates and other chordates exist in the head and are derived from neural crest cells, epidermal (neurogenic) placodes, and the muscularized hypomere during embryogenesis. Furthermore, neural crest and placodes contribute to the formation of special sense organs and other neural structures which may be homologous to portions of the epidermal nerve plexus of protochordates. Thus, many of the defining or characteristic features of vertebrates became concentrated in the head and reflect the fact that the evolution of vertebrates is associated with a shift from passive to active modes of predation.

Despite nearly 150 years of study, key questions about neural crest cells still remain. For example, what was the prototype of the neural crest cell and when did a neural crest cell really become a neural crest cell? The first challenge to answering such a question is to delineate the critical properties that define a neural crest cell. These include: (i) an origin at the neural plate border, or the junction between neural ectoderm and nonneural ectoderm; (ii) multipotency; (iii) formation via an epithelial-to-mesenchymal transition and acquisition of polarity and migratory ability; and (iv) regulation by a conserved gene regulatory network. These features must be considered individually and collectively in any analysis of the evolutionary origins of neural crest cells (Munoz & Trainor, 2015).

Numerous model systems including, but not limited to *Amphioxus*, jawless basal vertebrates such as lamprey and hagfish, as well as numerous species of amphibians, fish, avians, and rodents, have all been employed to explore the origins, behavior, contributions, and potential of neural crest cells. Neural crest cells have consequently been shown to contribute to nearly all tissues within a vertebrate organism, giving rise to neurons, glia, Schwann cells, cartilage, bone, smooth muscles, adipocytes, and melanocytes, among many cell types and tissues (Fig. 1). Neural crest cells are also integral to organization of the vertebrate brain and have possibly enhanced its growth in vertebrates (Creuzet, Martinez, & Le Douarin, 2006; Le Douarin, Brito, & Creuzet, 2007; Le Douarin, Couly, & Creuzet, 2012). Neural crest cells have evolved to provide specific advantages suited to an animal's environment, especially through contributions that improved metabolism, circulation, and respiration (Green & Bronner, 2013; Green, Simoes-Costa, & Bronner, 2015; Landacre, 1921; Lievre, 1974; Mongera et al., 2013). However, one of the most significant accomplishments of neural crest cells has been to the evolution of the "new head" which includes hinged jaws, special sensory organs, and neural circuitry. These predominately neural crest cell-derived structures facilitated vertebrates becoming predatory, shifting away from the filtration-feeding lifestyle of their *Amphioxus*-like ancestors (Gans & Northcutt, 1983; Northcutt & Gans, 1983).

Within the head and face, neural crest cells can differentiate into chondrocytes that form cartilage, osteoblasts that generate bone, and odontoblasts that produce teeth and scale dentine (Chai et al., 2000; Couly, Coltey, & Le Douarin, 1993; Jiang, Iseki, Maxson, Sucov, & Morriss-Kay, 2002). Interestingly, only the dentine-secreting cells are considered to be an exclusively neural crest cell derivative. Each of the other cell types can also originate from within mesoderm (Hall, 2009; Hall & Gillis, 2013). This suggests that dentine may be a definitive marker for the presence of neural crest cells in fossils and extant organisms, underpinning the idea that the first mineralized skeletal tissues of the vertebrate subphylum were of neural crest cell origin. Fossilized early vertebrates, such as pteraspidomorphs

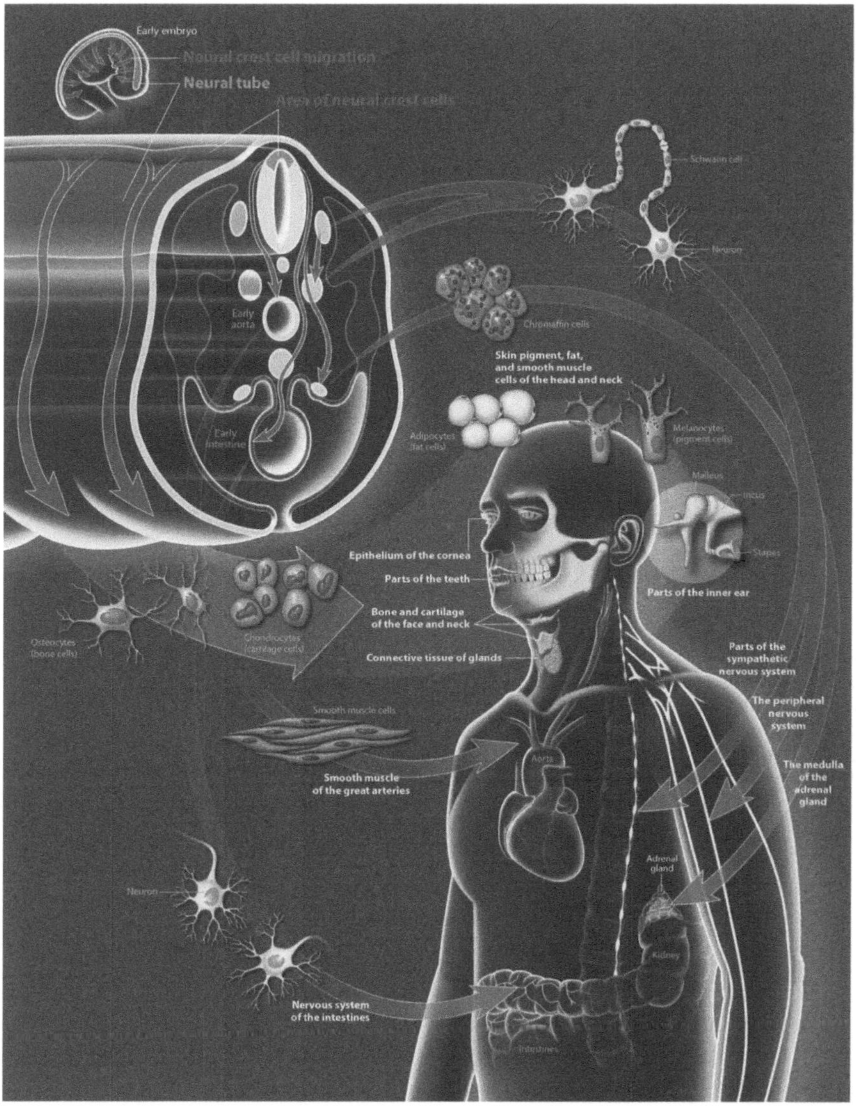

Figure 1 Neural crest cells. Schematic representation of neural crest cell derivation from the neural tube (green) and their migration via specific pathways (purple arrows) to give rise to numerous distinct cell types throughout the human body. (A color version of this figure is located at the end of this chapter.)

agnathans found in the Burgess Shale, exhibit dentine-derived ossified external armor (Janvier, 1996; Le Douarin & Dupin, 2012; Smith, 1991). However, as vertebrate evolution proceeded, gnathostomes developed an endoskeleton comprised somite-derived cartilage and bone. Consequently, the putatively neural crest cell-derived exoskeleton, which was presumably rendered obsolete, was lost, such that only the neural crest cell-derived skull, facial bones, and cartilages remained (Janvier, 2011). Interestingly, while trunk neural crest cells generally no longer contribute to skeletal formation in avians and mammals, they still maintain a latent ability to produce mesenchymal derivatives including skeletal tissues under appropriate *in vitro* conditions (Calloni, Glavieux-Pardanaud, Le Douarin, & Dupin, 2007; Coelho-Aguiar, Le Douarin, & Dupin, 2013; Ido & Ito, 2006; McGonnell & Graham, 2002; Nakamura & Ayer-le Lievre, 1982). Furthermore, extant Chelonians (turtles, terrapins, tortoises) possess an extensive trunk bony exoskeleton and appear to have retained the ancestral neural crest-exoskeleton condition in the form of trunk neural crest cell contributions to the nuchal bones and plastron of the carapace (Cebra-Thomas et al., 2007, 2013; Clark et al., 2001; Gilbert, Bender, Betters, Yin, & Cebra-Thomas, 2007).

The origin of neural crest cells and therefore by default the origins of vertebrates has been debated based primarily on shared structural homology between primitive vertebrates and protochordates (urochordates and cephalochordates), which had separated by the early Cambrian period (Dupret, Sanchez, Goujet, Tafforeau, & Ahlberg, 2014; Gai, Donoghue, Zhu, Janvier, & Stampanoni, 2011; Hall & Gillis, 2013; Mallatt & Chen, 2003). Comparative embryology suggests that cephalochordates are the sister clade to vertebrates and *Amphioxus* for example has been used extensively as a model to explore the origins of vertebrates (Bertrand & Escriva, 2011; Delsuc, Brinkmann, Chourrout, & Philippe, 2006; Holland, 2013; Holland et al., 2008). However, decades of extensive work in *Amphioxus* have not revealed the presence of bona fide primitive neural crest cell-like cells by either morphology or molecular markers (Hall & Gillis, 2013; Yu, 2010).

Perhaps, it is highly unlikely that neural crest cell-like cells will be found in *Amphioxus* given the lack of typical vertebrate neural crest cell derivatives such as peripheral pigment cells, dentine-, bone-, and cartilage-forming cells. Although dorsal root nerves in *Amphioxus* are ensheathed by Schwann cell-like glial cells (Bone, 1960; Peters, 1963), the origin of these cells is ectodermal. Furthermore, numerous invertebrate animals also possess peripheral glial cells despite a complete absence of neural crest cell-like cells (Coles & Abbott, 1996; Hall & Gillis, 2013). This suggests that the glial cell fate of ectodermal cells evolved prior to the divergence of neural crest cells from ectoderm.

The missing evolutionary link could perhaps be explained by the recent discovery that urochordates appear genomically more closely related to vertebrates than cephalochordates (Delsuc et al., 2006). Consistent with this idea, it was recently provocatively proposed that rudimentary neural crest cell-like cells may exist in at least one genera of urochordates, the mangrove tunicate *Ciona intestinalis* (Abitua, Wagner, Navarrete, & Levine, 2012). Blastomere *a9.49* found near the neural plate border of *C. intestinalis* expresses various neural plate border markers and perhaps some neural crest cell specification signals (Abitua et al., 2012; Imai, Levine, Satoh, & Satou, 2006; Jeffery et al., 2008; Squarzoni, Parveen, Zanetti, Ristoratore, & Spagnuolo, 2011). Furthermore, this blastomere gives rise to the gravity sensing otolith, and the melanocytes in light-sensing tissues (Nishida & Satoh, 1989). Interestingly, ectopic expression of *Twist* endows the *a9.49* blastomere lineage with migratory ability. This suggests that a gene regulatory network of the type that might be expected in rudimentary neural crest cells was indeed already present in this lineage and furthermore that it evolved before the divergence of urochordates and vertebrates. More importantly, the identification and characterization of this neural crest cell-like lineage that gives rise to pigment cells in a urochordate suggests that the vertebrate neural crest cell specification gene regulatory network was partially achieved by co-opting existing differentiation networks together with an EMT gene regulatory network to endow rudimentary neural crest cell-like cells with migratory ability. It still remains to be determined whether more neural crest cell-like cells will be identified in this subphyla of diverse extant urochordates, which would imply that neural crest cells began to evolve in the Precambrian period.

Thus, there is still much to be learned about how neural crest cells evolved from primitive neural crest cell-like cells. The identification of additional neural crest cell-like cells and variation between the cells, their derivatives and active pathways, will promote our understanding of the steps that were necessary for neural crest cell formation prior the divergence of vertebrates from protochordates. Furthermore, this knowledge may also facilitate a better understanding of neural crest cell formation in mammals. Although we understand many of the signals required for neural crest cell formation in avian and aquatic species, the counterparts underpinning mammalian neural crest cell formation remain to be functionally determined. What is clear, however, is that the origins of neural crest cells are closely linked to evolution of the vertebrate lineage.

In the twenty-first century, much of our focus now revolve around the contributions of neural crest cells to congenital disorders and diseases, which are collectively termed neurocristopathies. The ability to do so has been largely facilitated by recent advances in our comprehension of the roles of neural crest cells in vertebrate evolution and development. Understanding the true genetic and cellular etiology and pathogenesis of individual neurocristopathies offers the potential for developing therapeutic avenues for their clinical prevention. What is clear is that depending on the phase of neural crest cell development that is disrupted, whether it be formation, migration, or differentiation, very different malformations or syndromes can arise (Jones & Trainor, 2004; Noack Watt & Trainor, 2014; Walker & Trainor, 2006). For instance, within the head and face, perturbed neural crest cell formation is associated with Treacher Collins syndrome (Dixon et al., 2006; Jones et al., 2008), whereas perturbed neural crest cell migration and/or differentiation is associated with craniosynostosis (Merrill et al., 2006; Reardon et al., 1994). Within the trunk, perturbed vagal neural crest cell formation can lead to total intestinal aganglionosis, while altered neural crest cell migration, survival, or differentiation result in Hirschsprung disease (Butler Tjaden & Trainor, 2013). Numerous studies have uncovered and characterized the pluripotent stem cell-like characteristics of neural crest cells during embryogenesis and their persistence into adulthood. Thus, there is tremendous excitement in the potential to repair the affected neural crest cell process during embryogenesis, as well as for neural crest cells to be used in tissue engineering, surgery, and regenerative medicine.

Studies of neural crest cell development, evolution, and disease are notable not just for their fundamental contributions to our understanding of basic developmental biology and clinical disease, but also because they serve as an important reminder that every organism is a model organism. From sea quirts to *Amphioxus*, lampreys and hagfish, frogs, fish, birds, mice, and rats, as well as humans, each of these organisms has played an important role in furthering our understanding of Developmental Biology. It is important, however, not to get hung up or overly focused on working with one organism only. Organisms that may have had held a prominent place in Developmental Biology studies for historical reasons, may not always remain the most ideal model organisms. Every organism has its advantages and disadvantages as a model for study in a laboratory setting, and it is more important to ask the right questions and then determine and choose the right organisms with which to answer those questions. Evidence for this is particularly true in the neural crest cell field from evolutionary studies in *Amphioxus*, tunicates and lampreys, lineage tracing and transplantation in amphibians and avians, time-lapse imaging in fish, conditional spatiotemporal genetic alterations in mice, and genomic and epidemiological studies in humans. This illustrates the need for education and instruction across multiple organisms, which is not always easy to achieve during the normal course of one's training in Developmental Biology. However, this is where unique courses such as the MBL Embryology Course come to the fore, which for over 120 years has highlighted for eager students the endless diversity of organisms with which to study the breadth of Developmental Biology.

In summary, this essay set out to capture the classic principles of Developmental Biology and recent advances in our understanding of the roles of neural crest cells in vertebrate evolution and development. The classical definition of Developmental Biology appears to be too narrow and its contribution to the Stem Cell, Regeneration, and Tissue Bioengineering fields among others remains under appreciated and under recognized. In Developmental Biology, everything changes with time and everything is also context dependent. The fundamental knowledge that Developmental Biology has provided has, however, directly influenced our understanding of vertebrate development and evolution, as well as the pathogenesis of congenital disorders, and set the stage for tissue engineering and regenerative medicine to treat disease. This is particularly true with respect to neural crest cells. Thus, fundamental investments in basic Developmental Biology have provided the basis for translational discoveries and applications and will continue to so. As the essays in this celebratory issue attest, the concept of "Developmental Biology" has undergone as much change as the field itself and must continue to evolve to the capture the mechanisms that give rise to the immense diversity of living organisms.

ACKNOWLEDGMENTS

I greatly appreciate the contribution of Mark Miller who was solely responsible for the artwork in Fig. 1 as it originally appeared in the Stowers Report Spring 2014. Research in the Trainor Laboratory is supported by the Stowers Institute for Medical Research and the National Institute of Dental and Craniofacial Research (DE 016082).

REFERENCES

Abitua, P. B., Wagner, E., Navarrete, I. A., & Levine, M. (2012). Identification of a rudimentary neural crest in a non-vertebrate chordate. *Nature*, *492*(7427), 104–107. http://dx.doi.org/10.1038/nature11589.

Andres, G. (1946). Uber Induction und Entwicklung von Kopforganen aus Unkenektoderm im Molch (Epidermis, Plakoden and Derivate der Nemalleiste). *Revue Suisse de Zoologie*, *53*, 502–510.

Andres, G. (1949). Untersuchungen an chimaren von Triton und Bombinator. *Genetics*, *24*, 387–534.

Barrallo-Gimeno, A., & Nieto, M. A. (2005). The Snail genes as inducers of cell movement and survival: Implications in development and cancer. *Development*, *132*(14), 3151–3161. http://dx.doi.org/10.1242/dev.01907.

Bertrand, S., & Escriva, H. (2011). Evolutionary crossroads in developmental biology: Amphioxus. *Development*, *138*(22), 4819–4830. http://dx.doi.org/10.1242/dev.066720.

Bone, Q. (1960). The central nervous system in amphioxus. *The Journal of Comparative Neurology*, *115*(1), 27–64. http://dx.doi.org/10.1002/cne.901150105.

Bronner-Fraser, M., & Fraser, S. (1988). Cell lineage analysis reveals multipotency of some avian neural crest cells. *Nature*, *335*, 161–164.

Bronner-Fraser, M., & Fraser, S. (1989). Developmental potential of avian trunk neural crest cells in situ. *Neuron*, *3*(6), 755–766.

Butler Tjaden, N. E., & Trainor, P. A. (2013). The developmental etiology and pathogenesis of Hirschsprung disease. *Translational Research*, *162*(1), 1–15. http://dx.doi.org/10.1016/j.trsl.2013.03.001.

Calloni, G. W., Glavieux-Pardanaud, C., Le Douarin, N. M., & Dupin, E. (2007). Sonic Hedgehog promotes the development of multipotent neural crest progenitors endowed with both mesenchymal and neural potentials. *Proceedings of the National Academy of Sciences of The United States of America*, *104*(50), 19879–19884. http://dx.doi.org/10.1073/pnas.0708806104.

Cebra-Thomas, J. A., Betters, E., Yin, M., Plafkin, C., McDow, K., & Gilbert, S. F. (2007). Evidence that a late-emerging population of trunk neural crest cells forms the plastron bones in the turtle *Trachemys scripta*. *Evolution & Development*, *9*(3), 267–277. http://dx.doi.org/10.1111/j.1525-142X.2007.00159.x.

Cebra-Thomas, J. A., Terrell, A., Branyan, K., Shah, S., Rice, R., Gyi, L., et al. (2013). Late-emigrating trunk neural crest cells in turtle embryos generate an osteogenic ectomesenchyme in the plastron. *Developmental Dynamics*, *242*(11), 1223–1235. http://dx.doi.org/10.1002/dvdy.24018.

Chai, Y., Jiang, X., Ito, Y., Bringas, P., Jr., Han, J., Rowitch, D. H., et al. (2000). Fate of the mammalian cranial neural crest during tooth and mandibular morphogenesis. *Development*, *127*(8), 1671–1679.

Chibon, P. (1964). Analysis by the method of nuclear labelling with tritiated thymidine of derivatives of the cephalic neural crest in the Urodele Pleurodeles waltlii Michah. *Comptes Rendus Hebdomadaires des Séances de l'Académie des Sciences*, *259*, 3624–3627.

Chibon, P. (1965). Radioautographic study after labelling with tritiated thymidine of the derivatives of the truncal neural crest in the urodelan amphibian Pleurodeles waltlii Michah. *Comptes Rendus Hebdomadaires des Séances de l'Académie des Sciences Série D*, *261*(25), 5645–5648.

Chibon, P. (1967). Nuclear labelling by tritiated thymidine of neural crest derivatives in the amphibian Urodele Pleurodeles waltlii Michah. *Journal of Embryology and Experimental Morphology*, *18*(3), 343–358.

Clark, K., Bender, G., Murray, B. P., Panfilio, K., Cook, S., Davis, R., et al. (2001). Evidence for the neural crest origin of turtle plastron bones. *Genesis*, *31*(3), 111–117.

Coelho-Aguiar, J. M., Le Douarin, N. M., & Dupin, E. (2013). Environmental factors unveil dormant developmental capacities in multipotent progenitors of the trunk neural crest. *Developmental Biology*, *384*(1), 13–25. http://dx.doi.org/10.1016/j.ydbio.2013.09.030.

Cohen, S., Levi-Montalcini, R., & Hamburger, V. (1954). A nerve growth-stimulating factor isolated from sarcom as 37 and 180. *Proceedings of the National Academy of Sciences of the United States of America*, *40*(10), 1014–1018.

Coles, J. A., & Abbott, N. J. (1996). Signalling from neurones to glial cells in invertebrates. *Trends in Neurosciences*, *19*(8), 358–362.

Couly, G. F., Coltey, P. M., & Le Douarin, N. M. (1993). The triple origin of skull in higher vertebrates: A study in quail-chick chimeras. *Development, 117*, 409–429.

Creuzet, S. E., Martinez, S., & Le Douarin, N. M. (2006). The cephalic neural crest exerts a critical effect on forebrain and midbrain development. *Proceedings of the National Academy of Sciences of the United States of America, 103*(38), 14033–14038. http://dx.doi.org/10.1073/pnas.0605899103.

Delsuc, F., Brinkmann, H., Chourrout, D., & Philippe, H. (2006). Tunicates and not cephalochordates are the closest living relatives of vertebrates. *Nature, 439*(7079), 965–968.

Diaz, R. E., & Trainor, P. A. (2015). Clefting, skeletal reduction and skeletal 're-evolution' of mesopodial elements drive the evolution and radiation of Chameleons. *BMC Evolutionary Biology, 15*, 184–208.

Dixon, J., Jones, N. C., Sandell, L. L., Jayasinghe, S. M., Crane, J., Rey, J. P., et al. (2006). Tcof1/Treacle is required for neural crest cell formation and proliferation deficiencies that cause craniofacial abnormalities. *Proceedings of the National Academy of Sciences of the United States of America, 103*(36), 13403–13408. http://dx.doi.org/10.1073/pnas.0603730103.

Driesch, H. (1892). Entwicklungsmechanisme Studien. I. Der Werth der beiden ersten Furchungszellen in der Echinodermentwicklung. Experimentelle Erzeugen von Theil und Doppelbildung. *Zeitschrift für wissenschaftliche Zoologie, 53*, 160–184.

Dupret, V., Sanchez, S., Goujet, D., Tafforeau, P., & Ahlberg, P. E. (2014). A primitive placoderm sheds light on the origin of the jawed vertebrate face. *Nature, 507*(7493), 500–503. http://dx.doi.org/10.1038/nature12980.

Gai, Z., Donoghue, P. C., Zhu, M., Janvier, P., & Stampanoni, M. (2011). Fossil jawless fish from China foreshadows early jawed vertebrate anatomy. *Nature, 476*(7360), 324–327. http://dx.doi.org/10.1038/nature10276.

Gans, C., & Northcutt, R. (1983). Neural crest and the origin of vertebrates: A new head. *Science, 220*, 268–274.

Gans, C., & Wever, E. G. (1975). The amphisbaenian ear: *Blanus cinereus* and *Diplometopon zarudnyi*. *Proceedings of the National Academy of Sciences of the United States of America, 72*(4), 1487–1490.

Gilbert, S. F., Bender, G., Betters, E., Yin, M., & Cebra-Thomas, J. A. (2007). The contribution of neural crest cells to the nuchal bone and plastron of the turtle shell. *Integrative and Comparative Biology, 47*(3), 401–408. http://dx.doi.org/10.1093/icb/icm020.

Green, S. A., & Bronner, M. E. (2013). Gene duplications and the early evolution of neural crest development. *Seminars in Cell & Developmental Biology, 24*(2), 95–100. http://dx.doi.org/10.1016/j.semcdb.2012.12.006.

Green, S. A., Simoes-Costa, M., & Bronner, M. E. (2015). Evolution of vertebrates as viewed from the crest. *Nature, 520*(7548), 474–482. http://dx.doi.org/10.1038/nature14436.

Hall, B. K. (2009). *The neural crest and neural crest cells in vertebrate development and evolution* (2nd ed.). New York: Springer.

Hall, B. K., & Gillis, J. A. (2013). Incremental evolution of the neural crest, neural crest cells and neural crest-derived skeletal tissues. *Journal of Anatomy, 222*(1), 19–31. http://dx.doi.org/10.1111/j.1469-7580.2012.01495.x.

Hay, E. D. (1963). Recent studies of embryonic induction. *New England Journal of Medicine, 268*, 1114–1122. http://dx.doi.org/10.1056/NEJM196305162682006.

Hay, E. D. (1995). An overview of epithelio-mesenchymal transformation. *Acta Anatomica (Basel), 154*(1), 8–20.

His, W. (1868). *Untersuchungen uber die erste Analge des Wirbeltierleibes. Die erste Entwicklung des Hunchens im Ei*. Leipzig: F.C.W. Vogel.

Holland, L. Z. (2013). Evolution of new characters after whole genome duplications: Insights from amphioxus. *Seminars in Cell & Developmental Biology, 24*(2), 101–109. http://dx.doi.org/10.1016/j.semcdb.2012.12.007.

Holland, L. Z., Albalat, R., Azumi, K., Benito-Gutierrez, E., Blow, M. J., Bronner-Fraser, M., et al. (2008). The amphioxus genome illuminates vertebrate origins and cephalochordate biology. *Genome Research, 18*(7), 1100–1111. http://dx.doi.org/10.1101/gr.073676.107.

Horstadius, S. (1950). *The neural crest.* London: Oxford University Press.

Ido, A., & Ito, K. (2006). Expression of chondrogenic potential of mouse trunk neural crest cells by FGF2 treatment. *Developmental Dynamics, 235*(2), 361–367. http://dx.doi.org/10.1002/dvdy.20635.

Imai, K. S., Levine, M., Satoh, N., & Satou, Y. (2006). Regulatory blueprint for a chordate embryo. *Science, 312*(5777), 1183–1187. http://dx.doi.org/10.1126/science.1123404.

Janvier, P. (1996). *Early vertebrates.* Oxford/New York: Clarendon Press/Oxford University Press.

Janvier, P. (2011). Comparative anatomy: All vertebrates do have vertebrae. *Current Biology, 21*(17), R661–663. http://dx.doi.org/10.1016/j.cub.2011.07.014.

Jeffery, W. R., Chiba, T., Krajka, F. R., Deyts, C., Satoh, N., & Joly, J. S. (2008). Trunk lateral cells are neural crest-like cells in the ascidian *Ciona intestinalis*: Insights into the ancestry and evolution of the neural crest. *Developmental Biology, 324*(1), 152–160. http://dx.doi.org/10.1016/j.ydbio.2008.08.022.

Jiang, X., Iseki, S., Maxson, R. E., Sucov, H. M., & Morriss-Kay, G. M. (2002). Tissue origins and interactions in the mammalian skull vault. *Developmental Biology, 241*, 106–116.

Jones, N. C., Lynn, M. L., Gaudenz, K., Sakai, D., Aoto, K., Rey, J. P., et al. (2008). Prevention of the neurocristopathy Treacher Collins syndrome through inhibition of p53 function. *Nature Medicine, 14*(2), 125–133. http://dx.doi.org/10.1038/nm1725.

Jones, N. C., & Trainor, P. A. (2004). The therapeutic potential of stem cells in the treatment of craniofacial abnormalities. *Expert Opinion on Biological Therapy, 4*(5), 645–657.

Kalluri, R., & Weinberg, R. A. (2009). The basics of epithelial-mesenchymal transition. *Journal of Clinical Investigation, 119*(6), 1420–1428. http://dx.doi.org/10.1172/JCI39104.

Landacre, F. L. (1921). The fate of the neural crest in the head of the urodeles. *Journal of Comparative Neurology, 33*, 1–43.

Le Douarin, N. M. (1973). A Feulgen-positive nucleolus. *Experimental Cell Research, 77*(1), 459–468.

Le Douarin, N., & Barq, G. (1969). Use of Japanese quail cells as "biological markers" in experimental embryology. *Comptes Rendus Hebdomadaires des Séances de l'Académie des Sciences Série D, 269*(16), 1543–1546.

Le Douarin, N. M., Brito, J. M., & Creuzet, S. (2007). Role of the neural crest in face and brain development. *Brain Research Reviews, 55*(2), 237–247. http://dx.doi.org/10.1016/j.brainresrev.2007.06.023.

Le Douarin, N. M., Couly, G., & Creuzet, S. E. (2012). The neural crest is a powerful regulator of pre-otic brain development. *Developmental Biology, 366*(1), 74–82. http://dx.doi.org/10.1016/j.ydbio.2012.01.007.

Le Douarin, N. M., & Dupin, E. (2012). The neural crest in vertebrate evolution. *Current Opinion in Genetics & Development, 22*(4), 381–389. http://dx.doi.org/10.1016/j.gde.2012.06.001.

Le Douarin, N., & Kalcheim, C. (1999). *The neural crest* (2nd ed.). Cambridge, UK: Cambridge University Press.

Lee, G., & Studer, L. (2010). Induced pluripotent stem cell technology for the study of human disease. *Nature Methods,* 7(1), 25–27. http://dx.doi.org/10.1038/nmeth.f.283.

Levi-Montalcini, R., Meyer, H., & Hamburger, V. (1954). In vitro experiments on the effects of mouse sarcomas 180 and 37 on the spinal and sympathetic ganglia of the chick embryo. *Cancer Research, 14*(1), 49–57.

Lievre, C. L. (1974). Role of mesectodermal cells arising from the cephalic neural crest in the formation of the branchial arches and visceral skeleton. *Journal of Embryology and Experimental Morphology, 31*(2), 453–477.

Mallatt, J., & Chen, J. Y. (2003). Fossil sister group of craniates: Predicted and found. *Journal of Morphology*, *258*(1), 1–31. http://dx.doi.org/10.1002/jmor.10081.
McBride, W. G. (1963). The teratogenic action of drugs. *Medical Journal of Australia*, *2*, 689–692.
McBride, W. G. (1976). Studies of the etiology of thalidomide dysmorphogenesis. *Teratology*, *14*(1), 71–87. http://dx.doi.org/10.1002/tera.1420140110.
McBride, W. G. (1977). Thalidomide embryopathy. *Teratology*, *16*(1), 79–82. http://dx.doi.org/10.1002/tera.1420160113.
McBride, W. G. (1978). Teratogenic action of thalidomide. *Lancet*, *1*(8078), 1362.
McBride, W. G. (1981). Another, late thalidomide abnormality. *Lancet*, *2*(8242), 368.
McGonnell, I. M., & Graham, A. (2002). Trunk neural crest has skeletogenic potential. *Current Biology*, *12*(9), 767–771.
Merrill, A. E., Bochukova, E. G., Brugger, S. M., Ishii, M., Pilz, D. T., Wall, S. A., et al. (2006). Cell mixing at a neural crest-mesoderm boundary and deficient ephrin-Eph signaling in the pathogenesis of craniosynostosis. *Human Molecular Genetics*, *15*(8), 1319–1328. http://dx.doi.org/10.1093/hmg/ddl052.
Mongera, A., Singh, A. P., Levesque, M. P., Chen, Y. Y., Konstantinidis, P., & Nusslein-Volhard, C. (2013). Genetic lineage labeling in zebrafish uncovers novel neural crest contributions to the head, including gill pillar cells. *Development*, *140*(4), 916–925. http://dx.doi.org/10.1242/dev.091066.
Munoz, W. A., & Trainor, P. A. (2015). Neural crest cell evolution: How and when did a neural crest cell become a neural crest cell. In P. A. Trainor (Ed.), *Neural crest and placodes: Vol. 111* (pp. 3–16). New York: Elsevier.
Nakamura, H., & Ayer-le Lievre, C. S. (1982). Mesectodermal capabilities of the trunk neural crest of birds. *Journal of Embryology and Experimental Morphology*, *70*, 1–18.
Nieto, M. A. (2009). Epithelial-mesenchymal transitions in development and disease: Old views and new perspectives. *International Journal of Developmental Biology*, *53*(8–10), 1541–1547. http://dx.doi.org/10.1387/ijdb.072410mn.
Nishida, H., & Satoh, N. (1989). Determination and regulation in the pigment cell lineage of the ascidian embryo. *Developmental Biology*, *132*(2), 355–367.
Noack Watt, K. E., & Trainor, P. A. (2014). Neurocristopathies: Disorders of neural crest cell development. In P. A. Trainor (Ed.), *Neural crest cells: Evolution development and disease*. New York: Elsevier.
Noden, D. M. (1975). An analysis of migratory behavior of avian cephalic neural crest cells. *Developmental Biology*, *42*(1), 106–130.
Noden, D. M. (1983). The role of the neural crest in patterning of avian cranial skeletal, connective, and muscle tissues. *Developmental Biology*, *96*(1), 144–165.
Noden, D. M. (1984). Craniofacial development: New views on old problems. *Anatomical Record*, *208*(1), 1–13. http://dx.doi.org/10.1002/ar.1092080103.
Noden, D. M. (1986). Origins and patterning of craniofacial mesenchymal tissues. *Journal of Craniofacial Genetics and Developmental Biology, Supplement*, *2*, 15–31.
Northcutt, R. G., & Gans, C. (1983). The genesis of neural crest and epidermal placodes: A reinterpretation of vertebrate origins. *Quarterly Review of Biology*, *58*(1), 1–28.
Opie, I. A., & Opie, P. (1951). *The Oxford dictionary of nursery rhymes*. Oxford: Clarendon Press.
Peters, A. (1963). The structure of the dorsal root nerves of amphioxus an electron microscope study. *The Journal of Comparative Neurology*, *121*(2), 287–304. http://dx.doi.org/10.1002/cne.901210210.
Reardon, W., Winter, R. M., Rutland, P., Pulleyn, L. J., Jones, B. M., & Malcolm, S. (1994). Mutations in the fibroblast growth factor receptor 2 gene cause Crouzon syndrome. *Nature Genetics*, *8*(1), 98–103.
Schilling, T. F., & Kimmel, C. B. (1994). Segment and cell type lineage restrictions during pharyngeal arch development in the zebrafish embryo. *Development*, *120*(3), 483–494.

Smith, M. M. (1991). Putative skeletal neural crest cells in early late ordovician vertebrates from Colorado. *Science*, *251*(4991), 301–303. http://dx.doi.org/10.1126/science.251.4991.301.

Spemann, H., & Mangold, H. (2001). Induction of embryonic primordia by implantation of organizers from a different species. 1923. *International Journal of Developmental Biology*, *45*(1), 13–38.

Squarzoni, P., Parveen, F., Zanetti, L., Ristoratore, F., & Spagnuolo, A. (2011). FGF/MAPK/Ets signaling renders pigment cell precursors competent to respond to Wnt signal by directly controlling Ci-Tcf transcription. *Development*, *138*(7), 1421–1432. http://dx.doi.org/10.1242/dev.057323.

Trainor, P. A. (Ed.), (2014). *Neural crest cells: Evolution, development and disease*. New York: Elsevier.

Thiery, J. P., Acloque, H., Huang, R. Y., & Nieto, M. A. (2009). Epithelial-mesenchymal transitions in development and disease. *Cell*, *139*(5), 871–890. http://dx.doi.org/10.1016/j.cell.2009.11.007.

Trainor, P. A., & Tam, P. P. (1995). Cranial paraxial mesoderm and neural crest cells of the mouse embryo: Co-distribution in the craniofacial mesenchyme but distinct segregation in branchial arches. *Development*, *121*(8), 2569–2582.

Trainor, P. A., Tan, S. S., & Tam, P. P. (1994). Cranial paraxial mesoderm: Regionalisation of cell fate and impact on craniofacial development in mouse embryos. *Development*, *120*(9), 2397–2408.

Trelstad, R. L., Hay, E. D., & Revel, J. D. (1967). Cell contact during early morphogenesis in the chick embryo. *Developmental Biology*, *16*(1), 78–106.

Vega, S., Morales, A. V., Ocana, O. H., Valdes, F., Fabregat, I., & Nieto, M. A. (2004). Snail blocks the cell cycle and confers resistance to cell death. *Genes and Development*, *18*(10), 1131–1143. http://dx.doi.org/10.1101/gad.294104.

Wagner, G. (1949). Die Bedeutung der Neualleiste fur die Kpfgestaltung der Amphibienlarven. *Revue Suisse de Zoologie*, *56*, 519–620.

Wagner, G. (1959). Untersuchungen an Bombinator-Triton-Chimareren. *Roux Arch Entwicklungsmech*, *151*, 136–158.

Walker, M. B., & Trainor, P. A. (2006). Craniofacial malformations: Intrinsic vs extrinsic neural crest cell defects in Treacher Collins and 22q11 deletion syndromes. *Clinical Genetics*, *69*(6), 471–479. http://dx.doi.org/10.1111/j.0009-9163.2006.00615.x.

Weston, J. A. (1963). A radioautographic analysis of the migration and localization of trunk neural crest cells in the chick. *Developmental Biology*, *6*, 279–310.

Weston, J. A. (1970). The migration and differentiation of neural crest cells. *Advances in Morphogenesis*, *8*, 41–114.

Wiens, J. J., Brandley, M. C., & Reeder, T. W. (2006). Why does a trait evolve multiple times within a clade? Repeated evolution of snakelike body form in squamate reptiles. *Evolution*, *60*(1), 123–141.

Ye, X., Tam, W. L., Shibue, T., Kaygusuz, Y., Reinhardt, F., Ng Eaton, E., et al. (2015). Distinct EMT programs control normal mammary stem cells and tumour-initiating cells. *Nature*. http://dx.doi.org/10.1038/nature14897.

Yu, J. K. (2010). The evolutionary origin of the vertebrate neural crest and its developmental gene regulatory network—Insights from amphioxus. *Zoology (Jena, Germany)*, *113*(1), 1–9. http://dx.doi.org/10.1016/j.zool.2009.06.001.

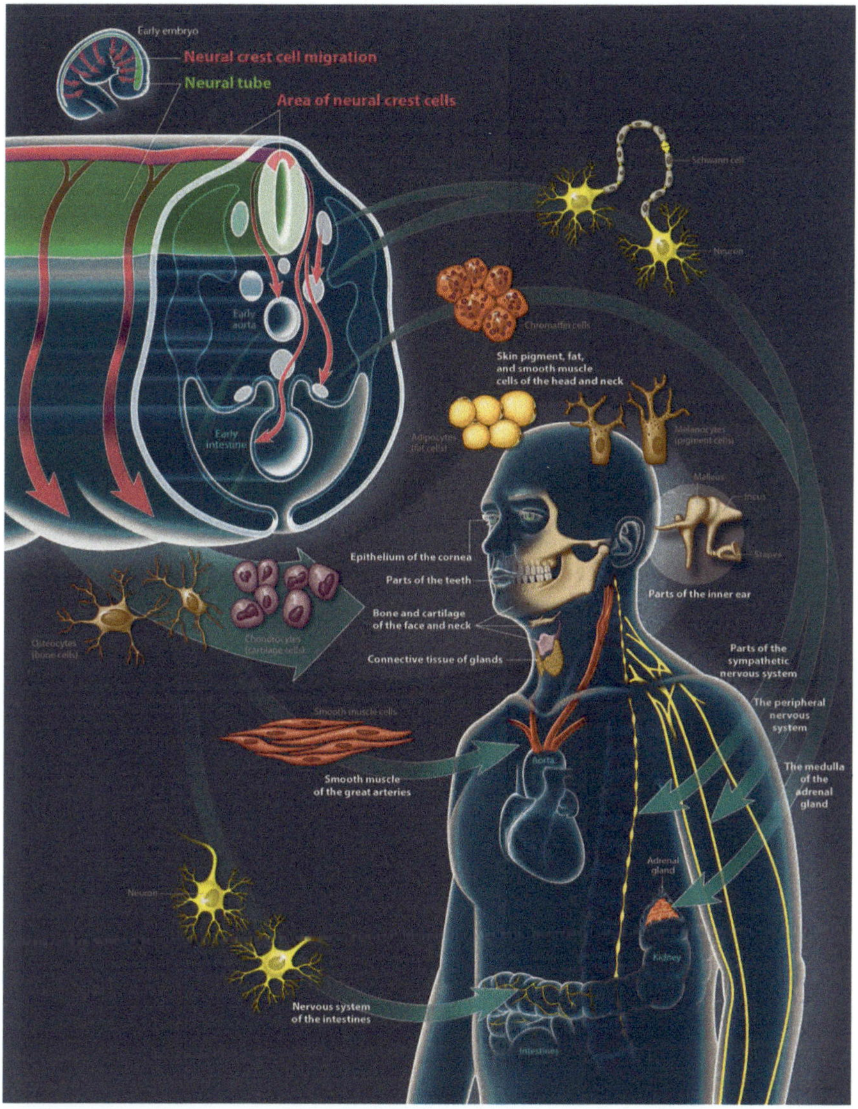

Paul A. Trainor, Figure 1 Neural crest cells. Schematic representation of neural crest cell derivation from the neural tube (green) and their migration via specific pathways (purple arrows) to give rise to numerous distinct cell types throughout the human body.

CHAPTER THIRTY-ONE

A Bespoke Coat for Eggs: Getting Ready for Fertilization

Paul M. Wassarman[1], Eveline S. Litscher
Department of Developmental and Regenerative Biology, Icahn School of Medicine at Mount Sinai, New York, USA
[1]Corresponding author: e-mail address: paul.wassarman@mssm.edu

Contents

Abstract

All eggs have an extracellular coat (EC) that plays unique roles during reproduction and development. ECs are designed to protect eggs and support their growth, regulate fertilization of eggs, and protect early embryos. ECs of mammalian and nonmammalian eggs consist of only a few proteins that are closely related to one another. All these proteins possess regulatory elements and a structural domain responsible for processing and assembly of the proteins into ECs. They also possess regions responsible for their functional roles during and after fertilization. Our essay addresses these and other aspects of EC biology.

Essay

Bespoke coat—A custom-made coat to fit a particular individual.

1. INTRODUCTION

For more than three decades, we have investigated various aspects of mouse zona pellucida (ZP) proteins in our laboratory, from their synthesis, structure, and assembly into a ZP, to their roles during oogenesis and

Current Topics in Developmental Biology, Volume 117
ISSN 0070-2153
http://dx.doi.org/10.1016/bs.ctdb.2015.10.018

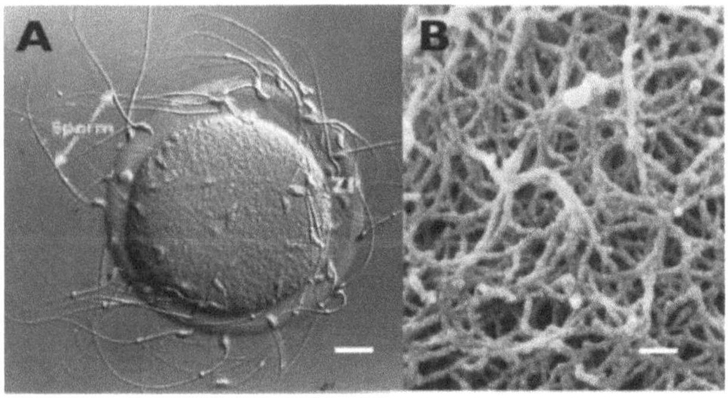

Figure 1 Light and electron micrographs of the mouse ZP. (A) Light micrograph of sperm bound to the mouse egg's ZP. Bar ≃13 μm. (B) Scanning electron micrograph of the mouse egg's ZP. Bar ≃200 nm. *Reproduced with permission from Wassarman (2008).* (A color version of this figure is located at the end of this chapter.)

fertilization. In this essay, we describe some major features of these unique proteins that make up the mammalian egg's extracellular coat (EC) or ZP and that participate directly in interactions between eggs and sperm during fertilization.

All mammalian eggs, including eggs from monotremes, marsupials, and placental mammals, are completely surrounded by a relatively thick EC or ZP (Fig. 1). During fertilization, sperm must bind to and then penetrate through the ZP to reach and fuse with the egg's plasma membrane. The ZP is laid down during oogenesis when oocytes, precursors of unfertilized eggs that are arrested in meiosis, undergo extensive growth to become unfertilized eggs. Each mouse oocyte grows from about 12 to 80 μm in diameter, about a 300-fold increase in volume, during a 2–3 week period. A ZP appears first as an extremely thin layer overlying the growing oocyte's plasma membrane and then increases in thickness to a terminal width of about 6–7 μm (contains about 3.5 ng of protein) around fully grown oocytes. All ZP are composed of long interconnected fibrils (Fig. 1) constructed of proteins held together by noncovalent bonds. The egg's ZP performs unique functions before, during, and after fertilization.

2. ZP AND VE PROTEINS

In view of the unique functions of the egg's EC as compared to the somatic cell's EC, it was not a surprise in 1980 when we found that the mouse egg's ZP consisted of three novel proteins that we named ZP1, 2, and 3 based on their apparent molecular weights (Bleil & Wassarman, 1980a; Wassarman, 1988a). All three mouse ZP proteins are glycosylated to different extents with asparagine- (N-) and serine/threonine- (O-) linked oligosaccharides; some of the oligosaccharides are sialylated and sulfated making the proteins relatively acidic (Greve, Salzmann, Roller, & Wassarman, 1982; Kinloch, Roller, Fimiani, Wassarman, & Wassarman, 1988; Liu, Litscher, & Wassarman, 1997; Salzmann, Greve, Roller, & Wassarman, 1983). Subsequently a fourth protein closely related to ZP1, called ZP4, was found in the ZP of human eggs (Hughes & Barratt, 1999). ZP4 is a pseudogene in mice.

The EC or vitelline envelope (VE) surrounding fish, reptilian, amphibian, and bird eggs also consists of a limited number of proteins, all closely related to ZP1–4 (Litscher & Wassarman, 2007, 2014, 2015). For example, comparison of ZP protein orthologs from human eggs with those from trout (33%), frog, chicken, and possum (43–51%), mouse, rat, cow, pig, and dog (64–69%), and macaca and chimpanzee (93–99%) reveals a high percentage average identity (33–99%) in overlapping sequences.

3. EVOLUTION OF ZP GENES

It is likely that egg-coat proteins are derived from a common ancestral gene, possibly *ZP3* (Litscher & Wassarman, 2014, 2015). This gene duplicated several times hundreds-of-millions of years ago and gave rise to 3–4 genes in fish (*ZP1*, *ZP3*, *ZPax*, and variants of *ZP1* and *ZP3*), 4–5 genes in amphibians (*ZP2–4*, *ZPd*, *ZPax*), 6 genes in birds (*ZP1–4*, *ZPd*, *ZPax*), and 3–4 genes (*ZP1–4*) in mammals. *ZP1–4* are found in mammals and other vertebrates, *ZPd* in amphibians and birds, and *ZPax* in fish, amphibians, and birds. The absence of *ZPax* and *ZPd* in mammals suggests that these genes were lost during evolution; that is, recent *ZP* gene evolution may have occurred by "gene death." In this context, *ZPax* is present as a nonfunctional pseudogene in cows, macacas, chimpanzees, and human beings.

4. ZP DOMAIN

In 1992, a structural element, called the ZP domain (ZPD), was identified in all ZP proteins by pattern-based sequence analysis (Bork & Sander, 1992). To date a ZPD has been found in hundreds of other kinds of proteins that have diverse functions and are present in a wide variety of organisms, from jellyfish to human beings (Jovine, Darie, Litscher, & Wassarman, 2005; Litscher & Wassarman, 2014, 2015). The latter suggests that the ZPD, which consists of approximately 260 amino acids (aa), has been conserved through at least 600 million years of evolution. Since it has been proposed that a yeast mating protein adopts a three-dimensional fold identical to one found in the ZPD of ZP proteins (Swanson et al., 2011), it is possible that the ZPD has been conserved for an even longer period of evolution.

The ZPD is composed of two subdomains, called ZP-N (about 120 aa) and ZP-C (about 130 aa) (Fig. 2), each with four conserved cysteine residues present as intramolecular disulfides. It is likely that the presence of additional cysteine residues and different disulfide bonding patterns in ZP proteins from some animals cause their polypeptides to adopt conformations that could alter the specificity of EC assembly in these species. A comparison of ZPD sequences of mammalian ZP proteins (ZP1–4), from platypus to human beings, reveals that the average percent identity (about 73%) and percent similarity (about 90%) is quite high (Litscher & Wassarman, 2015).

ZP-N is a conserved module for polymerization of ZP proteins (Jovine et al., 2002, 2005; Jovine, Janssen, Litscher, & Wassarman, 2006), can serve as an independent structural domain in the absence of ZP-C, and is found in single or multiple copies at the N-terminus of ZP1–4 (Fig. 2). ZP-N

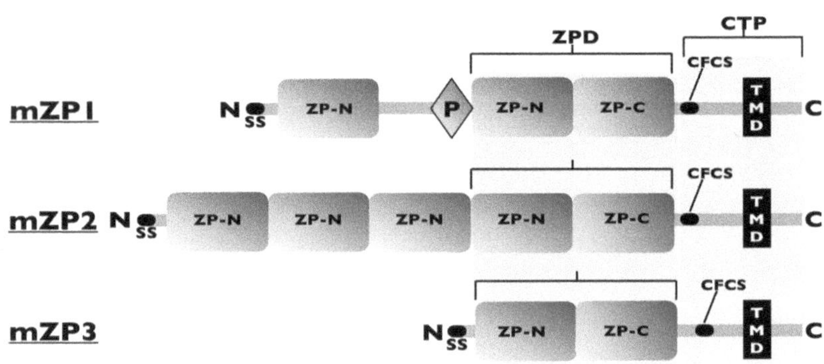

Figure 2 Schematic representation of the organization of mZP1, 2, and 3. In each case, the polypeptide contains an SS at the N-terminus, a ZPD, and a CFCS, TMD, and CT at the C-terminal propeptide (CTP). mZP1 also has a trefoil (P) domain adjacent to the ZPD and an extra ZP-N subdomain close to the N-terminus of the polypeptide. mZP2 has three extra ZP-N subdomains between the ZPD and N-terminus of the polypeptide. mZP3, the smallest of the three proteins, consists primarily of a ZPD. *Reproduced with permission from Litscher and Wassarman (2015).*

and ZP-C are always present in one order, ZP-N→ZP-C, and not the other, ZP-C→ZP-N, although the ZP-N subdomain can be found by itself. ZP-N and ZP-C share a common immunoglobulin (Ig)-like topology with β-sheet arrangements symmetrical to each other (Han et al., 2010; Monné & Jovine, 2011). In general, regions adjoining the ZPD contribute functional diversification that may be caused by selective pressure related to specific functions. For example, there are two clusters of sites N- and C-terminal to the ZPD of mouse ZP3 that are apparently under positive Darwinian selection (Claw & Swanson, 2012; Swanson, Yang, Wolfner, & Aquadro, 2001).

ZPD proteins are structural components of egg coats and animal tissues. They serve as receptors, mechanotransducers, and antimicrobials and often play important roles during signaling, differentiation, and morphogenesis (e.g., Chung, Zhu, Han, & Kernan, 2001; Fernandes et al., 2010; Heiman & Shaham, 2009; Plaza, Chanut-Delalande, Fernandes, Wassarman, & Payre, 2010). ZPD proteins play an important role in organizing and shaping highly specialized apical structures in epithelial cells and participate in functioning of the senses, taste and smell. Included among ZPD proteins are TGF-β receptor III (betaglycan), uromodulin, tectorin-α and -β, endoglin, vomeroglandin, hensin, cuticlins, oikosins, and mesoglein. Mutations in ZPD proteins can lead to severe human pathologies, including deafness, vascular and renal disease, cancer, and possibly infertility. It has been estimated that there are more than 10 ZPD proteins in nematodes (*C. elegans*), more than 20 in flies (*D. melanogaster*), and more than 100 in birds.

5. SYNTHESIS AND POLYMERIZATION OF ZP PROTEINS

In mammals and amphibians, ZP proteins are synthesized only in the ovary by oocytes and/or follicle cells, whereas in fish and birds ZP proteins are synthesized in the ovary and/or liver (Litscher & Wassarman, 2014, 2015). In mice it was found that ZP proteins originate from the growing oocyte rather than from surrounding follicle cells (Bleil & Wassarman, 1980b). Single-copy genes encoding ZP1–3 are expressed at very high levels by female mice during oocyte growth and are regulated by *cis*-acting DNA elements located close to the transcription start-site of *ZP* genes (Lira, Kinloch, Mortillo, & Wassarman, 1990; Lira, Schickler, & Wassarman, 1993). For example, ZP3 messenger-RNA is first detected when mouse oocytes begin to grow and accumulates to as much as 250,000–300,000 copies per oocyte in the terminal stages of growth. However, the level of ZP3 messenger-RNA falls dramatically to about 5000 copies per ovulated egg and to undetectable levels (less than 1000 copies) in fertilized eggs (Roller, Kinloch, Hiraoka, Li, & Wassarman, 1989).

Each nascent precursor ZP polypeptide has a signal-sequence (SS) at the N-terminus, a consensus furin cleavage site (CFCS) and transmembrane domain (TMD) near the C-terminus, and a short C-terminal tail (CT); ZP1 and 4 also have a trefoil (P) domain (Wassarman, 2008; Fig. 2). Nascent ZP proteins are packaged into relatively large intracellular secretory vesicles and incorporated into only the innermost layer of the thickening ZP (Qi, Williams, & Wassarman, 2002). Mature, secreted ZP proteins that polymerize into ECs lack the SS, CFCS, TMD, and CT due to proteolytic processing of the precursors (Litscher, Qi, & Wassarman, 1999; Williams & Wassarman, 2001). Two hydrophobic peptides (HP), an internal HP (IHP) located within the ZPD and an external HP (EHP) located between the CFCS and TMD, ensure that polymerization of ZP proteins occurs outside, not inside, growing oocytes (Jovine, Qi, Williams, Litscher, & Wassarman, 2004). Cleavage at the CFCS releases the EHP from the IHP and ZP proteins are activated for assembly into an EC (Fig. 3).

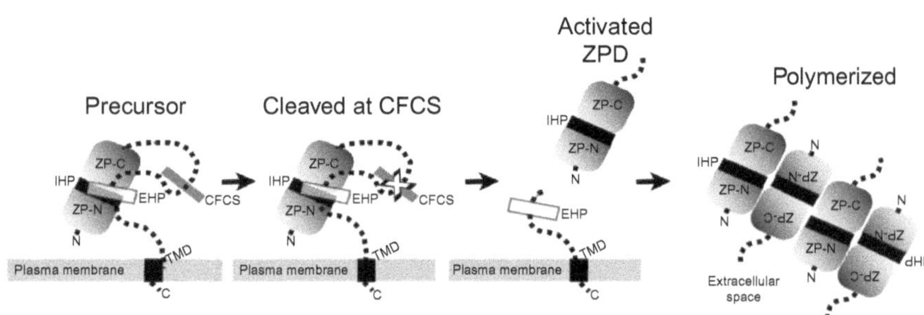

Figure 3 Schematic representation of a general mechanism for assembly of nascent ZP proteins. In all ZPD precursor proteins (precursor), the ZPD is followed by a C-terminal propeptide (CTP) that contains a basic cleavage site, such as a CFCS, an EHP, and, in most cases, a TMD or glycosylphosphatidylinositol- (GPI-) anchor site. Precursors do not polymerize within the cell either as a result of direct interaction between the EHP and IHP or because they adopt a conformation dependent on the presence of both hydrophobic patches. C-terminal processing at the CFCS by a pro-protein convertase (cleaved at CFCS) leads to dissociation of mature proteins from the EHP and activation of the ZPD (activated ZPD) for assembly (polymerization) into fibrils and matrices. *Reproduced with permission from Litscher and Wassarman (2015).*

6. STRUCTURE OF THE ZP

ZP1–4 are structural proteins. Nascent ZP proteins assemble/polymerize in the extracellular space into long interconnected fibrils that exhibit a structural repeat (Greve & Wassarman, 1985; Wassarman, 2008). A ZP2:ZP3 dimer is present every 14 nm or so along the fibrils that make up the matrix and fibrils are interconnected by ZP1/ZP4. Mutant female mice carrying a single ZP3 allele possess a ZP and reproduce normally, but their ZP is less than one-half the thickness of the wild-type ZP (Wassarman, Qi, & Litscher, 1997). On the other hand, homozygous-knockout female mice devoid of either ZP2 or ZP3 fail to construct a ZP and are infertile (Liu et al., 1996; Rankin et al., 1996, 2001). The latter is due to a paucity of growing oocytes in ovaries that leads to a scarcity of ovulated eggs in oviducts. In the absence of a ZP, the formation of gap junctions between oocytes and surrounding follicle cells is apparently reduced, thereby compromising transfer of nutrients, metabolites, and other molecules to growing oocytes and resulting in fewer growing oocytes and ovulated eggs (Wassarman & Litscher, 2012). Presumably the ZP helps to stabilize gap junctions under normal conditions of oocyte growth and follicle expansion. Both ZP2 and ZP3 must be present during oocyte growth in order to assemble a ZP that supports both oogenesis and fertility.

7. ZP PROTEINS AND FERTILIZATION

The notion that the egg's ZP harbors species-restricted receptors for sperm dates back more than a century. In the 1970s, it was found that solubilized ZP preparations from unfertilized hamster and mouse eggs inhibited binding of hamster sperm to ovulated eggs *in vitro* (Gwatkin, 1977; Gwatkin & Williams, 1976). This suggested that solubilized ZP preparations from unfertilized eggs contained so-called "sperm receptors" that bound to sperm and prevented them from binding to ovulated eggs *in vitro*.

Results of experiments carried out in our laboratory strongly suggest that unfertilized egg ZP3 and ZP2 serve as "primary" and "secondary" sperm receptors, respectively, during fertilization in mice (Bleil, Greve, & Wassarman, 1988; Bleil & Wassarman, 1980c, 1986; Wassarman, 1987, 1988b, 1990, 1999; Wassarman et al., 1985; Wassarman & Litscher, 2008). Binding of sperm to eggs is inhibited *in vitro* when sperm are exposed to nanomolar concentrations of ZP3 purified from unfertilized egg ZP (Table 1); on the other hand, binding is not inhibited by purified ZP1 or ZP2 (Bleil & Wassarman, 1980c). When sperm bind to ZP3 they undergo the acrosome reaction, a membrane fusion event that enables sperm to penetrate the ZP and fuse with the egg's plasma membrane (Arnoult, Zeng, & Florman, 1996; Bleil & Wassarman, 1983; Jungnickel, Sutton, Wang, & Florman, 2007). Acrosome-reacted sperm bind to ZP2 (Bleil et al., 1988; Bleil & Wassarman, 1986). However, in response to fertilization of eggs by a single sperm both ZP2 and ZP3 are inactivated as sperm receptors; the former by limited proteolytic cleavage and the latter by an unknown mechanism. This is consistent with the finding that preparations of solubilized ZP from two-cell mouse embryos have no effect on binding of sperm to eggs (Bleil & Wassarman, 1980c) and do not induce sperm to undergo the acrosome reaction *in vitro* (Bleil & Wassarman, 1983). This accounts for the general observation that mammalian sperm bind to the ZP of unfertilized eggs, but not to the ZP of fertilized eggs.

Table 1 Inhibitors of Sperm-Binding to Mouse Eggs *In Vitro*

mZP3[a] (Bleil & Wassarman, 1980b)
mZP3 glycopeptides[b] (Florman, Bechtol, & Wassarman, 1984)
mZP3 55 kD glycopeptide[c] (Litscher & Wassarman, 1996)
EC-mZP3[d] (Kinloch, Mortillo, Stewart, & Wassarman, 1991)
EC-ZP3[h1–5/m6–8][e] (Kinloch, Sakai, & Wassarman, 1995)
EC-hZP3/m7[f] (Williams, Litscher, Jovine, & Wassarman, 2006)
EC-huIgG(Fc)/mZP3(7,8)[g] (Williams et al., 2006)
mZP3 O-linked oligosaccharides[h] (Florman & Wassarman, 1985)
Biantennary oligosaccharide-α-Gal[i] (Litscher et al., 1995)
Tetraantennary oligosaccharides-α/β-Gal[j] (Litscher et al., 1995)
Lewis X (Le[x])-containing glycans[k] (Johnston et al., 1998)

[a]mZP3 purified from unfertilized egg ZP.
[b]Mixture of mZP3 glycopeptides from pronase digests.
[c]mZP3 glycopeptide (≃55 kD MW) purified from papain digests.
[d]mZP3 secreted by embryonal carcinoma (EC) cells transfected with *mZP3*.
[e]Hamster ZP3 secreted by embryonal carcinoma (EC) cells transfected with *hZP3* in which *hZP3* exons 6–8 were replaced with *mZP3* exons 6–8.
[f]Hamster ZP3 secreted by embryonal carcinoma (EC) cells transfected with *hZP3* in which *hZP3* exon-7 was replaced with *mZP3* exon-7.
[g]Human IgG(Fc) chimeric protein secreted by embryonal carcinoma (EC) cells transfected with *huIgG (Fc)* to which *mZP3* exons 7 and 8 were fused.
[h]Oligosaccharides released by mild alkaline reduction of mZP3.
[i]Biantennary blood group I-related dodecasaccharide with Gal in α-linkage at the nonreducing end.
[j]Tetraantennary blood group I-related tetra- and octa-decasaccharides with Gal in α- or β-linkage at the nonreducing end.
[k]Tetrasaccharides with Gal in α- or β-linkage at the nonreducing end with a Fuc residue and trisaccharides lacking the Fuc residue.

Binding of sperm to ZP3 is dependent on a portion of the protein's polypeptide encoded by exon-7, the polypeptide's C-terminal region, and O-linked carbohydrate (e.g., Bleil & Wassarman, 1988; Florman & Wassarman, 1985; Han et al., 2010; Johnston et al., 1998; Kinloch et al., 1995; Litscher et al., 1995; Williams et al., 2006; Table 1). This is a situation similar to binding of sperm from marine animals and amphibians to eggs, sexual agglutination in budding yeast, and binding of bacteria, animal viruses, and other pathogens to their cellular hosts. Antibodies directed against polypeptide encoded by exon-7 inhibit binding of sperm to eggs *in vivo* (Millar et al., 1989) and *in vitro* (Rosiere & Wassarman, 1992); the former causing long-lasting infertility. On the other hand, binding of acrosome-reacted sperm to ZP2 is dependent solely on the protein's polypeptide. Limited proteolysis of ZP2 following fertilization inactivates it as a secondary sperm receptor (Avella, Baibakov, & Dean, 2014; Bleil, Beall, & Wassarman, 1981) and also results in a substantial change in the solubility of the ZP, making it a much more resilient or "harder" structure (Moller & Wassarman, 1989). The change in solubility of the ZP following fertilization is accompanied by an approximately two- to fourfold increase in its stiffness and viscosity (Kim & Kim, 2013).

ZP3 from different mammalian eggs exhibits a high degree of divergence in its C-terminal region, encoded by exon-7, compared to other regions of the protein, perhaps to ensure some species-specificity during fertilization (Claw & Swanson, 2012; Swanson et al., 2001; Wassarman & Litscher, 1995; Williams et al., 2006). This would be analogous to the accelerated evolution of EC proteins of mollusk eggs, accessory gland proteins of flies, and agglutinin proteins of yeast that are thought to be responsible for facilitating speciation.

8. FINAL COMMENTS

For nearly four decades, characterization of EC (ZP and VE) genes and proteins, from mollusks and fish to human beings, has provided vital information about EC synthesis, structure, evolution, and function. Much of this information has been obtained over the years by the application of impressive technological advances in biochemistry, molecular genetics, protein crystallography, genomics, and proteomics to research on EC genes and proteins.

Today, it is clear that the proteins that constitute ECs surrounding eggs from all mammalian and many nonmammalian species are closely related to one another and have certain structural features in common (e.g., a ZPD), and that genes encoding EC proteins may be derived from a common ancestral gene, possibly *ZP3*. Recently, it has been proposed that ZP proteins have features or properties similar to those of amyloids that self-aggregate and form cross-β sheet fibrillar structures (Egge, Muthusubramanian, & Cornwall, 2015; Louros, Iconomidou, Giannelou, & Hamodrakas, 2013); a proposal that possibly bears on the molecular basis of ZP assembly and structure, and on evolution of ZP proteins.

A pathway has been described for proteolytic processing and assembly/polymerization of nascent mammalian ZP proteins (Jovine et al., 2005; Jovine, Qi, Williams, Litscher, & Wassarman, 2007) and a high-resolution, three-dimensional structure has been determined for at least one ZP protein, chicken ZP3 (Han et al., 2010). The latter provided considerable information, not only about the overall structure of ZP3 but also about certain of its features (e.g., oligosaccharide) that help to mediate sperm–egg interaction. In this context, it is likely that high-resolution structures for ZP1 and ZP2 will become available in the near future and recent advances in cryo-electron microscopy soon may permit high-resolution analyses of intact ZP.

However, despite considerable progress numerous questions remain unanswered, especially with respect to the role of EC proteins in fertilization. Perhaps, key among these questions are (1) what is the molecular basis of species-specific/restricted fertilization and (2) is there a single pathway or are there redundant pathways that ultimately lead to fusion of sperm and egg to form a zygote. The latter question is particularly relevant since attempts to identify sperm molecules that support binding of mammalian sperm to eggs have proved to be particularly ambiguous. Hopefully, answers to these and other questions will be forthcoming in the not-too-distant future and should extend our knowledge of the process of fertilization.

ACKNOWLEDGMENTS

This essay is dedicated to the memory of Jeffrey David Bleil (1952–2014), a most valued colleague and close friend who carried out pioneering research on the mouse egg's zona pellucida and mammalian fertilization. We thank the graduate students, postdoctoral fellows, and research technicians who contributed to the research on ZP proteins and fertilization carried out in our laboratory at Harvard Medical School, Roche Institute of Molecular Biology, and Icahn School of Medicine at Mount Sinai. Our research was supported in part by the National Institutes of Health (NICHD, HD-35105) and Hoffmann-LaRoche, Inc.

REFERENCES

Arnoult, C., Zeng, Y., & Florman, H. M. (1996). ZP3-dependent activation of sperm cation channels regulates acrosomal secretion during mammalian fertilization. *Journal of Cell Biology*, *134*, 637–645.

Avella, M. A., Baibakov, B., & Dean, J. (2014). A single domain of the ZP2 zona pellucida protein mediates gamete recognition in mice and humans. *Journal of Cell Biology*, *205*, 801–809.

Bleil, J. D., Beall, C. F., & Wassarman, P. M. (1981). Mammalian sperm-egg interaction: Fertilization of mouse eggs triggers modification of the major zona pellucida glycoprotein, ZP2. *Developmental Biology*, *86*, 189–197.

Bleil, J. D., Greve, J. M., & Wassarman, P. M. (1988). Identification of a secondary sperm receptor in the mouse egg zona pellucida: Role in maintenance of binding of acrosome-reacted sperm to eggs. *Developmental Biology*, *128*, 376–385.

Bleil, J. D., & Wassarman, P. M. (1980a). Structure and function of the zona pellucida: Identification and characterization of the proteins of the mouse oocyte's zona pellucida. *Developmental Biology*, *76*, 185–202.

Bleil, J. D., & Wassarman, P. M. (1980b). Synthesis of zona pellucida proteins by denuded and follicle-enclosed mouse oocytes during culture in vitro. *Proceedings of the National Academy of Sciences of the United States of America*, *77*, 1029–1033.

Bleil, J. D., & Wassarman, P. M. (1980c). Mammalian sperm-egg interaction: Identification of a glycoprotein in mouse egg zonae pellucidae possessing receptor activity for sperm. *Cell*, *20*, 873–882.

Bleil, J. D., & Wassarman, P. M. (1983). Mammalian sperm-egg interaction: Sequence of events and induction of the acrosome reaction by a zona pellucida glycoprotein. *Developmental Biology*, *95*, 317–324.

Bleil, J. D., & Wassarman, P. M. (1986). Autoradiographic visualization of the mouse egg's sperm receptor bound to sperm. *Journal of Cell Biology*, *102*, 1363–1371.

Bleil, J. D., & Wassarman, P. M. (1988). Galactose at the nonreducing terminus of O-linked oligosaccharides of mouse egg zona pellucida glycoprotein ZP3 is essential for the glycoprotein's sperm receptor activity. *Proceedings of the National Academy of Sciences of United States of America*, *85*, 6778–6782.

Bork, P., & Sander, C. (1992). A large domain common to sperm receptors (Zp2 and Zp3) and TGF-beta type III receptor. *FEBS Letters*, *300*, 237–240.

Chung, Y. D., Zhu, J., Han, Y., & Kernan, M. J. (2001). nompA encodes a PNS-specific, ZP domain protein required to connect mechanosensory dendrites to sensory structures. *Neuron*, *29*, 415–428.

Claw, K. G., & Swanson, W. J. (2012). Evolution of the egg: New findings and challenges. *Annual Review of Genomics and Human Genetics*, *13*, 109–125.

Egge, N., Muthusubramanian, A., & Cornwall, G. A. (2015). Amyloid properties of the mouse egg zona pellucida. *PLOS One*, *10*, e0129907.

Fernandes, I., Chanut-Delalande, H., Ferrer, P., Latapie, Y., Waltzer, L., Affolter, M., et al. (2010). Zona pellucida domain proteins remodel the apical compartment for localized cell shape changes. *Developmental Cell*, *18*, 64–76.

Florman, H. M., Bechtol, K. B., & Wassarman, P. M. (1984). Enzymatic dissection of the functions of the mouse egg's receptor for sperm. *Developmental Biology*, *106*, 243–255.

Florman, H. M., & Wassarman, P. M. (1985). O-Linked oligosaccharides of mouse egg ZP3 account for its sperm receptor activity. *Cell*, *41*, 313–324.

Greve, J. M., Salzmann, G. S., Roller, R. J., & Wassarman, P. M. (1982). Biosynthesis of the major zona pellucida glycoprotein secreted by oocytes during mammalian oogenesis. *Cell*, *31*, 749–759.

Greve, J. M., & Wassarman, P. M. (1985). Mouse egg extracellular coat is a matrix of interconnected filaments possessing a structural repeat. *Journal of Molecular Biology, 181*, 253–264.

Gwatkin, R. B. L. (1977). *Fertilization mechanisms in man and mammals*. New York, NY: Plenum. 161 pp.

Gwatkin, R. B. L., & Williams, D. T. (1976). Receptor activity of the solubilized hamster and mouse zona pellucida before and after the zona reaction. *Journal of Reproduction and Fertility, 49*, 55–59.

Han, L., Monné, M., Okumura, H., Schwend, T., Cherry, A. L., Flot, D., et al. (2010). Insights into egg-coat assembly and egg-sperm interaction from the X-ray structure of full-length ZP3. *Cell, 143*, 1–12.

Heiman, M. G., & Shaham, S. (2009). DEX-1 and DYF-7 establish sensory dendrite length by anchoring dendritic tips during cell migration. *Cell, 137*, 344–355.

Hughes, D. C., & Barratt, C. L. R. (1999). Identification of the true orthologue of the mouse ZP1 gene: Evidence for greater complexity in the mammalian zona pellucida. *Biochimica et Biophysica Acta, 1447*, 303–306.

Johnston, D. S., Wright, W. W., Shaper, J. H., Hokke, C. H., Van den Eijnden, D. H., & Joziasse, D. H. (1998). Murine sperm-zona binding, a fucosyl residue is required for a high affinity sperm-binding ligand. *Journal of Biological Chemistry, 273*, 1888–1895.

Jovine, L., Darie, C. C., Litscher, E. S., & Wassarman, P. M. (2005). Zona pellucida domain proteins. *Annual Review of Biochemistry, 74*, 83–114.

Jovine, L., Janssen, W. G., Litscher, E. S., & Wassarman, P. M. (2006). The PLAC1-homology region of the ZP domain is sufficient for protein polymerization. *BMC Biochemistry, 7*, 11.

Jovine, L., Qi, H., Williams, Z., Litscher, E., & Wassarman, P. M. (2002). The ZP domain is a conserved module for protein polymerization. *Nature Cell Biology, 4*, 457–461.

Jovine, L., Qi, H., Williams, Z., Litscher, E. S., & Wassarman, P. M. (2004). A duplicated motif controls assembly of zona pellucida domain proteins. *Proceedings of the National Academy of Sciences of United States of America, 101*, 5922–5927.

Jovine, L., Qi, H., Williams, Z., Litscher, E. S., & Wassarman, P. M. (2007). Features that affect secretion and assembly of zona pellucida glycoproteins during mammalian oogenesis. In S. K. Gupta, K. Koyama, & J. F. Murray (Eds.), *Gamete biology* (pp. 187–201). UK: Nottingham University Press.

Jungnickel, M. K., Sutton, K. A., Wang, Y., & Florman, H. M. (2007). Phosphoinositide-dependent pathways in mouse sperm are regulated by egg ZP3 and drive the acrosome reaction. *Developmental Biology, 304*, 116–126.

Kim, J., & Kim, J. (2013). Viscoelastic characterization of mouse zona pellucida. *IEEE Transactions of Biomedical Engineering, 60*, 569–575.

Kinloch, R. A., Mortillo, S., Stewart, C. L., & Wassarman, P. M. (1991). Embryonal carcinoma cells transfected with ZP3 genes differentially glycosylate similar polypeptides and secrete active mouse sperm receptor. *Journal of Cell Biology, 115*, 655–664.

Kinloch, R. A., Roller, R. J., Fimiani, C. M., Wassarman, D. A., & Wassarman, P. M. (1988). Primary structure of the mouse sperm receptor's polypeptide chain determined by genomic cloning. *Proceedings of the National Academy of Sciences of United States of America, 85*, 6409–6413.

Kinloch, R. A., Sakai, Y., & Wassarman, P. M. (1995). Mapping the mouse ZP3 combining site for sperm by exon swapping and site-directed mutagenesis. *Proceedings of the National Academy of Sciences of United States of America, 92*, 263–267.

Lira, S. A., Kinloch, R. A., Mortillo, S., & Wassarman, P. M. (1990). An upstream region of the mouse ZP3 gene directs expression of firefly luciferase specifically to growing oocytes in transgenic mice. *Proceedings of the National Academy of Sciences of United States of America, 87*, 7215–7219.

Lira, S. A., Schickler, M., & Wassarman, P. M. (1993). Cis-acting DNA elements involved in oocyte-specific expression of mouse sperm receptor gene mZP3 are located close to the gene's transcription start site. *Molecular Reproduction and Development, 36*, 494–499.

Litscher, E. S., Juntunen, K., Seppo, A., Penttila, L., Niemela, R., Renkonen, O., et al. (1995). Oligosaccharide constructs with defined structures that inhibit binding of mouse sperm to unfertilized eggs *in vitro*. *Biochemistry, 34*, 4662–4669.

Litscher, E. S., Qi, H., & Wassarman, P. M. (1999). Mouse zona pellucida glycoproteins mZP2 and mZP3 undergo carboxy-terminal proteolytic processing in growing oocytes. *Biochemistry, 38*, 12280–12287.

Litscher, E. S., & Wassarman, P. M. (1996). Characterization of a mouse ZP3-derived glycopeptide, gp55, that exhibits sperm receptor and acrosome reaction-inducing activity *in vitro*. *Biochemistry, 35*, 3980–3986.

Litscher, E. S., & Wassarman, P. M. (2007). Egg extracellular coat proteins: From fish to mammals. *Histology and Histopathology, 22*, 337–347.

Litscher, E. S., & Wassarman, P. M. (2014). Evolution, structure, and synthesis of vertebrate egg-coat proteins. *Trends in Developmental Biology, 8*, 65–76.

Litscher, E. S., & Wassarman, P. M. (2015). *A guide to zona pellucida domain proteins*. Hoboken, NJ: John Wiley and Sons. 208 pp.

Liu, C., Litscher, E. S., Mortillo, S., Sakai, Y., Kinloch, R. A., Stewart, C. L., et al. (1996). Targeted disruption of the mZP3 gene results in production of eggs lacking a zona pellucida and infertility in female mice. *Proceedings of the National Academy of Sciences of United States of America, 93*, 5431–5436.

Liu, C., Litscher, E. S., & Wassarman, P. M. (1997). Zona pellucida glycoprotein mZP3 bioactivity is not dependent on the extent of glycosylation of its polypeptide or on sulfation and sialylation of its oligosaccharides. *Journal of Cell Science, 110*, 745–752.

Louros, N. N., Iconomidou, V. A., Giannelou, P., & Hamodrakas, S. J. (2013). Structural analysis of peptide-analogues of human zona pellucida ZP1 protein with amyloidogenic properties: Insights into mammalian zona pellucida formation. *PLOS One, 8*, e73258.

Millar, S. E., Chamow, S. M., Baur, A. W., Oliver, C., Robey, F., & Dean, J. (1989). Vaccination with a synthetic zona pellucida peptide produces long-term contraception in female mice. *Science, 246*, 935–938.

Moller, C. C., & Wassarman, P. M. (1989). Characterization of a proteinase that cleaves zona pellucida glycoprotein ZP2 following activation of mouse eggs. *Developmental Biology, 132*, 103–112.

Monné, M., & Jovine, L. (2011). A structural view of egg coat architecture and function in fertilization. *Biology of Reproduction, 85*, 661–669.

Plaza, S., Chanut-Delalande, H., Fernandes, I., Wassarman, P. M., & Payre, F. (2010). From A to Z: Apical structures and zona pellucida-domain proteins. *Trends in Cell Biology, 20*, 524–532.

Qi, H., Williams, Z., & Wassarman, P. M. (2002). Secretion and assembly of zona pellucida glycoproteins by growing mouse oocytes microinjected with epitope-tagged cDNAs for mZP2 and mZP3. *Molecular Biology of the Cell, 13*, 530–541.

Rankin, T., Familiari, M., Lee, E., Ginsburg, A., Dwyer, N., Blanchette-Mackie, J., et al. (1996). Mice homozygous for an insertional mutation in the ZP3 gene lack a zona pellucida and are infertile. *Development, 122*, 2903–2910.

Rankin, T., O'Brien, M., Lee, E., Wigglesworth, K., Eppig, J., & Dean, J. (2001). Defective zonae pellucidae in Zp2-null mice disrupt folliculogenesis, fertility, and development. *Development, 128*, 1119–1126.

Roller, R. J., Kinloch, R. A., Hiraoka, B. Y., Li, S. S.-L., & Wassarman, P. M. (1989). Gene expression during mammalian oogenesis and early embryogenesis: Quantification of three messenger-RNAs abundant in fully grown mouse oocytes. *Development, 106*, 251–261.

Rosiere, T. K., & Wassarman, P. M. (1992). Identification of a region of mouse zona pellucida glycoprotein mZP3 that possesses sperm receptor activity. *Developmental Biology, 154*, 309–317.

Salzmann, G. S., Greve, J. M., Roller, R. J., & Wassarman, P. M. (1983). Biosynthesis of the sperm receptor during oogenesis in the mouse. *EMBO Journal, 2*, 1451–1456.

Swanson, W. J., Aagaard, J. E., Vacquier, V. D., Monné, M., Al Hosseini, H., & Jovine, L. (2011). The molecular basis of sex: Linking yeast to human. *Molecular Biology and Evolution, 28*, 1963–1966.

Swanson, W. J., Yang, Z., Wolfner, M. F., & Aquadro, C. F. (2001). Positive Darwinian selection drives the evolution of several female reproductive proteins in mammals. *Proceedings of the National Academy of Sciences of United States of America, 98*, 2509–2514.

Wassarman, P. M. (1987). The biology and chemistry of fertilization. *Science, 235*, 553–560.

Wassarman, P. M. (1988a). Zona pellucida glycoproteins. *Annual Review of Biochemistry, 57*, 415–442.

Wassarman, P. M. (1988b). Fertilization in mammals. *Scientific American, 255*, 78–84.

Wassarman, P. M. (1990). Profile of a mammalian sperm receptor. *Development, 108*, 1–19.

Wassarman, P. M. (1999). Mammalian fertilization: Molecular aspects of gamete adhesion. *Cell, 96*, 175–183.

Wassarman, P. M. (2008). Zona pellucida glycoproteins. *Journal of Biological Chemistry, 283*, 24285–24289.

Wassarman, P. M., Bleil, J. D., Florman, H. M., Greve, J. M., Roller, R. J., Salzmann, G. S., et al. (1985). The mouse egg's sperm receptor: What is it and how does it work? *Cold Spring Harbor Symposium on Quantitative Biology, 50*, 11–18.

Wassarman, P. M., & Litscher, E. S. (1995). Sperm-egg recognition mechanisms in mammals. *Current Topics in Developmental Biology, 30*, 1–19.

Wassarman, P. M., & Litscher, E. S. (2008). Mammalian fertilization: The egg's multifunctional zona pellucida. *International Journal of Developmental Biology, 52*, 665–676.

Wassarman, P. M., & Litscher, E. S. (2012). Influence of the zona pellucida of the mouse egg on folliculogenesis and fertility. *International Journal of Developmental Biology, 56*, 833–839.

Wassarman, P. M., Qi, H., & Litscher, E. S. (1997). Mutant female mice carrying a single mZP3 allele produce eggs with a thin zona pellucida, but reproduce normally. *Proceedings of the Royal Society: Biological Sciences, 264*, 323–328.

Williams, Z., Litscher, E. S., Jovine, L., & Wassarman, P. M. (2006). Polypeptide encoded by mouse *ZP3* exon-7 is necessary and sufficient for sperm binding *in vitro*. *Journal of Cellular Physiology, 207*, 30–39.

Williams, Z., & Wassarman, P. M. (2001). Secretion of mouse ZP3, the sperm receptor, requires cleavage of its polypeptide at a consensus furin cleavage site. *Biochemistry, 40*, 929–937.

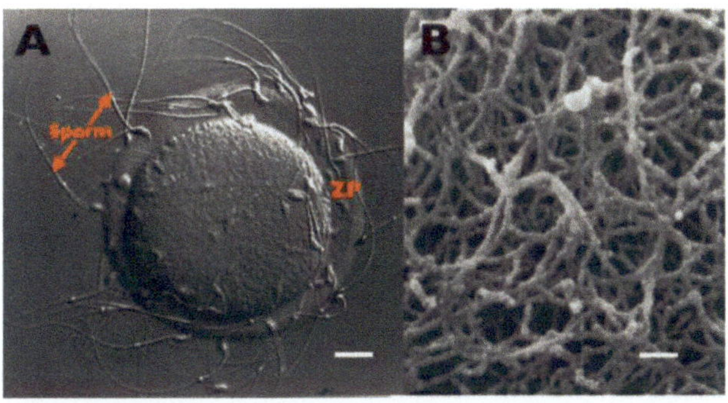

Paul M. Wassarman and Eveline S. Litscher, Figure 1 Light and electron micrographs of the mouse ZP. (A) Light micrograph of sperm bound to the mouse egg's ZP. Bar $\simeq$13 μm. (B) Scanning electron micrograph of the mouse egg's ZP. Bar $\simeq$200 nm. *Reproduced with permission from Wassarman (2008).*

CHAPTER THIRTY-TWO

Germ Line Mechanics—And Unfinished Business

Gary M. Wessel[1]
Department of Molecular and Cellular Biology, Brown University, Providence, Rhode Island, USA
[1]Corresponding author: e-mail address: rhet@brown.edu

Contents

Abstract

Primordial germ cells are usually made early in the development of an organism. These are the *mother of all stem cells* that are necessary for propagation of the species, yet use highly diverse mechanisms between organisms. How they are specified, and when and where they form, are central to developmental biology. Using diverse organisms to study this development is illuminating for understanding the mechanics these cells use in this essential function and for identifying the breadth of evolutionary changes that have occurred between species. This essay emphasizes how echinoderms may contribute to the patchwork quilt of our understanding of germ line formation during embryogenesis.

1. INTRODUCTION

The field of reproductive biology is as old a biological science as they come. Our fascination in the process of reproduction, the application to agriculture, the impact on society and the family unit, and perhaps the

Current Topics in Developmental Biology, Volume 117
ISSN 0070-2153
http://dx.doi.org/10.1016/bs.ctdb.2015.11.030

enormous variations in how plants, animals, and microbes reproduce has captivated the attention of scientists for centuries. In 1966, at the birth of the first volume of *Current Topics in Developmental Biology*, we saw the field of reproduction very differently than we do in 2016. We have indeed experienced great progress in the field during this time, and much of it is captured in the pages of this series. This progress has formed great intersections with all aspects of biology and society; think, for example, of the Nobel Prize in 2010 to Robert Edwards for his work in development of human *in vitro* fertilization (Steptoe & Edwards, 1978).

It is of great fun to page through the early volumes of the Current Topics series and to see how investigators thought, what the pressing questions of the time were, and how they were approached. It is not surprising that these topics were priorities in the early volumes of CTDB. Not only were they important topics in the field, but they were key for how the gametes functioned, and eventually for how the germ line was established. One of the original editors of CTDB was Alberto Monroy, a leader in molecular biology at the time and in dissecting translational mechanisms in transitions of eggs to embryos, largely in the sea urchin. His interest and role in this field of germ line and gamete function was highly impactful.

2. IN CONTEXT

The germ line is that lineage of cells that eventually forms the eggs and sperm in the adult. The precursor germ line cells, the primordial germ cells, are formed in the early embryo and will eventually migrate to the developing gonad, form the germ line stem cells, and in the adult, make the gametes(the eggs or sperm). What is clear in 2016 is that animals have tinkered enormously with the mechanisms to make the primordial germ cells in the embryo. Perhaps the strongest, and certainly the earliest impacts of germ line mechanics was revealed in *Drosophila* with the wonderful genetic screens that resulted in the identification of many germ line mechanisms, factors, and determinants—molecules which are sufficient to initiate a germ line fate in the early embryo (Cohen, 1995; Nusslein-Volhard & Wieschaus, 1980; Roush, 1995; for current reviews on this biology, see Slaidina & Lehmann, 2014; Weil, 2014). *Drosophila* determines its germ line in the posterior region of the syncytial embryo by accumulation of factors scaffolded by oskar—a highly divergent germ line determinant in Diptera that is essential for germ line formation (Lehmann & Nusslein-Volhard, 1986). This type of germ line formation is largely dependent on the localization of oskar in the posterior pole, followed by aggregation of many diverse functional elements that lead to a germ line fate (Lehmann & Nusslein-Volhard, 1986; Little, Sinsimer, Lee, Wieschaus, & Gavis, 2015; Lynch et al., 2011; Ryu & Macdonald, 2015; Yang et al., 2015). Cytoplasmic localization of factors is the main strategy for germ line formation in *Drosophila*, *C. elegans*, zebrafish, and *Xenopus*, among many other animals (Extavour & Akam, 2003).

But wait! This strategy of germ line determination based on localized material in the egg and early embryo appears to be a highly derived character in development (Extavour & Akam, 2003). Studies in the mouse embryo show an extreme opposite approach to germ line formation; intercellular signaling between different sites of the embryo initiates new gene expression that leads to chromatin modifications, the results of which lead to PGCs. This type of determination in the germ line is now referred to as an inductive mechanism, to contrast it from an acquired, or inherited mechanism. The most is known for these mechanics in mouse, in which the general region of the embryo wherein the germ line forms will otherwise develop into mesodermal fates (McLaren, 1998; Tam & Zhou, 1996). Yet, a small cohort of eventually as many as 40 cells receive a combination of BMP and Wnt signals from adjacent regions of the embryo that "rescues" them from a somatic cell fate and redirects them to a germ line fate. This inductive mechanism to make primordial germ cells is distinct from the localized cytoplasmic materials seen in, e.g., *Drosophila* and appears to be a basally derived strategy for germ line formation. An inductive mechanism is used by a wide variety of organisms, not just mammals, and includes representatives from all major phyla, including salamanders, some arthropods, and cnidarian (Extavour & Akam, 2003). Although the mechanisms of this inductive process are not as well described for most animals as in the mouse, the fact that many animals only make a germ line in adults strongly suggests that those animals also use inductive mechanisms, if only for the argument that it is hard to imagine how a cytoplasmic deposit in the egg and early embryo could have been passed on through enormous numbers of cell divisions and end up in an adult cell to become the germ line. The parsimonious explanation for such animals is that at some point in the adult, a stem-like cell is stimulated by intercellular signaling to acquire a germ line fate that may lead directly to gamete development, or to a precursor cell that eventually develops directly into a gamete.

The formation of a mammalian germ line precursor cell (primordial germ cell) can now even be recapitulated *in vitro* beginning with iPS cells (Hayashi, de Sousa Lopes, & Surani, 2007). Beginning with a pluripotent cell, investigators can sequentially stimulate these cells to divert their fate from a somatic cell fate to one of a germ line precursor cell. Although complete development to functional gamete is now been recapitulated, the later steps of gamete formation must occur via transplant back into a host mouse. It is likely only a matter of time though before the complete process of pluripotent stem cell-to-functional gamete is possible in a dish—especially with the recent successful engineering of 3-D organoids of a variety of tissue types *in vitro*, and of biofunctional matrices made synthetically. For more thorough reviews of the mechanisms, and the variations, in germ line determination between animals, the reader is referred instead to several significant references (Ewen-Campen, Schwager, & Extavour, 2010; Paksa & Raz, 2015; Rangan, DeGennaro, & Lehmann, 2008; Seydoux & Braun, 2006; Strome & Updike, 2015; Surani, 2015; Williamson & Lehmann, 1996).

The emphasis of this essay is to look forward. Here, I will address several important areas for future research in germ line determination mechanisms, with an emphasis of how especially echinoderms may serve as an important model system for answering these questions.

Echinoderms have recently become an exciting model organism for understanding mechanisms in germ line determination (Wessel, Brayboy, et al., 2014; Wessel, Fresques, Kiyomoto, Yajima, & Zazueta, 2014). The animals yield millions of oocytes and eggs, develop readily *in vitro* with excellent access to observation, enormous genome resources are available (echinobase.org), and a rich literature of gene regulatory networks each add to our understanding of the embryonic context for the reproductive cells and their precursors. At the birth of CTDB 50 years ago, sea urchins were primary sources of cells for many landmark observations. These include revealing mRNA biosynthesis and polyadenylation, regulation of mRNA translation, complexity of the genome and transcribed mRNAs, mechanisms of cell and tissue morphogenesis, egg activation and signal transduction mechanisms, and cell-type specific transcriptional mechanics. This rich history adds to the utility of sea urchins currently for studying the germ line determination mechanisms (Figs. 1 and 2).

Four important directions in germ line determination mechanisms are posed here with how echinoderms might be particularly relevant to these studies.

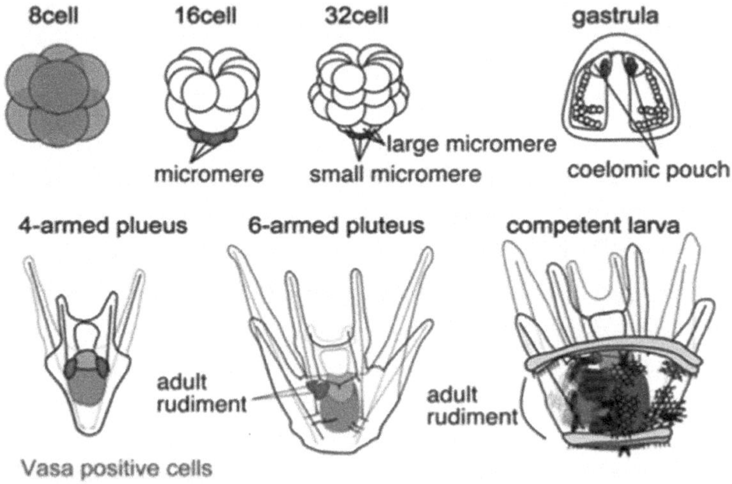

Figure 1 Diagram of the development of a sea urchin. Early development yields Vasa-positive cells (shown in red, beginning with a uniform Vasa-positive egg and early embryo). At the 32-cell stage, the sMics are uniquely Vasa-positive. These cells move into the coelom during gastrulation, segregate into the left and right coelomic pouches, and expand to contribute to the germ cells of the adult, and likely also to some somatic cells of the rudiment. *From Wessel, Brayboy, et al. (2014), John Wiley & Son.* (A color version of this figure is located at the end of this chapter.)

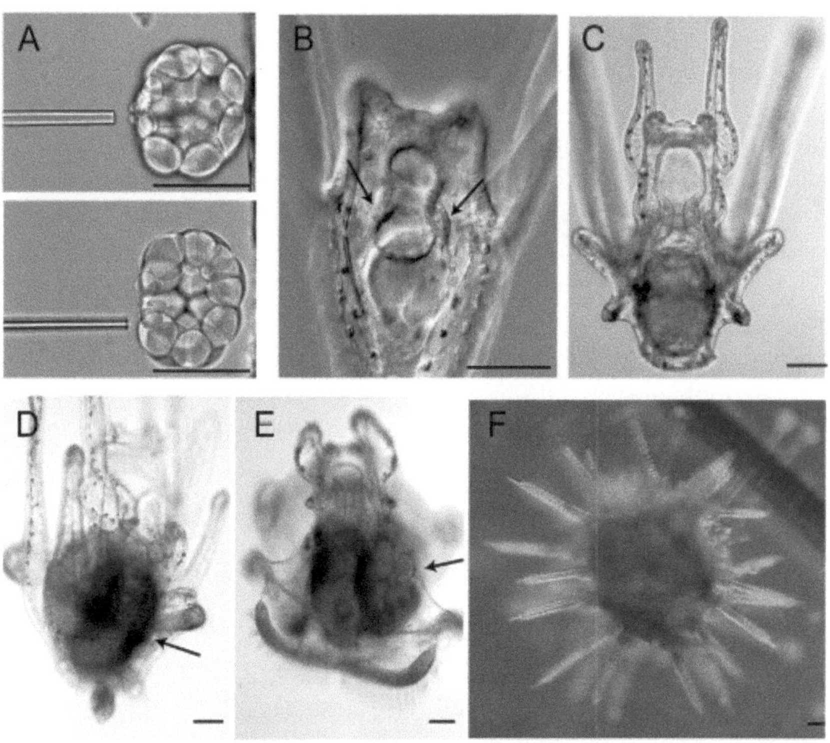

Figure 2 The sMics are dispersible for development but are required for gamete development. (A) Method of sMic removal with a glass pipette. The resulting embryos develop into a larva (B)–(E), form an adult rudiment (D, E, arrow in E), and metamorphos into an adult (F) that does not produce gametes. Scale bars, 80 μm. *From Yajima and Wessel (2011), Company of Biologists.* (A color version of this figure is located at the end of this chapter.)

3. IS NUAGE, OR GERMINAL GRANULES, ESSENTIAL FOR ANIMALS USING ACQUIRED FACTORS FOR GERM LINE DETERMINATION?

In animals utilizing cytoplasmically localized materials to determine their germ line, distinct aggregates of proteins and mRNAs are seen, and appear important, if not essential for germ line determination. A popular example of this structure/function relationship is in *Drosophila* in which the polar granules are formed in the posterior. Transplanting this cytoplasm containing the polar granules to the anterior pole of an embryo is sufficient to form pole cells (the primordial germ cells in this animal) in the anterior pole (Illmensee & Mahowald, 1974; Mahowald, Illmensee, & Turner, 1976). Indeed, even introduction of the oskar protein in the opposite end of the embryo, the anterior, is sufficient to initiate new pole cell formation ectopically in this organism (Ephrussi & Lehmann, 1992; Wharton & Struhl, 1991). Further, germinal granules likely form following oskar introduction, suggesting these structures as a necessary prerequisite for the germ line determination mechanism. Many animals have similar such aggregates important for acquired germ line formation.

Sea urchins also appear to utilize localized determination mechanisms for germ line formation (Voronina et al., 2008; Yajima & Wessel, 2012). The major difference in sea urchins though is that no germinal granules have ever been reported. Sea urchin eggs and early embryos were prime subjects of examination for electron microscopy studies in the era of the birth of the CTDB series. Innumerable reports of egg structures that included cytoplasmic organelles, nuclear structures, cell surface changes in fertilization etc., were wonderfully documented in sea urchins, yet no germinal granule-aggregate looking structures have ever been reported to become localized in the primordial germ cells (see, however, Yakovlev, 2015).

However, perinuclear granules are seen with Vasa antibodies, and to be induced by Vasa overexpression. This later approach of Vasa–GFP expression may be one way to identify if aggregates are actually present. Perinuclear granules are apparent following introduction of Vasa–GFP into the embryo (and when appropriately mutated so it is better able to accumulate outside of the germ line); it is clear that all cells are capable of accumulating perinuclear granular structures as seen in other organisms whose germ lines are determined by acquired cytoplasmic factors. The possible explanation for these observations is that either (a) all cells of the early sea urchin embryo already have perinuclear granules not otherwise apparent by electron microscopy, or

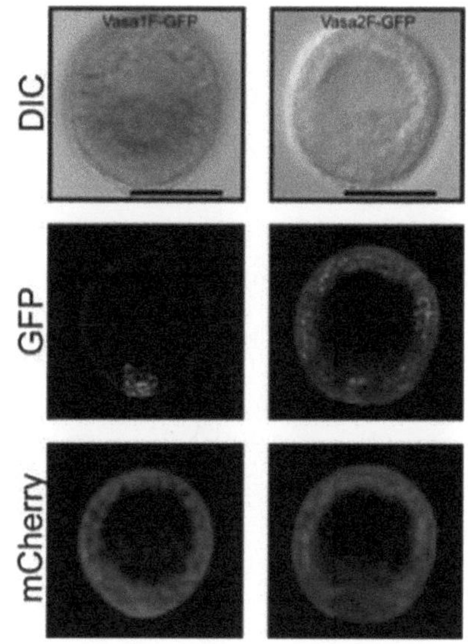

Figure 3 Vasa in perinuclear granules throughout the embryo. The *S. purpuratus* Vasa open reading frame is sufficient for specific small micromere enrichment of Vasa protein (left column), whereas deletion of 141 N-terminal amino acids causes broad Vasa expression throughout the embryo in perinuclear granules. Scale bar, 50 μm. *From Eric A. Gustafson, Mamiko Yajima, Celina E. Juliano, Gary M. Wessel, 2011, Post-translational regulation by gustavus contributes to selective Vasa protein accumulation in multipotent cells during embryogenesis, Developmental Biology, 349:2, 440-450, Elsevier.* (A color version of this figure is located at the end of this chapter.)

(b) they do not have such granules, and Vasa overexpression, induces formation of new perinuclear granules. This may be analogous to the example of oskar misexpression in the anterior region of the *Drosophila* embryo in which a nucleating stimulus is essential for aggregate formation, although this would be a surprising function for Vasa. The sea urchin embryos overexpressing Vasa have not been grown to sexual maturity, so it is not clear what impact, if any, may result in germ line functions with potentially induced perinuclear granule phenotype Fig. 3.

4. THE BROAD UTILITY OF GERM LINE FACTORS IN SOMATIC CELLS AND CANCER

Some of the first factors identified to be involved in germ line determination were identified genetically (Ephrussi, Dickinson, & Lehmann, 1991; Ephrussi & Lehmann, 1992; Gavis & Lehmann, 1992; Irish,

Lehmann, & Akam, 1989; Lasko & Ashburner, 1988). A mutation in the RNA-binding protein nanos, in *Drosophila* resulted in both posterior patterning defects (by virtue of its early regulation in *hunchback* translation) and in germ line viability. Nanos was subsequently found to be limited in its expression to the germ line. The RNA helicase Vasa was similarly identified and found in flies to be selectively expressed in the germ line. At that time (early- to mid-1980s), some transcription factors were discovered to be developmental determinants, that is, the fate of the cell changed by expression of an orthologous protein. This concept was highly popularized by the discovery of, e.g., myoD, myogenin, and their role in skeletal muscle formation (Edmondson & Olson, 1989; Tapscott et al., 1988; Wright, Sassoon, & Lin, 1989). It was tempting then to look for the same type of "determinant function" in the germ line mechanism.

What we now know, however, is that in most animals, the germ line factors are more general factors used broadly in development, and even into adulthood. An early report of this conclusion was in the mud snail, *Ilyanassa*, in which Vasa and Nanos were found in the 4d lineage, a cell that gives rise to many different mesoderm derivatives, as well as to the germ line (Juliano, Swartz, & Wessel, 2010; Swartz, Chan, & Lambert, 2008). In the *Ilyanassa* embryo though, it was clearly functioning much more broadly than in making a germ line stem cell, and instead was used in somatic progenitor cells for many different cell types (Rabinowitz, Chan, Kingsley, Duan, & Lambert, 2008; Swartz et al., 2008). Subsequent to that finding, other animal embryos were examined with similar results. The sea urchin embryo was of particular interest. Vasa expression was present throughout the early embryo, in cells, for example, that gave rise only to somatic ectoderm, or in larval tissues of the ectoderm (Yajima & Wessel, 2011, 2015), and recently reported even in tissues of the adult tube feet (Reinardy, Emerson, Manley, & Bodnar, 2015). Moreover, when Vasa expression was knocked down in the embryos, the cell cycle of those embryos was highly disorganized (Yajima & Wessel, 2011). Upon closer inspection, Vasa was found to accumulate on the spindle during mitosis, and potentially be involved in localized translation of messages important for cell cycle completion. Indeed, in this embryo, Vasa appeared to bind to most, if not all, mRNAs and knocking down Vasa caused broad translational cessation.

At about the same time, reports began to arrive that documented Vasa and other germ line factors are involved in tumor formation and progression. The most startling example was in flies in which one of the polycomb repressor groups was knocked out, causing a malignant brain tumor-like phenotype (Janic, Mendizabal, Llamazares, Rossell, & Gonzalez, 2010). Within these tumorigenic cells were high levels of Vasa, Nanos, Piwi, and other factors normally restricted to the germ line in this animal. On its own, this is interesting, but not exciting, since after all tumor cells are messed up and they may aberrantly express many different genes without posing a functional output. The key though was that when the investigators knocked out one or another of the "germ line factors," the tumors regressed, demonstrating not just aberrant expression, but functional contribution to the tumor phenotype. A significant percentage (estimated at ~10–15%) of some human tumor types also express one or another "germ line factor," the functionality of which is not yet known (Lasko, 2013).

Our rich dataset of "germ line factors" remains enigmatic. Vasa, for example, when it is present in a cell, seems essential to that cell's general functionality. Clearly, though not all cells have Vasa expression, but when they do, it seems essential for viability of that cell. It is not clear what Vasa may be doing in the balance of the cell's translational machinery, but the biochemical role of this factor, as well as the other major germ line factors, needs to be teased apart. The approach in these cases will likely not be completely served by knocking genes out of the genome, even conditionally. Instead, *in vitro* biochemical assays for functionality may lend an important perspective to the role these factors play both in the germ line for those animals in which it is highly restricted, and for a host of other organisms in which these same factors appear to be more broadly functional. Moreover, the role of germ line factors in tumorigenesis is both paradigm shifting and clinically important. While the sea urchin and other echinoderms may not be direct models for tumorigenesis, the tractability of these many embryos to such experimentation is highly attractive.

5. CONSEQUENCES OF EARLY GERM LINE DETERMINATION

It is strongly concluded from many studies that the ancestral mechanism for germ line determination is by inductive cell interactions and that an acquired mechanism is a derived state. This conclusion is based on broad phylogenetic analysis integrated into the developmental morphogenesis mechanism of the animal (Extavour & Akam, 2003). The derived character, acquisition, seems to have been evolutionarily tested and adopted throughout each major taxon, suggesting that it occurs broadly, is not constrained by any of the developmental mechanisms currently in place in those embryos, and that there must be some selection to transition from the inductive mechanism into the acquisition mode. It also argues that the transition might be more readily accomplished than perhaps thought just a decade ago.

As for mechanism(s) of the transition, it is likely that the acquired and inductive mechanisms are a continuum of characters (Juliano & Wessel, 2010). It is comfortable to speak of one versus the other, black versus white, but in actuality, these features are likely many shades of gray. An important point to this issue is what is the selection, if any, for having an acquired mechanism of germ line determination? Since the ancestral state is an inductive event, an acquired mechanism must in some way be favorable to fitness to have occurred so frequently and so broadly.

Recent research suggests that one selection mechanism may be that embryos using acquired mechanics are able to diversify their developmental program more rapidly, and thereby capitalize on new developmental niches to outcompete other organisms and increase their own fitness (Johnson & Alberio, 2015). Since their germ line is segregated early, sometimes very early (*C. elegans*), and even just prior to cellularization (*Drosophila*), any changes in the genome for a new developmental mechanism or feature would still be passed on to the next generation for testing because the germ line is already established early in development. An inductive mechanism necessarily occurs later in development, may be constrained by the positioning of the various signaling and receiving cells, and may therefore be limited in what developmental changes might be possible, while still retaining an ability to specify their germ line.

The sea urchin appears to set aside its germ line extremely early in development, at the fifth cell division (Wessel, Brayboy, et al., 2014). Even though these cells do not seem to have germinal granules, or nuage, or perinuclear granules early in development, the vast amount of data suggesting an acquired germ line mechanism in these organisms is overwhelming. Importantly, the other taxa of echinoderms, which include sea stars, crinoids, brittle stars, and sea cucumbers, each appear instead to make their germ line following gastrulation, after the assemblage of thousands of cells and following significant amounts of major tissue morphogenesis (Wessel, Brayboy, et al., 2014). Although the detailed mechanism of germ line determination in these other, non-echinoid echinoderms is not yet clear, they instead appear to use inductive mechanisms.

A further consequence of early germ line formation is that in many cases, the primordial germ cells are made so early, that the embryo has yet to begin formation of a gonadal precursor. In fact in *Drosophila, C. elegans*, and in sea urchins, no cells of the embryo are even migrating at the time of primordial germ cell formation. As a consequence, the primordial germ cells "wait" for the rest of the embryo to catch up. For this mechanism, at least some germ cells in various organisms demonstrate that they become quiescent. In sea urchins, for example, the PGCs are quiescent in terms of cell division, transcription, translation, motility, and even mitochondrial activity. Indeed, these cells appear to be a wonderful model for the quiescent stem cell phenotype that is also seen in neuronal stem cells, and hematopoietic stem cells of the mammal (Cheung & Rando, 2013; Nakamura-Ishizu, Takizawa, & Suda, 2014). Thus, with the readily accessible PGCs of the sea urchin, their ability to replicate the *in vivo* condition *in vitro*, and the strong manipulations possible in these cells, they may serve as an excellent model for general quiescent stem cell characteristics.

6. DIVERSITY IN MECHANISMS OF GERM LINE FORMATION

Echinodermata has five well-defined clades, Crinoidea (sea lilies and feather stars), Ophiuroidea (basket stars and brittle stars), Asteroidea (starfishes), Echinoidea (sea urchins, sand dollars, and sea biscuits), and Holothuroidea (sea cucumbers). Sea urchins appear to be the only clade in this phylum that uses acquired germ line determination mechanisms. All other echinoderms appear instead to specify their germ line late in development, likely necessitating inductive mechanisms. Sea stars have been the most intensively studied clade outside of echinoids, and the site of germ line formation appears to be within the posterior enterocoele. This structure buds off the endoderm following gastrulation and becomes progressively well defined by mRNAs of germ line factors (e.g., Vasa, Nanos, Piwi) and the absence of somatic markers (e.g., Blimp). Thus, although less is known about this group of animals in general than in the sea urchin, this phylum lends itself well to understanding both the acquired mechanism as seen in flies, roundworms, and fish, as well as the inductive mechanism seen in, for example, mammals. With the many manipulations feasible in sea stars, one may be able to more readily dissect inductive mechanisms for germ line determination and contribute to those studies which instead rely largely on genetic manipulations.

7. FINAL COMMENTS

No one animal is ideal, or sufficient, to test for mechanisms of germ line determination mechanisms. Some animals are beautifully situated for genetic manipulations, or for transplantations, or visualizations, or biochemical approaches, but no animal offers all the important features necessary for efficient and rapid progress. This essay points to several important goals in the field and how echinoderms might be particularly useful as a model organism for reaching these goals. An important part of study in this area, as could be argued for much of developmental biology, is a reliance of comparative analysis; how a patchwork of observations in diverse organisms may reveal information not generally accessible with only one or another experimental approach. Genetic applications over the past several decades have made enormous progress in this and many fields of the discipline, and now adding perspectives from different approaches, and different organisms less intensively documented to this point, should be of great benefit to the advancement of the field as a whole. Change is inevitable in biology, especially in the area of reproduction, and embracing these changes and differences is essential for real progress.

ACKNOWLEDGMENTS

Fifty years of *Current Topics in Developmental Biology*!—Happy Golden Anniversary! You just keep looking better, and healthier too! Your volumes keep reproducing and like a germ line stem cell—you retain your potential for immortality! Congratulations to you! I look forward to another 50 of the same! The author thanks Paul Wassarman for the opportunity to be involved in this project and to members of PRIMO for rich discussions and discoveries. The author acknowledges the NIH (HD028152 and HD075561) for research support. This work is dedicated to the memory of Betty (March 15, 1929–August 21, 2015), and Bob (December 30, 1923–September 6, 2015) Wessel. They were great people, special parents, and are missed dearly.

REFERENCES

Cheung, T. H., & Rando, T. A. (2013). Molecular regulation of stem cell quiescence. *Nature Reviews. Molecular Cell Biology*, *14*, 329–340.

Cohen, B. (1995). Nobel committee rewards pioneers of development studies in fruitflies. *Nature*, *377*, 465.

Edmondson, D. G., & Olson, E. N. (1989). A gene with homology to the myc similarity region of MyoD1 is expressed during myogenesis and is sufficient to activate the muscle differentiation program. *Genes & Development*, *3*, 628–640.

Ephrussi, A., Dickinson, L. K., & Lehmann, R. (1991). Oskar organizes the germ plasm and directs localization of the posterior determinant nanos. *Cell*, *66*, 37–50.

Ephrussi, A., & Lehmann, R. (1992). Induction of germ cell formation by oskar. *Nature*, *358*, 387–392.

Ewen-Campen, B., Schwager, E. E., & Extavour, C. G. (2010). The molecular machinery of germ line specification. *Molecular Reproduction and Development*, *77*, 3–18.

Extavour, C. G., & Akam, M. (2003). Mechanisms of germ cell specification across the metazoans: Epigenesis and preformation. *Development*, *130*, 5869–5884.

Gavis, E. R., & Lehmann, R. (1992). Localization of nanos RNA controls embryonic polarity. *Cell*, *71*, 301–313.

Hayashi, K., de Sousa Lopes, S. M., & Surani, M. A. (2007). Germ cell specification in mice. *Science*, *316*, 394–396.

Illmensee, K., & Mahowald, A. P. (1974). Transplantation of posterior polar plasm in Drosophila. Induction of germ cells at the anterior pole of the egg. *Proceedings of the National Academy of Sciences of the United States of America*, *71*, 1016–1020.

Irish, V., Lehmann, R., & Akam, M. (1989). The Drosophila posterior-group gene nanos functions by repressing hunchback activity. *Nature*, *338*, 646–648.

Janic, A., Mendizabal, L., Llamazares, S., Rossell, D., & Gonzalez, C. (2010). Ectopic expression of germline genes drives malignant brain tumor growth in Drosophila. *Science*, *330*, 1824–1827.

Johnson, A. D., & Alberio, R. (2015). Primordial germ cells: The first cell lineage or the last cells standing? *Development*, *142*, 2730–2739.

Juliano, C. E., Swartz, S. Z., & Wessel, G. M. (2010). A conserved germline multipotency program. *Development*, *137*, 4113–4126.

Juliano, C., & Wessel, G. (2010). Versatile germline genes. *Science*, *329*, 640–641.

Lasko, P. (2013). The DEAD-box helicase Vasa: Evidence for a multiplicity of functions in RNA processes and developmental biology. *Biochimica et Biophysica Acta*, *1829*, 810–816.

Lasko, P. F., & Ashburner, M. (1988). The product of the Drosophila gene Vasa is very similar to eukaryotic initiation factor-4A. *Nature*, *335*, 611–617.

Lehmann, R., & Nusslein-Volhard, C. (1986). Abdominal segmentation, pole cell formation, and embryonic polarity require the localized activity of oskar, a maternal gene in Drosophila. *Cell*, *47*, 141–152.

Little, S. C., Sinsimer, K. S., Lee, J. J., Wieschaus, E. F., & Gavis, E. R. (2015). Independent and coordinate trafficking of single Drosophila germ plasm mRNAs. *Nature Cell Biology*, *17*, 558–568.

Lynch, J. A., Ozuak, O., Khila, A., Abouheif, E., Desplan, C., & Roth, S. (2011). The phylogenetic origin of oskar coincided with the origin of maternally provisioned germ plasm and pole cells at the base of the Holometabola. *PLoS Genetics*, *7*, e1002029.

Mahowald, A. P., Illmensee, K., & Turner, F. R. (1976). Interspecific transplantation of polar plasm between Drosophila embryos. *The Journal of Cell Biology*, *70*, 358–373.

McLaren, A. (1998). Germ cells and germ cell transplantation. *The International Journal of Developmental Biology*, *42*, 855–860.

Nakamura-Ishizu, A., Takizawa, H., & Suda, T. (2014). The analysis, roles and regulation of quiescence in hematopoietic stem cells. *Development*, *141*, 4656–4666.

Nusslein-Volhard, C., & Wieschaus, E. (1980). Mutations affecting segment number and polarity in Drosophila. *Nature*, *287*, 795–801.

Paksa, A., & Raz, E. (2015). Zebrafish germ cells: Motility and guided migration. *Current Opinion in Cell Biology*, *36*, 80–85.

Rabinowitz, J. S., Chan, X. Y., Kingsley, E. P., Duan, Y., & Lambert, J. D. (2008). Nanos is required in somatic blast cell lineages in the posterior of a mollusk embryo. *Current Biology: CB*, *18*, 331–336.

Rangan, P., DeGennaro, M., & Lehmann, R. (2008). Regulating gene expression in the Drosophila germ line. *Cold Spring Harbor Symposia on Quantitative Biology*, *73*, 1–8.

Reinardy, H. C., Emerson, C. E., Manley, J. M., & Bodnar, A. G. (2015). Tissue regeneration and biomineralization in sea urchins: Role of notch signaling and presence of stem cell markers. *PLoS One*, *10*, e0133860.

Roush, W. (1995). Nobel prizes: Fly development work bears prize-winning fruit. *Science*, *270*, 380–381.

Ryu, Y. H., & Macdonald, P. M. (2015). RNA sequences required for the noncoding function of oskar RNA also mediate regulation of oskar protein expression by bicoid stability factor. *Developmental Biology*, *407*(2), 211–223.

Seydoux, G., & Braun, R. E. (2006). Pathway to totipotency: Lessons from germ cells. *Cell*, *127*, 891–904.

Slaidina, M., & Lehmann, R. (2014). Translational control in germline stem cell development. *The Journal of Cell Biology*, *207*, 13–21.

Steptoe, P. C., & Edwards, R. G. (1978). Birth after the reimplantation of a human embryo. *Lancet*, *2*, 366.

Strome, S., & Updike, D. (2015). Specifying and protecting germ cell fate. *Nature Reviews. Molecular Cell Biology*, *16*, 406–416.

Surani, M. A. (2015). Human germline: A new research frontier. *Stem Cell Reports*, *4*, 955–960.

Swartz, S. Z., Chan, X. Y., & Lambert, J. D. (2008). Localization of Vasa mRNA during early cleavage of the snail Ilyanassa. *Development Genes and Evolution*, *218*, 107–113.

Tam, P. P., & Zhou, S. X. (1996). The allocation of epiblast cells to ectodermal and germ-line lineages is influenced by the position of the cells in the gastrulating mouse embryo. *Developmental Biology*, *178*, 124–132.

Tapscott, S. J., Davis, R. L., Thayer, M. J., Cheng, P. F., Weintraub, H., & Lassar, A. B. (1988). MyoD1: A nuclear phosphoprotein requiring a Myc homology region to convert fibroblasts to myoblasts. *Science*, *242*, 405–411.

Voronina, E., Lopez, M., Juliano, C. E., Gustafson, E., Song, J. L., Extavour, C., et al. (2008). Vasa protein expression is restricted to the small micromeres of the sea urchin, but is inducible in other lineages early in development. *Developmental Biology*, *314*, 276–286.

Weil, T. T. (2014). mRNA localization in the Drosophila germline. *RNA Biology*, *11*, 1010–1018.

Wessel, G. M., Brayboy, L., Fresques, T., Gustafson, E. A., Oulhen, N., Ramos, I., et al. (2014). The biology of the germ line in echinoderms. *Molecular Reproduction and Development*, *81*, 679–711.

Wessel, G. M., Fresques, T., Kiyomoto, M., Yajima, M., & Zazueta, V. (2014). Origin and development of the germ line in sea stars. *Genesis*, *52*, 367–377.

Wharton, R. P., & Struhl, G. (1991). RNA regulatory elements mediate control of Drosophila body pattern by the posterior morphogen nanos. *Cell*, *67*, 955–967.

Williamson, A., & Lehmann, R. (1996). Germ cell development in Drosophila. *Annual Review of Cell and Developmental Biology*, *12*, 365–391.

Wright, W. E., Sassoon, D. A., & Lin, V. K. (1989). Myogenin, a factor regulating myogenesis, has a domain homologous to MyoD. *Cell*, *56*, 607–617.

Yajima, M., & Wessel, G. M. (2011). The DEAD-box RNA helicase Vasa functions in embryonic mitotic progression in the sea urchin. *Development*, *138*, 2217–2222.

Yajima, M., & Wessel, G. M. (2012). Autonomy in specification of primordial germ cells and their passive translocation in the sea urchin. *Development*, *139*, 3786–3794.

Yajima, M., & Wessel, G. M. (2015). Essential elements for translation: The germ line factor Vasa functions broadly in somatic cells. *Development*, *142*, 1960–1970.

Yakovlev, K. V. (2015). Localization of germ plasm-related structures during sea urchin oogenesis. *Developmental Dynamics*. http://dx.doi.org/10.1002/dvdy.24348 [Epub ahead of print].

Yang, N., Yu, Z., Hu, M., Wang, M., Lehmann, R., & Xu, R. M. (2015). Structure of Drosophila oskar reveals a novel RNA binding protein. *Proceedings of the National Academy of Sciences of the United States of America*, *112*, 11541–11546.

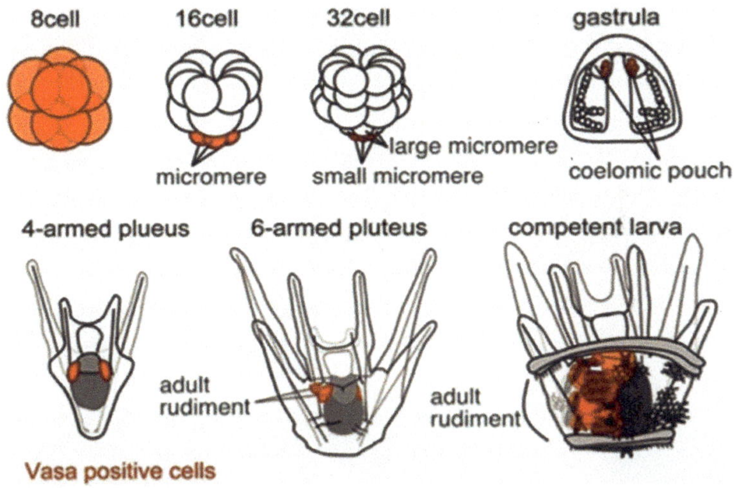

Gary M. Wessel, Figure 1 Diagram of the development of a sea urchin. Early development yields Vasa-positive cells (shown in red, beginning with a uniform Vasa-positive egg and early embryo). At the 32-cell stage, the sMics are uniquely Vasa-positive. These cells move into the coelom during gastrulation, segregate into the left and right coelomic pouches, and expand to contribute to the germ cells of the adult, and likely also to some somatic cells of the rudiment. *From Wessel, Brayboy, et al. (2014), John Wiley & Son.*

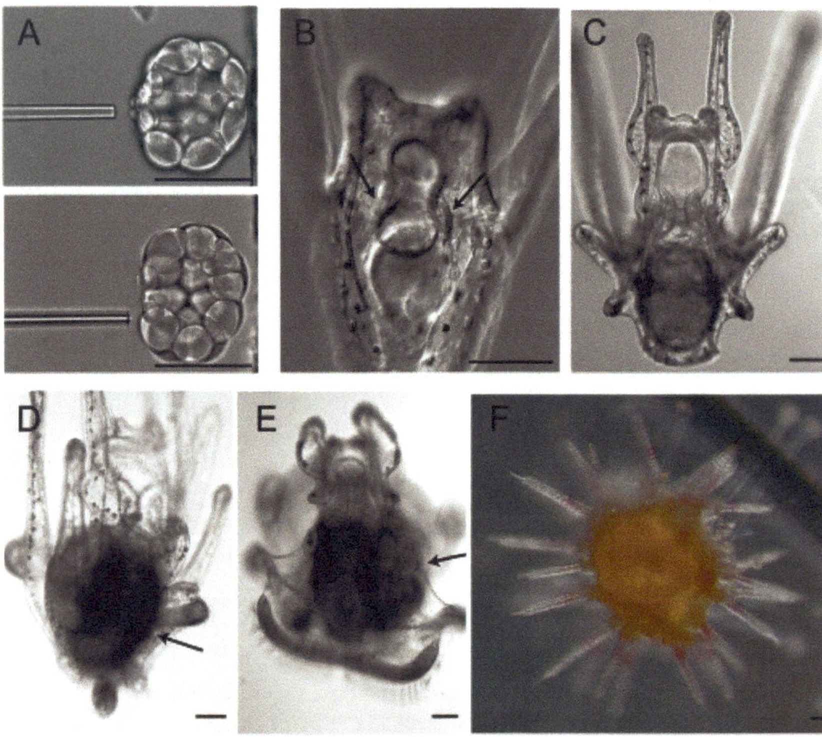

Gary M. Wessel, Figure 2 The sMics are dispersible for development but are required for gamete development. (A) Method of sMic removal with a glass pipette. The resulting embryos develop into a larva (B)–(E), form an adult rudiment (D, E, arrow in E), and metamorphos into an adult (F) that does not produce gametes. Scale bars, 80 μm. *From Yajima and Wessel (2011), Company of Biologists.*

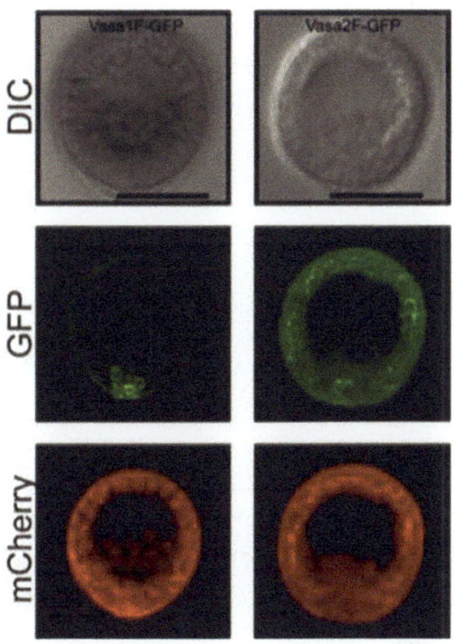

Gary M. Wessel, Figure 3 Vasa in perinuclear granules throughout the embryo. The *S. purpuratus* Vasa open reading frame is sufficient for specific small micromere enrichment of Vasa protein (left column), whereas deletion of 141 N-terminal amino acids causes broad Vasa expression throughout the embryo in perinuclear granules. Scale bar, 50 μm. *From Eric A. Gustafson, Mamiko Yajima, Celina E. Juliano, Gary M. Wessel, 2011, Post-translational regulation by gustavus contributes to selective Vasa protein accumulation in multipotent cells during embryogenesis, Developmental Biology, 349:2, 440-450, Elsevier.*

CHAPTER THIRTY-THREE

Positional Information and Cell Fate Determination in the Early *Drosophila* Embryo

Eric Wieschaus[1]
Molecular Biology Department, HHMI, Princeton University, Princeton, New Jersey, USA
[1]Corresponding author: e-mail address: efw@princeton.edu

Contents

Abstract

During early development in *Drosophila*, the spatial information of maternal gradients is translated into discrete transcriptional states determining cell fate. Information transfer depends on reproducibility of the gradients themselves, as well as the ability of cells to accurately measure and utilize morphogen concentrations in biologically meaning ways. In the following essay, these issues are discussed in context of the Bicoid gradient.

In biology, spatial patterns build on previously existing patterns. This is a reflection of the fundamental principle of all life, which is not carried in specific molecules but in the self-propagating relationship between those molecules. At any given moment, those relationships are defined by and in their spatial distributions. When the great German biologists of the nineteenth century, Schleiden, Schwann, and Virchow proposed that all cells are derived from previously existing cells, they did not mean that the molecules of life are passed on from one generation to the next, but the cellular pattern of those molecules in one generation provides a template for the next. Similarly, when a fully synthetic version of a bacterial genome is used to replace a bacterial host's own genome (Gibson et al., 2010), it does not "create life," but simply takes over the life and organization of the host bacteria. In this

Current Topics in Developmental Biology, Volume 117
ISSN 0070-2153
http://dx.doi.org/10.1016/bs.ctdb.2015.11.020

view, life arose once, most likely through an extraordinary chance event, but once that organization existed, the pattern could be past on and give rise to the extraordinary diversity of life on earth.

In a bacteria or any single celled organism, the pattern that is transferred between generations is essentially the life-giving internal organization of the cell. Before each cell division, cellular components are duplicated and then partitioned by biological machines to daughter cells that are often essentially identical to the mother cell. In a multicellular organism, the internal cell organization of individual cells is also passed from one cell cycle to the next, but more interestingly, complex patterns of cells and organs are passed between generations. Because spatial patterns transit between a single celled stage (the fertilized egg) and the functional multicellular organism, the transfer is necessarily more complex than in cell division. It is a *sequence* of orders, each of which can be fundamentally different from the preceding organization. The egg is not a miniature blueprint of the mother, for example, and the gastrula is not a printout of the spatial organization of molecules in the unfertilized egg. It is the sequential changes in organization that make development such an interesting problem in biology, and pose the central problem for scientists interested in information content and the transfer of information from one stage to the next.

It is that information transfer that is this subject of this review. There are two basic questions, how much information is present in each stage (what is the grain or resolution of the relevant information), and how the information present at one stage "maps" onto the information and spatial distributions present at the next. In this essay, I will focus on the early *Drosophila* embryo, which has played central role in research studies on this problem over the past 30 years.

One of the principle advantages of the *Drosophila* embryo for studies on pattern formation is its rapid development and the relatively simple geometry of its spatial organization. Following fertilization, the single celled embryo (the fertilized egg) undergoes a sequence of rapid nuclear replications. Unlike the cleavage divisions in most other embryos, these nuclear mitoses are not accompanied by cell divisions, such that the embryo remains a single cell with multiple nuclei (the syncytial blastoderm). After about 2 h and 13 rounds of such division, the number of nuclei on the surface of this single cell has reached about 6000 and the divisions cease. Prior to this stage, transcription of new RNAs has been minimal and the embryo has relied on RNAs and proteins supplied by the mother. Once the nuclear replications have paused, cell membranes partition the nuclei on the surface of the embryo into individual cells. Although low levels of spatially patterned transcription are observed in earlier stages, it is during cycles 13 and 14 that the protein levels from these newly transcribed genes reach their maximal levels. Since the activity of genes (the interaction of transcription factors with their target enhancers) is concentration dependent, it is likely that most of the relevant interactions between patterning genes occur at these stages. Cellularization stages during cycle 14 are characterized by particular rabid and complex changes in the pattern of gene expression. By the completion of cellularization, the spatial pattern of gene expression along the anterior–posterior axis of the embryo is sufficiently reproducible and detailed to distinguish each of the 100 individual cells along the anterior–posterior axis. The cell identities established at this stage are maintained throughout the remainder of development and although specific gene expression patterns change, it is possible to map blastoderm cells with particular gene expression patterns to distinct structures and cell fates in the larval epidermis.

Extensive genetic analyses in the 1980s identified three maternal systems that provide the initial inputs for spatial pattern along the anterior–posterior axis of the embryo (Nüsslein-Volhard, Frohnhöfer, & Lehmann, 1987; Schüpbach & Wieschaus, 1986). In each system, positional information is ultimately encoded in the concentration of specific transcription factors whose distributions are graded and could in principle provide positional information in the region of the egg where they have their most profound effect. The graded distributions in the three systems are generated by remarkably diverse cell biological and molecular mechanisms. The transcription factor that operates in the "anterior" system relies on the anterior localization during oogenesis of an mRNA encoding the transcription factor Bicoid (Driever & Nüsslein-Volhard, 1988a, 1988b, 1989). The posterior system operates through posteriorly localized translational control of a uniform maternal mRNA encoding the Hb transcription factor (Gavis & Lehmann, 1992). The distribution of the transcription factor in the terminal system (Capicua) is governed by a receptor tyrosine kinase Torso activated at the anterior and posterior poles of the egg (Jiménez, Guichet, Ephrussi, & Casanova, 2000). This review focuses on Bicoid which has been the system most intensely studied at the quantitative level.

1. ORIGIN AND FORMATION OF THE BICOID GRADIENT

In Driever and Nusslein-Volhard's original description of Bcd's role in embryonic patterning (1989), the Bicoid protein gradient was modeled using a point source (maternally supplied bcdRNA) located at the anterior tip of the egg. Translation of this RNA and subsequent diffusion of the protein was thought to give rise to the protein gradient that supplied positional information to nuclei along the anterior–posterior axis of the embryo. One attractive feature of a mechanism based on diffusion is that away from the source, the distribution of the protein will depend only on the diffusion rate and the rate that the protein is degraded (Crick, 1970; Wolpert, 1969).

Modeling the Bcd gradient using a point source is an obvious simplification that facilitates the mathematical treatment. Bcd is initially tightly anchored to the anterior-most cortex during oogenesis, but at fertilization, it moves away from the cortex to form a somewhat diffuse cloud of transcripts at the anterior end of the egg (Weil, Parton, Davis, & Gavis, 2008). This diffuse distribution has some graded features and raised the question of whether the final protein distribution might not be a direct reflection of the RNA distribution, with little role for protein diffusion from anterior source (Spirov et al., 2009). This is extremely unlikely. The most detailed study of bcdRNA distribution (Little, Tkačik, Kneeland, Wieschaus, & Gregor, 2011) utilized fluorescent *in situ* hybridization and probes capable of detecting and counting single RNA molecules. These studies conclude that 90% of the RNA is located in the anterior 20% of the egg and that the total distribution cannot be brought into a match with the more broadly distributed protein. Bcd protein can be detected out to 80% EL and under certain genotypes can pattern the posterior abdominal segments formed in that region.

Because it is difficult to compare RNA and protein gradients directly, the more telling indication of the role of diffusion in establishing the shape of protein gradient are experiments (Little et al., 2011) in which Bcd protein was marked with fluorescent tags (Venus, EGFP, RFG) that differ in the times required for their maturation. In such experiments, the fluorescent probe itself effectively acts as a timer for when each Bicoid molecule was translated. Fusion proteins with rapidly maturing forms (Venus tagged) display steep gradients that mirror the total protein gradient seen in antibody staining, whereas the slow-maturing forms, which only allow fusion proteins to be detected after they have had ample time to diffuse from the site of translation, show gradients that are much flatter. If the proteins do not move from where they are made, fusion proteins that have different maturation times might be detected at different levels, but the shapes of the gradients should be identical. The fact that fast-maturing Bcd fusion proteins give sharp gradient, whereas slow-maturing fusion of the same size give flat gradients is therefore the strongest demonstration that the RNA and protein gradients differ and that diffusion plays an essential role in establishing the spatial distribution of the protein.

The diffuse nature of the RNA distribution and the fact that it changes over time does make modeling the Bcd protein gradient more complicated. This is particularly true if the total amount of maternally supplied RNA is relatively constant, as shown by Petkova, Little, Liu, and Gregor (2014), but if the distribution is more variable or more diffuse in different embryos. The same amount of RNA will give a sharper peak of protein if it is more tightly localized that if it is more diffuse, but this applies only to the zone where protein is being made. Outside this zone, the protein distribution is substantially independent of the distribution of the source RNA and depends only on the protein levels at the margin of the synthesis zone and the rates of diffusion and degradation. Variability in the RNA distributions may explain a feature of most attempts to quantify absolute concentration of Bcd protein along the anterior–posterior axis of the embryo, namely, that the greatest variability from one embryo to the next is measured at the anterior pole of the egg. Values in the middle region of the egg are much more comparable between embryos (Gregor, Tank, Wieschaus, & Bialek, 2007). High variability at the anterior end of the embryos is especially problematic when the anterior peak values are used to anchor normalization of the gradient. This was a common procedure in earlier characterizations of Bcd gradient using antibodies to detect the protein. These experiments (e.g., Houchmandzadeh, Wieschaus, & Leibler, 2002) provide information on the shapes of gradients, but by anchoring normalization at the anterior pole, they effectively spread the variability initially concentrated at the anterior end over the entire length of the embryo.

Beyond the technical issues encountered in their analysis, the diffuseness of bcdRNA points to a fundamental feature of diffusion gradients in all biological systems, namely that their information content is not high at the source. On the other hand, as long as the synthesis capacity of the source is constant, the distribution of the source itself is not important for information to cells outside the source. Thus, under most conditions, given the role of diffusion and degradation in smoothing out local irregularities, gradient information will only get better and more reliable the farther it is from the source.

2. THE INFORMATION CONTENT OF THE BICOID GRADIENT

One of the attractive features of a gradient is that the concentration profile of a single molecule like Bicoid can in principle be used to define a very large number of cell states. This arises from the continuous incremental changes that characterize a diffusible gradient. On the other hand, the ability of the embryo to use a gradient to define cell fates depends on two features: (1) how reliable the gradient is, that is, how reproducible and therefore informative the concentration gradient is about position and (2) the ability of the cells that lie along the gradient to measure [and distinguish] subtle difference in concentrations to determine cell fate. A gradient may be extremely reproducible from one embryo to the next, such that an experimenter can always distinguish neighboring cells based on their concentration, but this would be of little value to the embryo if the cells themselves were only able to detect a 10-fold difference in concentration. Both the reliability/reproducibility of the Bicoid gradient itself and the sensitivity of the response have been the subject of intense investigation over the past 10 years and general picture is beginning to emerge.

Based on measurement of fluorescently labeled fusion proteins, the Bcd gradient is remarkable reproducible between embryos, particularly in regions posterior to the RNA source. At 48% EL, in cells that would form the posterior boundary of HB expression, the measured standard deviation in intensity is less than 10% of the mean value in that region (Gregor et al., 2007). This variability is roughly of the same magnitude as the measured difference in concentration between two neighboring cells and suggests that cells have sufficient information to correctly distinguish their positions from that of their anterior and posterior immediate neighbors based on Bcd concentration alone.

The reading of the gradient and its translation into specific cell fates occurs through its transcriptional activation of specific target genes. Bcd binding has been detected in the vicinity of approximately 600–1000 genes (Berman et al., 2002). It is not known how many of these binding sites function as enhancers that control the expression of specific downstream genes. At least 15% of the ~5000 *Drosophila* genes expressed at the blastoderm stage have nonuniform anterior–posterior patterns (Combs & Eisen, 2013). In RNAseq-based profiling of thin slices of microtome sectioned embryos, proper RNA levels in the posterior for specific genes are perturbed upon removal of Bcd, suggesting that directly or indirectly Bcd influences gene expression pattern along the entire length of the embryo.

In the simplest case, the anterior–posterior ordering of Bcd target gene expression would depend on the affinity of the protein of each gene's enhancers (Driever & Nüsslein-Volhard, 1989; Gao & Finkelstein, 1998;

Struhl, Struhl, & Macdonald, 1989). In posterior regions where concentrations are low, only genes whose enhancers have the highest affinity for Bcd would be capable of binding and thus be expressed. Genes expressed in the anterior regions of the embryo might be expected to have lower affinities such that they are only expressed where Bcd levels are high. Achieving a precision in gene expression patterns that preserves the information content of the Bcd gradient may require an unparalleled subtlety in different binding affinities at enhancers. A first test of those affinities is supplied at the level of gap gene expression. What is known is that already at that stage, the levels and spatial pattern of expression of the four major gap genes have achieve a precision in expression that mimics the information content of the Bcd gradient and effectively defines individual cells long the anterior–posterior axis (Dubuis, Tkacik, Wieschaus, Gregor, & Bialek, 2013). Whether that precision reflects an enhancer based direct readout of Bcd concentrations is not known.

3. A DIRECT RELATIONSHIP BETWEEN BCD CONCENTRATION AND FINAL PATTERN?

Can Bcd be read with sufficient accuracy in a direct readout to produce the precision observed at the level of gap and pair-rule expression? Based the small number of Bcd molecules per nucleus measured using Bcd fusions, our own speculative calculations for the PII enhancer of Hb (Gregor et al., 2007) concluded that there was not enough time during the cleavage stages and early cycle 14 to accurately sample the concentration, i.e., the accuracy of the HB response to Bcd concentration exceeds the theoretical physical limits achievable in a rapid direct response of a receptor (the enhancer) and to a diffusible ligand (Berg & Purcell, 1977). Based on those calculations, we proposed that alternate mechanisms (e.g., spatial averaging between adjacent nuclei) were needed. Recognition of the physical limits in cartoon models points to the value in accurate measurements of physical parameters in evaluation their feasibility. Subsequent studies have demonstrated the extent of movement of Hb RNAs between adjacent nuclei (Little, Tikhonov, & Gregor, 2013), as well as providing new estimates for the absolute numbers of Bcd molecules per nucleus (Drocco, Grimm, Tank, & Wieschaus, 2011) and the rates with which they diffuse (Abu-Arish, Porcher, Czerwonka, Dostatni, & Fradin, 2010).

Accuracy can be improved by increasing the number of independent readings an enhancer has access to in evaluating Bcd concentration.

Bcd is in fact read multiple times. For example, in addition to its role as a transcriptional activator, Bcd also functions as a negative translational regulator of maternal Caudal RNA, such that the two genes ultimately have expression patterns that form complementary gradients. The enhancers of many early acting genes, especially those expressed in the posterior of the embryo, have binding sites for both Bcd and Cad and this means that their enhancers read Bcd concentration twice, once directly by Bcd binding and once indirectly by binding Caudal. Gap genes also regulate each other. This cross regulation provides a second opportunity to get multiple reading of the Bcd gradient in a given cell. If the Kr enhancer misreads Bcd and would express Kr in cells that it should not, chances are that the independent reading of Bcd concentration by the Hb enhancer will drive Hb protein expression in that cell to repress Kr, thereby correcting the initial reading error. Pair-rule gene expression represents the final step in the pattern. In many cases, all the earlier genes (Bcd, Cad, multiple gap genes) bind to the enhancers of the pair-rule genes, giving multiple readings of the information. As pointed out in Tikhonov, Little, and Gregor (2015), it is the *simultaneous* availability of independent readings of the gradient that contributes to the precision of the final readout. This is very different view of the classical segmentation sequence, in which the segmental hierarchy was viewed as a temporal sequence of information transfer. In that view, each step away from the information-rich Bcd gradient has the potential to introduce reading errors that are propagated to the next layer (much like the classic children's telephone/whispering game). For that reason the classical hierarchical model tended to regard the information content of the maternal gradients as a transient generalized cues that establish broad distributions, subsequently elaborated by interaction between the gap genes and pair-rule genes to achieve a final precise cell by cell distinctions at the completion of cellularization.

4. HOW MUCH POSITIONAL INFORMATION IS RELEVANT?

The compartmental model developed by García-Bellido and his colleagues in the 1970s provides one of the historically best-developed views for how position information at the blastoderm stage could be used (García-Bellido, 1975). In this model, initial positional inhomogeneities establish large domains (=polyclones) corresponding to segments or compartments. Once established, the cells within a compartment are developmentally equivalent until a subsequent binary decision results in further subdivision. The developmental equivalence means that within a compartment, size and growth are easily regulated and compartments are thought to function as units of grow control. Subsequent subdivisions in each compartment are thought to be governed by interactions and signaling within the compartment and across compartment boundaries (Lawrence, 1987; Lawrence & Struhl, 1996), and there is no fixed relationship between positional differences within a compartment and the final fates of cells. The major challenge for the compartmental model is to explain how, despite the regulative nature of the compartment, subtle deviations in pattern at the blastoderm stage are preserved throughout development. Each pair-rule gene, for example, is expressed in a subset of the compartment or segment, and when a given pair-rule gene is mutant, a corresponding subset of the pattern within the segment anlage at the blastoderm stage is deleted (Gergen, Coulter, & Wieschaus, 1986). Unlike what would be expected in the simple compartment model, those pattern defects, however, are not corrected or regulated for during subsequent development. The fact that it was possible to predict the overlapping domains of the various pair-rule genes viewing only the deletion patterns in the final cuticle 24 h later argues that the details of cell fate decision within the compartment may be maintained and have precise effects on final pattern.

Clonal analysis suggests that by the blastoderm stage, segments and anterior–posterior compartments within those segments have been established as distinct, polarized cell lineage groups. These distinctions, and the associated polarity, would demand at least three distinct cell types within each segment. The existence of three distinct cell fates is compatible with the observation that in the *Drosophila* blastoderm, each segment primordial is 3–4 cells wide, and the complexity of the expression patterns at that stage is sufficient to determine rows of single cells, which may, for example, express *eve* and *runt* but not *paired*. Analyses of genetic mosaics suggest that mutant patterns at the blastoderm stage (e.g., runt; Gergen & Wieschaus, 1986) are autonomously propagated to the final cuticle pattern, arguing for minimal regulation within segments during postblastoderm development. There would be at least three meaningful states per segment primordium, and thus as many such cell fates as there are cells along the anterior–posterior axis of the embryo.

An estimate of three meaningful states per segment at the blastoderm stage is perhaps conservative, since the number of pair-rule genes and their graded expression levels actually allow a much more extensive combinatorial range of cell states. In fact, although the cells in a single row along the anterior–posterior axis are more similar to each other than they are to immediately adjacent cells in rows to the anterior or posterior, they are not identical. Within each row, there is variability in absolute levels of the different pair-rule genes that cannot be related to DV patterning. In models where each AP stripe is assumed to define a single cell fate, this variability is assumed to have no long-term consequences on subsequent behavior (to be sub-threshold). In strict compartmental models, such variability will of course be irrelevant once the cells are assigned to a polyclone and become developmentally equivalent. It is possible however, that subtle differences in pair-rule gene expression pattern have consequences on cell behavior during gastrulation and ultimately affects the kinds of pattern cells make. The discrepancy between the number of cells and the number of potential cell fates in part arises from the focus on the blastoderm. Immediately after that stage, the embryo undergoes germ-band extension. Cells intercalate between each other and the length of the germ band and the width of each segment expands 2.5-fold. The intercalary behavior of cells depends on the pair-rule expression pattern and more specifically on the juxtaposition of cells with differences in pair-rule gene expression. In early models for germ-band extension (Gergen et al., 1986), variability in pair-rule gene expression within single stripes leads to adhesive preferences that cell drive intercalation. Subsequent experiments confirmed the role of pair-rule gene expression differences in driving germ-band extension (Irvine & Wieschaus, 1994; Zallen & Wieschaus, 2004) but have focused more on their role in polarizing myosin accumulation to AP interfaces (Bertet, Sulak, & Lecuit, 2004). In all these models, it is the juxtaposition of different pair-rule expression states, rather than any particular state, that polarizes cell surfaces to drive the intercalary behavior. The required differences between cells can be subtler than single stripes of identity and could in principle incorporate repositioning cells of subtly different identities within single stripes during germ-band extension. The germ-band extension models provide an example of a more general model for using positional information in which the sequence of positional values, rather than any particular cell state is the essential outcome of the information transfer.

5. CONCLUSIONS

The previous discussion of positional information has emphasized a view where initial information-rich maternal gradients are directly translated into complex patterns of zygotic gene expression in which most of the initial information is preserved. The view requires that the number of meaningful states is large at all stages of development. It therefore contrasts with traditional hierarchical or compartmental strategies where initially simple patterns evolve to greater complexity. Information-rich transfers of the former type may be particularly useful in organisms or stages of development that develop rapidly with little growth. In flies where a relatively rapid embryonic period is followed by a longer larval growth stage, it is perhaps not surprising that development makes sequential use of both strategies.

The essay has also focused rather specifically on the Bicoid gradient. Like all model systems, the early *Drosophila* embryo is rather specialized. The specific molecules whose concentrations have evolved to carry and transfer positional information transfer are not the same in other organisms. Bicoid itself is recently evolved and not conserved outside the higher Diptera (Stauber, Jäckle, & Schmidt-Ott, 1999). The cellular context in which information is transferred and interpreted, i.e., the syncytial blastoderm, is also specialized. This may not be a problem. Historically, peculiarities of *Drosophila* development have provided simplified environments for identifying components in the patterning process and for their subsequent functional analyses. It seems to me that those same simple features will continue to facilitate investigations of information transfer and cell fate determination, revealing rules that are ultimately applicable to all organisms.

REFERENCES

Abu-Arish, A., Porcher, A., Czerwonka, A., Dostatni, N., & Fradin, C. (2010). High mobility of bicoid captured by fluorescence correlation spectroscopy: Implication for the rapid establishment of its gradient. *Biophysical Journal*, *99*(4), L33–L35.

Berg, H. C., & Purcell, E. M. (1977). Physics of chemoreception. *Biophysical Journal*, *20*, 193–219.

Berman, B. P., Nibu, Y., Pfeiffer, B. D., Tomancak, P., Celniker, S. E., Levine, M., et al. (2002). Exploiting transcription factor binding site clustering to identify cis-regulatory modules involved in pattern formation in the Drosophila genome. *Proceedings of the National Academy of Sciences of the United States of America*, *99*(2), 757–762.

Bertet, C., Sulak, L., & Lecuit, T. (2004). Myosin-dependent junction remodelling controls planar cell intercalation and axis elongation. *Nature*, *429*(6992), 667–671.

Combs, P. A., & Eisen, M. (2013). Sequencing mRNA from cryo-sliced Drosophila embryos to determine genome-wide spatial patterns of gene expression. *PLoS One*, *8*(8), e71820.

Crick, F. (1970). Diffusion in embryogenesis. *Nature*, *225*, 420–422.

Driever, W., & Nüsslein-Volhard, C. (1988a). A gradient of bicoid protein in Drosophila embryos. *Cell*, *54*, 83–93.

Driever, W., & Nüsslein-Volhard, C. (1988b). The bicoid protein determines position in the Drosophila embryo. *Cell*, *54*, 95–104.

Driever, W., & Nüsslein-Volhard, C. (1989). The bicoid protein is a positive regulator of hunchback transcription in the early Drosophila embryo. *Nature*, *337*, 138–143.

Drocco, J. A., Grimm, O., Tank, D. W., & Wieschaus, E. (2011). Measurement and perturbation of morphogen lifetime: Effects on gradient shape. *Biophysical Journal, 101*(8), 1807–1815.

Dubuis, J. O., Tkacik, G., Wieschaus, E. F., Gregor, T., & Bialek, W. (2013). Positional information, in bits. *Proceedings of the National Academy of Sciences of the United States of America, 110*, 16301–16308.

Gao, Q., & Finkelstein, R. (1998). Targeting gene expression to the head: The Drosophila orthodenticle gene is a direct target of the Bicoid morphogen. *Development, 125*, 4185–4193.

García-Bellido, A. (1975). Genetic control of wing disc development in Drosophila. *Ciba Foundation Symposium, 29*, 161–182.

Gavis, E. R., & Lehmann, R. (1992). Localization of nanos RNA controls embryonic polarity. *Cell, 71*(2), 301–313.

Gergen, J. P., Coulter, D., & Wieschaus, E. F. (1986). Segmental pattern and blastoderm cell identities. In J. G. Gall (Ed.), *Gametogenesis and the early embryo* (pp. 195–220). New York: Alan R. Liss, Inc.

Gergen, J. P., & Wieschaus, E. (1986). Dosage requirements for runt in the segmentation of Drosophila embryos. *Cell, 25*, 289–299.

Gibson, D. G., Glass, J. I., Lartigue, C., Noskov, V. N., Chuang, R. Y., Algire, M. A., et al. (2010). Creation of a bacterial cell controlled by a chemically synthesized genome. *Science, 329*(5987), 52–56.

Gregor, T., Tank, D. W., Wieschaus, E. F., & Bialek, W. (2007). Probing the limits to positional information. *Cell, 130*, 153–164.

Houchmandzadeh, B., Wieschaus, E., & Leibler, S. (2002). Establishment of developmental precision and proportions in the early Drosophila embryo. *Nature, 415*, 798–802.

Irvine, K. D., & Wieschaus, E. (1994). Cell intercalation during *Drosophila* germband extension and its regulation by pair-rule segmentation genes. *Development, 120*, 827–841.

Jiménez, G., Guichet, A., Ephrussi, A., & Casanova, J. (2000). Relief of gene repression by torso RTK signaling: Role of capicua in Drosophila terminal and dorsoventral patterning. *Genes & Development, 14*(2), 224–231.

Lawrence, P. A. (1987). Pair-rule genes: Do they paint stripes or draw lines? *Cell, 51*(6), 879–880.

Lawrence, P. A., & Struhl, G. (1996). Morphogens, compartments, and pattern: Lessons from Drosophila? *Cell, 85*(7), 951–961.

Little, S. C., Tikhonov, M., & Gregor, T. (2013). Precise developmental gene expression arises from globally stochastic transcriptional activity. *Cell, 154*, 789–800.

Little, S. C., Tkačik, G., Kneeland, T. B., Wieschaus, E. F., & Gregor, T. (2011). The formation of the Bicoid morphogen gradient requires protein movement from anteriorly localized mRNA. *PLoS Biology, 9*(3), e1000596.

Nüsslein-Volhard, C., Frohnhöfer, H. G., & Lehmann, R. (1987). Determination of anteroposterior polarity in Drosophila. *Science, 238*(4834), 1675–1681.

Petkova, M. D., Little, S. C., Liu, F., & Gregor, T. (2014). Maternal origins of developmental reproducibility. *Current Biology, 24*(11), 1283–1288.

Schüpbach, T., & Wieschaus, E. (1986). Maternal-effect mutations altering the anterior-posterior pattern of the Drosophila embryo. *Roux's Archives of Developmental Biology, 195*, 302–317.

Spirov, A., Fahmy, K., Schneider, M., Frei, E., Noll, M., & Baumgartner, S. (2009). Formation of the bicoid morphogen gradient: An mRNA gradient dictates the protein gradient. *Development, 136*(4), 605–614.

Stauber, M., Jäckle, H., & Schmidt-Ott, U. (1999). The anterior determinant bicoid of Drosophila is a derived Hox class 3 gene. *Proceedings of the National Academy of Sciences of the United States of America, 69*(7), 3786–3789.

Struhl, G., Struhl, K., & Macdonald, P. M. (1989). The gradient morphogen bicoid is a concentration-dependent transcriptional activator. *Cell*, *57*, 1259–1273.

Tikhonov, M., Little, S. C., & Gregor, T. (2015). Only accessible information is useful: Insights from gradient-mediated patterning. arXiv:1501.07342.

Weil, T. T., Parton, R., Davis, I., & Gavis, E. R. (2008). Changes in bicoid mRNA anchoring highlight conserved mechanisms during the oocyte-to-embryo transition. *Current Biology*, *18*(14), 1055–1061.

Wolpert, L. (1969). Positional information and the spatial pattern of cellular differentiation. *Journal of Theoretical Biology*, *25*, 1–47.

Zallen, J. A., & Wieschaus, E. (2004). Patterned gene expression directs bipolar planar polarity in Drosophila. *Developmental Cell*, *6*(3), 343–355.

CHAPTER THIRTY-FOUR

Segment Identity and Cell Segregation in the Vertebrate Hindbrain

Megan Addison, David G. Wilkinson[1]
The Francis Crick Institute, Mill Hill Laboratory, London, United Kingdom
[1]Corresponding author: e-mail address: david.wilkinson@crick.ac.uk

Contents

Abstract

The subdivision of tissues into sharply demarcated regions with distinct and homogenous identity is an essential aspect of embryonic development. Along the anteroposterior axis of the vertebrate nervous system, this involves signaling which induces spatially restricted expression of transcription factors that specify regional identity. The spatial expression of such transcription factors is initially imprecise, with overlapping expression of genes that specify distinct identities, and a ragged border at the interface of adjacent regions. This pattern becomes sharpened by establishment of mutually exclusive expression of transcription factors, and by cell segregation that underlies formation of a straight border. In this review, we discuss studies of the vertebrate hindbrain which have revealed how discrete regional identity is established, the roles of Eph–ephrin signaling in cell segregation and border sharpening, and how cell identity and cell segregation are coupled.

Current Topics in Developmental Biology, Volume 117
ISSN 0070-2153
http://dx.doi.org/10.1016/bs.ctdb.2015.10.019

1. INTRODUCTION

In many tissues, the generation of a complex organization of cell types involves an initial subdivision into domains along one or several body axes, each with a distinct identity. At the molecular level, regional identity is determined by specific transcription factors which have spatially restricted expression within the tissue induced by intercellular signaling. An extensively studied example is the patterning along the anteroposterior axis of the vertebrate neural epithelium, in which graded signals induce transcription factors that define distinct regions of the brain and spinal cord. At early stages, the borders of adjacent regions are imprecise, and then are sharpened, reflecting that correct tissue organization requires the formation of discrete territories (Batlle & Wilkinson, 2012; Dahmann, Oates, & Brand, 2011). The sharpening of borders involves two processes (Kiecker & Lumsden, 2005; Fig. 1). First, cells at the prospective border of adjacent regions may initially coexpress transcription factors that specify distinct identities, likely reflecting an intrinsic imprecision in the interpretation of graded signals. Such overlaps can be resolved through cross-repression that establishes mutually exclusive expression of transcription factors. Second, once distinct identities are established, the border of adjacent regions is initially ragged due to the spatial imprecision in cell specification, as well as cell intermingling that occurs during tissue growth and morphogenetic movements. The ragged interface is then straightened by segregation of the cells with distinct identity. Direct evidence for an imprecise pattern generated by noisy inductive signaling being sharpened by cell segregation has been obtained by *in vivo* imaging studies of the ventral neural tube in zebrafish (Xiong et al., 2013). This implies that mechanisms that underlie cell segregation are upregulated in concert with, or downstream of, regional specification. Understanding of how sharp regional domains are formed thus requires elucidation of interconnected processes: the establishment of mutually exclusive expression of transcription factors that specify regional identity, cell segregation and border sharpening, and the coupling between regional identity and cell segregation. In this review, we discuss current understanding of these mechanisms in the segmentation of the vertebrate hindbrain.

2. VERTEBRATE HINDBRAIN SEGMENTATION

During development of the vertebrate hindbrain, the neuroepithelium becomes subdivided into seven morphological units, known as rhombomeres

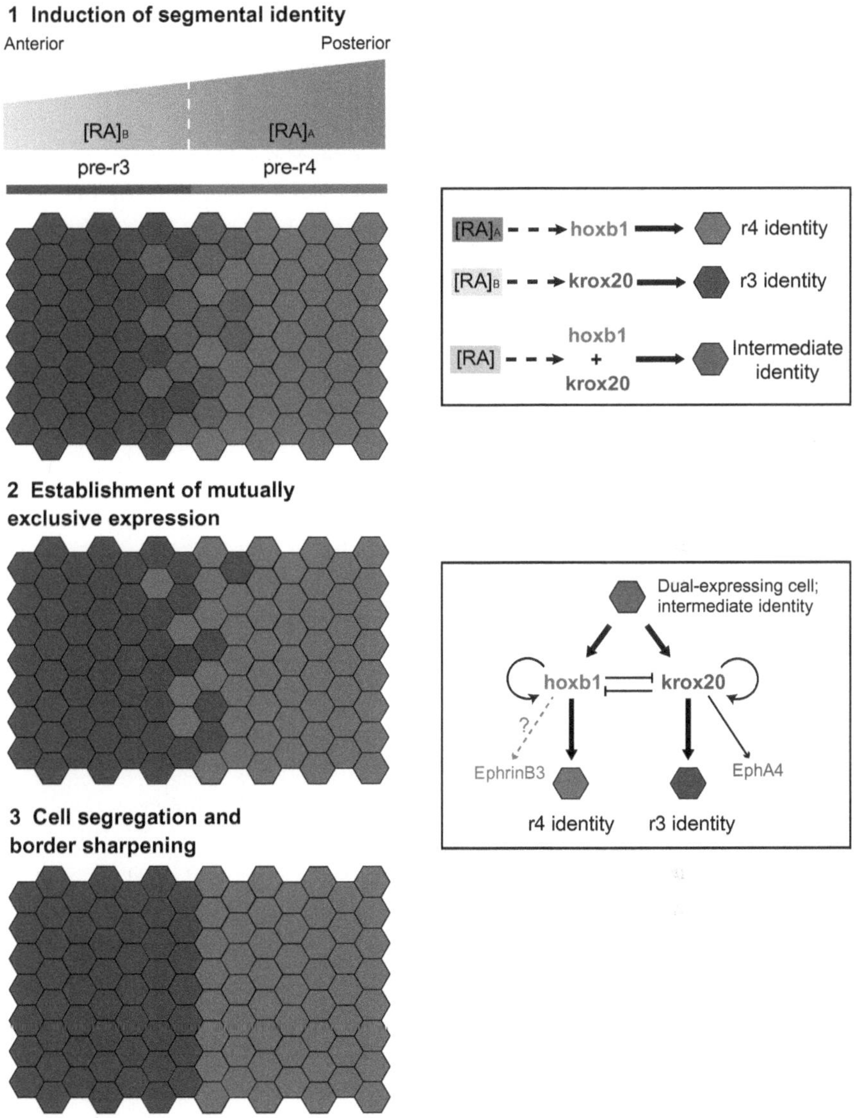

Figure 1 Border formation during hindbrain segmentation. The left-hand side depicts the progressive sharpening of segment borders during hindbrain development, illustrated for the r3/r4 border. During induction of segmental identity, graded retinoic acid (RA) upregulates expression of *hoxb1* and *krox20*, with some cells at the prospective border expressing both genes (top right). *hoxb1* and *krox20* expression becomes mutually exclusive through cross-repression, thus resolving the identity of those cells which initially express both genes (bottom right). Autoregulatory loops maintain the expression of *hoxb1* and *krox20*. Krox20 upregulates EphA4 in r3, while unknown factors (perhaps directly or indirectly downstream of *hoxb1*) upregulate ephrinB3 in r4. Signaling of EphA4 and ephrinB3, and of other Eph–ephrin pairs, underlies cell segregation that sharpens the segment borders. (A color version of this figure is located at the end of this chapter.)

(r1–r7). The subdivision of the hindbrain underlies the segmental patterning of neuronal organization (Clarke & Lumsden, 1993; Hanneman, Trevarrow, Metcalfe, Kimmel, & Westerfield, 1988; Lumsden & Keynes, 1989), as well as specification of branchial neural crest cells and establishment of their migratory pathways (Birgbauer, Sechrist, Bronner-Fraser, & Fraser, 1995; Lumsden, Sprawson, & Graham, 1991). At the molecular level, anteroposterior identity is determined by segmentally expressed *Hox* genes which have overlapping and nested expression patterns in the hindbrain (Tumpel, Wiedemann, & Krumlauf, 2009). For example, *Hoxb1* (*hoxb1a* in zebrafish) confers regional identity to r4 (Bell, Wingate, & Lumsden, 1999; McClintock, Kheirbek, & Prince, 2002; Rohrschneider, Elsen, & Prince, 2007; Studer, Lumsden, Ariza-McNaughton, Bradley, & Krumlauf, 1996). Specific *Hox* genes together with other transcription factors are required for segmentation of the hindbrain. For example, *Krox20* is essential for the formation of r3 and r5 (Schneider-Maunoury, Seitanidou, Charnay, & Lumsden, 1997; Schneider-Maunoury et al., 1993; Swiatek & Gridley, 1993; Voiculescu et al., 2001), while *MafB* is required for formation of r5 in mouse (Manzanares, Trainor, et al., 1999), and for segmentation to form r5 and r6 in zebrafish (Moens, Yan, Appel, Force, & Kimmel, 1996). As discussed in detail later, sharp borders form at the interface of adjacent rhombomeres, across which cell intermingling is prevented through signaling by Eph receptors and ephrins. In parallel, specialized cells with distinct cellular and molecular properties are induced to form at the interfaces of rhombomeres (Guthrie & Lumsden, 1991; Heyman, Faissner, & Lumsden, 1995; Lumsden & Keynes, 1989). There is a lower proliferation of boundary cells and accumulation of extracellular matrix which could in principle contribute to maintenance of sharp borders, but it has been found that cell intermingling is still restricted after blocking of boundary cell formation in chick (Nittenberg et al., 1997). There is increasing evidence that hindbrain boundary cells function as signaling centers that regulate segmental gene expression in chick (Sela-Donenfeld, Kayam, & Wilkinson, 2009), and the patterning of cell differentiation within rhombomeres in zebrafish (Gonzalez-Quevedo, Lee, Poss, & Wilkinson, 2010; Terriente, Gerety, Watanabe-Asaka, Gonzalez-Quevedo, & Wilkinson, 2012).

3. SEGMENT IDENTITY

3.1 Induction of Segmental Identity in the Vertebrate Hindbrain

The induction of different transcription factors along the anteroposterior axis of the hindbrain involves a combination of transient posteriorizing signals, including graded retinoic acid, and Wnt and Fgf proteins (Hernandez, Putzke, Myers, Margaretha, & Moens, 2007; Kudoh, Wilson, & Dawid, 2002; White, Nie, Lander, & Schilling, 2007). Retinoic acid is synthesized in the paraxial mesoderm posterior to the hindbrain and is necessary for the specification of hindbrain segments posterior to the r3/4 border. Domains of *Hox* expression here are patterned by differential responses to graded levels of retinoic acid along the anteroposterior axis. Retinoic acid can induce the expression of certain *Hox* genes directly by activating RAR/RXR heterodimers at retinoic acid response elements (RAREs) in regulatory regions, for example, of members of the *Hox1* and *Hox4* paralogous groups (Marshall et al., 1994; Nolte, Amores, Nagy Kovacs, Postlethwait, & Featherstone, 2003). The expression of other posterior retinoic acid-responsive *hox* genes, including *Hox3* genes, is regulated indirectly by retinoic acid; for example, *Hoxa3* expression in r5 and r6 is regulated by *MafB*, which is itself regulated in part by retinoic acid (via *Vhnf1*) (Hernandez, Rikhof, Bachmann, & Moens, 2004; Manzanares, Cordes, et al., 1999). More posterior retinoic acid-responsive *Hox* genes require higher levels of retinoic acid for their induction and are expressed later than more anterior retinoic acid-responsive *Hox* genes, which are induced at lower levels of retinoic acid (Dupe & Lumsden, 2001; Maves & Kimmel, 2005). However, the spatial patterning of *Hox* gene expression along the anteroposterior axis by retinoic acid is not purely achieved by differential sensitivities to retinoic acid, as the RAREs of different *Hox* genes have been found to have comparable sensitivities (Nolte et al., 2003). It has also been shown that *Hox* genes expressed in more posterior hindbrain domains do not require longer exposures to retinoic acid for their induction (Hernandez et al., 2007; Maves & Kimmel, 2005).

The mechanisms by which the graded retinoic acid activity required for hindbrain segmentation is generated and maintained are not completely understood. The concentration of retinoic acid along the anteroposterior axis of the hindbrain is influenced both spatially and temporally by members of the Cyp26 family of cytochrome p450 enzymes—Cyp26a1, b1 and c1—which degrade retinoic acid. In the absence of Cyp26 enzymes, the anteroposterior patterning of the hindbrain is disrupted (Hernandez et al., 2007). Furthermore, exogenous retinoic acid applied over a 20-fold concentration range and at a variety of developmental stages is sufficient to correctly pattern the hindbrain of embryos that are depleted of endogenous retinoic acid, and Cyp26 enzymes are required for this rescue (Dupe & Lumsden, 2001; Hernandez et al., 2007; Maves & Kimmel, 2005). *Cyp26a1* is induced by retinoic acid itself and has an important role in the negative feedback of retinoic acid signaling (White et al., 2007). This self-enhanced degradation allows retinoic acid-induced patterning to be robust despite variations in retinoic acid levels. *Cyp26b1* and *cyp26c1* are segmentally and dynamically expressed during hindbrain segmentation and refine retinoic acid responsiveness within the hindbrain through localized retinoic acid degradation (White & Schilling, 2008). It is not clear whether segmentally expressed Cyp26 enzymes generate a step-wise gradient of retinoic acid. In mouse, use of a RARE–lacZ reporter has revealed distinct boundaries of retinoic acid responsiveness, which move during hindbrain segmentation (Rossant, Zirngibl, Cado, Shago, & Goguere, 1991; Sirbu, Gresh, Barra, & Duester, 2005). However, a RARE-eYFP transgenic zebrafish line has revealed a smooth gradient of retinoic acid responsiveness, though only at later stages of hindbrain segmentation (White et al., 2007). More recently, a smooth gradient of retinoic acid in the zebrafish hindbrain has been detected by visualization of unbound intracellular retinoic acid using FRET-based genetically encoded retinoic acid probes (Shimozono, Iimura, Kitaguchi, Higashijima, & Miyawaki, 2013).

Fgfs have important roles in hindbrain patterning, in part by inhibiting retinoic acid-mediated *cyp26a1* upregulation (White et al., 2007). This enables Fgfs to influence the retinoic acid gradient within the hindbrain, which has been suggested to help couple growth of the hindbrain with a corresponding expansion of the retinoic acid gradient (Schilling, Nie, & Lander, 2012; White et al., 2007). In combination with retinoic acid, Fgfs and Wnts also initiate the expression of certain posterior genes, including *hoxb1b* (posteriorly from pre-r4) and *vhnf1*, (in pre-r5–r6) (Kudoh et al., 2002). Later in hindbrain segmentation, fgf3 and fgf8 are upregulated by Hox1 proteins in r4, and are involved in the patterning of surrounding rhombomeres (Hernandez et al., 2004; Marin & Charnay, 2000; Maves, Jackman, & Kimmel, 2002; Walshe, Maroon, McGonnell, Dickson, & Mason, 2002; Waskiewicz, Rikhof, & Moens, 2002; Wiellette & Sive, 2003). For example, Fgf signals from r4 cooperate with *vhnf1* and retinoic acid to drive initiation of *mafB* expression in r5 and r6 (Hernandez et al., 2004), and Fgfs also contribute to induction of *krox20* expression (Labalette et al., 2015).

3.2 Establishment of Mutually Exclusive Segmental Identity

The expression domains of segmentally expressed transcription factors initially have diffuse borders, in which some cells coexpress factors that confer conflicting identities and thus have an intermediate identity. For example in zebrafish, some cells at the r3/r4 and r4/r5 interfaces express both *krox20* and *hoxb1a*, which confer opposing identities; likewise, ectopic *krox20*-expressing cells within r4 can also contain *hoxb1a* transcripts (Zhang et al., 2012). This raises the question of how one or the other gene is downregulated in order to establish and maintain a single identity.

There have been detailed investigations of gene regulatory interactions in the specification of r3, r4, and r5 in the hindbrain. Following initiation of *krox20* expression, an autoregulatory loop amplifies and maintains *krox20* expression in r3 and r5 (Bouchoucha et al., 2013; Chomette, Frain, Cereghini, Charnay, & Ghislain, 2006; Giudicelli, Taillebourg, Charnay, & Gilardi-Hebenstreit, 2001). Likewise, *Hoxb1* upregulates its own expression in r4 via an autoregulatory loop (Pöpperl et al., 1995). Mutually exclusive expression is established since Krox20 and Hoxb1 repress each other: ectopic overexpression of *krox20* in r4 causes downregulation of *hoxb1* expression (Giudicelli et al., 2001), while *hoxb1* represses *krox20* expression indirectly via activation of Nlz factors (Labalette et al., 2015). These interactions create a bistable switch, which enables cells to adopt an exclusive identity and commit to a particular fate following transient inputs to identity specification by morphogens (Fig. 1). This generates sharp spatial transitions in identity despite shallow and noisy morphogen gradients (Bouchoucha et al., 2013). Computational modeling has demonstrated that such a bistable switch regulating *krox20* and *hoxb1* expression, in combination with noise in the retinoic acid gradient, can contribute to border sharpening in the hindbrain by driving the refinement of gene expression domains via noise-induced switching of cell identity (Schilling et al., 2012; Zhang et al., 2012).

3.3 Plasticity in Cell Identity

Intermingling of cells occurs during hindbrain development, driven by cell division and morphogenetic rearrangements such as convergent extension. This is revealed by the dispersal of the progeny of individual cells in the chick hindbrain, which when labeled at early stages can contribute to adjacent rhombomeres, but when labeled after morphological boundary formation are restricted to single segments (Fraser, Keynes, & Lumsden, 1990). This raises the question of how sharp gene expression boundaries are established despite some cell movement between segments at early stages. There is evidence that cells are capable of switching their identity due to plasticity in *Hox* expression status. In mouse embryos, transpositions of small numbers of cells between rhombomeres have found that cells will change *Hox* expression (Trainor & Krumlauf, 2000). Likewise, in zebrafish, cells can alter their *hox* expression following transplantation between rhombomeres at early stages, when border sharpening is occurring (Kemp, Cooke, & Moens, 2009; Schilling, Prince, & Ingham, 2001). Transplantations at later stages, when segmental gene expression domains are sharp, revealed a progressive loss of plasticity in cell identity (Schilling et al., 2001); by these stages, there is no longer cell mixing between segments (Calzolari, Terriente, & Pujades, 2014), and thus switching is not required during normal development. Cells transplanted to more posterior regions of the hindbrain have been found to exhibit increased plasticity compared to posterior-to-anterior transplantations (Grapin-Botton, Bonnin, McNaughton, Krumlauf, & Le Douarin, 1995; Itasaki, Sharpe, Morrison, & Krumlauf, 1996; Schilling et al., 2001), which may reflect the importance of posteriorizing signals in regulating *Hox* expression. Intriguingly, cell plasticity depends on the size of the group of transplanted cells, with larger groups maintaining their original *Hox* expression at a different position along the anteroposterior axis (Schilling et al., 2001; Trainor & Krumlauf, 2000). A failure to switch identity can account for the groups of *krox20*-expressing cells present in even-numbered segments following increased intermingling due to dominant negative blocking of Eph receptors (Xu, Alldus, Holder, & Wilkinson, 1995). These observations suggest that community effects have a role in the maintenance of segmental gene expression. The finding that *Krox20* nonautonomously induces its own expression in adjacent cells (Giudicelli et al., 2001) suggests a model in which it upregulates a signal which establishes and maintains homogenous identity within groups of cells.

It is not clear from these experiments the extent to which cell plasticity contributes to the establishment and maintenance of distinct segmental identity. Lineage analysis in chick reveals cell intermingling at early stages in hindbrain segmentation (Fraser et al., 1990), but the extent to which this occurs has not been quantitated. Use of Cre-mediated recombination in mouse to permanently label cells of the hindbrain that have expressed the r3/r5 marker, *Krox20*, has found some labeled cells in even-numbered rhombomeres (Voiculescu et al., 2001). This is consistent with cells dynamically expressing *Krox20*, and suggests that switching of cell identity occurs. However, such labeling could reflect resolution of initial overlapping expression, for example of *Krox20* and *Hoxb1*, rather than intermingling of cells that have established an r3/r5 identity. In zebrafish, cells expressing stable fluorescent reporters of r3 and r5 cell identity have not been detected in adjacent segments once sharpening of gene expression domains is complete (Calzolari et al., 2014). This suggests that in zebrafish there is little, if any, intermingling between segments and cell identity switching in normal development. However, it is possible that early intermingling is not detected since insufficient levels of the transgenic fluorescent reporter have accumulated in cells with transient r3/r5 identity.

4. CELL SEGREGATION

4.1 Restriction of Cell Intermingling Across Segment Borders

During early hindbrain development, the initially ragged expression limits of molecular determinants of rhombomere identity become straight (Irving, Nieto, DasGupta, Charnay, & Wilkinson, 1996; Oxtoby & Jowett, 1993). This sharpening is presumably driven by cell segregation mechanisms that underlie the restriction of intermingling between hindbrain segments (Calzolari et al., 2014; Fraser et al., 1990; Jimenez-Guri et al., 2010; Mathis, Sieur, Voiculescu, Charnay, & Nicolas, 1999). Transplantation experiments in chick showed that the inhibition of cell mixing involves cell properties that differ most prominently between odd versus even-numbered rhombomeres (Guthrie, Prince, & Lumsden, 1993). There is much evidence that interactions between Eph receptor tyrosine kinases and ephrins have a major role in this restriction of cell mixing. Eph receptors and ephrins comprise a cell contact-dependent signaling system, which upon binding and clustering transduce signals bidirectionally (Klein, 2012). In mouse, several Eph receptors are expressed in r3 and r5, and ephrinB proteins that they bind are expressed in even-numbered segments (Xu, Mellitzer, & Wilkinson, 2000). There is a different situation in zebrafish (Chan et al., 2001; Cooke et al., 2001; Xu et al., 1995), in which Eph–ephrin pairs that have high affinity have complementary expression, but not simply in odd versus even-numbered segments: expression of EphA4 and ephrinB3 occurs in r3/r5 and r2/r4/r6, respectively; expression of EphB4a and ephrinb2a in r5/r6 and r2/r4/r7, respectively.

The effects of blocking or knockdown of EphA4a and ephrinB2a reveal that Eph–ephrin signaling is required to inhibit cell intermingling and sharpen segment borders in zebrafish (Cooke, Kemp, & Moens, 2005; Kemp et al., 2009; Xu et al., 1995). The restriction of cell mixing is mediated, at least in part, by Eph–ephrin pairs with complementary expression that are interacting at the border (Mellitzer, Xu, & Wilkinson, 1999; Xu, Mellitzer, Robinson, & Wilkinson, 1999). Eph–ephrin signaling could act through one or several mechanisms that have been shown to drive cell segregation in other tissues: differential cell–cell adhesion, cortical tension, and cell repulsion (Batlle & Wilkinson, 2012; Cayuso, Xu, & Wilkinson, 2015; Fagotto, 2014; Fagotto, Winklbauer, & Rohani, 2014). A recent study has shown that there is an Eph–ephrin-dependent accumulation of actomyosin cables at segment borders, which can potentially generate increased cortical tension that restricts cell intermingling (Calzolari et al., 2014). Since the formation of actomyosin cables occurs several hours after border sharpening, cortical tension may play a late role in boundary stabilization. Intriguingly, mosaic knockdown of EphA4 or of ephrinB2a in zebrafish was found to lead to segregation of the knockdown cells to the borders of the segments expressing the corresponding Eph receptor or ephrin (Cooke et al., 2005; Kemp et al., 2009). The mechanism(s) that drives segregation of cells with decreased Eph or ephrin expression within segments is not known, but in principle could involve differential adhesion, tension, or repulsion. This suggests that EphA4 and ephrinB2 activity within segments contributes to cell segregation, presumably acting in parallel with bidirectional signaling at the segment borders to drive sharpening.

4.2 Coupling Between Cell Identity and Cell Segregation

The finding that border sharpening requires complementary expression of interacting Eph receptors and ephrins raises the question of how their segmental expression in the hindbrain is regulated. One possibility is that this is mediated by the transcription factors that underlie segmentation and anteroposterior identity in the hindbrain. Indeed, *mafB* is required to upregulate *EphB4a* and suppress *ephrinB2a* expression in r5 and r6, such that this Eph–ephrin pair is in complementary domains (Cooke et al., 2001; Hernandez et al., 2004). Likewise, *Hoxa2* is required for *EphA7* expression in r3 (Taneja et al., 1996). However, due to the cross-regulation that occurs between *mafB*, *Krox20*, and *Hox* genes, it is not known from these studies which of the transcription factors are the direct regulator of Eph receptor or ephrin expression. Evidence has been obtained by analysis of enhancer elements of the *EphA4* gene showing that Krox20 directly regulates *EphA4* expression in r3 and r5 (Theil et al., 1998). Since Krox20 also directly regulates expression of specific *Hox* genes (Tumpel et al., 2009), there is thus a coupling of anteroposterior identity and cell segregation. Direct regulation has also been revealed in studies showing that *EphA2* expression in r4 is upregulated by *Hoxa1* and *Hoxb1* (Studer et al., 1998) through binding to Hox/Pbx binding sites in an *EphA2* enhancer (Chen & Ruley, 1998). However, the roles of EphA2 in the hindbrain are not known.

A recent study has given further insights into the relationships between *Hox* genes and cell segregation in the hindbrain (Prin, Serpente, Itasaki, & Gould, 2014). The anterior limit of *Hoxb4* and *Hoxd4* expression is at the r6/r7 boundary, which is absent in double *Hoxb4/Hoxd4* mutants, as seen by loss of molecular markers of boundary cells and of the morphological constriction. Furthermore, boundary formation in more anterior regions of the hindbrain is disrupted when *Hox4* genes are ectopically expressed. Importantly, it was found that mosaic ectopic expression of *Hox4* genes leads to segregation of *Hox4+* and *Hox4–* cells. This segregation is accompanied by formation of an apical constriction in whichever of the two populations is in the minority, and alterations in the apical surface area of cells 3–4 diameters away from the *Hox4+/Hox4–* border. These observations are consistent with a tension-based mechanism, which may involve Eph–ephrin signaling since ectopic expression of *ephrinB2* creates similar constrictions. Furthermore, *Hox4* genes underlie the posterior limit of *EphA7* expression at the r6/r7 border by downregulating its expression in r7, albeit it is not known whether *EphA7* is involved in boundary formation. Taken together, these findings reveal that *Hox4* genes underlie formation of the r6/r7 boundary and regulate mechanisms that drive cell segregation. Since cell segregation also occurs following ectopic expression of *Hoxa2* and *Hoxa3*, this role may be a general feature of *Hox* genes (Prin et al., 2014). Evidence that *Hox* genes act through regulation of Eph–ephrin expression is currently circumstantial, and it is possible that cell segregation also involves other *Hox* targets (Prin et al., 2014). The results of overexpression experiments show that *krox20* has a dominant role in cell segregation in r3 and r5 through regulation of EphA4, which overrides the control of cell segregation by *Hox* genes in these segments. *Hox* genes may thus act in the other rhombomeres to promote border formation.

5. SUMMARY AND FUTURE PERSPECTIVES

Major progress has been made in elucidating the signaling and gene regulatory networks that underlie segmental expression of key transcription factors in hindbrain patterning, including *Hox* genes, *krox20*, and *mafB*. These involve positive feedback and mutual repression that establishes complementary segmental expression. Recent studies have focussed on cross-repression of *hoxb1* and *krox20* at the borders of r4, and it is important to elucidate the regulatory relationships that establish discrete cell identities across the other boundaries in the hindbrain. Another important question is whether such networks account for cell identity switching following transplantation between segments, and which may also occur during cell intermingling in normal development. Presumably, the ectopic cells are reading a different level of graded retinoic acid and other signals at the new anteroposterior position. Another possibility is that ectopic cells come under the influence of signaling from their new neighbors, as suggested by the lack of identity switching when groups of cells are transplanted. How does the gene regulatory network respond to such changes in the signaling environment? In particular, how does altered signaling overcome positive feedback loops that maintain *krox20* and *hoxb1* gene expression? Do such loops and/or other mechanisms underlie the decreasing ability of cells to switch identity at later stages of hindbrain development?

There has been less progress in understanding the coupling between segment identity and cell segregation. It is clear that *Hox* genes, *krox20*, and *mafB* are important regulators which lie upstream of Eph receptor and ephrin expression in the hindbrain. However, for the Eph receptors and ephrins known to underlie cell segregation in the hindbrain, the only direct link shown thus far is the upregulation of *EphA4* by *krox20*. It will, in particular, be interesting to uncover whether *Hox* genes are direct regulators of the relevant Eph receptors and ephrins, and how the gene regulatory network in the hindbrain establishes the complementary expression of Eph–ephrin pairs that is required for cell segregation.

ACKNOWLEDGMENTS

We are grateful to Alex Gould, Fabrice Prin, Jordi Cayuso, and Qiling Xu for their valuable comments on this chapter. Work in the author's lab is funded by the Francis Crick Institute.

REFERENCES

Batlle, E., & Wilkinson, D. G. (2012). Molecular mechanisms of cell segregation and boundary formation in development and tumorigenesis. *Cold Spring Harbor Perspectives in Biology*, *4*, a008227.

Bell, E., Wingate, R., & Lumsden, A. (1999). Homeotic transformation of rhombomere identity after localized *Hoxb1* misexpression. *Science*, *284*, 21682171.

Birgbauer, E., Sechrist, J., Bronner-Fraser, M., & Fraser, S. (1995). Rhombomeric origin and rostrocaudal reassortment of neural crest cells revealed by intravital microscopy. *Development*, *121*, 935–945.

Bouchoucha, Y. X., Reingruber, J., Labalette, C., Wassef, M. A., Thierion, E., Desmarquet-Trin Dinh, C., et al. (2013). Dissection of a Krox20 positive feedback loop driving cell fate choices in hindbrain patterning. *Molecular Systems Biology, 9*, 690.

Calzolari, S., Terriente, J., & Pujades, C. (2014). Cell segregation in the vertebrate hindbrain relies on actomyosin cables located at the interhombomeric boundaries. *The EMBO Journal, 33*, 686–701.

Cayuso, J., Xu, Q., & Wilkinson, D. G. (2015). Mechanisms of boundary formation by Eph receptor and ephrin signaling. *Developmental Biology, 401*, 122–131.

Chan, J., Mably, J. D., Serluca, F. C., Chen, J. N., Goldstein, N. B., Thomas, M. C., et al. (2001). Morphogenesis of prechordal plate and notochord requires intact Eph/ephrin B signaling. *Developmental Biology, 234*, 470–482.

Chen, J., & Ruley, H. E. (1998). An enhancer element in the EphA2 (Eck) gene sufficient for rhombomere-specific expression is activated by HOXA1 and HOXB1 homeobox proteins. *The Journal of Biological Chemistry, 273*, 24670–24675.

Chomette, D., Frain, M., Cereghini, S., Charnay, P., & Ghislain, J. (2006). Krox20 hindbrain cis-regulatory landscape: Interplay between multiple long-range initiation and autoregulatory elements. *Development, 133*, 1253–1262.

Clarke, J. D., & Lumsden, A. (1993). Segmental repetition of neuronal phenotype sets in the chick embryo hindbrain. *Development, 118*, 151–162.

Cooke, J. E., Kemp, H. A., & Moens, C. B. (2005). EphA4 is required for cell adhesion and rhombomere-boundary formation in the zebrafish. *Current Biology, 15*, 536–542.

Cooke, J., Moens, C., Roth, L., Durbin, L., Shiomi, K., Brennan, C., et al. (2001). Eph signalling functions downstream of Val to regulate cell sorting and boundary formation in the caudal hindbrain. *Development, 128*, 571–580.

Dahmann, C., Oates, A. C., & Brand, M. (2011). Boundary formation and maintenance in tissue development. *Nature Reviews Genetics, 12*, 43–55.

Dupe, V., & Lumsden, A. (2001). Hindbrain patterning involves graded responses to retinoic acid signalling. *Development, 128*, 2199–2208.

Fagotto, F. (2014). The cellular basis of tissue separation. *Development, 141*, 3303–3318.

Fagotto, F., Winklbauer, R., & Rohani, N. (2014). Ephrin-Eph signaling in embryonic tissue separation. *Cell Adhesion & Migration, 8*, 308–326.

Fraser, S., Keynes, R., & Lumsden, A. (1990). Segmentation in the chick embryo hindbrain is defined by cell lineage restrictions. *Nature, 344*, 431–435.

Giudicelli, F., Taillebourg, E., Charnay, P., & Gilardi-Hebenstreit, P. (2001). Krox-20 patterns the hindbrain through both cell-autonomous and non cell-autonomous mechanisms. *Genes & Development, 15*, 567–580.

Gonzalez-Quevedo, R., Lee, Y., Poss, K. D., & Wilkinson, D. G. (2010). Neuronal regulation of the spatial patterning of neurogenesis. *Developmental Cell, 18*, 136–147.

Grapin-Botton, A., Bonnin, M. A., McNaughton, L. A., Krumlauf, R., & Le Douarin, N. M. (1995). Plasticity of transposed rhombomeres: Hox gene induction is correlated with phenotypic modifications. *Development, 121*, 2707–2721.

Guthrie, S., & Lumsden, A. (1991). Formation and regeneration of rhombomere boundaries in the developing chick hindbrain. *Development, 112*, 221–229.

Guthrie, S., Prince, V., & Lumsden, A. (1993). Selective dispersal of avian rhombomere cells in orthotopic and heterotopic grafts. *Development, 118*, 527–538.

Hanneman, E., Trevarrow, B., Metcalfe, W. K., Kimmel, C. B., & Westerfield, M. (1988). Segmental pattern of development of the hindbrain and spinal cord of the zebrafish embryo. *Development, 103*, 49–58.

Hernandez, R. E., Putzke, A. P., Myers, J. P., Margaretha, L., & Moens, C. B. (2007). Cyp26 enzymes generate the retinoic acid response pattern necessary for hindbrain development. *Development, 134*, 177–187.

Hernandez, R. E., Rikhof, H. A., Bachmann, R., & Moens, C. B. (2004). vhnf1 integrates global RA patterning and local FGF signals to direct posterior hindbrain development in zebrafish. *Development*, *131*, 4511–4520.
Heyman, I., Faissner, A., & Lumsden, A. (1995). Cell and matrix specialisations of rhombomere boundaries. *Developmental Dynamics*, *204*, 301–315.
Irving, C., Nieto, M. A., DasGupta, R., Charnay, P., & Wilkinson, D. G. (1996). Progressive spatial restriction of Sek-1 and Krox-20 gene expression during hindbrain segmentation. *Developmental Biology*, *173*, 26–38.
Itasaki, N., Sharpe, J., Morrison, A., & Krumlauf, R. (1996). Reprogramming Hox expression in the vertebrate hindbrain: Influence of paraxial mesoderm and rhombomere transposition. *Neuron*, *16*, 487–500.
Jimenez-Guri, E., Udina, F., Colas, J. F., Sharpe, J., Padron-Barthe, L., Torres, M., et al. (2010). Clonal analysis in mice underlines the importance of rhombomeric boundaries in cell movement restriction during hindbrain segmentation. *PLoS One*, *5*, e10112.
Kemp, H. A., Cooke, J. E., & Moens, C. B. (2009). EphA4 and EfnB2a maintain rhombomere coherence by independently regulating intercalation of progenitor cells in the zebrafish neural keel. *Developmental Biology*, *327*, 313–326.
Kiecker, C., & Lumsden, A. (2005). Compartments and their boundaries in vertebrate brain development. *Nature Review Neuroscience*, *6*, 553–564.
Klein, R. (2012). Eph/ephrin signalling during development. *Development*, *139*, 4105–4109.
Kudoh, T., Wilson, S. W., & Dawid, I. B. (2002). Distinct roles for Fgf, Wnt and retinoic acid in posteriorizing the neural ectoderm. *Development*, *129*, 4335–4346.
Labalette, C., Wassef, M. A., Desmarquet-Trin Dinh, C., Bouchoucha, Y. X., Le Men, J., Charnay, P., et al. (2015). Molecular dissection of segment formation in the developing hindbrain. *Development*, *142*, 185–195.
Lumsden, A., & Keynes, R. (1989). Segmental patterns of neuronal development in the chick hindbrain. *Nature*, *337*, 424–428.
Lumsden, A., Sprawson, N., & Graham, A. (1991). Segmental origin and migration of neural crest cells in the hindbrain region of the chick embryo. *Development*, *113*, 1281–1291.
Manzanares, M., Cordes, S., Ariza-McNaughton, L., Sadl, V., Maruthainar, K., Barsh, G., et al. (1999). Conserved and distinct roles of kreisler in regulation of the paralogous Hoxa3 and Hoxb3 genes. *Development*, *126*, 759–769.
Manzanares, M., Trainor, P. A., Nonchev, S., Ariza-McNaughton, L., Brodie, J., Gould, A., et al. (1999). The role of kreisler in segmentation during hindbrain development. *Developmental Biology*, *211*, 220–237.
Marin, F., & Charnay, P. (2000). Hindbrain patterning: FGFs regulate Krox20 and mafB/kr expression in the otic/preotic region. *Development*, *127*, 4925–4935.
Marshall, H., Studer, M., Pöpperl, H., Aparicio, S., Kuroiwa, A., Brenner, S., et al. (1994). A conserved retinoic acid response element required for early expression of the homeobox gene *Hoxb-1*. *Nature*, *370*, 567–571.
Mathis, L., Sieur, J., Voiculescu, O., Charnay, P., & Nicolas, J. F. (1999). Successive patterns of clonal cell dispersion in relation to neuromeric subdivision in the mouse neuroepithelium. *Development*, *126*, 4095–4106.
Maves, L., Jackman, W., & Kimmel, C. B. (2002). FGF3 and FGF8 mediate a rhombomere 4 signaling activity in the zebrafish hindbrain. *Development*, *129*, 3825–3837.
Maves, L., & Kimmel, C. B. (2005). Dynamic and sequential patterning of the zebrafish posterior hindbrain by retinoic acid. *Developmental Biology*, *285*, 593–605.
McClintock, J. M., Kheirbek, M. A., & Prince, V. E. (2002). Knockdown of duplicated zebrafish hoxb1 genes reveals distinct roles in hindbrain patterning and a novel mechanism of duplicate gene retention. *Development*, *129*, 2339–2354.
Mellitzer, G., Xu, Q., & Wilkinson, D. G. (1999). Eph receptors and ephrins restrict cell intermingling and communication. *Nature*, *400*, 77–81.

Moens, C. B., Yan, Y.-L., Appel, B., Force, A. G., & Kimmel, C. B. (1996). valentino: A zebrafish gene required for normal hindbrain segmentation. *Development, 122*, 3981–3990.

Nittenberg, R., Patel, K., Joshi, Y., Krumlauf, R., Wilkinson, D. G., Brickell, P. M., et al. (1997). Cell movements, neuronal organisation and gene expression in hindbrains lacking morphological boundaries. *Development, 124*, 2297–2306.

Nolte, C., Amores, A., Nagy Kovacs, E., Postlethwait, J., & Featherstone, M. (2003). The role of a retinoic acid response element in establishing the anterior neural expression border of Hoxd4 transgenes. *Mechanisms of Development, 120*, 325–335.

Oxtoby, E., & Jowett, T. (1993). Cloning of the zebrafish *krox-20* gene (*krx-20*) and its expression during hindbrain development. *Nucleic Acids Research, 21*, 1087–1095.

Pöpperl, H., Bienz, M., Studer, M., Chan, S. K., Aparicio, S., Brenner, S., et al. (1995). Segmental expression of Hoxb-1 is controlled by a highly conserved autoregulatory loop dependent upon exd/pbx. *Cell, 81*, 1031–1042.

Prin, F., Serpente, P., Itasaki, N., & Gould, A. P. (2014). Hox proteins drive cell segregation and non-autonomous apical remodelling during hindbrain segmentation. *Development, 141*, 1492–1502.

Rohrschneider, M. R., Elsen, G. E., & Prince, V. E. (2007). Zebrafish Hoxb1a regulates multiple downstream genes including prickle1b. *Developmental Biology, 309*, 358–372.

Rossant, J., Zirngibl, R., Cado, D., Shago, M., & Goguere, V. (1991). Expression of retinoic acid response element-hsp *lacZ* transgene defines specific domains of transcriptional activity during mouse embryogenesis. *Genes & Development, 5*, 1333–1344.

Schilling, T. F., Nie, Q., & Lander, A. D. (2012). Dynamics and precision in retinoic acid morphogen gradients. *Current Opinion in Genetics & Development, 22*, 562–569.

Schilling, T. F., Prince, V., & Ingham, P. W. (2001). Plasticity in zebrafish hox expression in the hindbrain and cranial neural crest. *Developmental Biology, 231*, 201–216.

Schneider-Maunoury, S., Seitanidou, T., Charnay, P., & Lumsden, A. (1997). Segmental and neuronal architecture of the hindbrain of Krox-20 mouse mutants. *Development, 124*, 1215–1226.

Schneider-Maunoury, S., Topilko, P., Seitanidou, T., Levi, G., Cohen-Tannoudji, M., Pournin, S., et al. (1993). Disruption of Krox-20 results in alteration of rhombomeres 3 and 5 in the developing hindbrain. *Cell, 75*, 1199–1214.

Sela-Donenfeld, D., Kayam, G., & Wilkinson, D. G. (2009). Boundary cells regulate a switch in the expression of FGF3 in hindbrain rhombomeres. *BMC Developmental Biology, 9*, 16.

Shimozono, S., Iimura, T., Kitaguchi, T., Higashijima, S., & Miyawaki, A. (2013). Visualization of an endogenous retinoic acid gradient across embryonic development. *Nature, 496*, 363–366.

Sirbu, I. O., Gresh, L., Barra, J., & Duester, G. (2005). Shifting boundaries of retinoic acid activity control hindbrain segmental gene expression. *Development, 132*, 2611–2622.

Studer, M., Gavalas, A., Marshall, H., Ariza-McNaughton, L., Rijli, F. M., Chambon, P., et al. (1998). Genetic interactions between Hoxa1 and Hoxb1 reveal new roles in regulation of early hindbrain patterning. *Development, 125*, 1025–1036.

Studer, M., Lumsden, A., Ariza-McNaughton, L., Bradley, A., & Krumlauf, R. (1996). Altered segmental identity and abnormal migration of motor neurones in mice lacking Hoxb-1. *Nature, 384*, 630–634.

Swiatek, P. J., & Gridley, T. (1993). Perinatal lethality and defects in hindbrain development in mice homozygous for a targeted mutation of the zinc finger gene Krox-20. *Genes & Development, 7*, 2071–2084.

Taneja, R., Thisse, B., Rijli, F. M., Thisse, C., Bouillet, P., Dolle, P., et al. (1996). The expression pattern of the mouse receptor tyrosine kinase gene MDK1 is conserved through evolution and requires Hoxa-2 for rhombomere-specific expression in mouse embryos. *Developmental Biology, 177*, 397–412.

Terriente, J., Gerety, S. S., Watanabe-Asaka, T., Gonzalez-Quevedo, R., & Wilkinson, D. G. (2012). Signalling from hindbrain boundaries regulates neuronal clustering that patterns neurogenesis. *Development*, *139*, 2978–2987.

Theil, T., Frain, M., Gilardi-Hebenstreit, P., Flenniken, A., Charnay, P., & Wilkinson, D. G. (1998). Segmental expression of the EphA4 (Sek-1) receptor tyrosine kinase in the hindbrain is under direct transcriptional control of Krox-20. *Development*, *125*, 443–452.

Trainor, P., & Krumlauf, R. (2000). Plasticity in mouse neural crest cells reveals a new patterning role for cranial mesoderm. *Nature Cell Biology*, *2*, 96–102.

Tumpel, S., Wiedemann, L. M., & Krumlauf, R. (2009). Hox genes and segmentation of the vertebrate hindbrain. *Current Topics in Developmental Biology*, *88*, 103–137.

Voiculescu, O., Taillebourg, E., Pujades, C., Kress, C., Buart, S., Charnay, P., et al. (2001). Hindbrain patterning: Krox20 couples segmentation and specification of regional identity. *Development*, *128*, 4967–4978.

Walshe, J., Maroon, H., McGonnell, I. M., Dickson, C., & Mason, I. (2002). Establishment of hindbrain segmental identity requires signaling by FGF3 and FGF8. *Current Biology*, *12*, 1117–1123.

Waskiewicz, A. J., Rikhof, H. A., & Moens, C. B. (2002). Eliminating zebrafish pbx proteins reveals a hindbrain ground state. *Developmental Cell*, *3*, 723–733.

White, R. J., Nie, Q., Lander, A. D., & Schilling, T. F. (2007). Complex regulation of cyp26a1 creates a robust retinoic acid gradient in the zebrafish embryo. *PLoS Biology*, *5*, e304.

White, R. J., & Schilling, T. F. (2008). How degrading: Cyp26s in hindbrain development. *Developmental Dynamics*, *237*, 2775–2790.

Wiellette, E. L., & Sive, H. (2003). vhnf1 and Fgf signals synergize to specify rhombomere identity in the zebrafish hindbrain. *Development*, *130*, 3821–3829.

Xiong, F., Tentner, A. R., Huang, P., Gelas, A., Mosaliganti, K. R., Souhait, L., et al. (2013). Specified neural progenitors sort to form sharp domains after noisy Shh signaling. *Cell*, *153*, 550–561.

Xu, Q., Alldus, G., Holder, N., & Wilkinson, D. G. (1995). Expression of truncated Sek-1 receptor tyrosine kinase disrupts the segmental restriction of gene expression in the Xenopus and zebrafish hindbrain. *Development*, *121*, 4005–4016.

Xu, Q., Mellitzer, G., Robinson, V., & Wilkinson, D. G. (1999). In vivo cell sorting in complementary segmental domains mediated by Eph receptors and ephrins. *Nature*, *399*, 267–271.

Xu, Q., Mellitzer, G., & Wilkinson, D. G. (2000). Roles of Eph receptors and ephrins in segmental patterning. *Philosophical Transactions of the Royal Society of London. Series B: Biological Sciences*, *355*, 993–1002.

Zhang, L., Radtke, K., Zheng, L., Cai, A. Q., Schilling, T. F., & Nie, Q. (2012). Noise drives sharpening of gene expression boundaries in the zebrafish hindbrain. *Molecular Systems Biology*, *8*, 613.

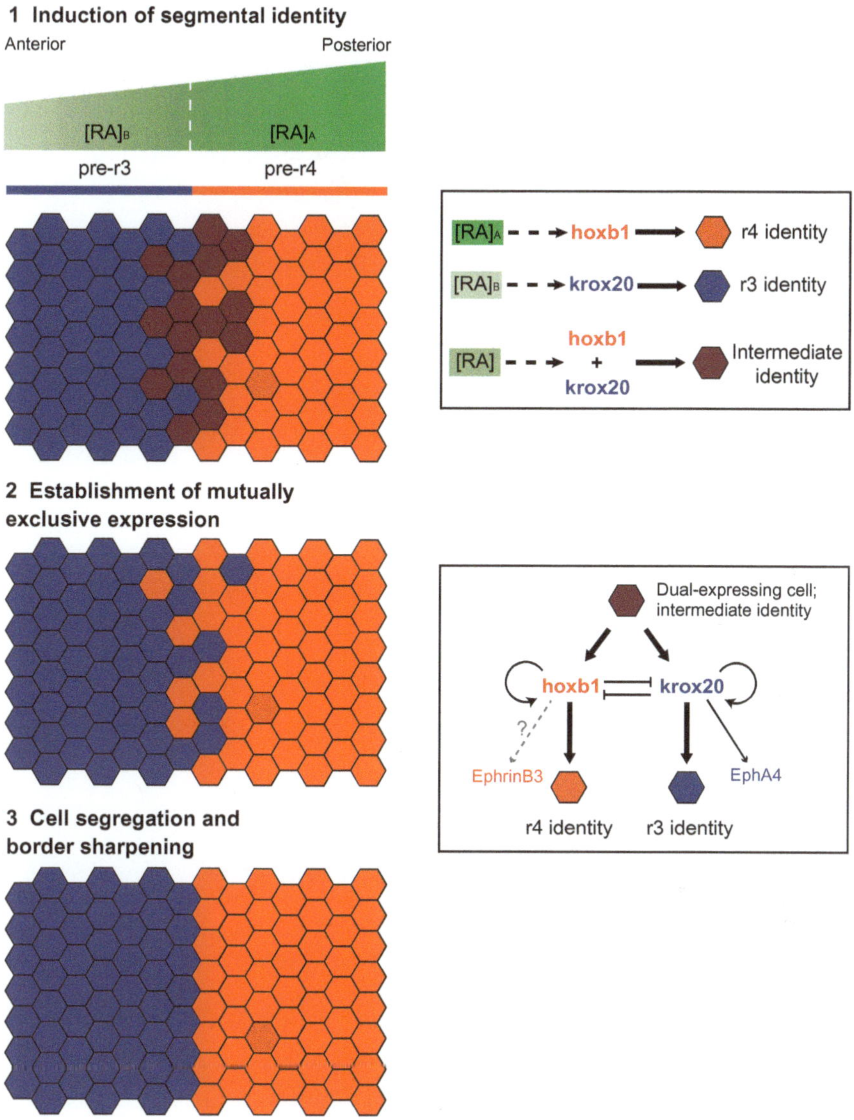

Megan Addison and David G. Wilkinson, Figure 1 Border formation during hindbrain segmentation. The left-hand side depicts the progressive sharpening of segment borders during hindbrain development, illustrated for the r3/r4 border. During induction of segmental identity, graded retinoic acid (RA) upregulates expression of *hoxb1* and *krox20*, with some cells at the prospective border expressing both genes (top right). *hoxb1* and *krox20* expression becomes mutually exclusive through cross-repression, thus resolving the identity of those cells which initially express both genes (bottom right). Autoregulatory loops maintain the expression of *hoxb1* and *krox20*. Krox20 upregulates EphA4 in r3, while unknown factors (perhaps directly or indirectly downstream of *hoxb1*) upregulate ephrinB3 in r4. Signaling of EphA4 and ephrinB3, and of other Eph–ephrin pairs, underlies cell segregation that sharpens the segment borders.

CHAPTER THIRTY-FIVE

Positional Information and Pattern Formation

Lewis Wolpert[1]
Cell and Developmental Biology, University College London, London, United Kingdom
[1]Corresponding author: e-mail address: l.wolpert@ucl.ac.uk

Abstract

The concept of positional information proposes that cells acquire positional values as in a coordinate system, which they interpret by developing in particular ways to give rise to spatial patterns. Some of the best evidence for positional information comes from regeneration experiments, and the patterning of the leg and antenna in *Drosophila*, and the vertebrate limb. Central problems are how positional information is set up, how it is recorded, and then how it is interpreted by the cells. A number of models have been proposed for the setting up of positional gradients, and most are based on diffusion of a morphogen and its interactions with extracellular molecules; however, diffusion may not be reliable mechanism. There are also mechanisms based on timing. There is no good evidence for the quantitative aspects of any of the proposed gradients and details how they are set up. The way in which a signaling gradient regulates differential gene expression in a concentration-dependent manner also raises several technical and quite difficult issues. A key feature of positional information being the basis for pattern formation is that there is no prepattern in the embryo.

Positional information is a concept that relates to pattern formation, that is, the development of spatial organization in the embryo that results from cells differentiating at specific positions. The concept of positional information proposes that cells acquire positional identities as in a coordinate system, and then interpret their positions to give rise to spatial patterns (Wolpert, 2011). It had its origin over 100 years ago, when Hans Driesch was working on sea urchin development. He found that if he separated the cells of the embryo at the two-cell stage, each developed into a small but normal larva. This implies that their patterning involves a mechanism for whose pattern is size invariant. He also claimed—mistakenly—that if any region of the early embryo were removed, the embryo still developed normally. He, therefore,

Current Topics in Developmental Biology, Volume 117
ISSN 0070-2153
http://dx.doi.org/10.1016/bs.ctdb.2015.11.008

concluded that there was a coordinate system that specified the position of each cell so that it could develop in the correct manner, but this was a mystical process. He was of course wrong, but his ideas helped give rise to the French flag mechanism for patterning, the pattern of flags being the same no matter their size (Wolpert, 1969). This is the basis of positional information which requires that the cells have positional values as in a coordinate system. However, few models take change of size into account as there is a scaling effect, which most models of morphogens do not take into account. There are several proposed mechanisms such as two diffusible molecules that emanate from opposing poles and define the activation profile through their ratio. A model based on expansion–repression by Ben-Zvi and Barkai (2010) suggests that scaling emerges as a natural consequence of a feedback topology.

Positional information implies that cells have a graded set of values that varies continuously along an axis, and under the influence of the local positional information, a cell obtains a positional value which they interpret their position by developing in particular ways. One of the main mechanisms proposed for setting up positional information is based on gradients of morphogens which are diffusible molecules produced by the cells, whose concentration specifies position. This was first proposed by Crick (1970) because the lengths of the proposed gradients, such as those in the vertebrate limb and insect imaginal discs, are quite small, around 50 cells long (Wolpert, 1969). The mechanisms for specifying position are discussed later.

Quite simple experiments that show that cells have positional values came from regeneration experiments that show intercalation of missing regions of invertebrate limbs (French, Bryant, & Bryant, 1976). There seems to be a set of positional values along the tibia of the cockroach leg since if a portion of the tibia is removed, the missing region will be replaced by intercalary regeneration. More striking, when a proximal cut tibia is grafted on to a more distal site making the tibia longer, intercalation makes the tibia even longer by intercalating the missing positional values. Another clear example of intercalation is from the salamander limb, when a distal regeneration blastema is grafted in place of a proximal one, and the missing regions are intercalated to form a normal limb.

An important example providing evidence for positional information came from the antenna and leg of *Drosophila*. If the Hox gene Antennapedia, which is normally expressed in parasegments 4 and 5 of the *Drosophila* embryo, is expressed in the head region, the antenna develops as a leg. Clones of Antennapedia cells in the antenna disc develop as leg cells, and the type of developed leg cells depends on their position along the proximodistal axis; if, for example, they are at the tip, they will develop as a claw (Postlethwait & Schneiderman, 1971). This strongly suggests that the positional values along the leg and antenna are the same, but because of the Hox genes, the cells interpret their positional values differently. The molecular basis of these positional values is not known, and the downstream action of the Hox genes that controls this process is also not understood.

An important model for pattern formation involving repeated structures is the model which is different from positional information. The Turing model generates a number of peaks in diffusible morphogens (Turing, 1952). There are two key components, an activator which is self stimulating, and an inhibitor which inhibits the activator but which is stimulated by it. There is thus a reaction–diffusion system. The model has been applied to a number of systems such as digit formation. The molecular nature of the morphogens has still to be reliably identified, but there is one encouraging case, the transverse ridges of the mammalian palate. Sonic hedgehog (Shh) and fibroblast growth factor (FGF) seem to play a key role (Economou et al., 2012). More important is the recent development of a realistic two-dimensional simulation of digit patterning by Raspopovic, Marcon, Russo, and Sharpe (2014).

Regeneration of amphibian limbs (Brockes & Gates, 2014) provides some of the best evidence for cells having positional values, and regeneration requires the specification of new positional values. There is good evidence for a continuous set of positional values along the newt limb from intercalation experiments (Pescitelli & Stocum, 1981). Of major significance, in the newt limb, a membrane molecule, Prod1, is graded from one end of the limb to the other, the proximodistal axis, and can be considered to be the molecular basis of positional information in the limb (Kumar, Gates, & Brockes, 2007). Retinoic acid can proximalise the positional values in the blastema of a regenerating limb by increasing the concentration of Prod1. If the limb is amputated at its distal end and then treated with retinoic acid, a whole limb will then regenerate from a cut at the distal end; the treatment increases the concentration of Prod1 so that it has a proximal positional value. This is excellent evidence for graded positional values. Further evidence comes from removing a blastema from a regenerating limb cut at a proximal level and replacing it with a distal blastema, intercalation takes place and the limb regenerates. Regeneration of an amputated limb requires the generation of new distal positional values. The mechanism is not known. According to Nachtrab, Kikuchi, Tornini, and Poss (2013), bone-forming cells in a regenerating appendage like the fin of zebrafish maintain expression throughout life of transcription factor genes that can influence the anteroposterior pattern. In the axolotl, McCusker and Gardiner (2013) found that cells from early stage blastemas as well as cells at the tip of late-stage blastemas are developmentally labile such that their positional identity is reprogrammed by interactions with more proximal cells with stable positional information. In contrast, cells from the adjacent, more proximal stump tissues as well as the basal region of late-bud blastemas are positionally stable.

There is also evidence for positional gradients in hydra which has a strong capacity for regeneration, when the head end is removed. The head end produces an inhibitor of head formation, and this is graded down the body. There is also a gradient in what can be thought of as positional values. When the inhibitor falls below a related level of this gradient following head removal, a new head forms. Grafting a head to the basal end of a hydra and then determining how long it took for the inhibitor to prevent head regeneration when the host head was removed, together with computer simulation, suggested a diffusion coefficient of 2×10^{-7} cm^2/s (Wolpert, Clarke, & Horrnbruch, 1972). This is remarkably similar to Crick's (1970) initial conjecture for setting up gradients.

Central problems are how positional information is set up, how it is recorded, and then how it is interpreted by the cells. One simple example is where there is at one end a source of the morphogen and at the other end a sink, both with fixed concentrations, and this results in the formation of a linear gradient. If there is just a localized source and breakdown of the morphogen, the diffusion will give rise to a gradient with an exponential form. Meinhardt (2009) has shown that primary pattern formation is possible if, and only if, a locally restricted self-enhancing reaction is coupled with an antagonistic reaction that acts on a longer range. A prototype of such a pattern-forming reaction consists of a short-ranging substance, the activator, which promotes—directly or indirectly—its own production, and also regulates the synthesis of its rapidly diffusing antagonist, the inhibitor. However, most models for setting up gradients do not consider this requirement.

Morphogen gradients that provide positional information are claimed to exist in many developing systems and are reviewed in detail by Rogers and Schier (2011). They point out that the prevailing model of morphogen gradient formation is that the morphogen is produced from a localized source, diffuses through tissues, and is cleared. But they ask why do the measured diffusion coefficients of morphogens differ by four orders of magnitude? There is still much to be understood. A very different mechanism for setting up a gradient is that of planar polarity (Strutt, 2009). The mechanism is based not on a morphogen but on cell-to-cell interactions involving the membrane component Frizzled at the junction, where there is asymmetry.

Eldar, Rosin, Shilo, and Barka (2003) described how having a morphogen that stimulates its own degradation can provide a powerful way to build gradients that are highly robust to variation in the level of morphogen production at a source. They claim that exponential profiles cannot, at the same time, buffer fluctuations in morphogen production rate and define long-range gradients. To comply with both requirements, morphogens should decay rapidly close to their source but at a significantly slower rate over most of the field. Morphogens could enhance their own degradation, leading to a higher degradation rate close to their source through reciprocal interactions between the morphogen and its receptor. To generate steady-state gradients, morphogens must be degraded (Wartlick, Kicheva, & González-Gaitán, 2009).

Mechanisms based on positional information have been used to explain vertebrate limb development, particularly the chick limb (Towers, Wolpert, & Tickle, 2012). Limb pattern is specified along three principal axes: anteroposterior (thumb to little finger), dorsoventral (back of hand to palm), and proximodistal (shoulder to finger tips). Digit pattern across the anteroposterior axis is a classic example of a signaling gradient that specifies positional values. Hox genes are specified in the limb. The specification of the three digits, 4,3, and 2, across the anteroposterior axis of the chick wing involves cells responding to different levels of a signal, Shh produced by the polarizing region a group of mesenchyme cells at the posterior margin of the limb bud. Specification of digits procedes in a step-wise fashion from anterior to posterior (Tickle, Summerbell, & Wolpert, 1975). The most posterior digit is exposed to the highest concentration of Shh. If Shh is applied to the anterior of the limb bud the pattern of digits 432234. There is recent evidence that Shh does not diffuse but is carried on particles and transported by cell extensions, cytonemes (Sanders, Llagostera, & Barna, 2013). The digits themselves are thought to be specified by a Turing mechanism. Progressive reduction in distal Hox genes together with a Turing mechanism results in progressively more severe polydactyly, displaying thinner and densely packed digits (Sheth et al., 2012).

In contrast to a signaling mechanism, specification of proximodistal positional values in the limb bud has been proposed to involve a timing mechanism linked to growth (Summerbell, Lewis, & Wolpert, 1973). Beneath the apical ridge at the tip of the limb bud is the progress zone and the proliferating cells push cells out of the zone. The cells measure the time they spend in the zone, the longer the time, the more distal the structures they give rise to. The progress zone model can readily explain why X-irradiation of early chick wing buds leads to absence of proximal structures. X-irradiation leads to extensive mesenchyme cell death, and as cells spend longer in the progress zone in order to repopulate it, they would acquire distal positional values, while fewer cells leave the zone and acquire proximal positional values (Wolpert, Tickle, & Sampford, 1979). Nevertheless, there are criticisms of the model (Galloway, Delgado, Ros, & Tabin, 2009). A molecular clock has been proposed by Sheeba, Andrade, and Palmeirim (2014) to integrate proximal–distal and anterior–posterior limb patterning, involving positional specification along both the axes. Dorsoventral patterning is less studied and involves signals from dorsal and ventral ectoderm.

Evidence for a morphogen role of FGF in zebrafish development mainly comes from patterning of the dorsoventral axis and the brain. It is claimed to involve extracellular diffusion and endocytosis (Bökel & Brand, 2013). Ramel and Hill (2013) argue that the establishment of the BMP activity gradient in early zebrafish embryos is determined by graded expression of the BMP ligands and suggest that there are different ways in which BMP gradients are set up. Xu, Houssin, Ferri-Lagneau, Thisse, and Thisse (2014) showed that setting up two gradients of BMP and Nodal in uncommitted cells of the zebrafish blastula animal pole caused the development of a vertebrate embryo.

There is evidence for positional signaling in the early development of *Drosophila*, but it is not clear whether related mechanism is involved in other systems, or whether these are specific for the fly. A gradient along the anteroposterior axis of the protein bicoid is set up early. The mechanism is special as it is preceded by a gradient in bicoid mRNA in the cortex probably transported by microtubules. However, the role of bicoid as providing positional information has been questioned (Chen, Xu, Mei, Yu, & Small, 2012). There are also gradients in hunchback and caudal which are thought to pattern the early embryo by activating genes at specific positions. It is again not clear whether positional information involved in the patterning of the dorsoventral axis of *Drosophila* is typical. A gradient in the intranuclear dorsal protein, with its high point in the ventral region, divides the dorsoventral axis into different regions with different genes being expressed. However, two studies used quantitative imaging to measure the distribution of the dorsal protein along the dorsoventral axis of the early *Drosophila* embryo (Bothma, Levine, & Boettiger, 2010). One claimed that the region of graded nuclear dorsal in the ventral region does not extend to the dorsal half of the embryo and that the expression boundaries of key genes in this region are not determined by dorsal, whereas another reported a smoothly changing nuclear concentration gradient that spans nearly the entire circumference of the embryo. A recent study claimed that pattern formation by graded signals depends on uniformly expressed transcriptional activators, such as Zelda (Kanodia et al., 2012).

In *Xenopus*, chordin protein secreted by the dorsal Spemann organizer was found to diffuse along a narrow region that separates the ectoderm from the anterior endoderm and mesoderm and forms a smooth gradient that encircles the embryo. Plouhinec, Zakin, Moriyama, and De Robertis (2013) suggest that both germ layers read their positional information coordinately from this morphogen gradient. Gurdon et al. (1999) found that cells respond to activin in a concentration-related way irrespective of their degree of contact with other cells. Therefore, interaction among cells is not required for the interpretation of morphogen concentration, at least in the case of the early genes studied here. They concluded that isolated blastula cells can sense and respond individually to activin by expressing genes in a concentration-dependent way.

Most of the signals specifying position discussed so far are based on diffusion of the morphogen, an exception being those based on time. However, we (Kerszberg & Wolpert, 2007) have argued that diffusion is not a reliable mechanism for specifying position as the concentration changes with time. A recent discovery has shown that a quite different mechanism is probably involved. This mechanism is based on cell extensions, specialized filopodia—cytonemes—transporting the morphogen to cells, and similarly being able to receive it from other cytonemes. In the fly wing disc, cytonemes extend to the midline where decapentaplegic (Dpp) is expressed and probably carry Dpp to this region (Ramírez-Weber & Kornberg, 1999; Roy, Huang, Liu, & Kornberg, 2014). Also, as mentioned earlier, in the developing chick limb, there is evidence for cytonemes transporting Hedgehog from the polarising region (Sanders et al., 2013).

A study by Lei, Wang, Song, Nie, and Wan (2013) have developed a model for developing a gradient. "Key elements of this mechanism are rapid turnover of morphogen to nonreceptors of neighboring cells resulting in significant degradation and transport of nonreceptor–morphogen complexes, the latter moving downstream through a 'bucket brigade' process."

Müller et al. (2013) in a detailed review have examined morphogen movement and discuss evidence for and against each transport model, focusing on the morphogens Nodal, FGF and Dpp. Based on their analysis, they conclude that most of the available data are consistent with diffusion models, together with interactions with extracellular molecules. They recognize that there is still much research that needs to be done. Another review by Restrepo, Zartman, and Basler (2014) focuses on DPP and its role in patterning the *Drosophila* wing imaginal disc. DPP is secreted from a stripe of cells at the boundary between the anterior and posterior compartments, and they suggest that there are two types of mechanism for its movement, those based on diffusion and those based on active transport. In their view, diffusion-based mechanisms are involved.

Positional information also has a role in plants. Gradients of auxin are required for tissue patterning within the embryo and the root of plants (Benková, Ivanchenko, Friml, Shishkova, & Dubrovsky, 2009). Graded auxin distribution seems to be crucial for cell specification within the root meristem, and the cells seem to react flexibly to changes in the auxin level by modulation of their developmental fate. Computer studies have shown that this is a plausible mechanism for the generation of the auxin distribution patterns in the root of plants. Mironova et al. (2010) demonstrated that the reflected flow mechanism, which relies on the presence of positive and negative regulations between auxin and expression of its carriers not only provides self-organization of the observed auxin distribution in the root but can also explain much of the positional information in root patterning. Benkovics and Timmermans (2014) propose that small RNAs are involved in signaling in plants. A new mechanism for specifying positional information in both plants and animals has been proposed by Hernández-Hernández, Rueda, Caballero, Alvarez-Buylla, and Benítez (2014) and is based on mechanical forces such as tension and compression.

The function of a positional gradient is to specify position and how this is recorded and appropriate genes turned on and off is a central problem. Ashe and Briscoe (2006) reviewed how morphogens act as graded positional cues and raises several mechanistic issues, such as how responding cells perceive and interpret the concentration-dependent information to generate precise patterns of gene expression and cell differentiation in developing tissues. The mechanisms of gene regulation by morphogen signaling must provide a means to translate small differences in signal strength into a threshold response. Thresholds are found when the modifying enzymes are saturated by their protein substrates in conditions of zero-order ultrasensitivity (Goldbetter & Wolpert, 1990).

The most detailed model of how gradients can be interpreted comes from how Shh signaling from the notochord assigns the positional identities of distinct neuronal subtype progenitors throughout the ventral neural tube (Dessaud et al., 2010). Assays of intracellular signal transduction and gene expression indicate that the duration as well as level of signaling is critical for morphogen interpretation. Progenitors of the ventral neuronal subtypes are established sequentially, with progressively more ventral identities requiring correspondingly higher levels and longer periods of Shh signaling. Moreover, cells remain sensitive to changes in Shh signaling for an extended time, reverting to antecedent identities if signaling levels fall below a threshold. Thus, the duration of signaling is important not only for the assignment but also for the refinement and maintenance of positional identity. In addition, the data suggest a dynamic model for ventral neural tube patterning in which positional information corresponds to the time integral of Shh signaling. This suggests an alternative to conventional models of morphogen action that rely solely on the level of signaling. The dynamic nature of Shh-mediated pattern formation is emphasized by the requirement for continued Shh signaling to maintain the more ventral progenitor identities even after the gene expression profiles that define these progenitor domains have been induced. Cohen, Briscoe, and Blassberg (2013) argue, that is, that cell identity results from the combined input of Shh signaling, uniformly expressed neural factors, and the cross-regulatory network of downstream Shh target genes. They provide evidence that "Shh morphogen interpretation in the neural tube is an emergent property of its downstream gene regulatory network. Cells transform the extracellular gradient of Shh into a dynamic profile of intracellular Gli activity that engages a transcriptional circuit, the regulatory logic of which is responsible for the generation of the characteristic temporal and spatial patterns of gene expression. This mechanism offers a powerful strategy to achieve the characteristic precision and robustness of morphogen-mediated pattern formation."

While there are many encouraging studies, we still do not know the molecular basis of positional information with the exception of Prod1, nor do we have convincing evidence of how positional values are specified or interpreted. Even the role of diffusion in morphogens is unclear.

REFERENCES

Ashe, H. L., & Briscoe, J. (2006). The interpretation of morphogen gradients. *Development, 133*, 385–394.

Benková, E., Ivanchenko, M. G., Friml, J., Shishkova, S., & Dubrovsky, J. G. (2009). A morphogenetic trigger: Is there an emerging concept in plant developmental biology? *Trends in Plant Science, 14*(4), 189–193.

Benkovics, A. H., & Timmermans, M. C. P. (2014). Developmental patterning by gradients of mobile small RNAs. *Current Opinion in Genetics & Development, 27*, 83–91.

Ben-Zvi, D., & Barkai, N. (2010). Scaling of morphogen gradients by an expansion–repression integral feedback control. *Proceedings of the National Academy of Sciences of the United States of America, 107*, 6924–6929.

Bökel, C., & Brand, M. (2013). Generation and interpretation of FGF morphogen gradients in vertebrates. *Current Opinion in Genetics & Development, 23*, 415–422.

Bothma, J. P., Levine, M., & Boettiger, A. (2010). Morphogen gradients: Limits to signaling or limits to measurement? *Current Biology, 20*(5), R232–R234.

Brockes, J. P., & Gates, P. B. (2014). Mechanisms underlying vertebrate limb regeneration: Lessons from the salamander. *Biochemical Society Transactions, 242*, 625–630.

Chen, H., Xu, Z., Mei, C., Yu, D., & Small, S. (2012). A system of repressor gradients spatially organizes the boundaries of Bicoid-dependent target genes. *Cell, 147*(3), 618–629.

Cohen, M., Briscoe, J., & Blassberg, R. (2013). Morphogen interpretation: The transcriptional logic of neural tube patterning. *Current Opinion in Genetics & Development, 23*, 423–428.

Crick, F. H. C. (1970). Diffusion in embryogenesis. *Nature, 225*, 420.

Dessaud, E., Ribes, V., Balaskas, N., Yang, L. L., Pierani, A., Kicheva, A., et al. (2010). Dynamic assignment and maintenance of positional identity in the ventral neural tube by the morphogen sonic hedgehog. *PLoS Biology, 8*(6), e1000382.

Economou, A. E., Ohazama, A., Porntaveetus, T., Sharpe, P. T., Kondo, S., Basson, M. A., et al. (2012). Periodic stripe formation by a Turing-mechanism operating at growth zones in the mammalian palate. *Science, 338*, 1476–1480.

Eldar, A., Rosin, D., Shilo, B. Z., & Barkai, N. (2003). Self-enhanced ligand degradation underlies robustness of morphogen gradients. *Developmental Cell, 5*, 635–646.

French, V., Bryant, P. J., & Bryant, S. V. (1976). Pattern regulation in epimorphic fields. *Science, 193*, 969–981.

Galloway, J. L., Delgado, I., Ros, M. A., & Tabin, C. J. (2009). A reevaluation of X-irradiation-induced phocomelia and proximodistal limb patterning. *Nature, 460*, 400–404.

Goldbetter, A., & Wolpert, L. (1990). Covalent modification of proteins as a threshold mechanism in development. *Journal of Theoretical Biology, 142*, 243–250.

Gurdon, J. B., Standley, H., Dyson, S., Butler, K., Langon, T., Ryan, K., et al. (1999). Single cells can sense their position in a morphogen gradient. *Development, 126*, 5309–5317.

Hernández-Hernández, V., Rueda, D., Caballero, L., Alvarez-Buylla, E. R., & Benítez, M. (2014). Mechanical forces as information: An integrated approach to plant and animal development. *Frontiers in Plant Science, 5*, 265.

Kanodia, J. S., Liang, H., Kim, Y., Li, B., Lu, H., Rushlow, C. A., et al. (2012). Pattern formation by graded and uniform signals in the early *Drosophila* embryo. *Biophysical Journal, 102*, 427–433.

Kerszberg, M., & Wolpert, L. (2007). Specifying positional information in the embryo: Looking beyond morphogens. *Cell, 130*, 205–209.

Kumar, A., Gates, P. B., & Brockes, J. P. (2007). Positional identity of adult stem cells in salamander limb regeneration. *Comptes Rendus Biologies, 330*(6–7), 485–490.

Lei, J., Wang, D., Song, Y., Nie, Q., & Wan, F. Y. (2013). Robustness of morphogen gradients with "Bucket Brigade" transport through membrane-associated non-receptors. *Discrete and Continuous Dynamical Systems. Series B, 1*, 18.

McCusker, C. D., & Gardiner, D. M. (2013). Positional information is reprogrammed in blastema cells of the regenerating limb of the axolotl (*Ambystoma mexicanum*). *PLoS One, 28*(9), e77064.

Meinhardt, H. (2009). Models for the generation and interpretation of gradients. *Cold Spring Harbor Perspectives in Biology, 1*(4), a001362.

Mironova, V. V., Omelyanchuk, N. A., Yosiphon, G., Fadeev, S. I., Kolchanov, N. A., Mjolsness, E., et al. (2010). A plausible mechanism for auxin patterning along the developing root. *BMC Systems Biology, 4*, 98.

Müller, P., Rogers, K. W., Yu, S. R., Brand, M., & Schier, A. F. (2013). Morphogen transport. *Development, 140*, 1621–1638.

Nachtrab, G., Kikuchi, K., Tornini, V. A., & Poss, K. D. (2013). Transcriptional components of anteroposterior positional information during zebrafish fin regeneration. *Development, 140*, 3754–3764.

Pescitelli, M. J., & Stocum, D. L. (1981). Nonsegmental organization of positional information in regenerating Ambystoma limbs. *Developmental Biology, 82*, 69–85.

Plouhinec, J. L., Zakin, L., Moriyama, Y., & De Robertis, E. M. (2013). Chordin forms a self-organizing morphogen gradient in the extracellular space between ectoderm and mesoderm in the Xenopus embryo. *Proceedings of the National Academy of Sciences of the United States of America, 110*, 20372–20379.

Postlethwait, J. H., & Schneiderman, H. A. (1971). Pattern formation and determination in the antenna of the homoeotic mutant Antennapedia of Drosophila melanogaster. *Developmental Biology, 25*, 606–640.

Ramel, M., & Hill, C. S. (2013). The ventral to dorsal BMP activity gradient in the early zebrafish embryo is determined by graded expression of BMP ligands. *Developmental Biology, 378*, 170–182.

Ramírez-Weber, F. A., & Kornberg, T. B. (1999). Cytonemes: Cellular processes that project to the principal signaling center in Drosophila imaginal discs. *Cell, 5*, 599–607.

Raspopovic, J., Marcon, L., Russo, L., & Sharpe, J. (2014). Modeling digits. Digit patterning is controlled by a Bmp-Sox9-Wnt Turing network modulated by morphogen gradients. *Science, 345*, 566–570.

Restrepo, S., Zartman, J. J., & Basler, K. (2014). Coordination of patterning and growth by the morphogen DPP. *Current Biology, 24*, R245–R255.

Rogers, K. W., & Schier, A. F. (2011). Morphogen gradients: From generation to interpretation. *Annual Review of Cell and Developmental Biology, 27*, 377–407.

Roy, S., Huang, H., Liu, S., & Kornberg, T. B. (2014). Cytoneme-mediated contact-dependent transport of the Drosophila decapentaplegic signaling protein. *Science, 343*, 1244624.

Sanders, T. A., Llagostera, E., & Barna, M. (2013). Specialized filopodia direct long-range transport of SHH during vertebrate tissue patterning. *Nature, 497*, 628–632.

Sheeba, C. J., Andrade, R. P., & Palmeirim, I. (2014). Limb patterning: From signaling gradients to molecular oscillations. *Journal of Molecular Biology, 426*, 780–784.

Sheth, R., Marcon, L., Bastida, M. F., Junco, M., Quintana, L., Dahn, R., et al. (2012). Hox genes regulate digit patterning by controlling the wavelength of a Turing-type mechanism. *Science, 338*, 1476–1480.

Strutt, D. (2009). Gradients and the specification of planar polarity in the insect cuticle. *Cold Spring Harbor Perspectives in Biology, 1*(5), a000489.

Summerbell, D., Lewis, J. H., & Wolpert, L. (1973). Positional information in chick limb morphogenesis. *Nature, 244*, 492–496.

Tickle, C., Summerbell, D., & Wolpert, L. (1975). Positional signalling and specification of digits in chick limb morphogenesis. *Nature, 254*, 199–202.

Towers, M., Wolpert, L., & Tickle, C. (2012). Gradients of signalling in the developing limb. *Current Opinion in Cell Biology, 24*, 181–187.

Turing, A. M. (1952). The chemical basis of morphogenesis. *Philosophical Transactions of the Royal Society of London, 237*, 37–72.

Wartlick, O., Kicheva, A., & González-Gaitán, M. (2009). Morphogen gradient formation. *Cold Spring Harbor Perspectives in Biology, 1*(3), a001255. http://dx.doi.org/10.1101/cshperspect.a001255.

Wolpert, L. (1969). Positional information and the spatial pattern of cellular differentiation. *Journal of Theoretical Biology, 25*, 1–47.

Wolpert, L. (2011). Positional information and patterning revisited. *Journal of Theoretical Biology, 269*, 359–365.

Wolpert, L., Clarke, M. R. B., & Hornbruch, A. (1972). Positional signalling along hydra. *Nature*, *239*, 101–105.

Wolpert, L., Tickle, C., & Sampford, M. (1979). The effect of cell killing by X-irradiation on pattern formation in the chick limb. *Journal of Embryology and Experimental Morphology*, *50*, 175–193.

Xu, P. F., Houssin, N., Ferri-Lagneau, K. F., Thisse, B., & Thisse, C. (2014). Construction of a vertebrate embryo from two opposing vertebrate gradients. *Science*, *344*, 87–89.

CHAPTER THIRTY-SIX

The Birth of Animal Development: Multicellularity and the Germline

Hugh R. Woodland[1]
School of Life Sciences, University of Warwick, Coventry, United Kingdom
[1]Corresponding author: e-mail address: h.r.woodland@warwick.ac.uk

Contents

Abstract

The evolution of multicellular animals has been attributed to many kinds of selective advantage; here I suggest that the evolution of somatic cells to feed and protect the germline was central to the appearance of animals. This would have been driven by selection for extreme anisogamy—the evolution of sperm and egg. Evidence is adduced from the germline stem cells of simple animals (defining germline as any cell that normally produces the next generation via the sexual process) and from the control circuitry ubiquitous in animal germlines. With the soma and its elaboration came animal development, as we understand it.

1. INTRODUCTION

This volume celebrates a series which has charted the progress of developmental biology over 50 years. Here I wish to discuss how the organisms that are the raw material of the subject, animals, came into existence.

Current Topics in Developmental Biology, Volume 117
ISSN 0070-2153
http://dx.doi.org/10.1016/bs.ctdb.2015.10.020

While single-celled organisms may be very complicated, and therefore show developmental processes, the overwhelming proportion of papers published in *Current Topics* consider the diversification of cells and the formation of patterns in multicellular animals and plants. Development, in the sense usually meant here, is therefore the cyclical appearance of complexity in the life cycles of multicellular organisms. Here I propose to discuss the appearance of multicellularity in animals—the birth of animal development.

Where is the raw evidence? The fossil record does not help, since we are dealing with events deep in the Precambrian and with organisms unlikely to leave fossils, so we must look at extant species. Sponges have the simplest body forms of all known animals, barely at the tissue level of organization, but they must already be more complicated than the very first animals. The same is truer of animals fully at the tissue level of organization, Ctenophora, Cnidaria, and Placozoa (*Trichoplax*). While these animals are relatively simple, their genomes are enormously complicated and include most of the families of genes critical for the development of the most complex animals. On the one hand, it has been questioned whether *Trichoplax*, with only five cell types, has an undiscovered complex life cycle stage (Miller & Ball, 2008; Srivastava et al., 2008), on the other it is questioned whether the complicated developmental toolkits of sponges and cnidarians indicate that they are simplified forms, possibly even with more mobile, bilaterally symmetrical ancestors (Erwin, 2009). Of course, it is possible that it takes more genetic complexity to produce even a simple animal than one might suspect. It is dangerous to draw definite conclusions about the complexity of ancestral animals from current forms. For example, the life style of current Cnidaria is to trap swimming animals that are members of groups that emerged later in evolutionary time, mostly crustaceans or vertebrates. To do this, they evolved the complex stinging cells, cnidoblasts, from which they get their name. This means that the cnidarians that were the ancestors of all Eumetazoa (except maybe ctenophores and placozoans) were likely to have been rather different from modern forms. This again highlights our uncertainty about the morphology of the earliest animals and the extent to which modern descendants represent them. Nevertheless, modern forms may still provide clues to the metazoan ancestors, especially as regards their development.

Multicellularity has evolved several times. A spectacular example is found in the photosynthetic Volvocales, which form a beautiful series from single cells to multicellularity (Herron & Michod, 2008; Kirk, 2005; Nozaki et al., 2000). It is remarkable that so many intermediate levels of complexity still exist, showing that increased multicellularity is not necessarily synonymous with success. However, such organisms may shed limited light on animal evolution because they are phototrophs. It is likely that animals, which actively ingest animal and plant food, would be subject to rather different selective forces. The evolutionary forces working on various phototrophic multicellular algae and their land derivatives, terrestrial plants, are also likely to be different. For this reason, the discussion here is restricted to animals.

Here I argue that an important selection pressure generating multicellular animals was on sexual reproduction, producing a multicellular soma feeding and protecting an anisogametic germline. At these early stages of evolution, the processes by which cells built animal bodies would have been similar in sexual reproduction (embryonic development from the single cell produced by gamete fusion—the zygote), asexual reproduction by budding or fragmentation, and in regeneration from parts of the animal body. All depended on a pluripotent germline stem cell system, as they do in simple animals today.

2. THE ORIGINS OF ANIMALS

Recent phylogenetic analyses have actually placed ctenophores as the sister group to all other animals, even though they are more complex than sponges and Placozoa (Whelan, Kocot, Moroz, & Halanych, 2015; Fig. 1). Clearly, they have pursued their own history of the evolution of complexity—for example, their neurons develop by a different genetic

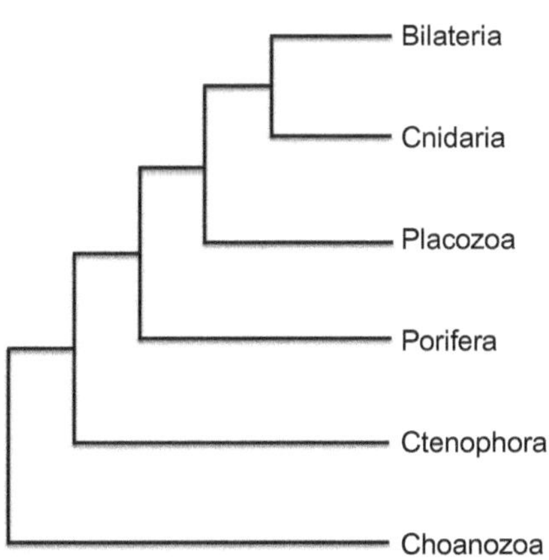

Figure 1 Animal phylogeny, according to a recent molecular analysis (Whelan et al., 2015).

program from those of other animals (Moroz et al., 2014). For this reason, I will restrict my discussion to the sponge/Placozoa/Cnidaria/Bilateria clade of animals. And there is every reason to believe that the first animal was somewhat sponge-like, with choanocytes, reflecting animals' single-celled sister group, the choanoflagellates (King et al., 2008; Ruiz-Trillo, Roger, Burger, Gray, & Lang, 2008). Indeed, typical molecular analyses suggest that sponges are a sister group to the other complex animals (Philippe et al., 2009; Whelan et al., 2015). While Ctenophores seem to be the sister group to all animals, they are so complicated that they shed little light on the common ancestor. Whether the last common ancestor (LCA) looked like a sponge or an even simpler blastula-like larva is unclear, and it is not proposed to discuss this here (see Nielsen, 2008). Rather, we are concerned with the argument that the germline and stem cells of the first animals were crucial in the initial evolution of multicellularity, and the way in which selection operated on these cellular systems. Simpler Metazoa do provide clues about this from common features of their germlines and stem cells.

Various sources of multicellular animals among protistan groups have been proposed historically, but current molecular evidence clearly points to Choanozoa as their closest sister group. How similar the common ancestor was to a modern choanoflagellate, perhaps a colonial form (Nielsen, 2008), is a matter for speculation. Historically, the key morphological clue to the choanozoan origin of animals was that sponges possess choanocytes, which are similar to single-celled choanoflagellates. Choanocytes are cells that move water through sponges and are involved in feeding and intracellular digestion. Choanoflagellates and choanocytes are very similar, with a "collar" (choana-) of microvilli surrounding a single flagellum. Somewhat similar collared cells are found as sensory neurons or protonephridia in other invertebrates, but these are thought to be evolved convergently (Nielsen, 2001). While it is possible that choanocytes and choanoflagellates are also similar because of convergent evolution, the great similarity of these cells most likely indicates a common origin (Maldonado, 2004). Crucially, molecular data strongly support the position of choanoflagellates as the sister group to animals (Cavalier-Smith & Chao, 2003; Dunn et al., 2008; King et al., 2008; Shalchian-Tabrizi et al., 2008). Although there are extant colonial choanoflagellates, none are seriously proposed to be an ancestor of animals. The best known is *Proterospongia*, which in the original species *P. haeckeli* exists as a colony of flagellated cells embedded in a matrix, with the flagellae pointing outward, not inward, as in a sponge. However, there has been little study of this genus. In the matrix are to be found nonflagellated amoeboid cells, which divide and form flagellated cells, so they might be a stem cell population comparable to the archeocytes of sponges. Even more obscure is *Diaphanoeca*, which forms small colonies like a chamber of a sponge, with the choanocytes pointing inward (Nielsen, 2008; Thomsen, 1982). Both the forms mentioned are free swimming, but other colonial choanoflagellates are sessile (e.g., *Sphaeroeca*), and it is possible that the metazoan ancestor was sedentary. These colonial choanoflagellates illustrate the plausibility of the theory that a choanoflagellate-like organism might have made the very first steps to colonial and multicellular animal life.

While our scenario envisages multicellularity arising via coloniality, origins have also been proposed via syncytial forms and symbiosis. The former envisages an ancestry in ciliates, which is now excluded by molecular data, and there is no evidence for the latter. On the other hand, colonial flagellated forms are commonplace. But why would coloniality have evolved and why would one line have progressed to multicellular animals?

3. SEX AND THE ORIGINS OF MULTICELLULARITY

I start from the proposition that animals were ancestrally sexual and arose from sexual protists. This is hardly controversial since the molecular processes of meiosis are so conserved across eukaryotes. It is believed that the earliest protists were sexual (Dacks & Roger, 1999). Until recently, sex was unknown in Choanozoa, but very recently sex by mild anisogamy has been discovered in one choanoflagellate, *Salpingoeca rosetta* (Levin & King, 2013).

Sex in animals invariably involves dimorphic gametes, large immotile eggs, and tiny, motile sperm. Both sorts of germline cells are dependent on nonreproductive sister somatic cells for growth and protection. One might imagine this need not be so in a phototroph, indeed photosynthetic protists like diatoms and dinoflagellates do show anisogamy, or in a parasite bathed in nutrients. But for free-living unicellular forms, the evolution of a large, female gamete would have costs. Movement would be compromised, as would defense and feeding in an animal hunting and ingesting macroscopic food. The generation of macrogametes would be most efficient if they could concentrate solely on multiplication and growth. Thus, the evolution of large macrogametes in free-living forms, feeding on other organisms, would be expected to be accompanied by selection for altruistic sister cells that fed and protected the reproductive macrogamete. If there were continued selection for enlargement of the macrogamete, there would be the essential coevolution of a supporting multicellular soma. In Volvocales, anisogamy increased as colonial complexity increased (Nozaki, Yamada, Takahashi, Matsuzaki, & Nakada, 2014; Randerson & Hurst, 2001), suggesting a causal link.

But why would macrogametes evolve? An obvious reason is to lay down food reserves for inimical conditions, a time often triggering sexual reproduction in organisms that alternate sexual and asexual reproduction. The food reserves both permit a prolonged period of dormancy during the unfavorable period and fuel rapid growth and/or subsequent asexual reproduction when conditions improve. Other reasons for anisogamy have also been proposed (Dacks & Roger, 1999). For example, that it maximizes the number of zygotes (Levitan, 1996; Parker, Smith, & Baker, 1972), it optimizes nuclear/cytoplasmic compatibility (Cosmides & Tooby, 1981), or it reduces parasitism (Hurst, 1990). Whichever factors were involved, the argument for the appearance of trophic sister cells, the nascent soma, would remain valid.

What of the parallel microgametes? The first step in the series described would have been an increase in investment of food reserves in isogametes. The consequences of increased gamete size would be not only decreased efficiency of feeding, but decreased mobility. But gametes need mobility to meet and fuse, so the development of macrogametes produces parallel selection for motile microgametes. This would introduce a rapidly evolving situation, since small microgametes involve little individual investment by the parent; hence, larger numbers can be afforded. In turn, this would introduce intense competition between the microgametes, which we can now call the male line. Sperm competition is well known to produce rapid molecular evolution (Torgerson, Kulathinal, & Singh, 2002; Turner, Chuong, & Hoekstra, 2008). Any feeding adaptations of microgametes would compromise efficiency in fertilization, so inevitably selection would favor diversification into specialized fertilizing sperm and helper cells to feed and protect them, again the soma.

Thus, the somatic cells would feed and protect the larger macrogametes and enable the evolution of the animal egg. This feeding relationship has been shown to be favored in *Volvox* under high nutrient conditions (Koufopanou & Bell, 1993). In animals, arguably the strongest initial selection would be on accessory cells to support the production of microgametes, driven by sperm competition. However, male somatic cells need to be little different from those supporting the female germline; indeed, it is quite likely that the first animals would have been hermaphrodites. In sponges and

Cnidaria, the gonads show little specialization, being basically accumulations of germ cells within spaces in the soma. There are no follicle cells with tight associations with the oocytes. Thus, possessing male and female gametes would need little if any dimorphism in the soma of ancestral metazoans. Meanwhile, there would be strong selection for integration of the soma to form a well-organized colony, rapidly moving to the multicellular organism grade. It is easy to see how delineation of a soma would permit its further diversification accompanied by more advanced cell adhesion, metabolic specialization and so on, that is true multicellularity.

A well-established explanation for the origins of multicellularity is that there was an advantage in the specialization of cells in a colony for metabolic or structural functions. This would be comparable to the forces producing diversification of individuals within a colonial cnidarian, like the Portuguese Man of War. What is proposed here is basically a variant of this idea, suggesting the very first specialization was into reproductive and somatic cells. Once the soma was relieved of its sexual reproductive function, it is likely that the ease with which further cell diversity could be produced would be increased. Moreover germlines are by definition immortal, in contrast to known somas (see below). This is presumably because immortality is expensive and would detract from the reproductive capacity of the germline in the soma combination. Thus, it would rapidly be lost from nonreproductive somatic cell lineages and the colony/multicellular organism would reproduce sexually more efficiently than a wholly immortal equivalent.

4. WHAT IS THE GERMLINE?

At this point, it is useful to consider exactly what is meant by the "germline." Recent definitions often adopt a stem cell perspective, defining it as germ cells and stem cells that can only form germ cells. This gives rise to the conclusion that the germline is discontinuous in many animals, such as mammals (Strome & Updike, 2015). Solana has discussed shortcomings of this idea when applied to simple animals (Solana, 2013). Mostly, the definition is perfectly workable when applied to taxa like Vertebrata, but even in these there may be issues. In *Xenopus*, transplanted migratory PGCs can differentiate into endoderm (Wylie, Holwill, Odriscoll, Snape, & Heasman, 1985) and cultured male mouse germline stem cells can form ES-like cells, which in turn can form any kind of somatic cell (Guan et al., 2006; Kanatsu-Shinohara et al., 2004). In addition, migratory derivatives of *Drosophila* pole cells normally form mid-gut cuprophilic cells (Poulson & Waterhouse, 1960). Most people would want to call these cells "germline." A way out of the dilemma is to suppose that PGCs are so poised that even brief culture can reprogram them. Thus in *Xenopus*, naturally mislocalized ectopic PGCs may persist as such in embryos (Ikenishi, Ohno, & Komiya, 2007) and mouse PGCs may not be reprogrammable by return to a normal embryo (Leitch et al., 2014). So without intervention (or mutation, as in germline cancers), gonad germline cells may or may not be able to make soma. A practical problem with this definition is that it is based on a negative assertion, so can never be definitively applied.

The original nineteenth century idea of the germline, initiated by Weismann (1892), is evolutionary in concept and based on two ideas. One is the cell theory, which asserted that cells only arise from other cells, so a line exists from every cell back to the original cell nearly 4 billion years ago. This is the germline (Keimbahn). In each generation, the soma is a dead end; the germline is potentially immortal. The other part of the concept was the wish to exclude Lamarckian inheritance by proposing the "Weismann barrier." The germline represents the only line of genetic inheritance; there is no route for mutations to cross from the soma to the germline. This definition was recently used by Lesch and Page (2012), for example. A modern interpretation of this idea is thoroughly discussed by Solana (2013). He calls any cells that naturally may form germ cells germline, and when they usually form either germ cells and/or soma they are germline primordial stem cells (PriSCs). These include the kind of pluripotent stem cells found in simple animals (see below), but the cells of the mouse inner cell mass would also be of this kind; even though most will form soma, any could form germ cells (so far as we know). Surprisingly, the PriSCs in simple animals that reproduce only asexually are called asexual germline, since they form any somatic tissue and generate the next generation by fission, budding, etc. Thus, they lie within the Weismann barrier—they are the only route for genetic change (or indeed stasis) to transmit to the next generation.

I will adopt this definition of the sexual germline—cells that can, under natural circumstances, form germ cells, and which are part of a potentially immortal lineage.

5. THE GERMLINE–SOMA IN SIMPLE ANIMALS

What is the germline in so-called basal animals, here called simple animals, and does it give any evidence for the proposal that the germline/soma split was involved in the evolution of multicellularity? The undesirability of calling any animals basal has recently been emphasized (Omland, Cook, & Crisp, 2008). All bifurcations in phylogenetic trees are symmetrical, but one branch is usually called basal (or early branching) because it does not lead to humans, or it has morphologically simpler animals, or it is species poor. Logically, there is no reason to call either branch basal, but more importantly there is no reason to believe *a priori* that the "basal" clade is more representative of the LCA of any two branches. Rather, characters must be looked at one by one, and their distribution in many branches of a clade examined. Where characters are widely spread among descendent clades, there is an increased probability that they were present in the most recent common ancestor. In this way, the likely arrangement of germline and soma in ancestral animals can be predicted.

The typical, much studied, arrangement of the germline and soma in simple animals is seen in platyhelminths like planaria. A pluripotent cell population, which generates soma and germline, is distributed through most of the adult body. In platyhelminths, these stem cells are called neoblasts. Conventionally, they are regarded as *somatic* cells which generate more soma, but from which the germline is also continuously derived throughout adult life. Thus, conventionally, the germline is discontinuous in these animals. But according to the definition used here, they are true germline cells. However, beyond our functional definition, which might be regarded as a semantic issue, animal germlines have molecular and structural fingerprints during much of their existence. One is the expression of a circuitry of characteristic genes, such as *piwi*, *vasa*, and *nanos*. Another is the presence of massive cytoplasmic structures concerned with storing mRNAs and associated with mitochondria (Balbiani bodies, chromatoid bodies, germ plasm; Fig. 2). Platyhelminth neoblasts express *vasa* and *nanos* and have typical chromatoid bodies. Thus, platyhelminth neoblasts are a disseminated population of germline stem cells, PriSCs according to Solana's nomenclature. They continuously generate new soma and germline, thus enabling asexual reproduction by division of the body and regeneration by fragments. Remarkably, in planaria, they are the only dividing cells, apart from later members of the germline itself. Moreover, and importantly, because the germline is immortal, barring accidents or deleterious mutation, these whole animals are also potentially immortal, the adult being continuously renewed from neoblasts (Galloni, 2012; Salo, 2006).

This situation is found throughout simple animals, for example, in annelids and Cnidaria, where the germline PriSCs are called interstitial cells. (Actually, within Cnidaria, interstitial cells are known only in Hydrozoa

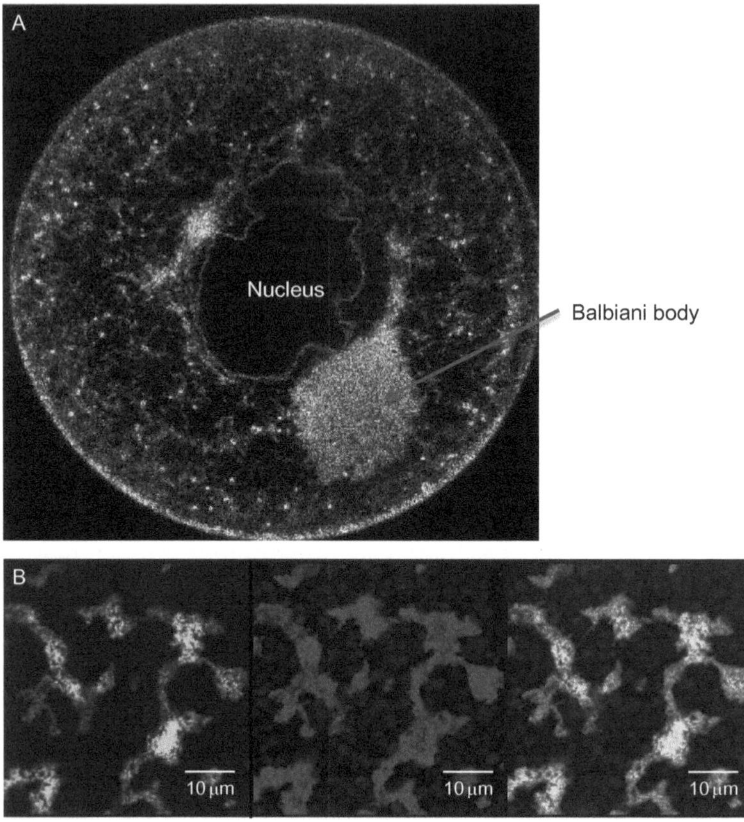

Figure 2 Balbiani bodies and germ plasm. (A) A previtellogenic *Xenopus* oocyte. The optical section from confocal imaging shows staining with antibodies to the Balbiani body/germ plasm protein Xpat (Machado et al., 2005). This forms particles distinct from RNPs. (B) Germ plasm in a living full-grown *Xenopus* oocyte showing: left, RNP particles, visualized with GFP-Xvelo fusion protein; center, mitochondria, visualized with TMRE; right, overlay (Nijjar & Woodland, 2013). In *Xenopus*, the germ plasm is derived directly from the Balbiani body, from which its mitochondria are derived. (A color version of this figure is located at the end of this chapter.)

and the stem cell situation in other cnidarians is uncertain (Gold & Jacobs, 2013). There is no way to know what the situation was in Pre-Cambrian Cnidaria. It seems reasonable to guess that there were stem cells with some attributes of interstitial cells, like chromatoid bodies.) This supports the view that it was the ancestral animal condition, and further strong support comes from observations that it applies to Acoela (Gschwentner, Ladurner, Nimeth, & Rieger, 2001). While acoel flatworms were long regarded as simple, even simplified platyhelminths, evidence is growing that they are the most similar known forms to the ancestral bilaterian, and constitute a sister group to all other Bilateria, although this remains controversial (Baguna, Martinez, Paps, & Riutort, 2008; Baguna & Riutort, 2004; Cook, Jimenez, Akam, & Salo, 2004; Egger et al., 2009; Giribet, Dunn, Edgecombe, & Rouse, 2007; Ruiz-Trillo, Riutort, Fourcade, Baguna, & Boore, 2004; Sempere, Martinez, Cole, Baguna, & Peterson, 2007).

In the more poorly studied sponges, germ cells can be derived from archeocytes, a totipotent stem cell-like population within the matrix of sponges, and also from differentiated choanocytes (Agata et al., 2006; Muller, 2006). Oocytes mostly arise from archeocytes and sperm from choanocytes, but archeocytes can also form choanocytes, and choanocytes may form archeocytes (Funayama, 2013). Thus, archeocytes and choanocytes would seem to be germline stem cells analogous to neoblasts and interstitial cells. Asexual reproduction via gemmules also depends on archeocytes (Agata et al., 2006; Muller, 2006), and gemmules produce sexually competent individuals from fragments of the body lacking gonads, so they must have germline stem cells. Archeocytes have the molecular characteristics of germline stem cells, for example, expressing *nanos*- and *vasa*-like genes (mentioned in Funayama, 2013), and Piwi RNAs are expressed in archeocytes and choanocytes (Funayama, Nakatsukasa, Mohri, Masuda, & Agata, 2010). They also contain perinuclear, granular, Balbiani/chromatoid structures, similar to those in the germline of other animals (Isaeva, Akhmadieva, Aleksandriova, & Shukaliuk, 2009).

The ubiquity of neoblast/interstitial cell/archeocyte PriSCs in simple animals strongly suggests that the LCA of all Metazoa had a similar, dispersed pluripotent stem cell lineage, containing germ plasm/Balbiani body/chromatoid body granular structures. Given that choanocytes are joint germline stem cells in sponges, it is hard not to link this directly with the putative choanoflagellate ancestor. So perhaps we see a relic of the flagellate ancestor that produced somatic sisters to aid its reproduction; rather like the queen bee producing sterile daughters to nurture its germline-bearing offspring.

6. WHY DID ANIMALS LOSE THEIR UBIQUITOUS NEOBLAST-LIKE CELLS?

The next two sections are something of an aside from the argument developed so far, but expand on several issues raised above. First, from the point of view of survival, neoblast-type PriSCs seem like a good adaptation! They enable regeneration in the face even of enormous damage and fuel asexual reproduction that is much more efficient in the short term than the sexual process. So why were they lost in more complex animals?

Complex animals develop from a zygote via a highly stereotyped program of embryonic development, controlled by complex regulatory circuits, often involving inputs from stored maternal RNAs or proteins. Simple animals may also reproduce by fragmentation or budding, and may regenerate new organs or regenerate entirely from tiny fragments. Morgan found that a planarian can regenerate the complete animal from as little as a 1/267th of its body (Morgan, 1898). The evolution of complexity largely involved the evolution of control circuits—genetic and signaling. It was control complexity, rather than the toolkit, which evolved during animal evolution, since the basic controlling molecular families were already present in animals like cnidarians (Erwin, 2009; Rokas, 2008). It is relatively easy to imagine how mutation and selection could change these circuits to generate more and more complexity. It is much harder to see how these changes could have simultaneously applied to the contexts of embryonic development, with maternal inputs, asexual reproduction, and regeneration, particularly since the last may have many different starting points. So in clades of animals where increased complexity was favored, it is likely that regeneration and asexual reproduction would have been lost. With it went somatic immortality, but efficiency of reproduction in a purely sexual animal must have outweighed this.

7. HOW DO GERMLINES ACHIEVE IMMORTALITY?

Are simple animals really potentially immortal? Proving this requires rather long-term experiments! A study of *Hydra vulgaris* showed no evidence of aging over 4 years; rates of rapid cell proliferation were maintained, and it was argued that this period of time was sufficient to detect aging in such a small animal if it showed normal mortality (Martinez, 1998). Another study did show "degeneration" in *Hydra oligactis* when it reproduced sexually at high rates (Yoshida, Fujisawa, Hwang, Ikeo, & Gojobori, 2006), but it has been argued that this is a consequence of the load associated with high rates of gamete production, and it is not seen in other species (Bosch, 2009). Indeed, if animals in this group are maintained in the absence of sexual reproduction, aging is absent (discussed by Martinez and Bridge, 2012). Life times of thousands of years have been estimated in sponges (McMurray, Blum, & Pawlik, 2008), even more than 10,000 years (Gatti, 2002; Jochum, Wang, Vennemann, Sinha, & Mueller, 2012). They continually replace their somas from archeocytes, even at the rate of two thirds of their cellular compliment per day (De Goeij et al., 2009). Thus, barring accidents, many "basal" animals are not subject to aging (Bosch, 2009; Salo, 2006). How germline stem cells prevent aging, beyond retaining telomerase, is unknown.

There is another issue here relating to *Hydra*. It is an unusual cnidarian containing three independent stem cell compartments, endoderm, ectoderm, and interstitial cells (which form all other cells and the germline); these have never been observed to interconvert (Hobmayer et al., 2012). So, if *Hydra* is truly immortal, the two epithelial cell lineages must also be potentially immortal. *vasa* has been shown to be expressed in the ectodermal lineage (Mochizuki, Nishimiya-Fujisawa, & Fujisawa, 2001) and Piwi proteins are found in perinuclear granules in both epithelia (Juliano et al., 2014; Lim, Anand, Nishimiya-Fujisawa, Kobayashi, & Kai, 2014). This suggests that chromatoid bodies may exist in the ectoderm and endoderm, and therefore could protect against aging.

It is outside the scope of this essay to discuss all the causes of cellular aging and hence its prevention (Petralia, Mattson, & Yao, 2014), but one aspect is potentially relevant to our discussion so far—mitochondrial aging. Since mitochondria have no sex, how do they evade progressive degradation by Müller's ratchet? Maybe the Balbiani and chromatoid bodies are relevant. These structures are a ubiquitous feature of the germline, except for short periods in the PriSCs of some early developments, so what are they for?

A major role of Balbiani and chromatoid bodies is undoubtedly in storing mRNAs. Most animals have oocytes that lay down large stores of RNAs for development, because in gametogenesis and early development there are periods when transcriptional control is impractical; in spermatogenesis because of nuclear condensation; and in oogenesis and embryonic cleavage because of the unfavorable nuclear cytoplasmic ratio and early transcriptional inactivity. In these cases, Balbiani bodies may contain a proportion of the RNAs that later become localized, predominantly in germ plasm and hence later in PGCs. Balbiani and chromatoid bodies typically also contain a number of molecules concerned with splicing and RNA regulation (Kloc, Bilinski, & Etkin, 2004; Kloc & Etkin, 2005). Such observations have led to the conclusion that the prime role of both bodies is in regulating RNA metabolism. Since the chromatoid bodies of the neoblast-type cells of basal animals contain stored RNAs, this role may have been important in the first Balbiani and chromatoid bodies. Because they share roles in control of RNA metabolism and hence some relevant molecular components, Balbiani/chromatoid bodies may have originally evolved from P-bodies (Gallo, Munro, Rasoloson, Merritt, & Seydoux, 2008; Kotaja et al., 2006; Standart & Minshall, 2008).

Why do flatworm neoblasts have a chromatoid body if it is concerned with storing RNAs? Planaria can undergo one round of regeneration when transcription is blocked with actinomycin D, which might mean that the chromatoid bodies store enough mRNA for this regeneration (Coward, 1968; Gabriel & Lemoigne, 1971). Perhaps, the chromatoid bodies are important in allowing regeneration under starvation conditions. This may have been important in their origin, but is it important enough for chromatoid bodies to have been retained for over half a billion years?

If the function of the Balbiani body is to store RNA, how can it be that the Balbiani body of the mouse does not do so (Pepling, 2010; Pepling, Wilhelm, O'Hara, Gephardt, & Spradling, 2007)? Similarly, the Balbiani body is transitory in *Drosophila* oogenesis and has no role in storing maternal RNAs (Cox & Spradling, 2003). This strongly suggests that translational control and mRNA storage are not its only functions. Since it is specific to all germlines, the only immortal cell lineage, could there be a role in immortality? Balbiani, chromatoid bodies, and germ plasm contain concentrations of mitochondria (Fig. 2). Cox and Spradling (2003) and Pepling et al. (2007) have suggested that the Balbiani body selects healthy mitochondria and were able to test this in *Drosophila*. Here, the germline acquires mitochondria in two phases, initially via the Balbiani body and later at the time of nurse cell dumping. The first phase is dependent on the protein Milton which permitted Cox and Spradling (2006) to disrupt this phase in *milton* mutants. Viable, fertile flies resulted, suggesting that, if mitochondrial selection occurs, it is not essential. However, a reduction in fitness and a long-term effect on aging cannot be excluded. It is intriguing that Balbiani body mitochondria are different from those generally dispersed in oocyte cytoplasm of both *Drosophila* and *Xenopus* (Cox & Spradling, 2003; Kogo et al., 2011; Mignotte, Tourte, & Mounolou, 1987). In the fly, they differ in their reactivity to human cytochrome c_1 oxidase antibodies and in *Xenopus*, they have reduced cytochrome oxidase activity and replication. It is clear that mitochondria are selected during oogenesis (Hill, Chen, & Xu, 2014), and evidence that Balbiani bodies have a role in maintaining mitochondrial quality has recently been reviewed by Zhou, Wang, Wang, Schatten, and Zhang (2010). Pathways of mitochondrial surveillance are known, so it would be worth looking for these in the Balbiani body. If this turns out to be true, we may have discovered a key property of the LCA of all animals, related to immortality.

8. PREVIOUS EXPLANATIONS OF MULTICELLULARITY

There have been several hypotheses to explain how multicellularity arose (Rokas, 2008), and it has to be said from the outset that they need not be mutually exclusive. Cell specialization (division of labor) may have been advantageous, just like specialized individuals in a colonial animal. Another attractive idea, because it is supported by experimental evidence, might be described as the big mouthful hypothesis, which proposes that in a world of unicellular predators it was of advantage for prey to become multicellular and hence larger. Boraas, Seale, and Boxhorn (1998) found that culturing single-celled *Chlorella* with the predator *Ochromonas* led to the appearance of colonies with as many as eight cells within 10–20 generations and they were the predominant form of the prey within 100 generations. It is quite conceivable that such a process was the kind of selection that favored the very first simple, possibly colonial forms, but less likely that it produced complex integrated animal structures. It certainly could have been an adjunct to the mechanism suggested here.

Another idea is that multicellularity provided food for hard times (Kerszberg & Wolpert, 1998). When food was short, some cells could be used as a reserve for the others. Alternatively, the ancestor is envisaged to be a flagellate in which there is a conflict between motility and cell division. This is because the centriole organizes the flagellum and division spindle, and cannot do both at the same time. This scenario envisages the ancestral animal as a planktonic form. This is plausible, but by no means certain; for example, there are more sessile, colonial choanoflagellates than free-swimming forms. However, the absence of a flagellum would also compromise feeding.

The important point is that all these mechanisms have plausible advantages and therefore may have all played a role in the evolution of multicellularity. However, it is argued here that the fundamental imperative of reproduction, which was partly or wholly sexual, means that the germline/soma evolution would have been particularly important in the first steps to multicellularity.

Wolpert and Szathmary (2002) have argued that an essential early step in animal evolution was the evolution of the egg. It was argued that a multicellular organism which reproduced asexually by subdivision of the animal, or budding would suffer evolutionary inertia because any inherited morphological change would require the simultaneous mutation of many cells. It is hard to argue with this logic, at least in the context of an asexually reproducing ancestor, but the animal ancestor must also have reproduced sexually, since this character goes back to early eukaryotes. Thus, before multicellularity the ancestral life cycle would certainly have involved germ cells and a zygote, i.e., the egg preceded multicellularity. Of course, strictly speaking the egg is a macrogamete, but in the only known sexual choanoflagellate, *S. rosetta*, there is already anisogamy, as well as a multicellular/colonial stage (Levin & King, 2013).

9. PROTECTING THE GERMLINE

The only thing that matters for an organism, from an evolutionary perspective, is successful reproduction. For well-discussed, but still controversial reasons, sex is an essential attribute of eukaryotes, and when sex is an obligatory part of the life cycle, it is essential that germline cells must not irreversibly become somatic. This is achieved by a battery of repressive circuitry, operating at different developmental stages (Solana, 2013; Strome & Updike, 2015). One ubiquitous system includes *vasa* and *nanos*. These are expressed in all developing germ cells and they are constituents of germ plasm in animals that predetermine their germline by cytoplasmic localization in the egg. In vertebrates which have it, the germ plasm derives from the oocyte Balbiani body (Wilk, Bilinski, Dougherty, & Kloc, 2005).

I propose here that the first appearance of multicellular animals involved the evolution of somatic sister cells to feed and protect the germline. This would entail producing positive genetic circuitry to differentiate these cells from the reproductive line. Equally, there would need to be the evolution of circuitry to protect the germline from somatic differentiation, which would be a dead end. One could imagine that the preexistence of this circuitry would preadapt the unicellular organism to develop the positive circuits promoting somatic development. Initially, it could have protected against mutation and transposable elements (which strongly promote mutation), and it is notable that germ plasm repressive circuitry has this function (discussed in Seydoux & Schedl, 2001; Solana, 2013).

The repressive circuits are highly conserved, but known positive mechanisms directing germline development are not. Even in recent evolutionary times, mice and humans have co-opted different essential transcription factors to promote germ cell differentiation, Brachyury in the former, Sox17 in the latter (Aramaki et al., 2013; Tang et al., 2015). The conservation of repressive circuits emphasizes their particular importance. So the ideas expressed here would be supported if precursors could be found in Choanozoa.

10. CONCLUSIONS

The "smoking gun" for the hypothesis presented here would come from the discovery of an extremely simple organism close in form and ancestry to the first animal. Clearly, there is little prospect of this happening. The only alternative is to look at the common characters of the simplest extant animals. We have argued that, since all have similar totipotent germline stem cells (PriSCs) that continually generate somatic cells and germline, this must have been a feature of their LCA, in which the germline

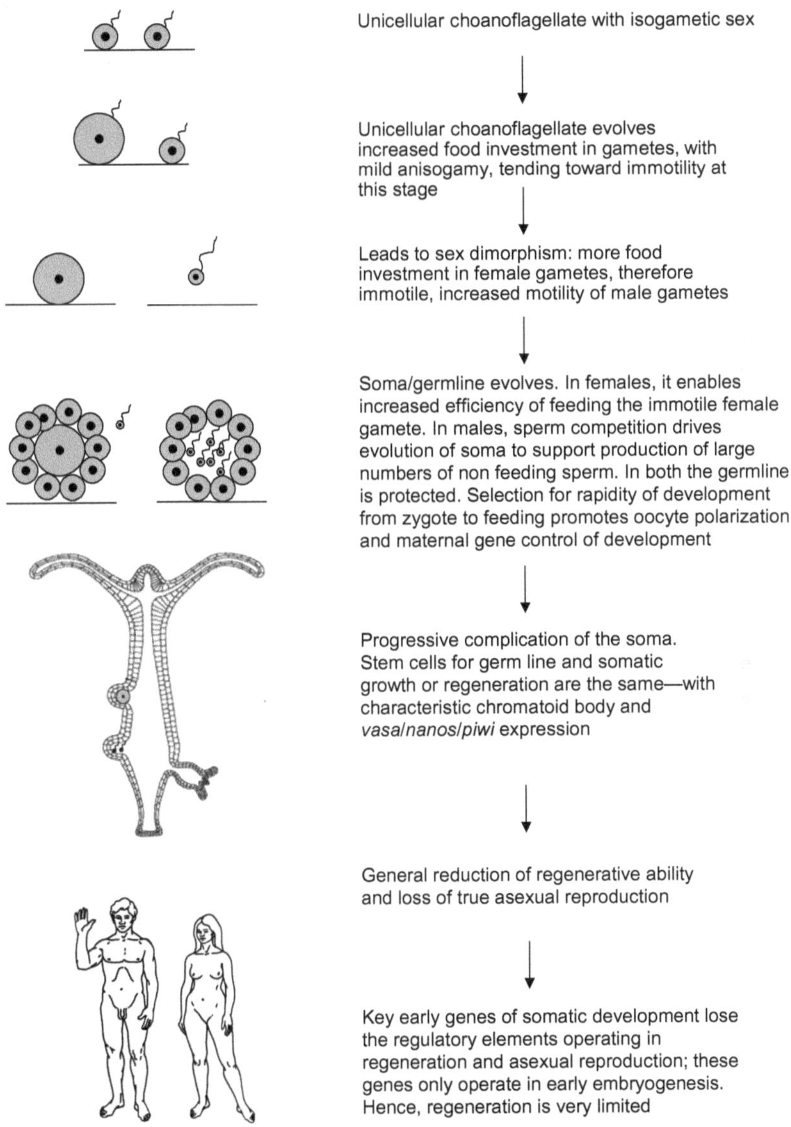

Figure 3 A scheme for the evolution of the animal germline from unicellular ancestors to complex animals. (Within the term "asexual reproduction," we exclude parthenogenetic processes in which eggs are produced via meiosis, but do not need fertilization.)

generated the soma for its own protection and propagation. Further support for this view might arise from increased understanding of the germline of sponges, where a choanoflagellate-like cell acts as germline. There might be other insights from further study of choanoflagellates, in particular colonial forms.

It is not surprising that this essay has been very speculative (Fig. 3). Analyzing the relative importance of causation chains so long in the past is obviously problematic. One is reminded that economists still argue about the causation of the Great Depression. There is no doubt that deflation, bank failure, money supply contraction, and so on were involved, but their primacy is very controversial. If there can be argument about such recent events, it is not surprising that we can only speculate about events over half a billion years ago.

ACKNOWLEDGMENTS

I am grateful for very helpful comments from John Gerhart on an early precursor of this chapter, and to Karuna Sampath for comments on the current version.

REFERENCES

Agata, K., Nakajima, E., Funayama, N., Shibata, N., Saito, Y., & Umesono, Y. (2006). Two different evolutionary origins of stem cell systems and their molecular basis. *Seminars in Cell & Developmental Biology*, *17*, 503–509.

Aramaki, S., Hayashi, K., Kurimoto, K., Ohta, H., Yabuta, Y., Iwanari, H., et al. (2013). A mesodermal factor, T, specifies mouse germ cell fate by directly activating germline determinants. *Developmental Cell*, *27*, 516–529.

Baguna, J., Martinez, P., Paps, J., & Riutort, M. (2008). Back in time: A new systematic proposal for the Bilateria. *Philosophical Transactions of the Royal Society of London. Series B: Biological Sciences*, *363*, 1481–1491.

Baguna, J., & Riutort, M. (2004). The dawn of bilaterian animals: The case of acoelomorph flatworms. *Bioessays*, *26*, 1046–1057.

Boraas, M. E., Seale, D. B., & Boxhorn, J. E. (1998). Phagotrophy by a flagellate selects for colonial prey: A possible origin of multicellularity. *Evolutionary Ecology*, *12*, 153–164.

Bosch, T. C. G. (2009). Hydra and the evolution of stem cells. *Bioessays*, *31*, 478–486.

Cavalier-Smith, T., & Chao, E. E. (2003). Phylogeny of choanozoa, apusozoa, and other protozoa and early eukaryote megaevolution. *Journal of Molecular Evolution*, *56*, 540–563.

Cook, C. E., Jimenez, E., Akam, M., & Salo, E. (2004). The Hox gene complement of acoel flatworms, a basal bilaterian clade. *Evolution & Development*, *6*, 154–163.

Cosmides, L. M., & Tooby, J. (1981). Cytoplasmic inheritance and intragenomic conflict. *Journal of Theoretical Biology*, *89*, 83–129.

Coward, S. J. (1968). Effects of actinomycin D on regeneration give evidence of sequential gene activation. *Nature*, *219*, 1257–1258.

Cox, R. T., & Spradling, A. C. (2003). A Balbiani body and the fusome mediate mitochondrial inheritance during Drosophila oogenesis. *Development*, *130*, 1579–1590.

Cox, R. T., & Spradling, A. C. (2006). Milton controls the early acquisition of mitochondria by Drosophila oocytes. *Development, 133*, 3371–3377.
Dacks, J., & Roger, A. J. (1999). The first sexual lineage and the relevance of facultative sex. *Journal of Molecular Evolution, 48*, 779–783.
De Goeij, J. M., De Kluijver, A., Van Duyl, F. C., Vacelet, J., Wijffels, R. H., De Goeij, A., et al. (2009). Cell kinetics of the marine sponge Halisarca caerulea reveal rapid cell turnover and shedding. *Journal of Experimental Biology, 212*, 3892–3900.
Dunn, C. W., Hejnol, A., Matus, D. Q., Pang, K., Browne, W. E., Smith, S. A., et al. (2008). Broad phylogenomic sampling improves resolution of the animal tree of life. *Nature, 452*, 745–749.
Egger, B., Steinke, D., Tarui, H., De Mulder, K., Arendt, D., Borgonie, G., et al. (2009). To be or not to be a flatworm: The acoel controversy. *PLoS One, 4*, e5502.
Erwin, D. H. (2009). Early origin of the bilaterian developmental toolkit. *Philosophical Transactions of the Royal Society of London. Series B: Biological Sciences, 364*, 2253–2261.
Funayama, N. (2013). The stem cell system in demosponges: Suggested involvement of two types of cells: Archeocytes (active stem cells) and choanocytes (food-entrapping flagellated cells). *Development Genes and Evolution, 223*, 23–38.
Funayama, N., Nakatsukasa, M., Mohri, K., Masuda, Y., & Agata, K. (2010). *Piwi* expression in archeocytes and choanocytes in demosponges: Insights into the stem cell system in demosponges. *Evolution & Development, 12*, 275–287.
Gabriel, A., & Lemoigne, A. (1971). Action of actinomycin-D on cellular differentiation during regeneration in planarians. 1. Morphological, histological and ultrastructural studies on regeneration capacity in presence of antibiotic. *Zeitschrift Fur Zellforschung Und Mikroskopische Anatomie, 115*, 426–441.
Gallo, C. M., Munro, E., Rasoloson, D., Merritt, C., & Seydoux, G. (2008). Processing bodies and germ granules are distinct RNA granules that interact in *C. elegans* embryos. *Developmental Biology, 323*, 76–87.
Galloni, M. (2012). Global irradiation effects, stem cell genes and rare transcripts in the planarian transcriptome. *The International Journal of Developmental Biology, 56*, 103–116.
Gatti, S. (2002). The role of sponges in high-Antarctic carbon and silicon cycling—A modelling approach. *Berichte zur Polar- und Meeresforschung, 434*, 1–124.
Giribet, G., Dunn, C. W., Edgecombe, G. D., & Rouse, G. W. (2007). A modern look at the animal tree of life. *Zootaxa, 1668*, 61–79.
Gold, D. A., & Jacobs, D. K. (2013). Stem cell dynamics in Cnidaria: are there unifying principles? *Development Genes and Evolution, 223*, 53–66.
Gschwentner, R., Ladurner, P., Nimeth, K., & Rieger, R. (2001). Stem cells in a basal bilaterian—S-phase and mitotic cells in *Convolutriloba longifissura* (Acoela, Platyhelminthes). *Cell and Tissue Research, 304*, 401–408.
Guan, K., Nayernia, K., Maier, L. S., Wagner, S., Dressel, R., Lee, J. H., et al. (2006). Pluripotency of spermatogonial stem cells from adult mouse testis. *Nature, 440*, 1199–1203.
Herron, M. D., & Michod, R. E. (2008). Evolution of complexity in the volvocine algae: Transitions in individuality through Darwin's eye. *Evolution, 62*, 436–451.
Hill, J. H., Chen, Z., & Xu, H. (2014). Selective propagation of functional mitochondrial DNA during oogenesis restricts the transmission of a deleterious mitochondrial variant. *Nature Genetics, 46*, 389–392.
Hobmayer, B., Jenewein, M., Eder, D., Eder, M. K., Glasauer, S., Gufler, S., et al. (2012). Stemness in Hydra—A current perspective. *The International Journal of Developmental Biology, 56*, 509–517.
Hurst, L. D. (1990). Parasite diversity and the evolution of diploidy, multicellularity and anisogamy. *Journal of Theoretical Biology, 144*, 429–443.
Ikenishi, K., Ohno, T., & Komiya, T. (2007). Ectopic germline cells in embryos of *Xenopus laevis*. *Development, Growth & Differentiation, 49*, 561–570.

Isaeva, V. V., Akhmadieva, A. V., Aleksandriova, Y. N., & Shukaliuk, A. I. (2009). Morphofunctional organization of reserve stem cells providing for asexual and sexual reproduction of invertebrates. *Russian Journal of Developmental Biology, 40*, 57–68.

Jochum, K. P., Wang, X., Vennemann, T. W., Sinha, B., & Mueller, W. E. G. (2012). Siliceous deep-sea sponge *Monorhaphis chuni*: A potential paleoclimate archive in ancient animals. *Chemical Geology, 300*, 143–151.

Juliano, C. E., Reich, A., Liu, N., Gotzfried, J., Zhong, M., Uman, S., et al. (2014). PIWI proteins and PIWI-interacting RNAs function in Hydra somatic stem cells. *Proceedings of the National Academy of Sciences of the United States of America, 111*, 337–342.

Kanatsu-Shinohara, M., Inoue, K., Lee, J., Yoshimoto, M., Ogonuki, N., Miki, H., et al. (2004). Generation of pluripotent stem cells from neonatal mouse testis. *Cell, 119*, 1001–1012.

Kerszberg, M., & Wolpert, L. (1998). The origin of metazoa and the egg: A role for cell death. *Journal of Theoretical Biology, 193*, 535–537.

King, N., Westbrook, M. J., Young, S. L., Kuo, A., Abedin, M., Chapman, J., et al. (2008). The genome of the choanoflagellate Monosiga brevicollis and the origin of metazoans. *Nature, 451*, 783–788.

Kirk, D. L. (2005). A twelve-step program for evolving multicellularity and a division of labor. *Bioessays, 27*, 299–310.

Kloc, M., Bilinski, S., & Etkin, L. D. (2004). The Balbiani body and germ cell determinants: 150 years later. In G. P. Schatten (Ed.), *Current topics in developmental biology: Vol. 59* (pp. 1–36). San Diego: Elsevier Inc.

Kloc, M., & Etkin, L. D. (2005). RNA localization mechanisms in oocytes. *Journal of Cell Science, 118*, 269–282.

Kogo, N., Tazaki, A., Kashino, Y., Morichika, K., Orii, H., Mochii, M., et al. (2011). Germline mitochondria exhibit suppressed respiratory activity to support their accurate transmission to the next generation. *Developmental Biology, 349*, 462–469.

Kotaja, N., Bhattacharyya, S. N., Jaskiewicz, L., Kimmins, S., Parvinen, M., Filipowicz, W., et al. (2006). The chromatoid body of male germ cells: Similarity with processing bodies and presence of Dicer and microRNA pathway components. *Proceedings of the National Academy of Sciences of the United States of America, 103*, 2647–2652.

Koufopanou, V., & Bell, G. (1993). Soma and germ—An experimental approach using *Volvox*. *Proceedings of the Royal Society of London. Series B: Biological Sciences, 254*, 107–113.

Leitch, H. G., Okamura, D., Durcova-Hills, G., Stewart, C. L., Gardner, R. L., Matsui, Y., et al. (2014). On the fate of primordial germ cells injected into early mouse embryos. *Developmental Biology, 385*, 155–159.

Lesch, B. J., & Page, D. C. (2012). Genetics of germ cell development. *Nature Reviews. Genetics, 13*, 781–794.

Levin, T. C., & King, N. (2013). Evidence for sex and recombination in the choanoflagellate *Salpingoeca rosetta*. *Current Biology, 23*, 2176–2180.

Levitan, D. R. (1996). Effects of gamete traits on fertilization in the sea and the evolution of sexual dimorphism. *Nature, 382*, 153–155.

Lim, R. S., Anand, A., Nishimiya-Fujisawa, C., Kobayashi, S., & Kai, T. (2014). Analysis of Hydra PIWI proteins and piRNAs uncover early evolutionary origins of the piRNA pathway. *Developmental Biology, 386*, 237–251.

Machado, R. J., Moore, W., Hames, R., Houliston, E., Chang, P., King, M. L., et al. (2005). *Xenopus* Xpat protein is a major component of germ plasm and may function in its organisation and positioning. *Developmental Biology, 287*, 289–300.

Maldonado, M. (2004). Choanoflagellates, choanocytes, and animal multicellularity. *Invertebrate Biology, 123*, 1–22.

Martinez, D. E. (1998). Mortality patterns suggest lack of senescence in hydra. *Experimental Gerontology, 33*, 217–225.

Martinez, D. E., & Bridge, D. (2012). Hydra, the everlasting embryo, confronts aging. *The International Journal of Developmental Biology, 56*, 479–487.

McMurray, S. E., Blum, J. E., & Pawlik, J. R. (2008). Redwood of the reef: Growth and age of the giant barrel sponge *Xestospongia muta* in the Florida Keys. *Marine Biology, 155*, 159–171.

Mignotte, F., Tourte, M., & Mounolou, J. C. (1987). Segregation of mitochondria in the cytoplasm of *Xenopus vitellogenic* oocytes. *Biology of the Cell, 60*, 97–102.

Miller, D. J., & Ball, E. E. (2008). Animal evolution: Trichoplax, trees, and taxonomic turmoil. *Current Biology, 18*, R1003–R1005.

Mochizuki, K., Nishimiya-Fujisawa, C., & Fujisawa, T. (2001). Universal occurrence of the vasa-related genes among metazoans and their germline expression in Hydra. *Development Genes and Evolution, 211*, 299–308.

Morgan, T. H. (1898). Experimental studies of the regeneration of *Planaria maculata. Archiv für Entwicklungsmechanik der Organismen, 7*, 364–397.

Moroz, L. L., Kocot, K. M., Citarella, M. R., Dosung, S., Norekian, T. P., Povolotskaya, I. S., et al. (2014). The ctenophore genome and the evolutionary origins of neural systems. *Nature, 510*, 109–114.

Muller, W. E. G. (2006). The stem cell concept in sponges (Porifera): Metazoan traits. *Seminars in Cell & Developmental Biology, 17*, 481–491.

Nielsen, C. (2001). *Animal evolution: Interrelationships of the living phyla.* Oxford, New York: Oxford University Press. 563 p.

Nielsen, C. (2008). Six major steps in animal evolution: Are we derived sponge larvae? *Evolution & Development, 10*, 241–257.

Nijjar, S., & Woodland, H. R. (2013). Protein interactions in *Xenopus* germ plasm RNP particles. *PLoS One, 8*, e80077.

Nozaki, H., Misawa, K., Kajita, T., Kato, M., Nohara, S., & Watanabe, M. M. (2000). Origin and evolution of the colonial volvocales (Chlorophyceae) as inferred from multiple, chloroplast gene sequences. *Molecular Phylogenetics and Evolution, 17*, 256–268.

Nozaki, H., Yamada, T. K., Takahashi, F., Matsuzaki, R., & Nakada, T. (2014). New "missing link" genus of the colonial volvocine green algae gives insights into the evolution of oogamy. *BMC Evolutionary Biology, 14*, 37.

Omland, K. E., Cook, L. G., & Crisp, M. D. (2008). Tree thinking for all biology: The problem with reading phylogenies as ladders of progress. *Bioessays, 30*, 854–867.

Parker, G. A., Smith, V. G. F., & Baker, R. R. (1972). Origin and evolution of gamete dimorphism and male–female phenomenon. *Journal of Theoretical Biology, 36*, 529–553.

Pepling, M. E. (2010). A novel maternal mRNA storage compartment in mouse oocytes. *Biology of Reproduction, 82*, 807–808.

Pepling, M. E., Wilhelm, J. E., O'Hara, A. L., Gephardt, G. W., & Spradling, A. C. (2007). Mouse oocytes within germ cell cysts and primordial follicles contain a Balbiani body. *Proceedings of the National Academy of Sciences of the United States of America, 104*, 187–192.

Petralia, R. S., Mattson, M. P., & Yao, P. J. (2014). Aging and longevity in the simplest animals and the quest for immortality. *Ageing Research Reviews, 16*, 66–82.

Philippe, H., Derelle, R., Lopez, P., Pick, K., Borchiellini, C., Boury-Esnault, N., et al. (2009). Phylogenomics revives traditional views on deep animal relationships. *Current Biology, 19*, 706–712.

Poulson, D. F., & Waterhouse, D. F. (1960). Experimental studies on pole cells and mid-gut differentiation in Diptera. *Australian Journal of Biological Sciences, 13*, 541–567.

Randerson, J. P., & Hurst, L. D. (2001). A comparative test of a theory for the evolution of anisogamy. *Proceedings of the Royal Society of London. Series B: Biological Sciences, 268*, 879–884.

Rokas, A. (2008). The origins of multicellularity and the early history of the genetic toolkit for animal development. *Annual Review of Genetics, 42*, 235–251.

Ruiz-Trillo, I., Riutort, M., Fourcade, H. M., Baguna, J., & Boore, J. L. (2004). Mitochondrial genome data support the basal position of Acoelomorpha and the polyphyly of the Platyhelminthes. *Molecular Phylogenetics and Evolution, 33*, 321–332.

Ruiz-Trillo, I., Roger, A. J., Burger, G., Gray, M. W., & Lang, B. F. (2008). A phylogenomic investigation into the origin of metazoa. *Molecular Biology and Evolution, 25*, 664–672.

Salo, E. (2006). The power of regeneration and the stem-cell kingdom: Freshwater planarians (Platyhelminthes). *Bioessays, 28*, 546–559.

Sempere, L. F., Martinez, P., Cole, C., Baguna, J., & Peterson, K. J. (2007). Phylogenetic distribution of microRNAs supports the basal position of acoel flatworms and the polyphyly of Platyhelminthes. *Evolution & Development, 9*, 409–415.

Seydoux, G., & Schedl, T. (2001). The germline in *C. elegans*: Origins, proliferation, and silencing. *International Review of Cytology, 203*, 139–185.

Shalchian-Tabrizi, K., Minge, M. A., Espelund, M., Orr, R., Ruden, T., Jakobsen, K. S., et al. (2008). Multigene phylogeny of Choanozoa and the origin of animals. *PLoS One, 3*, e2098.

Solana, J. (2013). Closing the circle of germline and stem cells: The Primordial Stem Cell hypothesis. *Evodevo, 4*, 2.

Srivastava, M., Begovic, E., Chapman, J., Putnam, N. H., Hellsten, U., Kawashima, T., et al. (2008). The Trichoplax genome and the nature of placozoans. *Nature, 454*, 955–960.

Standart, N., & Minshall, N. (2008). Translational control in early development: CPEB, P-bodies and germinal granules. *Biochemical Society Transactions, 36*, 671–676.

Strome, S., & Updike, D. (2015). Specifying and protecting germ cell fate. *Nature Reviews. Molecular Cell Biology, 16*, 406–416.

Tang, W. W. C., Dietmann, S., Irie, N., Leitch, H. G., Floros, V. I., Bradshaw, C. R., et al. (2015). A unique gene regulatory network resets the human germline epigenome for development. *Cell, 161*, 1453–1467.

Thomsen, H. A. (1982). Planktonic choanoflagellates from Disko Bugt, West Greenland, with a survey of the marine nanoplankton of the area. *Meddelelser om Grønland, Bioscience, 8*, 1–35.

Torgerson, D. G., Kulathinal, R. J., & Singh, R. S. (2002). Mammalian sperm proteins are rapidly evolving: Evidence of positive selection in functionally diverse genes. *Molecular Biology and Evolution, 19*, 1973–1980.

Turner, L. M., Chuong, E. B., & Hoekstra, H. E. (2008). Comparative analysis of testis protein evolution in rodents. *Genetics, 179*, 2075–2089.

Weismann, A. J. F. (1892). *Das Keimplasma: eine Theorie der Vererbung*. Jena: Fischer.

Whelan, N. V., Kocot, K. M., Moroz, L. L., & Halanych, K. M. (2015). Error, signal, and the placement of Ctenophora sister to all other animals. *Proceedings of the National Academy of Sciences of the United States of America, 112*, 5773–5778.

Wilk, K., Bilinski, S., Dougherty, M. T., & Kloc, M. (2005). Delivery of germinal granules and localized RNAs via the messenger transport organizer pathway to the vegetal cortex of *Xenopus* oocytes occurs through directional expansion of the mitochondrial cloud. *The International Journal of Developmental Biology, 49*, 17–21.

Wolpert, L., & Szathmary, E. (2002). Multicellularity: Evolution and the egg. *Nature, 420*, 745.

Wylie, C. C., Holwill, S., Odriscoll, M., Snape, A., & Heasman, J. (1985). Germ plasm and germ-cell determination in *Xenopus laevis* as studied by cell transplantation analysis. *Cold Spring Harbor Symposia on Quantitative Biology, 50*, 37–43.

Yoshida, K., Fujisawa, T., Hwang, J. S., Ikeo, K., & Gojobori, T. (2006). Degeneration after sexual differentiation in hydra and its relevance to the evolution of aging. *Gene, 385*, 64–70.

Zhou, R. R., Wang, B., Wang, J., Schatten, H., & Zhang, Y. Z. (2010). Is the mitochondrial cloud the selection machinery for preferentially transmitting wild-type mtDNA between generations? Rewinding Muller's ratchet efficiently. *Current Genetics, 56*, 101–107.

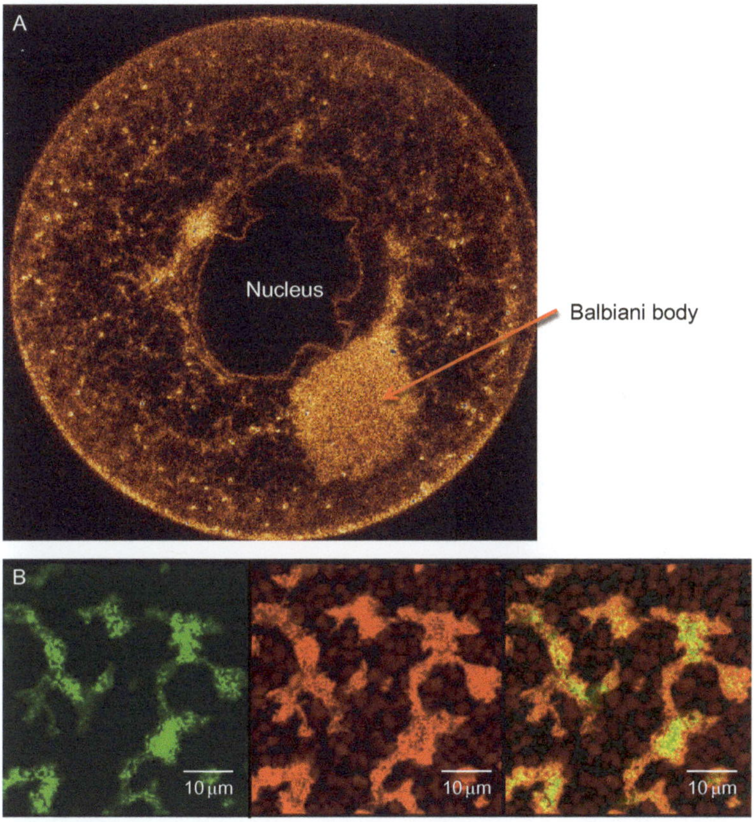

Hugh R. Woodland, Figure 2 Balbiani bodies and germ plasm. (A) A previtellogenic *Xenopus* oocyte. The optical section from confocal imaging shows staining with antibodies to the Balbiani body/germ plasm protein Xpat (Machado et al., 2005). This forms particles distinct from RNPs. (B) Germ plasm in a living full-grown *Xenopus* oocyte showing: left, RNP particles, visualized with GFP-Xvelo fusion protein; center, mitochondria, visualized with TMRE; right, overlay (Nijjar & Woodland, 2013). In *Xenopus*, the germ plasm is derived directly from the Balbiani body, from which its mitochondria are derived.

CHAPTER THIRTY-SEVEN

Pli Selon Pli: Mechanochemical Feedback and the Morphogenetic Role of Contractility at Cadherin Cell–Cell Junctions

Bipul R. Acharya, Alpha S. Yap[1]

Division of Cell Biology and Molecular Medicine, Program in Membrane Interface Biology, Institute for Molecular Bioscience, The University of Queensland, Brisbane, Queensland, Australia

[1]Corresponding author: e-mail address: a.yap@imb.uq.edu.au

Contents

Abstract

Cellular contractility, driven by actomyosin networks coupled to cadherin cell–cell adhesion junctions, is a major determinant of cellular rearrangement during morphogenesis. It now emerges that contractility arises as the emergent property of a mechanochemical feedback system that encompasses the signals that regulate contractility and the elements of the actomyosin network itself.

Current Topics in Developmental Biology, Volume 117
ISSN 0070-2153
http://dx.doi.org/10.1016/bs.ctdb.2015.10.021

1. INTRODUCTION

The development of multicellular organisms and tissues operates at many levels. Their constituent cells integrate many molecular mechanisms that allow them to serve as the building blocks of tissues. Yet interactions between cells also generate complex, emergent properties at the tissue level. Thus, a comprehensive understanding of development presents the great challenge of understanding how these different (molecular, cellular, tissue) levels are integrated, despite the substantial differences in length and timescales at which each level operates.

The challenges, and our progress in responding to them, are well illustrated by the classical cadherins, whose founding member, E-cadherin, was first discovered by Vestweber and Kemler as a protein necessary for compaction of the early mouse embryo (Vestweber & Kemler, 1984, 1985) and by Takeichi as a mediator of adhesion between cultured cells (Takeichi, 1977). Since then, the tissue-specific expression of different classical cadherins has been comprehensively characterized, and their contributions to development, morphogenesis, and organogenesis have been established in a range of invertebrate and vertebrate model systems (Tepass, Truong, Godt, Ikura, & Peifer, 2000). In parallel, we have learnt much about how cadherins operate and are regulated at the molecular and cellular levels (Niessen, Leckband, & Yap, 2011; Takeichi, 2014). In this essay, we will discuss how some recent insights into the cellular mechanisms of cadherin biology can influence the way we think about the processes that drive epithelial morphogenesis. We focus on how adhesion cooperates with the actin cytoskeleton.

2. CADHERINS AND THE CONTRACTILE CYTOSKELETON

We have long believed that cadherins cooperate, functionally and physically, with the actin cytoskeleton. Much of the early evidence for this derived from cell culture, such as the observation that inhibitors of F-actin integrity disrupt cadherin adhesion (Jaffe et al., 1990). Cellular and biochemical studies further suggested that cadherins could interact physically with actin filaments (Nagafuchi & Takeichi, 1988; Ozawa, Ringwald, & Kemler, 1990). Indeed, many actin-binding proteins (ABPs) have now been identified as capable of interacting with classical cadherins (Ratheesh & Yap, 2012), and recent screens using proximity-biotinylation techniques have identified a host of other potential cadherin-associated ABPs (Guo et al., 2014; Van Itallie et al., 2014).

An important advance in understanding the functional impact of cadherin–actin cooperation has come from the *in vivo* demonstration that mechanical forces exerted by the actin cytoskeleton are transmitted to cadherin junctions. In epithelia, this can take the form of pulsatile movements of junctions that are driven by pulsatile contractions of the cytoskeleton (Martin, Kaschube, & Wieschaus, 2009; Munjal, Philippe, Munro, & Lecuit, 2015; Rauzi, Lenne, & Lecuit, 2010; Roh-Johnson et al., 2012; Wu et al., 2014) or sustained contractile tension that is found in the junctions (Ratheesh et al., 2012; Wu et al., 2014). These different patterns of contractility often distinguish different morphogenetic processes and are associated with different actomyosin networks that associate with junctions. Pulsatile contraction is commonly seen with apical constriction and invagination (Martin et al., 2009; Roh-Johnson et al., 2012) and is driven by two-dimensional (2D) actomyosin networks found at the apical poles of cells (often called medial–apical networks) that undergo periodic condensation (Martin et al., 2009; Munjal et al., 2015). In contrast, sustained junctional tension is associated with processes such as neighbor exchange, boundary formation, and tissue homeostasis (Fernandez-Gonzalez, Simoes Sde, Roper, Eaton, & Zallen, 2009; Monier, Pelissier-Monier, Brand, & Sanson, 2009; Wu et al., 2014); it is attributed to circumferential actomyosin bundles that run parallel to, and associate with, cadherin junctions, especially the apically positioned zonulae adherente (ZA). However, these two modes of contractile organization can coexist within the same cell (Rauzi et al., 2010; Wu et al., 2014; Fig. 1).

This fact—that cadherin junctions are sites where the cytoskeleton generates force—carries several important implications. First, this mechanical connection implies that there must be physical connections between the cadherins and the actin cytoskeleton. Second, since contractile forces are generated by nonmuscle Myosin II (NMII) acting on actin filaments, cadherins interact with an actomyosin contractile apparatus. This is further supported by the observation that mechanical stress exerted on cadherins can cause the cytoskeleton to stiffen (Bazellieres et al., 2015; le Duc et al., 2010), a process that requires both F-actin and myosin. Third, it alters the way we approach the biology of cell–cell adhesion: rather than principally being a mechanism that passively resists detachment forces to preserve tissue integrity, cadherins serve to generate tensile forces by coupling contractile cortices of neighboring cells together (Maitre et al., 2012). Furthermore,

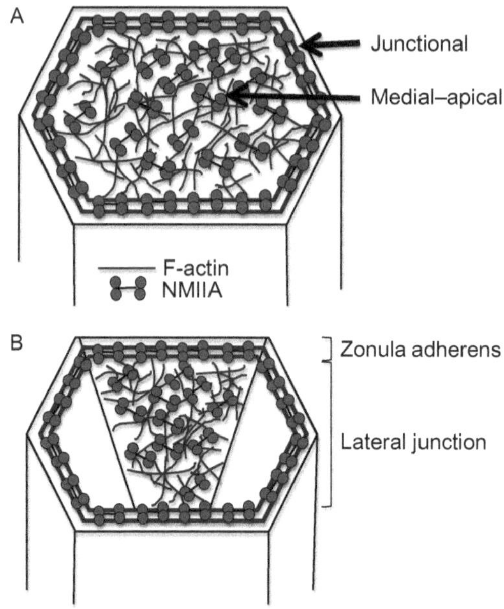

Figure 1 Patterns of actomyosin contractility at epithelial cell–cell junctions. (A) Apical actomyosin networks include two-dimensional medial–apical networks and junctional actomyosin bundles, as found in epithelial cells undergoing apical constriction. (B) In polarized epithelia, apical bundles at the zonula adherens (ZA) can coexist with two-dimensional networks found at the lateral junctions, below the ZA. (A color version of this figure is located at the end of this chapter.)

cadherins can promote assembly and homeostasis of the contractile apparatus itself, while actomyosin feeds back to promote adhesion (Shewan et al., 2005; Smutny et al., 2010).

Taken together, these observations suggest that at the cellular level the functional interrelationship between cadherin and actomyosin represents an integrated cellular system. As we shall now describe, considerable recent progress has been made in characterizing the elements that contribute to cadherin–actin cooperation. In the process, we have begun to discover feedback interactions between these elements that integrate them into a complex network whose emergent properties have implications for understanding morphogenesis.

3. THE ELEMENTS OF THE SYSTEM

Integrating cadherin adhesion with contractility involves several components. These include the physical association of adhesion with actomyosin, the building of a contractile actomyosin apparatus that is linked to cadherin, and signaling pathways which coordinate these individual elements.

3.1 Molecular Basis of Cadherin–F-Actin Binding

At a basic level, junctional contractility arises from a physical association of cadherin adhesion complexes with actomyosin. To date, the mechanical coupling of contractility to cadherin adhesion appears to be mediated by the F-actin elements of the actomyosin apparatus; there is no evidence that NMII interacts directly with the cadherin complex. Three broad pathways are available to couple cadherins to F-actin. First, cadherins may associate with actin filaments through α-catenin, a ubiquitous component of the core cadherin–catenin complex. Although free α-catenin can directly bind F-actin (Pokutta, Drees, Takai, Nelson, & Weis, 2002; Rimm, Koslov, Kebriaei, Cianci, & Morrow, 1995), its role as a linker between cadherins and F-actin became controversial when it proved difficult to reconstitute F-actin binding when α-catenin was incorporated into a minimal cadherin/catenin complex *in vitro* (Drees, Pokutta, Yamada, Nelson, & Weis, 2005; Yamada, Pokutta, Drees, Weis, & Nelson, 2005). This conundrum has been recently resolved by the demonstration that minimal cadherin–catenin complexes will bind F-actin when under force (Buckley et al., 2014), whereas the earlier reconstitution studies were performed in solution.

Second, α-catenin can recruit other ABPs to the cadherin complex (Abe & Takeichi, 2008). This is exemplified by vinculin, which bears multiple F-actin-binding sites (Ziegler, Liddington, & Critchley, 2006) and can bind directly to α-catenin itself (Choi et al., 2012). Junctional vinculin is increased when force is stimulated and decreased when forces are relaxed (Leerberg et al., 2014; Peng, Cuff, Lawton, & DeMali, 2010; Yonemura, Wada, Watanabe, Nagafuchi, & Shibata, 2010). This appears to reflect an ability of force to reveal cryptic vinculin-binding sites that are present in α-catenin (Yao et al., 2014; Yonemura et al., 2010). Thus, force appears to play an integral role in controlling how α-catenin may mediate interactions between cadherins and actin filaments.

Finally, other proteins can mediate physical association between cadherins and actin filaments. For example, Myosin VI is an actin-binding motor that can directly interact with the E-cadherin cytoplasmic tail and is found at mature cell–cell junctions (Maddugoda, Crampton, Shewan, & Yap, 2007). However, Myosin VI is recruited to E-cadherin relatively late in the process of junctional biogenesis, where it is necessary for assembly of the ZA and organization of its associated actin cytoskeleton through vinculin. Together, these observations reinforce the notion that cadherins associate with F-actin through multiple pathways. The ability of α-catenin to directly bind F-actin is likely to constitute a ubiquitous mechanism that operates when cadherins come under force to link them with cortical actin (Chen et al., 2015). This may represent an elemental mechanism that can then be built upon by the force-dependent association of proteins like vinculin to α-catenin and the context-dependent recruitment of additional ABPs, such as Myosin VI.

3.2 Incorporating Myosin II

Cellular contractility in nonmuscle cells is generated by the motor action of NMII exerted upon F-actin networks (Murrell, Oakes, Lenz, & Gardel, 2015). Consistent with this central cell-biological role, NMII is found decorating the medial–apical and junctional F-actin networks that are thought to exert contractile forces on cell–cell junctions (Mason, Tworoger, & Martin, 2013; Smutny et al., 2010). Considering that cadherin-bound actin filaments are the scaffolds upon which actomyosin is assembled, the incorporation of NMII at junctions involves two parallel processes.

First, NMII must be activated. A dominant pathway involves the phosphorylation of the regulatory light chain of myosin through pathways mediated by kinases such as myosin light chain kinase (MLCK) and Rho-dependent kinase (ROCK) (Heissler & Manstein, 2013). This is thought to promote F-actin binding, mini-filament assembly, and hence formation of an effective contractile unit. Consistent with this, both MLCK and ROCK support junctional contractility and the recruitment of NMII to the junctional cytoskeleton (Smutny et al., 2010) and to medial–apical networks (Mason et al., 2013; Munjal et al., 2015).

Second, activated NMII must interact with F-actin networks. Of note, cadherin junctions are sites of dynamic actin assembly, initiated by actin nucleators such as Arp2/3 and formins, and modulated by postnucleation regulators such as Ena/VASP proteins and α-actinin-4 (Ratheesh & Yap, 2012). Their contribution to recruitment of NMII is illustrated by evidence that inhibiting Arp2/3 or its upstream activator, WAVE2, impaired the junctional recruitment of NMII and reduced junctional tension (Verma et al., 2012). This implies that the assembly of actin filaments at the junctional cortex serves to recruit activated NMII to the junctions, thereby allowing the contractile apparatus to assemble. However, the role of nucleators such as Arp2/3 carries a potential paradox. Arp2/3 generates branched actin networks that are thought to be poorly fitted for either stable NMII binding or contractile force generation. In part, this is because NMII tends to cause severing of branched networks, thereby turning over the actin scaffolds necessary for stable NMII localization (Reymann et al., 2012). Despite this, the cell clearly has developed strategies that compensate for this potential paradox, which may include stabilization of nascent filaments (Kovacs et al., 2011; Wu et al., 2014) and reorganization of actin networks to allow them to better sustain contractile stress (Murrell et al., 2015). While many mechanistic details have yet to be resolved, these observations carry the important implication that regulation of junctional F-actin (kinetics, organization) may constitute a pathway to influence contractility that operates orthogonal to that of the classical mode of NMII activation.

3.3 Upstream Regulation by Cell Signaling

It will be apparent from the discussion above that cell signaling plays a central role in coordinating the many elements of cadherin–actin cooperation. Key signals include canonical regulators of the actin cytoskeleton, notably the GTPases Rac and Rho, as well as many protein and lipid kinases which are found at cell–cell junctions (Niessen et al., 2011). Rather than focusing on the many details of the mechanisms, we would highlight the following concepts.

First, cadherin adhesion often plays an important role in establishing the junctional signaling pathways that control actomyosin. For example, depleting cadherin impairs the ability of cells to sustain both Rac and Rho signaling (Kovacs, Ali, McCormack, & Yap, 2002; Lampugnani et al., 2002; Priya, Yap, & Gomez, 2013). This dependence can reflect different paradigms for signal activation. In some instances, cadherins appear to behave like adhesion-activated signaling receptors. This is exemplified by Rac, whose activity is acutely stimulated by cadherin adhesion (Kovacs et al., 2002). In other cases, cadherins act more indirectly, as is seen with Rho. E-cadherin is necessary for Rho signaling to be established at cell–cell junctions (Priya et al., 2013), but acute ligation of cadherins can actually downregulate Rho, through feedback from Rac (Noren, Arthur, & Burridge, 2003; Noren, Niessen, Gumbiner, & Burridge, 2001). Here, cadherin adhesion appears to serve less as a ligand-activated receptor than as a membrane platform that is necessary for signaling zones to be established and maintained.

Second, key signals, such as Rac and Rho, reflect networks of upstream molecular activators and inhibitors, whose balanced action determines the expression of signaling. For example, Rho signaling at cadherin junctions is activated by exchange factors (GEFs), such as Ect2 (Ratheesh et al., 2012), and also potentially inhibited by GTPase-activating proteins (GAPs) such as Myosin XIA (Omelchenko & Hall, 2012) and p190 RhoGAP (Priya et al., 2015). The spatial control of GEFs is important to initiate Rho signaling (Ratheesh et al., 2012), and the complementary inhibition of GAPs can play an important role in sustaining signaling (Priya et al., 2015).

4. MECHANOCHEMICAL FEEDBACK AMONG THE ELEMENTS OF THE CADHERIN–ACTOMYOSIN SYSTEM

We conventionally think of the above elements of cadherin–actomyosin cooperativity as reflecting linear pathways that potentially work in parallel to one another. However, it is increasingly evident that these pathways are linked to ultimately form a network of chemical signals and mechanical effectors. One of the most recent examples involves Rho and NMII.

As noted above, Rho (Rho A in vertebrates and Rho1 in *Drosophila*) is a canonical activator of NMII that is necessary for the various forms of contractility that impinge on cell–cell junctions. Thus, inhibiting Rho blocked the pulsatile contractility that is seen in the medial–apical actomyosin networks during *Drosophila* germband extension (Munjal et al., 2015) and that is found at the lateral junctions between mammalian epithelial cells (Wu et al., 2014). Inhibiting Rho, or its upstream activators, also degrades the sustained junctional tension of the ZA, which is associated with peri-junctional actomyosin bundles (Fernandez-Gonzalez et al., 2009; Ratheesh et al., 2012). In all these instances, the impact of Rho is mediated by ROCK. Indeed, ligation of E-cadherin alone is sufficient to activate NMII in a ROCK-dependent fashion (Shewan et al., 2005). Together, these observations are consistent with a canonical linear pathway that leads from Rho to NMII via ROCK.

The close functional link between the elements of the Rho–ROCK–NMII pathway is emphasized by the coaccumulation of these components at sites of contractility. This is especially striking in pulsatile medial–lateral networks in *Drosophila*. Here, not only do GTP-Rho and ROCK coaccumulate with actomyosin, but they also condense with the contracting medial–apical networks (Munjal et al., 2015; Vasquez, Tworoger, & Martin, 2014), suggesting that the dynamic spatial distribution of these signals is closely linked with that of NMII. Similarly, if less dramatically, these signals concentrate at the contractile ZA of polarized epithelial cells along with actomyosin (Priya et al., 2015, 2013; Ratheesh et al., 2012; Simoes Sde, Mainieri, & Zallen, 2014).

At first, these patterns of coaccumulation fit well with a conventional linear signaling pathway. Surprisingly, however, not only does inhibiting Rho–ROCK degrade contractility, but directly inhibiting NMII itself also perturbs upstream Rho–ROCK signaling. Blocking contractility in gastrulating *Drosophila* embryonic cells, both prevented the pulsatile condensation of actomyosin seen in the medial–apical networks of these cells and also displaced GTP-Rho and ROCK from the apical cortex (Munjal et al., 2015). Similarly, inhibiting NMII directly in cultured mammalian epithelial cells reduced GTP-Rho and ROCK at the apical ZA of these cells (Priya et al., 2015). Together, these findings indicate that NMII can feedback to support Rho–ROCK signaling at cellular cortices.

Importantly, both these instances reflect modes of mechanochemical feedback. In the pulsatile medial–apical networks of *Drosophila* germband cells, Rho and ROCK appear to undergo advection, being concentrated by a process of active bulk flow in the cellular cortex when the medial–apical actomyosin networks contracted and condensed. Such concentration may promote the interaction of Rho activators, thereby supporting positive feedback (Howard, Grill, & Bois, 2011). Consistent with this, inhibition of pulsatile contractility, which blocks advection, also decreased the amplitude of Rho and ROCK pulses (Munjal et al., 2015). In the case of the mammalian ZA, the mechanism involved anchorage of ROCK by physical association with NMII. When anchored at the ZA, ROCK then antagonized the cortical localization of a Rho inhibitor, p190B RhoGAP, to thereby support Rho signaling (Priya et al., 2015). Computational modeling predicted that in this network the local concentration of NMII would stabilize zones of activated Rho by preventing its premature inactivation. These observations add to the increasing realization that mechanochemical factors influence contractility at junctions. Indeed, tension can stabilize junctional NMII itself (Fernandez-Gonzalez et al., 2009), which might now be predicted to feedback upstream to affect Rho signaling.

5. EMERGENT PROPERTIES OF MECHANOCHEMICAL NETWORKS

A common consequence of networked control systems is the generation of emergent properties, features that are not predictable from analysis of the individual components but arise from their interactions at the higher-order network level. Could feedback from NMII to Rho yield emergent properties that are morphogenetically relevant? Here, an intriguing question is how different patterns of contractility—pulsed contractions associated with 2D networks versus the more sustained contractility of junctional bundles—can be generated from the same principal components (Rho, ROCK, NMII, and F-actin). Some potential answers to these questions lie in considering how mechanochemical feedback acts in these different contexts.

5.1 Generating Contractile Pulsatility

By its very nature, pulsatility in actomyosin networks involves cycles of contraction/condensation and relaxation. While these coincide with pulsed fluctuations in Rho signaling found at the same sites as these actomyosin networks, contractile pulsatility is not simply an entrained response to those changes in Rho, as inhibiting contractility downregulates Rho signaling itself (Munjal et al., 2015). Instead, feedback within the Rho–NMII network contributes to generating pulses. One form of feedback occurs during the contraction phase when mechanochemical feedback by advection can promote local Rho signaling, thereby potentiating contractility (Munjal et al., 2015).

Such mechanochemical feedback also affects the relaxation phase. One mechanism to induce contractile relaxation is inactivation of the NMII motor itself. This is mediated by myosin phosphatase, which is necessary for pulsatility to occur (Mason et al., 2013; Munjal et al., 2015). Interestingly, NMII was necessary for the cortical localization of myosin phosphatase, which also coaccumulated with condensed actomyosin networks when they contracted (Mason et al., 2013; Munjal et al., 2015). This suggests that actomyosin contractility promotes the local recruitment of a mechanism (myosin phosphatase) that leads to its inactivation, a sequence which would be predicted to promote pulsatile behavior.

This mechanism for myosin inactivation is complemented by other changes in the contractile apparatus. Thus, advection declined during the transition from contraction to relaxation (Munjal et al., 2015), which constitutes a mechanism for negative feedback that is predicted to be necessary for disassembly of pulsed contractions. This was attributable to the increased density of actomyosin networks that developed during condensation, which can increase friction to oppose advection. In addition, NMII turnover increased (Munjal et al., 2015), suggesting that its association with actin networks became less stable. This can reflect inactivation of NMII as well the potential for stress to sever actin filaments and promote their turnover (Wu et al., 2014). Indeed, blocking contractility stabilized F-actin at lateral epithelial cell–cell junctions, regions of adhesion where pulsatile contraction also occurs (Wu et al., 2014). Therefore, rather than simply reflecting entrained responses to upstream signals, many factors, encompassing regulatory signals and the structural (F-actin) elements of the contractile apparatus, collaborate to drive the assembly and disassembly phases of contractile pulsatility.

5.2 Generating Sustained Tension

How, then, can a system capable of generating pulsed contractions be converted into one that generates sustained tension? One possibility is that signaling by upstream activators of NMII behaves in a sustained, rather than a pulsatile, fashion (Munjal et al., 2015). Indeed, this is found at the apical ZA of mammalian epithelial cells, where GTP-Rho forms a defined micron-sized signaling zone whose intensity and size is relatively stable over timescales (10 s of minutes) that are much longer than the dynamics of its constituent molecules (Priya et al., 2015) or the cycles associated with pulsatile actomyosin networks (~100 s) (Rauzi et al., 2010; Wu et al., 2014). This zone depends on NMII and reflects the myosin-based feedback network described above. Of note, computational analysis of the feedback network predicted that the system was capable of generating bistable outcomes (Priya et al., 2015), meaning that once NMII was activated and localized to the junctional cortex, the feedback network that it set up could stably sustain signaling within the population of molecules found at the ZA zone. Sustained Rho signaling can then reflect the impact of the mechanochemical feedback network implemented by NMII.

Computational analysis further predicted that surface stabilization of NMII plays an important role in establishing a stable Rho zone (Priya et al., 2015). Consistent with this, NMII was more stable at the ZA, where the stable Rho zone forms, than elsewhere at cell–cell junctions

(Priya et al., 2015). This differential cortical stability of NMII may be a response to Rho signaling (Priya et al., 2015), but it can also be influenced by the actin networks with which it interacts (Wu et al., 2014). For example, F-actin is more stable at the ZA than elsewhere at cell–cell junctions, an effect of the actin regulator, N-WASP (Kovacs et al., 2011; Wu et al., 2014), and this is necessary for strong junctional tension to be generated at the ZA (Wu et al., 2014). Potentially, regulators of F-actin that promote association of NMII may then influence NMII–Rho feedback at junctions.

5.3 Morphogenetic Implications

Ultimately, an important question is how these cellular properties of the cadherin–actomyosin system help us understand their impact on morphogenesis. Potential insight comes from considering the mechanical impact of pulsatile versus sustained contractile systems. Pulsatile contractile systems essentially contain the capacity to dissipate the stresses that they generate during their contractile phases. Their ability to support apical constriction is thought to arise from a rachet-like mechanism which allows the constricted adhesive junctions to be stabilized, i.e., to fail to relax when the medial–apical networks relax (Martin & Goldstein, 2014). Such relaxation may be important to coordinate apical constriction across populations of cells that undergo tissue invaginations (Vasquez et al., 2014). The mechanism responsible for rachet-like stabilization is not yet clear but coincides with the accumulation of actomyosin at junctions (Rauzi et al., 2010). As these are sites of sustained junctional tension, it is tempting to speculate that this contributes to generating the rachet.

In contrast, the pulsatile networks found at the lateral junctions of polarized epithelial cells (i.e., below the apical ZA; Fig. 1) do not appear to be coupled to a rachet-like mechanism: the cadherin clusters with which they are associated also relax when the condensed actomyosin networks dissipate and relax. This limits the contractile tension that can be generated at these regions of junctions (Wu et al., 2014). One consequence of reducing junctional tension in these lateral regions may be to minimize mechanical barriers to hydrodynamic flow between the cytoplasmic compartments of cells, something that is also important for apical constriction to occur during development (He, Doubrovinski, Polyakov, & Wieschaus, 2014).

6. CONCLUDING COMMENTS

Overall, these recent findings indicate that distinct contractile outputs can be generated by the core actomyosin apparatus, depending on different settings within the mechanochemical feedback network that encompasses Rho, NMII, and actin at cell–cell junctions. It seems likely that we have only begun to appreciate the versatility that can be generated by this Rho-actomyosin network.

However, a key outstanding question is the role that the cadherin system may play in this network. To date, adhesion has principally been treated as the mechanism that mechanically integrates the contractile cortices of cells together (Maitre et al., 2012). This is also consistent with current focus on understanding the molecular mechanisms that physically couple cadherin adhesion complexes to F-actin. However, as we have discussed, cadherin adhesion also supports the biogenesis of junctional actomyosin, through processes such as actin assembly (Verma et al., 2012) and Rho signaling (Ratheesh et al., 2012). This active feedback from adhesion to contractility might be predicted to influence morphogenetic outcomes by tuning contractile outcomes. Further, junctional actomyosin promotes the accumulation and stability of cadherins at junctions and can influence cadherin clustering to promote adhesion (Priya et al., 2013; Smutny et al., 2010; Wu, Kanchanawong, & Zaidel-Bar, 2015). It is tempting to speculate that this might affect mechanical coupling between cells and even the degree of friction that adhesion presents as cells slide across each other during morphogenetic events such as neighbor exchange. If so, then, the physically coupled cadherin system would constitute another active feedback element in the mechanochemical network that allows cellular contractility to drive morphogenesis.

ACKNOWLEDGMENTS

We thank all our lab colleagues for feedback and endlessly stimulating conversations. Our work is funded by the Human Frontiers Science Program, The Australian Research Council, and the National Health and Medical Research Council of Australia (1044041, 1037320, and 1067405).

REFERENCES

Abe, K., & Takeichi, M. (2008). EPLIN mediates linkage of the cadherin catenin complex to F-actin and stabilizes the circumferential actin belt. *Proceedings of the National Academy of Sciences of the United States of America*, *105*, 13–19.

Bazellieres, E., Conte, V., Elosegui-Artola, A., Serra-Picamal, X., Bintanel-Morcillo, M., Roca-Cusachs, P., et al. (2015). Control of cell-cell forces and collective cell dynamics by the intercellular adhesome. *Nature Cell Biology*, *17*, 409–420.

Buckley, C. D., Tan, J. L., Anderson, K. L., Hanein, D., Volkmann, N., Weis, W. I., et al. (2014). The minimal cadherin-catenin complex binds to actin filaments under force. *Science*, *346*, 1254211.

Chen, C. S., Hong, S., Indra, I., Sergeeva, A. P., Troyanovsky, R. B., Shapiro, L., et al. (2015). alpha-Catenin-mediated cadherin clustering couples cadherin and actin dynamics. *The Journal of Cell Biology*, *210*, 647–661.

Choi, H. J., Pokutta, S., Cadwell, G. W., Bobkov, A. A., Bankston, L. A., Liddington, R. C., et al. (2012). alphaE-catenin is an autoinhibited molecule that coactivates vinculin. *Proceedings of the National Academy of Sciences of the United States of America*, *109*, 8576–8581.

Drees, F., Pokutta, S., Yamada, S., Nelson, W. J., & Weis, W. I. (2005). Alpha-catenin is a molecular switch that binds E-cadherin-beta-catenin and regulates actin-filament assembly. *Cell*, *123*, 903–915.

Fernandez-Gonzalez, R., Simoes Sde, M., Roper, J. C., Eaton, S., & Zallen, J. A. (2009). Myosin II dynamics are regulated by tension in intercalating cells. *Developmental Cell*, *17*, 736–743.

Guo, Z., Neilson, L. J., Zhong, H., Murray, P. S., Zanivan, S., & Zaidel-Bar, R. (2014). E-cadherin interactome complexity and robustness resolved by quantitative proteomics. *Science Signaling*, *7*, rs7.

He, B., Doubrovinski, K., Polyakov, O., & Wieschaus, E. (2014). Apical constriction drives tissue-scale hydrodynamic flow to mediate cell elongation. *Nature*, *508*, 392–396.

Heissler, S. M., & Manstein, D. J. (2013). Nonmuscle myosin-2: Mix and match. *Cellular and Molecular Life Sciences*, *70*, 1–21.

Howard, J., Grill, S. W., & Bois, J. S. (2011). Turing's next steps: The mechanochemical basis of morphogenesis. *Nature Reviews. Molecular Cell Biology*, *12*, 400–406.

Jaffe, S. H., Friedlander, D. R., Matsuzaki, F., Crossin, K. L., Cunningham, B. A., & Edelman, G. M. (1990). Differential effects of the cytoplasmic domains of cell adhesion molecules on cell aggregation and sorting-out. *Proceedings of the National Academy of Sciences of the United States of America*, *87*, 3589–3593.

Kovacs, E. M., Ali, R. G., McCormack, A. J., & Yap, A. S. (2002). E-cadherin homophilic ligation directly signals through Rac and PI3-kinase to regulate adhesive contacts. *The Journal of Biological Chemistry*, *277*, 6708–6718.

Kovacs, E. M., Verma, S., Ali, R. G., Ratheesh, A., Hamilton, N. A., Akhmanova, A., et al. (2011). N-WASP regulates the epithelial junctional actin cytoskeleton through a non-canonical post-nucleation pathway. *Nature Cell Biology*, *13*, 934–943.

Lampugnani, M. G., Zanetti, A., Breviario, F., Balconi, G., Orsenigo, F., Corada, M., et al. (2002). VE-cadherin regulates endothelial actin activating Rac and increasing membrane association of Tiam. *Molecular Biology of the Cell*, *13*, 1175–1189.

le Duc, Q., Shi, Q., Blonk, I., Sonnenberg, A., Wang, N., Leckband, D., et al. (2010). Vinculin potentiates E-cadherin mechanosensing and is recruited to actin-anchored sites within adherens junctions in a myosin II-dependent manner. *The Journal of Cell Biology*, *189*, 1107–1115.

Leerberg, J. M., Gomez, G. A., Verma, S., Moussa, E. J., Wu, S. K., Priya, R., et al. (2014). Tension-sensitive actin assembly supports contractility at the epithelial zonula adherens. *Current Biology*, *24*, 1689–1699.

Maddugoda, M. P., Crampton, M. S., Shewan, A. M., & Yap, A. S. (2007). Myosin VI and vinculin cooperate during the morphogenesis of cadherin cell-cell contacts in mammalian epithelial cells. *The Journal of Cell Biology*, *178*, 529–540.

Maitre, J. L., Berthoumieux, H., Krens, S. F., Salbreux, G., Julicher, F., Paluch, E., et al. (2012). Adhesion functions in cell sorting by mechanically coupling the cortices of adhering cells. *Science*, *338*, 253–256.

Martin, A. C., & Goldstein, B. (2014). Apical constriction: Themes and variations on a cellular mechanism driving morphogenesis. *Development*, *141*, 1987–1998.

Martin, A. C., Kaschube, M., & Wieschaus, E. F. (2009). Pulsed contractions of an actin-myosin network drive apical constriction. *Nature*, *457*, 495–499.

Mason, F. M., Tworoger, M., & Martin, A. C. (2013). Apical domain polarization localizes actin-myosin activity to drive ratchet-like apical constriction. *Nature Cell Biology*, *15*, 926–936.

Monier, B., Pelissier-Monier, A., Brand, A. H., & Sanson, B. (2009). An actomyosin-based barrier inhibits cell mixing at compartmental boundaries in Drosophila embryos. *Nature Cell Biology*, *12*, 60–65.

Munjal, A., Philippe, J. M., Munro, E., & Lecuit, T. (2015). A self-organized biomechanical network drives shape changes during tissue morphogenesis. *Nature*, *524*, 351–355.

Murrell, M., Oakes, P. W., Lenz, M., & Gardel, M. L. (2015). Forcing cells into shape: The mechanics of actomyosin contractility. *Nature Reviews. Molecular Cell Biology*, *16*, 486–498.

Nagafuchi, A., & Takeichi, M. (1988). Cell binding function of E-cadherin is regulated by the cytoplasmic domain. *The EMBO Journal*, *7*, 3679–3684.

Niessen, C., Leckband, D., & Yap, A. S. (2011). Tissue organization by classical cadherin adhesion molecules: Dynamic molecular and cellular mechanisms of morphogenetic regulation. *Physiological Reviews*, *91*, 691–731.

Noren, N. K., Arthur, W. T., & Burridge, K. (2003). Cadherin engagement inhibits RhoA via p190RhoGAP. *The Journal of Biological Chemistry*, *278*, 13615–13618.

Noren, N. K., Niessen, C. M., Gumbiner, B. M., & Burridge, K. (2001). Cadherin engagement regulates Rho family GTPases. *The Journal of Biological Chemistry*, *276*, 33305–33308.

Omelchenko, T., & Hall, A. (2012). Myosin-IXA regulates collective epithelial cell migration by targeting RhoGAP activity to cell-cell junctions. *Current Biology*, *22*, 278–288.

Ozawa, M., Ringwald, M., & Kemler, R. (1990). Uvomorulin-catenin complex formation is regulated by a specific domain in the cytoplasmic region of the cell adhesion molecule. *Proceedings of the National Academy of Sciences of the United States of America*, *87*, 4246–4250.

Peng, X., Cuff, L. E., Lawton, C. D., & DeMali, K. A. (2010). Vinculin regulates cell-surface E-cadherin expression by binding to beta-catenin. *Journal of Cell Science*, *123*, 567–577.

Pokutta, S., Drees, F., Takai, Y., Nelson, W. J., & Weis, W. I. (2002). Biochemical and structural definition of the l-afadin- and actin-binding sites of alpha-catenin. *The Journal of Biological Chemistry*, *277*, 18868–18874.

Priya, R., Gomez, G. A., Budnar, S., Verma, S., Cox, H. L., Hamilton, N. A., et al. (2015). Feedback regulation through myosin II confers robustness on RhoA signalling at E-cadherin junctions. *Nature Cell Biology*, *17*, 1282–1293.

Priya, R., Yap, A. S., & Gomez, G. A. (2013). E-cadherin supports steady-state Rho signaling at the epithelial zonula adherens. *Differentiation*, *86*, 133–140.

Ratheesh, A., Gomez, G. A., Priya, R., Verma, S., Kovacs, E. M., Jiang, K., et al. (2012). Centralspindlin and alpha catenin regulate Rho signalling at the epithelial zonula adherens. *Nature Cell Biology*, *14*, 818–828.

Ratheesh, A., & Yap, A. S. (2012). A bigger picture: Classical cadherins and the dynamic actin cytoskeleton. *Nature Reviews. Molecular Cell Biology*, *13*, 673–679.

Rauzi, M., Lenne, P. F., & Lecuit, T. (2010). Planar polarized actomyosin contractile flows control epithelial junction remodelling. *Nature*, *468*, 1110–1114.

Reymann, A. C., Boujemaa-Paterski, R., Martiel, J. L., Guerin, C., Cao, W., Chin, H. F., et al. (2012). Actin network architecture can determine myosin motor activity. *Science*, *336*, 1310–1314.

Rimm, D. L., Koslov, E. R., Kebriaei, P., Cianci, C. D., & Morrow, J. S. (1995). α_1(E)-catenin is an actin-binding and -bundling protein mediating the attachment of F-actin to the membrane adhesion complex. *Proceedings of the National Academy of Sciences of the United States of America*, *92*, 8813–8817.

Roh-Johnson, M., Shemer, G., Higgins, C. D., McClellan, J. H., Werts, A. D., Tulu, U. S., et al. (2012). Triggering a cell shape change by exploiting preexisting actomyosin contractions. *Science*, *335*, 1232–1235.
Shewan, A. M., Maddugoda, M., Kraemer, A., Stehbens, S. J., Verma, S., Kovacs, E. M., et al. (2005). Myosin 2 is a key Rho kinase target necessary for the local concentration of E-cadherin at cell-cell contacts. *Molecular Biology of the Cell*, *16*, 4531–4532.
Simoes Sde, M., Mainieri, A., & Zallen, J. A. (2014). Rho GTPase and Shroom direct planar polarized actomyosin contractility during convergent extension. *The Journal of Cell Biology*, *204*, 575–589.
Smutny, M., Cox, H. L., Leerberg, J. M., Kovacs, E. M., Conti, M. A., Ferguson, C., et al. (2010). Myosin II isoforms identify distinct functional modules that support integrity of the epithelial zonula adherens. *Nature Cell Biology*, *12*, 696–702.
Takeichi, M. (1977). Functional correlation between cell adhesive properties and some cell surface proteins. *The Journal of Cell Biology*, *75*, 464–474.
Takeichi, M. (2014). Dynamic contacts: Rearranging adherens junctions to drive epithelial remodelling. *Nature Reviews. Molecular Cell Biology*, *15*, 397–410.
Tepass, U., Truong, K., Godt, D., Ikura, M., & Peifer, M. (2000). Cadherins in embryonic and neural morphogenesis. *Nature Reviews. Molecular Cell Biology*, *1*, 91–100.
Van Itallie, C. M., Tietgens, A. J., Aponte, A., Fredriksson, K., Fanning, A. S., Gucek, M., et al. (2014). Biotin ligase tagging identifies proteins proximal to E-cadherin, including lipoma preferred partner, a regulator of epithelial cell-cell and cell-substrate adhesion. *Journal of Cell Science*, *127*, 885–895.
Vasquez, C. G., Tworoger, M., & Martin, A. C. (2014). Dynamic myosin phosphorylation regulates contractile pulses and tissue integrity during epithelial morphogenesis. *The Journal of Cell Biology*, *206*, 435–450.
Verma, S., Han, S. P., Michael, M., Gomez, G. A., Yang, Z., Teasdale, R. D., et al. (2012). A WAVE2-Arp2/3 actin nucleator apparatus supports junctional tension at the epithelial zonula adherens. *Molecular Biology of the Cell*, *23*, 4601–4610.
Vestweber, D., & Kemler, R. (1984). Rabbit antiserum against a purified surface glycoprotein decompacts mouse preimplantation embryos and reacts with specific adult tissues. *Experimental Cell Research*, *152*, 169–178.
Vestweber, D., & Kemler, R. (1985). Identification of a putative cell adhesion domain of uvomorulin. *The EMBO Journal*, *4*, 3393–3398.
Wu, S. K., Gomez, G. A., Michael, M., Verma, S., Cox, H. L., Lefevre, J. G., et al. (2014). Cortical F-actin stabilization generates apical-lateral patterns of junctional contractility that integrate cells into epithelia. *Nature Cell Biology*, *16*, 167–178.
Wu, Y., Kanchanawong, P., & Zaidel-Bar, R. (2015). Actin-delimited adhesion-independent clustering of E-cadherin forms the nanoscale building blocks of adherens junctions. *Developmental Cell*, *32*, 139–154.
Yamada, S., Pokutta, S., Drees, F., Weis, W. I., & Nelson, W. J. (2005). Deconstructing the cadherin-catenin-actin complex. *Cell*, *123*, 889–901.
Yao, M., Qiu, W., Liu, R., Efremov, A. K., Cong, P., Seddiki, R., et al. (2014). Force-dependent conformational switch of alpha-catenin controls vinculin binding. *Nature Communications*, *5*, 4525.
Yonemura, S., Wada, Y., Watanabe, T., Nagafuchi, A., & Shibata, M. (2010). alpha-Catenin as a tension transducer that induces adherens junction development. *Nature Cell Biology*, *12*, 533–542.
Ziegler, W. H., Liddington, R. C., & Critchley, D. R. (2006). The structure and regulation of vinculin. *Trends in Cell Biology*, *16*, 453–460.

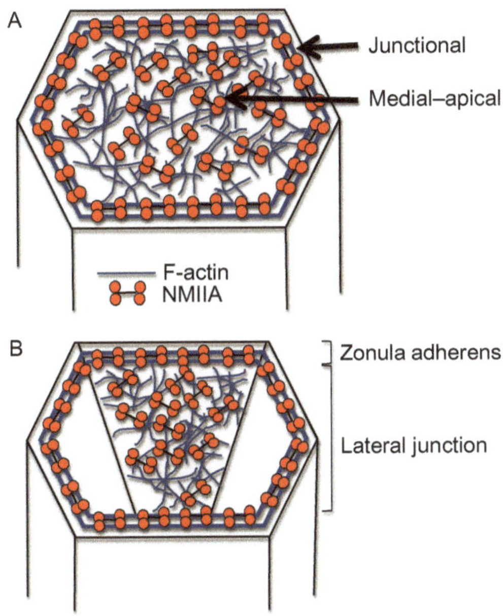

Bipul R. Acharya and Alpha S. Yap, Figure 1 Patterns of actomyosin contractility at epithelial cell–cell junctions. (A) Apical actomyosin networks include two-dimensional medial–apical networks and junctional actomyosin bundles, as found in epithelial cells undergoing apical constriction. (B) In polarized epithelia, apical bundles at the zonula adherens (ZA) can coexist with two-dimensional networks found at the lateral junctions, below the ZA.

CHAPTER THIRTY-EIGHT

From Endoderm to Liver Bud: Paradigms of Cell Type Specification and Tissue Morphogenesis

Kenneth S. Zaret[1]
Institute for Regenerative Medicine, Epigenetics Program, Department of Cell and Developmental Biology, Perelman School of Medicine, University of Pennsylvania, Philadelphia, Pennsylvania, USA
[1]Corresponding author: e-mail address: zaret@upenn.edu

Contents

Abstract

The early specification, rapid growth and morphogenesis, and conserved functions of the embryonic liver across diverse model organisms have made the system an experimentally facile paradigm for understanding basic regulatory mechanisms that govern cell differentiation and organogenesis. This essay highlights concepts that have emerged from studies of the discrete steps of foregut endoderm development into the liver bud, as well as from modeling the steps via embryonic stem cell differentiation. Such concepts include understanding the chromatin basis for the competence of progenitor cells to develop into specific lineages; the importance of combinatorial signaling from different sources to induce cell fates; the impact of inductive signaling on preexisting chromatin states; the ability of separately specified domains of cells to merge into a common tissue; and the marked cell biological dynamics, including interactions with the developing vasculature, which establish the initial morphogenesis and patterning of a tissue. The principles gleaned from these studies, focusing on the 2 days it takes for the endoderm to develop into a liver bud, should be instructive for many other organogenic systems and for manipulating tissues in regenerative contexts for biomedical purposes.

Current Topics in Developmental Biology, Volume 117
ISSN 0070-2153
http://dx.doi.org/10.1016/bs.ctdb.2015.12.015

The liver is the first tissue to be specified in the endoderm germ layer and its well-defined location, relative cellular homogeneity, and rapid growth have made it easy to observe and analyze the underlying regulatory and morphogenetic processes. With few exceptions, these processes are remarkably well conserved in embryos of the mouse, chick, fish, and frog, allowing the best of each experimental model to be employed toward the common goal of understanding mechanisms of liver development. Together, the model systems have allowed dynamics in chromatin, signaling, and cell interactions to be studied in ways that have eluded other tissues that develop in a more dispersed fashion, that are composed of more cell types, or that grow more slowly.

On the other hand, the internal location of the liver has caused it to be excluded from most large-scale screens of genes affecting development, because such screens typically have been based on external visual phenotypes, such as changes in embryo or animal morphology or movement. But with the advent of reverse genetics in the 1990s and RNA interference technologies in the 2000s, where genes of interest could be knocked out in mice or knocked down in chick, fish, and frogs, diverse genes have been found to perturb liver development, more than adequately providing definitive genetic insight into the steps of liver organogenesis. This essay is divided into fundamental concepts of developmental biology that have been investigated with the liver system and the gaps that remain, and is intended to be useful for people thinking about development, stem cell differentiation, and regenerative biology in general. The essay is not intended to be a comprehensive review of the field, but rather is focused on questions that have driven thinking and research in the area and problems that remain to be solved in the future.

1. DISTINCT EMBRYONIC AND ADULT FUNCTIONS FOR THE SAME TISSUE

In the adult, the liver receives nutrient-laden blood from the intestine, after a meal, and the hepatocytes in the liver absorb and metabolize the nutrients for the rest of the body. Hepatocytes also store glucose, in the form of glycogen, and release it upon energy need. Hepatocytes secrete various serum proteins, such as albumin, which act as blood carriers for fatty acids and diverse other molecules. Hepatocytes also secrete bile into ducts that connect to the gall bladder; bile released from the gall bladder, into the gastrointestinal tract, aids in the digestion of fats. Finally, since the liver is the first tissue to receive ingested compounds via the bloodstream, the liver expresses an extensive array of proteins that detoxify harmful substances. Indeed, the liver is well known for its regenerative capacity (Miyajima, Tanaka, & Itoh, 2014; Zaret & Grompe, 2008), likely having evolved because the organ is the first line of defense when the body absorbs toxins. Given the unusually diverse functions of the liver and its coverage of so many metabolic pathways, it is no surprise that hepatocytes express a large number of genes and that liver function is necessary for adult survival.

Yet, as best we know, few if any of the above functions are operative in the embryo or fetus, despite that the liver is the first endodermal tissue to develop and it grows very large, very rapidly. The liver is, however, the major site of fetal hematopoiesis. Fetal hematopoietic stem cells (HSCs) are specified from the ventral aorta near the developing gonad and mesonephros (Medvinsky & Dzierzak, 1996). The HSCs migrate to the emerging liver, starting at days 10.5–12.5 of mouse embryo gestation (E10.5–12.5), and develop into the various hematopoietic lineages. Blood cells rapidly constitute up to 60% of the liver cell mass during fetal life (Paul, Conkie, & Freshney, 1969). Fortunately, for studies of hepatocytes themselves at these stages, simple mincing of the fetal livers releases the blood cells and enriches for hepatocytes remaining within the liver matrix. Shortly after birth, the hematopoietic system migrates to the bone marrow.

In summary, the liver has markedly distinct functions before and after birth. It is clear that both of these functions are elicited by hepatocytes, the primary functional cell (or, "parenchymal cell") of the liver. Various adult liver genes, such as *albumin* and *ttr* (for transthyretin, another secreted serum protein), are expressed at the onset of liver development, during the birth of hepatoblasts from the ventral foregut endoderm (Cascio & Zaret, 1991) at about E8.5 in the mouse. During the period of E8.5–10, the liver buds from the endoderm, and by E10.5 the liver develops into a separate organ into which the hematopoietic cells can invade. The constituent hepatoblasts are bipotential, differentiating mostly into hepatocytes and partially into cholangiocytes (bile duct cells) from E14.5 to E16.5, with the hepatocytes continuing to support hematopoiesis. In response to signals prior to birth, hepatocytes induce new mRNAs as they mature from supporting hematopoiesis to preparing for metabolic action on ingested food (Kamiya, Kinoshita, & Miyajima, 2001). Thus, hepatocytes exhibit a marked, stage-specific change in their transcriptional program in the perinatal period. Yet the genes and networks necessary for the hepatic support of fetal hematopoiesis are an understudied area, and how the hematopoietic support is shut off in postnatal hepatocytes is simply unknown. Significantly, liver bud emergence and the fetal and adult functions of the liver are critically dependent upon contacts with endothelial cells and the vasculature (LeCouter et al., 2003; Matsumoto, Yoshitomi, Rossant, & Zaret, 2001). Thus, a vascular influence is a common theme for all stages of liver development, as it is for many other tissues (Red-Horse, Crawford, Shojaei, & Ferrara, 2007).

2. PIONEER FACTORS AS A BASIS FOR DEVELOPMENTAL COMPETENCE

Of the three germ layers that arise during gastrulation, endoderm, ectoderm, and mesoderm, only the endoderm is competent to induce the liver. This conclusion was initially drawn from careful embryo tissue dissection studies of the chick (Fukuda-Taira, 1981; Le Douarin, 1975) and later affirmed in studies of mouse (Gualdi et al., 1996; Houssaint, 1980) and frog (Zorn & Wells, 2007) embryos. While there is much focus presently on the pluripotent states in embryonic stem (ES) cells, the embryonic epiblast, and primordial germ cells, for all other progenitor and stem cells at all other developmental stages, multipotency is the norm, not pluripotency. This raises the question of how multipotent progenitor or stem cells gain the competence to develop into certain kinds of descendant cell types and not others. Since the aforementioned tissue dissection studies showed that isolated endodermal cells retain the competence to be induced into liver *in vitro*, such competence must be manifest in a cell-intrinsic manner. At the chromosomal level, it suggests "priming" of genes for endoderm descendant fates, repression of genes of ectodermal and mesodermal fates, or both.

Since the *albumin* gene is activated at the moment of hepatic induction and remains a highly liver-specific gene throughout life, it was a useful model for understanding the mechanisms of liver gene control (Zaret, 1999). Furthermore, mouse embryonic endoderm dissection studies showed that dorsal–posterior endoderm, but not ectoderm, outside the foregut domain where the liver normally is induced, was silent for the *albumin* gene but competent to induce it *in vitro* from the E8.5–12.5 stages (Bossard & Zaret, 1998; Gualdi et al., 1996). More recent studies showed that posterior endoderm at the corresponding stages in zebrafish is also competent for hepatic induction (Shin, Lee, Poss, & Stainier, 2011). By E14.5 in the mouse, the dorsal–posterior endoderm loses such competence to induce *albumin*, as the cells initiate intestinal differentiation (Bossard & Zaret, 2000).

In vivo footprinting of the *albumin* enhancer in such uninduced, but competent, endoderm from E9.5 to E12.5 showed that essential binding sites for the endodermal transcription factors FoxA1 and FoxA2 (formerly HNF3A and HNF3B) and GATA4/6 are occupied, and that by E14.5, when competence is lost, the binding sites are no longer occupied (Bossard & Zaret, 1998, 2000; Gualdi et al., 1996). In liver buds at E9.5, when the *albumin* gene turns on, nearby binding sites become occupied (Gualdi et al., 1996); such sites are known to be essential for enhancer activity (DiPersio, Jackson, & Zaret, 1991; Jackson et al., 1993; Liu, DiPersio, & Zaret, 1991). Taking the data together, it suggests that the FoxA and GATA4/6 factors impart the competence of the *albumin* gene to be activated in the endoderm and that the absence of the factors in ectoderm and the posterior–distal endoderm, after E14.5, explains a lack of competence for the *albumin* gene to be induced in those tissues (Zaret, 1999).

Indeed, genetic ablation of the unlinked FoxA1 and FoxA2 genes in the mouse foregut endoderm results in a lack of induction of hepatoblasts in the endoderm, confirming the necessity of the factors for hepatic competence (Lee, Sund, Behr, Herrera, & Kaestner, 2005). At the onset of endoderm formation, during gastrulation, FoxA2 and then FoxA1 are first expressed (Ang et al., 1993; Monaghan, Kaestner, Grau, & Schütz, 1993; Sasaki & Hogan, 1993), and germline FoxA2 homozygous deletion results in embryonic lethality, due to the defects in endoderm and the other tissues in which the factor is expressed (Ang & Rossant, 1994; Hannenhalli & Kaestner, 2009; Weinstein et al., 1994). GATA4 and GATA6 are also expressed in the foregut endoderm and necessary for liver development in both mice and zebrafish (Holtzinger & Evans, 2005; Watt, Zhao, Li, & Duncan, 2007; Zhao et al., 2005).

Contemporaneously with these studies, the crystal structure of the FoxA DNA-binding domain revealed it to have an overall fold that is similar to that seen for linker histone (Clark, Halay, Lai, & Burley, 1993; Ramakrishnan, Finch, Graziano, Lee, & Sweet, 1993) and FoxA was then shown to be able to bind its target DNA sequence on a nucleosome *in vitro* (Cirillo et al., 1998) and *in vivo* (Chaya, Hayamizu, Bustin, & Zaret, 2001; Li, Schug, Tuteja, White, & Kaestner, 2011). Purified FoxA protein, and to a lesser extent purified GATA4 protein, could bind to target sequences in a compacted nucleosome array *in vitro*, and such binding is sufficient to make the underlying nucleosome accessible to different nucleases (Cirillo et al., 2002). This "chromatin opening" *in vitro* was a model for the action of the factors on transcriptionally silent chromatin in the embryonic endoderm.

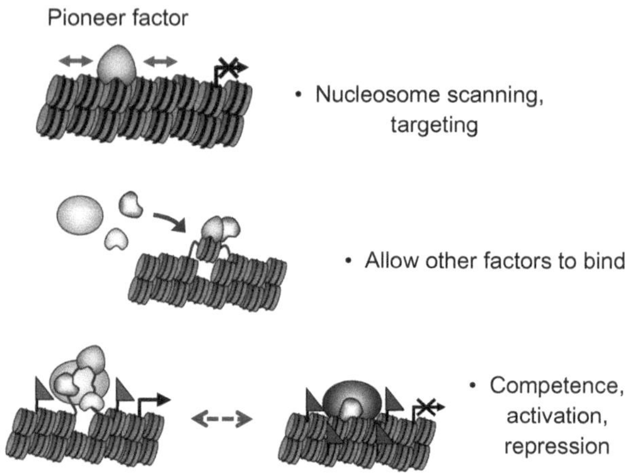

Figure 1 Chromatin binding by pioneer factors and consequences for gene expression. *From Zaret and Grompe (2008). Reprinted with permission from AAAS.* (A color version of this figure is located at the end of this chapter.)

Thus, the factors were called "pioneer factors" for their ability to be the first to bind a silent chromatin region and initiate gene regulatory events (Cirillo et al., 2002) (Fig. 1).

Recent studies using mouse and human ES cells differentiating to endoderm and hepatic lineages provided further support for the concept that FoxA proteins act as pioneer factors in the endoderm. First, genome-wide studies showed that when FoxA2 is first expressed in endoderm derived from ES cells, it occupies sites of many genes that are expressed in endoderm as well as silent liver genes, including at the *albumin* enhancer, prior to the differentiating the cells to a hepatic fate *in vitro* (Li et al., 2011; Xu et al., 2012). Thus, FoxA2 operates genome wide as was seen for the single site monitored by the original *in vivo* footprinting studies. Second, similar, more extensive studies showed that FoxA2 also occupies silent genes in endoderm for non-liver endodermal fates and helps impart the competence to activate those fates (Wang, Yue, et al., 2015).

It is now appreciated that different pioneer factors work in diverse developmental and gene regulatory contexts (Budry et al., 2012; Carroll et al., 2005; Menet, Pescatore, & Rosbash, 2014; Zaret & Carroll, 2011). Sherwood and colleagues showed that it is possible to mine genomic chromatin accessibility data to discover new pioneer factors by their motif occurrence at sites that become opened (accessible) during a developmental transition (Sherwood et al., 2014). These authors were able to distinguish between "pioneers" that initiate gene regulatory events and "settler" transcription factors that perform secondary steps of gene activation. While most FoxA pioneer factor binding in mammalian cells seems to occur at distal enhancers, recent studies in *Caenorhabditis elegans* reveal that pioneer factors can directly lead to RNA polymerase recruitment at promoters (Hsu et al., 2015). Yet pioneer factors such as FoxA can also recruit corepressors such as GRG/Groucho, causing the local closing of chromatin structure (Sekiya & Zaret, 2007), as well as bind with repressive transcription factors that in turn suppress gene activity (Watts et al., 2011). Thus, depending upon the genetic context, pioneer factor binding can enable developmental competence, transcriptional activation, or transcriptional repression (Fig. 1).

Pioneer factor activity is now being appreciated to be crucial for transcription factors that can reprogram mammalian cell fate, such as during reprogramming fibroblasts to induced pluripotent stem cells (Soufi, Donahue, & Zaret, 2012), fibroblasts to neurons (Wapinski et al., 2013), and B cells to macrophages (van Oevelen et al., 2015). The basis for pioneer activity appears to reside in the ability of a transcription factor to recognize its DNA motif, or a partial DNA motif, on the surface of a nucleosome, which can in turn enable cooperatively with other factors (Soufi et al., 2015). In summary, studies of hepatic competence in the endoderm have been useful for revealing generally applicable principles by which cell fate changes can be enabled during developmental transitions and transdifferentiation (Iwafuchi-Doi & Zaret, 2014).

3. INDEPENDENT PROGENITOR CELL SOURCES FOR ONE CELL TYPE

Various internal organs are constructed from separate progenitor domains. For example, the pancreas arises from two buds of tissue from the endoderm, from the ventral foregut endoderm, adjacent to the liver domain, and from the dorsal endoderm (Zaret, 2008). Each bud gives rise to distinct portions of the adult pancreas, and both descendant sections are composed of acinar, duct, and endocrine cells, indicating common final genetic programs. Yet the induction of the ventral and dorsal pancreatic buds is elicited by different signals and employs only partially overlapping sets of transcription factor genes (Zaret, 2008). More sophisticated genetic lineage labeling studies, along with dye labeling, show that heart develops from two distinct progenitor domains of mesoderm, though here the different descendant domains have different functions in heart structure (Abu-Issa, Waldo, & Kirby, 2004).

When studying the location of the endodermal origins of the liver, it can be confounding to compare fate maps derived from fish, where the early endoderm first forms a rod of cells, to fate maps from frog, chick, and mouse embryos, where the early endoderm first forms an epithelial sheet of cells (Chalmers & Slack, 2000; Horne-Badovinac et al., 2001; Kirby et al., 2003; Lawson, Meneses, & Pedersen, 1986; Lawson & Pedersen, 1987; Tam, Khoo, Wong, Tsang, & Behringer, 2004; Warga & Nusslein-Volhard, 1999). Yet for either type of endoderm development, dye labeling studies have found that paired lateral domains of endoderm converge to form a liver bud (Chalmers & Slack, 2000; Rosenquist, 1971; Tremblay & Zaret, 2005; Warga & Nusslein-Volhard, 1999). More detailed studies of mouse embryos found an additional site of cells that contribute to the liver bud: those emerging from the ventral midline of the endoderm lip (VMEL) (Tremblay & Zaret, 2005), where the "lip" corresponds to where the sheet of endoderm folds into the anterior intestinal portal, which will make the foregut. The VMEL cells have been seen in chick and mouse as a cluster of perhaps 10 cells that migrate rostrally along the ventral midline as the foregut closes off (Kirby et al., 2003; Tam et al., 2004; Tremblay & Zaret, 2005). Further studies showed that the VMEL cells are a multipotent cell population that contributes to the ventral midline cells within the thyroid, liver, and ventral pancreas bud (Angelo, Guerrero-Zayas, & Tremblay, 2012). In summary, the liver bud develops from three endodermal cell sources: paired lateral domains which become fated directly to become the liver, and VMEL cells which, when labeled, give rise to descendants in the liver and other endodermal tissues (Fig. 2).

We do not yet know how each of the descendants of the lateral and medial (VMEL) sources of liver bud cells may contribute to different spatial or functional domains of the adult liver. The multipotency of the VMEL progenitors is intriguing and it would be particularly interesting if VMEL descendants in the liver possess a different regenerative capacity than the bulk of the tissue, the latter of which seems to be generated by the lateral liver domains. Clearly, genetic lineage tracing is needed whereby the lateral and VMEL progenitors can be marked and their descendant cells followed through organogenesis. Lineage tracing has been used to provide evidence for a KDR-positive progenitor of a subset of liver cells; KDR expression has been thought to be restricted to mesodermal cells (Goldman et al., 2013). The relationship of the early KDR+ liver progenitors and the VMEL and lateral progenitors is unclear and represents another area for further research.

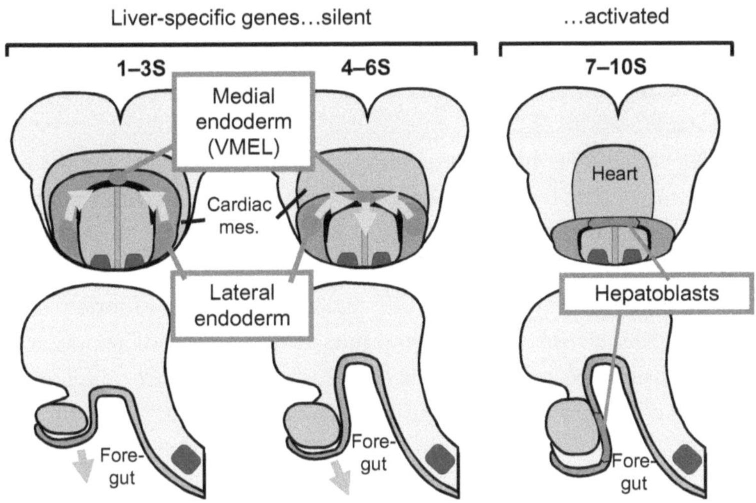

Figure 2 Location of three hepatic progenitor domains (pink) within the ventral foregut endoderm (green). Arrows depict cell movement during foregut closure. (A color version of this figure is located at the end of this chapter.)

4. COMBINATORIAL INDUCTIVE SIGNALING AND ALTERNATIVE FATE CONTROL

Diverse experimental approaches have revealed that the hepatic fate is induced in the foregut endoderm by combinatorial signaling from different overlying mesoderm-derived cell populations. The classic chick embryo tissue recombination studies of Le Douarin showed that overlying cardiac mesoderm induces the hepatic fate in the endoderm (Le Douarin, Houssaint, Jotereau, & Belo, 1975). Later studies of mouse endoderm explants confirmed cardiac induction and showed that cardiac fibroblast growth factor (FGF) signaling is required for hepatogenesis (Jung, Zheng, Goldfarb, & Zaret, 1999). FGF signaling was subsequently shown to induce hepatogenesis in the frog (Chen et al., 2003) and fish (Shin et al., 2007). Recent studies using FGF signaling inhibitors with whole mouse embryos indicate that FGF signaling controls the anterior portion of the liver bud, with the inhibitor not affecting the posterior portion of the bud (Wang, Rhee, Palaria, & Tremblay, 2015). Detailed studies with different FGF pathway inhibitors showed that downstream MAPK signaling appears most important for hepatic gene induction, whereas PI3K signaling appears most important for hepatic endoderm outgrowth (Calmont et al., 2006). Genetic expression of Spry2, a dominant FGF signaling inhibitor, in the mouse foregut endoderm of live mouse embryos inhibited hepatic differentiation, with "escaper" cells that did not induce Spry2 eventually generating a liver (Calmont et al., 2006). Indeed, studies with chimeric embryos have shown that the hepatic bud is highly regenerative in the context of early hepatic failure and can select for rare cells without a defect, for outgrowth (Bort, Signore, Tremblay, Barbera, & Zaret, 2006).

Further studies with mouse embryonic ventral foregut endoderm demonstrated that different amounts of FGF signaling result in different fate outcomes. Very low or no FGF input results in induction of pancreas genes, low-to-medium levels of FGF result in induction of hepatic genes, and high levels of FGF result in induction of lung genes (Deutsch, Jung, Zheng, Lóra, & Zaret, 2001; Serls, Doherty, Parvatiyar, Wells, & Deutsch, 2005). A gap remains in understanding how different threshold levels of FGF signaling result in different cell fate outcomes in the ventral foregut endoderm. Expression of the transcription factor vHNF1 in the endoderm is necessary for hepatic induction, apparently by allowing responsiveness to FGF (Lokmane et al., 2008). It is not yet known how vHNF1 coordinates with other known endodermal transcription factors in the FoxA and GATA families.

The initial studies of FGF signaling in mouse endoderm involved a small observational detail with high importance for studying developmental events *in vitro*. Endoderm fragments that contained a few mesenchymal cells were able to attach to the substratum and respond to FGF by inducing hepatic genes, whereas fragments that lacked mesenchymal cells failed to attach and were difficult to maintain in culture (Gualdi et al., 1996; Jung et al., 1999). Subsequent studies showed that the mesenchymal cells express BMP4 and that indeed BMP4 from rare septum transversum mesenchyme cells that "contaminated" the endoderm cultures have an important role, *in vitro* and *in vivo*, in inducing the hepatic fate (Rossi, Dunn, Hogan, & Zaret, 2001) (Fig. 3). Later in development, the septum transversum mesenchyme cells give rise to the mesothelial capsule that surrounds the liver and to stellate cells, the latter of which get activated upon liver damage (Asahina, Zhou, Pu, & Tsukamoto, 2011).

Furthermore, BMP signaling suppresses the pancreatic fate in the endoderm while allowing the hepatic fate, as seen in both mouse and zebrafish (Chung, Shin, & Stainier, 2008; Deutsch et al., 2001; Shin et al., 2007). A beautiful lineage labeling study in zebrafish soaked the embryos in a dye that required laser activation for fluorescence; by activating the flour in selected cells, they could track descendant cells live in developing

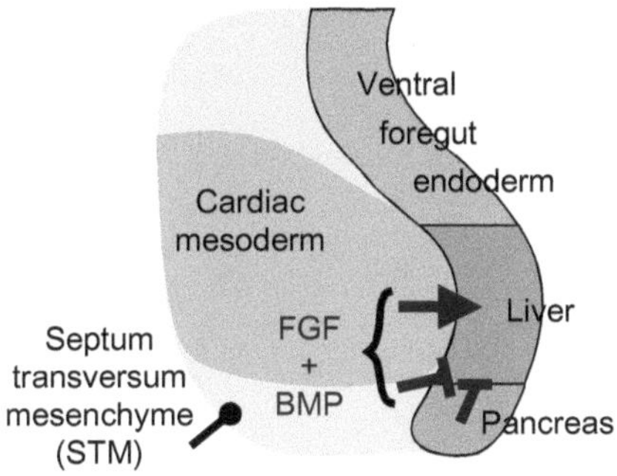

Figure 3 Combinatorial FGF and BMP signaling induces the liver fate (pink) while repressing the pancreas fate (dark green) within the endoderm (pale green). (A color version of this figure is located at the end of this chapter.)

embryos (Chung et al., 2008). Using this method, the authors unequivocally proved the existence of a bipotential endodermal cell that could generate hepatic or pancreatic descendants, with a role for BMP signaling in driving the hepatic fate at the expense of the pancreatic fate (Chung et al., 2008). Additional complexity was revealed when prostaglandin signaling was found to modulate the initial BMP role in hepatic induction (Nissim et al., 2014). Thus, an elaborate network of combinatorial signals from two mesoderm-derived tissues, the cardiac mesoderm and the septum transversum mesenchyme, precisely governs the induction of the liver in the endoderm.

An analysis of the signal transduction pathways activated in response to the FGF and BMP signals in the foregut endoderm revealed two unexpected features. First, pERK phosphorylation in the endoderm, downstream of FGF, appears in the lateral liver progenitors prior to appearing in the VMEL progenitors (Calmont et al., 2006). An hour or so later, pERK phosphorylation appears in the VMEL progenitors as well. Conversely, phosphorylation of SMAD1,5,8, downstream of BMP, appears first in the VMEL progenitors and, then an hour or so later, in the lateral liver progenitors (Wandzioch & Zaret, 2009). These observations suggest that while both the lateral and VMEL liver progenitors receive FGF and BMP signaling, they do so in a different order. The role of such order in affecting gene expression patterns within the cells, or effects on descendant cell capabilities, is unknown. In addition, the transcription factor Hex appears to affect the movement of endoderm cells with respect to the overlying mesoderm tissues, thereby affecting the timing by which the endoderm receives signals for the liver versus pancreas fates (Bort, Martinez-Barbera, Beddington, & Zaret, 2004).

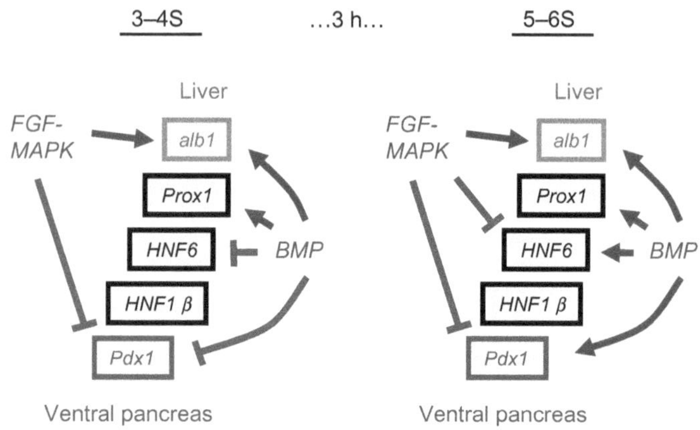

Figure 4 Dynamics in inductive signaling networks during hepatic and ventral pancreatic specification within the endoderm. (A color version of this figure is located at the end of this chapter.)

The second unexpected feature of the hepatic versus pancreatic induction network, with regard to BMP signaling, is that it is dynamic. That is, as noted above, at the time of hepatic induction, BMP signaling is positive for the liver fate and inhibitory for the pancreatic fate. But hours after the hepatic versus pancreatic decision is made, BMP signaling then promotes early ventral pancreas development, instead of inhibiting it (Wandzioch & Zaret, 2009) (Fig. 4). Differences in FGF and TGFbeta signal responsiveness were also seen at the different time points (Wandzioch & Zaret, 2009). It is remarkable that the cell could change so rapidly with regard to its interpretation of an inductive signal; how such is achieved remains to be determined. Also, these findings underscore the importance of precisely defining and staging the cell population under analysis.

A discrepancy between liver induction in the fish and the mouse has been seen for canonical Wnt signaling. In zebrafish, Wnt signaling appears to be necessary for hepatic induction (Ober, Verkade, Field, & Stainier, 2006), whereas in the frog (Xenopus) and mouse, it appears that Wnt signaling needs to be repressed for the initial hepatic induction (Finley, Tennessen, & Shawlot, 2003; McLin, Rankin, & Zorn, 2007). EpCAM signaling may derepress Wnt to allow hepatic induction in the fish, while a signal for repressing Wnt in the mouse, to allow hepatic induction, could be from local endothelial cells (Gouon-Evans et al., 2006; Han et al., 2002). The absence of early Wnt in frogs allows the induction of the transcription factor Hex (McLin et al., 2007), the latter of which is required to be expressed in the endoderm for liver development (Bort et al., 2004; Martinez-Barbera et al., 2000), whereas in the fish, Hex expression is dependent upon Wnt. A previous review described differences between fish and amniote gut development that can explain the disparity in Wnt signaling between the species (Zaret, 2008).

Yet immediately after hepatic induction in the mouse and frog, it seems that Wnt signaling is positive for hepatic bud outgrowth and further differentiation (Monga et al., 2003; Monga, Pediaditakis, Mule, Stolz, & Michalopoulos, 2001; Ueno et al., 2007). Also, a detailed transcriptomic study using RNA-seq at early developmental stages revealed dynamics in noncanonical Wnt signaling-related genes during the hepatic–pancreatic decision (Rodriguez-Seguel et al., 2013). Recent studies discovered a population of hepatocytes that surround the central vein of the adult liver, driven by canonical Wnt signaling from adjacent endothelial cells, that may have a distinct role in homeostatic self-renewal (Wang, Zhao, Fish, Logan, & Nusse, 2015). Clearly, there is much to learn about the influences of Wnt and other signals in the endoderm and how such signals may be interpreted differently during hepatic induction, hepatic bud outgrowth, and later stages of development and homeostasis of the liver. A comprehensive analysis of transcription factor expression dynamics in embryos should help illuminate how the liver and other endodermal organs are controlled (Sherwood, Chen, & Melton, 2009).

5. CONNECTING INDUCTIVE SIGNALING TO COMPETENT STATES OF CHROMATIN AND CELL FATE CONTROL

The signals that induce a hepatic fate and suppress a pancreatic fate within the ventral foregut endoderm must connect to networks within cells that impact gene expression in very discrete way, enabling one genetic program while repressing another. Inductive signaling acts upon pioneer factors that have established competent states in progenitors, as discussed above, but pioneer factors also operate in the context of chromatin, which contains nucleosomal histones with diverse covalent modifications and attendant enzymes and binding complexes that, in turn, help modify gene activity and affect the persistence of gene expression states.

To address this issue, a superovulation protocol to isolate many early-stage mouse embryos per litter, along with a method for isolating mouse ventral foregut endoderm cells by fluorescence-activated cell sorting (Gadue et al., 2009), allowed the development of chromatin immunoprecipitation technology from 10,000 to 20,000 sorted endoderm cells and their immediate hepatoblast descendants (Xu et al., 2011). The undifferentiated endoderm and hepatoblast cells were screened for differences in histone modifications, reflective of different chromatin states, at representative liver-specific gene regulatory elements, pancreas-specific gene regulatory elements, and a regulatory element active in both cells (GAPDH promoter). Notably, the target genes of the liver and pancreas regulatory elements were silent in the endoderm, and only the liver genes become activated in the hepatoblasts. These studies revealed a "prepattern" of chromatin states in the endoderm by which the liver-specific elements lack histone H3 acetylation, normally indicative of an active state, and lack H3K27me3, normally indicative of a repressed state. By contrast, the pancreas-specific elements contain both H3 acetylation and H3K27me3, reflecting a novel "bivalent" state of inactive chromatin in the endoderm. During hepatic induction, the pattern does not change at the pancreas elements, but the liver elements gain H3 acetylation. Further studies indicated that the gain of H3 acetylation in the hepatoblasts is due to Smad4, apparently downstream of BMP signaling, which in turn recruits the histone acetyltransferase P300; and knockouts of genes encoding Smad4 and P300, or an inhibitor of P300, suppress hepatic induction and expand pancreatic induction (Xu et al., 2011). In addition, the study showed that genetic inactivation of Ezh2, one of the enzymes that elicits the repressive H3K27me3 mark, or treating embryo fragments with an Ezh2 inhibitor (Xu et al., 2014), enhances the yield of pancreatic progenitors (Fig. 5). These studies revealed how inductive signaling can connect to preconditioned states at target gene chromatin, though much remains to be learned about the exact regulatory mechanisms involved.

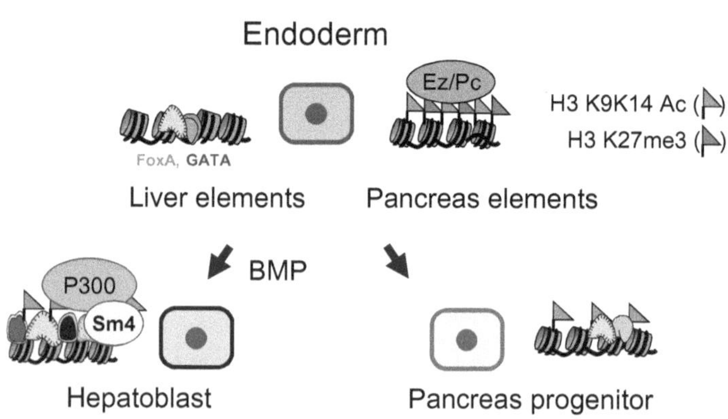

Figure 5 Prepattern in chromatin states differs between selected liver and pancreas regulatory sequences in the undifferentiated endoderm, and changes elicited by BMP signaling. (A color version of this figure is located at the end of this chapter.)

Despite that the aforementioned chromatin analysis yielded clues about histone modifiers that could be genetically confirmed to affect the liver–pancreas fate choice, a far more extensive, genome-wide study of chromatin states in ES cells and their hepatic and pancreatic descendants failed to observe a prepattern across all liver and pancreas genes (Loh et al., 2014). This could reflect differences in gene curation, differences in cell culture versus embryos (e.g., see Xu et al., 2014), or a simple absence of consistent features genome wide. Regardless, the study availed of the much greater cell populations that can be obtained from ES cultures and provided a wealth of data regarding the chromatin transitions and gene expression changes that accompany the lineage bifurcations discussed above (Loh et al., 2014). Notably, they discovered a variety of chromatin states at enhancers, prior to their activation during cell fate commitment, which reflect a previously unappreciated complexity with which the diverse inductive signals described above must integrate with genomic sites in chromatin.

6. ONSET OF MORPHOGENESIS

Two aspects of bud emergence figure prominently in our understanding of the postspecification phase of early liver development: tissue morphogenesis and an increased rate of cell division. The first discernable morphogenetic change in the hepatic endoderm, after being induced, is a transition from a columnar morphology to a pseudostratified morphology, with concomitant interkinetic nuclear migration during cell division (Bort et al., 2006) (Fig. 6, compare 10–11S with 18–20S panels). This transition is dependent upon the transcription factor Hex (Bort et al., 2006), which as mentioned above also affects the earlier movement of the endoderm against the overlying mesoderm. Next, the nascent hepatoblasts emerge into the stromal environment and continue generating the liver bud (Bort et al., 2006) (Fig. 6, 27–30S). During this transition, the transcription factor Prox1 promotes a breakdown in cell–cell contacts that allow the liver bud cells to emerge (Sosa-Pineda, Wigle, & Oliver, 2000). The Klf6 transcription factor also promotes hepatic bud outgrowth (Zhao et al., 2010). The induction of the transcription factor Tbx3 suppresses the growth suppressing gene p19Arf, thereby allowing enhanced replication of the hepatoblasts and their emergence into a bud (Suzuki et al., 2008). Finally, after the initial emergence of the liver bud, as the differentiating hepatocytes reepithelialize,

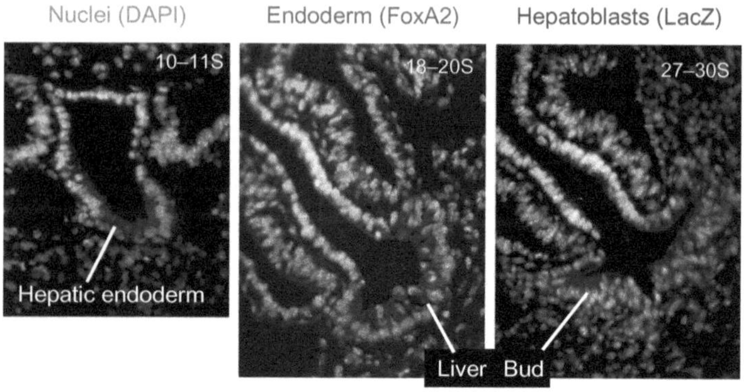

Figure 6 Morphogenesis of the liver bud. The three panels depict the columnar epithelium (10–11S) of newly specified hepatoblasts, the pseudostratified epithelium of the thickened hepatic endoderm (hepatoblast) domain (18–20S), and the hepatoblasts beginning to migrate into the stroma to generate the liver bud (27–30S). (A color version of this figure is located at the end of this chapter.)

the transcription factor HNF4 plays an important role (Parviz et al., 2003). As can be seen, bud outgrowth is complex and diverse factors control the process.

Interestingly, when the bud emergence phase is blocked by a homozygous mutation in the Hex gene, within a day or two the hepatoblasts switch from expressing liver genes to become gut-like cells (Bort et al., 2006). Thus, if the liver program is not stabilized by some means, after a few days the cells switch to a gut fate. It will be important to understand the gene regulatory transitions within the cells during this period that normally stabilize a hepatic fate.

By the time of mouse hepatic bud emergence, the endoderm has moved caudally, during gut closure, away from the cardiac region. The emergent hepatic cells enter the stromal environment, which contains septum transversum mesenchyme cells and endothelial cells. While later, the endothelial cells will form the vasculature upon which hepatic function so intimately depends, at the initial stage, prior to vascular formation, the endothelial cells are clearly nonvascular and yet still send an important growth signal to the hepatocytes (Matsumoto et al., 2001). In the fish, endothelial signals control the polarity of the nascent hepatocytes, again affecting early bud emergence (Sakaguchi, Sadler, Crosnier, & Stainier, 2008). Thus, once again, coordinate signaling from multiple environmental cell types modulates liver development. Interestingly, endothelial cell signals affect adult liver homeostasis, as noted above, as well as hepatocyte regeneration after injury (Ding et al., 2010; LeCouter et al., 2003). It will be important to determine how the endothelial cell population comatures with hepatocytes so that the two interdependent cell types coordinate their developmental state and function.

7. THE FUTURE OF PROGRAMMING FROM EMBRYONIC STEM CELLS AND OTHER CELL TYPES

While studying developmental biology, such as of the liver, is an interesting scientific endeavor, in and of itself, there are three major reasons for applying such knowledge to make new hepatocytes from other cell sources. First, hepatocytes are essential for drug toxicity testing, so if we can make hepatocytes from different human genetic backgrounds, e.g., by first making iPS cells from blood or skin cell biopsies (Takahashi & Yamanaka, 2006) and then generating hepatocytes from them, it could be possible to provide far more specificity with regard to predicting toxic behaviors of chemicals in different human populations. Second, the same technology could be applied to study hepatocytes from different genetic disease backgrounds that affect liver function (Merkle & Eggan, 2013). And third, given the paucity of liver donors, there is an imperative to generate large quantities of hepatocytes for transplantation (Bhatia, Underhill, Zaret, & Fox, 2014).

Some have argued that adult hepatocytes have uniquely adult features that will be difficult to replicate using principles of developmental biology (Dor & Stanger, 2007) and, to date, there are no protocols that are reproducible in different labs for readily generating fully mature hepatocytes from stem cells. However, application of the signaling inducers for embryonic hepatogenesis has been invaluable for generating hepatic cells from ES cells (see Zaret & Grompe, 2008). And iPS differentiation to hepatoblasts, followed by coculture of the cells with mesenchyme and endothelial cells, leads to a liver bud like morphology that has better transplantation outcomes than the hepatoblasts alone (Takebe et al., 2013). Despite how many years it has taken the field to get this far, as noted in each section above, there are many features yet to learn about liver development. Indeed, despite even more years of research on adult hepatocyte culture, there is not yet a common protocol for expanding the cell populations *in vitro*. In the end, it seems likely that a better understanding of developmental events, coupled with a better understanding of adult hepatocyte function, will bring us to a point where liver biology and function can be controlled and replicated at will.

REFERENCES

Abu-Issa, R., Waldo, K., & Kirby, M. L. (2004). Heart fields: One, two or more? *Developmental Biology*, *272*, 281–285.

Ang, S.-L., & Rossant, J. (1994). *HNF-3b* is essential for node and notochord formation in mouse development. *Cell*, *78*, 561–574.

Ang, S.-L., Wierda, A., Wong, D., Stevens, K. A., Cascio, S., Rossant, J., et al. (1993). The formation and maintenance of the definitive endoderm lineage in the mouse: Involvement of HNF3/*forkhead* proteins. *Development*, *119*, 1301–1315.

Angelo, J. R., Guerrero-Zayas, M. I., & Tremblay, K. D. (2012). A fate map of the murine pancreas buds reveals a multipotent ventral foregut organ progenitor. *PLoS One*, *7*, e40707.

Asahina, K., Zhou, B., Pu, W. T., & Tsukamoto, H. (2011). Septum transversum-derived mesothelium gives rise to hepatic stellate cells and perivascular mesenchymal cells in developing mouse liver. *Hepatology*, *53*, 983–995.

Bhatia, S. N., Underhill, G. H., Zaret, K. S., & Fox, I. J. (2014). Cell and tissue engineering for liver disease. *Science Translational Medicine*, *6*, 245sr2.

Bort, R., Martinez-Barbera, J. P., Beddington, R. S., & Zaret, K. S. (2004). *Hex* homeobox gene-dependent tissue positioning is required for organogenesis of the ventral pancreas. *Development*, *131*, 797–806.

Bort, R., Signore, M., Tremblay, K., Barbera, J. P., & Zaret, K. S. (2006). Hex homeobox gene controls the transition of the endoderm to a pseudostratified, cell emergent epithelium for liver bud development. *Developmental Biology*, *290*, 44–56.

Bossard, P., & Zaret, K. S. (1998). GATA transcription factors as potentiators of gut endoderm differentiation. *Development*, *125*, 4909–4917.

Bossard, P., & Zaret, K. S. (2000). Repressive and restrictive mesodermal interactions with gut endoderm: Possible relation to Meckel's diverticulum. *Development*, *127*, 4915–4923.

Budry, L., Balsalobre, A., Gauthier, Y., Khetchoumian, K., L'Honore, A., Vallette, S., et al. (2012). The selector gene Pax7 dictates alternate pituitary cell fates through its pioneer action on chromatin remodeling. *Genes and Development*, *26*, 2299–2310.

Calmont, A., Wandzioch, E., Tremblay, K. D., Minowada, G., Kaestner, K. H., Martin, G. R., et al. (2006). An FGF response pathway that mediates hepatic gene induction in embryonic endoderm cells. *Developmental Cell*, *11*, 339–348.

Carroll, J. S., Liu, X. S., Brodsky, A. S., Li, W., Meyer, C. A., Szary, A. J., et al. (2005). Chromosome-wide mapping of estrogen receptor binding reveals long-range regulation requiring the forkhead protein FoxA1. *Cell*, *122*, 33–43.

Cascio, S., & Zaret, K. S. (1991). Hepatocyte differentiation initiates during endodermal–mesenchymal interactions prior to liver formation. *Development*, *113*, 217–225.

Chalmers, A. D., & Slack, J. M. (2000). The Xenopus tadpole gut: Fate maps and morphogenetic movements. *Development*, *127*, 381–392.

Chaya, D., Hayamizu, T., Bustin, M., & Zaret, K. S. (2001). Transcription factor FoxA (HNF3) on a nucleosome at an enhancer complex in liver chromatin. *The Journal of Biological Chemistry*, *276*, 44385–44389.

Chen, Y., Jurgens, K., Hollemann, T., Claussen, M., Ramadori, G., & Pieler, T. (2003). Cell-autonomous and signal-dependent expression of liver and intestine marker genes in pluripotent precursor cells from Xenopus embryos. *Mechanisms of Development*, *120*, 277–288.

Chung, W. S., Shin, C. H., & Stainier, D. Y. (2008). Bmp2 signaling regulates the hepatic versus pancreatic fate decision. *Developmental Cell*, *15*, 738–748.

Cirillo, L., Lin, F. R., Cuesta, I., Jarnik, M., Friedman, D., & Zaret, K. (2002). Opening of compacted chromatin by early developmental transcription factors HNF3 (FOXA) and GATA-4. *Molecular Cell*, *9*, 279–289.

Cirillo, L. A., McPherson, C. E., Bossard, P., Stevens, K., Cherian, S., Shim, E.-Y., et al. (1998). Binding of the winged-helix transcription factor HNF3 to a linker histone site on the nucleosome. *The EMBO Journal*, *17*, 244–254.

Clark, K. L., Halay, E. D., Lai, E., & Burley, S. K. (1993). Co-crystal structure of the HNF3/*fork head* DNA recognition motif resembles histone H5. *Nature*, *364*, 412–420.

Deutsch, G., Jung, J., Zheng, M., Lóra, J., & Zaret, K. S. (2001). A bipotential precursor population for pancreas and liver within the embryonic endoderm. *Development*, *128*, 871–881.

Ding, B. S., Nolan, D. J., Butler, J. M., James, D., Babazadeh, A. O., Rosenwaks, Z., et al. (2010). Inductive angiocrine signals from sinusoidal endothelium are required for liver regeneration. *Nature*, *468*, 310–315.

DiPersio, C. M., Jackson, D. A., & Zaret, K. S. (1991). The extracellular matrix coordinately modulates liver transcription factors and hepatocyte morphology. *Molecular and Cellular Biology*, *11*, 4405–4414.

Dor, Y., & Stanger, B. Z. (2007). Regeneration in liver and pancreas: Time to cut the umbilical cord? *Science's STKE*, *2007*, pe66.

Finley, K. R., Tennessen, J., & Shawlot, W. (2003). The mouse secreted frizzled-related protein 5 gene is expressed in the anterior visceral endoderm and foregut endoderm during early post-implantation development. *Gene Expression Patterns*, *3*, 681–684.

Fukuda-Taira, S. (1981). Hepatic induction in the avian embryo: Specificity of reactive endoderm and inductive mesoderm. *Journal of Embryology and Experimental Morphology*, *63*, 111–125.

Gadue, P., Gouon-Evans, V., Cheng, X., Wandzioch, E., Zaret, K. S., Grompe, M., et al. (2009). Generation of monoclonal antibodies specific for cell surface molecules expressed on early mouse endoderm. *Stem Cells*, *27*, 2103–2113.

Goldman, O., Han, S., Sourisseau, M., Dziedzic, N., Hamou, W., Corneo, B., et al. (2013). KDR identifies a conserved human and murine hepatic progenitor and instructs early liver development. *Cell Stem Cell*, *12*, 748–760.

Gouon-Evans, V., Boussemart, L., Gadue, P., Nierhoff, D., Koehler, C. I., Kubo, A., et al. (2006). BMP-4 is required for hepatic specification of mouse embryonic stem cell-derived definitive endoderm. *Nature Biotechnology*, *24*, 1402–1411.

Gualdi, R., Bossard, P., Zheng, M., Hamada, Y., Coleman, J. R., & Zaret, K. S. (1996). Hepatic specification of the gut endoderm in vitro: Cell signaling and transcriptional control. *Genes and Development*, *10*, 1670–1682.

Han, H., Bearss, D. J., Browne, L. W., Calaluce, R., Nagle, R. B., & Von Hoff, D. D. (2002). Identification of differentially expressed genes in pancreatic cancer cells using cDNA microarray. *Cancer Research*, *62*, 2890–2896.

Hannenhalli, S., & Kaestner, K. H. (2009). The evolution of Fox genes and their role in development and disease. *Nature Reviews. Genetics*, *10*, 233–240.

Holtzinger, A., & Evans, T. (2005). Gata4 regulates the formation of multiple organs. *Development*, *132*, 4005–4014.

Horne-Badovinac, S., Lin, D., Waldron, S., Schwarz, M., Mbamalu, G., Pawson, T., et al. (2001). Positional cloning of heart and soul reveals multiple roles for PKC lambda in zebrafish organogenesis. *Current Biology*, *11*, 1492–1502.

Houssaint, E. (1980). Differentiation of the mouse hepatic primordium. I. An analysis of tissue interactions in hepatocyte differentiation. *Cell Differentiation*, *9*, 269–279.

Hsu, H.-T., Chen, H.-M., Yang, Z., Wang, J., Lee, N. K., Burger, A., et al. (2015). Recruitment of RNA polymerase II by the pioneer transcription factor PHA-4. *Science*, *348*, 1372–1376.

Iwafuchi-Doi, M., & Zaret, K. S. (2014). Pioneer transcription factors in cell reprogramming. *Genes and Development*, *28*, 2679–2692.

Jackson, D. A., Rowader, K. E., Stevens, K., Jiang, C., Milos, P., & Zaret, K. S. (1993). Modulation of liver-specific transcription by interactions between hepatocyte nuclear factor 3 and nuclear factor 1 binding DNA in close apposition. *Molecular and Cellular Biology, 13*, 2401–2410.

Jung, J., Zheng, M., Goldfarb, M., & Zaret, K. S. (1999). Initiation of mammalian liver development from endoderm by fibroblast growth factors. *Science, 284*, 1998–2003.

Kamiya, A., Kinoshita, T., & Miyajima, A. (2001). Oncostatin M and hepatocyte growth factor induce hepatic maturation via distinct signaling pathways. *FEBS Letters, 492*, 90–94.

Kirby, M. L., Lawson, A., Stadt, H. A., Kumiski, D. H., Wallis, K. T., McCraney, E., et al. (2003). Hensen's node gives rise to the ventral midline of the foregut: Implications for organizing head and heart development. *Developmental Biology, 253*, 175–188.

Lawson, K. A., Meneses, J. J., & Pedersen, R. A. (1986). Cell fate and cell lineage in the endoderm of the presomite mouse embryo, studied with an intracellular tracer. *Developmental Biology, 115*, 325–339.

Lawson, K. A., & Pedersen, R. A. (1987). Cell fate, morphogenetic movement and population kinetics of embryonic endoderm at the time of germ layer formation in the mouse. *Development, 101*, 627–652.

LeCouter, J., Moritz, D. R., Li, B., Phillips, G. L., Liang, X. H., Gerber, H. P., et al. (2003). Angiogenesis-independent endothelial protection of liver: Role of VEGFR-1. *Science, 299*, 890–893.

Le Douarin, N. M. (1975). An experimental analysis of liver development. *Medical Biology, 53*, 427–455.

Le Douarin, N. M., Houssaint, E., Jotereau, F. V., & Belo, M. (1975). Origin of hemopoietic stem cells in embryonic bursa of Fabricius and bone marrow studied through interspecific chimeras. *Proceedings of the National Academy of Sciences of the United States of America, 72*, 2701–2705.

Lee, C. S., Sund, N. J., Behr, R., Herrera, P. L., & Kaestner, K. H. (2005). Foxa2 is required for the differentiation of pancreatic alpha-cells. *Developmental Biology, 278*, 484–495.

Li, Z., Schug, J., Tuteja, G., White, P., & Kaestner, K. H. (2011). The nucleosome map of the mammalian liver. *Nature Structural and Molecular Biology, 18*, 742–746.

Liu, J. K., DiPersio, C. M., & Zaret, K. S. (1991). Extracellular signals that regulate liver transcription factors during hepatic differentiation in vitro. *Molecular and Cellular Biology, 11*, 773–784.

Loh, K. M., Ang, L. T., Zhang, J., Kumar, V., Ang, J., Auyeong, J. Q., et al. (2014). Efficient endoderm induction from human pluripotent stem cells by logically directing signals controlling lineage bifurcations. *Cell Stem Cell, 14*, 237–252.

Lokmane, L., Haumaitre, C., Garcia-Villalba, P., Anselme, I., Schneider-Maunoury, S., & Cereghini, S. (2008). Crucial role of vHNF1 in vertebrate hepatic specification. *Development, 135*, 2777–2786.

Martinez-Barbera, J. P., Clements, M., Thomas, P., Rodriguez, T., Meloy, D., Kioussis, D., et al. (2000). The homeobox gene *hex* is required in definitive endodermal tissues for normal forebrain, liver and thyroid formation. *Development, 127*, 2433–2445.

Matsumoto, K., Yoshitomi, H., Rossant, J., & Zaret, K. S. (2001). Liver organogenesis promoted by endothelial cells prior to vascular function. *Science, 294*, 559–563.

McLin, V. A., Rankin, S. A., & Zorn, A. M. (2007). Repression of Wnt/beta-catenin signaling in the anterior endoderm is essential for liver and pancreas development. *Development, 134*, 2207–2217.

Medvinsky, A., & Dzierzak, E. (1996). Definitive hematopoiesis is autonomously initiated by the AGM region. *Cell, 86*, 897–906.

Menet, J. S., Pescatore, S., & Rosbash, M. (2014). CLOCK:BMAL1 is a pioneer-like transcription factor. *Genes and Development, 28*, 8–13.

Merkle, F. T., & Eggan, K. (2013). Modeling human disease with pluripotent stem cells: From genome association to function. *Cell Stem Cell, 12*, 656–668.
Miyajima, A., Tanaka, M., & Itoh, T. (2014). Stem/progenitor cells in liver development, homeostasis, regeneration, and reprogramming. *Cell Stem Cell, 14*, 561–574.
Monaghan, A. P., Kaestner, K. H., Grau, E., & Schütz, G. (1993). Postimplantation expression patterns indicate a role for the mouse forkhead/HNF-3a, b, and g genes in determination of the definitive endoderm, chordamesoderm and neuroectoderm. *Development, 119*, 567–578.
Monga, S. P., Monga, H. K., Tan, X., Mule, K., Pediaditakis, P., & Michalopoulos, G. K. (2003). Beta-catenin antisense studies in embryonic liver cultures: Role in proliferation, apoptosis, and lineage specification. *Gastroenterology, 124*, 202–216.
Monga, S. P., Pediaditakis, P., Mule, K., Stolz, D. B., & Michalopoulos, G. K. (2001). Changes in WNT/beta-catenin pathway during regulated growth in rat liver regeneration. *Hepatology, 33*, 1098–1109.
Nissim, S., Sherwood, R. I., Wucherpfennig, J., Saunders, D., Harris, J. M., Esain, V., et al. (2014). Prostaglandin E2 regulates liver versus pancreas cell-fate decisions and endodermal outgrowth. *Developmental Cell, 28*, 423–437.
Ober, E. A., Verkade, H., Field, H. A., & Stainier, D. Y. (2006). Mesodermal Wnt2b signalling positively regulates liver specification. *Nature, 442*, 688–691.
Parviz, F., Matullo, C., Garrison, W. D., Savatski, L., Adamson, J. W., Ning, G., et al. (2003). Hepatocyte nuclear factor 4alpha controls the development of a hepatic epithelium and liver morphogenesis. *Nature Genetics, 34*, 292–296.
Paul, J., Conkie, D., & Freshney, R. I. (1969). Erythropoietic cell population changes during the hepatic phase of erythropoiesis in the fetal mouse. *Cell and Tissue Kinetics, 2*, 283–294.
Ramakrishnan, V., Finch, J. T., Graziano, V., Lee, P. L., & Sweet, R. M. (1993). Crystal structure of globular domain of histone H5 and its implications for nucleosome binding. *Nature, 362*, 219–224.
Red-Horse, K., Crawford, Y., Shojaei, F., & Ferrara, N. (2007). Endothelium–microenvironment interactions in the developing embryo and in the adult. *Developmental Cell, 12*, 181–194.
Rodriguez-Seguel, E., Mah, N., Naumann, H., Pongrac, I. M., Cerda-Esteban, N., Fontaine, J. F., et al. (2013). Mutually exclusive signaling signatures define the hepatic and pancreatic progenitor cell lineage divergence. *Genes and Development, 27*, 1932–1946.
Rosenquist, G. C. (1971). The location of the pregut endoderm in the chick embryo at the primitive streak stage as determined by radioautographic mapping. *Developmental Biology, 26*, 323–335.
Rossi, J. M., Dunn, N. R., Hogan, B. L., & Zaret, K. S. (2001). Distinct mesodermal signals, including BMPs from the septum transversum mesenchyme, are required in combination for hepatogenesis from the endoderm. *Genes and Development, 15*, 1998–2009.
Sakaguchi, T. F., Sadler, K. C., Crosnier, C., & Stainier, D. Y. (2008). Endothelial signals modulate hepatocyte apicobasal polarization in zebrafish. *Current Biology, 18*, 1565–1571.
Sasaki, H., & Hogan, B. L. M. (1993). Differential expression of multiple fork head related genes during gastrulation and pattern formation in the mouse embryo. *Development, 118*, 47–59.
Sekiya, T., & Zaret, K. S. (2007). Repression by Groucho/TLE/Grg proteins: Genomic site recruitment generates compacted chromatin in vitro and impairs activator binding in vivo. *Molecular Cell, 28*, 291–303.
Serls, A. E., Doherty, S., Parvatiyar, P., Wells, J. M., & Deutsch, G. H. (2005). Different thresholds of fibroblast growth factors pattern the ventral foregut into liver and lung. *Development, 132*, 35–47.

Sherwood, R. I., Chen, T. Y., & Melton, D. A. (2009). Transcriptional dynamics of endodermal organ formation. *Developmental Dynamics*, *238*, 29–42.
Sherwood, R. I., Hashimoto, T., O'Donnell, C. W., Lewis, S., Barkal, A. A., van Hoff, J. P., et al. (2014). Discovery of directional and nondirectional pioneer transcription factors by modeling DNase profile magnitude and shape. *Nature Biotechnology*, *32*, 171–178.
Shin, D., Lee, Y., Poss, K. D., & Stainier, D. Y. (2011). Restriction of hepatic competence by Fgf signaling. *Development*, *138*, 1339–1348.
Shin, D., Shin, C. H., Tucker, J., Ober, E. A., Rentzsch, F., Poss, K. D., et al. (2007). Bmp and Fgf signaling are essential for liver specification in zebrafish. *Development*, *134*, 2041–2050.
Sosa-Pineda, B., Wigle, J. T., & Oliver, G. (2000). Hepatocyte migration during liver development requires Prox1. *Nature Genetics*, *25*, 254–255.
Soufi, A., Donahue, G., & Zaret, K. S. (2012). Facilitators and impediments of the pluripotency reprogramming factors' initial engagement with the genome. *Cell*, *151*, 994–1004.
Soufi, A., Garcia, M. F., Jaroszewicz, A., Osman, N., Pellegrini, M., & Zaret, K. S. (2015). Pioneer transcription factors target partial DNA motifs on nucleosomes to initiate reprogramming. *Cell*, *161*, 555–568.
Suzuki, A., Sekiya, S., Buscher, D., Izpisua Belmonte, J. C., & Taniguchi, H. (2008). Tbx3 controls the fate of hepatic progenitor cells in liver development by suppressing p19ARF expression. *Development*, *135*, 1589–1595.
Takahashi, K., & Yamanaka, S. (2006). Induction of pluripotent stem cells from mouse embryonic and adult fibroblast cultures by defined factors. *Cell*, *126*, 663–676.
Takebe, T., Sekine, K., Enomura, M., Koike, H., Kimura, M., Ogaeri, T., et al. (2013). Vascularized and functional human liver from an iPSC-derived organ bud transplant. *Nature*, *499*, 481–484.
Tam, P. P., Khoo, P. L., Wong, N., Tsang, T. E., & Behringer, R. R. (2004). Regionalization of cell fates and cell movement in the endoderm of the mouse gastrula and the impact of loss of Lhx1(Lim1) function. *Developmental Biology*, *274*, 171–187.
Tremblay, K. D., & Zaret, K. S. (2005). Distinct populations of endoderm cells converge to generate the embryonic liver bud and ventral foregut tissues. *Developmental Biology*, *280*, 87–99.
Ueno, S., Weidinger, G., Osugi, T., Kohn, A. D., Golob, J. L., Pabon, L., et al. (2007). Biphasic role for Wnt/beta-catenin signaling in cardiac specification in zebrafish and embryonic stem cells. *Proceedings of the National Academy of Sciences of the United States of America*, *104*, 9685–9690.
van Oevelen, C., Collombet, S., Vicent, G., Hoogenkamp, M., Lepoivre, C., Badeaux, A., et al. (2015). C/EBPalpha activates pre-existing and de novo macrophage enhancers during induced pre-B cell transdifferentiation and myelopoiesis. *Stem Cell Reports*, *5*, 232–247.
Wandzioch, E., & Zaret, K. S. (2009). Dynamic signaling network for the specification of embryonic pancreas and liver progenitors. *Science*, *324*, 1707–1710.
Wang, J., Rhee, S., Palaria, A., & Tremblay, K. D. (2015). FGF signaling is required for anterior but not posterior specification of the murine liver bud. *Developmental Dynamics*, *244*, 431–443.
Wang, A., Yue, F., Li, Y., Xie, R., Harper, T., Patel, N. A., et al. (2015). Epigenetic priming of enhancers predicts developmental competence of hESC-derived endodermal lineage intermediates. *Cell Stem Cell*, *16*, 386–399.
Wang, B., Zhao, L., Fish, M., Logan, C. Y., & Nusse, R. (2015). Self-renewing diploid Axin2(+) cells fuel homeostatic renewal of the liver. *Nature*, *524*, 180–185.

Wapinski, O. L., Vierbuchen, T., Qu, K., Lee, Q. Y., Chanda, S., Fuentes, D. R., et al. (2013). Hierarchical mechanisms for direct reprogramming of fibroblasts to neurons. *Cell*, *155*, 621–635.

Warga, R. M., & Nusslein-Volhard, C. (1999). Origin and development of the zebrafish endoderm. *Development*, *126*, 827–838.

Watt, A. J., Zhao, R., Li, J., & Duncan, S. A. (2007). Development of the mammalian liver and ventral pancreas is dependent on GATA4. *BMC Developmental Biology*, *7*, 37.

Watts, J. A., Zhang, C., Klein-Szanto, A. J., Kormish, J. D., Fu, J., Zhang, M. Q., et al. (2011). Study of FoxA pioneer factor at silent genes reveals Rfx-repressed enhancer at Cdx2 and a potential indicator of esophageal adenocarcinoma development. *PLoS Genetics*, *7*, e1002277.

Weinstein, D. C., Ruiz i Altaba, A., Chen, W. S., Hoodless, P., Prezioso, V. R., Jessell, T. M., et al. (1994). The winged-helix transcription factor *HNF-3b* is required for notochord development in the mouse embryo. *Cell*, *78*, 575–588.

Xu, C. R., Cole, P. A., Meyers, D. J., Kormish, J., Dent, S., & Zaret, K. S. (2011). Chromatin "prepattern" and histone modifiers in a fate choice for liver and pancreas. *Science*, *332*, 963–966.

Xu, C. R., Li, L. C., Donahue, G., Ying, L., Zhang, Y. W., Gadue, P., et al. (2014). Dynamics of genomic H3K27me3 domains and role of EZH2 during pancreatic endocrine specification. *The EMBO Journal*, *33*, 2157–2170.

Xu, C., Lu, X., Chen, E. Z., He, Z., Uyunbilig, B., Li, G., et al. (2012). Genome-wide roles of FoxA2 in directing liver specification. *Journal of Molecular Cell Biology*, *4*, 420–422.

Zaret, K. (1999). Developmental competence of the gut endoderm: Genetic potentiation by GATA and HNF3/fork head proteins. *Developmental Biology*, *209*, 1–10.

Zaret, K. S. (2008). Genetic programming of liver and pancreas progenitors: Lessons for stem-cell differentiation. *Nature Reviews. Genetics*, *9*, 329–340.

Zaret, K. S., & Carroll, J. S. (2011). Pioneer transcription factors: Establishing competence for gene expression. *Genes and Development*, *25*, 2227–2241.

Zaret, K. S., & Grompe, M. (2008). Generation and regeneration of cells of the liver and pancreas. *Science*, *322*, 1490–1494.

Zhao, X., Monson, C., Gao, C., Gouon-Evans, V., Matsumoto, N., Sadler, K. C., et al. (2010). Klf6/copeb is required for hepatic outgrowth in zebrafish and for hepatocyte specification in mouse ES cells. *Developmental Biology*, *344*, 79–93.

Zhao, R., Watt, A. J., Li, J., Luebke-Wheeler, J., Morrisey, E. E., & Duncan, S. A. (2005). GATA6 is essential for embryonic development of the liver but dispensable for early heart formation. *Molecular and Cellular Biology*, *25*, 2622–2631.

Zorn, A. M., & Wells, J. M. (2007). Molecular basis of vertebrate endoderm development. *International Review of Cytology*, *259*, 49–111.

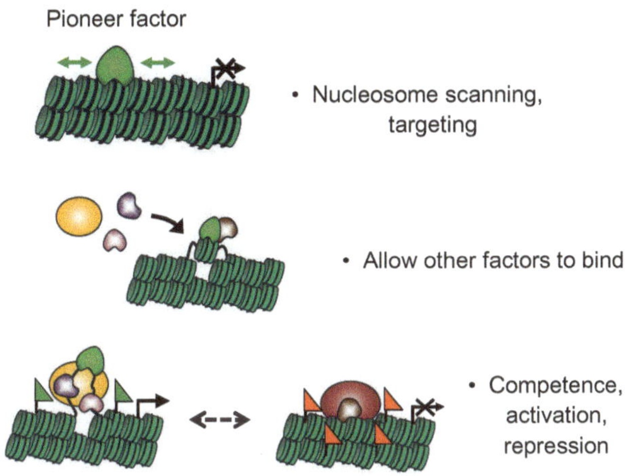

Kenneth S. Zaret, Figure 1 Chromatin binding by pioneer factors and consequences for gene expression. *From Zaret and Grompe (2008). Reprinted with permission from AAAS.*

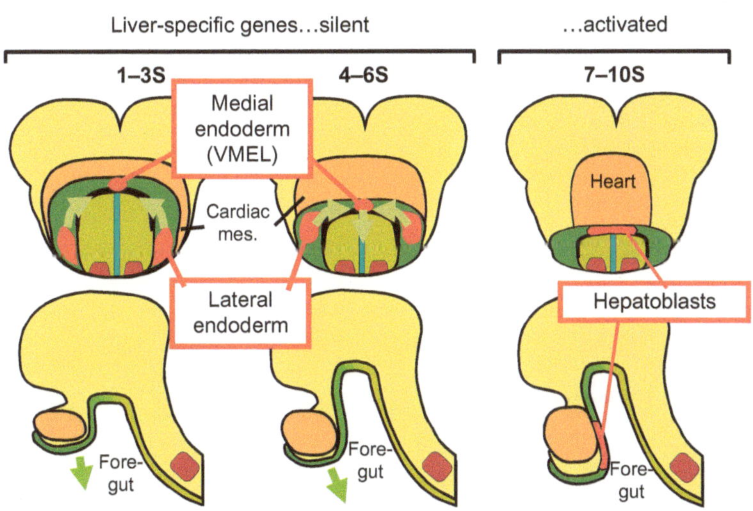

Kenneth S. Zaret, Figure 2 Location of three hepatic progenitor domains (pink) within the ventral foregut endoderm (green). Arrows depict cell movement during foregut closure.

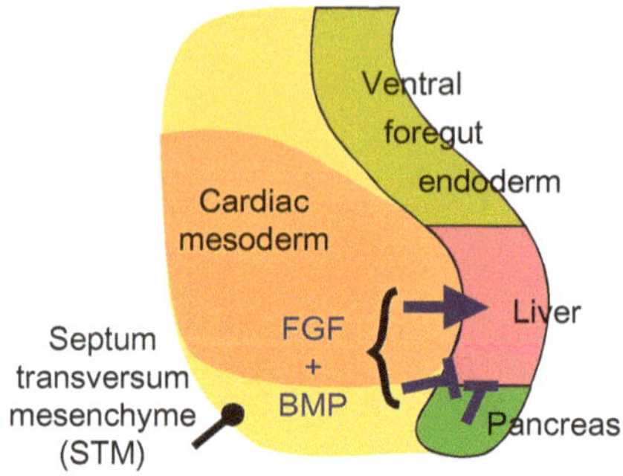

Kenneth S. Zaret, Figure 3 Combinatorial FGF and BMP signaling induces the liver fate (pink) while repressing the pancreas fate (dark green) within the endoderm (pale green).

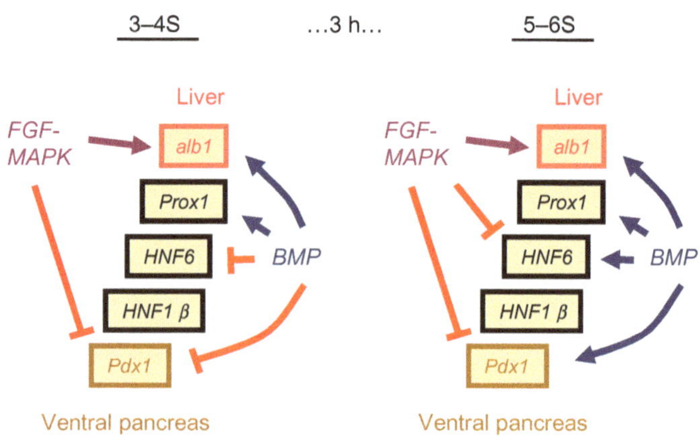

Kenneth S. Zaret, Figure 4 Dynamics in inductive signaling networks during hepatic and ventral pancreatic specification within the endoderm.

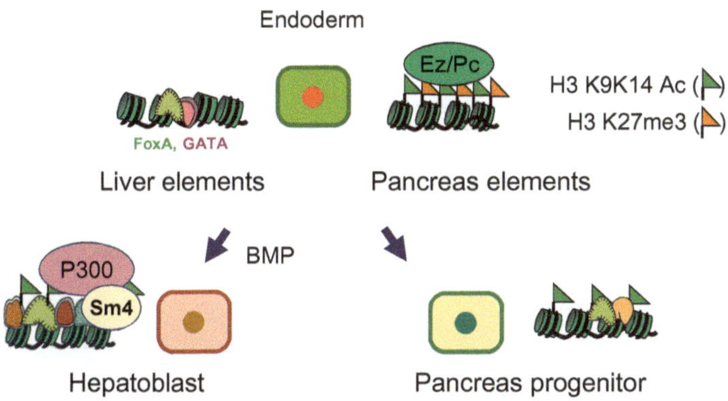

Kenneth S. Zaret, Figure 5 Prepattern in chromatin states differs between selected liver and pancreas regulatory sequences in the undifferentiated endoderm, and changes elicited by BMP signaling.

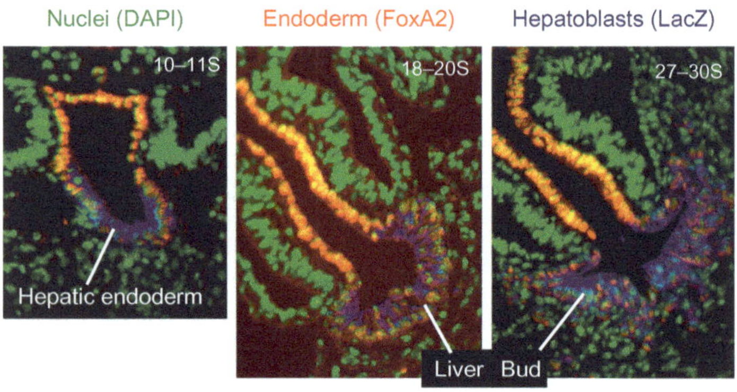

Kenneth S. Zaret, Figure 6 Morphogenesis of the liver bud. The three panels depict the columnar epithelium (10–11S) of newly specified hepatoblasts, the pseudostratified epithelium of the thickened hepatic endoderm (hepatoblast) domain (18–20S), and the hepatoblasts beginning to migrate into the stroma to generate the liver bud (27–30S).

CHAPTER THIRTY-NINE

The Acquisition of Cell Fate in Mouse Development: How Do Cells First Become Heterogeneous?

Sarah J.L. Graham, Magdalena Zernicka-Goetz[1]
Department of Physiology, Development and Neuroscience, University of Cambridge, Cambridge, United Kingdom
[1]Corresponding author: e-mail address: mz205@cam.ac.uk

Contents

Abstract

Specification of cell fate in the early mouse embryo involves the generation of heterogeneities between cells in polarity, position, gene expression, and protein activity. Identifying when and how these differences first become established represents an important challenge for understanding the roles of stochastic and cell history-dependent mechanisms in mammalian development.

1. INTRODUCTION

Mammalian early development involves the transition of a totipotent fertilized egg (zygote) into an organized embryo comprising different lineages of cells that will contribute to either extraembryonic tissues or to the fetus itself (Fig. 1). Totipotency is lost over the first few days of development as cell fate and potential become progressively restricted and individual cells

Current Topics in Developmental Biology, Volume 117
ISSN 0070-2153
http://dx.doi.org/10.1016/bs.ctdb.2015.11.021

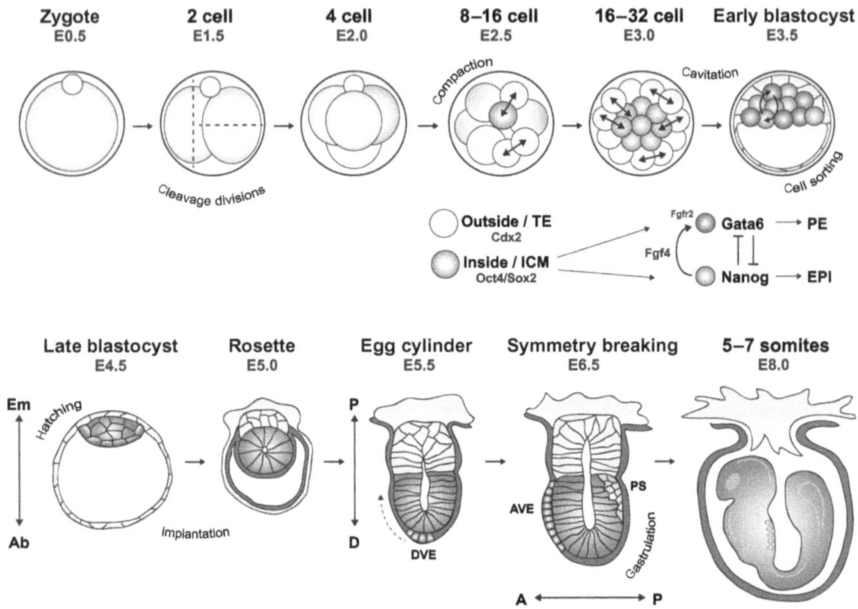

Figure 1 Early mouse embryo development. Through a series of cleavage divisions the zygote forms a mature blastocyst by E4.5, which then implants into the uterus. By this stage three different cell lineages have been specified, the extraembryonic trophectoderm (TE; yellow) and primitive endoderm (PE; red) and the embryonic epiblast (EPI; blue). These lineages begin to be separated by the formation of outside and inside cells: Outside cells form the TE while inside cells form the pluripotent inner cell mass (ICM; purple). ICM cells will be further segregated into PE and EPI cells. The location of the ICM within the embryo determines the embryonic–abembryonic (Em–Ab) axis of the embryo, which is translated into the proximal–distal (P–D) axis following implantation. The fetal EPI lineage is organized into a cup-shaped epithelium through the formation of a polarized rosette and the opening of a central lumen during peri-implantation stages. The anterior–posterior (A–P) axis is determined by the migration of the distal visceral endoderm (DVE, derived from the PE) to form the anterior visceral endoderm (AVE) on one side of the egg cylinder. Primitive streak (PS) formation on the posterior side of the embryo marks the onset of gastrulation. The contribution of early blastomeres to either the TE or ICM is biased by early heterogeneities between cells. (A color version of this figure is located at the end of this chapter.)

are directed toward either pluripotency or differentiation. The critical mechanisms driving these cell fate decisions are difficult to study because the mammalian embryo is regulative, that is, it can compensate in situations where cells are either lost or gained (Hillman, Sherman, & Graham, 1972; Tarkowski, 1959, 1961), and therefore multiple complementary mechanisms exist to modulate cell fate. Preimplantation mouse embryo development culminates with the formation of a blastocyst at embryonic day 4.5 (E4.5), a small hollow ball of cells that contains the precursors of both embryonic and extraembryonic cell lineages and is capable of hatching out of the protective *zona pellucida* and implanting into the maternal uterine wall. The blastocyst is generated from the zygote through a series of cleavages, which divide the embryo into separate cells, blastomeres, without increasing the total embryo size. Following embryo implantation, the embryonic lineage of the blastocyst, the epiblast (EPI), grows rapidly, transitioning from an apolar mass of cells into a polarized epithelium, which forms an elongated egg cylinder surrounding a central lumen. These EPI cells are the progenitors of the fetus and their development has been modeled *in vitro* using embryonic stem cells (derived from the blastocyst) and epiblast stem cells (EpiSCs; derived from the egg cylinder). Extraembryonic cell lineages are critical for supporting and patterning the EPI throughout this time. The outer surface of the blastocyst is trophectoderm (TE), the first epithelial layer to be specified, which mediates embryo implantation and later forms the embryonic components of the placenta. The EPI is positioned on the interior of one side of the ball of TE, covered by the epithelial primitive endoderm (PE), which will later give rise to the yolk sac. Signals from the derivatives of the TE and PE provide important spatial information to the developing EPI, resulting in the specification of the anterior–posterior (A–P) axis and triggering the onset of gastrulation with the formation of the primitive streak on the posterior side of the egg cylinder (Fig. 1). Thus, in the mouse embryo, symmetry is initially broken with the specification of embryonic and extraembryonic cell lineages before implantation, but axial polarity is established later. The correct specification and organization of the three preimplantation lineages is essential for successful development and how cell fate is established in the early mammalian embryo is an important question for understanding the very first steps of life. The techniques available for the study of early mouse development are limited by the very small amounts of material that can be collected from the tiny embryos; however, recent advances in single-cell techniques are vastly improving our ability to understand cell fate decisions on a molecular level. Improvements in imaging and computational analyses are also allowing us to follow the movements of individual cells as they adopt different fates. Traditionally, these cell-tracing studies were limited to preimplantation development, when the embryo can be cultured free-floating in a dish, but now the study of peri-implantation and early postimplantation development is possible with new techniques for the *in vitro* culture and imaging of embryos throughout implantation stages (Bedzhov & Zernicka-Goetz, 2014; Morris, Grewal, et al., 2012). This essay will discuss how different cell lineages become specified in the preimplantation mouse embryo. This knowledge is useful not just for understanding early development, but it also provides important insights into the mechanisms that govern the regulation of pluripotency and differentiation in mammalian cells.

2. FORMATION OF THE TE AND INNER CELL MASS

The first cell fate decision involves the creation of two distinct cell populations that differ in both position and fate: the extraembryonic TE on the outside of the embryo and the pluripotent inner cell mass (ICM) on the inside. These two cell lineages can be clearly identified by the 32-cell stage (E3.5), by which time the maturing TE has formed a single layer epithelium with intercellular tight junctions creating a water-tight seal that mediates the formation of a fluid-filled blastocyst cavity (Smith & McLaren, 1977). The location of the cavity determines the embryonic–abembryonic (Em–Ab) axis of the embryo, with the ICM on the embryonic side. Prior to the eight-cell stage, each cell in the embryo occupies a comparative position, having an exposed surface on the outside of the embryo and minimal adhesion to other blastomeres at areas of cell–cell contact. At the mid eight-cell stage the embryo undergoes a process called compaction, in which the blastomeres become flattened in morphology and adhere tightly together to form a morula (Ziomek & Johnson, 1980). As the embryo compacts, each cell becomes polarized, with an external apical domain, characterized by the localization of aPKC and Par6, and a basolateral domain with E-cadherin localized to regions of direct cell–cell contact (Plusa et al., 2005; Vestweber, Gossler, Boller, & Kemler, 1987; Vinot et al., 2005; Ziomek & Johnson, 1980). This apical–basal polarization results in the blastomeres then being able to divide in different orientations: either more perpendicular to the apical–basal axis so that two polarized daughter cells are positioned on the outside of the embryo (symmetric division), or more along the apical–basal axis so that one apolar daughter cell is directed toward the inside of the embryo (asymmetric division). Two major rounds of asymmetric divisions, at the 8- to 16-cell and 16- to 32-cell transitions, generate an inside cell population that will form the ICM (Fleming, 1987; Johnson & Ziomek, 1981; Morris et al., 2010). Inside cells can also be generated through the apical constriction and "engulfment" of cells originally positioned on the outside (Samarage et al., 2015). How common this mechanism is for ICM formation is not clear, as discrepancies between studies may reflect differences in the methods of classifying cells as inside or outside and divisions as symmetric or asymmetric, as well as the methods of cell tracking and differences between genetic strains of mice (Anani, Bhat, Honma-Yamanaka, Krawchuk, & Yamanaka, 2014; McDole, Xiong, Iglesias, & Zheng, 2011; Morris, Guo, & Zernicka-Goetz, 2012; Samarage et al., 2015; Watanabe, Biggins, Tannan, & Srinivas, 2014; Yamanaka, Lanner, & Rossant, 2010).

The development of each lineage relies on the expression of specific transcription factors, several of which are initially expressed in all cells of the embryo, before becoming restricted to a specific lineage. The TE transcriptional network includes *Cdx2* (Niwa et al., 2005; Strumpf et al., 2005), *Gata3* (Ralston et al., 2010), and *Eomes* (Russ et al., 2000). Embryos that lack zygotic expression of *Cdx2* have severe defects in TE differentiation, are unable to downregulate the expression of pluripotency markers in outside cells, and fail to form a functional TE epithelium, resulting in the collapse of the blastocyst cavity (Strumpf et al., 2005). *Cdx2−/−* embryos are, however, able to initiate cavity formation, possibly due to the presence of maternal *Cdx2* mRNA. Major zygotic genome activation occurs at the two-cell stage in mouse embryos and prior to this time, the embryo is reliant on maternal mRNAs and proteins inherited from the egg (Bachvarova & De Leon, 1980; Braude, Pelham, Flach, & Lobatto, 1979; Flach, Johnson, Braude, Taylor, & Bolton, 1982). When RNAi approaches were used to deplete all *Cdx2* transcripts from the zygote stage, an earlier detrimental effect on development was observed than in the zygotic knockout in one study (Jedrusik et al., 2010), but not another (Wu et al., 2010). Initial analysis of maternal/zygotic (MZ) *Cdx2* knockout mouse embryos (Blij, Frum, Akyol, Fearon, & Ralston, 2012) suggested that maternal Cdx2 was not important for development; however, more recent analysis of these mice has supported a role of maternal Cdx2 in the morula to blastocyst transition (Jedrusik, Cox, Wicher, Glover, & Zernicka-Goetz, 2015). Gata3 acts in parallel to Cdx2 to regulate TE differentiation (Ralston et al., 2010), while Eomes is important later in TE development, acting together with Cdx2 and Elf5 to maintain the extraembryonic ectoderm (ExE) in the proximal region of the egg cylinder (Donnison et al., 2005; Russ et al., 2000; Strumpf et al., 2005). Activation of the TE transcriptional network is determined by the activity of the master transcriptional regulator Tead4, which directly regulates the expression of *Cdx2* and *Gata3* (Nishioka et al., 2008; Ralston et al., 2010; Yagi et al., 2007). Without Tead4 expression, embryos completely lack TE development, appearing as abnormal morulae that are unable to form cavities (Nishioka et al., 2008; Yagi et al., 2007).

Inside cells, in contrast, activate an ICM-specific network of pluripotency-associated genes, including *Oct4* (Nichols et al., 1998; Palmieri, Peter, Hess, & Scholer, 1994), *Nanog* (Mitsui et al., 2003), and *Sox2* (Avilion et al., 2003). Maternal Oct4 is present in the zygote, but is rapidly degraded, with zygotic expression of Oct4 protein first apparent in all cells at the eight-cell stage (Palmieri et al., 1994). *Oct4−/−* embryos form blastocyst-like structures; however, the inside cells are unable to develop into ICM cells capable of forming either EPI or PE (Le Bin et al., 2014; Nichols et al., 1998). Nanog and Sox2 are both essential for the development of the EPI (Avilion et al., 2003; Mitsui et al., 2003). The TE and ICM transcriptional networks each act to establish a specific cell fate and to repress the alternative. For example, Cdx2 both positively regulates its own expression and inhibits Oct4 transcription factor activity, while Oct4 in return has a reciprocal inhibitory effect on the expression of TE genes (Niwa et al., 2005). The two fates are, however, not fixed until after the 32-cell stage, until this point the fate of some cells is flexible and can be diverted in response to experimental influence or perturbations (Handyside, 1978; Spindle, 1978; Suwińska, Czołowska, Ożdżeński, & Tarkowski, 2008; Tarkowski, Suwińska, Czołowska, & Ożdżeński, 2010).

3. WHAT CONTROLS THE FIRST CELL FATE DECISION?

The first cell fate decision is influenced by position, polarity, and gene expression, with the relationship between each of these factors the subject of much recent research. Cell position has been demonstrated to be an important driver of fate, as cells translocated in the embryo will adopt the appropriate fate for their new position (Handyside, 1978; Hillman et al., 1972; Rossant & Lis, 1979; Spindle, 1978). The importance of polarity is more recently becoming understood, as the molecular links between position, polarity and cell fate are being uncovered. The master regulator that controls the first cell fate decision, Tead4, is expressed uniformly between inside and outside cells at the 16-cell stage (Nishioka et al., 2008), yet its activity is restricted to the outside polarized cells through the action of the Hippo pathway (Fig. 2). Tead4 transcription factor activity requires the homologous transcriptional coactivators, Yap and Taz (Nishioka et al., 2009), which are regulated by the Hippo pathway kinases Lats1/2 that act to sequester Yap and Taz in the cytoplasm and therefore inhibit Tead4 transcriptional activity (Nishioka et al., 2009). In outside polarized cells, the Hippo pathway component Angiomotin (Amot) is restricted to the apical domain where it is

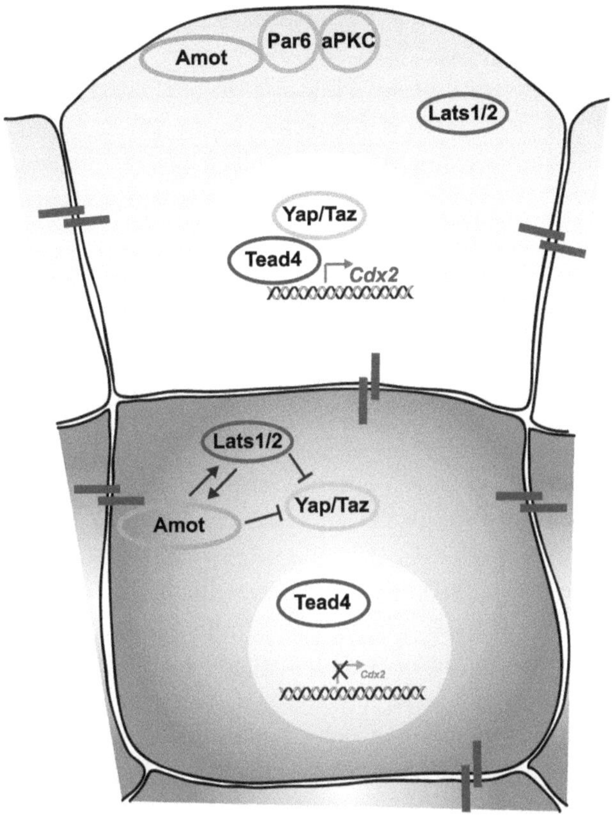

Figure 2 Polarity and position in the first cell fate decision. Tead4 activity is differentially regulated in outside and inside cells due to differences in polarity and Hippo pathway signaling. In outside polarized cells, Angiomotin (Amot) is restricted to the apical domain, allowing Yap/Taz to enter the nucleus and facilitate upregulation of Cdx2 expression by Tead4. In inside apolar cells, Amot is localized at adherens junctions (gray bars), where it acts to sequester Yap/Taz in the cytoplasm, through both the activation of the Hippo pathway kinases Lats1/2 and through Hippo-independent binding to Yap/Taz, thereby inhibiting Tead4 activity and Cdx2 expression. (A color version of this figure is located at the end of this chapter.)

unable to activate Hippo signaling, allowing Yap and Taz to enter the nucleus (Hirate et al., 2013; Leung & Zernicka-Goetz, 2013). In inside apolar cells, Amot is localized to adherens junctions where it inhibits Yap/Taz nuclear localization through both the activation of Lats1/2 and through Hippo-independent binding to Yap and Taz (Chan et al.,

2011; Hirate et al., 2013; Leung & Zernicka-Goetz, 2013; Paramasivam, Sarkeshik, Yates, Fernandes, & McCollum, 2011). When Amot expression is artificially downregulated, Yap becomes localized to the nuclei of inside cells, leading to activation of the TE transcriptional program and expression of Cdx2 in the ICM (Leung & Zernicka-Goetz, 2013).

The formation of outside and inside cells seems inevitable in a compact embryo with only a certain amount of surface area; however, the decision of each cell to contribute to the TE or ICM is not random, but influenced by both polarity and the expression of lineage-specific genes. Polarity itself is suggested to be more important than position for regulating Hippo pathway activity and cell fate at the 16-cell stage, with apolar outside cells showing cytoplasmic phospho-Yap expression and subsequently being engulfed by neighboring polar cells (Anani et al., 2014). Disruption of apical–basal polarity through the inhibition of Tfap2c and the Rho–ROCK pathway during preimplantation development results in the activation of Hippo signaling, directing cells toward an ICM fate (Cao et al., 2015; Kono, Tamashiro, & Alarcon, 2014). Furthermore, eight-cell stage blastomeres with reduced levels of aPKC or Par3 preferentially give rise to inside cells, both through an increased frequency of asymmetric cell divisions and through the internalization of outside daughter cells that are less polarized than their neighbors (Plusa et al., 2005). In half embryos, which contain half the normal complement of cells due to blastomere separation at the two-cell stage, inside cells are predominantly generated by engulfment rather than asymmetric cell divisions (Morris, Guo, et al., 2012). In these embryos, all of the blastomeres divide symmetrically at the four- to eight-cell transition (equivalent to the 8- to 16-cell transition in whole embryos); however, a proportion of the outside cells generated do not have apical aPKC and these less-polarized cells become internalized to contribute to the ICM (Morris, Guo, et al., 2012). Cell polarization also influences the fate of daughter cells of asymmetric divisions through differential inheritance of *Cdx2* mRNA. At the eight-cell stage, *Cdx2* mRNA becomes localized to the apical domain, such that when cells divide asymmetrically only the outside cell inherits the transcripts (Skamagki, Wicher, Jedrusik, Ganguly, & Zernicka-Goetz, 2013). This asymmetric inheritance of *Cdx2* mRNA is not critical for promoting the TE fate in outside cells, but instead has an important role in inhibiting Cdx2 expression in inside cells. If the apical transcript localization is disrupted, inside cells express Cdx2 and are eliminated by apoptosis (Skamagki et al., 2013). The relationship between Cdx2 and polarity appears to be mutually reinforcing, cells that are more polarized express higher levels of Cdx2, while increasing Cdx2 levels enhances cell polarity and increases the frequency of symmetric cell divisions (Jedrusik et al., 2008; Ralston & Rossant, 2008). Together, the data published so far suggest that differential polarity of blastomeres is the most critical factor for specifying cell fate in the first cell fate decision. The exact molecular mechanism through which polarization regulates Amot activity remains an important unanswered question, but it most likely involves the inhibition of Amot by Par6 and aPKC at the apical domain (Hirate et al., 2013).

4. FORMATION OF THE EPI AND PE

In the second cell fate decision, ICM cells are specified to develop into either EPI or PE. Progenitors of these two lineages can be first identified in the early blastocyst by the expression of Nanog in EPI precursors and Gata6 in PE precursors (Chazaud, Yamanaka, Pawson, & Rossant, 2006). These precursor cells are initially found in a mixed "salt and pepper"-like distribution throughout the ICM, but are then sorted into their correct positions through a combination of active cell migration, positional effects, and selective apoptosis (Meilhac et al., 2009; Morris et al., 2010; Plusa, Piliszek, Frankenberg, Artus, & Hadjantonakis, 2008; Saiz, Grabarek, Sabherwal, Papalopulu, & Plusa, 2013). By E4.5 the polarized epithelial PE forms a single layer adjacent to the blastocyst cavity, covering the apolar pluripotent EPI. The EPI is the precursor cell lineage of the future fetus and is characterized by the expression of *Nanog* (Chambers et al., 2003; Mitsui et al., 2003) and *Sox2* (Avilion et al., 2003). During implantation stages the EPI changes from a "naïve" pluripotent state to a "primed" pluripotent state, in preparation for the generation of different germ layers at gastrulation. This change in potency is associated with a dramatic morphogenetic change. Between E4.5 and E5.5 the EPI is completely transformed, from an apolar mass of cells into a single-layer polarized epithelium surrounding a central lumen. Studies in embryoid bodies initially suggested that this transformation was mediated by apoptosis of inner EPI cells and subsequent polarization and epithelialization (Coucouvanis & Martin, 1995); however, recent work has demonstrated that the EPI first forms a polarized rosette-like structure and that the formation and expansion of the lumen is independent of cell death (Bedzhov & Zernicka-Goetz, 2014). The trigger for rosette formation appears to be the production of extracellular matrix (ECM) proteins by the extraembryonic lineages surrounding the EPI. This important event can be modeled *in vitro* by placing ES cells in a matrix of ECM proteins, where they will form a polarized rosette and a central lumen, recapitulating the peri-implantation morphogenesis of the EPI (Bedzhov & Zernicka-Goetz, 2014).

Development of the PE is driven by an endodermal transcriptional network that includes *Gata6* (Koutsourakis, Langeveld, Patient, Beddington, & Grosveld, 1999; Schrode, Saiz, Di Talia, & Hadjantonakis, 2014), *Gata4* (Molkentin, Lin, Duncan, & Olson, 1997), platelet-derived growth factor receptor alpha (*PDGFRα*; Plusa et al., 2008), *Sox17* (Niakan et al., 2010), and *Sox7* (Artus, Piliszek, & Hadjantonakis, 2011). Gata6 is the earliest marker of the PE and *Gata6−/−* embryos completely lack PE formation, with the entire ICM containing Nanog-positive EPI cells (Bessonnard et al., 2014; Chazaud et al., 2006; Schrode et al., 2014). Sox17 acts downstream of Gata6 and is important for promoting PE differentiation (Morris et al., 2010; Niakan et al., 2010), while Gata4 is involved later in development (Molkentin et al., 1997). Sox7 is expressed in the late PE, once the cells have been positioned on the surface of the ICM (Artus et al., 2011). The survival of developing PE cells is dependent upon both PDGF (Artus, Kang, Cohen-Tannoudji, & Hadjantonakis, 2013) and bone morphogenetic protein (BMP) signaling (Graham et al., 2014). Disruption of either of these pathways results in the loss of PE precursor cells, most likely due to the elimination of cells that are unable to differentiate appropriately (Artus et al., 2013; Graham et al., 2014). Differentiation of the PE into a polarized epithelium has been demonstrated to be important for both the organization of the PE on the ICM surface and the survival of PE cells. If aPKC activity is blocked during blastocyst formation, preventing proper polarization, PE precursor cells are unable to maintain their position at the surface of the ICM, or to express later PE markers, and are eliminated by apoptosis (Saiz et al., 2013). Once the PE epithelium has been formed it gives rise to two extraembryonic tissues, the visceral endoderm that covers the elongating egg cylinder and is essential for axis formation, and the parietal endoderm that migrates along the mural TE and forms the endodermal component of the parietal yolk sac.

5. WHAT CONTROLS THE SECOND CELL FATE DECISION?

Segregation of the PE and EPI lineages is regulated by the Fibroblast growth factor (Fgf) signaling pathway. Nanog-positive EPI precursor cells express the Fgf4 ligand, while Gata6-positive PE precursor cells upregulate expression of the Fgfr2 receptor (Frankenberg et al., 2011; Guo et al., 2010; Kurimoto et al., 2006; Ohnishi et al., 2014). If Fgf signaling is blocked in the embryo, all cells of the ICM are specified to become EPI and express Nanog (Frankenberg et al., 2011; Yamanaka et al., 2010). Fgf signaling is therefore required to commit cells to the PE fate program, though it is unable to trigger PE development in *Gata6−/−* embryos (Kang, Piliszek, Artus, & Hadjantonakis, 2013; Schrode et al., 2014). The presence of Nanog-positive EPI cells is itself important for PE specification. Both the EPI and PE fail to develop in *Nanog−/−* embryos (Messerschmidt & Kemler, 2010), although the development of the PE in these embryos can be rescued by the application of exogenous Fgf4 (Frankenberg et al., 2011). But how are the initial Nanog- and Gata6-positive cells specified? The "salt and pepper" distribution of EPI and PE precursors in the early ICM appears random; however, cell-tracking experiments have identified a cell fate bias dependent upon the time of cell internalization. ICM cells that are internalized earlier, at the 8- to 16-cell transition, are more likely to give rise to the EPI while cells internalized later, at the 16- to 32-cell transition, are biased toward forming PE (Krupa et al., 2014; Morris et al., 2010). The existence of this cell fate bias has been challenged (Yamanaka et al., 2010). However, when the number of cells generated in each wave of asymmetric divisions is taken into account, discrepancies between results can be resolved (Morris, Graham, Jedrusik, & Zernicka-Goetz, 2013). Importantly, the strength of this bias is related to the unique ICM composition of each embryo and is most apparent when both division waves contribute to the ICM in a similar proportion (Krupa et al., 2014; Morris et al., 2013).

It has been proposed that random fluctuations in gene expression can explain ICM lineage specification (Ohnishi et al., 2014). In this model, random heterogeneous expression of lineage-specific genes is gradually refined to create two separate cell populations, through the positive feedback and mutual repression of each ICM cell fate program. Generating inside cells in sequential waves of cell internalizations, however, means that a heterogeneous ICM cell population can be created and provides a mechanism by which three different preimplantation lineages can be specified in a situation where there are only two different positional states: inside and outside. An internalization wave-dependent bias requires that inside cells generated at different times are inherently unique and that their differences are influential for the second cell fate decision. This model, however, does not exclude stochastic influences and early ICM cells do retain flexibility in cell fate potential. An important difference that has been identified is the expression of Fgf4 and Fgfr2 in ICM cells generated in the first and second waves of asymmetric divisions, respectively (Krupa et al., 2014; Morris et al., 2013). At the 16-cell stage inside cells (internalized during the 8- to 16-cell transition) upregulate Fgf4, while outside cells upregulate Fgfr2 and the inside progeny of these cells (internalized during the 16- to 32-cell transition) therefore inherit high levels of Fgfr2, biasing their development toward the PE lineage (Morris et al., 2013; Fig. 3A). The effect of ICM composition on the strength of this bias is most likely due to the relative levels of Fgf "producers" and "responders" in each embryo, depending on how many cells are internalized during each round of divisions. As demonstrated by *Fgf4+/−* heterozygotes, the level of Fgf signaling in the embryo has a dose-dependent effect on ICM cell fate specification (Kang et al., 2013; Krawchuk, Honma-Yamanaka, Anani, & Yamanaka, 2013). If there are many early-internalized inside cells, there will be high Fgf4 levels, stimulating PE differentiation in cells with relatively low Fgfr2 expression. In contrast, if there are few early-internalized inside cells, there will be low Fgf4 levels and only the later-internalized cells with high Fgfr2 expression will form PE. This context-dependent regulation of the bias ensures an equal balance of PE and EPI cells are specified, regardless of the number of inside cells generated in each division wave (Fig. 3B). Deep mRNA sequencing analysis of the first inside and outside cells has further suggested that the internalization time-dependent bias may be enhanced by the production of BMP ligands in early-internalized ICM cells and the expression of BMP receptors in the progenitors of later-internalized cells (Graham et al., 2014; Fig. 3C). It is not yet known if the later-internalized ICM cells inherit the BMP receptors from their outside "parents," and the extent to which ICM cells internalized at different times differ on a molecular level is an important question that warrants further investigation.

6. EARLY HETEROGENEITIES THAT BIAS CELL FATE

The regulative nature of preimplantation mouse development implies that early blastomeres have the potential to contribute to any cell lineage, and therefore to both embryonic and extraembryonic tissues. It was initially assumed that each cell in the embryo was therefore "equal" in potential and that cell fate decisions were guided by external factors—positional or signaling information—or by chance. This view, however, is changing as studies are increasingly discovering heterogeneities between cells that influence cell fate decisions. The first suggestions that early blastomeres have biased

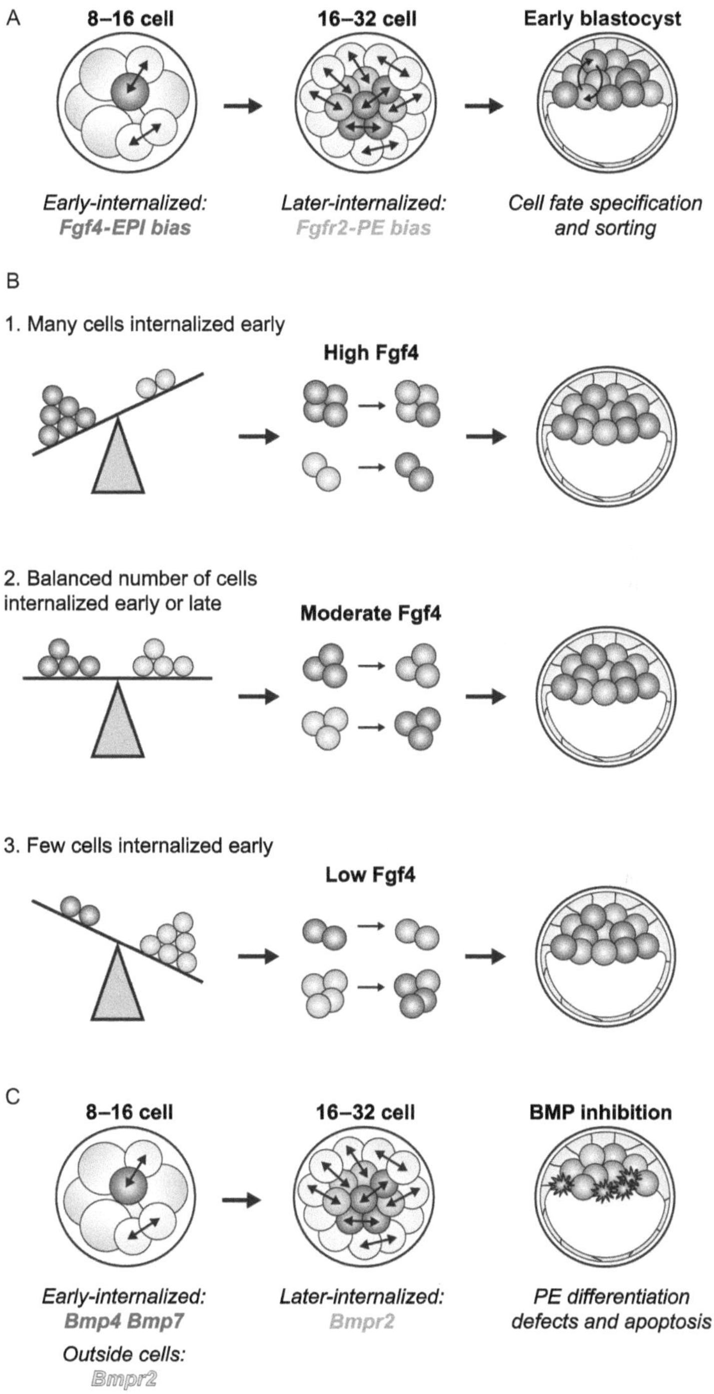

Figure 3 See figure legend on next page.

potential came from lineage tracing studies that tracked the fate of two-cell stage blastomeres and found that the embryonic cells of the blastocyst are predominantly derived from one blastomere (Fujimori, Kurotaki, Miyazaki, & Nabeshima, 2003; Gardner, 2001; Piotrowska, Wianny, Pedersen, & Zernicka-Goetz, 2001). These findings were challenged by studies in which the majority of labeled two-cell stage blastomeres were not biased in their contribution along the Em–Ab axis of the blastocyst (Alarcón & Marikawa, 2003; Chróścicka, Komorowski, & Maleszewski, 2004; Motosugi, Bauer, Polanski, Solter, & Hiiragi, 2005). These discrepancies between findings are still not resolved, but more recent single-cell microarray and mRNA sequencing of two-cell stage blastomeres have indicated that these cells are heterogeneous in gene expression (Biase, Cao, & Zhong, 2014; Roberts, Katayama, Magnuson, Falduto, & Torres, 2011).

In contrast to the controversial data from two-cell stage embryos, at the four-cell stage, specific molecular heterogeneities have been identified that can influence cell fate and developmental potential (Fig. 4A and B). The first molecular difference that was identified between four-cell stage blastomeres of the same embryo was the methylation of particular arginine residues of Histone H3 (Torres-Padilla, Parfitt, Kouzarides, & Zernicka-Goetz, 2007). The abundance of the methyltransferase Carm1, which methylates Histone H3R26, was also found to be heterogeneous between the cells, correlating with Histone H3R26 methylation levels. Overexpression of Carm1 in specific cells resulted in increased Histone H3R26 methylation, upregulation of the pluripotency-associated genes *Nanog* and *Sox2*, and a

Figure 3—Cont'd Internalization time bias in the second cell fate decision. (A) ICM cells are internalized at different times, with those internalized earlier (purple) expressing Fgf4 and being biased toward forming EPI (blue) and those internalized later (pink) expressing Fgfr2 and being biased to form PE (red). (B) The strength of the internalization time bias depends on the composition of the ICM and the local levels of Fgf signaling. 1. When there are many early-internalized cells there are high levels of Fgf4 and some early-internalized cells are specified to become PE. 2. When the numbers of early- and late-internalized cells are roughly equal the bias is most apparent, as there is an equal number of Fgf "producers" and "responders." 3. When there are few early-internalized cells there are low levels of Fgf4 and some later-internalized cells are specified to form EPI. All three scenarios have the same outcome, producing a balanced number of EPI and PE cells in the ICM. (C) The internalization time bias may be enhanced by BMP signaling. At the 16-cell stage, BMP ligands (*Bmp4* and *7*) are expressed in early-internalized ICM cells and BMP receptor 2 (*Bmpr2*) is expressed in the outside progenitors of later-internalized cells. BMP signaling is critical for PE differentiation and survival, as evidenced by the apoptosis of PE cells when BMP signaling is inhibited. (A color version of this figure is located at the end of this chapter.)

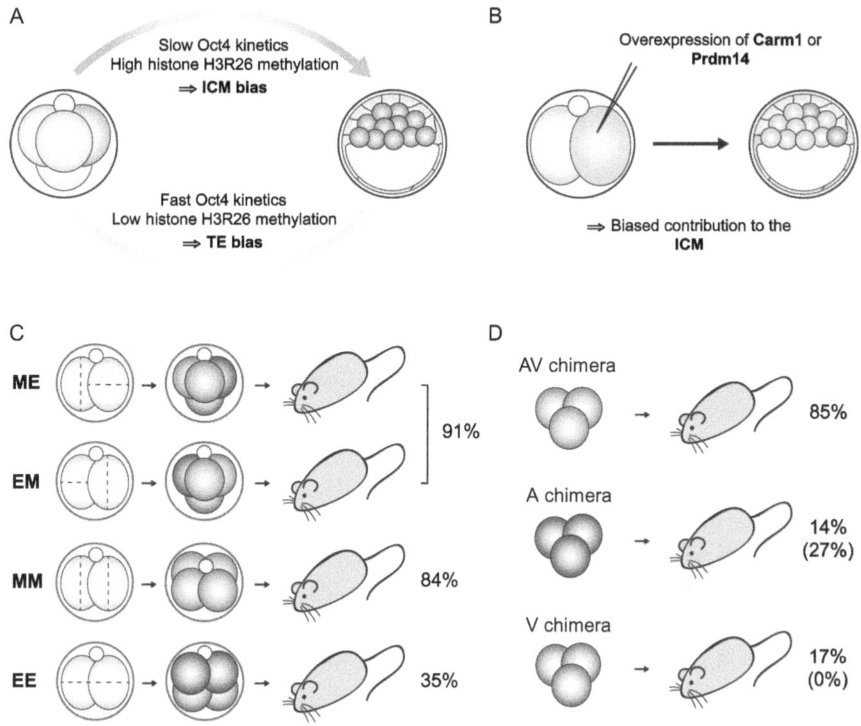

Figure 4 Early heterogeneities and developmental potential. (A) Early heterogeneities at the four-cell stage relate to lineage contribution in the blastocyst. (B) Overexpression of Carm1 or Prdm14 results in increased Histone H3R26 methylation and a biased contribution to the pluripotent ICM. (C) ME, EM, and MM embryos all have high developmental potential. EE embryos, which contain only A and V blastomeres at the four-cell stage, have reduced potential with only 35% of transferred embryos developing to term. (D) Chimeras made from "like" blastomeres from four-cell stage ME embryos have different developmental potentials. AV chimeras develop with normal success, while the majority of A and V chimeras fail to develop to term. Percentages in brackets represent the proportion of chimeras that developed when cell rotation was excluded, indicating the extremely compromised development of V blastomeres. (A color version of this figure is located at the end of this chapter.)

dramatically increased contribution of the cell's progeny to the ICM (Torres-Padilla et al., 2007). Subsequently, differential expression of another epigenetic modifier Prdm14, which interacts with Carm1, was also identified in four-cell stage blastomeres (Burton et al., 2013). Prmd14 is initially expressed in all early four-cell stage blastomeres, but by the late four-cell stage its expression has been restricted to only two cells through an unknown mechanism. As with Carm1, overexpression of Prdm14 increased Histone

H3R26 methylation and biased cells toward a pluripotent fate (Burton et al., 2013). How this epigenetic modification can be linked to cell fate and lineage specification is not completely understood, but possibly involves differential accessibility of specific DNA-binding sites for lineage-determining transcription factors. An example of such a transcription factor is Oct4 and differences in Oct4 DNA-binding kinetics between four-cell blastomeres have been demonstrated using a fluorescence decay after photoactivation assay (Plachta, Bollenbach, Pease, Fraser, & Pantazis, 2011). Using this assay, cells were found to cluster into two groups: one having faster kinetics and a smaller immobile fraction of Oct4 and the other having slower kinetics and a larger immobile fraction, indicating increased binding of the transcription factor to DNA-binding sites. These two groups of cells had different fates, with those having faster Oct4 kinetics dividing more symmetrically and contributing to the TE, and those with slower Oct4 kinetics dividing asymmetrically to contribute to the ICM (Plachta et al., 2011). Together, these results suggest that the pluripotent ICM lineage is predominantly formed from specific cells at the four-cell stage and that these cells already display unique properties before individual lineages can be identified. That some four-cell stage blastomeres preferentially contribute to either the TE or ICM lineages has further been demonstrated using Rainbow transgenic mice, in which each cell and its progeny are labeled in a different fluorescent color using a noninvasive recombination technique (Tabansky et al., 2013). In these mice, cells labeled at the four-cell stage were found to have a biased contribution to either the TE or ICM in 30% of embryos, which could not be explained by a stochastic mechanism (Tabansky et al., 2013).

How do these early differences between four-cell stage blastomeres affect later division orientations and contributions to the TE or ICM? At the eight-cell stage, just before the generation of inside and outside cells, Cdx2 is heterogeneously expressed (Dietrich & Hiiragi, 2007; Jedrusik et al., 2008; Ralston & Rossant, 2008). Given that Cdx2 expression enhances polarization and increases the frequency of symmetric cell divisions (Ajduk, Biswas Shivhare, & Zernicka-Goetz, 2014; Jedrusik et al., 2008), this heterogeneity provides a potential mechanism by which cells can be biased in their fate contributions. The transcription factor Tfap2C has recently been found to be required for the initial zygotic expression of Cdx2 at very early stages (Cao et al., 2015); however, a clear mechanistic link between the heterogeneities observed at the four-cell stage and Cdx2 expression at the eight-cell stage has not yet been shown. One likely possibility is that it reflects the activity of pluripotency-related genes downstream of Carm1 and Histone H3R26 methylation that act to inhibit Cdx2 expression.

7. HOW DOES HETEROGENEITY FIRST ARISE?

One of the most intriguing and difficult to answer questions of early development is whether mammalian embryos exhibit any prepatterning, as in the embryos of most other model systems, or if their development is entirely governed by stochastic mechanisms. Trying to answer this question in the mouse embryo is particularly difficult as regulative developmental mechanisms mean that cell fate is flexible, at least at early stages, and therefore how important a certain factor is for cell fate decisions can depend on context. Furthermore, key differences between cells that influence fate may not be in gene expression but rather in the localization or activity of particular transcription factors or epigenetic modifiers, which can be difficult to identify. *Oct4* is an example of a gene that is uniformly expressed at the four-cell stage, yet it has dramatically different protein activity in different blastomeres, which correlates with cell fate (Plachta et al., 2011).

A model in which asymmetries between cells arise as a consequence of stochastic differences in gene expression has been proposed (Dietrich & Hiiragi, 2007). In this model, random differences in gene expression between blastomeres are refined as the embryo develops and cells adopt different positions, encountering different environments and positional cues. All cleavage-stage blastomeres would therefore have equal developmental potential and the same capability to contribute to any cell lineage. This "random model" makes sense when you consider how regulative mouse development is; however, key experiments have found innate biases in cells that relate to their division history and the animal–vegetal (A–V) axis of the egg. When the zygote divides, this first cleavage division is typically along the A–V axis (with the animal pole marked by the second attached polar body). The next round of cleavages, at the two- to four-cell transition, can be classified as either meridional (M; along the A–V axis) or equatorial (E; perpendicular to the A–V axis). Daughter cells of M divisions inherit both animal and vegetal portions of the blastomere (AV cells), while E divisions generate separate animal (A) and vegetal (V) cells. The majority of embryos have sequential divisions through each axis to form a tetrahedral four-cell stage embryo (Bischoff, Parfitt, & Zernicka-Goetz, 2008; Gardner, 2002; Piotrowska-Nitsche & Zernicka-Goetz, 2005). Much less common are flat four-cell embryos, which are generated by two M or two E divisions. Four different kinds of four-cell stage embryos can therefore exist depending on their division orientation and order: ME, EM, MM, and EE (Fig. 4C). Interestingly, although ME, EM, and MM embryos have normal developmental potential, EE embryos are developmentally compromised and the majority fail to develop past implantation (Piotrowska-Nitsche & Zernicka-Goetz, 2005), indicating that cell division pattern is in some way linked to developmental success. These EE embryos contain two A cells and two V cells at the four-cell stage, while all other embryo types contain at least two AV blastomeres. To determine whether the different kinds of four-cell stage blastomeres exhibited different developmental characteristics, chimeras were generated that were composed only of A, V, or AV blastomeres isolated from tetrahedral ME embryos (Morris, Guo, et al., 2012; Piotrowska-Nitsche, Perea-Gomez, Haraguchi, & Zernicka-Goetz, 2005). AV chimeras developed normally, but A and V chimeras, containing only cells that inherited the animal or vegetal parts of the egg, had very reduced developmental potential when transferred to recipient mothers (Piotrowska-Nitsche et al., 2005; Fig. 4D). Identifying A and V blastomeres can be difficult, as they can occasionally rotate such that the cell that occupies the more animal position is actually formed from the vegetal portion of the egg, and vice versa. When chimeras were made from A and V cells in which this rotation had been controlled for, it was found that V cells in particular had compromised development, with no chimeras comprising truly vegetal V cells developing to term (Piotrowska-Nitsche et al., 2005). Cell fate tracking of individual cells within ME embryos discovered that A and V blastomeres contribute much less frequently to the pluripotent ICM than AV blastomeres (Piotrowska-Nitsche & Zernicka-Goetz, 2005). The dramatically restricted potential of V blastomeres correlates with Carm1 expression and Histone H3R26 methylation levels, both of which are lowest in the V blastomere of ME embryos (Torres-Padilla et al., 2007).

Why do A and V blastomeres from ME embryos exhibit reduced developmental potential and a biased contribution to the TE? The most obvious possibility is that this is related to the A–V axis of the egg. The embryo can develop normally when either the animal or vegetal pole of the zygote is removed (Zernicka-Goetz, 1998), therefore nothing essential for development is localized at either pole. However, the possibility remains that the poles either lack or contain something that biases cell fate. Even if this is the case, it cannot be the sole reason why the A and V blastomeres of ME embryos are developmentally compromised, as these blastomeres in

EM embryos, which are generated by the first of the two- to four-cell cleavage divisions, do not show such restricted fate (Piotrowska-Nitsche et al., 2005; Piotrowska-Nitsche & Zernicka-Goetz, 2005). This suggests that division order also influences developmental potential, and that the progeny of the first and second dividing cell may have different properties. These combined factors appear to have the most effect in ME embryos, in which A and V cells are separated in the second two- to four-cell cleavage, as these embryos show the most evidence of pattern when looking at the contributions of each cell to the blastocyst (Bischoff et al., 2008). Indeed, in ME embryos, the eight-cell stage progeny of the second-diving E blastomere have the highest Cdx2 expression, while in EM embryos the progeny of the first-dividing E blastomere do not show this enhanced Cdx2 expression (Jedrusik et al., 2008). The potential influence of other factors such as the sperm entry point (Piotrowska & Zernicka-Goetz, 2001) and epigenetic asymmetries (Takaoka & Hamada, 2014) also need to be considered, and a comprehensive explanation for how cell fate bias can be generated by cell history remains elusive. Understanding these biases in the mammalian embryo and determining how influential they are for early development represents a considerable challenge for the future. It is an important challenge to take, however, as it will help us to understand the origins of our own symmetry breaking events and the mechanisms by which a minimal guiding force can lead to tissue diversification.

ACKNOWLEDGMENTS

We thank the Wellcome Trust for supporting the work in our laboratory and Meng Zhu for comments on the manuscript.

REFERENCES

Ajduk, A., Biswas Shivhare, S., & Zernicka-Goetz, M. (2014). The basal position of nuclei is one pre-requisite for asymmetric cell divisions in the early mouse embryo. *Developmental Biology*, *392*, 133–140.

Alarcón, V. B., & Marikawa, Y. (2003). Deviation of the blastocyst axis from the first cleavage plane does not affect the quality of mouse postimplantation development. *Biology of Reproduction*, *69*, 1208–1212.

Anani, S., Bhat, S., Honma-Yamanaka, N., Krawchuk, D., & Yamanaka, Y. (2014). Initiation of Hippo signaling is linked to polarity rather than to cell position in the pre-implantation mouse embryo. *Development*, *141*, 2813–2824.

Artus, J., Kang, M., Cohen-Tannoudji, M., & Hadjantonakis, A.-K. (2013). PDGF signaling is required for primitive endoderm cell survival in the inner cell mass of the mouse blastocyst. *Stem Cells*, *31*, 1932–1941.

Artus, J., Piliszek, A., & Hadjantonakis, A.-K. (2011). The primitive endoderm lineage of the mouse blastocyst: Sequential transcription factor activation and regulation of differentiation by Sox17. *Developmental Biology*, *350*, 393–404.
Avilion, A. A., Nicolis, S. K., Pevny, L. H., Perez, L., Vivian, N., & Lovell-Badge, R. (2003). Multipotent cell lineages in early mouse development depend on SOX2 function. *Genes & Development*, *17*, 126–140.
Bachvarova, R., & De Leon, V. (1980). Polyadenylated RNA of mouse ova and loss of maternal RNA in early development. *Developmental Biology*, *74*, 1–8.
Bedzhov, I., & Zernicka-Goetz, M. (2014). Self-organizing properties of mouse pluripotent cells initiate morphogenesis upon implantation. *Cell*, *156*, 1032–1044.
Bessonnard, S., De Mot, L., Gonze, D., Barriol, M., Dennis, C., Goldbeter, A., et al. (2014). Gata6, Nanog and Erk signaling control cell fate in the inner cell mass through a tristable regulatory network. *Development*, *141*, 3637–3648.
Biase, F. H., Cao, X., & Zhong, S. (2014). Cell fate inclination within 2-cell and 4-cell mouse embryos revealed by single-cell RNA sequencing. *Genome Research*, *24*, 1787–1796.
Bischoff, M., Parfitt, D. E., & Zernicka-Goetz, M. (2008). Formation of the embryonic-abembryonic axis of the mouse blastocyst: Relationships between orientation of early cleavage divisions and pattern of symmetric/asymmetric divisions. *Development*, *135*, 953–962.
Blij, S., Frum, T., Akyol, A., Fearon, E., & Ralston, A. (2012). Maternal Cdx2 is dispensable for mouse development. *Development*, *139*, 3969–3972.
Braude, P., Pelham, H., Flach, G., & Lobatto, R. (1979). Post-transcriptional control in the early mouse embryo. *Nature*, *282*, 102–105.
Burton, A., Muller, J., Tu, S., Padilla-Longoria, P., Guccione, E., & Torres-Padilla, M.-E. (2013). Single-cell profiling of epigenetic modifiers identifies PRDM14 as an inducer of cell fate in the mammalian embryo. *Cell Reports*, *5*, 687–701.
Cao, Z., Carey, T. S., Ganguly, A., Wilson, C. A., Paul, S., & Knott, J. G. (2015). Transcription factor AP-2γ induces early Cdx2 expression and represses HIPPO signaling to specify the trophectoderm lineage. *Development*, *142*, 1606–1615.
Chambers, I., Colby, D., Robertson, M., Nichols, J., Lee, S., Tweedie, S., et al. (2003). Functional expression cloning of Nanog, a pluripotency sustaining factor in embryonic stem cells. *Cell*, *113*, 643–655.
Chan, S. W., Lim, C. J., Chong, Y. F., Pobbati, A. V., Huang, C., & Hong, W. (2011). Hippo pathway-independent restriction of TAZ and YAP by angiomotin. *The Journal of Biological Chemistry*, *286*, 7018–7026.
Chazaud, C., Yamanaka, Y., Pawson, T., & Rossant, J. (2006). Early lineage segregation between epiblast and primitive endoderm in mouse blastocysts through the Grb2-MAPK pathway. *Developmental Cell*, *10*, 615–624.
Chróścicka, A., Komorowski, S., & Maleszewski, M. (2004). Both blastomeres of the mouse 2-cell embryo contribute to the embryonic portion of the blastocyst. *Molecular Reproduction and Development*, *68*, 308–312.
Coucouvanis, E., & Martin, G. R. (1995). Signals for death and survival: A two-step mechanism for cavitation in the vertebrate embryo. *Cell*, *83*, 279–287.
Dietrich, J. E., & Hiiragi, T. (2007). Stochastic patterning in the mouse pre-implantation embryo. *Development*, *134*, 4219–4231.
Donnison, M., Beaton, A., Davey, H. W., Broadhurst, R., L'Huillier, P., & Pfeffer, P. L. (2005). Loss of the extraembryonic ectoderm in Elf5 mutants leads to defects in embryonic patterning. *Development*, *132*, 2299–2308.
Flach, G., Johnson, M. H., Braude, P. R., Taylor, R. A., & Bolton, V. N. (1982). The transition from maternal to embryonic control in the 2-cell mouse embryo. *The EMBO Journal*, *1*, 681–686.

Fleming, T. P. (1987). A quantitative analysis of cell allocation to trophectoderm and inner cell mass in the mouse blastocyst. *Developmental Biology, 119*, 520–531.

Frankenberg, S., Gerbe, F., Bessonnard, S., Belville, C., Pouchin, P., Bardot, O., et al. (2011). Primitive endoderm differentiates via a three-step mechanism involving Nanog and RTK signaling. *Developmental Cell, 21*, 1005–1013.

Fujimori, T., Kurotaki, Y., Miyazaki, J., & Nabeshima, Y. (2003). Analysis of cell lineage in two- and four-cell mouse embryos. *Development, 130*, 5113–5122.

Gardner, R. L. (2001). Specification of embryonic axes begins before cleavage in normal mouse development. *Development, 128*, 839–847.

Gardner, R. L. (2002). Experimental analysis of second cleavage in the mouse. *Human Reproduction, 17*, 3178–3189.

Graham, S. J. L., Wicher, K. B., Jedrusik, A., Guo, G., Herath, W., Robson, P., et al. (2014). BMP signalling regulates the pre-implantation development of extra-embryonic cell lineages in the mouse embryo. *Nature Communications, 5*, 5667.

Guo, G., Huss, M., Tong, G. Q., Wang, C., Li Sun, L., Clarke, N. D., et al. (2010). Resolution of cell fate decisions revealed by single-cell gene expression analysis from zygote to blastocyst. *Developmental Cell, 18*, 675–685.

Handyside, A. H. (1978). Time of commitment of inside cells isolated from preimplantation mouse embryos. *Journal of Embryology and Experimental Morphology, 45*, 37–53.

Hillman, N., Sherman, M. I., & Graham, C. (1972). The effect of spatial arrangement on cell determination during mouse development. *Journal of Embryology and Experimental Morphology, 28*, 263–278.

Hirate, Y., Hirahara, S., Inoue, K., Suzuki, A., Alarcon, V. B., Akimoto, K., et al. (2013). Polarity-dependent distribution of angiomotin localizes Hippo signaling in preimplantation embryos. *Current Biology, 23*, 1181–1194.

Jedrusik, A., Bruce, A. W., Tan, M. H., Leong, D. E., Skamagki, M., Yao, M., et al. (2010). Maternally and zygotically provided Cdx2 have novel and critical roles for early development of the mouse embryo. *Developmental Biology, 344*, 66–78.

Jedrusik, A., Cox, A., Wicher, K. B., Glover, D. M., & Zernicka-Goetz, M. (2015). Maternal-zygotic knockout reveals a critical role of Cdx2 in the morula to blastocyst transition. *Developmental Biology, 398*, 147–152.

Jedrusik, A., Parfitt, D.-E., Guo, G., Skamagki, M., Grabarek, J. B., Johnson, M. H., et al. (2008). Role of Cdx2 and cell polarity in cell allocation and specification of trophectoderm and inner cell mass in the mouse embryo. *Genes & Development, 22*, 2692–2706.

Johnson, M. H., & Ziomek, C. A. (1981). The foundation of two distinct cell lineages within the mouse morula. *Cell, 24*, 71–80.

Kang, M., Piliszek, A., Artus, J., & Hadjantonakis, A.-K. (2013). FGF4 is required for lineage restriction and salt-and-pepper distribution of primitive endoderm factors but not their initial expression in the mouse. *Development, 140*, 267–279.

Kono, K., Tamashiro, D. A. A., & Alarcon, V. B. (2014). Inhibition of RHO–ROCK signaling enhances ICM and suppresses TE characteristics through activation of hippo signaling in the mouse blastocyst. *Developmental Biology, 394*, 142–155.

Koutsourakis, M., Langeveld, A., Patient, R., Beddington, R., & Grosveld, F. (1999). The transcription factor GATA6 is essential for early extraembryonic development. *Development, 126*, 723–732.

Krawchuk, D., Honma-Yamanaka, N., Anani, S., & Yamanaka, Y. (2013). FGF4 is a limiting factor controlling the proportions of primitive endoderm and epiblast in the ICM of the mouse blastocyst. *Developmental Biology, 384*, 65–71.

Krupa, M., Mazur, E., Szczepańska, K., Filimonow, K., Maleszewski, M., & Suwińska, A. (2014). Allocation of inner cells to epiblast vs primitive endoderm in the mouse embryo is biased but not determined by the round of asymmetric divisions (8→16- and 16→32-cells). *Developmental Biology, 385*, 136–148.

Kurimoto, K., Yabuta, Y., Ohinata, Y., Ono, Y., Uno, K. D., Yamada, R. G., et al. (2006). An improved single-cell cDNA amplification method for efficient high-density oligonucleotide microarray analysis. *Nucleic Acids Research, 34*, e42.

Le Bin, G. C., Muñoz-Descalzo, S., Kurowski, A., Leitch, H., Lou, X., Mansfield, W., et al. (2014). Oct4 is required for lineage priming in the developing inner cell mass of the mouse blastocyst. *Development, 141*, 1001–1010.

Leung, C. Y., & Zernicka-Goetz, M. (2013). Angiomotin prevents pluripotent lineage differentiation in mouse embryos via Hippo pathway-dependent and -independent mechanisms. *Nature Communications, 4*, 2251.

McDole, K., Xiong, Y., Iglesias, P. A., & Zheng, Y. (2011). Lineage mapping the pre-implantation mouse embryo by two-photon microscopy, new insights into the segregation of cell fates. *Developmental Biology, 355*, 239–249.

Meilhac, S. M., Adams, R. J., Morris, S. A., Danckaert, A., Le Garrec, J. F., & Zernicka-Goetz, M. (2009). Active cell movements coupled to positional induction are involved in lineage segregation in the mouse blastocyst. *Developmental Biology, 331*, 210–221.

Messerschmidt, D. M., & Kemler, R. (2010). Nanog is required for primitive endoderm formation through a non-cell autonomous mechanism. *Developmental Biology, 344*, 129–137.

Mitsui, K., Tokuzawa, Y., Itoh, H., Segawa, K., Murakami, M., Takahashi, K., et al. (2003). The homeoprotein Nanog is required for maintenance of pluripotency in mouse epiblast and ES cells. *Cell, 113*, 631–642.

Molkentin, J. D., Lin, Q., Duncan, S. A., & Olson, E. N. (1997). Requirement of the transcription factor GATA4 for heart tube formation and ventral morphogenesis. *Genes & Development, 11*, 1061–1072.

Morris, S. A., Graham, S. J. L., Jedrusik, A., & Zernicka-Goetz, M. (2013). The differential response to Fgf signalling in cells internalized at different times influences lineage segregation in preimplantation mouse embryos. *Open Biology, 3*, 130104.

Morris, S. A., Grewal, S., Barrios, F., Patankar, S. N., Strauss, B., Buttery, L., et al. (2012). Dynamics of anterior-posterior axis formation in the developing mouse embryo. *Nature Communications, 3*, 673.

Morris, S. A., Guo, Y., & Zernicka-Goetz, M. (2012). Developmental plasticity is bound by pluripotency and the Fgf and Wnt signaling pathways. *Cell Reports, 2*, 756–765.

Morris, S. A., Teo, R. T. Y., Li, H., Robson, P., Glover, D. M., & Zernicka-Goetz, M. (2010). Origin and formation of the first two distinct cell types of the inner cell mass in the mouse embryo. *Proceedings of the National Academy of Sciences of the United States of America, 107*, 6364–6369.

Motosugi, N., Bauer, T., Polanski, Z., Solter, D., & Hiiragi, T. (2005). Polarity of the mouse embryo is established at blastocyst and is not prepatterned. *Genes & Development, 19*, 1081–1092.

Niakan, K. K., Ji, H., Maehr, R., Vokes, S. A., Rodolfa, K. T., Sherwood, R. I., et al. (2010). Sox17 promotes differentiation in mouse embryonic stem cells by directly regulating extraembryonic gene expression and indirectly antagonizing self-renewal. *Genes & Development, 24*, 312–326.

Nichols, J., Zevnik, B., Anastassiadis, K., Niwa, H., Klewe-Nebenius, D., Chambers, I., et al. (1998). Formation of pluripotent stem cells in the mammalian embryo depends on the POU transcription factor Oct4. *Cell, 95*, 379–391.

Nishioka, N., Inoue, K., Adachi, K., Kiyonari, H., Ota, M., Ralston, A., et al. (2009). The hippo signaling pathway components Lats and Yap pattern Tead4 activity to distinguish mouse trophectoderm from inner cell mass. *Developmental Cell, 16*, 398–410.

Nishioka, N., Yamamoto, S., Kiyonari, H., Sato, H., Sawada, A., Ota, M., et al. (2008). Tead4 is required for specification of trophectoderm in pre-implantation mouse embryos. *Mechanisms of Development, 125*, 270–283.

Niwa, H., Toyooka, Y., Shimosato, D., Strumpf, D., Takahashi, K., Yagi, R., et al. (2005). Interaction between Oct3/4 and Cdx2 determines trophectoderm differentiation. *Cell, 123*, 917–929.

Ohnishi, Y., Huber, W., Tsumura, A., Kang, M., Xenopoulos, P., Kurimoto, K., et al. (2014). Cell-to-cell expression variability followed by signal reinforcement progressively segregates early mouse lineages. *Nature Cell Biology, 16*, 27–37.

Palmieri, S. L., Peter, W., Hess, H., & Scholer, H. R. (1994). Oct-4 transcription factor is differentially expressed in the mouse embryo during establishment of the first two extraembryonic cell lineages involved in implantation. *Developmental Biology, 166*, 259–267.

Paramasivam, M., Sarkeshik, A., Yates, J. R., 3rd., Fernandes, M. J., & McCollum, D. (2011). Angiomotin family proteins are novel activators of the LATS2 kinase tumor suppressor. *Molecular Biology of the Cell, 22*, 3725–3733.

Piotrowska, K., Wianny, F., Pedersen, R. A., & Zernicka-Goetz, M. (2001). Blastomeres arising from the first cleavage division have distinguishable fates in normal mouse development. *Development, 128*, 3739–3748.

Piotrowska, K., & Zernicka-Goetz, M. (2001). Role for sperm in spatial patterning of the early mouse embryo. *Nature, 409*, 517–521.

Piotrowska-Nitsche, K., Perea-Gomez, A., Haraguchi, S., & Zernicka-Goetz, M. (2005). Four-cell stage mouse blastomeres have different developmental properties. *Development, 132*, 479–490.

Piotrowska-Nitsche, K., & Zernicka-Goetz, M. (2005). Spatial arrangement of individual 4-cell stage blastomeres and the order in which they are generated correlate with blastocyst pattern in the mouse embryo. *Mechanisms of Development, 122*, 487–500.

Plachta, N., Bollenbach, T., Pease, S., Fraser, S. E., & Pantazis, P. (2011). Oct4 kinetics predict cell lineage patterning in the early mammalian embryo. *Nature Cell Biology, 13*, 117–123.

Plusa, B., Frankenberg, S., Chalmers, A., Hadjantonakis, A.-K., Moore, C. A., Papalopulu, N., et al. (2005). Downregulation of Par3 and aPKC function directs cells towards the ICM in the preimplantation mouse embryo. *Journal of Cell Science, 118*, 505–515.

Plusa, B., Piliszek, A., Frankenberg, S., Artus, J., & Hadjantonakis, A. K. (2008). Distinct sequential cell behaviours direct primitive endoderm formation in the mouse blastocyst. *Development, 135*, 3081–3091.

Ralston, A., Cox, B. J., Nishioka, N., Sasaki, H., Chea, E., Rugg-Gunn, P., et al. (2010). Gata3 regulates trophoblast development downstream of Tead4 and in parallel to Cdx2. *Development, 137*, 395–403.

Ralston, A., & Rossant, J. (2008). Cdx2 acts downstream of cell polarization to cell-autonomously promote trophectoderm fate in the early mouse embryo. *Developmental Biology, 313*, 614–629.

Roberts, R. M., Katayama, M., Magnuson, S. R., Falduto, M. T., & Torres, K. E. O. (2011). Transcript profiling of individual twin blastomeres derived by splitting two-cell stage murine embryos. *Biology of Reproduction, 84*, 487–494.

Rossant, J., & Lis, W. T. (1979). Potential of isolated mouse inner cell masses to form trophectoderm derivatives in vivo. *Developmental Biology, 70*, 255–261.

Russ, A. P., Wattler, S., Colledge, W. H., Aparicio, S. A., Carlton, M. B., Pearce, J. J., et al. (2000). Eomesodermin is required for mouse trophoblast development and mesoderm formation. *Nature, 404*, 95–99.

Saiz, N., Grabarek, J. B., Sabherwal, N., Papalopulu, N., & Plusa, B. (2013). Atypical protein kinase C couples cell sorting with primitive endoderm maturation in the mouse blastocyst. *Development*, *140*, 4311–4322.

Samarage, C. R., White, M. D., Álvarez, Y. D., Fierro-González, J. C., Henon, Y., Jesudason, E. C., et al. (2015). Cortical tension allocates the first inner cells of the mammalian embryo. *Developmental Cell*, *34*, 435–447.

Schrode, N., Saiz, N., Di Talia, S., & Hadjantonakis, A.-K. (2014). GATA6 levels modulate primitive endoderm cell fate choice and timing in the mouse blastocyst. *Developmental Cell*, *29*, 454–467.

Skamagki, M., Wicher, K. B., Jedrusik, A., Ganguly, S., & Zernicka-Goetz, M. (2013). Asymmetric localization of Cdx2 mRNA during the first cell-fate decision in early mouse development. *Cell Reports*, *3*, 442–457.

Smith, R., & McLaren, A. (1977). Factors affecting the time of formation of the mouse blastocoele. *Journal of Embryology and Experimental Morphology*, *41*, 79–92.

Spindle, A. I. (1978). Trophoblast regeneration by inner cell masses isolated from cultured mouse embryos. *Journal of Experimental Zoology*, *203*, 483–489.

Strumpf, D., Mao, C.-A., Yamanaka, Y., Ralston, A., Chawengsaksophak, K., Beck, F., et al. (2005). Cdx2 is required for correct cell fate specification and differentiation of trophectoderm in the mouse blastocyst. *Development*, *132*, 2093–2102.

Suwińska, A., Czołowska, R., Ożdżeński, W., & Tarkowski, A. K. (2008). Blastomeres of the mouse embryo lose totipotency after the fifth cleavage division: Expression of Cdx2 and Oct4 and developmental potential of inner and outer blastomeres of 16- and 32-cell embryos. *Developmental Biology*, *322*, 133–144.

Tabansky, I., Lenarcic, A., Draft, R. W., Loulier, K., Keskin, D. B., Rosains, J., et al. (2013). Developmental bias in cleavage-stage mouse blastomeres. *Current Biology*, *23*, 21–31.

Takaoka, K., & Hamada, H. (2014). Origin of cellular asymmetries in the pre-implantation mouse embryo: A hypothesis. *Philosophical Transactions of the Royal Society of London. Series B, Biological Sciences*, *369*. (article ID: 20130536).

Tarkowski, A. K. (1959). Experiments on the development of isolated blastomers of mouse eggs. *Nature*, *184*, 1286–1287.

Tarkowski, A. K. (1961). Mouse chimeras developed from fused eggs. *Nature*, *190*, 857–860.

Tarkowski, A. K., Suwińska, A., Czołowska, R., & Ożdżeński, W. (2010). Individual blastomeres of 16- and 32-cell mouse embryos are able to develop into foetuses and mice. *Developmental Biology*, *348*, 190–198.

Torres-Padilla, M. E., Parfitt, D. E., Kouzarides, T., & Zernicka-Goetz, M. (2007). Histone arginine methylation regulates pluripotency in the early mouse embryo. *Nature*, *445*, 214–218.

Vestweber, D., Gossler, A., Boller, K., & Kemler, R. (1987). Expression and distribution of cell adhesion molecule uvomorulin in mouse preimplantation embryos. *Developmental Biology*, *124*, 451–456.

Vinot, S., Le, T., Ohno, S., Pawson, T., Maro, B., & Louvet-VallÈe, S. (2005). Asymmetric distribution of PAR proteins in the mouse embryo begins at the 8-cell stage during compaction. *Developmental Biology*, *282*, 307–319.

Watanabe, T., Biggins, J. S., Tannan, N. B., & Srinivas, S. (2014). Limited predictive value of blastomere angle of division in trophectoderm and inner cell mass specification. *Development*, *141*, 2279–2288.

Wu, G., Gentile, L., Fuchikami, T., Sutter, J., Psathaki, K., Esteves, T. C., et al. (2010). Initiation of trophectoderm lineage specification in mouse embryos is independent of Cdx2. *Development*, *137*, 4159–4169.

Yagi, R., Kohn, M. J., Karavanova, I., Kaneko, K. J., Vullhorst, D., DePamphilis, M. L., et al. (2007). Transcription factor TEAD4 specifies the trophectoderm lineage at the beginning of mammalian development. *Development*, *134*, 3827–3836.

Yamanaka, Y., Lanner, F., & Rossant, J. (2010). FGF signal-dependent segregation of primitive endoderm and epiblast in the mouse blastocyst. *Development*, *137*, 715–724.
Zernicka-Goetz, M. (1998). Fertile offspring derived from mammalian eggs lacking either animal or vegetal poles. *Development*, *125*, 4803–4808.
Ziomek, C. A., & Johnson, M. H. (1980). Cell surface interaction induces polarization of mouse 8-cell blastomeres at compaction. *Cell*, *21*, 935–942.

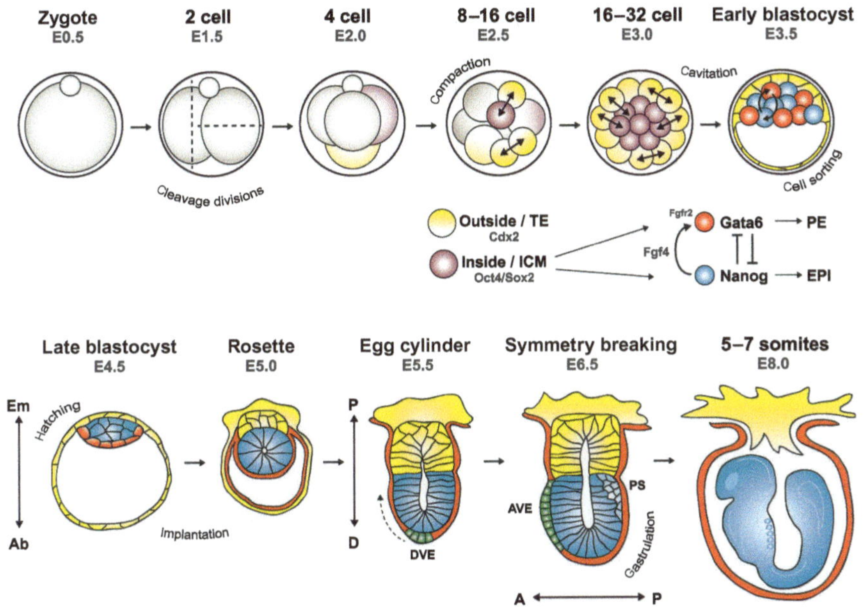

Sarah J.L. Graham and Magdalena Zernicka-Goetz, Figure 1 Early mouse embryo development. Through a series of cleavage divisions the zygote forms a mature blastocyst by E4.5, which then implants into the uterus. By this stage three different cell lineages have been specified, the extraembryonic trophectoderm (TE; yellow) and primitive endoderm (PE; red) and the embryonic epiblast (EPI; blue). These lineages begin to be separated by the formation of outside and inside cells: Outside cells form the TE while inside cells form the pluripotent inner cell mass (ICM; purple). ICM cells will be further segregated into PE and EPI cells. The location of the ICM within the embryo determines the embryonic–abembryonic (Em–Ab) axis of the embryo, which is translated into the proximal–distal (P–D) axis following implantation. The fetal EPI lineage is organized into a cup-shaped epithelium through the formation of a polarized rosette and the opening of a central lumen during peri-implantation stages. The anterior–posterior (A–P) axis is determined by the migration of the distal visceral endoderm (DVE, derived from the PE) to form the anterior visceral endoderm (AVE) on one side of the egg cylinder. Primitive streak (PS) formation on the posterior side of the embryo marks the onset of gastrulation. The contribution of early blastomeres to either the TE or ICM is biased by early heterogeneities between cells.

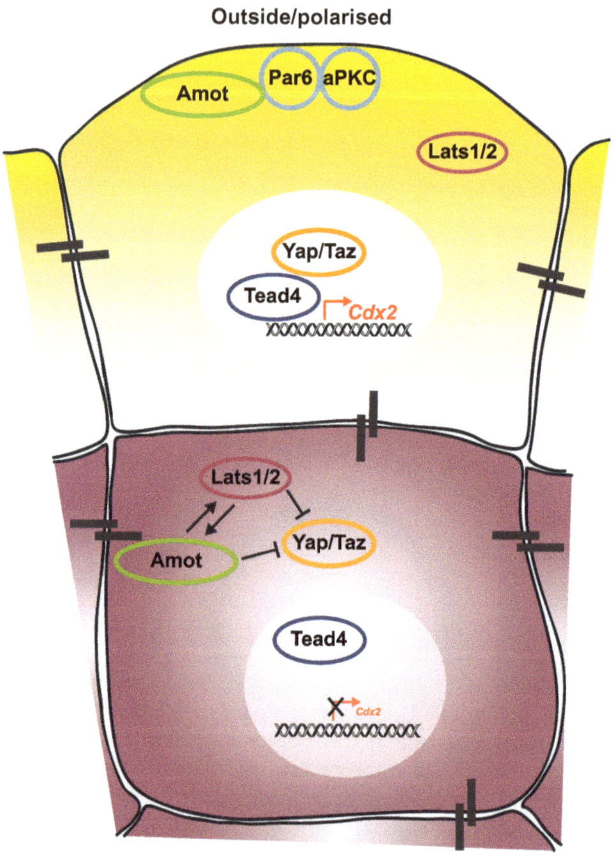

Sarah J.L. Graham and Magdalena Zernicka-Goetz, Figure 2 Polarity and position in the first cell fate decision. Tead4 activity is differentially regulated in outside and inside cells due to differences in polarity and Hippo pathway signaling. In outside polarized cells, Angiomotin (Amot) is restricted to the apical domain, allowing Yap/Taz to enter the nucleus and facilitate upregulation of Cdx2 expression by Tead4. In inside apolar cells, Amot is localized at adherens junctions (gray bars), where it acts to sequester Yap/Taz in the cytoplasm, through both the activation of the Hippo pathway kinases Lats1/2 and through Hippo-independent binding to Yap/Taz, thereby inhibiting Tead4 activity and Cdx2 expression.

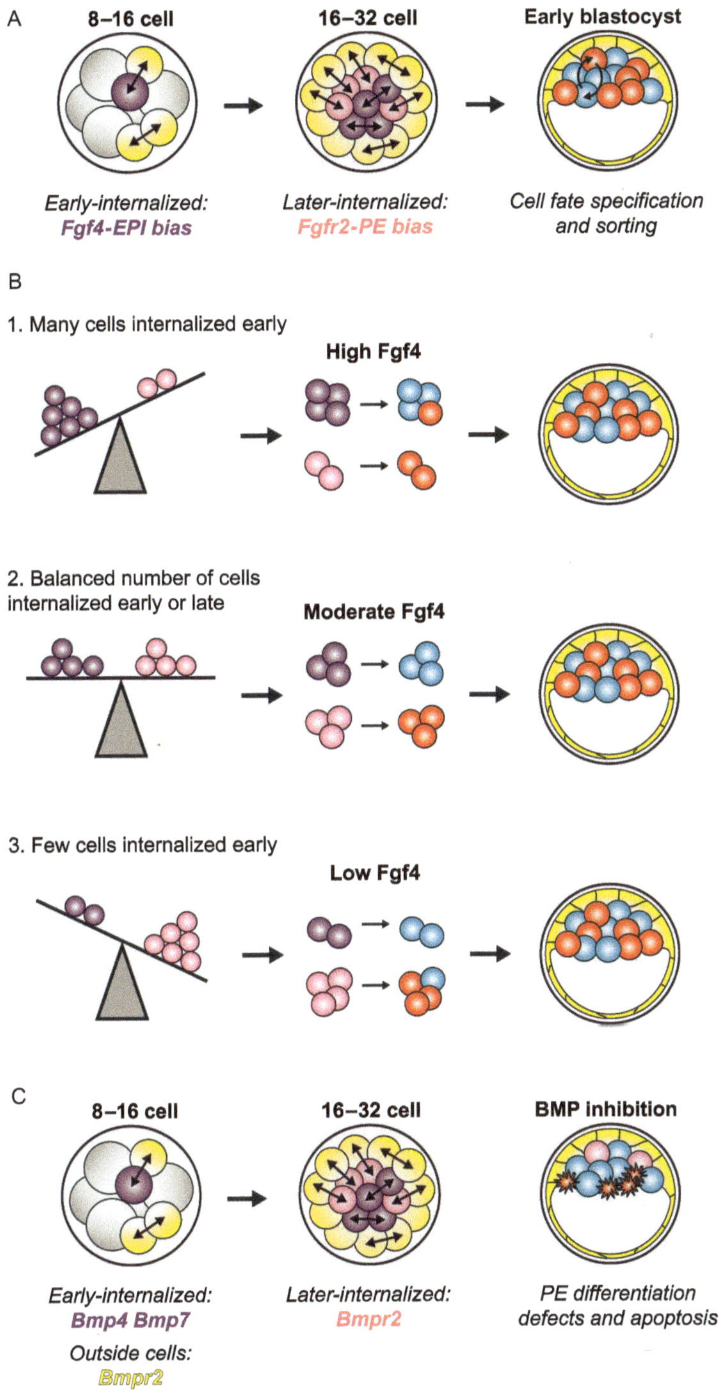

Sarah J.L. Graham and Magdalena Zernicka-Goetz, Figure 3 See figure legend on next page.

Sarah J.L. Graham and Magdalena Zernicka-Goetz, Figure 3 Internalization time bias in the second cell fate decision. (A) ICM cells are internalized at different times, with those internalized earlier (purple) expressing Fgf4 and being biased toward forming EPI (blue) and those internalized later (pink) expressing Fgfr2 and being biased to form PE (red). (B) The strength of the internalization time bias depends on the composition of the ICM and the local levels of Fgf signaling. 1. When there are many early-internalized cells there are high levels of Fgf4 and some early-internalized cells are specified to become PE. 2. When the numbers of early- and late-internalized cells are roughly equal the bias is most apparent, as there is an equal number of Fgf "producers" and "responders." 3. When there are few early-internalized cells there are low levels of Fgf4 and some later-internalized cells are specified to form EPI. All three scenarios have the same outcome, producing a balanced number of EPI and PE cells in the ICM. (C) The internalization time bias may be enhanced by BMP signaling. At the 16-cell stage, BMP ligands (*Bmp4* and *7*) are expressed in early-internalized ICM cells and BMP receptor 2 (*Bmpr2*) is expressed in the outside progenitors of later-internalized cells. BMP signaling is critical for PE differentiation and survival, as evidenced by the apoptosis of PE cells when BMP signaling is inhibited.

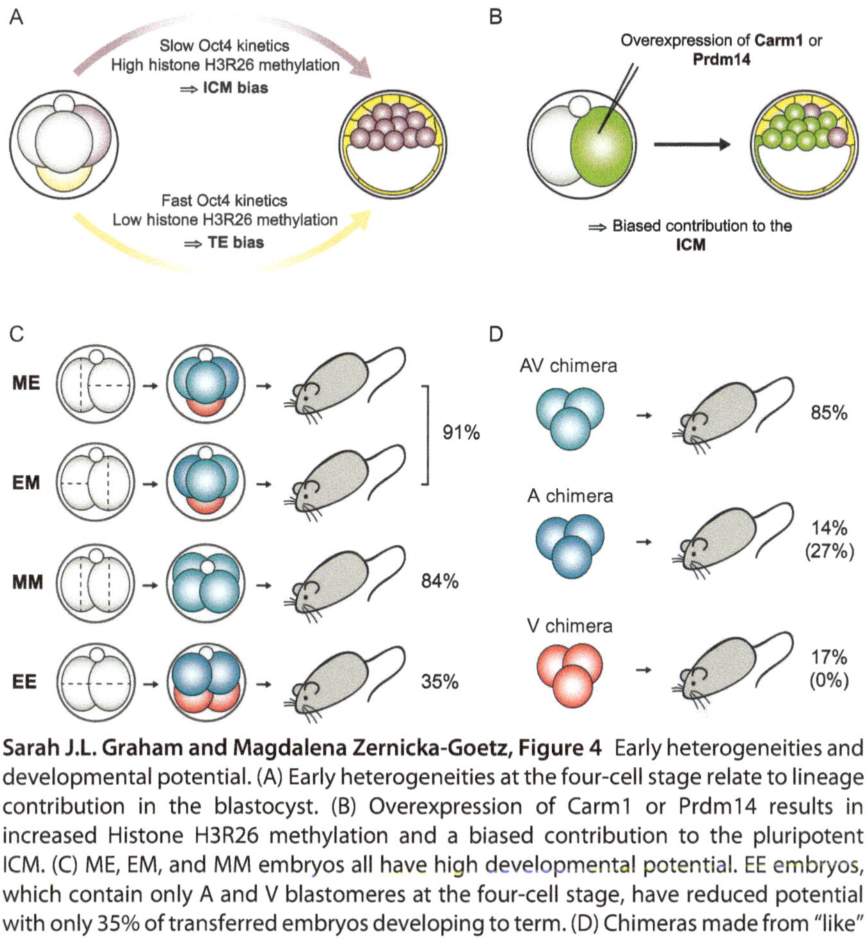

Sarah J.L. Graham and Magdalena Zernicka-Goetz, Figure 4 Early heterogeneities and developmental potential. (A) Early heterogeneities at the four-cell stage relate to lineage contribution in the blastocyst. (B) Overexpression of Carm1 or Prdm14 results in increased Histone H3R26 methylation and a biased contribution to the pluripotent ICM. (C) ME, EM, and MM embryos all have high developmental potential. EE embryos, which contain only A and V blastomeres at the four-cell stage, have reduced potential with only 35% of transferred embryos developing to term. (D) Chimeras made from "like" blastomeres from four-cell stage ME embryos have different developmental potentials. AV chimeras develop with normal success, while the majority of A and V chimeras fail to develop to term. Percentages in brackets represent the proportion of chimeras that developed when cell rotation was excluded, indicating the extremely compromised development of V blastomeres.

INDEX

Note: Page numbers followed by "*f*" indicate figures, and "*t*" indicate tables.

D

F

M

T

Kolibri 360 GmbH
Ferdinand-Jühlke-Straße 7
99095 Erfurt, Deutschland
produktsicherheit@kolibri360.de